Astrocytes in CNS Disorders

Astrocytes in CNS Disorders

Editors

Michael Brenner
Vladimir Parpura

MDPI

Basel • Beijing • Wuhan • Barcelona • Belgrade • Novi Sad • Cluj • Manchester

Editors

Michael Brenner
Department of Neurobiology
The University of Alabama
at Birmingham
Birmingham, AL
United States

Vladimir Parpura
International Translational
Neuroscience Research Institute
Zhejiang Chinese
Medical University
Hangzhou
P.R. China

Editorial Office
MDPI AG
Grosspeteranlage 5
4052 Basel, Switzerland

This is a reprint of articles from the Special Issue published online in the open access journal *Cells* (ISSN 2073-4409) (available at: https://www.mdpi.com/journal/cells/special_issues/Astrocytes_CNS).

For citation purposes, cite each article independently as indicated on the article page online and as indicated below:

Lastname, A.A.; Lastname, B.B. Article Title. *Journal Name* **Year**, *Volume Number*, Page Range.

ISBN 978-3-7258-2297-3 (Hbk)
ISBN 978-3-7258-2298-0 (PDF)
doi.org/10.3390/books978-3-7258-2298-0

Cover image courtesy of Vladimir Parpura
Indirect immunochemistry of glial fibrillary acidic protein (GFAP) in a cultured rat astrocyte isolated from the visual cortex of a 2-day-old pup.

Contents

About the Editors ix

Preface xi

Michael Brenner and Vladimir Parpura
The Role of Astrocytes in CNS Disorders: Historic and Contemporary Views
Reprinted from: *Cells* **2024**, *13*, 1388, doi:10.3390/cells13161388 1

Tracy L. Hagemann, Sierra Coyne, Alder Levin, Liqun Wang, Mel B. Feany and Albee Messing
STAT3 Drives GFAP Accumulation and Astrocyte Pathology in a Mouse Model of Alexander Disease
Reprinted from: *Cells* **2023**, *12*, 978, doi:10.3390/cells12070978 14

Jodie H. K. Man, Charlotte A. G. H. van Gelder, Marjolein Breur, Daniel Okkes, Douwe Molenaar, Sophie van der Sluis, et al.
Cortical Pathology in Vanishing White Matter
Reprinted from: *Cells* **2022**, *11*, 3581, doi:10.3390/cells11223581 33

Juao-Guilherme Rosa, Katherine Hamel, Carrie Sheeler, Ella Borgenheimer, Stephen Gilliat, Alyssa Soles, et al.
Spatial and Temporal Diversity of Astrocyte Phenotypes in Spinocerebellar Ataxia Type 1 Mice
Reprinted from: *Cells* **2022**, *11*, 3323, doi:10.3390/cells11203323 52

Jordi Duran
Role of Astrocytes in the Pathophysiology of Lafora Disease and Other Glycogen Storage Disorders
Reprinted from: *Cells* **2023**, *12*, 722, doi:10.3390/cells12050722 68

Sanghoon Kim, Edward Pajarillo, Ivan Nyarko-Danquah, Michael Aschner and Eunsook Lee
Role of Astrocytes in Parkinson's Disease Associated with Genetic Mutations and Neurotoxicants
Reprinted from: *Cells* **2023**, *12*, 622, doi:10.3390/cells12040622 80

Igor C. Fontana, Miriam Scarpa, Mona-Lisa Malarte, Filipa M. Rocha, Sira Ausellé-Bosch, Marina Bluma, et al.
Astrocyte Signature in Alzheimer's Disease Continuum through a Multi-PET Tracer Imaging Perspective
Reprinted from: *Cells* **2023**, *12*, 1469, doi:10.3390/cells12111469 99

Chiara F. Valori, Claudia Sulmona, Liliana Brambilla and Daniela Rossi
Astrocytes: Dissecting Their Diverse Roles in Amyotrophic Lateral Sclerosis and Frontotemporal Dementia
Reprinted from: *Cells* **2023**, *12*, 1450, doi:10.3390/cells12111450 111

Shotaro Michinaga, Shigeru Hishinuma and Yutaka Koyama
Roles of Astrocytic Endothelin ET_B Receptor in Traumatic Brain Injury
Reprinted from: *Cells* **2023**, *12*, 719, doi:10.3390/cells12050719 138

Nika Gržeta Krpan, Anja Harej Hrkać, Tamara Janković, Petra Dolenec, Elena Bekyarova, Vladimir Parpura and Kristina Pilipović
Chemically Functionalized Single-Walled Carbon Nanotubes Prevent the Reduction in Plasmalemmal Glutamate Transporter EAAT1 Expression in, and Increase the Release of Selected Cytokines from, Stretch-Injured Astrocytes in Vitro
Reprinted from: *Cells* **2024**, *13*, 225, doi:10.3390/cells13030225 . . . . **153**

Terese A. Garcia, Carrie R. Jonak and Devin K. Binder
The Role of Aquaporins in Spinal Cord Injury
Reprinted from: *Cells* **2023**, *12*, 1701, doi:10.3390/cells12131701 . . . . **171**

Maria Stefania Brignone, Angela Lanciotti, Antonio Michelucci, Cinzia Mallozzi, Serena Camerini, Luigi Catacuzzeno, et al.
The CaMKII/MLC1 Axis Confers Ca^{2+}-Dependence to Volume-Regulated Anion Channels (VRAC) in Astrocytes
Reprinted from: *Cells* **2022**, *11*, 2656, doi:10.3390/cells11172656 . . . . **187**

Katharina Everaerts, Pawan Thapaliya, Nils Pape, Simone Durry, Sara Eitelmann, Eleni Roussa, et al.
Inward Operation of Sodium-Bicarbonate Cotransporter 1 Promotes Astrocytic Na^+ Loading and Loss of ATP in Mouse Neocortex during Brief Chemical Ischemia
Reprinted from: *Cells* **2023**, *12*, 2675, doi:10.3390/cells12232675 . . . . **212**

Osama Al-Dalahmah, Alexander A. Sosunov, Yu Sun, Yang Liu, Nacoya Madden, E. Sander Connolly, et al.
The Matrix Receptor CD44 Is Present in Astrocytes throughout the Human Central Nervous System and Accumulates in Hypoxia and Seizures
Reprinted from: *Cells* **2024**, *13*, 129, doi:10.3390/cells13020129 . . . . **237**

Patrycja K. Kruk, Karolina Nader, Anna Skupien-Jaroszek, Tomasz Wójtowicz, Anna Buszka, Gabriela Olech-Kochańczyk, et al.
Astrocytic CD44 Deficiency Reduces the Severity of Kainate-Induced Epilepsy
Reprinted from: *Cells* **2023**, *12*, 1483, doi:10.3390/cells12111483 . . . . **261**

Peter Bedner and Christian Steinhäuser
Role of Impaired Astrocyte Gap Junction Coupling in Epileptogenesis
Reprinted from: *Cells* **2023**, *12*, 1669, doi:10.3390/cells12121669 . . . . **283**

Maja Potokar, Robert Zorec and Jernej Jorgačevski
Astrocytes Are a Key Target for Neurotropic Viral Infection
Reprinted from: *Cells* **2023**, *12*, 2307, doi:10.3390/cells12182307 . . . . **295**

Maria Angeliki S. Pavlou, Kartikeya Singh, Srikanth Ravichandran, Rashi Halder, Nathalie Nicot, Cindy Birck, et al.
Transcriptional and Chromatin Accessibility Profiling of Neural Stem Cells Differentiating into Astrocytes Reveal Dynamic Signatures Affected under Inflammatory Conditions
Reprinted from: *Cells* **2023**, *12*, 948, doi:10.3390/cells12060948 . . . . **314**

Ali Sepehrinezhad, Fin Stolze Larsen, Rezan Ashayeri Ahmadabad, Ali Shahbazi and Sajad Sahab Negah
The Glymphatic System May Play a Vital Role in the Pathogenesis of Hepatic Encephalopathy: A Narrative Review
Reprinted from: *Cells* **2023**, *12*, 979, doi:10.3390/cells12070979 . . . . **342**

Benjamin M. Siemsen, Adam R. Denton, Jeffrey Parrila-Carrero, Kaylee N. Hooker, Eilish A. Carpenter, Meagan E. Prescot, et al.
Heroin Self-Administration and Extinction Increase Prelimbic Cortical Astrocyte–Synapse Proximity and Alter Dendritic Spine Morphometrics That Are Reversed by N-Acetylcysteine
Reprinted from: *Cells* **2023**, *12*, 1812, doi:10.3390/cells12141812 . . . . **361**

Kateryna Murlanova, Yan Jouroukhin, Ksenia Novototskaya-Vlasova, Shovgi Huseynov, Olga Pletnikova, Michael J. Morales, et al.
Loss of Astrocytic μ Opioid Receptors Exacerbates Aversion Associated with Morphine Withdrawal in Mice: Role of Mitochondrial Respiration
Reprinted from: *Cells* **2023**, *12*, 1412, doi:10.3390/cells12101412 . . . . **394**

Xiaolu Zhang, Alyssa Wolfinger, Xiaojun Wu, Rawan Alnafisah, Ali Imami, Abdul-rizaq Hamoud, et al.
Gene Enrichment Analysis of Astrocyte Subtypes in Psychiatric Disorders and Psychotropic Medication Datasets
Reprinted from: *Cells* **2022**, *11*, 3315, doi:10.3390/cells11203315 . . . . **412**

Ying Zhu, Rui Wang, Anthony C. Pappas, Philip Seifert, Andrej Savol, Ruslan I. Sadreyev, et al.
Astrocytes in the Optic Nerve Are Heterogeneous in Their Reactivity to Glaucomatous Injury
Reprinted from: *Cells* **2023**, *12*, 2131, doi:10.3390/cells12172131 . . . . **434**

About the Editors

Michael Brenner

Michael Brenner received his Ph.D. in Biochemistry from the University of California, Berkeley. He served on the faculty of Harvard College and Temple University, and was a Research Scientist at the National Institutes of Health before joining the University of Alabama at Birmingham in 1998, where he is presently Professor Emeritus of Neurobiology. His research investigated both the transcriptional regulation of the gene encoding glial fibrillary acidic protein (GFAP), an intermediate filament protein used as a marker for astrocytes, and GFAP's biological role. Findings from the transcriptional studies include identification of the GFAP mRNA and protein start sites and analysis of the GFAP promoter. Studies using site-directed mutation of GFAP promoter-driven mouse transgenes identified multiple interacting sites affecting transcription, revealing a complex regulatory system for the gene. Several of the promoter constructs developed are extensively used by other laboratories for directing gene expression to astrocytes. These include gfa2, which extends 2.2kb upstream of the mRNA start site, and the smaller 681 bp gfaABC1D promoter, which is useful for viral delivery systems with limited DNA capacity. GFAP promoters developed in the Brenner laboratory have been used by more than 500 laboratories in 32 countries. The investigation of the biological role of GFAP revealed that its absence in mice renders them hypersensitive to traumatic spinal cord injury, indicating a novel role for GFAP in structural support. The laboratory also discovered that mutations within the coding sequence of the GFAP gene are responsible for nearly all cases of Alexander disease, a rare but often fatal neurodegenerative disorder of humans. This was the first identification of a genetic disorder attributable to a primary defect in astrocytes. It made a simple diagnostic genetic test for the disease possible and provided a therapeutic target. Dr. Brenner has written 103 peer-reviewed publications and 16 book chapters.

Vladimir Parpura

Vladimir Parpura, MD, PhD, MAE, holds both a medical degree, awarded from the University of Zagreb in Croatia, and a doctorate in Neuroscience and Zoology received from Iowa State University. He held faculty appointments at the Department of Zoology and Genetics, Iowa State University, U.S.A; the Department of Cell Biology and Neuroscience, University of California Riverside, U.S.A; the Department of Biotechnology, University of Rijeka, Croatia; and the Department of Neurobiology, University of Alabama at Birmingham, U.S.A. He is presently Distinguished Professor and Director of the International Translational Neuroscience Research Institute, Zhejiang Chinese Medical University, Hangzhou, P.R. China, as well as a Cheung Kong Scholar and Kun Peng Action Expert. Parpura has been elected/inducted as a Member of Academia Europaea (MAE) and of Dana Alliance for Brain Initiatives, a corresponding member of the Slovenian Academy of Sciences and Arts, a Fellow of The American Association for the Advancement of Science and of the American Physiological Society. He received the McNulty Civitan Scientist Award. Parpura was the 25th President of the American Society for Neurochemistry and he served on the Council of the International Society for Neurochemistry. Parpura was awarded the Honorary Professor title at the University of Rijeka, and the journal *Neurochemical Research* published a Special Issue in his honor. Parpura discovered the astrocyte–neuron glutamate-mediated signaling pathway, i.e., gliotransmission. This led to the concept of the tripartite synapse, whereby astrocytes, by releasing a gliotransmitter, can modulate synaptic transmission and plasticity.

Subsequently, gliotransmission has proven to be important for biological processes, including sleep, respiration, learning, memory, gut motility and secretion. Parpura has been interfacing neuroscience with nanoscience/nanotechnology, neurochemistry, synthetic biology and biomedical engineering.

Preface

Over the past 30 years, astrocytes have emerged from a supporting role to co-starring with neurons and oligodendrocytes as players in CNS disorders. Dysfunction of astrocytes is now implicated in disorders once thought to be solely of neuronal origin, such as epilepsy, amyotrophic lateral sclerosis, and depression. It is also implicated in disorders previously ascribed solely to oligodendrocytes, like multiple sclerosis, vanishing white matter disease, and megalencephalic leukoencephalopathy with subcortical cysts. In some instances, the disease cause is a defect originating in astrocytes. The first and perhaps most definitive example is Alexander disease, in which coding mutations in the GFAP gene can cause massive dysmyelination and perturb multiple neuronal circuits. In other instances, induced disruption of normal astrocyte functions, such as glutamate or potassium transport, exacerbates the disease process. The purpose of this Special Issue is to illustrate the broad-ranging involvement of astrocytes in CNS disorders. Understanding their role in a variety of CNS disorders will be informative both for revealing normal interactions between astrocytes and other CNS cell types, and for identifying therapeutic targets.

Michael Brenner and Vladimir Parpura
Editors

cells

MDPI

Editorial

The Role of Astrocytes in CNS Disorders: Historic and Contemporary Views

Michael Brenner [1,*] and Vladimir Parpura [2,*]

1 Department of Neurobiology, The University of Alabama at Birmingham, Birmingham, AL 35294, USA
2 International Translational Neuroscience Research Institute, Zhejiang Chinese Medical University, Hangzhou 310053, China
* Correspondence: michaelb@uab.edu (M.B.); vlad@zcmu.edu.cn (V.P.)

1. Introduction

This Special Issue of *Cells* presents a collection of 22 published, peer-reviewed articles on the theme of "Astrocytes in CNS Disorders," including 9 reviews of the evidence implicating astrocytes in the etiology of specific disorders, and 13 original research papers providing such evidence. As discussed in these papers, astrocyte dysfunction is now implicated in disorders once thought to be solely of neuronal origin, such as epilepsy [1,2] and amyotrophic lateral sclerosis [3]. It is also implicated in disorders previously ascribed solely to oligodendrocytes, like vanishing white matter disease [4] and megalencephalic leukoencephalopathy with subcortical cysts [5]. In some instances, the cause of the disorder is a defect originating in astrocytes. The first and perhaps most definitive example is Alexander disease, in which coding mutations in the *GFAP* gene cause massive dysmyelination and perturb multiple neuronal circuits. In other instances, induced disruption of normal astrocytic functions, such as glutamate transport or potassium flux, exacerbates the disorder. Although the identification of specific astrocyte-mediated mechanisms contributing to disorders as described in these papers is a relatively new endeavor, the likelihood that astrocytes have a role in CNS disorders was recognized by the earliest investigators of this cell type. Herein, we briefly review this historical context, followed by introductions to the current contributions.

2. A Brief History of Astrocytes and Their Role in CNS Disorders

Astrocytes are a molecularly and morphologically diverse subtype of neuroglia in the central nervous system (CNS) [6]. The first published observation of neuroglia was in 1858 by Rudolf Ludwig Karl Virchow [7], a German physician and pathologist, who noted the mass present between neurons and referred to it as "neuroglia" (*nervenkitt*). Virchow considered this mass to be acellular, serving to hold neurons in place. A cellular component of neuroglia was recognized in 1872, by Moritz Jastrowitz, a German internist and psychiatrist, who described spider-like glial cells (*spinnenähnliche gliazellen*) [8]. In 1893, British pathologist William Lloyd Andriezen and Swiss anatomist Albert von Kölliker subdivided neuroglia (presumably astrocytes) based on their morphology [9,10]. Andriezen described *fibrous* cells as having processes that are "extremely long, smooth contoured, and of uniform calibre"; whereas *protoplasmic* glial cell processes are much shorter in length, "exhibit great variations of calibre" and have the "striking feature" of a "shaggy granular contour." This latter feature likely equates to the recently described "bushy" morphology of astrocytes (e.g., ref. [11]). Andriezen also observed that the fibrous cells tended to be present in white matter, whereas the protoplasmic cells were generally present in gray matter. Shortly thereafter, in 1895, Hungarian anatomist and histologist Michael von Lenhossék coined the term "astrocytes" (*astrocyten*) for these stellate neuroglia; the term was rooted in the ancient Greek words *astron* (star) and *kytos* (container) [12].

Citation: Brenner, M.; Parpura, V. The Role of Astrocytes in CNS Disorders: Historic and Contemporary Views. *Cells* **2024**, *13*, 1388. https://doi.org/10.3390/cells13161388

Received: 22 July 2024
Accepted: 29 July 2024
Published: 20 August 2024

That astrocytes, like neurons, derive from the neuroectoderm was demonstrated in 1889 by Wilhelm His, a Swiss-born anatomist and cardiologist, and the inventor of a microtome [13,14]. Study of astrocytes was facilitated by development of a specific stain in 1913 by Santiago Ramón y Cajal, a Spanish pathologist and histologist with a keen interest in the CNS [15]. We know today that his gold sublimation technique targets the astrocyte cytoskeleton, in particular, intermediate filaments, the major component of which is glial fibrillary acidic protein (GFAP), a commonly used astrocyte marker [16]. The gold sublimate method is tedious (a recipe is available on pp. 323–324 of ref. [17]) and has been replaced by the simpler and more reproducible method of immunostaining against GFAP (e.g., ref. [18]). Nonetheless, by using the gold sublimate method to analyze the adult human hippocampus, Cajal inferred a proliferating ability of adult astrocytes, as he observed *"astrocitos gemelos"*, which translates to twin astrocytes; i.e., pairs of astrocytes held together by their somata [15].

As morphological studies advanced in the late 19th century and the beginning of the 20th century, speculations of astrocyte function also emerged. Camillo Golgi, an Italian biologist and pathologist, speculated in 1895 that they serve "for distributing the nutritive material"; i.e., they have a metabolic role [19]. Ernesto Lugaro, an Italian psychiatrist, expanded on this idea in 1907, suggesting that astrocytes may take up and metabolize neuroactive substances at the synapse. He wrote:

> "Elsewhere I have exposed the arguments that can make us to think that the actions carried out at the level of the neuronal articulations [synapses], between neuronal terminals and dendrites and cellular bodies of subsequent neurons, are of chemical nature. Every nervous ending would undergo in the moment of the excitation a chemical modification and this chemical modification would act as a stimulus on the other neuron. If that were [true], the interneuronal articulations would be a center of active chemical exchanges; and one would therefore comprise the infiltrating of protoplasmic "tufted" extensions of neuroglia in all the nearby free interstices, in order to perhaps pick and instantaneously to fix even the smallest product of refusal" (page 294 of ref. [20]; translation in [21]).

Experimental confirmation of Lugaro's speculation awaited an elegant study performed a century later demonstrating that astrocytes in culture responded to glutamatergic neurotransmission by uptake of glutamate via their plasma membrane glutamate transporters, a process which contributes to the termination of synaptic events [22]. We know today that 80% of glutamate removal from the extracellular space in the CNS is done by astrocytes, while the remaining 20% is taken up by neurons [23]. A metabolic cross-talk between astrocytes and neurons was also discovered, whereby astrocytes convert glutamate to glutamine [24], which is then released to the extracellular space to be taken up by neurons, which convert the glutamine to glutamate and GABA. Furthermore, Lugaro's concept that three synaptic compartments (pre- and post-synaptic neuronal elements as well as the glial element) constitute a functional synapse found further theoretical traction in the so-called *triad* [25]. The concept of the "tripartite synapse" expanded the synaptic role of astrocytes from serving solely for glutamate removal to being an active participant by the finding that "gliotransmission"; e.g., glutamate released from astrocytes in a Ca^{2+}-dependent manner [26], can modulate synaptic activity [27,28].

Although discovery of Ca^{2+}-dependent glutamate release from astrocytes is a relatively new finding, the possibility that astrocytes had secretory properties was raised over a hundred years ago by Hans Held in Germany and Jean Nageotte in France (for a review, see [29]). Held observed dark molybdenum-hematoxylin-stained granular inclusions in processes of marginal subpial astrocytes in 1909 [30]. Cytosolic granules in glial cells were also reported by several microanatomists of that era (1906–1916), including Alzheimer, Eisath, Fieandt, Cajal, Achucarro and Rio Hortega; with Alzheimer hypothesizing that these granules are evidence for glial secretion mediating metabolic support for neurons (for a review and references, see [29]). In 1910, Nageotte provided the first spatio-temporal description of secretion from gray matter neuroglia, which were likely astrocytes. It should be noted, however, that he misidentified the secretory granules as mitochondrial fragments. He wrote (pages 1068–1069 of ref. [31], translated by Vedrana Montana, Zhejiang Chinese Medical University and Mathieu Lesort, The University of Alabama at Birmingham, with the parenthetical addition for clarification):

> "The facts that I have just observed seem to shed new light on the physiology of neuroglial cells, not only of those which associate closely to neuronal cells and deserve the name of satellite cells, but also, and especially, of cells which are in connection with the vasculature walls.
>
> Indeed, I was able to present evidence of robust active secretion phenomena in the protoplasm of these cells in rabbit and guinea pig. This observation is especially visible within the protoplasmic extensions which cross the empty space created by the retraction of tissues around the vascular walls, on which they [neuroglial cell extensions] attach using an enlarged foot.
>
> In a previous note, I have described the mitochondria that exist in these protoplasmic extensions, and I have shown that many, and maybe all of the granulations located in the gray matter outside the protoplasm of neuronal cells, in reality belong to the neuroglia. Today, I am poised to follow the evolution that occurs within these granulations and to show their progressive transformation into secretion grains. These phenomena are exactly similar to those described by Altmann in the glandular cells; the observed granulations are of three types: 1. excessively small round grains that, by the Altmann method, color themselves in intense red; 2. more voluminous grains, with clear centers; 3. grains that do not take color with fuchsin. The last ones are slightly smaller than the most voluminous red grains. All intermediates exist between these three types, which represent the successive phases of the transformation of mitochondria into secretion grains.
>
> Using Benda's method, the smallest grains color themselves in blue, the largest ones in purple and the secretion grains in red."

More modern and direct evidence for astrocyte transmitter release came from demonstration that primary cultured astrocytes can release taurine [32]. We know today that astrocytes can release many chemical transmitters, neuromodulators, hormones, and metabolic, trophic and plastic factors, under various conditions and by several different mechanisms (reviewed in [33]). For instance, glutamate can be released from astrocytes: (i) through plasmalemmal channels like anion channels [34], hemichannels formed from unpaired connexons [35], or pore-forming $P2X_7$ ionotropic purinergic receptors [36]; (ii) through transporters, such as the reversal of uptake via plasma membrane excitatory amino acid transporters [37], exchange via the cystine-glutamate antiporter [38] or through organic anion transporters [39]; and (iii) via Ca^{2+}-dependent exocytosis [26].

That dynamic neuronal-glial interactions underlie brain functions was postulated as long ago as the late 19th century. Carl Ludwig Schleich, a German surgeon, suggested that variations in the volume of glial cells represented the mechanism for control of neuronal communication; swollen glial cells inhibited neuronal communication, and communication was facilitated when glia shrink [40]. Cajal suggested that astrocytes act as a switch between active and passive states in neuronal networks; retraction of astroglial processes allows information to flow during wakefulness, whereas expansion of astroglial processes disrupts interneuronal connectivity, thus inducing sleep [41,42]. Variants of these ideas have since been experimentally demonstrated; for example, in the supraoptic nucleus of the hypothalamus the retraction of astrocytic processes during lactation or chronic dehydration allows for an increased formation of dendrodendritic electrical synapses between magnocellular neurons, resulting in synchronization of neurons in discharging action potentials (reviewed in [43]).

From the above postulated and proven properties of astrocytes it was to be expected that their faulty functioning would contribute to or even cause CNS disorders. Indeed, they have been implicated in various CNS pathologies from the onset of their description. Virchow thought that neuroglia represent "one of the most frequent seats of morbid change" [7], while Andriezen considered astrocytes to "exhibit a morbid hypertrophy in pathological conditions" [9]. We now introduce the 22 papers in this Special Issue that fulfill this expectation.

3. Overview of the Special Issue Papers

Leukodystrophies are a group of heritable neurological diseases affecting primarily the white matter. Alexander disease (AxD) is a rare, often fatal leukodystrophy of humans, in which the primary deficit in astrocytes affects oligodendrocytes and neurons. Dominant gain-of-function mutations within the coding sequence of the *GFAP* gene are responsible for nearly all cases of Alexander disease [44]. Using molecular genetics and a pre-clinical mouse model of AxD, Hagemann et al. [45] demonstrate that signal transducer and activator of transcription (STAT)-3 in astrocytes plays a role in AxD pathology by promoting GFAP accumulation and aggregation, one of the hallmarks of AxD histopathology. Preventing STAT-3 activation could therefore be therapeutic for AxD in human patients.

Vanishing white matter (VWM) disease is another heritable and progressive leukodystrophy which mainly affects the CNS. In addition to loss of white matter, cortical atrophy is present in long-term patients. VWM is unusual in its progression, as stressors such as infection, mild head trauma or other injuries, can lead to periods of rapid and severe deterioration [46]. The genetic defect may occur in any of the five housekeeping genes that encode individual subunits of eukaryotic translation initiation factor eIF2B (*EIF2B1*, *EIF2B2*, *EIF2B3*, *EIF2B4* and *EIF2B5*) [47]. Surprisingly, although the defect is in a general housekeeping gene, oligodendrocytes and astrocytes are predominantly affected among all cell types. In this Special Issue, Man et al. [4] analyzed the cortex of the middle frontal gyrus of VWM patients for protein composition using mass spectrometry and for astrocyte morphology using immunochemistry. Of the several pathways identified by gene ontology analysis, 14 proteins, all downregulated in VWM, were associated with mitochondrial activity; 10 of the genes encoding these 14 proteins are expressed at least five times more strongly in astrocytes than in neurons or oligodendrocytes. Further implicating cortical astrocytes in VWM, they appeared immature, unreactive, and morphologically less complex than astrocytes from control patients. These results are the first description of molecular and pathological changes in the frontal cortex of VWM patients.

Spinocerebellar ataxia type 1 (SCA1) is an inherited progressive neurodegenerative disease. Its cause is expansion in the *ataxin1 (ATXN1)* gene of the triplet nucleotide repeat cytosine-adenine-guanine (CAG), which encodes glutamine. An expansion of the CAG tract to at least 39 repeats results in a fully penetrant pathogenic allele [48]. Normal ataxin-1 is a DNA-binding protein with unknown function. Expression of the ataxin-1 gene occurs in neurons, astrocytes and microglia. Both of the glial cell types become activated, likely by signals from distressed neurons [49], with reactive astrocytes having been shown to contribute to pathogenesis of SCA1 [50]. Using a knock-in *Atxn1*$^{154Q/2Q}$ mouse model of SCA1, in this Special Issue Rosa et al. [51] report spatial and temporal variation in the reactivity of astrocytes and their expression of neuroprotective genes in the four clinically relevant brain regions: cerebellum, brainstem, hippocampus and motor cortex. Changes were observed early in the disease progression, before neuronal loss. Regional variability in astrocyte homeostatic functions correlated with subsequent pathology, implicating astrocytes in the region- and stage-specific pathogenesis of SCA1.

Glycogen is a metabolically dynamic polymer that serves as a storage form for glucose [52]. In the brain it is primarily located in astrocytes, where it is mobilized for such functions as memory formation [53], protection from hypoxia [54] and glutamatergic gliotransmission [55]. Three glycogen storage diseases affect the CNS: adult polyglucosan body disease, which generally presents in the fifth or sixth decade of life, and Lafora disease and polyglucosan body myopathy 1 with or without immunodeficiency, which occur in children. Each of these disorders is caused by accumulation of insufficiently branched glycogen, which becomes sequestered in aggregates referred to as Lafora bodies or *corpora amylacea*. Duran [56] reviews what is known about the genetic defect in each of these three disorders, and describes the experimental evidence for a primary role for astrocytes in disease development. Perhaps widening the relevance of these glycogen storage disorders, Duran points out that similar aggregates also accumulate in several other CNS diseases, such as Alzheimer's and Parkinson's diseases, and temporal lobe epilepsy.

A defining feature of Parkinson's disease (PD) is the degeneration of dopaminergic neurons in the basal ganglia leading to movement disorders. Although variants of several genes have been associated with PD, about 90% of cases are idiopathic, suggesting involvement of environmental toxins. Nevertheless, studies of the genetic variants have been pursued with the expectation that they will provide clues to causative environmental factors. In this Special Issue, Kim et al. [57] review results from PD models induced by either genetic variants or neurotoxins that point to astrocyte dysfunction as contributing to dopaminergic neuron loss. Greater attention to this role of astrocytes in PD may lead to development of more effective therapeutics.

Alzheimer's disease (AD) is characterized by a progressive brain parenchyma atrophy leading to severe dementia. At the histopathological level there is a presence of both senile plaques, which consist of extracellular deposits of β-amyloid protein, and intraneuronal accumulation of neurofibrillary tangles, which are composed of tau-protein filaments [58,59]. Astrocyte involvement in AD is suggested by studies in transgenic animal models that find that astrocytes become atrophic at the very early stages of pathology and then become reactive prior to the formation of plaques [60–62]. It is also suggested by AD model mice having enhanced Ca^{2+} signaling. That this enhanced signaling can be induced by neuronally-derived soluble Aβ injected into wild type mice in the absence of plaques suggests that it is an early event in AD development [63]. In this Special Issue, Fontana et al. [64] review and correlate results from clinical positron emission tomography (PET) analyses that detect astrocyte reactivity, the presence of Aβ plaques and tau tangles. These findings were also correlated with those for cerebrospinal fluid (CSF) and blood markers of astrogliosis and AD. The authors observe two waves of astrogliosis; the first is associated with an early increase in monoamine oxidase B that precedes Aβ plaque formation by up to 10 years, and the second is associated with a later increase in GFAP, which correlates with plaque deposition. These observations raise the possibility that the early activation phase of astrocytes may contribute to the progression of AD. They also raise the promise that monitoring markers of early astrogliosis may permit detection of AD in its very early stages.

Like AD, amyotrophic lateral sclerosis (ALS, also known as Lou Gehrig's disease) is a progressive, fatal neurological disorder involving the death of neurons—in this case, motor neurons. In addition to motor deficits, about 15% of ALS patients develop behavioral and cognitive disorders, leading to co-diagnosis of frontotemporal dementia (FTD). Conversely, about 20% of FTD patients are co-diagnosed with ALS. In this Special Issue Valori et al. [3] review data demonstrating astrocyte reactivity as an early event in both ALS and FTD, with the resulting alteration of astrocyte functions perhaps contributing to disease progression. Data reviewed include transcriptome analyses of patients and of mouse models, and detection of astrocyte reactivity by PET imaging and fluid biomarkers, similar to those described in the above-mentioned Fontana et al. [64] AD review. Like for AD, appreciation of the early role of astrocytes in ALS and FTD may facilitate early diagnosis and be the basis for therapeutic intervention.

Two papers in this Special Issue explore the role of astrocytes in neurotrauma—injury to the CNS resulting from an abrupt external physical assault. Tissue directly impacted by traumatic brain injury (TBI) is irretrievably lost, and adjacent (penumbral) tissue is at risk due to hypoperfusion, excitotoxicity, neuroinflammation and edema. Saving this penumbral tissue is thus the focus of medical intervention. Astrocytes respond to neurotrauma by reactive astrogliosis that has both detrimental and supportive effects on the outcome of TBI. In this Special Issue, Michinaga et al. [65] review the primacy of the astrocyte endothelin receptor B (ET_B-R) in these astrocyte responses. Levels of both astrocytic ET_B-R and endothelin-1 secretion are increased directly by TBI by an unknown mechanism. The resultant activation of the ET_B-R triggers astrocyte reactivity, leading to secretion of cytokines and proteinases that disrupt the blood-brain barrier (BBB), contributing to both edema and neuroinflammation in the acute stage of injury. Studies in animal models suggest that drugs inhibiting the ET_B-R may be therapeutic in the early stage of injury. However, in the recovery stage of TBI, the astrocytic ET_B-R mediates release of neurotrophic factors that promote repair of the damaged brain parenchyma. Thus, in this later phase of the injury process, stimulation of ET_B-R activity may be therapeutic.

As mentioned above, excitotoxicity is one of the pathological processes detrimental to the penumbra following TBI. A major contributing factor is the downregulation of glutamate uptake in reactive astrocytes [66]. Prevention of such loss could thus have therapeutic value. In this Special Issue, Gržeta Krpan et al. [67] use an in vitro simulated (by stretch) TBI model to demonstrate that colloidal single-walled carbon nanotubes derivatized with polyethylene glycol can prevent the injury-induced reduction in the level of plasmalemmal excitatory amino acid transporter 1 in astrocytes. A possible additional therapeutic benefit of this nanomaterial is that it caused an increase in the release of cytokines from injured astrocytes that were mainly anti-inflammatory. These might ameliorate the effects of the excess of proinflammatory cytokines present in TBI that contribute to disruption of the BBB, brain edema and neuroinflammation.

Other papers in this Special Issue focus on mechanisms of edema in CNS injury. Garcia et al. [68] review experimental evidence for a role for aquaporin 4 (APQ-4) in spinal cord injury. Particularly relevant to causality are studies performed with general and astrocyte-specific AQP-4 knockout mice. Following an injury that disrupts the blood-spinal cord barrier (BSCB), and thus produces vasogenic edema, the AQP-4 knockout mice are compromised in resolution of this edema. In contrast, following an injury producing cytotoxic edema, in which there is no compromise of the BSCB but swelling occurs in astrocytes due to water uptake, edema and its sequelae are ameliorated in the AQP-4 knockout mice. These results can be rationalized by the function of AQP-4 as a water channel. In cytotoxic edema, water enters astrocytes through the channel, and thus its absence attenuates the edema. In vasogenic edema, the compromised BSCB allows water to freely flow into the parenchyma along with proteins and other solutes, but once the BSCB heals, AQP-4 provides the channel for dissipating the edema. The insights provided by this review suggest that matching the timing and direction of AQP-4 manipulation to the type of traumatic CNS injury may provide clinical benefit.

Cytotoxic edema is counteracted by activation of volume-regulated anion channels (VRAC), whose release of chloride anion from the cell provides a driving force for export of water through the AQP-4 channels. In their research paper in this Special Issue, Brignone et al. [5] investigate the mechanism linking cell swelling to VRAC activation by building on observations from patients and animal models of megalencephalic leukoencephalopathy with subcortical cysts (MLC), a fatal genetic disorder marked by astrocytic edema. Mutations of several proteins cause this disease, but they occur primarily in MLC-1 [69], which is predominantly present in astrocytic endfeet. Brignone et al. employ a variety of techniques, including pull-down assays, electrophysiology and intracellular calcium ion imaging, to parse the pathway linking MLC-1 to edema. Using the glioblastoma cell line U251 transfected with wild type or mutant MLC-1, they demonstrate that MLC-1 potentiates the hypoosmotic-induced outward chloride current, and that this capacity requires phosphorylation of MLC-1 by Ca^{2+}/Calmodulin-dependent protein kinase II (CaMKII). They also present evidence that influx of calcium ions into the cells, possibly through TRPV4 channels, is required for the activation of CaMKII. However, it does not do so directly, but instead through stimulation of calcium ion release from internal stores. Whether it is possible to manipulate this pathway to prevent or resolve cytotoxic edema for clinical benefit for MLC patients and for patients with other causes of cytotoxic edema remains to be determined.

Ischemia in the CNS results in accumulation of lactic acid and Na^+ in astrocytes. Lactic acid accumulation, and hence acidosis, occurs from the anerobic metabolism of residual glucose and stored glycogen. Na^+ accumulation occurs when the eventual depletion of these energy sources leads to decreased intracellular ATP, resulting in the failure of the Na^+/K^+-ATPase to pump Na^+ from the cells. Both acidosis and Na^+ accumulation have detrimental effects: acidosis can result in irreversible damage leading to necrosis [70], while Na^+ accumulation inhibits glutamate uptake and increases Ca^{2+} uptake [71]. The astrocyte sodium-bicarbonate cotransporter 1 (NBCe1), which cotransports Na^+ and HCO_3^- across the plasma membrane, affects both these changes, but in opposite directions. In this Special Issue, Everaerts et al. [72] determined the direction of net flux through the astrocytic NBCe1 under ischemic conditions. They exposed hippocampal slice cultures from wild type mice or mice deficient for NBCe1 to a 2 min chemical ischemia, using 2-deoxyglucose to inhibit glycolysis and azide to inhibit oxidative phosphorylation. Assays with fluorescent indicators for Na^+, pH and ATP revealed that NBCe1 operated in the forward mode under these conditions; that is, there was net import of Na^+ and HCO_3^- into the cell, ameliorating the ischemia-induced acidification, but exacerbating the increase in intracellular Na^+. Awaiting further investigation is whether the same direction of net flux occurs during ischemia in animals, and whether manipulation of NBCe1 activity could be of clinical benefit in humans experiencing ischemia.

CD44 in a transmembrane glycoprotein that interacts with multiple binding partners at both its N-terminal extracellular domain and its C-terminal intracellular domain. A primary extracellular binding partner is hyaluronic acid, which is important for cell adhesion and perhaps has a role in axon guidance. Intracellular binding partners include several signaling proteins, such as Src kinase and focal adhesion kinase, which may regulate intracellular Ca^{2+} levels (reviewed in ref. [73]). Given its multiple binding partners and thus likely functions, it's surprising that CD44 null mice are grossly normal; however, targeted analyses have uncovered defects in behavior, memory and regulation of inflammation. In the limited regions of the mature human CNS surveyed, CD44 has been found primarily on a subclass of astrocytes that reside in subpial and subependymal regions and extend long, unbranched, processes; it is not found on bushy, protoplasmic astrocytes. In this Special Issue, Al-Dalahmah et al. [74] confirm and extend these findings in a comprehensive analysis of the entire human CNS. Curiously, they find that in some CNS regions neurons are surrounded by CD44-positive astrocyte processes, whereas in other regions they are not. They also observe that although protoplasmic astrocytes do not express CD44 under normal conditions, they do in hypoxic and epileptic tissue from human patients and in vivo rat models of hypoxia and seizures. These findings provide basic information about CD44 expression in the CNS which should facilitate deciphering its functions.

In this Special Issue, Kruk et al. [2] investigated what functional ramifications epilepsy-induced CD44 expression might have by examining the effect of inhibiting astrocytic CD44 expression in the hippocampus of mice prior to kainate-induced seizures. Hippocampal-specific knockdown of CD44 was accomplished by injecting adeno-associated viruses expressing *cre* driven by a *GFAP* promoter into the hippocampus of mice carrying a floxed CD44 gene. A significant reduction in the number of behavioral seizures in the CD44 knockdown mice compared to controls was observed, as well as a reduction in astrocyte reactivity as measured by GFAP expression. Structural analyses revealed fewer contacts between astrocyte processes and synapses in the CD44 deficient mice compared to controls, and prevention of the loss of dendritic spines produced by KA-induced seizures that occurs in controls. These results, as well as others discussed in their paper, suggest CD44 as a therapeutic target for ameliorating seizures.

The findings of Kruk et al. [2] regarding CD44 implicate astrocytes as contributors to epileptogenesis. As reviewed by Bedner and Steinhäuser [1] in this Special Issue, astrocytes might also contribute to epileptogenesis through alterations in connexin proteins and the gap junctions they form. Relevant to epilepsy, gap junctions link groups of astrocytes to facilitate K^+ buffering, which is critical for preventing the accumulation of K^+ around active synapses that could lead to depolarization and hyperactivity. Consistent with K^+ buffering being critical in epilepsy, gap junction connectivity is lost in animal models of temporal lobe epilepsy, and this loss precedes onset of behavioral seizures. Surprisingly, however, knockout of the primary astrocytic connexins, Cx30 and Cx43, does not cause seizures despite eliminating gap junction connectivity. Bedner and Steinhäuser explain that this likely reflects the multiple roles of connexins and gap junctions. Absence of some of these functions, like K^+ buffering, could initiate or exacerbate seizures; whereas absence of others, like the provision of energy substrates to the neural network, could lower neuronal activity. The authors also review the possible relevance of connexin functions unrelated to gap junctions, including their formation of hemichannels that expose the astrocyte cytoplasm directly to the interstitial fluid. Their commentary on these complexities concludes with suggestions for determining causality, and for the potential for targeting connexins for epilepsy treatment.

In addition to TBI and ischemia, viral infection can be another source of chronic inflammation in the CNS. Many of the CNS viruses, including Zika, herpes, coronavirus and HIV, replicate in astrocytes [75,76]. In their review in this Special Issue, Potokar et al. [77] describe the particular properties of astrocytes that cause them to be welcoming viral hosts. These include their physical location in the CNS, metabolism, and array of viral receptors. Treatment of CNS viral infections will require appreciation of the role of astrocytes in their replication.

As noted above, inflammation accompanies many insults to the CNS, including TBI, ischemia, and viral infection. In addition to producing a direct deleterious effect on the injured tissue, inflammation can have an insidious indirect impact by disrupting the restorative effect of regeneration by differentiation of neural stem cells. In this Special Issue, Pavlou et al. [78] investigate the effects of inflammation on the differentiation of astrocytes from neural stem cells. Changes in gene expression and chromatin accessibility during the differentiation of astrocytes from murine neurospheres in the absence and presence of tumor necrosis factor-alpha (TNF) were compared. Their results indicate that the presence of TNF directs the stem cells along a reactive and incomplete astrocyte differentiation pathway, resulting in compromised astrocyte function. Transcription factors key to astrocyte differentiation in the presence and absence of TNF were also identified. These findings are paralleled by the restriction in chromatin accessibility that occurs during astrocyte differentiation being less extensive in the presence of TNF. Understanding the mechanisms and effects of inflammation on neural progenitor cell differentiation could lead to greater success for the use of neural progenitor cells to treat CNS disorders that are accompanied by inflammation.

Hyperammonemic encephalopathy is a clinical condition displaying both mental and motor manifestations. Coma occurs in its terminal stage, associated with fatal brain edema [79–82]. The disorder is defined by an elevated level of blood ammonia, most commonly caused by a deficiency in the urea cycle or liver failure, the latter being referred to as hepatic encephalopathy. In the brain, ammonia is metabolized by the astrocyte-specific enzyme glutamine synthetase, which joins ammonia to glutamate to form glutamine [83–85]. However, if CNS ammonia levels rise above the capacity of the glutamine synthetase to detoxify it, multiple astrocyte functions become compromised, including regulation of glutamate, K^+, Na^+, Ca^{2+} and pH, leading to death if unchecked [86–90]. In this Special Issue, Sepehrinezhad et al. [91] review another function of astrocytes that could contribute to the pathology of hepatic encephalopathy—its role in the glymphatic system. This system serves to remove potentially toxic metabolic products and misfolded proteins from the brain through the flow of CSF into and through the brain parenchyma to eventually drain into the lymph [92]. This flow is driven by the pressure difference between pulsating arterioles and the brain interstitial fluid, and is facilitated by water flow through astrocytic AQP4 channels. Sepehrinezhad et al. [91] review the evidence for reduced flow through the glymphatic system in hepatic encephalopathy. Among the possible causes are astrocytic swelling, which reduces the arterial/brain parenchyma pressure differential, and mislocalization of AQP4 channels from astrocytic endfeet to the cell body.

Alterations in astrocyte function and morphology are associated with multiple addictive disorders (reviewed in ref. [93]). In this Special Issue, Siemsen et al. [94] report observations of astrocytic changes associated with the remodeling of the connectivity of cortical prelimbic neurons to the nucleus accumbens that is considered responsible for relapse of heroin addiction. Previous studies from their lab and others have focused on the nucleus accumbens, where heroin self-administration in rats was found to produce downregulation of glutamate uptake by astrocytes and decreased contact between astrocyte processes and synapses, with the net result of increased glutamate spillover to perisynaptic targets. Administration of the antioxidant N-acetylcysteine prevented both of these astrocytic changes and drug-seeking relapse, suggesting a causal relationship [95–97]. In their present report, Siemsen et al. [94] performed similar studies of astrocytes at the cortical prelimbic end of the neuronal connection. Changes in astrocytes were also observed in this region, but differed from those in the nucleus accumbens—there was little if any reduction in glutamate uptake, and an increase, rather than a decrease, in the association of astrocytic processes with synapses. In addition, heroin self-administration resulted in the astrocytic processes becoming longer and more highly branched. All of these changes were prevented by treatment with N-acetylcysteine, raising the possibility that astrocyte remodeling in the prelimbic cortex as well as in the nucleus accumbens contributes to drug addiction.

Another pathway by which astrocytes could affect drug addiction is via their opioid receptors. Murlanova et al. [98] explore this possibility in this Special Issue by studying mice in which one of the alleles encoding the μ-opioid receptor 1 has been knocked out specifically in astrocytes. Several effects of this heterozygous knockout are noted, including a greater increase in the morphine-induced locomotor activity, a greater and longer-lived conditioned place aversion following morphine withdrawal, and an increased rate of oxygen consumption by astrocytes cultured from untreated or morphine-injected knockout mice compared to wild type controls. These findings suggest that the astrocyte μ-opioid receptor 1 serves to tamp down morphine-induced behaviors, perhaps by decreasing the rate of astrocytic oxidative phosphorylation.

Multiple studies have reported alterations of astrocytes in psychiatric disorders, but often with conflicting findings. In the Introduction to their paper in this Special Issue, Zhang et al. [99] summarize these prior observations and point out the variables that could cause the discrepancies, including heterogeneity among astrocytes and effects of treatment interventions. They then take these variables into account in an analysis of gene expression in several psychiatric diseases. For example, when the gene expression data from patients with schizophrenia or depression were compared to the expression profiles that define several astrocyte subgroups, the expression data matched the profiles of some astrocyte subgroups, but not others, indicating a role for these astrocyte subsets in the diseases. Proteins encoded by genes with increased expression identified biological pathways that might contribute to disease phenotype. Among these were several protein kinases, raising the possibility that their targeting could be therapeutic. As a cautionary note, experiments using rats to test the expression effects of psychotropic drugs under controlled conditions revealed discrepancies with data from humans, calling into question the use of rodent models to study human psychiatric disease.

Glaucoma is an insidious eye disorder affecting more than 2% of the world's population aged 40 or older [100]. It involves damage to the optic nerve axons and death of retinal ganglion cells (RGCs) resulting in progressive reduction of the visual field. The primary cause is increased intraocular pressure (IOP) due to defective drainage of aqueous humor inside the eye. Death of the RGCs is believed to be caused by physical stress exerted by the IOP on their axons after they gather together to form the optic nerve. The region believed particularly sensitive to this stress is the astrocytic glial lamina, through which the optic nerve passes as it exits the eye. Since this region is primarily composed of astrocytes, and it contains an increased number of reactive astrocytes in glaucoma, astrocytes are considered a likely contributor to the disorder. In this Special Issue, Zhu et al. [101] use a mouse model of glaucoma to identify changes in astrocytes that are linked to glaucoma. They note that astrocytes in the glial lamina of older mice (40 weeks old) show some of the same changes found in younger mice (13 weeks old) treated to produce elevated IOP, yet do not have glaucoma. Thus, to sharpen their discrimination of changes specific to glaucoma, they compare gene expression profiles of their young glaucoma model mice to both age-matched and 40-week-old controls. Although about 80% of the expression differences between the young treated and untreated mice are also present in the comparison to normal older mice, the glaucoma mice uniquely increased expression of genes involved with process formation and extension. This suggested the possibility of increased phagocytosis of axonal mitochondria undergoing degeneration, a process normally present in the astrocytic glial lamina. Double fluorescent labeling of axonal mitochondria and astrocytes indeed revealed an increase in this activity. Surprisingly, for both this activity and gene expression, astrocytes with the same properties were observed in both the treated and untreated mice, with only the proportion of activated astrocytes increased. The authors discuss the possibility that this reflects two distinct but interconvertible populations of astrocytes, and that there might be clinical benefit to regulating their interconversion.

Together, the 22 papers in this Special Issue describe 22 different ways in which astrocytes are implicated in CNS disorders. This is probably just the tip of the iceberg, but a tip worth recognizing.

Author Contributions: M.B. and V.P. contributed equally to all aspects of producing this Editorial. All authors have read and agreed to the published version of the manuscript.

Funding: VP's work is supported by the Cheung Kong (Yangtze River) Scholar Program, Ministry of Education, China and by the Kun Peng Action of Zhejiang Province, China.

Acknowledgments: We thank all the contributors to this Special Issue.

Conflicts of Interest: The authors declare no conflict of interest.

References

1. Bedner, P.; Steinhäuser, C. Role of impaired astrocyte gap junction coupling in epileptogenesis. *Cells* **2023**, *12*, 1669. [CrossRef]
2. Kruk, P.K.; Nader, K.; Skupien-Jaroszek, A.; Wójtowicz, T.; Buszka, A.; Olech-Kochańczyk, G.; Wilczynski, G.M.; Worch, R.; Kalita, K.; Włodarczyk, J.; et al. Astrocytic CD44 deficiency reduces the severity of kainate-induced epilepsy. *Cells* **2023**, *12*, 1483. [CrossRef]
3. Valori, C.F.; Sulmona, C.; Brambilla, L.; Rossi, D. Astrocytes: Dissecting their diverse roles in amyotrophic lateral sclerosis and frontotemporal dementia. *Cells* **2023**, *12*, 1450. [CrossRef]
4. Man, J.H.K.; van Gelder, C.; Breur, M.; Okkes, D.; Molenaar, D.; van der Sluis, S.; Abbink, T.; Altelaar, M.; van der Knaap, M.S.; Bugiani, M. Cortical pathology in vanishing white matter. *Cells* **2022**, *11*, 3581. [CrossRef] [PubMed]
5. Brignone, M.S.; Lanciotti, A.; Michelucci, A.; Mallozzi, C.; Camerini, S.; Catacuzzeno, L.; Sforna, L.; Caramia, M.; D'Adamo, M.C.; Ceccarini, M.; et al. The CaMKII/MLC1 axis confers Ca^{2+}-dependence to volume-regulated anion channels (VRAC) in astrocytes. *Cells* **2022**, *11*, 2656. [CrossRef]
6. Endo, F.; Kasai, A.; Soto, J.S.; Yu, X.; Qu, Z.; Hashimoto, H.; Gradinaru, V.; Kawaguchi, R.; Khakh, B.S. Molecular basis of astrocyte diversity and morphology across the CNS in health and disease. *Science* **2022**, *378*, eadc9020. [CrossRef]
7. Virchow, R. *Die Cellularpathologie in Ihrer Begründung auf Physiologische and Pathologische Gewebelehre. Zwanzig Vorlesungen Gehalten Während der Monate Februar, März und April 1858 im Pathologischen Institut zu Berlin*, 1st ed.; August Hirschwald: Berlin, Germany, 1858; p. 440.
8. Jastrowitz, M. Studien über die encephalitis und myelitis des ersten kindesalters. *Arch. Psychiatr. Nervenkrankh.* **1872**, *3*, 162–213. [CrossRef]
9. Andriezen, W.L. The neuroglia elements of the brain. *Br. Med. J.* **1893**, *2*, 227–230. [CrossRef] [PubMed]
10. Kölliker, A. *Handbuch der Gewebelchre des Menschen*; Wilhelm Engelmann: Leipzig, Germany, 1893.
11. Baldwin, K.T.; Murai, K.K.; Khakh, B.S. Astrocyte morphology. *Trends Cell Biol.* **2024**, *34*, 547–565. [CrossRef] [PubMed]
12. Lenhossék, M.V. *Der Feinere bau des Nervensystems im Lichte Neuester Forschung*, 2nd ed.; Fischer's Medicinische Buchhandlung H. Kornfield: Berlin, Germany, 1895.
13. His, W. Die Formentwickelung des Menschlichen Vorderhirns vom Ende des Ersten bis Zum Beginn des Dritten Moknats. *Abh. Kgl. Sachs. Ges. Wissensch. Math. Phys. Kl* **1889**, *15*, 673–736.
14. His, W. Die Neuroblasten und Deren Entstehung im Embryonalen Mark. *Abh. Kgl. Sachs. Ges. Wissensch. Math. Phys. Kl* **1889**, *15*, 311–372.
15. Ramón y Cajal, S. Un nuevo proceder para la impregnación de la neuroglía. *Bol. Soc. Esp. Biol.* **1913**, *II*, 104–108.
16. Kimelberg, H.K. The problem of astrocyte identity. *Neurochem. Int.* **2004**, *45*, 191–202. [CrossRef] [PubMed]

17. Kiernan, J.A. *Histological & Histochemical Methods: Theory & Practice*, 2nd ed.; Pergamon Press: Oxford, UK, 1990.
18. Gottipati, M.K.; Bekyarova, E.; Brenner, M.; Haddon, R.C.; Parpura, V. Changes in the morphology and proliferation of astrocytes induced by two modalities of chemically functionalized single-walled carbon nanotubes are differentially mediated by glial fibrillary acidic protein. *Nano Lett.* **2014**, *14*, 3720–3727. [CrossRef] [PubMed]
19. Golgi, C. *Sulla Fina Anatomia Degli Organi Centrali del Sistema Nervoso*; Calderini e Figlio: Reggio-Emila, Italy, 1895.
20. Lugaro, E. Sulle funzioni della nevroglia. *Riv. Patol. Nerv. E Ment.* **1907**, *12*, 225–233.
21. Parpura, V.; Verkhratsky, A. Astrocytes revisited: Concise historic outlook on glutamate homeostasis and signaling. *Croat. Med. J.* **2012**, *53*, 518–528. [CrossRef]
22. Mennerick, S.; Zorumski, C.F. Glial contributions to excitatory neurotransmission in cultured hippocampal cells. *Nature* **1994**, *368*, 59–62. [CrossRef]
23. Danbolt, N.C. Glutamate uptake. *Prog. Neurobiol.* **2001**, *65*, 1–105.
24. Martinez-Hernandez, A.; Bell, K.P.; Norenberg, M.D. Glutamine synthetase: Glial localization in brain. *Science* **1977**, *195*, 1356–1358. [CrossRef]
25. Kettenmann, H.; Faissner, A.; Trotter, J. Neuron-glia interactions in homeostasis and degeneration. In *Comprehensive Human Physiology*; Greger, R., Windhorst, U., Eds.; Springer: Berlin/Heidelberg, Germany, 1996; Volume I, pp. 533–543.
26. Parpura, V.; Basarsky, T.A.; Liu, F.; Jeftinija, K.; Jeftinija, S.; Haydon, P.G. Glutamate-mediated astrocyte-neuron signalling. *Nature* **1994**, *369*, 744–747. [CrossRef]
27. Araque, A.; Parpura, V.; Sanzgiri, R.P.; Haydon, P.G. Glutamate-dependent astrocyte modulation of synaptic transmission between cultured hippocampal neurons. *Eur. J. Neurosci.* **1998**, *10*, 2129–2142. [CrossRef] [PubMed]
28. Araque, A.; Parpura, V.; Sanzgiri, R.P.; Haydon, P.G. Tripartite synapses: Glia, the unacknowledged partner. *Trends Neurosci.* **1999**, *22*, 208–215. [CrossRef] [PubMed]
29. Glees, P. *Neuroglia Morphology and Function*; Blackwell: Oxford, UK, 1955.
30. Held, H. Über die neuroglia marginalis der menschlichen grosshirnrinde. *Monatschr. Psychol. Neurol.* **1909**, *26*, 360–416. [CrossRef]
31. Nageotte, J. Phenomenes de secretion dans le protoplasma des cellules nevrogliques de la substance grise. *C. R. Soc. Biol.* **1910**, *68*, 1068–1069.
32. Shain, W.G.; Martin, D.L. Activation of beta-adrenergic receptors stimulates taurine release from glial cells. *Cell. Mol. Neurobiol.* **1984**, *4*, 191–196. [CrossRef] [PubMed]
33. Verkhratsky, A.; Matteoli, M.; Parpura, V.; Mothet, J.P.; Zorec, R. Astrocytes as secretory cells of the central nervous system: Idiosyncrasies of vesicular secretion. *EMBO J.* **2016**, *35*, 239–257. [CrossRef]
34. Pasantes-Morales, H.; Schousboe, A. Volume regulation in astrocytes: A role for taurine as an osmoeffector. *J. Neurosci. Res.* **1988**, *20*, 503–509.
35. Cotrina, M.L.; Lin, J.H.; Alves-Rodrigues, A.; Liu, S.; Li, J.; Azmi-Ghadimi, H.; Kang, J.; Naus, C.C.; Nedergaard, M. Connexins regulate calcium signaling by controlling ATP release. *Proc. Natl. Acad. Sci. USA* **1998**, *95*, 15735–15740. [CrossRef]
36. Duan, S.; Anderson, C.M.; Keung, E.C.; Chen, Y.; Chen, Y.; Swanson, R.A. $P2X_7$ receptor-mediated release of excitatory amino acids from astrocytes. *J. Neurosci.* **2003**, *23*, 1320–1328. [CrossRef]
37. Szatkowski, M.; Barbour, B.; Attwell, D. Non-vesicular release of glutamate from glial cells by reversed electrogenic glutamate uptake. *Nature* **1990**, *348*, 443–446. [CrossRef]
38. Warr, O.; Takahashi, M.; Attwell, D. Modulation of extracellular glutamate concentration in rat brain slices by cystine-glutamate exchange. *J. Physiol.* **1999**, *514 Pt 3*, 783–793. [CrossRef] [PubMed]
39. Rosenberg, P.A.; Knowles, R.; Knowles, K.P.; Li, Y. Beta-adrenergic receptor-mediated regulation of extracellular adenosine in cerebral cortex in culture. *J. Neurosci.* **1994**, *14*, 2953–2965. [CrossRef]
40. Schleich, C.L. *Schmerzlose Operationen: Örtliche Betäubung Mit Indiffrenten Flüssigkeiten. Psychophysik des Natürlichen und Künstlichen Schlafes*; Julius Springer: Berlin/Heidelberg, Germany, 1894; p. 256.
41. Ramón y Cajal, S. *Algunas Conjeturas Sobre el Mechanismoanatomico de la Ideacion, Asociacion y Atencion*; Imprenta y Libreria de Nicolas Moya: Madrid, Spain, 1895.
42. Ramón y Cajal, S. Contribution a la connaissance de la nevroglia cerebrale et cerebeleuse dans la paralyse generale progressive. *Trab. Lab. Investig. Biol. Univ. Madr.* **1925**, *23*, 157–216.
43. Hatton, G.I. Glial-neuronal interactions in the mammalian brain. *Adv. Physiol. Educ.* **2002**, *26*, 225–237. [CrossRef]
44. Brenner, M.; Johnson, A.B.; Boespflug-Tanguy, O.; Rodriguez, D.; Goldman, J.E.; Messing, A. Mutations in GFAP, encoding glial fibrillary acidic protein, are associated with Alexander disease. *Nat. Genet.* **2001**, *27*, 117–120. [CrossRef]
45. Hagemann, T.L.; Coyne, S.; Levin, A.; Wang, L.; Feany, M.B.; Messing, A. STAT3 drives GFAP accumulation and astrocyte pathology in a mouse model of Alexander disease. *Cells* **2023**, *12*, 978. [CrossRef] [PubMed]
46. van der Knaap, M.S.; Pronk, J.C.; Scheper, G.C. Vanishing white matter disease. *Lancet Neurol.* **2006**, *5*, 413–423. [CrossRef] [PubMed]
47. Scali, O.; Di Perri, C.; Federico, A. The spectrum of mutations for the diagnosis of vanishing white matter disease. *Neurol. Sci.* **2006**, *27*, 271–277. [CrossRef]
48. Nethisinghe, S.; Pigazzini, M.L.; Pemble, S.; Sweeney, M.G.; Labrum, R.; Manso, K.; Moore, D.; Warner, J.; Davis, M.B.; Giunti, P. PolyQ tract toxicity in SCA1 is length dependent in the absence of CAG repeat interruption. *Front. Cell. Neurosci.* **2018**, *12*, 200. [CrossRef]

49. Cvetanovic, M.; Ingram, M.; Orr, H.; Opal, P. Early activation of microglia and astrocytes in mouse models of spinocerebellar ataxia type 1. *Neuroscience* **2015**, *289*, 289–299. [CrossRef]
50. Kim, J.H.; Lukowicz, A.; Qu, W.; Johnson, A.; Cvetanovic, M. Astroglia contribute to the pathogenesis of spinocerebellar ataxia Type 1 (SCA1) in a biphasic, stage-of-disease specific manner. *Glia* **2018**, *66*, 1972–1987. [CrossRef] [PubMed]
51. Rosa, J.G.; Hamel, K.; Sheeler, C.; Borgenheimer, E.; Gilliat, S.; Soles, A.; Ghannoum, F.J.; Sbrocco, K.; Handler, H.P.; Rainwater O.; et al. Spatial and temporal diversity of astrocyte phenotypes in spinocerebellar ataxia type 1 mice. *Cells* **2022**, *11*, 3323 [CrossRef] [PubMed]
52. Obel, L.F.; Müller, M.S.; Walls, A.B.; Sickmann, H.M.; Bak, L.K.; Waagepetersen, H.S.; Schousboe, A. Brain glycogen-new perspectives on its metabolic function and regulation at the subcellular level. *Front. Neuroenerg.* **2012**, *4*, 3. [CrossRef]
53. Suzuki, A.; Stern, S.A.; Bozdagi, O.; Huntley, G.W.; Walker, R.H.; Magistretti, P.J.; Alberini, C.M. Astrocyte-neuron lactate transport is required for long-term memory formation. *Cell* **2011**, *144*, 810–823. [CrossRef]
54. López-Ramos, J.C.; Duran, J.; Gruart, A.; Guinovart, J.J.; Delgado-García, J.M. Role of brain glycogen in the response to hypoxia and in susceptibility to epilepsy. *Front. Cell. Neurosci.* **2015**, *9*, 431. [CrossRef]
55. Montana, V.; Flint, D.; Waagepetersen, H.S.; Schousboe, A.; Parpura, V. Two metabolic fuels, glucose and lactate, differentially modulate exocytotic glutamate release from cultured astrocytes. *Neurochem. Res.* **2021**, *46*, 2551–2579.
56. Duran, J. Role of astrocytes in the pathophysiology of Lafora disease and other glycogen storage disorders. *Cells* **2023**, *12*, 722 [CrossRef] [PubMed]
57. Kim, S.; Pajarillo, E.; Nyarko-Danquah, I.; Aschner, M.; Lee, E. Role of astrocytes in Parkinson's disease associated with genetic mutations and neurotoxicants. *Cells* **2023**, *12*, 622. [CrossRef] [PubMed]
58. Braak, H.; de Vos, R.A.; Jansen, E.N.; Bratzke, H.; Braak, E. Neuropathological hallmarks of Alzheimer's and Parkinson's diseases *Prog. Brain Res.* **1998**, *117*, 267–285.
59. Selkoe, D.J. Alzheimer's disease: Genes, proteins, and therapy. *Physiol. Rev.* **2001**, *81*, 741–766. [CrossRef]
60. Olabarria, M.; Noristani, H.N.; Verkhratsky, A.; Rodriguez, J.J. Concomitant astroglial atrophy and astrogliosis in a triple transgenic animal model of Alzheimer's disease. *Glia* **2010**, *58*, 831–838. [CrossRef] [PubMed]
61. Rodriguez, J.J.; Verkhratsky, A. Neuroglial roots of neurodegenerative diseases? *Mol. Neurobiol.* **2011**, *43*, 87–96. [CrossRef]
62. Fuller, S.; Munch, G.; Steele, M. Activated astrocytes: A therapeutic target in Alzheimer's disease? *Expert Rev. Neurother.* **2009**, *9*, 1585–1594. [CrossRef]
63. Kelly, P.; Sanchez-Mico, M.V.; Hou, S.S.; Whiteman, S.; Russ, A.; Hudry, E.; Arbel-Ornath, M.; Greenberg, S.M.; Bacskai, B.J. Neuronally derived soluble Abeta evokes cell-wide astrocytic calcium dysregulation in absence of amyloid plaques in vivo. *J. Neurosci.* **2023**, *43*, 4926–4940. [CrossRef]
64. Fontana, I.C.; Scarpa, M.; Malarte, M.L.; Rocha, F.M.; Ausellé-Bosch, S.; Bluma, M.; Bucci, M.; Chiotis, K.; Kumar, A.; Nordberg, A. Astrocyte signature in Alzheimer's disease continuum through a multi-PET tracer imaging perspective. *Cells* **2023**, *12*, 1469. [CrossRef]
65. Michinaga, S.; Hishinuma, S.; Koyama, Y. Roles of astrocytic endothelin ET_B receptor in traumatic brain injury. *Cells* **2023**, *12*, 719. [CrossRef]
66. Rao, V.L.; Başkaya, M.K.; Doğan, A.; Rothstein, J.D.; Dempsey, R.J. Traumatic brain injury down-regulates glial glutamate transporter (GLT-1 and GLAST) proteins in rat brain. *J. Neurochem.* **1998**, *70*, 2020–2027. [PubMed]
67. Gržeta Krpan, N.; Harej Hrkać, A.; Janković, T.; Dolenec, P.; Bekyarova, E.; Parpura, V.; Pilipović, K. Chemically functionalized single-walled carbon nanotubes prevent the reduction in plasmalemmal glutamate transporter EAAT1 expression in, and increase the release of selected cytokines from, stretch-injured astrocytes in vitro. *Cells* **2024**, *13*, 225. [CrossRef]
68. Garcia, T.A.; Jonak, C.R.; Binder, D.K. The role of aquaporins in spinal cord injury. *Cells* **2023**, *12*, 1701. [CrossRef] [PubMed]
69. Bosch, A.; Estévez, R. Megalencephalic leukoencephalopathy: Insights into pathophysiology and perspectives for therapy. *Front. Cell. Neurosci.* **2020**, *14*, 627887. [CrossRef]
70. Kraig, R.P.; Petito, C.K.; Plum, F.; Pulsinelli, W.A. Hydrogen ions kill brain at concentrations reached in ischemia. *J. Cereb. Blood Flow Metab.* **1987**, *7*, 379–386. [CrossRef]
71. Kogure, T.; Kogure, K. Molecular and biochemical events within the brain subjected to cerebral ischemia (targets for therapeutical intervention). *Clin. Neurosci.* **1997**, *4*, 179–183.
72. Everaerts, K.; Thapaliya, P.; Pape, N.; Durry, S.; Eitelmann, S.; Roussa, E.; Ullah, G.; Rose, C.R. Inward operation of sodium-bicarbonate cotransporter 1 promotes astrocytic Na^+ loading and loss of ATP in mouse neocortex during brief chemical ischemia. *Cells* **2023**, *12*, 2675. [CrossRef]
73. Dzwonek, J.; Wilczynski, G.M. CD44: Molecular interactions, signaling and functions in the nervous system. *Front. Cell. Neurosci.* **2015**, *9*, 175. [CrossRef] [PubMed]
74. Al-Dalahmah, O.; Sosunov, A.A.; Sun, Y.; Liu, Y.; Madden, N.; Connolly, E.S.; Troy, C.M.; McKhann, G.M., 2nd; Goldman, J.E. The matrix receptor CD44 is present in astrocytes throughout the human central nervous system and accumulates in hypoxia and seizures. *Cells* **2024**, *13*, 129. [CrossRef] [PubMed]
75. Kempuraj, D.; Aenlle, K.K.; Cohen, J.; Mathew, A.; Isler, D.; Pangeni, R.P.; Nathanson, L.; Theoharides, T.C.; Klimas, N.G. COVID-19 and long COVID: Disruption of the neurovascular unit, blood-brain barrier, and tight junctions. *Neuroscientist* **2023**, *30*, 421–439. [CrossRef] [PubMed]

76. Mhatre, R.; Mahadevan, A.; Shankar, S.K. Biology of astrocytes in CNS infection. In *The Biology of Glial Cells: Recent Advances*; Patro, I., Seth, P., Patro, N., Tandon, P.N., Eds.; Springer: Singapore, 2022; pp. 171–198.
77. Potokar, M.; Zorec, R.; Jorgačevski, J. Astrocytes are a key target for neurotropic viral infection. *Cells* **2023**, *12*, 2307. [CrossRef] [PubMed]
78. Pavlou, M.A.S.; Singh, K.; Ravichandran, S.; Halder, R.; Nicot, N.; Birck, C.; Grandbarbe, L.; Del Sol, A.; Michelucci, A. Transcriptional and chromatin accessibility profiling of neural stem cells differentiating into astrocytes reveal dynamic signatures affected under inflammatory conditions. *Cells* **2023**, *12*, 948. [CrossRef]
79. Ede, R.J.; Williams, R.W. Hepatic encephalopathy and cerebral edema. *Semin. Liver Dis.* **1986**, *6*, 107–118. [CrossRef]
80. Felipo, V. Hepatic encephalopathy: Effects of liver failure on brain function. *Nat. Rev. Neurosci.* **2013**, *14*, 851–858. [CrossRef]
81. Link, A.; Kaplan, B.T.; Bohm, M. 21-jährige Patientin mit Reye-Syndrom nach grippalem Infekt. *Dtsch. Med. Wochenschr.* **2012**, *137*, 1853–1856. [CrossRef]
82. Rama Rao, K.V.; Jayakumar, A.R.; Norenberg, M.D. Brain edema in acute liver failure: Mechanisms and concepts. *Metab. Brain Dis.* **2014**, *29*, 927–936. [CrossRef] [PubMed]
83. Norenberg, M.D. The role of astrocytes in hepatic encephalopathy. *Neurochem. Pathol.* **1987**, *6*, 13–33. [CrossRef]
84. Rose, C.F.; Verkhratsky, A.; Parpura, V. Astrocyte glutamine synthetase: Pivotal in health and disease. *Biochem. Soc. Trans.* **2013**, *41*, 1518–1524. [CrossRef] [PubMed]
85. Albrecht, J.; Zielinska, M.; Norenberg, M.D. Glutamine as a mediator of ammonia neurotoxicity: A critical appraisal. *Biochem. Pharmacol.* **2010**, *80*, 1303–1308. [PubMed]
86. Obara-Michlewska, M.; Ruszkiewicz, J.; Zielinska, M.; Verkhratsky, A.; Albrecht, J. Astroglial NMDA receptors inhibit expression of Kir4.1 channels in glutamate-overexposed astrocytes in vitro and in the brain of rats with acute liver failure. *Neurochem. Int.* **2014**, *88*, 20–25. [CrossRef]
87. Rangroo Thrane, V.; Thrane, A.S.; Wang, F.; Cotrina, M.L.; Smith, N.A.; Chen, M.; Xu, Q.; Kang, N.; Fujita, T.; Nagelhus, E.A.; et al. Ammonia triggers neuronal disinhibition and seizures by impairing astrocyte potassium buffering. *Nat. Med.* **2013**, *19*, 1643–1648. [CrossRef]
88. Haack, N.; Dublin, P.; Rose, C.R. Dysbalance of astrocyte calcium under hyperammonemic conditions. *PLoS ONE* **2014**, *9*, e105832. [CrossRef] [PubMed]
89. Kelly, T.; Kafitz, K.W.; Roderigo, C.; Rose, C.R. Ammonium-evoked alterations in intracellular sodium and pH reduce glial glutamate transport activity. *Glia* **2009**, *57*, 921–934. [CrossRef]
90. Görg, B.; Morwinsky, A.; Keitel, V.; Qvartskhava, N.; Schrör, K.; Häussinger, D. Ammonia triggers exocytotic release of L-glutamate from cultured rat astrocytes. *Glia* **2010**, *58*, 691–705. [CrossRef]
91. Sepehrinezhad, A.; Stolze Larsen, F.; Ashayeri Ahmadabad, R.; Shahbazi, A.; Sahab Negah, S. The glymphatic system may play a vital role in the pathogenesis of hepatic encephalopathy: A narrative review. *Cells* **2023**, *12*, 979. [CrossRef]
92. Iliff, J.J.; Wang, M.; Liao, Y.; Plogg, B.A.; Peng, W.; Gundersen, G.A.; Benveniste, H.; Vates, G.E.; Deane, R.; Goldman, S.A.; et al. A paravascular pathway facilitates CSF flow through the brain parenchyma and the clearance of interstitial solutes, including amyloid β. *Sci. Transl. Med.* **2012**, *4*, 147ra111. [CrossRef] [PubMed]
93. Kruyer, A.; Scofield, M.D. Astrocytes in addictive disorders. *Adv. Neurobiol.* **2021**, *26*, 231–254.
94. Siemsen, B.M.; Denton, A.R.; Parrila-Carrero, J.; Hooker, K.N.; Carpenter, E.A.; Prescot, M.E.; Brock, A.G.; Westphal, A.M.; Leath, M.N.; McFaddin, J.A.; et al. Heroin self-administration and extinction increase prelimbic cortical astrocyte-synapse proximity and alter dendritic spine morphometrics that are reversed by N-acetylcysteine. *Cells* **2023**, *12*, 1812. [CrossRef] [PubMed]
95. Moussawi, K.; Zhou, W.; Shen, H.; Reichel, C.M.; See, R.E.; Carr, D.B.; Kalivas, P.W. Reversing cocaine-induced synaptic potentiation provides enduring protection from relapse. *Proc. Natl. Acad. Sci. USA* **2011**, *108*, 385–390. [CrossRef] [PubMed]
96. Reissner, K.J.; Gipson, C.D.; Tran, P.K.; Knackstedt, L.A.; Scofield, M.D.; Kalivas, P.W. Glutamate transporter GLT-1 mediates N-acetylcysteine inhibition of cocaine reinstatement. *Addict. Biol.* **2015**, *20*, 316–323. [CrossRef]
97. Kruyer, A.; Scofield, M.D.; Wood, D.; Reissner, K.J.; Kalivas, P.W. Heroin cue-evoked astrocytic structural plasticity at nucleus accumbens synapses inhibits heroin seeking. *Biol. Psychiatry* **2019**, *86*, 811–819. [CrossRef]
98. Murlanova, K.; Jouroukhin, Y.; Novototskaya-Vlasova, K.; Huseynov, S.; Pletnikova, O.; Morales, M.J.; Guan, Y.; Kamiya, A.; Bergles, D.E.; Dietz, D.M.; et al. Loss of astrocytic μ opioid receptors exacerbates aversion associated with morphine withdrawal in mice: Role of mitochondrial respiration. *Cells* **2023**, *12*, 1412. [CrossRef]
99. Zhang, X.; Wolfinger, A.; Wu, X.; Alnafisah, R.; Imami, A.; Hamoud, A.R.; Lundh, A.; Parpura, V.; McCullumsmith, R.E.; Shukla, R.; et al. Gene enrichment analysis of astrocyte subtypes in psychiatric disorders and psychotropic medication datasets. *Cells* **2022**, *11*, 3315. [CrossRef]
100. Tham, Y.C.; Li, X.; Wong, T.Y.; Quigley, H.A.; Aung, T.; Cheng, C.Y. Global prevalence of glaucoma and projections of glaucoma burden through 2040: A systematic review and meta-analysis. *Ophthalmology* **2014**, *121*, 2081–2090. [CrossRef]
101. Zhu, Y.; Wang, R.; Pappas, A.C.; Seifert, P.; Savol, A.; Sadreyev, R.I.; Sun, D.; Jakobs, T.C. Astrocytes in the optic nerve are heterogeneous in their reactivity to glaucomatous injury. *Cells* **2023**, *12*, 2131. [CrossRef]

cells

MDPI

Article

STAT3 Drives GFAP Accumulation and Astrocyte Pathology in a Mouse Model of Alexander Disease

Tracy L. Hagemann [1,*], Sierra Coyne [1], Alder Levin [1], Liqun Wang [2], Mel B. Feany [3] and Albee Messing [1,4]

1 Waisman Center, University of Wisconsin-Madison, Madison, WI 53705, USA; scoyne4@wisc.edu (S.C.); alevin5@wisc.edu (A.L.); amessing@wisc.edu (A.M.)
2 Wyss Institute, Harvard University, Boston, MA 02115, USA; liqun.wang@wyss.harvard.edu
3 Department of Pathology, Brigham and Women's Hospital, Harvard Medical School, Boston, MA 02115, USA; mel_feany@hms.harvard.edu
4 Department of Comparative Biosciences, School of Veterinary Medicine, University of Wisconsin-Madison, Madison, WI 53705, USA
* Correspondence: tlhagemann@wisc.edu

Abstract: Alexander disease (AxD) is caused by mutations in the gene for glial fibrillary acidic protein (GFAP), an intermediate filament expressed by astrocytes in the central nervous system. AxD-associated mutations cause GFAP aggregation and astrogliosis, and GFAP is elevated with the astrocyte stress response, exacerbating mutant protein toxicity. Studies in mouse models suggest disease severity is tied to *Gfap* expression levels, and signal transducer and activator of transcription (STAT)-3 regulates *Gfap* during astrocyte development and in response to injury and is activated in astrocytes in rodent models of AxD. In this report, we show that STAT3 is also activated in the human disease. To determine whether STAT3 contributes to GFAP elevation, we used a combination of genetic approaches to knockout or reduce STAT3 activation in AxD mouse models. Conditional knockout of *Stat3* in cells expressing *Gfap* reduced *Gfap* transactivation and prevented protein accumulation. Astrocyte-specific *Stat3* knockout in adult mice with existing pathology reversed GFAP accumulation and aggregation. Preventing STAT3 activation reduced markers of reactive astrocytes, stress-related transcripts, and microglial activation, regardless of disease stage or genetic knockout approach. These results suggest that pharmacological inhibition of STAT3 could potentially reduce GFAP toxicity and provide a therapeutic benefit in patients with AxD.

Keywords: GFAP; STAT3; astrocyte; Alexander disease; Rosenthal fibers; reactive gliosis

Citation: Hagemann, T.L.; Coyne, S.; Levin, A.; Wang, L.; Feany, M.B.; Messing, A. STAT3 Drives GFAP Accumulation and Astrocyte Pathology in a Mouse Model of Alexander Disease. *Cells* **2023**, *12*, 978. https://doi.org/10.3390/cells12070978

Academic Editors: Michael Brenner and Vladimir Parpura

Received: 30 December 2022
Revised: 18 March 2023
Accepted: 20 March 2023
Published: 23 March 2023

1. Introduction

Alexander disease (AxD) is a progressive and generally fatal disorder of the central nervous system (CNS), with a range of clinical phenotypes including cognitive and motor impairment and white matter deficits that are distributed throughout the neuroaxis [1]. AxD has been subclassified into two groups, with type I demonstrating frontal predominance and early onset, and type II presenting with lesions concentrated in the hindbrain and cervical spinal cord, less pronounced leukodystrophy and cognitive deficits, and onset at any age [2]. The hallmark pathological feature in both groups is the presence of Rosenthal fibers, cytoplasmic protein aggregates of astrocytes, particularly in perivascular, subpial, and subependymal locations.

Nearly all cases of AxD result from heterozygous missense mutations in the astrocyte intermediate filament, GFAP [3,4]. To better understand how mutations in GFAP cause disease, we generated mouse models with knock-in mutations causing single amino acid changes orthologous to common human mutations at the endogenous *Gfap* locus [5]. These models replicate many of the pathological characteristics found in the human disease. AxD mice spontaneously increase expression of *Gfap* above normal levels, form the characteristic

Rosenthal fibers within the cytoplasm of astrocytes, induce stress responses including Nrf2/ARE and small heat shock protein pathways, and activate microglia [6].

A major factor in the pathology induced by mutant GFAP in AxD is accumulation of GFAP protein to toxic levels. Even overexpression of wild-type GFAP causes protein aggregation, and at high enough levels, is lethal [7]. The cascade of stress response signaling pathways associated with protein misfolding leads to reactive gliosis with increased GFAP expression, which is in part due to transactivation of the *Gfap* promoter [8]. The transcription factor STAT3 regulates astrocyte development and astrogliosis and directly controls transcription of *Gfap* [9–13]. STAT3 binds to the *Gfap* promoter and is necessary to induce expression [13,14], and STAT3 activation has been shown to drive reactive gliosis in several models of acute injury [11,12,15] and neurodegeneration [16,17].

Although the disease process in AxD is initiated by expression of mutant protein, the resulting astrogliosis and transactivation of the *Gfap* promoter further promote protein accumulation and toxicity [8]. We have shown that STAT3 is phosphorylated in both mouse and rat models of AxD [18,19] and hypothesize that activation of STAT3 contributes to a positive feedback loop that drives increased GFAP expression, protein accumulation, and ultimately astrocyte dysfunction. Here, we used a combination of genetic approaches to manipulate STAT3 activation and clarify its involvement in chronic gliosis and to test whether blocking phosphorylation reduces astrocyte pathology in mouse models of the disease. Therapeutic approaches to target activation of the JAK/STAT pathway have been approved to treat inflammatory disorders, hematopoietic malignancies, and solid tumors [20,21], and inhibitors that cross the blood-brain barrier could potentially be repurposed to benefit patients with AxD.

2. Materials and Methods

2.1. Mice

All animal studies were approved by the College of Letters and Sciences and Vice Chancellor Office for Research and Graduate Education Animal Care and Use Committee within the University of Wisconsin-Madison. *Gfap*$^{+/\mathrm{R236H}}$ mice (*Gfap*$^{\mathrm{tm2Mes}}$) were maintained as heterozygotes congenic in either FVB/N-Tac or C57BL/6J genetic backgrounds [5]. *Stat3*$^{\mathrm{fl/fl}}$ mice (*Stat3*$^{\mathrm{tm2Aki}}$) [22], engineered with loxP sites flanking the exon coding for the Tyr-705 phosphorylation site, were congenic in C57BL/6J and maintained as homozygotes. Tg(Gfap-cre)73.12Mvs (C57BL/6J, JAX stock #012886) [23] and Tg(Aldh1l1-cre/ERT2)1Khakh (C57BL/6NJ, JAX stock #031008) [24] lines were maintained as hemizygotes. Ai14 Cre reporter mice (Gt(ROSA)26Sor$^{\mathrm{tm14(CAG\text{-}tdTomato)Hze}}$/J, C57BL/6J, JAX stock #007914) [25] were maintained as heterozygotes. Heterozygous constitutive *Stat3*$^{+/-}$ knockout mice were generated by crossing a nestin-promotor-driven Cre line (Tg(Nes-cre)1Kln, C57BL/6J, JAX stock #003771) [26] with *Stat3*$^{\mathrm{fl/fl}}$ mice for germline excision of the floxed allele. Cre excision removes the exon containing the Tyr-705 phosphorylation site while leaving the remaining coding sequence in frame and the translated protein intact. The genes for STAT3 and GFAP are located within a 2 megabase interval on mouse chromosome 11 and required homologous recombination between the two genes to combine the floxed *Stat3* allele with the R236H *Gfap* mutation. These mice were maintained by crossing *Gfap*$^{+/\mathrm{R236H}}$; *Stat3*$^{\mathrm{fl/fl}}$ mice with *Stat3*$^{\mathrm{fl/fl}}$ mice. Genotyping was performed by Transnetyx using real time PCR to analyze tail biopsy DNA.

2.2. Human Brain Tissue

Frozen frontal cortex from a 1-year-old AxD patient was obtained from the NICHD Brain and Tissue Bank for Developmental Disorders at the University of Maryland, Baltimore, MD, USA. Tissue was thawed and fixed in 4% paraformaldehyde overnight before paraffin embedding for immunolabeling.

2.3. Tamoxifen Induction

For inducible Cre activation in mice expressing the *Aldh1l1*-CreER^{T2} transgene, all mice were given intraperitoneal injections of 80 mg/kg tamoxifen (Sigma T5648, dissolved in corn oil at 20 mg/mL) daily for 5 days starting at approximately P56. Personnel administering the injections were blinded to genotypes. Tissues were collected for molecular and histological analysis at 3 months of age (~P90) as described below.

2.4. Tissue Collection

For molecular analysis of crosses including *Gfap*$^{+/R236H}$; *Stat3*$^{+/-}$ or *Gfap*$^{+/R236H}$; *Stat3*$^{-/fl}$; *Gfap*-Cre$^{Tg73.12}$ mice, animals were euthanized by CO_2 asphyxiation, and tissues were collected on ice and frozen immediately with dry ice and stored at −80 °C until processed. For experiments performed with crosses to generate *Gfap*$^{+/R236H}$; *Stat3*$^{fl/fl}$; *Gfap*-Cre$^{Tg73.12}$ mice, animals were first anesthetized with isoflurane and transcardially perfused with phosphate buffered saline (PBS, pH 7.4). Brains were collected and bisected at the midline: half was placed in 4% paraformaldehyde for histological analysis, and half was microdissected and frozen for molecular analysis.

For experiments performed with mice from the *Gfap*$^{+/R236H}$; *Stat3*$^{fl/fl}$; Aldh1l1-CreER^{T2}; Ai14 cross, animals that were positive for Ai14 were anesthetized with isoflurane and transcardially perfused with phosphate buffer, followed by 4% paraformaldehyde for histological analysis. Animals that were Ai14 negative were euthanized by CO_2 asphyxiation, and tissues were collected on ice and frozen immediately with dry ice and stored at −80 °C until processed.

2.5. RNA Extraction and Quantitative PCR

Tissues were homogenized in Trizol reagent (Thermo Fisher Scientific, Invitrogen, Waltham, MA, USA) for RNA analysis per the manufacturer's protocol. For quantitative PCR, 1 µg RNA was transcribed as cDNA using 1 µm random hexamers, 2 µm poly-(dT)15 primers and 500 nm (each) dNTP in a 20 µL reaction with Superscript III (Thermo Fisher Scientific, Invitrogen protocol). Reverse transcriptase reactions were diluted and incorporated into quantitative PCR at a ratio of 16 nl cDNA per µL reaction using SYBR Green Master Mix with 150 nM primers (Table 1) on an ABI ViiA7 Real-Time PCR System (Thermo Fisher Scientific, Applied Biosystems Inc.). Standards generated from purified amplicons were used to determine relative concentrations, and values were normalized to *Rn18s*.

Table 1. Primer sets for quantitative PCR.

Gene	Forward Primer	Reverse Primer
Gfap	CAA CGT TAA GCT AGC CCT GGA CAT	CTC ACC ATC CCG CAT CTC CAC AGT
Aif1	GCC AAA GCA GGG ATT TGC AG	GGG AAC CCC AAG TTT CTC CA
Ccl2	CTG GAG CAT CCA CGT GTT GG	CAT TCC TTC TTG GGG TCA GC
Ccnd1	GCG TAC CCT GAC ACC AAT CTC CT	CTT CGC ACT TCT GCT CCT CAC
Cxcl1	GCC ACA CTC AAG AAT GGT CG	ACC AGA CAG GTG CCA TCA GA
Il6	GAC TTC CAT CCA GTT GCC TTC T	AAG TAG GGA AGG CCG TGG TT
Lcn2	AGA CTT CCG GAG CGA TCA GT	TCT GAT CCA GTA GCG ACA GC
Nqo1	CGG TAT TAC GAT CCT CCC TCA ACA	AGC CTC TAC AGC AGC CTC CTT CAT
Socs3	AAG GCC GGA GAT TTC GCT TC	GGG AAA CTT GCT GTG GGT GA
Stat3	ACG AAA GTC AGG TTG CTG GT	GCT GCC GTT GTT AGA CTC CT
Thbs1	TGC CCT GCC CAC CAC GAT TCA	GCA TAG CCG GGC TTG CAC TCA CA
Rn18s	CGC CGC TAG AGG TGA AAT TCT	CGA ACC TCC GAC TTT CGT TCT

2.6. Protein Extraction and Western Analysis

Total protein lysates were prepared from frozen brain regions by homogenizing with a bead mill mixer (GenoGrinder, SPEX SamplePrep) with 50 mg tissue per ml 2% SDS, 50 mM, Tris-HCl (pH 7.4), 5 mM EDTA, 1 mM Pefabloc SC (Sigma-Aldrich, St. Louis, MO, USA) and Complete Protease Inhibitor Cocktail (Sigma-Aldrich, Roche). Samples were boiled for 20 min, and protein concentration was determined by bicinchoninic acid assay using the BCA Protein Assay kit (Thermo Fisher Scientific) with bovine serum albumin (BSA) as a standard. For western analysis, samples were diluted with lysis buffer to equivalent concentrations and 20–30 μg electrophoresed (125V at room temperature until bromophenol blue reached the bottom of the gel) on 26-well 10% Criterion Tris-HCl or TGX gels (BioRad, Hercules, CA, USA) before transferring to an Immobilon-FL membrane using the BioRad Criterion system. For most experiments, immunoblots were stained with REVERT Total Protein Stain (LI-COR Biosciences, Lincoln, NE, USA) and imaged with a LI-COR Odyssey for subsequent normalization (GAPDH was used for normalization otherwise). Membranes were blocked in SEA Blocking Buffer (Thermo Fisher Scientific, Pierce) before incubating with primary antibodies overnight at 4 °C. Immunoblots were washed before adding secondary antibodies and incubating for 2 h at room temperature, washing again, and imaging with a LI-COR Odyssey. Tris buffered saline (TBS, pH 7.5) with 0.05% Tween 20 was used for all antibody dilutions and washes. A final wash in PBS was performed before imaging. Antibodies are listed in Table 2.

Table 2. Antibodies used in study.

Assay	Antigen	Host	Mono/Poly	Source	Catalog No.	RRID	Dilution
ELISA	GFAP	mouse	mono SMI 26	BioLegend	837602	AB_2565380	1:1000
ELISA	GFAP	rabbit	poly	Dako/Agilent	Z0334	AB_10013382	1:5000
IF	GFAP	mouse	mono GA5	Millipore Sigma	G6171	AB_1840893	1:1000
IF	GFAP	rabbit	poly	Dako/Agilent	Z0334	AB_10013382	1:1000
IF	Iba1	rabbit	poly	Wako	019-19741	AB_839504	1:500
IF	p62/SQSTM1	mouse	mono 2C11	Abnova	H00008878-M01	AB_437085	1:100
IF	STAT3-pY705	rabbit	mono D3A7	Cell Signaling	9145	AB_2491009	1:200
Western	GAPDH	mouse	mono 6C5	Fitzgerald	10R-G109A	AB_1285808	1:10,000
Western	GAPDH	rabbit	poly	Abcam	Ab9485	AB_307275	1:2500
Western	p62/SQSTM1	mouse	mono 2C11	Abnova	H00008878-M01	AB_437085	1:5000
Western	STAT3	mouse	mono 124H6	Cell Signaling	9139S	AB_331757	1:5000
Western	STAT3-pY705	rabbit	mono D3A7	Cell Signaling	9145S	AB_2799407	1:5000

Assay	2nd Antigen	Host	Conjugate	Source	Catalog No.	RRID	Dilution
ELISA	rabbit-IgG	goat	HRP	Millipore Sigma	A6154	AB_258284	1:10,000
IF	mouse-IgG	goat	Alexa Fluor 568	Thermo-Invitrogen	A11031	AB_144696	1:500
IF	mouse-IgG	goat	Alexa Fluor 647	Thermo-Invitrogen	A32728	AB_2633277	1:500
IF	rabbit-IgG	goat	Alexa Fluor 488	Thermo-Invitrogen	A11034	AB_2576217	1:500
IF	mouse-IgG	donkey	Alexa Fluor 488	Thermo-Invitrogen	A21202	AB_141607	1:500
IF	rabbit-IgG	donkey	Alexa Fluor 546	Thermo-Invitrogen	A10040	AB_2534016	1:500
Western	mouse-IgG	goat	Alexa Fluor 680	Thermo-Invitrogen	A21057	AB_2535723	1:10,000
Western	mouse-IgG	goat	DyLight 800	Thermo Scientific	SA5-10176	AB_2556756	1:10,000
Western	rabbit-IgG	goat	Alexa Fluor 680	Thermo-Invitrogen	A21109	AB_2535758	1:10,000
Western	rabbit-IgG	goat	DyLight 800	Thermo Scientific	35571	AB_614947	1:10,000

2.7. GFAP ELISA

GFAP was quantified from total protein extracted with 2% SDS (described above) by a sandwich enzyme linked immunosorbent assay (ELISA) as described previously [8]. Briefly, microtiter plates (Nunc MaxiSorp, Thermo Fisher Scientific, Waltham, MA, USA) were coated with a cocktail of monoclonal GFAP antibodies diluted in PBS (SMI-26, see Table 2 for antibodies) and blocked with BLOTTO (5% milk in PBS) before adding protein samples and standards diluted in PBS with 1% BSA and 0.05% Tween 20 (Tw20). Plates were incubated at room temperature for 2 h and washed with PBS/0.05% Tw20 before adding rabbit anti-GFAP detection antibody (Dako Z0334) diluted in BLOTTO and incubated overnight at 4 °C. Plates were washed with PBS/0.05% Tw20 followed by incubation with secondary HRP conjugated goat anti-rabbit IgG diluted in BLOTTO at room temperature for 2 h. Plates were washed with PBS/0.05%Tw20 followed by PBS before adding SuperSignal ELISA Femto Substrate (Thermo Fisher Scientific, Pierce, Waltham, MA, USA) and analyzing with a GloRunner luminometer (Turner Biosystems).

2.8. Histological Analysis

Mouse brains were collected as described above, fixed in 4% paraformaldehyde overnight before cryoprotecting in graded concentrations of sucrose (10–30%), and sectioning on a sliding microtome at 40 μm. Sections were stored at −20 °C in 25% glycerol/25% ethylene glycol/0.1 M phosphate buffer (pH 7.4) as cryoprotectant. For labeling, tissue sections were rinsed with PBS, blocked and permeabilized with 5% normal goat or donkey serum, 0.5% Triton-X-100 (TX100) in PBS, and incubated in primary antibody (Table 2) diluted in 1% BSA, 0.3% TX100 in PBS for approximately 72 h at 4 °C. Sections were washed in PBS with 0.05% TX100 before incubating in secondary antibody (Table 2) diluted in 1% BSA, 0.3% TX100 in PBS overnight at 4 °C. Finally, sections were washed in PBS/0.05% TX100 and mounted with ProLong Gold mounting media with DAPI (Thermo Fisher Scientific, Invitrogen). For phophoY705-STAT3 immunofluorescence, sections were further permeabilized with 100% methanol for 10 min at −20 °C prior to adding block. Fluorescence images were collected with a Nikon A1R-HD confocal microscope system with equivalent settings for comparisons between genotypes.

2.9. Image Analysis

All fluorescent images were collected with identical settings at 1024 × 1024 pixel resolution for comparisons and quantification, and analysis performed on grayscale images from individual fluorescent channels. For signal intensity, the entire field of view was selected, and the mean pixel intensity measured with ImageJ/Fiji (NIH). Values are expressed as arbitrary units (a.u.) of fluorescence normalized to the control group. To quantify p62 as a representation of protein aggregation and Rosenthal fibers, gray-scale images were converted to RGB, and p62 positive puncta highlighted using the color threshold adjustment in Fiji (default thresholding method, dark background on) by setting the upper limit for brightness to separate high intensity signals from background in the histogram. The highlighted puncta were selected, and the measured area converted to a percentage of the total image area for comparisons.

3. Results

3.1. STAT3 Activation in AxD

Gfap$^{+/R236H}$ mouse models of AxD demonstrate GFAP accumulation and a marked astrocyte stress response. In addition to perivascular, periventricular, and subpial astrocytes, protein aggregation and increased GFAP expression occur predominantly in forebrain regions including hippocampus, corpus callosum and olfactory bulb [5,8]. As a marker of astrogliosis, changes in the *Gfap* transcript show a similar pattern with greater elevation in forebrain compared to hindbrain regions and spinal cord, as previously reported [8] (Figure 1A). STAT3 is known to transactivate *Gfap* expression in reactive gliosis [12,15], and transcriptome analysis of a mouse model of AxD overexpressing human GFAP also suggests JAK/STAT pathway activation [27]. To determine whether STAT3 activation correlates with GFAP elevation in mouse models of AxD, we performed western analysis of STAT3 and phosphorylated STAT3 (Tyr705) in different regions of the CNS and found that STAT3 expression and activation were more pronounced in forebrain regions compared to hindbrain regions and spinal cord, similarly to *Gfap* transcript (Figure 1B,C). Although STAT3 protein was elevated approximately 2-fold, expression levels did not account for the 15- to 40-fold increase in phospho-STAT3. Further analysis showed other genes regulated by STAT3, including *Il6*, *Socs3*, *Cxcl1*, *Thbs1*, *Ccnd1*, as well as *Stat3*, were also elevated in *Gfap*$^{+/\mathrm{R236H}}$ mice (Figure 1D). Immunofluorescence labeling demonstrated nuclear localization of activated pSTAT3 specifically in *Gfap*$^{+/\mathrm{R236H}}$ mouse astrocytes in both forebrain and hindbrain regions (Figure 1E). Importantly, activation of STAT3 was also apparent in brain tissue from a patient with AxD, with nuclear localization of pSTAT3 in astrocytes (Figure 1F). These results suggest that STAT3 contributes to astrocyte reactivity in AxD, but whether the transcription factor is necessary for GFAP accumulation is not clear.

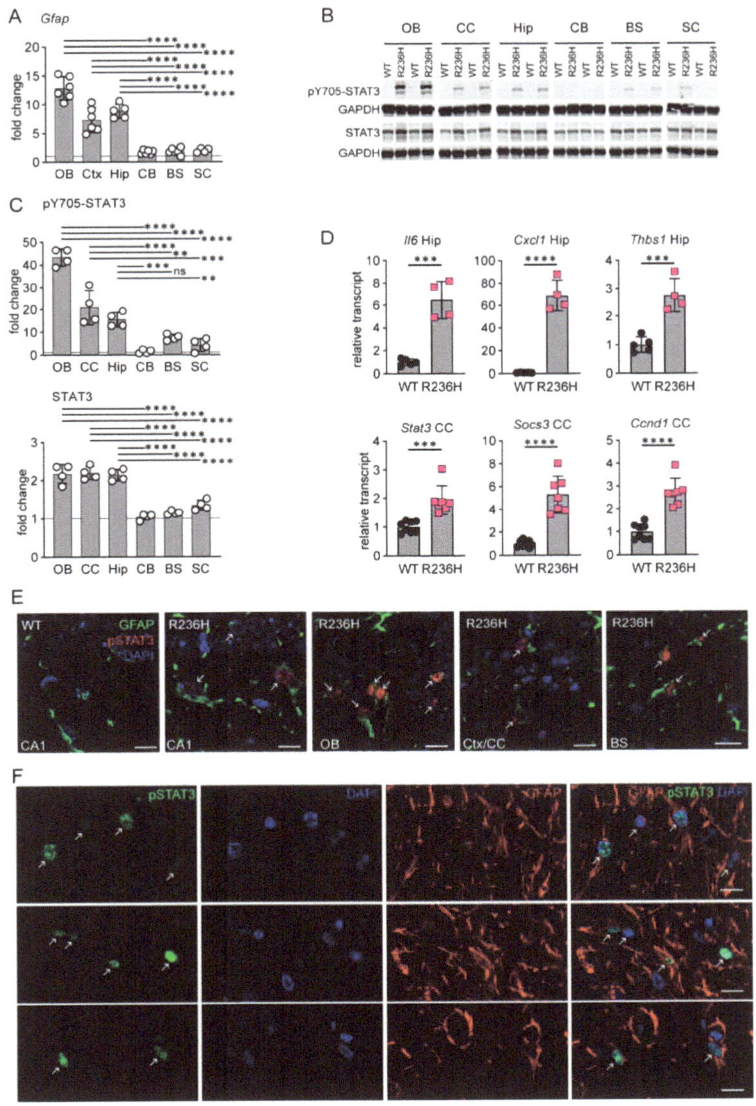

Figure 1. STAT3 activation in AxD. (**A**) Comparison of fold increase in *Gfap* transcript among different brain regions and spinal cord in *Gfap*$^{+/\mathrm{R236H}}$ mice as measured by qPCR. Results for olfactory bulb (OB), cortex (Ctx), hippocampus (Hip), cerebellum (CB), brainstem (BS) and cervical spinal cord (SC) are shown (males at 8 weeks, ANOVA with Sidak's post-tests, data reanalyzed from Figure 2 of

Jany et al., 2013 [8]). (**B**) Western analysis of pY705-STAT3 and total STAT3 in different brain regions and spinal cord of *Gfap*$^{+/R236H}$ (R236H) and wild-type (WT) mice (different animal in each lane) (**C**) Graphs represent fold change comparisons of R236H over WT for immunoblots as shown in (**B**) (males at 3 months, ANOVA with Sidak's post-tests) (**D**) Quantification of transcripts regulated by STAT3 or related to neuroinflammation in hippocampus (Hip) and corpus callosum (CC) in *Gfap*$^{+/R236H}$ mice. Relative values are normalized to *Rn18s* and expressed as a fold change compared to the wild-type average (males and females, Hip at 8 weeks, CC at 12 weeks, two-tailed *t*-test) (**E**) Immunolabeling of pY705-STAT3 (red) and GFAP (green) in *Gfap*$^{+/R236H}$ mouse hippocampus (CA1), olfactory bulb (OB), frontal cortex adjacent to the corpus callosum (Ctx/CC), and brainstem (BS). Left panel shows a wild-type (WT) mouse, and remaining panels represent *Gfap*$^{+/R236H}$ mice (R236H, males and females at 2 months). (**F**) Immunolabeling of pY705-STAT3 (green) and GFAP (red) in gray matter astrocytes from a 1-year-old patient with AxD. Arrows in (**E**,**F**) indicate pSTAT3 labeled nuclei. All mice are FVB/N. ** $p < 0.01$, *** $p < 0.001$, **** $p < 0.0001$ for all statistics. Scale bars = 10 µm.

3.2. Genetic Knockdown of STAT3 in AxD Model Mice

STAT3 is a pleiotropic transcription factor, and complete knockout by deletion of both alleles is lethal [28]. To further explore whether reduced STAT3 activation affects GFAP expression, we generated heterozygous constitutive knockout mice from the *Stat3*^{tm2Aki} strain [22] (see methods), and this line was crossed with *Gfap*$^{+/R236H}$ mice (Figure 2A). At 12 weeks of age, *Gfap*$^{+/R236H}$; *Stat3*$^{+/-}$ mice exhibited reduced pSTAT3 protein in hippocampus and olfactory bulb (Figure 2B), two regions known to have elevated GFAP in AxD mouse models. Quantitation of *Gfap* transcript and GFAP protein demonstrated only subtle and inconsistent differences (Figure 2C), suggesting that even at reduced levels, pSTAT3 activation is still sufficient to elevate *Gfap* expression, or that other pathways are involved in *Gfap* transactivation in AxD.

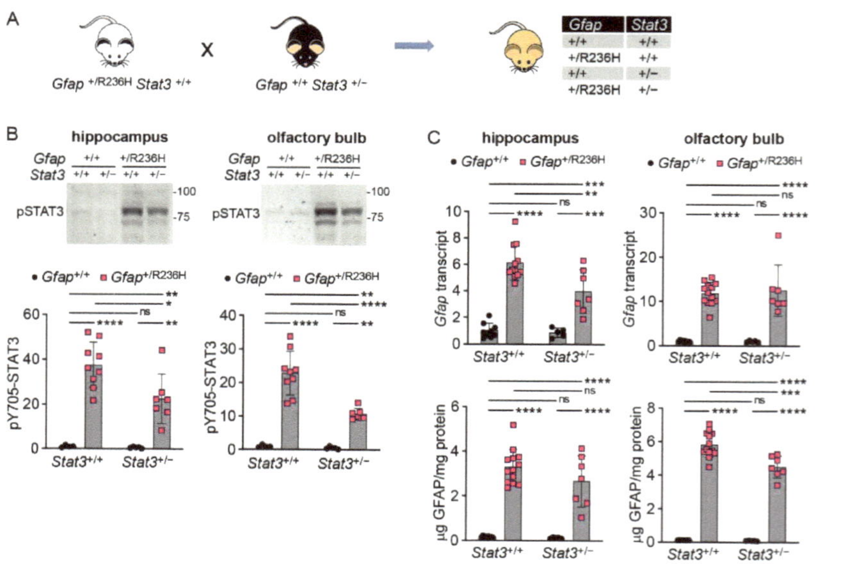

Figure 2. Genetic knockdown of *Stat3* in AxD model mice has marginal effects on *Gfap* expression. (**A**) Genetic cross to generate *Gfap*$^{+/R236H}$ mice with reduced STAT3 activation. (**B**,**C**) Quantitation of pY705-STAT3 by immunoblotting (**B**), and *Gfap* transcript by qPCR and protein by ELISA (**C**) in olfactory bulb and hippocampus from *Gfap*$^{+/R236H}$ mice and *Gfap*$^{+/+}$ littermate controls with heterozygous knockout of the STAT3-Y705 phosphorylation site. Values represent fold change relative to the average for *Gfap*$^{+/+}$; *Stat3*$^{+/+}$ mice. FVBB6F1 male and female mice were used at 12 weeks of age. Two-way ANOVA with Sidak's post-tests, * $p < 0.05$, ** $p < 0.01$, *** $p < 0.001$, **** $p < 0.0001$.

3.3. Conditional Knockout of STAT3 in Gfap Expressing Cells Prevents GFAP Elevation and Stress Response in AxD Model Mice

To specifically assess the role of STAT3 in *Gfap* expressing cells, we next used a transgenic line (Tg73.12) expressing Cre recombinase under the control of the mouse *Gfap* promoter [23], in combination with the floxed (fl) *Stat3* mice. The genes for GFAP and STAT3 are located within a 2 megabase interval on mouse chromosome 11. However, we were successful in combining the floxed *Stat3* allele with the R236H *Gfap* mutation, which required a homologous recombination event to generate a $Gfap^{+/R236H}$; $Stat3^{fl/fl}$ line. To facilitate complete STAT3 knockout in astrocytes, the *Gfap*-Cre-Tg73.12 line was crossed with heterozygous $Stat3^{+/-}$ knockout mice to produce Tg73.12; $Stat3^{+/-}$ mice. This line was crossed with the $Gfap^{+/R236H}$; $Stat3^{fl/fl}$ line, and animals carrying both knockout and floxed *Stat3* alleles (-/fl) were analyzed (Figure 3A, all mice congenic in C57BL/6J). At 8 weeks of age $Gfap^{+/R236H}$; $Stat3^{-/fl}$ mice negative for the Tg73.12 Cre transgene showed the typical increase in pSTAT3 activation, while those that were positive for Cre demonstrated pSTAT3 levels similar to mice with wild-type *Gfap* in both hippocampus and olfactory bulb (Figure 3B). Analysis of *Gfap* mRNA in the same mice demonstrated that conditional *Stat3* knockout in astrocytes reduced transcription to levels at or below the amount observed in $Gfap^{+/+}$ mice (Figure 3C). Quantification of *Lcn2* (lipocalin 2) and *Aif1* (Iba1) as markers of reactive astrocytes and microglia, respectively, also showed normalization in $Gfap^{+/R236H}$ mice with *Stat3* knockout (Figure 3C). Finally, GFAP protein measures showed reduction with *Stat3* knockout in $Gfap^{+/R236H}$ mice (Figure 3D), suggesting that protein accumulation is tied to elevated expression.

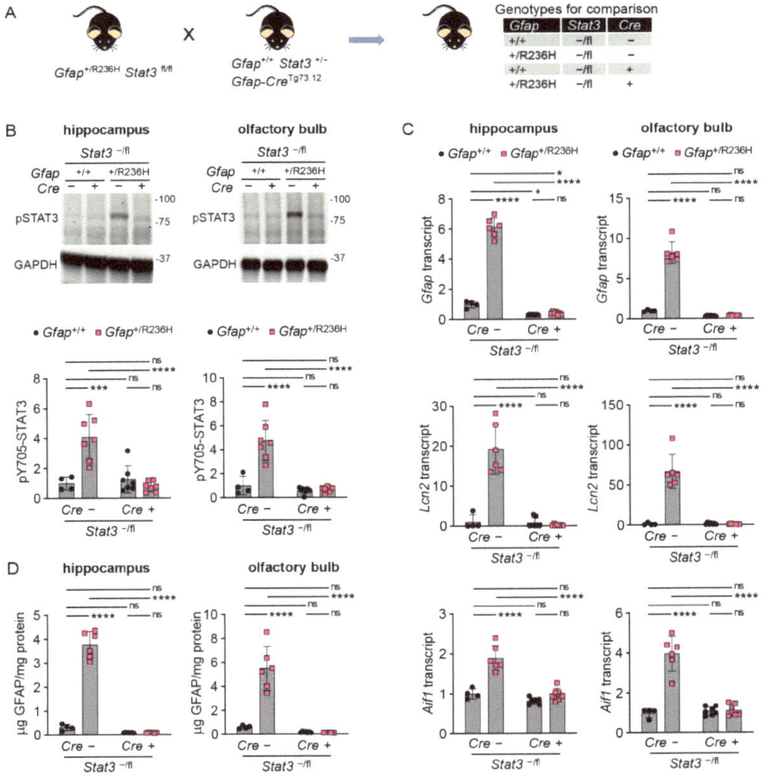

Figure 3. GFAP and stress markers in AxD model mice with conditional genetic knockout of *Stat3* in *Gfap* expressing cells at 8 weeks of age. (**A**) Genetic cross to generate $Gfap^{+/R236H}$ mice and $Gfap^{+/+}$

littermate controls with heterozygous knockout/floxed (-/fl) *Stat3* alleles and a *Gfap*-promoter-driven Cre transgene Tg73.12. (**B–D**) Quantitation of pY705-STAT3 protein by immunoblotting (**B**), *Gfap*, *Lcn2*, and *Aif1* transcripts by qPCR (**C**), and GFAP protein by ELISA (**D**) in hippocampus and olfactory bulb. Values in (**B**,**C**) are normalized to GAPDH and *Rn18s*, respectively, and represent fold change relative to the average for $Gfap^{+/+}$; $Stat3^{-/fl}$; Cre (-) mice. C57BL/6J males and females, two-way ANOVA with Sidak's post-tests, * $p < 0.05$, *** $p < 0.001$, **** $p < 0.0001$, ns = not significant.

To assess whether GFAP mutation might eventually cause protein accumulation with age, we performed similar experiments in older mice (12 weeks and 6 months), when pathology was increased in the model. For these experiments, we used mice carrying the floxed *Stat3* gene on both alleles (fl/fl) for a more exact comparison between full STAT3 expression and the targeted knockout condition (Figure 4A). At 12 weeks of age, we observed a reduction in phosphorylated STAT3 similar to that observed above at 8 weeks, and total STAT3 expression, which was also elevated in $Gfap^{+/R236H}$ mice, was normalized by blocking STAT3 activation (Figure 4B). GFAP protein was reduced by STAT3 inactivation in all brain regions analyzed and spinal cord (Figure 4C), with no significant difference between Cre positive $Gfap^{+/R236H}$ and $Gfap^{+/+}$ mice. Complete knockout of *Stat3* resulted in GFAP levels at or below those measured in wild-type animals.

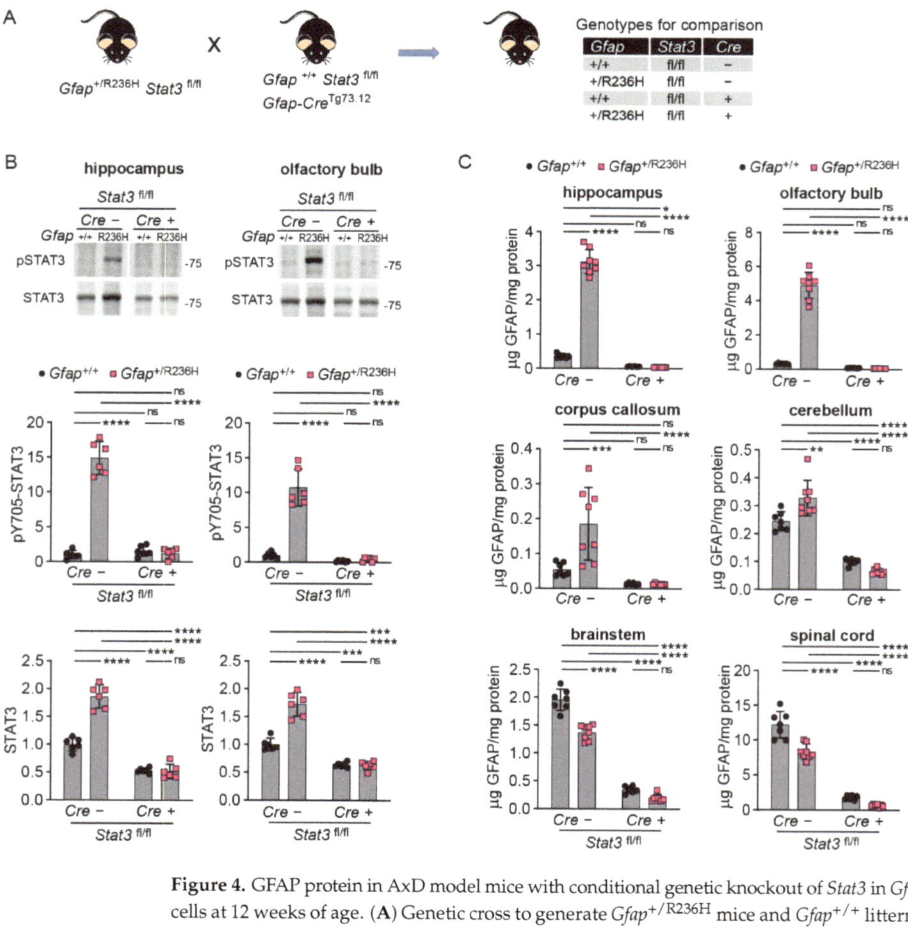

Figure 4. GFAP protein in AxD model mice with conditional genetic knockout of *Stat3* in *Gfap* expressing cells at 12 weeks of age. (**A**) Genetic cross to generate $Gfap^{+/R236H}$ mice and $Gfap^{+/+}$ littermate controls

with both *Stat3* alleles floxed (fl/fl) and a *Gfap*-promoter-driven Cre transgene (Tg73.12). (**B**,**C**) Quantitation of pY705-STAT3 and total STAT3 protein in olfactory bulb and hippocampus by immunoblotting (**B**), and GFAP protein in multiple brain regions and cervical spinal cord by ELISA (**C**). Values in (**B**) are normalized to total protein and represent fold change relative to the average for *Gfap*$^{+/+}$; *Stat3*$^{fl/fl}$; Cre (-) mice. C57BL/6J males and females, two-way ANOVA with Sidak's post-tests, * $p < 0.05$, ** $p < 0.01$, *** $p < 0.001$, **** $p < 0.0001$, ns = not significant.

To assess disrupted proteostasis and protein aggregation, we analyzed the autophagy cargo protein p62/SQSTM1 (sequestosome), which facilitates degradation of ubiquitinated proteins and colocalizes with Rosenthal fibers [29,30], and found that blocking STAT3 activation reduced p62 accumulation in *Gfap*$^{+/R236H}$ mice as shown by western analysis in Figure 5A. These results were confirmed by immunofluorescence in animals at 6 months of age, which exhibited no GFAP aggregation or p62 accumulation when STAT3 phosphorylation was prevented in astrocytes (Figure 5B,C). Finally, Iba1 immunolabeling showed reduced microglial expression, indicating the prevention of neuroinflammation (Figure 6).

3.4. Inducible STAT3 Knockout in Adult AxD Mouse Model Astrocytes Reverses GFAP Pathology and Normalizes Reactive Astrocytes and Microglia

To investigate whether genetic manipulation of STAT3 phosphorylation can affect GFAP accumulation in adult AxD model mice with established pathology [5,6], we took advantage of a tamoxifen-inducible astrocyte-specific CreER^{T2} line using the *Aldh1l1* promoter [24]. We first crossed the *Aldh1l1*-CreER^{T2} line to the Cre-reporter line Ai14 [25] and subsequently bred those mice with the Stat3$^{fl/fl}$ line. *Gfap*$^{+/R236H}$; *Stat3*$^{fl/fl}$ mice were then crossed to the CreER^{T2}; Ai14; *Stat3*$^{fl/fl}$ line. *Gfap*$^{+/R236H}$ and *Gfap*$^{+/+}$ animals that were either positive or negative for Cre were treated with tamoxifen for 5 days at 2 months of age, and tissues collected at 3 months of age (Figure 7A).

An initial assessment of efficacy for tamoxifen induction of Cre recombination in astrocytes using the Ai14 tdTomato reporter showed that 83% of GFAP positive astrocytes in hippocampus and 87% in olfactory bulb were also positive for tdTomato (364+/436 and 748+/858 astrocytes, respectively). We determined whether STAT3 was activated and found that *Gfap*$^{+/R236H}$; *Stat3*$^{fl/fl}$ mice (without Cre) showed the same increase in pSTAT3, whereas mice expressing Cre (*Stat3*$^{\Delta/\Delta}$) demonstrated pSTAT3 levels similar to those of *Gfap*$^{+/+}$ mice in both hippocampus and olfactory bulb (Figure 7B). Total STAT3 protein showed similar changes, although not as pronounced.

We next examined whether blocking STAT3 activation in adult AxD model mice could reverse GFAP accumulation. Transcript analysis demonstrated a significant reduction in *Gfap* expression in hippocampus and olfactory bulb (Figure 7C). Analysis of GFAP protein showed reductions in all brain regions examined except for cerebellum (Figure 7D). Spinal cord, which had decreased amounts of GFAP in *Gfap*$^{+/R236H}$ mice, also showed no further reduction in protein. *Lcn2* transcript was measured as a surrogate marker of astrogliosis, and expression was normalized with STAT3 inactivation (Figure 7E). Other transcripts regulated by STAT3, including *Socs3* and *Ccnd1*, were also decreased in hippocampus, and *Socs3* was reduced in olfactory bulb. Transcripts for small chemokines *Cxcl1* and *Ccl2* were reduced, suggesting a normalization of the neuroinflammatory response, and *Nqo1* (NAD(P)H quinone oxidoreductase), a target of the Nrf2/ARE antioxidant stress response pathway, was reduced in hippocampus but not olfactory bulb (Figure 7E).

To determine whether reduced GFAP expression led to reduced aggregation and improved proteostasis, we immunolabeled p62 positive aggregates as a measure of Rosenthal fibers and found that protein aggregation was reversed in *Gfap*$^{+/R236H}$ mice by blocking STAT3 activation (Figure 8A,B). Quantification of p62 by western analysis showed a similar reduction in hippocampus and olfactory bulb (Figure 8C).

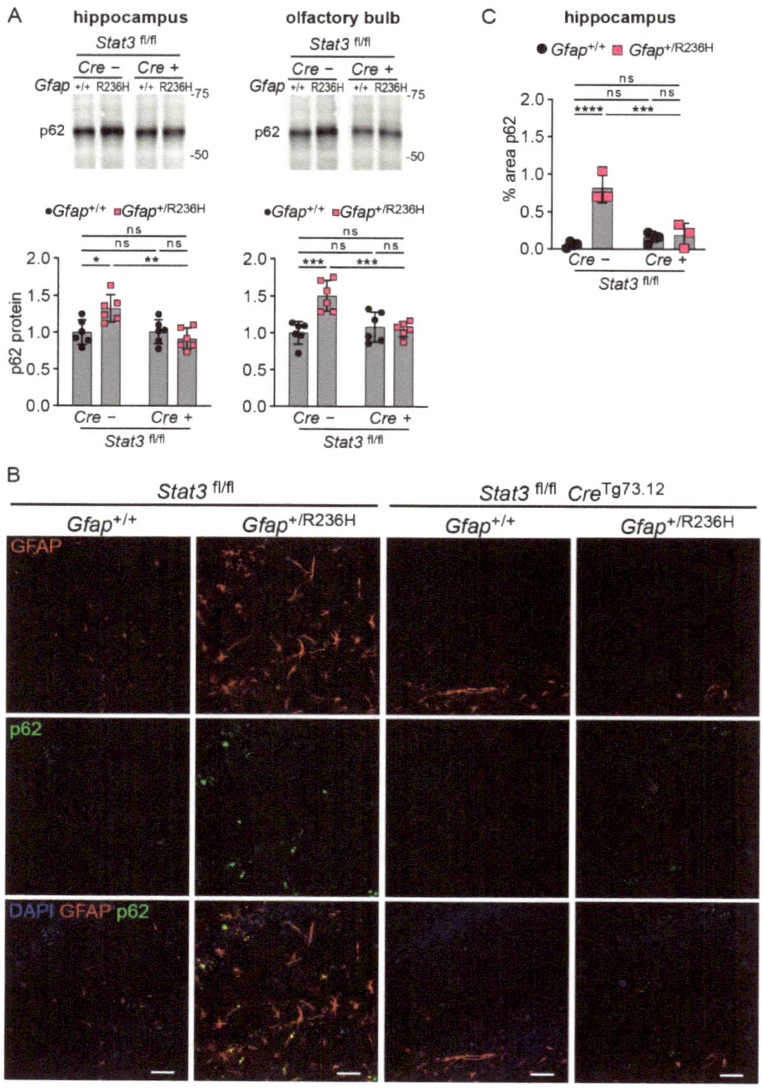

Figure 5. p62/SQSTM1 expression and protein aggregation in AxD model mice with conditional knockout of *Stat3* in *Gfap* expressing cells. (**A**) Quantification of p62/SQSTM1 protein in olfactory bulb and hippocampus from *Gfap*$^{+/R236H}$ mice and *Gfap*$^{+/+}$ littermate controls (3 months of age) collected from the same Stat3$^{fl/lf}$; Cre$^{Tg73.12}$ genetic cross as shown in Figure 4A. (**B**,**C**) Immunofluorescence labeling of p62 and GFAP in hippocampus (**B**), and quantification of percent area for p62 (**C**) in mice from the same cross at 6 months of age. Values in (**A**) are normalized to total protein and represent fold change relative to the average for *Gfap*$^{+/+}$; *Stat3*$^{fl/fl}$; Cre (-) mice. C57BL/6J males and females, two-way ANOVA with Sidak's post-tests, * $p < 0.05$, ** $p < 0.01$, *** $p < 0.001$, **** $p < 0.0001$, ns = not significant. Scale bars = 50 μm.

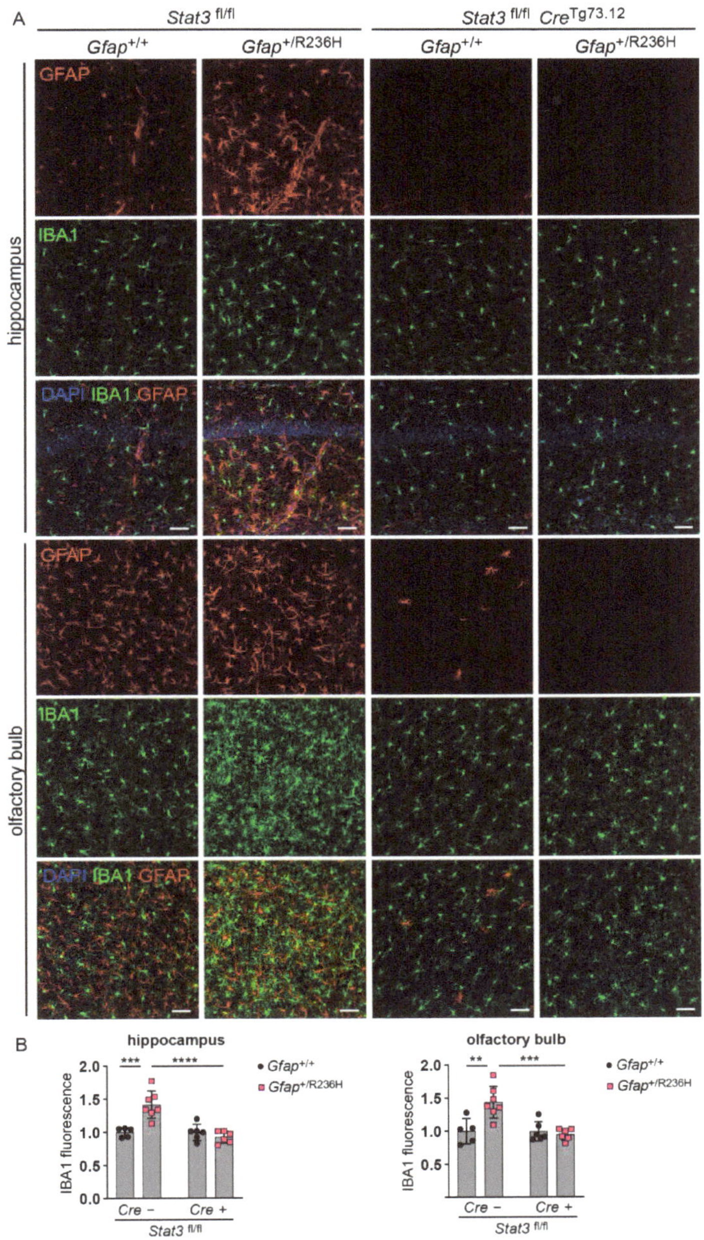

Figure 6. Iba1 expression in microglia from AxD model mice with conditional knockout of *Stat3* in *Gfap* expressing cells. (**A**,**B**) Immunofluorescence labeling (**A**) and quantification (**B**) of Iba1 in hippocampus and olfactory bulb from $Gfap^{+/R236H}$ mice and $Gfap^{+/+}$ littermate controls (6 months of age) collected from the same $Stat3^{fl/lf}$, $Cre^{Tg73.12}$ genetic cross as shown in Figure 4A. GFAP is also labeled for comparison. Values in (**B**) represent fold change relative to the average for $Gfap^{+/+}$; $Stat3^{fl/fl}$; Cre (-) mice. C57BL/6J males and females, two-way ANOVA with Sidak's post-tests, ** $p < 0.01$, *** $p < 0.001$, **** $p < 0.0001$, ns = not significant. Scale bars = 50 μm.

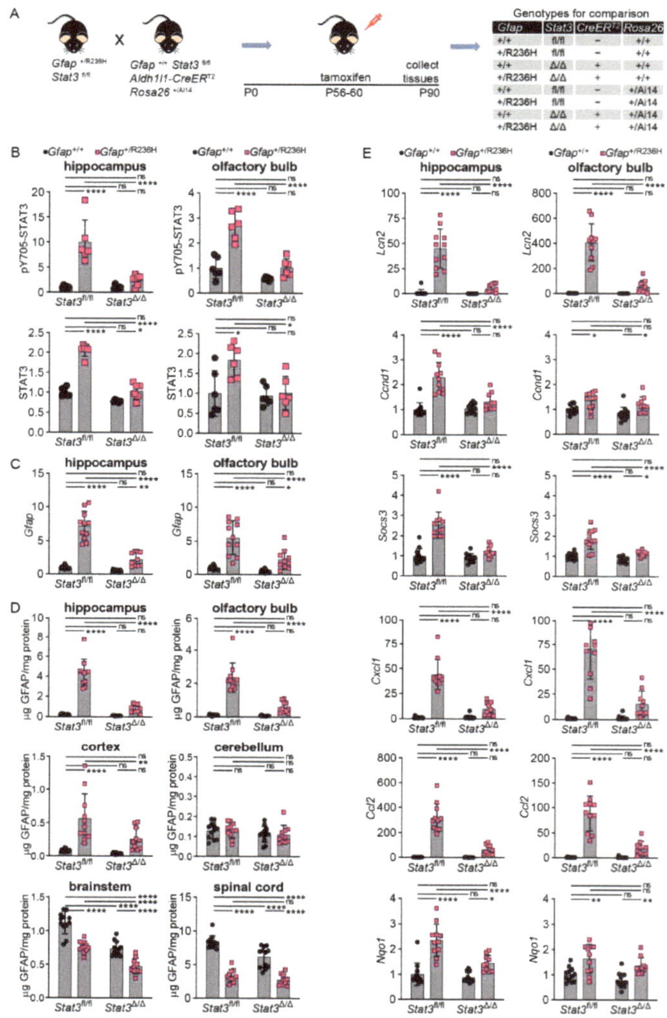

Figure 7. STAT3, GFAP, and other stress and immune related markers in AxD model mice after induced genetic knockout of *Stat3* in astrocytes. (**A**) Genetic cross to generate $Gfap^{+/R236H}$ mice and $Gfap^{+/+}$ littermate controls with both *Stat3* alleles floxed (fl/fl) and a tamoxifen-inducible *Aldh1l1*-promoter driven $CreER^{T2}$ transgene. All mice were treated with tamoxifen at 2 months of age for 5 days, and tissues collected at 3 months of age. Timeline in right panel shows targeted postnatal (P) days. *Stat3* alleles in animals expressing the *Aldh1l1*-$CreER^{T2}$ transgene are indicated as Δ ($Stat3^{\Delta/\Delta}$). (**B**,**C**) Quantitation of pY705-STAT3 and total STAT3 protein by immunoblotting (**B**), and *Gfap* transcript by qPCR (**C**) in hippocampus and olfactory bulb. (**D**) GFAP protein quantification in multiple brain regions and cervical spinal cord. (**E**) Quantification of transcripts reflecting reactive gliosis (*Lcn2*), STAT3 activation (*Ccnd1*, *Socs3*), neuroinflammation (*Cxcl1*, *Ccl2*), and oxidative stress (*Nqo1*) in hippocampus and olfactory bulb. Values in (**B**) are normalized to total protein and in (**C**,**E**) to *Rn18s*, and all represent fold change relative to the average for $Gfap^{+/+}$; $Stat3^{fl/fl}$; Cre (-) mice. C57BL/6J males and females, two-way ANOVA with Sidak's post-tests, * $p < 0.05$, ** $p < 0.01$, **** $p < 0.0001$, ns = not significant.

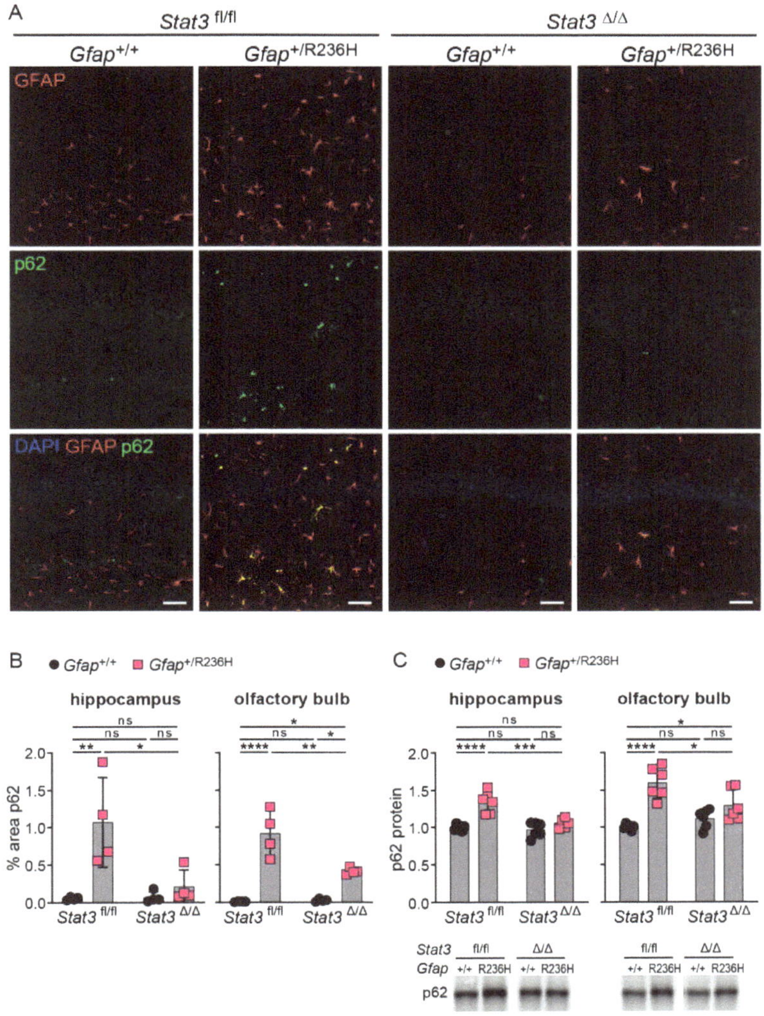

Figure 8. p62/SQSTM1 expression and protein aggregation in AxD model mice after inducing genetic knockout of *Stat3* in astrocytes. (**A**,**B**) Immunofluorescence labeling of p62 and GFAP in hippocampus (**A**), and quantitation of percent area (**B**) in hippocampus and olfactory bulb from $Gfap^{+/R236H}$ mice and $Gfap^{+/+}$ littermate controls (3 months of age) derived from the same $Stat3^{fl/fl}$, $CreER^{T2}$ genetic cross and tamoxifen treatment schedule as shown in Figure 7A. *Stat3* alleles in animals expressing the *Aldh1l1*-$CreER^{T2}$ transgene are indicated as Δ ($Stat3^{\Delta/\Delta}$). (**C**) Quantification of p62/SQSTM1 protein in olfactory bulb and hippocampus by western analysis of a separate group of mice from the same genetic cross and treatment described in (**A**,**B**). C57BL/6J males and females, two-way ANOVA with Sidak's post-tests, * $p < 0.05$, ** $p < 0.01$, *** $p < 0.001$, **** $p < 0.0001$, ns = not significant. Scale bars = 50 μm.

Finally, to assess the effects of STAT3 inhibition in astrocytes on neuroinflammation in general, we examined microglia activation by labeling with Iba1 and found that microglia appeared to return to a normal state with reduced Iba1 expression (Figure 9A,B). Transcript analysis of the gene for Iba1 (*Aif1*) also showed a reduction in hippocampus and olfactory bulb after blocking STAT3 activation in *Gfap*$^{+/R236H}$ mice (Figure 9C).

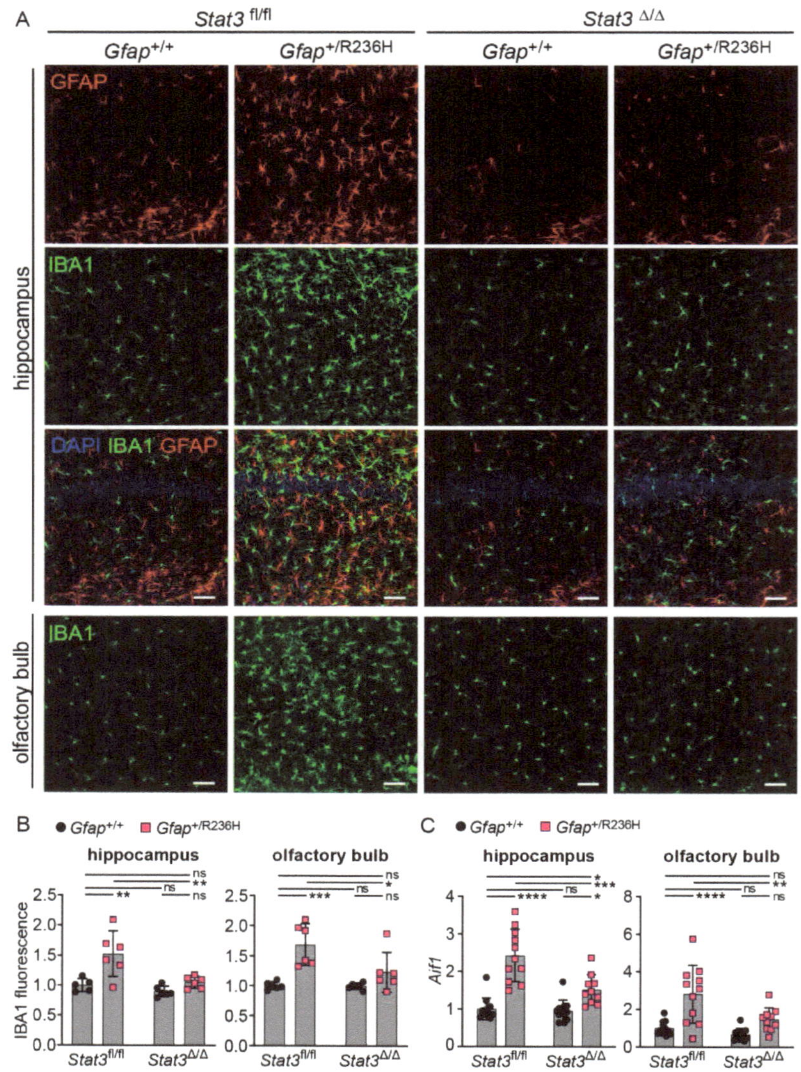

Figure 9. Microglial Iba1 expression in AxD model mice after inducible knockout of *Stat3* in astrocytes. (**A**,**B**) Immunofluorescence labeling (**A**) and quantification (**B**) of Iba1 in hippocampus and olfactory bulb from *Gfap*$^{+/R236H}$ mice and *Gfap*$^{+/+}$ littermate controls (3 months of age) derived from the same *Stat3*$^{fl/fl}$, CreER^{T2} genetic cross and tamoxifen treatment schedule as shown in Figure 7A. *Stat3* alleles in animals expressing the *Aldh1l1*-CreER^{T2} transgene are indicated as Δ (Stat3$^{Δ/Δ}$). GFAP is also shown in hippocampus for comparison. (**C**) Quantification of *Aif1* transcript (Iba1 gene, normalized to *Rn18s*) in olfactory bulb and hippocampus by qPCR. Values represent fold change relative to the average for *Gfap*$^{+/+}$; *Stat3*$^{fl/fl}$; Cre (-) mice. C57BL/6J males and females, two-way ANOVA with Sidak's post-tests, * $p < 0.05$, ** $p < 0.01$, *** $p < 0.001$, **** $p < 0.0001$, ns = not significant. Scale bars = 50 μm.

4. Discussion

GFAP expression and accumulation have been shown to correlate with AxD severity [8], and STAT3 is known to regulate GFAP transcription in development and in response to injury [14]. In mouse models of AxD, regions of the CNS with the greatest increase in GFAP expression also demonstrate significant STAT3 phosphorylation. In addition, we showed that STAT3 was activated in gray matter astrocytes in a patient with the disease. Here, we investigated whether blocking STAT3 phosphorylation could be used to manipulate *Gfap* expression and prevent or reverse protein accumulation in *Gfap*$^{+R236H}$ mice. Initial experiments with heterozygous *Stat3*$^{+/-}$ knockout mice achieved significant reduction in phosphorylated STAT3-Y705, but these differences were not sufficient to consistently reduce *Gfap* expression. In addition to STAT3, several other transcription factors and their binding sites in the *GFAP* promoter have been tested for their influence on basal expression and the injury response [14]. AP-1 in particular has been suggested to regulate *Gfap* elevation in different injury models [31], including a transgenic mouse model of AxD that overexpresses wild-type human GFAP [7], but whether AP-1 or other factors contribute to *Gfap* elevation in *Gfap*$^{+R236H}$ mice has not yet been tested.

To further investigate the role of STAT3 in reactive AxD astrocytes, we used a conditional knockout approach to target *Stat3* in *Gfap* expressing cells and found that *Stat3* knockout reduced GFAP levels and prevented the reactive astrocyte response, similarly to other studies in acute injury models [12,15] and models of neurodegeneration [16,32]. STAT3 is an essential regulator of astrocyte differentiation and maturation, and by prohibiting STAT3 activation during astrocyte development, GFAP transcript and protein were reduced even in *Gfap*$^{+/+}$ mice. AxD severity is tied to GFAP expression, and in these experiments, further transactivation of the *Gfap* promoter was prevented. Reduced GFAP suggests that mutant protein does not reach a toxic threshold and that astrocytes are able to maintain proteostasis and avoid a reactive stress response.

GFAP is also expressed in progenitor cells, and early activation of the *Gfap*-promoter-driven Cre transgene (Tg73.12) as described above leads to excision of floxed genes in both astrocytes and other differentiated cell types, including neurons [23]. To specifically evaluate the role of STAT3 activation in reactive astrocytes with existing GFAP accumulation and protein aggregation [5,6], we used an inducible CreER^{T2} line controlled by the *Aldh1l1* promoter [24]. Knocking down STAT3 in adult mice with established GFAP pathology reduced *Gfap* expression and the concomitant stress and neuroinflammatory response and reversed GFAP accumulation and aggregation. We cannot exclude the formal possibility that Cre itself has an effect independent of STAT3 [33]; however, we consider this unlikely given the lack of change in markers for astro- and microgliosis (e.g., *Lcn2*, *Aif1*) in the presence of Cre in *Gfap*$^{+/+}$ mice (Figures 3C, 7E and 9C). These results demonstrate that STAT3 is a key factor in *Gfap* elevation and GFAP accumulation, and that blocking STAT3 activation may provide a therapeutic target for AxD.

STAT3 activation is not necessarily detrimental, and as has been shown in other models of CNS injury or neurodegenerative disease, the consequences are context-dependent [34–40]. In Huntington's disease, for example, activation of JAK/STAT3 signaling improves proteostasis by increasing gene expression in proteasome and lysosomal pathways [39]. However, given the role of STAT3 in *Gfap* transactivation, inhibiting STAT3 in AxD would have the direct benefit of reducing levels of mutant GFAP and preventing protein misfolding and the resulting cellular toxicity. More extensive analyses of AxD tissues and controls will be needed to firmly establish the role of STAT3 in the human disorder.

Although we do not know the initiating events leading to STAT3 activation, our results support our hypothesis that activation of STAT3 contributes to a positive feedback loop that drives increased GFAP expression, protein accumulation, and ultimately astrocyte dysfunction. GFAP is one of the most highly expressed proteins in astrocytes, and a misfolded mutant form could disrupt a myriad of cellular processes, including vesicular transport and mitochondrial function, which could lead to oxidative stress and activation of MAP kinase pathways. Activation of JNK and p38 has been demonstrated in cellular models of AxD [41,42], and MAP kinases could activate AP-1 and NFκB, leading to increased IL-6 to activate the IL-6/JAK/STAT3 pathway.

In this study, we took advantage of the wealth of available genetic tools to manipulate STAT3 activation in our mouse models of AxD. *Gfap*$^{+R236H}$ mice recapitulate many of the pathological features of human disease and have been instrumental in investigating mechanisms and modifiers of the disease [43]; however, they have only subtle behavioral deficits and limited functional outcome measures [5,44]. For future studies to test whether inhibiting STAT3 pharmacologically has therapeutic value, we recently developed a rat model of AxD [19], which, in addition to STAT3 activation, exhibits motor deficits and other clinically relevant phenotypes for improved functional assessment. Inhibitors of the JAK/STAT pathway have been approved to treat inflammatory disorders, hematopoietic malignancies, and solid tumors [20,21], and JAK inhibitors have been shown to be effective in suppressing STAT3 activation in glioblastoma, experimental autoimmune encephalomyelitis (EAE) models of multiple sclerosis, and a model of Parkinson's disease [45–48]. The JAK inhibitor baricitinib was recently evaluated in clinical trials for Aicardi–Goutieres syndrome, an interferonopathy that also leads to white matter deficits, with results suggesting improved neurological function [49]. Although JAK inhibition can have adverse side effects [50,51], strategies to specifically target STAT3 with small molecules have been effective in preclinical models of Alzheimer's disease [32,52] and may provide alternatives for reducing GFAP pathology in AxD.

Author Contributions: Conceptualization and Methodology, T.L.H., L.W., M.B.F. and A.M.; Formal Analysis, T.L.H., S.C., A.L., L.W. and M.B.F.; Investigation, T.L.H., S.C., A.L. and L.W.; Resources, T.L.H., M.B.F. and A.M.; Writing—Original Draft Preparation, T.L.H. and A.M.; Writing—Review and Editing, T.L.H., L.W., M.B.F. and A.M.; Project Administration, T.L.H.; Funding Acquisition, T.L.H., M.B.F. and A.M. All authors have read and agreed to the published version of the manuscript.

Funding: This work was supported by grants from the NIH NINDS (NS093482 and NS110719 to T.L.H.) and NICHD (HD076892 to A.M., HD090256 and HD105353 core grants to the Waisman Center IDDRC), and by End Alexander disease and the Juanma Fund.

Institutional Review Board Statement: All animal studies were approved by the College of Letters and Sciences and Vice Chancellor Office for Research and Graduate Education Animal Care and Use Committee within the University of Wisconsin-Madison under protocol numbers G005568 (approved 8 September 2016), G006167 (approved 4 August 2019), and G006276 (approved 1 March 2020).

Informed Consent Statement: Not applicable.

Data Availability Statement: All relevant data are included in the current report.

Acknowledgments: We would like to thank Michael Sofroniew for sharing the floxed STAT3 line from Takeda et al., 1998, and Denice Springman and Jacob Loeffelholz for technical support.

Conflicts of Interest: The authors declare no conflict of interest. The funders had no role in the design of the study; in the collection, analyses, or interpretation of data; in the writing of the manuscript; or in the decision to publish the results.

References

1. Messing, A.; Brenner, M.; Feany, M.B.; Nedergaard, M.; Goldman, J.E. Alexander disease. *J. Neurosci.* **2012**, *32*, 5017–5023. [CrossRef] [PubMed]
2. Prust, M.; Wang, J.; Morizono, H.; Messing, A.; Brenner, M.; Gordon, E.; Hartka, T.; Sokohl, A.; Schiffmann, R.; Gordish-Dressman, H.; et al. GFAP mutations, age at onset, and clinical subtypes in Alexander disease. *Neurology* **2011**, *77*, 1287–1294. [CrossRef]
3. Brenner, M.; Johnson, A.B.; Boespflug-Tanguy, O.; Rodriguez, D.; Goldman, J.E.; Messing, A. Mutations in GFAP, encoding glial fibrillary acidic protein, are associated with Alexander disease. *Nat. Genet.* **2001**, *27*, 117–120. [CrossRef] [PubMed]
4. Brenner, M.; Goldman, J.E.; Quinlan, R.A.; Messing, A. Alexander Disease: A Genetic Disorder of Astrocytes. In *Astrocytes in Pathophysiology of the Nervous System*, 1st ed.; Parpura, V., Haydon, P.G., Eds.; Springer: Boston, MA, USA, 2009; pp. 591–648.
5. Hagemann, T.L.; Connor, J.X.; Messing, A. Alexander disease-associated glial fibrillary acidic protein mutations in mice induce Rosenthal fiber formation and a white matter stress response. *J. Neurosci.* **2006**, *26*, 11162–11173. [CrossRef] [PubMed]
6. Hagemann, T.L.; Jobe, E.M.; Messing, A. Genetic ablation of Nrf2/antioxidant response pathway in Alexander disease mice reduces hippocampal gliosis but does not impact survival. *PLoS ONE* **2012**, *7*, e37304. [CrossRef]

7. Messing, A.; Head, M.W.; Galles, K.; Galbreath, E.J.; Goldman, J.E.; Brenner, M. Fatal encephalopathy with astrocyte inclusions in GFAP transgenic mice. *Am. J. Pathol.* **1998**, *152*, 391–398.
8. Jany, P.L.; Hagemann, T.L.; Messing, A. GFAP expression as an indicator of disease severity in mouse models of Alexander disease. *ASN Neuro.* **2013**, *5*, art:e00109. [CrossRef]
9. Bonni, A.; Sun, Y.; Nadal-Vicens, M.; Bhatt, A.; Frank, D.A.; Rozovsky, I.; Stahl, N.; Yancopoulos, G.D.; Greenberg, M.E. Regulation of gliogenesis in the central nervous system by the JAK-STAT signaling pathway. *Science* **1997**, *278*, 477–483. [CrossRef]
10. Nakashima, K.; Yanagisawa, M.; Arakawa, H.; Kimura, N.; Hisatsune, T.; Kawabata, M.; Miyazono, K.; Taga, T. Synergistic signaling in fetal brain by STAT3-Smad1 complex bridged by p300. *Science* **1999**, *284*, 479–482. [CrossRef]
11. Sriram, K.; Benkovic, S.A.; Hebert, M.A.; Miller, D.B.; O'Callaghan, J.P. Induction of gp130-related cytokines and activation of JAK2/STAT3 pathway in astrocytes precedes up-regulation of glial fibrillary acidic protein in the 1-methyl-4-phenyl-1,2,3,6-tetrahydropyridine model of neurodegeneration: Key signaling pathway for astrogliosis in vivo? *J. Biol. Chem.* **2004**, *279*, 19936–19947. [CrossRef]
12. Herrmann, J.E.; Imura, T.; Song, B.; Qi, J.; Ao, Y.; Nguyen, T.K.; Korsak, R.A.; Takeda, K.; Akira, S.; Sofroniew, M.V. STAT3 is a critical regulator of astrogliosis and scar formation after spinal cord injury. *J. Neurosci.* **2008**, *28*, 7231–7243. [CrossRef] [PubMed]
13. Yeo, S.; Bandyopadhyay, S.; Messing, A.; Brenner, M. Transgenic analysis of GFAP promoter elements. *Glia* **2013**, *61*, 1488–1499. [CrossRef] [PubMed]
14. Brenner, M.; Messing, A. Regulation of GFAP Expression. *ASN Neuro.* **2021**, *13*, 1759091420981206. [CrossRef]
15. O'Callaghan, J.P.; Kelly, K.A.; VanGilder, R.L.; Sofroniew, M.V.; Miller, D.B. Early Activation of STAT3 Regulates Reactive Astrogliosis Induced by Diverse Forms of Neurotoxicity. *PLoS ONE* **2014**, *9*, e102003. [CrossRef] [PubMed]
16. Ben Haim, L.; Ceyzeriat, K.; Carrillo-de Sauvage, M.A.; Aubry, F.; Auregan, G.; Guillermier, M.; Ruiz, M.; Petit, F.; Houitte, D.; Faivre, E.; et al. The JAK/STAT3 Pathway Is a Common Inducer of Astrocyte Reactivity in Alzheimer's and Huntington's Diseases. *J. Neurosci.* **2015**, *35*, 2817–2829. [CrossRef] [PubMed]
17. Ceyzériat, K.; Ben Haim, L.; Denizot, A.; Pommier, D.; Matos, M.; Guillemaud, O.; Palomares, M.A.; Abjean, L.; Petit, F.; Gipchtein, P.; et al. Modulation of astrocyte reactivity improves functional deficits in mouse models of Alzheimer's disease. *Acta Neuropathol. Commun.* **2018**, *6*, 104. [CrossRef] [PubMed]
18. LaPash Daniels, C.M.; Paffenroth, E.; Austin, E.V.; Glebov, K.; Lewis, D.; Walter, J.; Messing, A. Lithium decreases glial fibrillary acidic protein in a mouse model of Alexander disease. *PLoS ONE* **2015**, *10*, e0138132. [CrossRef] [PubMed]
19. Hagemann, T.L.; Powers, B.; Lin, N.H.; Mohamed, A.F.; Dague, K.L.; Hannah, S.C.; Bachmann, G.; Mazur, C.; Rigo, F.; Olsen, A.L.; et al. Antisense therapy in a rat model of Alexander disease reverses GFAP pathology, white matter deficits, and motor impairment. *Sci. Transl. Med.* **2021**, *13*, eabg4711. [CrossRef]
20. Johnson, D.E.; O'Keefe, R.A.; Grandis, J.R. Targeting the IL-6/JAK/STAT3 signalling axis in cancer. *Nat. Rev. Clin. Oncol.* **2018**, *15*, 234–248. [CrossRef]
21. Tanaka, Y.; Luo, Y.; O'Shea, J.J.; Nakayamada, S. Janus kinase-targeting therapies in rheumatology: A mechanisms-based approach. *Nat. Rev. Rheumatol.* **2022**, *18*, 133–145. [CrossRef]
22. Takeda, K.; Kaisho, T.; Yoshida, N.; Takeda, J.; Kishimoto, T.; Akira, S. Stat3 activation is responsible for IL-6-dependent T cell proliferation through preventing apoptosis: Generation and characterization of T cell-specific Stat3-deficient mice. *J. Immunol.* **1998**, *161*, 4652–4660. [CrossRef] [PubMed]
23. Garcia, A.D.; Doan, N.B.; Imura, T.; Bush, T.G.; Sofroniew, M.V. GFAP-expressing progenitors are the principal source of constitutive neurogenesis in adult mouse forebrain. *Nat. Neurosci.* **2004**, *7*, 1233–1241. [CrossRef] [PubMed]
24. Srinivasan, R.; Lu, T.Y.; Chai, H.; Xu, J.; Huang, B.S.; Golshani, P.; Coppola, G.; Khakh, B.S. New Transgenic Mouse Lines for Selectively Targeting Astrocytes and Studying Calcium Signals in Astrocyte Processes In Situ and In Vivo. *Neuron* **2016**, *92*, 1181–1195. [CrossRef] [PubMed]
25. Madisen, L.; Zwingman, T.A.; Sunkin, S.M.; Oh, S.W.; Zariwala, H.A.; Gu, H.; Ng, L.L.; Palmiter, R.D.; Hawrylycz, M.J.; Jones, A.R.; et al. A robust and high-throughput Cre reporting and characterization system for the whole mouse brain. *Nat. Neurosci.* **2010**, *13*, 133–140. [CrossRef]
26. Tronche, F.; Kellendonk, C.; Kretz, O.; Gass, P.; Anlag, K.; Orban, P.C.; Bock, R.; Klein, R.; Schütz, G. Disruption of the glucocorticoid receptor gene in the nervous system results in reduced anxiety. *Nat. Genet.* **1999**, *23*, 99–103. [CrossRef]
27. Hagemann, T.L.; Gaeta, S.A.; Smith, M.A.; Johnson, D.A.; Johnson, J.A.; Messing, A. Gene expression analysis in mice with elevated glial fibrillary acidic protein and Rosenthal fibers reveals a stress response followed by glial activation and neuronal dysfunction. *Hum. Mol. Genet.* **2005**, *14*, 2443–2458. [CrossRef]
28. Takeda, K.; Noguchi, K.; Shi, W.; Tanaka, T.; Matsumoto, M.; Yoshida, N.; Kishimoto, T.; Akira, S. Targeted disruption of the mouse Stat3 gene leads to early embryonic lethality. *Proc. Natl. Acad. Sci. USA* **1997**, *94*, 3801–3804. [CrossRef]
29. Zatloukal, K.; Stumptner, C.; Fuchsbichler, A.; Heid, H.; Schnoelzer, M.; Kenner, L.; Kleinert, R.; Prinz, M.; Aguzzi, A.; Denk, H. p62 is a common component of cytoplasmic inclusions in protein aggregation diseases. *Am. J. Pathol.* **2002**, *160*, 255–263. [CrossRef]
30. Cohen-Kaplan, V.; Livneh, I.; Avni, N.; Cohen-Rosenzweig, C.; Ciechanover, A. The ubiquitin-proteasome system and autophagy: Coordinated and independent activities. *Int. J. Biochem. Cell Biol.* **2016**, *79*, 403–418. [CrossRef]
31. Brenner, M.; Messing, A.; Olsen, M.L. AP-1 and the injury response of the GFAP gene. *J. Neurosci. Res.* **2019**, *97*, 149–161. [CrossRef]

32. Reichenbach, N.; Delekate, A.; Plescher, M.; Schmitt, F.; Krauss, S.; Blank, N.; Halle, A.; Petzold, G.C. Inhibition of Stat3-mediated astrogliosis ameliorates pathology in an Alzheimer's disease model. *EMBO Mol. Med.* **2019**, *11*, e9665. [CrossRef]
33. Loonstra, A.; Vooijs, M.; Beverloo, H.B.; Allak, B.A.; van Drunen, E.; Kanaar, R.; Berns, A.; Jonkers, J. Growth inhibition and DNA damage induced by Cre recombinase in mammalian cells. *Proc. Natl. Acad. Sci. USA* **2001**, *98*, 9209–9214. [CrossRef] [PubMed]
34. Okada, S.; Nakamura, M.; Katoh, H.; Miyao, T.; Shimazaki, T.; Ishii, K.; Yamane, J.; Yoshimura, A.; Iwamoto, Y.; Toyama, Y.; et al. Conditional ablation of Stat3 or Socs3 discloses a dual role for reactive astrocytes after spinal cord injury. *Nat. Med.* **2006**, *12*, 829–834. [CrossRef] [PubMed]
35. Qin, H.; Yeh, W.I.; De Sarno, P.; Holdbrooks, A.T.; Liu, Y.; Muldowney, M.T.; Reynolds, S.L.; Yanagisawa, L.L.; Fox, T.H., 3rd; Park, K.; et al. Signal transducer and activator of transcription-3/suppressor of cytokine signaling-3 (STAT3/SOCS3) axis in myeloid cells regulates neuroinflammation. *Proc. Natl. Acad. Sci. USA* **2012**, *109*, 5004–5009. [CrossRef]
36. Hristova, M.; Rocha-Ferreira, E.; Fontana, X.; Thei, L.; Buckle, R.; Christou, M.; Hompoonsup, S.; Gostelow, N.; Raivich, G.; Peebles, D. Inhibition of Signal Transducer and Activator of Transcription 3 (STAT3) reduces neonatal hypoxic-ischaemic brain damage. *J. Neurochem.* **2016**, *136*, 981–994. [CrossRef]
37. Anderson, M.A.; Burda, J.E.; Ren, Y.; Ao, Y.; O'Shea, T.M.; Kawaguchi, R.; Coppola, G.; Khakh, B.S.; Deming, T.J.; Sofroniew, M.V. Astrocyte scar formation aids central nervous system axon regeneration. *Nature* **2016**, *532*, 195–200. [CrossRef] [PubMed]
38. Tyzack, G.E.; Hall, C.E.; Sibley, C.R.; Cymes, T.; Forostyak, S.; Carlino, G.; Meyer, I.F.; Schiavo, G.; Zhang, S.C.; Gibbons, G.M.; et al. A neuroprotective astrocyte state is induced by neuronal signal EphB1 but fails in ALS models. *Nat. Commun.* **2017**, *8*, 1164. [CrossRef]
39. Abjean, L.; Ben Haim, L.; Riquelme-Perez, M.; Gipchtein, P.; Derbois, C.; Palomares, M.A.; Petit, F.; Herard, A.S.; Gaillard, M.C.; Guillermier, M.; et al. Reactive astrocytes promote proteostasis in Huntington's disease through the JAK2-STAT3 pathway. *Brain* **2023**, *146*, 149–166. [CrossRef]
40. Burda, J.E.; O'Shea, T.M.; Ao, Y.; Suresh, K.B.; Wang, S.; Bernstein, A.M.; Chandra, A.; Deverasetty, S.; Kawaguchi, R.; Kim, J.H.; et al. Divergent transcriptional regulation of astrocyte reactivity across disorders. *Nature* **2022**, *606*, 557–564. [CrossRef]
41. Tang, G.; Yue, Z.; Talloczy, Z.; Hagemann, T.; Cho, W.; Messing, A.; Sulzer, D.L.; Goldman, J.E. Autophagy induced by Alexander disease-mutant GFAP accumulation is regulated by p38/MAPK and mTOR signaling pathways. *Hum. Mol. Genet.* **2008**, *17*, 1540–1555. [CrossRef]
42. Tang, G.; Xu, Z.; Goldman, J.E. Synergistic effects of the SAPK/JNK and the proteasome pathway on glial fibrillary acidic protein (GFAP) accumulation in Alexander disease. *J. Biol. Chem.* **2006**, *281*, 38634–38643. [CrossRef] [PubMed]
43. Hagemann, T.L. Alexander disease: Models, mechanisms, and medicine. *Curr. Opin. Neurobiol.* **2021**, *72*, 140–147. [CrossRef] [PubMed]
44. Hagemann, T.L.; Paylor, R.; Messing, A. Deficits in adult neurogenesis, contextual fear conditioning, and spatial learning in a Gfap mutant mouse model of Alexander disease. *J. Neurosci.* **2013**, *33*, 18698–18706. [CrossRef]
45. Liu, Y.; Holdbrooks, A.T.; De Sarno, P.; Rowse, A.L.; Yanagisawa, L.L.; McFarland, B.C.; Harrington, L.E.; Raman, C.; Sabbaj, S.; Benveniste, E.N.; et al. Therapeutic efficacy of suppressing the Jak/STAT pathway in multiple models of experimental autoimmune encephalomyelitis. *J. Immunol.* **2014**, *192*, 59–72. [CrossRef]
46. Qin, H.; Buckley, J.A.; Li, X.; Liu, Y.; Fox, T.H., 3rd; Meares, G.P.; Yu, H.; Yan, Z.; Harms, A.S.; Li, Y.; et al. Inhibition of the JAK/STAT Pathway Protects Against α-Synuclein-Induced Neuroinflammation and Dopaminergic Neurodegeneration. *J. Neurosci.* **2016**, *36*, 5144–5159. [CrossRef]
47. McFarland, B.C.; Ma, J.Y.; Langford, C.P.; Gillespie, G.Y.; Yu, H.; Zheng, Y.; Nozell, S.E.; Huszar, D.; Benveniste, E.N. Therapeutic potential of AZD1480 for the treatment of human glioblastoma. *Mol. Cancer Ther.* **2011**, *10*, 2384–2393. [CrossRef] [PubMed]
48. Groot, J.; Ott, M.; Wei, J.; Kassab, C.; Fang, D.; Najem, H.; O'Brien, B.; Weathers, S.P.; Matsouka, C.K.; Majd, N.K.; et al. A first-in-human Phase I trial of the oral p-STAT3 inhibitor WP1066 in patients with recurrent malignant glioma. *CNS Oncol.* **2022**, *11*, Cns87. [CrossRef]
49. Vanderver, A.; Adang, L.; Gavazzi, F.; McDonald, K.; Helman, G.; Frank, D.B.; Jaffe, N.; Yum, S.W.; Collins, A.; Keller, S.R.; et al. Janus Kinase Inhibition in the Aicardi-Goutières Syndrome. *N. Engl. J. Med.* **2020**, *383*, 986–989. [CrossRef]
50. Verstovsek, S.; Hoffman, R.; Mascarenhas, J.; Soria, J.C.; Bahleda, R.; McCoon, P.; Tang, W.; Cortes, J.; Kantarjian, H.; Ribrag, V. A phase I, open-label, multi-center study of the JAK2 inhibitor AZD1480 in patients with myelofibrosis. *Leuk. Res.* **2015**, *39*, 157–163. [CrossRef]
51. Scott, B.L.; Becker, P.S. JAK/STAT Pathway Inhibitors and Neurologic Toxicity: Above All Else Do No Harm? *JAMA Oncol.* **2015**, *1*, 651–652. [CrossRef]
52. Mehla, J.; Singh, I.; Diwan, D.; Nelson, J.W.; Lawrence, M.; Lee, E.; Bauer, A.Q.; Holtzman, D.M.; Zipfel, G.J. STAT3 inhibitor mitigates cerebral amyloid angiopathy and parenchymal amyloid plaques while improving cognitive functions and brain networks. *Acta Neuropathol. Commun.* **2021**, *9*, 193. [CrossRef] [PubMed]

cells

MDPI

Article

Cortical Pathology in Vanishing White Matter

Jodie H. K. Man [1,2,3], Charlotte A. G. H. van Gelder [4,5], Marjolein Breur [1,2,3], Daniel Okkes [1,2,3], Douwe Molenaar [6], Sophie van der Sluis [7], Truus Abbink [1,2,3], Maarten Altelaar [4,5], Marjo S. van der Knaap [1,2,3,8] and Marianna Bugiani [2,3,9,*]

1 Department of Child Neurology, Emma Children's Hospital, Amsterdam University Medical Centers, VU University Amsterdam, 1081 HV Amsterdam, The Netherlands
2 Amsterdam Leukodystrophy Center, Emma Children's Hospital, Amsterdam University Medical Centers, 1081 HV Amsterdam, The Netherlands
3 Molecular and Cellular Mechanisms, Amsterdam Neuroscience, 1081 HV Amsterdam, The Netherlands
4 Biomolecular Mass Spectrometry and Proteomics, Bijvoet Center for Biomolecular Research and Utrecht Institute for Pharmaceutical Sciences, University of Utrecht, 3584 CS Utrecht, The Netherlands
5 Netherlands Proteomics Center, 3584 CS Utrecht, The Netherlands
6 Department of Systems Bioinformatics, VU University Amsterdam, 1081 HV Amsterdam, The Netherlands
7 Department of Child and Adolescent Psychology and Psychiatry, Complex Trait Genetics, Amsterdam Neuroscience, VU University Medical Center, 1081 HV Amsterdam, The Netherlands
8 Department of Integrative Neurophysiology, Center for Neurogenomics and Cognitive Research, VU University Amsterdam, 1081 HV Amsterdam, The Netherlands
9 Department of Pathology, Amsterdam University Medical Centers, 1081 HV Amsterdam, The Netherlands
* Correspondence: m.bugiani@amsterdamumc.nl; Tel.: +31-6-48517239

Abstract: Vanishing white matter (VWM) is classified as a leukodystrophy with astrocytes as primary drivers in its pathogenesis. Magnetic resonance imaging has documented the progressive thinning of cortices in long-surviving patients. Routine histopathological analyses, however, have not yet pointed to cortical involvement in VWM. Here, we provide a comprehensive analysis of the VWM cortex. We employed high-resolution-mass-spectrometry-based proteomics and immunohistochemistry to gain insight into possible molecular disease mechanisms in the cortices of VWM patients. The proteome analysis revealed 268 differentially expressed proteins in the VWM cortices compared to the controls. A majority of these proteins formed a major protein interaction network. A subsequent gene ontology analysis identified enrichment for terms such as cellular metabolism, particularly mitochondrial activity. Importantly, some of the proteins with the most prominent changes in expression were found in astrocytes, indicating cortical astrocytic involvement. Indeed, we confirmed that VWM cortical astrocytes exhibit morphological changes and are less complex in structure than control cells. Our findings also suggest that these astrocytes are immature and not reactive. Taken together, we provide insights into cortical involvement in VWM, which has to be taken into account when developing therapeutic strategies.

Keywords: leukodystrophy; astrocytopathy; vanishing white matter; cortex; proteomics; astrocytes

Citation: Man, J.H.K.; van Gelder, C.A.G.H.; Breur, M.; Okkes, D.; Molenaar, D.; van der Sluis, S.; Abbink, T.; Altelaar, M.; van der Knaap, M.S.; Bugiani, M. Cortical Pathology in Vanishing White Matter. *Cells* **2022**, *11*, 3581. https://doi.org/10.3390/cells11223581

Academic Editors: Michael Brenner and Vladimir Parpura

Received: 25 July 2022
Accepted: 10 November 2022
Published: 12 November 2022

1. Introduction

Vanishing white matter (VWM) is one of the more prevalent leukodystrophies [1]. The disease most often has its onset in young children, but may occur at any age from the prenatal period to senescence [2,3]. Patients experience chronic neurological decline and are susceptible to stressors, including fever, infections and minor trauma [3–5]. These stressors may lead to episodes of rapid deterioration, which may result in a coma and sometimes death, but are most often followed by a partial recovery [3–5]. VWM is caused by bi-allelic mutations in any of the 5 genes encoding the subunits of the eukaryotic translation initiation factor 2B (eIF2B) [6,7]. eIF2B is a housekeeping factor crucial for the translation initiation of mRNA into proteins [8]. It also orchestrates the integrated stress response (ISR), a cell protective mechanism activated in response to cellular stress [9]. eIF2B in this way serves as a central regulator of the mRNA translation rate under normal and stress conditions [6,7]. The ISR is constitutively deregulated in VWM [10].

In VWM, the brain white matter is allegedly selectively affected, showing a lack of myelin and cystic degeneration [3–5]. There is meagre reactive gliosis and no glial scarring [3–5,11]. In the affected areas, astrocytes are immature and dysmorphic, and they overexpress the delta subunit of the glial fibrillary acidic protein (GFAPδ), indicating an abnormal intermediate filament composition [11]. Additionally, oligodendrocyte maturation is halted in a premyelinating state, explaining the lack of myelin [3–5,11]. Axons are lost in cavitated areas or appear thinned and swollen in more preserved tissue [12–14]. Recent evidence has pointed to astrocytes as primary drivers in the pathogenesis of VWM, with secondary defects in oligodendrocytes and axons [14–16]. Co-cultures of oligodendrocyte progenitor cells (OPCs) and astrocytes derived from wild-type and VWM mice showed that VWM astrocytes halt wild-type OPC maturation, whereas mutant OPCs mature normally when cultured with wild-type astrocytes [15]. Additionally, VWM astrocytes induce axonal pathology in myelinating culture systems containing wild-type mouse neurons [14]. On the basis of these findings, VWM is categorized under the astrocytopathies [5].

Magnetic resonance imaging (MRI) documents the progressive thinning of the cortices in long-surviving patients [3]. Post-mortem routine analyses, however, have not pointed to any cortical pathology so far [4,17,18]. This finding prompted us to explore the cortex in VWM in more depth using high-resolution-mass-spectrometry-based proteomics.

2. Materials and Methods

2.1. Patients

Post-mortem brain tissue from 4 genetically proven VWM patients was collected at the Amsterdam UMC location, VU university Amsterdam (Amsterdam, The Netherlands). All patients had a childhood onset of VWM. The controls included 4 subjects without confounding structural and neuropathological abnormalities obtained from the Netherlands Brain Bank (NBB, www.brainbank.nl). Cause of death of the latter was cardio-respiratory arrest, possibly due to arrhythmia. Demographic features of controls and patients are shown in Table 1. Tissue was collected within 6 h post-mortem. This study focused on the cortex of the middle frontal gyrus. Informed consent was obtained in all cases. The study was approved by the institutional review board of the Amsterdam UMC location, VU University Amsterdam (Amsterdam, The Netherlands) and conducted according to the declaration of Helsinki.

Table 1. Demographic features of controls and patients.

Subject	NBB Reference No.	Age at Death	Molecular Diagnosis
Control 1	2003–085	24 years	-
Control 2	2017–029	23 years	-
Control 3	2018–010	21 years	-
Control 4	2018–066	35 years	-
VWM 1	-	29 years	ε, Thr91Ala/Thr91Ala *
VWM 2	-	10 years	ε, Arg113His/Ala403Val *
VWM 3	-	6 years	ε, Thr91Ala/Val437Met *
VWM 4	-	12 years	ε, Thr91Ala/Ala403Val *

* Mutation: mutant eIF2B subunit and amino acid changes are indicated.

2.2. Fast Immunohistochemistry and Laser Capture Microdissection

Twenty μm thick frozen tissue sections were mounted on glass slides coated with polyethylene naphthalate, fixed in 100% ethanol for 20 min, and subsequently air-dried and rehydrated twice in sterile H_2O for 1 min. Next, tissue sections were incubated in toluidine (1% *w*/*v* in sterile H_2O) for 1 min at room temperature and washed in sterile H_2O 3 times for 1 min. Sections were then dehydrated in 100% ethanol twice for 3 min and air-dried. Laser capture microdissection was performed using a Leica LMD6500 system

(Leica Microsystems, Wetzlar, Germany). All cortical layers from the middle frontal gyrus were included. Microdissected tissue samples of 100 mm^3 were collected into adhesive caps (Zeiss, Göttingen, Germany) and stored at −80 °C until use.

2.3. In-Solution Protein Digestion

Microdissected samples were lysed, reduced and alkylated in lysis buffer (6 M guanidine hydrochloride, 5 mM tris(2-carboxyethyl)phosphine, 10 mM chloroacetamide and 100 mM Tris-HCl pH 8.0) in 50 mM ammonium bicarbonate with pH 8.0–8.5 for 10 min at 99 °C with mixing at 750 RPM. Samples were sonicated with 20 mg protein extraction beads (Diagenode, Seraing, Belgium) using a Bioruptor Plus (Diagenode, Seraing, Belgium) for 20 cycles with on/off pulses of 30 s. Protein (20 µg/sample) was then digested with Lys-C at an enzyme-to-substrate ratio of 1:100 at 37 °C for 4 h. Samples were next diluted to a final concentration of 2 M guanidine hydrochloride using 25 mM Tris-HCl (pH 8.0) and processed for overnight digestion with trypsin at an enzyme-to-substrate ratio of 1:100 at 37 °C. Peptide samples were then acidified using 10% formic acid with pH < 2.0. Finally, peptides were extracted using C18 cartridges (Agilent Technologies, Amstelveen, The Netherlands) and an automated AssayMAP Bravo Platform (Agilent Technologies, Amstelveen, The Netherlands) according to standard procedures. Eluted peptides were dried using a SpeedVac centrifuge at 37 °C and stored at −20 °C until analysis.

2.4. Mass Spectrometry Analysis

Peptide samples (1.5 µg/sample) were resuspended in 2% formic acid and analyzed using an Orbitrap Q-Exactive HF-X mass spectrometer (Thermo Fisher Scientific, Bleiswijk, The Netherlands) coupled to an UltiMate 3000 RSLCnano System (Thermo Fisher Scientific, Bleiswijk, The Netherlands). Peptides were loaded on a trap column (300 µm id × 5 mm, 5 µm particle size, reversed phase C18, Thermo Fisher Scientific, The Netherlands) at a flow rate of 10 µL/min in 100% solvent A (0.1% formic acid). Next, peptides were separated with an analytical column made in-house with solvent B (80% acetonitrile/0.1% formic acid) at a flow rate of 300 nL/min. The following gradient was used: 0–2.5 min with 92% solvent A; 2.5–157.5 min with 8% solvent B; 157.5–161 min with 35% solvent B; 161–165 min with 100% solvent B; and 165–175 min with 92% solvent A. Total run per sample was 175 min. Spectra were collected using a data-dependent acquisition method in which the top 15 precursors were selected from a full MS1 scan (m/z range = 315–1500, resolution = 60,000, target = 3×10^6 ions) for MS2 analysis with higher-collision dissociation (target ions = 1×10^5, max ion fill time = 50 milliseconds, isolation window = 1.4 m/z, normalized collision energy = 27%, resolution = 30,000). Precursor ions were excluded from fragmentation if charge state was unassigned or equal to 1, 6, 7, 8 or higher than 8. A dynamic exclusion of 16 s was included.

2.5. Mass Spectra Processing

RAW files were processed and searched against the Swiss-Prot human reference proteome database (20,431 entries, accessed on August 2019) using MaxQuant (version 1.6.6.0, Martinsried, Germany) [19]. Label-free quantification (LFQ) was performed by MaxLFQ with "match between runs" enabled, and minimal ratio count set at 2. False discovery rate (FDR) threshold was set to 1%. The minimum and maximum peptide length ranged from 7 to 25 amino acids. Carbamidomethylation of cysteine was set as fixed modification, and peptide N-terminal acetylation and methionine oxidation were set as variable modifications. A maximum of 2 missed trypsin cleavages was allowed. The maximum peptide and fragmentation mass tolerance was set at 20 ppm and 0.5 Da, respectively.

2.6. Differential Protein Expression Analysis

Statistical data analysis was conducted using R statistical software (version 1.3.959). Proteins identified as contaminants, only by site modification or in decoy reverse database were removed prior to analysis. Data were then log 2-transformed. Missing values were imputed based on a normal distribution in which the lowest detected LFQ intensity was set as mean and the average standard deviation of all detected proteins was set as standard deviation. Only proteins in which expression was detected in all subjects in either control or disease state were imputed. Differential protein expression analysis was carried out using the limma R package [20]. Statistical significance was set as p-value < 0.05 with Benjamini–Hochberg's (BH) test for multiple comparison adjustment. Data were then further filtered to only include proteins showing at least ±1.5-fold change (FC). Hierarchical clustering analysis was performed using the ComplexHeatmap R package [21]. LFQ intensities were z-scored and clustered using the Euclidean and average methods as distance and clustering measures, respectively. The transcriptome dataset from human cortical fetal astrocytes, mature astrocytes, neurons, oligodendrocytes, and microglia (GSE73721) [22] was used to generate cell type marker lists. These lists were then used to identify potential cell-type-enriched protein expression changes in the proteome dataset. In this context, the term enriched refers to the expression of a marker in a specific cell type relative to its expression across all other cell types. Only markers enriched >5-fold in a given cell type over all other cell types were included. A Fisher's exact test was performed to find significant cell type enrichment (p-value < 0.05).

2.7. Protein–Protein Interaction Network Analysis

Protein–protein interaction network was predicted with STRING (https://string-db.org/, accessed on 29 July 2022) and gene ontologies of the largest subnetwork were analyzed [23]. For the protein–protein interaction network, the confidence (score) cutoff and maximal additional interactions were set as 0.40 and 0, respectively. Data were visualized using Cytoscape (version 3.9.1, Seattle, WA, USA) [24]. The following gene ontology terms were included: cellular component, molecular function, biological processes, KEGG pathways, and REACTOME. Statistical significance was set as p-value < 0.05. For visualization, only clusters with more than 4 nodes are shown.

2.8. Immunohistochemical Staining

Five µm thick formalin-fixed paraffin-embedded tissue were deparaffinized, rehydrated and incubated in 0.3% (w/v) H_2O_2 in dH_2O for 30 min to block endogenous peroxidase activity. Heat-induced antigen retrieval was carried out in 10 mM citrate buffer (pH 6.0) or Tris/EDTA buffer (pH 9.0). Tissue was then stained according to standard protocols using the antibodies listed in Table 2. Immunopositivity was detected using 3,3′-diaminobenzidine chromogen or chromogen-bound Alexa Fluor®-labelled secondary antibodies. Hematoxylin or DAPI-Fluoromount-G® (ITK diagnostics, Uithoorn, The Netherlands) were used as counterstains. Images were taken using a Leica DM4000B light microscope or a Leica DM5000B fluorescence microscope.

Table 2. List of antibodies used for immunohistochemistry.

Antibody	Dilution	Vendor	Catalogue No.
GFAP	1:1000	DAKO	Z0334
GFAP	1:750	Merck Millipore	AB5541
Vimentin	1:2000	Clone V9 *	-
GFAPδ	1:500	†	-
Nestin	1:200	Millipore	MAB5623
GJA1	1:100	Sigma-Aldrich	C6219
GLUL	1:5000	ThermoFisher	PA5-81505
HRSP12	1:100	ThermoFisher	PA5-54622
ACTG1	1:100	Abcam	ab123034
TNC	1:200	Abcam	ab108930

Table 2. *Cont.*

Antibody	Dilution	Vendor	Catalogue No.
ACTB	1:1500	Abcam	ab8226
S100β	1:100	DAKO	Z311

* Manufactured at the department of Pathology, Amsterdam UMC, The Netherlands. † Manufactured at the Netherlands Institute for Neuroscience, Amsterdam, The Netherlands [25].

2.9. Astrocyte Morphology Analysis

Astrocyte morphology was assessed on GFAP-stained brain tissue using the ImageJ Sholl analysis plug-in. GFAP-positive astrocytes were manually traced. Sholl analysis was then performed, in which concentric rings with increasing sizes were placed around the astrocytes starting from the center of the cell body [26]. The distance between each concentric ring was 3 μm. The following parameters were analyzed: total number of intersections, maximum process length and process thickness.

3. Results

3.1. Proteomic Map of the VWM Cortex

To explore the proteome of the VWM cortex, high-resolution-mass-spectrometry was employed. We included the pan-cortical tissue of the middle frontal gyrus from 4 VWM subjects and 4 controls (see Table 1). In total, 3239 proteins were identified (Table S1). Of these proteins, 268 were differentially expressed (log 2FC at least ± 1.5 and BH adjusted p-value < 0.05) in VWM compared to the control (Table S2). The hierarchical clustering analysis of these proteins showed that the VWM samples were clearly separated from the control samples, forming distinct clusters, in which samples from biological replicates aggregated together. In total, 2 protein clusters were identified: cluster 1 contained 43 proteins with high expression in the VWM cortex; and cluster 2 contained 225 proteins with low expression in the VWM cortex (Figure 1 and Table S3).

Considering that our proteomics analysis was performed on the whole cortex of the frontal middle gyrus, the identified proteins may represent changes in multiple cell types. We therefore next examined cell-type-enriched protein expression changes in the cortices of VWM patients. In order to explore this, a transcriptomic dataset from human cortical fetal astrocytes, mature astrocytes, neurons, oligodendrocytes and microglia was analyzed and compared to our proteomic data [22]. Amongst the 43 upregulated proteins, 6 were enriched in fetal astrocytes, and 2 in neurons. Of the 225 downregulated proteins, 4 were enriched in fetal astrocytes, 33 in mature astrocytes, 16 in neurons, 10 in oligodendrocytes and 5 in microglia (Table S4). A Fisher's exact test revealed that amongst the overexpressed proteins, there was an enrichment in fetal astrocyte-enriched proteins ($p = 0.039$), while amongst the underexpressed proteins, there was an enrichment for mature astrocytes ($p = 4.5 \times 10^{-20}$), oligodendrocytes ($p = 2.3 \times 10^{-4}$) and neurons ($p = 0.017$).

3.2. Protein–Protein Interaction Network

To understand the organization of the differentially expressed proteins, we applied a protein–protein interaction network approach. This uncovered a protein network with one major connected component harboring 234 proteins, in which over- and under-expressed proteins clustered together (Figure 2A). We also observed several unconnected nodes, accounting for 34 proteins (Table S5A). Next, to gain insight into possible biological functions and pathways altered in the VWM cortex, a gene ontology enrichment analysis was performed on the differentially expressed proteins (Table S5B). In VWM cortices, differentially expressed proteins were mainly located in the extracellular exosomes, cytoplasm, vesicles and extracellular space (Figure 2B). Some of the network proteins were also enriched for gene ontology terms related to cellular metabolism, such as the oxidation–reduction process, oxidoreductase activity and NAD binding (Figure 2B). Consistently, we found an over-representation of terms for metabolic pathways, particularly glycolysis/gluconeogenesis, pyruvate metabolism, fatty acid degradation, the metabolism of amino acids and derivatives, the citric acid (TCA) cycle and respiratory electron transport body (Figure 2B).

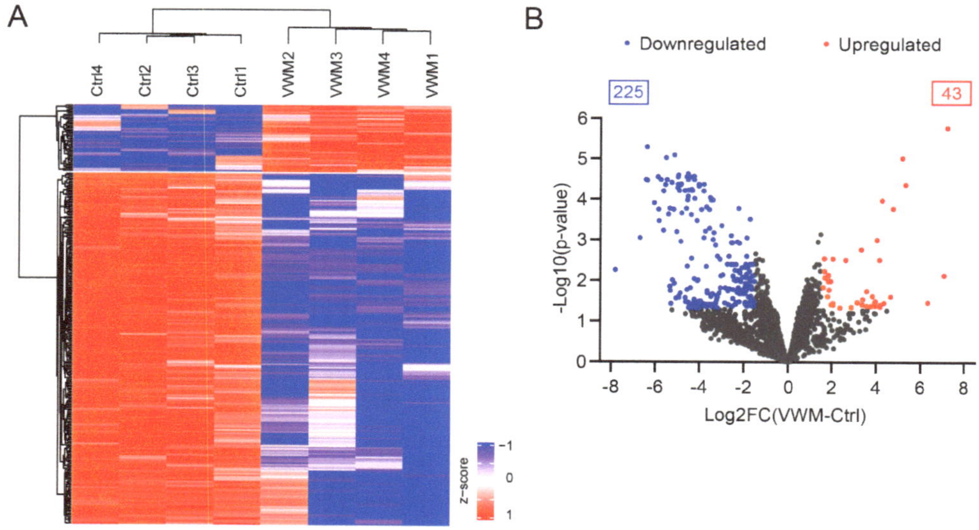

Figure 1. VWM cortex shows substantial differences from control cortex at the protein level. (**A**) Heat map of z-scored abundances (LFQ intensities) of the 268 significantly differentially expressed proteins in VWM cortex compared to control cortex. Hierarchical clustering revealed 2 separate expression clusters. (**B**) Differential protein expression in control versus VWM cortex. Volcano plot displaying significant differentially expressed proteins with at least ±1.5-fold change. Proteins downregulated and upregulated in VWM are colored in blue and red, respectively (adj. p-value < 0.05). Not significantly differentially expressed proteins (adj. p-value > 0.05) and proteins with fold changes between ±1.5 are colored in gray.

The cell type enrichment analysis of significantly differentially expressed proteins revealed a distinct cell type involvement in VWM cortices (Table S4). To further stratify the organization of the differentially expressed proteins from mixed cell populations, we repeated the analysis, only taking into account the proteins that we identified as enriched in different cell types. Among the 76 brain cell-type-enriched proteins, we found that 54 clustered together, forming a subnetwork (Figure 3A and Table S6A). A subsequent gene ontologies analysis revealed that these cell-type-enriched proteins were located in extracellular exosomes, vesicles, extracellular space, extracellular regions and neuron projections (Figure 3B and Table S6B). The protein network also showed enrichment for biological processes related to transport (Figure 3B and Table S6B). Notably, some proteins in this network were also enriched for gene ontology terms associated with cellular metabolism, especially oxidoreductase activity, with 12 out of 54 cell-type-enriched proteins overrepresented in that term. Amongst these 12 proteins, the majority were astrocyte-enriched, suggesting that cellular metabolism might be particularly affected in these cells in the VWM cortex (Table S6B). In summary, our findings suggest that the majority of the differentially expressed proteins physically or functionally interact with each other (Figures 2 and 3). The protein–protein interaction network analysis only revealed a single large subnetwork and no other specific clusters were evident.

3.3. Localization and Distribution of Differentially Expressed Proteins in VWM

To confirm our proteomic findings, we selected the top 3 up- and downregulated proteins in VWM compared to the control cortex (Table S2). The top 3 upregulated proteins were actin gamma 1 (ACTG1), tenascin C (TNC) and actin beta (ACTB). The top 3 downregulated proteins were glutamate-ammonia ligase (GLUL), heat-responsive protein 12 (HRSP12) and gap-junction alpha-1 protein (GJA1, also known as connexin-43). All proteins except for HRSP12 were part of the largest subnetwork (Figure 2A). Notably, based on the cell type enrichment analysis (Table S4), some proteins were enriched for fetal astrocytes (i.e., TNC) and mature astrocytes (i.e., GLUL, HRSP12 and GJA1). Neither ACTG1 and ACTB were assigned to a particular cell type (Table S4). To gain insight into the localization and distribution of these proteins in the cortex of VWM patients and the controls, immunohistochemistry was performed on all VWM patients and the controls (the results were comparable in each group).

We started with the top 3 upregulated proteins: ACTG1, TNC and ACTB. The immunohistochemistry against ACTG1 showed low expression in the control cortex and high expression in the VWM cortex. ACTG1 expression was found in between cell bodies, but also in neurons and in blood vessels (Figure 4A,B). TNC showed increased immunoreactivity in the cortex of VWM patients. In the control cortex, TNC expression was detected in only a few cells. Based on nuclear size and morphology, these are presumably glia and neurons. In the VWM cortex, TNC expression was predominantly extracellular (Figure 4C,D). Low immunoreactivity for ACTB was present in the control cortex. In the VWM cortex, ACTB immunoreactivity was increased, consistent with the differences observed in our proteomics findings. Remarkably, the expression was restricted to non-neuronal cells, possibly glia, based on morphology (Figure 4E,F).

Next, we looked into the expression patterns of the top 3 downregulated proteins in the VWM cortex: GLUL, HRSP12 and GJA1. GLUL was localized in astroglia-like cells in both the control and VWM cortices. The immunoreactivity in between cells, however, was decreased in the VWM cortex compared to the control cortex. Notably, in the control cortex, GLUL-positive cells were extensively branched, whereas in the VWM cortex, they were less complex in morphology (Figure 5A,B). HRSP12 expression was found in astrocyte-like cells in the control and VWM cortices. Overall immunoreactivity appeared less in the VWM cortex compared to the controls (Figure 5C,D). GJA1 was detected in cells with astrocyte morphology in both the control and VWM cortices. The immunoreactivity, however, was decreased in between cells in the VWM cortex. Notably, GJA1 and GLUL share similar expression patterns. In this, GJA1-positive cellular processes also appeared more complex in morphology in the control cortex compared to those in VWM (Figure 5E,F).

Notably, amongst the 6 validated proteins, 4 were found expressed in cells with glia morphology, particularly cells with astrocytic morphology and nuclear features. These proteins include ACTB, GLUL, HRSP12 and GJA1. We next set out to confirm their expression in astrocytes. Double-labeling with GFAP confirmed the expression of GLUL, HRSP12 and GJA1 in astrocytes in the control cortex (Figure 6), whereas ACTB was absent, consistent with the immunohistochemistry (Figure 4E). In VWM, the expression in astrocytes was only partly confirmed (Figure 6). The ACTB expression was clearly upregulated in the VWM cortex, with the expression restricted to cells (Figure 6B). These cells were, however, not positive for GFAP. Since cortical astrocytes might express GFAP below the level of antibody detection, we also stained for Ca^{2+}- and Zn^{2+}-binding protein S100β [27]. Double-labeling with S100β showed little co-expression with ACTB, indicating that ACTB is expressed mainly in other cells than astrocytes (Figure 6B). GLUL-positive cells co-localized with GFAP in VWM (Figure 6F), whereas HRSP12 and GJA1 expression was not detected (Figure 6G,H). Since HRSP12 and GJA1 were downregulated in the VWM cortex, the absence of these proteins might be explained by an antibody level below the detection threshold for immunofluorescence.

3.4. *Dysmorphic and Immature Astrocytes in the VWM Cortex*

Thus far, some of the proteins with the most prominent changes in expression in the VWM cortex were localized in astrocytes, especially GLUL, HRSP12 and GJA1 (Figures 5 and 6).

This, combined with the central role of astrocytes in VWM white matter pathology, prompted us to focus on this cell type in the cortex.

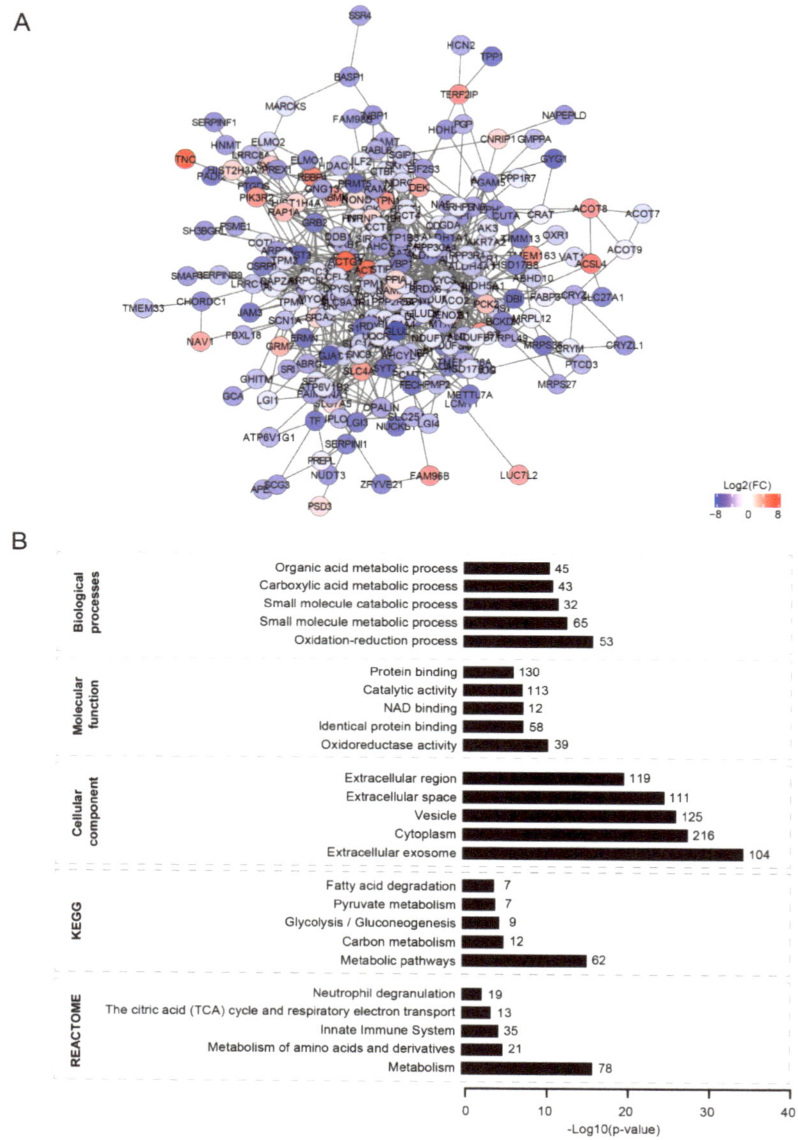

Figure 2. Protein–protein interaction network of differentially expressed proteins in VWM cortex compared to control cortex. (**A**) Protein–protein network of differentially expressed proteins in VWM cortex. The network consists of a single major component (largest subnetwork), harboring 234 of the differentially expressed proteins we identified. For visualization, only clusters with more than 4 nodes are shown. Red and blue nodes indicate upregulated and downregulated proteins, respectively. (**B**) Gene ontologies analysis of proteins clustered in the largest subnetwork. Graphs show the top 5 enriched terms (ranked by −log10 (p-value)) for each gene ontology. The number of proteins in each term is listed.

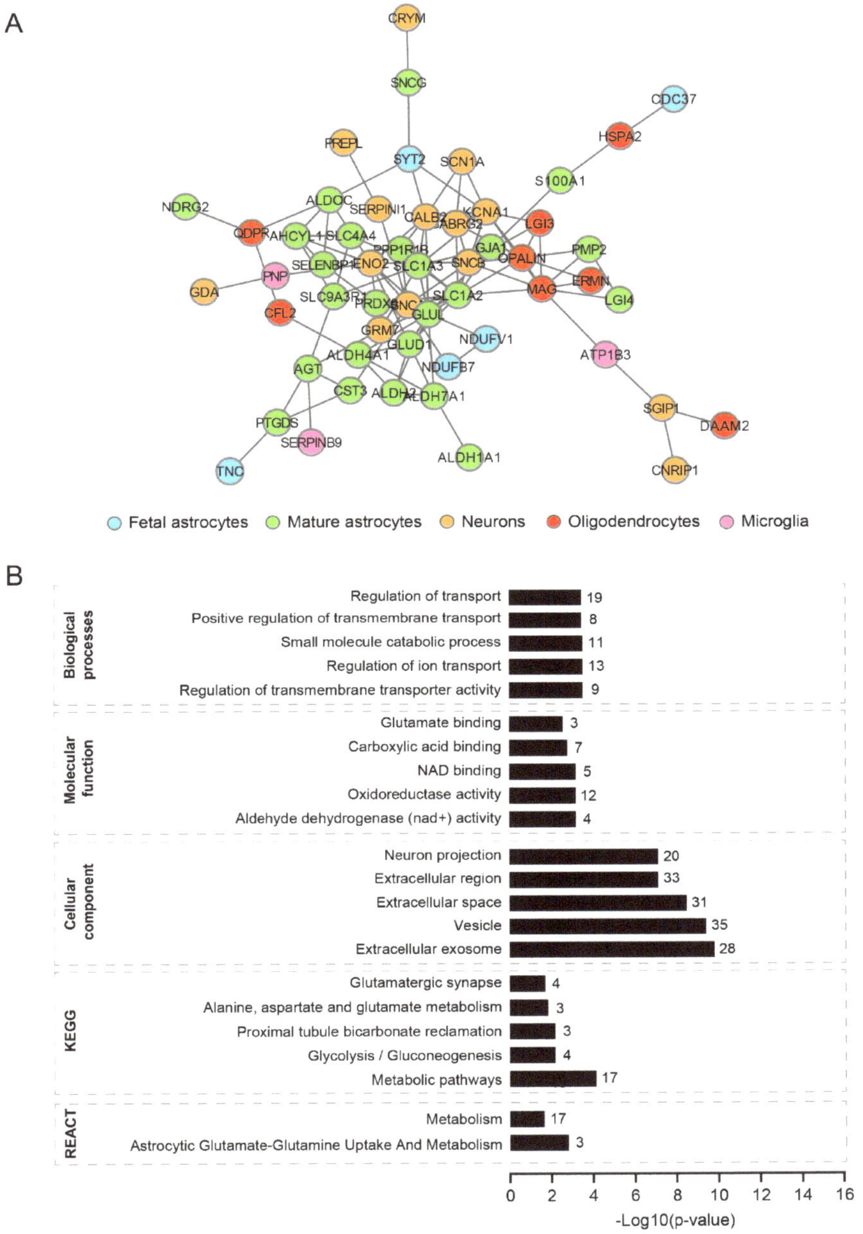

Figure 3. Protein–protein interaction network of differentially expressed proteins in different cell types. (**A**) Protein–protein network of differentially expressed and cell-type-enriched proteins in VWM cortex. The network consists of a large subnetwork, harboring 54 cell-type-enriched proteins. For visualization, only clusters with more than 4 nodes are shown. (**B**) Gene ontologies analysis of proteins clustered in the largest subnetwork. Graphs show the top 5 enriched terms (ranked by −log10 (*p*-value) for each gene ontology. The number of proteins in each term is listed. In (**B**), REACT is an abbreviation for REACTOME.

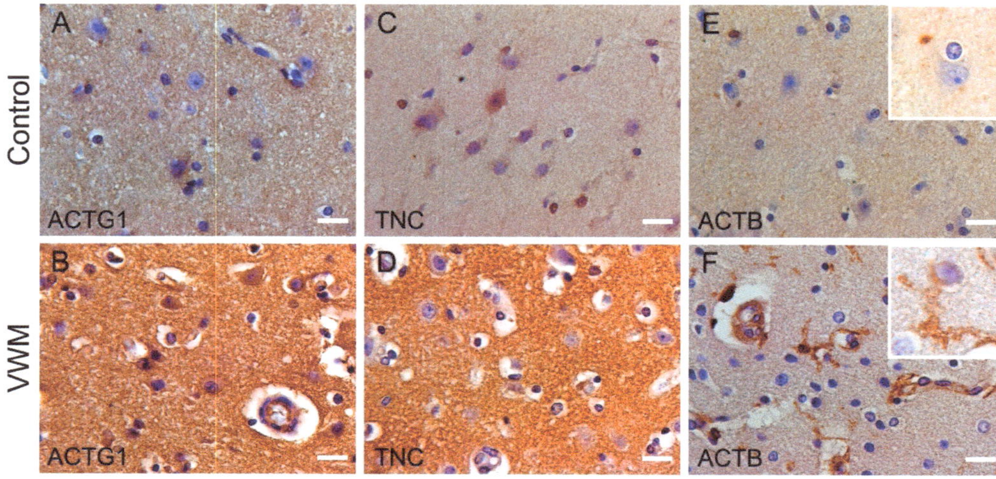

Figure 4. Expression pattern of ACTG1, TNC and ACTB in control and VWM cortices. (**A–F**) Immunohistochemistry against ACTG1, TNC and ACTB in control cortex (**upper panel**) and VWM cortex (**bottom panel**). Insets in (**E**,**F**) are higher magnification images, showing cellular immunoreactivity of ACTB in VWM cortex, but not in control cortex. Blue stain indicates nuclei. Scale bars: 20 μm. Depicted images are stains of cortical tissue of control 2 and VWM 1.

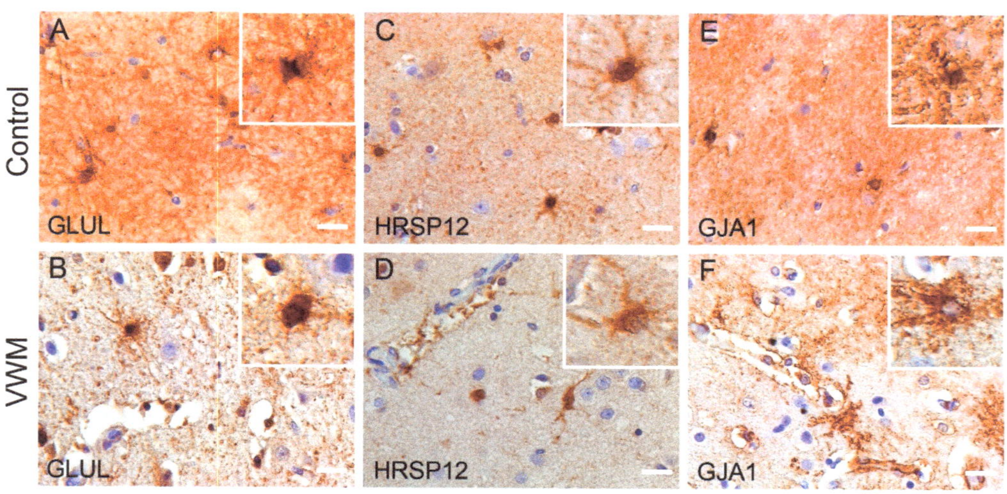

Figure 5. Expression pattern of GLUL, HRSP12 and GJA1 in control and VWM cortices. (**A–F**) Immunohistochemistry against GLUL, HRSP12 and GJA1 in control cortex (**upper panel**) and VWM cortex (**bottom panel**). Insets are higher magnification images, showing immunoreactive cells with astrocyte-like morphology. Blue stain indicates nuclei. Scale bars: 20 μm. Depicted images are stains of cortical tissue of control 4 and VWM 3.

To investigate whether cortical astrocytes in VWM show abnormalities, we performed a morphological analysis [26]. Typically, astrocytes in the cortex are highly branched with bushy processes [28], consistent with our control findings (Figure 7A). In the control, cortical astrocytes were only visible in layer I close to the glia limitans and around blood vessels. In the VWM cortex, astrocytes were detectable in all layers and were much less complex in morphology (data not shown). Although not as severely affected as the astrocytes in the VWM white matter (Figure S1), cortical astrocytes showed significantly less branched and shorter processes than those in controls (Figure 7B–D). No changes in process thickness were observed (Figure 7E).

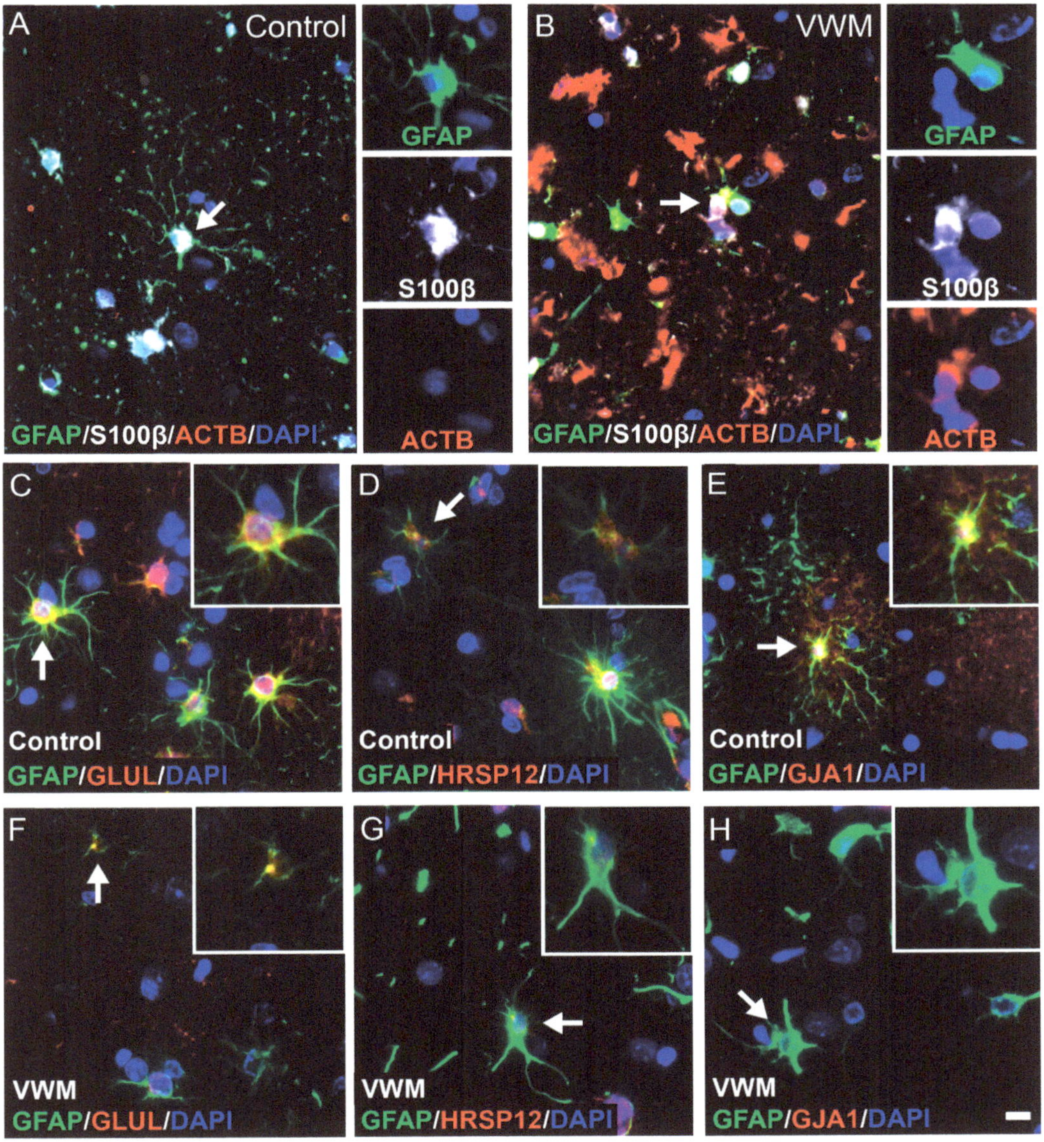

Figure 6. Cellular distribution of ACTB, GLUL, HRSP12 and GJA1 in control and VWM cortices. Triple stain against GFAP (green) and S100β (white) with ACTB (red) in control (**A**) and VWM (**B**) cortices. Double stain against GFAP (green) with GLUL, HRSP12 and GJA1 (red) in control (**C**–**E**) and VWM (**F**–**H**) cortices. Cells indicated by white arrows are displayed in insets. Blue stain indicates nuclei. Scale bar: 10 μm. Depicted images are stains of control 4 and VWM 2.

Specifically in VWM, and as in white matter, no reactive gliosis was seen in the cortex. VWM cortical astrocytes also co-expressed the intermediate filament protein vimentin (Figure 8A). Interestingly, VWM astrocytes throughout the cortex were also strongly immunopositive for GFAPδ, indicating a disrupted intermediate filament network (Figure 8B). Furthermore, they expressed the intermediate filament protein nestin, suggesting immaturity (Figure 8C). Very low, if any, GFAPδ and nestin expressions were found in the control cortical astrocytes (Figure 8D,E).

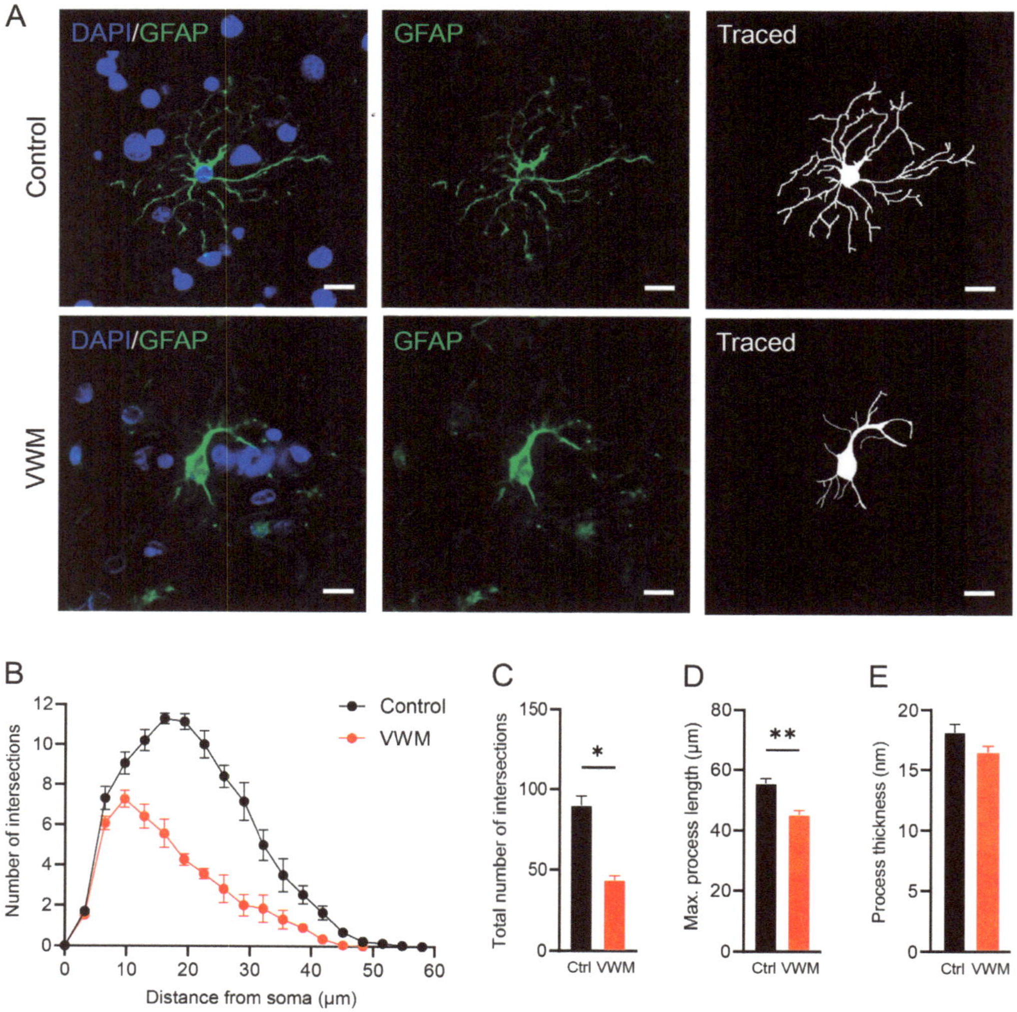

Figure 7. Astrocytes in the VWM cortex show reduced morphological complexity. (**A**) Immunostaining of control and VWM cortices showing GFAP-positive astrocytes (green) and their respective tracings. Blue stain indicates nuclei. Scale bars: 10 µm. Depicted images are stains of cortical tissue of control 1 and VWM 1. (**B**) Sholl diagram of astrocytes in the cortex of control and VWM brains. Diagram shows number of processes intersecting with concentric rings per 3 µm, starting from the cell body. (**C**) Total number of intersections. (**D**) Maximum process length. (**E**) Process thickness. * p-value < 0.05, ** p-value < 0.01.

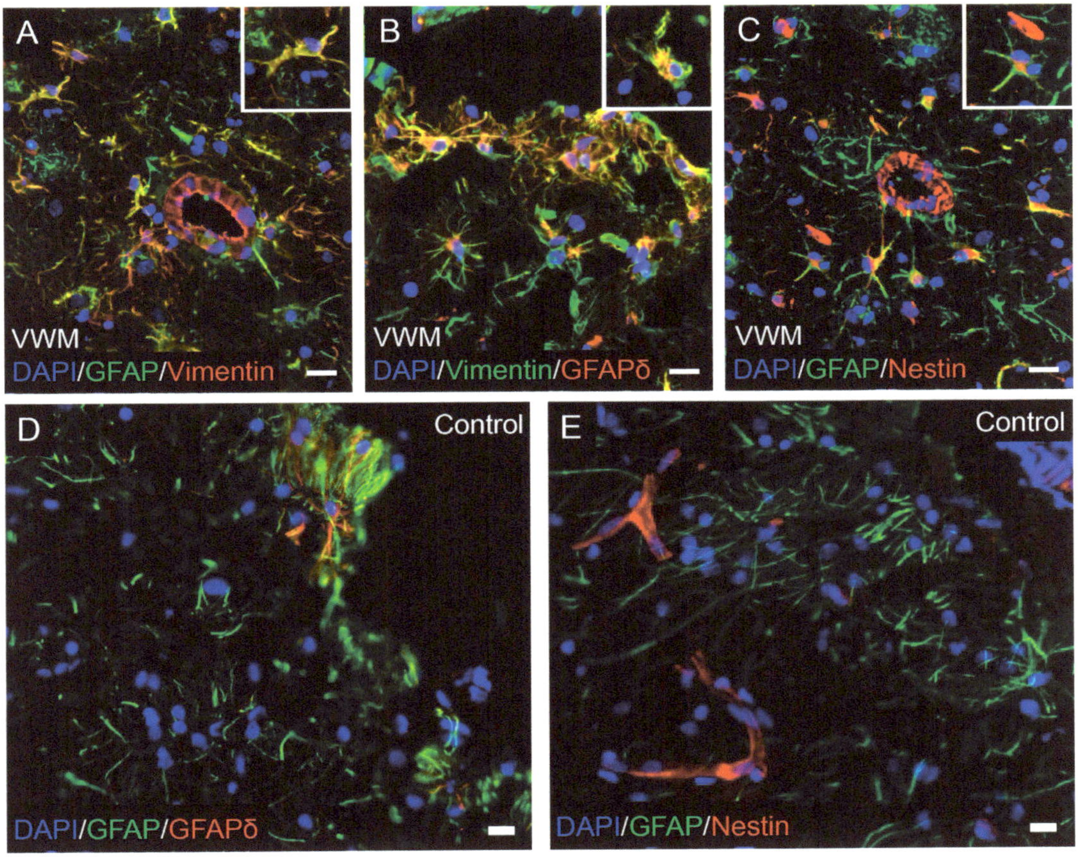

Figure 8. Astrocytes in the VWM cortex are immature and have a disrupted intermediate filament network. Double stain in VWM cortex against (**A**) GFAP (green) and vimentin (red), (**B**) vimentin (green) and GFAPδ (red), and (**C**) GFAP (green) and nestin (red). Insets show double-labeled cells. Double stain in control cortex for (**D**) GFAP (green) and GFAPδ (red), and (**E**) GFAP (green) and nestin (red). Blue stain indicates nuclei. Scale bars: 20 μm. Depicted images are stains of control 4 and VWM 3.

4. Discussion

In this study, we provide a comprehensive analysis of the VWM cortex. Our data show that the cortex, rather than being spared, is affected in VWM. The analysis of the cortices in the VWM patients at the protein level revealed substantial differences compared to those in the control subjects. Numerous differentially expressed proteins formed a large protein–protein interaction network. We also show that differentially expressed proteins enriched in different cell types were linked to each other. These findings indicate that the majority of differentially expressed proteins identified in our proteomics study physically or functionally interact with each other. The subsequent gene ontology analysis identified enrichment for various terms, including those associated with cellular metabolism. Our findings also hint at the likely astrocytic involvement in the VWM cortex. Some of the proteins with the most prominent changes were found in astrocytes. Additionally, a further analysis showed that VWM cortical astrocytes are immature and exhibit morphological changes resulting in a less complex structure.

To better understand cortical pathology in VWM, we first mapped the protein expression patterns using high-resolution-mass-spectrometry-based proteomics. We identified 268 proteins associated with VWM. The cell type enrichment analysis revealed that some of the higher-expressed proteins in the VWM cortex were most enriched in fetal astrocytes The set of lower-expressed proteins in the VWM cortex were enriched in mature astrocytes and, to a lesser extent, also oligodendrocytes and neurons. These findings suggest that multiple cell types contribute to or are affected in the VWM cortex. Moreover, it may also reflect changes in primary and secondary (e.g., cell autonomous and non-cell autonomous) cellular pathology. Notably, the enrichments found in the VWM cortex are consistent with the known pathology driving brain white matter degeneration of the disease, with astrocytes having a primary role and secondary effects on both oligodendrocytes and axons (neurons) [29]. This suggest that a similar cellular pathology could be present in the cortex.

Through the use of a protein–protein interaction network and a subsequent gene ontology term-enrichment analysis, we identified biological functions potentially altered in VWM cortices. In particular, there was enrichment in terms that were suggestive of altered cellular metabolism, particularly mitochondrial functions. Mitochondrial effects in astrocytes and oligodendrocytes have been described in experimental VWM mouse models [30–33]. This correlates with both glial cell types being predominantly affected in VWM [3,11]. We now add data suggesting that the mitochondrial function is changed in the cortices of VWM patients, hinting at a potential general defect in the VWM brain. Remarkably, an imaging phenocopy of VWM is mitochondrial leukodystrophy, due to respiratory chain complex II deficiency and *SDHA* mutations [34,35]. Both diseases feature diffuse white matter signal changes with cavitations in the cerebral white matter. It could be argued that a primary functional impairment of astrocytes is involved in both disorders, halting glia scarring. Currently, we do not know if mitochondrial changes are linked with the deregulated ISR in VWM gray matter astrocytes [10]; the ISR has been implicated in regulating several metabolic pathways [36].

Amongst the identified differentially expressed proteins, 6 proteins with the most prominent changes in expression were selected and validated by immunohistochemistry. Interestingly, the expression pattern of some of these proteins appear to coincide with that of astrocytes, in particular the lower-expressed proteins GLUL, HRSP12 and GJA1. GLUL and GJA1 are astrocytic proteins. GLUL is an enzyme involved in the conversion of glutamate and ammonium to glutamine, also known as glutamate metabolism. Glutamine plays an important role in cell metabolism, particularly in the tricarboxylic acid cycle and the biosynthesis of several amino acids, nucleotides and glutathione [37–40]. GJA1 (also known as connexin-43) is a transmembrane protein involved in the coupling of astrocytes to form domains within the so-called pan-glial syncytium [41–43]. Gap junctions allow the trafficking of molecules, including ions and metabolites, between cells [41–43]. HRSP12, on the other hand, is a reactive intermediate deaminase A protein that has not been associated with astrocytes before. Its functions are still unknown, but it is proposed to participate in cellular metabolism, such as 2-iminobutanoate deaminase activity and mRNA catabolic processes [44,45]. It should be noted that amongst the validated proteins, GLUL and HRSP12 relate to metabolic processes, supporting the possibility of altered cell metabolism. Other validated proteins (i.e., ACTG1 and TNC) were found in the extracellular matrix.

The fact that some of the most prominent changes were specifically observed in astrocytes suggests that cortical involvement in VWM may also be driven by changes in this cell type. The presence of dysmorphic astrocytes is a typical feature of VWM [3,11,15]. Astrocytes residing in the affected white matter usually have short blunt processes compared to the fine, long processes seen in the controls [3,11]. They remain immature and typically fail in mature astrocytic functions, including reactive gliosis and glial scarring [11]. Their intermediate filament composition is altered, with an absolute overexpression of GFAPδ, which could explain their dysmorphic appearance [11]. We now add to these findings and show that, although to a lesser degree than in the white matter, VWM cortical astrocytes are affected as well. In this, VWM cortical astrocytes exhibit abnormal morphology with less branched and shorter cellular processes. These astrocytes also appear immature, with an overexpression of GFAPδ, nestin and vimentin. The concurrent downregulation of nestin and vimentin is associated with astrocytic differentiation [46–48], supporting the possibility of an altered astrocytic maturation in VWM. Additionally, the cell type enrichment analysis showed that amongst the higher-expressed proteins in VWM, 6 were enriched in fetal astrocytes. Furthermore, amongst the lower-expressed proteins, 4 were enriched in fetal astrocytes and 33 were enriched in mature astrocytes. Maturation is a dynamic process, regulated by increasing the expression of some proteins and decreasing the expression of other proteins [49]. There is also no hint of reactive gliosis in the VWM cortex. The hallmarks of reactive gliosis are the induction of nestin and upregulation of GFAP and vimentin [49]. We found that VWM cortical astrocytes, such as reactive astrocytes, robustly express nestin and vimentin, but do not upregulate the *total* GFAP [11]. Additionally, in reactive astrocytes, GFAPδ represents only a small fraction of the total GFAP, not always immunohistochemically detectable, whereas GFAPδ expression is clearly visible in VWM cortical astrocytes. Moreover, VWM astrocytes are not hypertrophic. They do not show an increase in number nor in length of cellular processes, but rather the opposite [49]. All together, these data suggest immaturity rather than reactive gliosis. Our findings therefore suggest a potential role of astrocytes in both gray and white matter pathology in VWM. In line with this, a previous study using human-induced pluripotent stem cells further demonstrated astrocytic involvement in VWM, with white matter astrocytes showing higher vulnerability than gray matter astrocytes [50]. It remains unclear whether cortical astrocyte pathology relates to white matter pathology in VWM. In VWM, white matter astrocytes are severely abnormal. Nonetheless, the cortex has always reported to be normal [29]. Our new data show that gray matter astrocytes are also affected, although to a lesser degree. Astrocytes in the cortex, however, are completely different from white matter astrocytes [51]. The possibility of a knock-on effect cannot be excluded and warrants additional studies.

Some limitations apply to this work. Although VWM is one of the more prevalent leukodystrophies [1], it is still a very rare disease [2]. This results in the limited availability of human brain tissue. Our sample size was therefore restricted to 4 VWM cases. However, it should be noted that all patients included in this study had a childhood-onset form of VWM with comparable pathologies characterized by a lack of myelin, meager reactive gliosis and no glial scarring [3–5,11]. Some patients survive longer, while others die younger. The difference is in part explained by external factors (e.g., infections). Since there is also a limited availability of control human brain tissue of a high quality suitable for proteomics, especially from young healthy subjects, our sample size was restricted to 4 control cases. Notably, such tissue is extremely rare, especially that obtained within a short post-mortem period.

Another potential concern is the mismatch in ages between the patient and control groups with only 1 patient being of similar age. It is therefore possible that the changes in the abundance of certain proteins may be in part caused by aging. The substantial differences between the controls and the 29-year-old VWM patient and the fact that all patients cluster together, however, suggest that aging does not significantly impact the changes found.

The distribution and localization of several proteins identified by proteomics were mapped by immunohistochemistry. A validation by immunoblotting was not done because of several reasons. Firstly, the proteomics data indicate the relative levels of protein expression. Secondly, the tissue available was limited. Thirdly, the immunoblotting does not elucidate the regional expression. Although immunohistochemistry is not absolutely quantitative, it gives insight into the cell types expressing the proteins, and the location within the cortex these proteins are expressed.

The expression patterns of proteins selected for validation did not always coincide in tissue sections developed with diaminobenzidine chromogen or fluorochrome. In particular, strong cellular immunoreactivity for some proteins was detected with chromogen detection, whereas we were not able to confirm their expression with immunofluorescence. This might be due to the difference in the antibody detection method and threshold.

Another limitation is that our proteome analysis was performed on the whole tissue, and the data may represent changes in multiple cell types. To study the involvement of brain cell types in VWM, we applied a cell type enrichment analysis. This analysis is typically used on bulk transcriptomic datasets. There are currently no datasets available of single-cell-type proteomics of the human brain. For this reason, we compared our proteome dataset with a published human mRNA sequencing dataset [22]. It is, however, well known that the correlation between mRNA and protein levels is not straightforward, resulting in differences in specificity across cell types at the protein and mRNA levels [52,53]. The results should therefore be interpreted with caution. Future single-cell-type proteomics studies may further help us confirm our results.

5. Conclusions

In conclusion, our data support the notion that the frontal cortex is also affected in VWM. This has to be taken into account when developing therapeutic strategies.

Supplementary Materials: The following supporting information can be downloaded at: https://www.mdpi.com/article/10.3390/cells11223581/s1, Figure S1: Astrocytes show region-dependent vulnerability in VWM gray and white matter brain tissue; Table S1: List of all detected proteins in control and VWM cortex; Table S2: List of differentially expressed proteins in control versus VWM cortex; Table S3: Results obtained from hierarchical clustering analysis of significantly differentially expressed proteins; Table S4: List of cell-type-enriched differentially expressed proteins in VWM cortex; Table S5: Protein–protein interaction network analysis of significantly differentially expressed proteins; Table S6: Protein–protein interaction network analysis of significantly differentially expressed proteins in different cell types.

Author Contributions: Conceptualization, J.H.K.M., M.B. (Marianna Bugiani) and M.S.v.d.K.; methodology, J.H.K.M., C.A.G.H.v.G., M.B. (Marjolein Breur), D.O., D.M. and S.v.d.S.; software, J.H.K.M., D.O., D.M. and S.v.d.S.; validation, J.H.K.M., C.A.G.H.v.G., M.B. (Marjolein Breur) and D.O.; formal analysis, J.H.K.M. and M.B. (Marjolein Breur); investigation, J.H.K.M., C.A.G.H.v.G. and M.B. (Marjolein Breur); resources, M.A., M.B. (Marianna Bugiani) and M.S.v.d.K.; data curation, J.H.K.M.; visualization, J.H.K.M. and M.B. (Marjolein Breur); supervision, M.B. (Marianna Bugiani); project administration, M.B. (Marianna Bugiani) and M.S.v.d.K.; funding acquisition, M.S.v.d.K.; writing—original draft preparation, J.H.K.M. and M.B. (Marianna Bugiani); writing—review and editing, J.H.K.M., C.A.G.H.v.G., M.B. (Marjolein Breur), D.O., D.M., S.v.d.S., T.A., M.A., M.S.v.d.K. and M.B. (Marianna Bugiani). All authors have read and agreed to the published version of the manuscript.

Funding: The research was funded by the ZonMw TOP grant 91217006. S.v.d.S is supported by NWO Gravitation program BRAINSCAPES: A Roadmap from Neurogenetics to Neurobiology (NWO: 024.004.012). M.B. (Marianna Bugiani) is supported by the ZonMW VENI Grant 016.196.107.

Institutional Review Board Statement: The study was conducted in accordance with the Declaration of Helsinki and approved by the Institutional Review Board of the Amsterdam UMC location, VU University Amsterdam (2000.083, April 2000).

Informed Consent Statement: Informed consent was obtained from all subjects involved in the study by the Netherlands Brain Bank and the Amsterdam UMC location, VU University Amsterdam.

Data Availability Statement: The mass spectrometry proteomics data have been deposited into the ProteomeXchange Consortium via the PRIDE [54] partner repository with the dataset identifier PXD030831. Other datasets used during this study are available from the corresponding author on reasonable request.

Acknowledgments: We are grateful to all patients who participated in this study and their families, the Netherlands Brain Bank for providing the tissue of controls, and the autopsy assistants of the Amsterdam UMC. M.S.v.d.K. and M.B. (Marianna Bugiani) are members of the European Reference Network for Rare Neurological Disorders, project ID 739510.

Conflicts of Interest: The authors declare no conflict of interest. The funders had no role in the design of the study; in the collection, analyses, or interpretation of data; in the writing of the manuscript, or in the decision to publish the results.

References

1. van der Knaap, M.S.; Breiter, S.N.; Naidu, S.; Hart, A.A.; Valk, J. Defining and categorizing leukoencephalopathies of unknown origin: MR imaging approach. *Radiology* **1999**, *213*, 121–133. [CrossRef] [PubMed]
2. Hamilton, E.M.C.; van der Lei, H.D.W.; Vermeulen, G.; Gerver, J.A.M.; Lourenco, C.M.; Naidu, S.; Mierzewska, H.; Gemke, R.; de Vet, H.C.W.; Uitdehaag, B.M.J.; et al. Natural History of Vanishing White Matter. *Ann. Neurol.* **2018**, *84*, 274–288. [CrossRef] [PubMed]
3. van der Knaap, M.S.; Pronk, J.C.; Scheper, G.C. Vanishing white matter disease. *Lancet Neurol.* **2006**, *5*, 413–423. [CrossRef]
4. Bugiani, M.; Boor, I.; Powers, J.M.; Scheper, G.C.; van der Knaap, M.S. Leukoencephalopathy with vanishing white matter: A review. *J. Neuropathol. Exp. Neurol.* **2010**, *69*, 987–996. [CrossRef]
5. van der Knaap, M.S.; Bugiani, M. Leukodystrophies: A proposed classification system based on pathological changes and pathogenetic mechanisms. *Acta Neuropathol.* **2017**, *134*, 351–382. [CrossRef]
6. Leegwater, P.A.; Vermeulen, G.; Konst, A.A.; Naidu, S.; Mulders, J.; Visser, A.; Kersbergen, P.; Mobach, D.; Fonds, D.; van Berkel, C.G.; et al. Subunits of the translation initiation factor eIF2B are mutant in leukoencephalopathy with vanishing white matter. *Nat. Genet.* **2001**, *29*, 383–388. [CrossRef]
7. van der Knaap, M.S.; Leegwater, P.A.; Konst, A.A.; Visser, A.; Naidu, S.; Oudejans, C.B.; Schutgens, R.B.; Pronk, J.C. Mutations in each of the five subunits of translation initiation factor eIF2B can cause leukoencephalopathy with vanishing white matter. *Ann. Neurol.* **2002**, *51*, 264–270. [CrossRef]
8. Wortham, N.C.; Proud, C.G. eIF2B: Recent structural and functional insights into a key regulator of translation. *Biochem. Soc. Trans.* **2015**, *43*, 1234–1240. [CrossRef]
9. Proud, C.G. Regulation of eukaryotic initiation factor eIF2B. *Signal. Pathw. Transl.* **2001**, *26*, 95–114. [CrossRef]
10. Abbink, T.E.M.; Wisse, L.E.; Jaku, E.; Thiecke, M.J.; Voltolini-Gonzalez, D.; Fritsen, H.; Bobeldijk, S.; Ter Braak, T.J.; Polder, E.; Postma, N.L.; et al. Vanishing white matter: Deregulated integrated stress response as therapy target. *Ann. Clin. Transl. Neurol.* **2019**, *6*, 1407–1422. [CrossRef]
11. Bugiani, M.; Boor, I.; van Kollenburg, B.; Postma, N.; Polder, E.; van Berkel, C.; van Kesteren, R.E.; Windrem, M.S.; Hol, E.M.; Scheper, G.C.; et al. Defective glial maturation in vanishing white matter disease. *J. Neuropathol. Exp. Neurol.* **2011**, *70*, 69–82. [CrossRef]
12. Bruck, W.; Herms, J.; Brockmann, K.; Schulz-Schaeffer, W.; Hanefeld, F. Myelinopathia centralis diffusa (vanishing white matter disease): Evidence of apoptotic oligodendrocyte degeneration in early lesion development. *Ann. Neurol.* **2001**, *50*, 532–536. [CrossRef]
13. van der Knaap, M.S.; Kamphorst, W.; Barth, P.G.; Kraaijeveld, C.L.; Gut, E.; Valk, J. Phenotypic variation in leukoencephalopathy with vanishing white matter. *Neurology* **1998**, *51*, 540–547. [CrossRef]
14. Klok, M.D.; Bugiani, M.; de Vries, S.I.; Gerritsen, W.; Breur, M.; van der Sluis, S.; Heine, V.M.; Kole, M.H.P.; Baron, W.; van der Knaap, M.S. Axonal abnormalities in vanishing white matter. *Ann. Clin. Transl. Neurol.* **2018**, *5*, 429–444. [CrossRef]
15. Dooves, S.; Bugiani, M.; Postma, N.L.; Polder, E.; Land, N.; Horan, S.T.; van Deijk, A.L.; van de Kreeke, A.; Jacobs, G.; Vuong, C.; et al. Astrocytes are central in the pathomechanisms of vanishing white matter. *J. Clin. Investig.* **2016**, *126*, 1512–1524. [CrossRef]
16. Bugiani, M.; Postma, N.; Polder, E.; Dieleman, N.; Scheffer, P.G.; Sim, F.J.; van der Knaap, M.S.; Boor, I. Hyaluronan accumulation and arrested oligodendrocyte progenitor maturation in vanishing white matter disease. *Brain* **2013**, *136*, 209–222. [CrossRef]
17. Rodriguez, D.; Gelot, A.; della Gaspera, B.; Robain, O.; Ponsot, G.; Sarlieve, L.L.; Ghandour, S.; Pompidou, A.; Dautigny, A.; Aubourg, P.; et al. Increased density of oligodendrocytes in childhood ataxia with diffuse central hypomyelination (CACH) syndrome: Neuropathological and biochemical study of two cases. *Acta Neuropathol.* **1999**, *97*, 469–480. [CrossRef]
18. van der Knaap, M.S.; Barth, P.G.; Gabreels, F.J.; Franzoni, E.; Begeer, J.H.; Stroink, H.; Rotteveel, J.J.; Valk, J. A new leukoencephalopathy with vanishing white matter. *Neurology* **1997**, *48*, 845–855. [CrossRef]
19. Tyanova, S.; Temu, T.; Cox, J. The MaxQuant computational platform for mass spectrometry-based shotgun proteomics. *Nat. Protoc.* **2016**, *11*, 2301–2319. [CrossRef]
20. Ritchie, M.E.; Phipson, B.; Wu, D.; Hu, Y.; Law, C.W.; Shi, W.; Smyth, G.K. Limma powers differential expression analyses for RNA-sequencing and microarray studies. *Nucleic Acids Res.* **2015**, *43*, e47. [CrossRef]
21. Gu, Z.; Eils, R.; Schlesner, M. Complex heatmaps reveal patterns and correlations in multidimensional genomic data. *Bioinformatics* **2016**, *32*, 2847–2849. [CrossRef] [PubMed]
22. Zhang, Y.; Sloan, S.A.; Clarke, L.E.; Caneda, C.; Plaza, C.A.; Blumenthal, P.D.; Vogel, H.; Steinberg, G.K.; Edwards, M.S.; Li, G.; et al. Purification and Characterization of Progenitor and Mature Human Astrocytes Reveals Transcriptional and Functional Differences with Mouse. *Neuron* **2016**, *89*, 37–53. [CrossRef] [PubMed]
23. Szklarczyk, D.; Gable, A.L.; Nastou, K.C.; Lyon, D.; Kirsch, R.; Pyysalo, S.; Doncheva, N.T.; Legeay, M.; Fang, T.; Bork, P.; et al. The STRING database in 2021: Customizable protein–protein networks, and functional characterization of user-uploaded gene/measurement sets. *Nucleic Acids Res.* **2021**, *49*, D605–D612. [CrossRef] [PubMed]

24. Shannon, P.; Markiel, A.; Ozier, O.; Baliga, N.S.; Wang, J.T.; Ramage, D.; Amin, N.; Schwikowski, B.; Ideker, T. Cytoscape: A software environment for integrated models of biomolecular interaction networks. *Genome Res.* **2003**, *13*, 2498–2504. [CrossRef] [PubMed]
25. Middeldorp, J.; Boer, K.; Sluijs, J.A.; De Filippis, L.; Encha-Razavi, F.; Vescovi, A.L.; Swaab, D.F.; Aronica, E.; Hol, E.M. GFAPdelta in radial glia and subventricular zone progenitors in the developing human cortex. *Development* **2010**, *137*, 313–321. [CrossRef]
26. Ferreira, T.A.; Blackman, A.V.; Oyrer, J.; Jayabal, S.; Chung, A.J.; Watt, A.J.; Sjostrom, P.J.; van Meyel, D.J. Neuronal morphometry directly from bitmap images. *Nat. Methods* **2014**, *11*, 982–984. [CrossRef]
27. Lundgaard, I.; Osorio, M.J.; Kress, B.T.; Sanggaard, S.; Nedergaard, M. White matter astrocytes in health and disease. *Neuroscience* **2014**, *276*, 161–173. [CrossRef]
28. Garcia-Marin, V.; Garcia-Lopez, P.; Freire, M. Cajal's contributions to glia research. *Trends Neurosci.* **2007**, *30*, 479–487. [CrossRef]
29. Bugiani, M.; Vuong, C.; Breur, M.; van der Knaap, M.S. Vanishing white matter: A leukodystrophy due to astrocytic dysfunction. *Brain Pathol* **2018**, *28*, 408–421. [CrossRef]
30. Elroy-Stein, O. Mitochondrial malfunction in vanishing white matter disease: A disease of the cytosolic translation machinery. *Neural Regen. Res.* **2017**, *12*, 1610–1612. [CrossRef]
31. Herrero, M.; Daw, M.; Atzmon, A.; Elroy-Stein, O. The Energy Status of Astrocytes Is the Achilles' Heel of eIF2B-Leukodystrophy. *Cells* **2021**, *10*, 1858. [CrossRef]
32. Herrero, M.; Mandelboum, S.; Elroy-Stein, O. eIF2B Mutations Cause Mitochondrial Malfunction in Oligodendrocytes. *Neuromolecular Med.* **2019**, *21*, 303–313. [CrossRef]
33. Raini, G.; Sharet, R.; Herrero, M.; Atzmon, A.; Shenoy, A.; Geiger, T.; Elroy-Stein, O. Mutant eIF2B leads to impaired mitochondrial oxidative phosphorylation in vanishing white matter disease. *J. Neurochem.* **2017**, *141*, 694–707. [CrossRef]
34. Ghezzi, D.; Goffrini, P.; Uziel, G.; Horvath, R.; Klopstock, T.; Lochmuller, H.; D'Adamo, P.; Gasparini, P.; Strom, T.M.; Prokisch, H.; et al. SDHAF1, encoding a LYR complex-II specific assembly factor, is mutated in SDH-defective infantile leukoencephalopathy. *Nat. Genet.* **2009**, *41*, 654–656. [CrossRef]
35. Moroni, I.; Bugiani, M.; Bizzi, A.; Castelli, G.; Lamantea, E.; Uziel, G. Cerebral white matter involvement in children with mitochondrial encephalopathies. *Neuropediatrics* **2002**, *33*, 79–85. [CrossRef]
36. Baird, T.D.; Wek, R.C. Eukaryotic initiation factor 2 phosphorylation and translational control in metabolism. *Adv. Nutr.* **2012**, *3*, 307–321. [CrossRef]
37. Anlauf, E.; Derouiche, A. Glutamine synthetase as an astrocytic marker: Its cell type and vesicle localization. *Front. Endocrinol.* **2013**, *4*, 144. [CrossRef]
38. Jayakumar, A.R.; Norenberg, M.D. Glutamine Synthetase: Role in Neurological Disorders. *Adv. Neurobiol.* **2016**, *13*, 327–350. [CrossRef]
39. Rose, C.F.; Verkhratsky, A.; Parpura, V. Astrocyte glutamine synthetase: Pivotal in health and disease. *Biochem. Soc. Trans.* **2013**, *41*, 1518–1524. [CrossRef]
40. Yoo, H.C.; Yu, Y.C.; Sung, Y.; Han, J.M. Glutamine reliance in cell metabolism. *Exp. Mol. Med.* **2020**, *52*, 1496–1516. [CrossRef]
41. Orthmann-Murphy, J.L.; Abrams, C.K.; Scherer, S.S. Gap junctions couple astrocytes and oligodendrocytes. *J. Mol. Neurosci.* **2008**, *35*, 101–116. [CrossRef] [PubMed]
42. Spray, D.C.; Ye, Z.C.; Ransom, B.R. Functional connexin "hemichannels": A critical appraisal. *Glia* **2006**, *54*, 758–773. [CrossRef] [PubMed]
43. Theis, M.; Giaume, C. Connexin-based intercellular communication and astrocyte heterogeneity. *Brain Res.* **2012**, *1487*, 88–98. [CrossRef]
44. Irons, J.L.; Hodge-Hanson, K.; Downs, D.M. RidA Proteins Protect against Metabolic Damage by Reactive Intermediates. *Microbiol. Mol. Biol. Rev.* **2020**, *84*, e00024-20. [CrossRef] [PubMed]
45. Manjasetty, B.A.; Delbruck, H.; Pham, D.T.; Mueller, U.; Fieber-Erdmann, M.; Scheich, C.; Sievert, V.; Bussow, K.; Niesen, F.H.; Weihofen, W.; et al. Crystal structure of Homo sapiens protein hp14.5. *Proteins* **2004**, *54*, 797–800. [CrossRef]
46. Eliasson, C.; Sahlgren, C.; Berthold, C.H.; Stakeberg, J.; Celis, J.E.; Betsholtz, C.; Eriksson, J.E.; Pekny, M. Intermediate filament protein partnership in astrocytes. *J. Biol. Chem.* **1999**, *274*, 23996–24006. [CrossRef]
47. Shibuya, S.; Miyamoto, O.; Auer, R.N.; Itano, T.; Mori, S.; Norimatsu, H. Embryonic intermediate filament, nestin, expression following traumatic spinal cord injury in adult rats. *Neuroscience* **2002**, *114*, 905–916. [CrossRef]
48. Jurga, A.M.; Paleczna, M.; Kadluczka, J.; Kuter, K.Z. Beyond the GFAP-Astrocyte Protein Markers in the Brain. *Biomolecules* **2021**, *11*, 1361. [CrossRef]
49. Hol, E.M.; Pekny, M. Glial fibrillary acidic protein (GFAP) and the astrocyte intermediate filament system in diseases of the central nervous system. *Curr. Opin. Cell Biol.* **2015**, *32*, 121–130. [CrossRef]
50. Leferink, P.S.; Dooves, S.; Hillen, A.E.J.; Watanabe, K.; Jacobs, G.; Gasparotto, L.; Cornelissen-Steijger, P.; van der Knaap, M.S.; Heine, V.M. Astrocyte Subtype Vulnerability in Stem Cell Models of Vanishing White Matter. *Ann. Neurol.* **2019**, *86*, 780–792. [CrossRef]
51. Kohler, S.; Winkler, U.; Hirrlinger, J. Heterogeneity of Astrocytes in Grey and White Matter. *Neurochem. Res.* **2021**, *46*, 3–14. [CrossRef]
52. de Sousa Abreu, R.; Penalva, L.O.; Marcotte, E.M.; Vogel, C. Global signatures of protein and mRNA expression levels. *Mol. Biosyst.* **2009**, *5*, 1512–1526. [CrossRef]

53. Vogel, C.; Marcotte, E.M. Insights into the regulation of protein abundance from proteomic and transcriptomic analyses. *Nat. Rev. Genet.* **2012**, *13*, 227–232. [CrossRef]
54. Perez-Riverol, Y.; Csordas, A.; Bai, J.; Bernal-Llinares, M.; Hewapathirana, S.; Kundu, D.J.; Inuganti, A.; Griss, J.; Mayer, G.; Eisenacher, M.; et al. The PRIDE database and related tools and resources in 2019: Improving support for quantification data. *Nucleic Acids Res.* **2019**, *47*, D442–D450. [CrossRef]

cells

MDPI

Article

Spatial and Temporal Diversity of Astrocyte Phenotypes in Spinocerebellar Ataxia Type 1 Mice

Juao-Guilherme Rosa [1,2], Katherine Hamel [1], Carrie Sheeler [1], Ella Borgenheimer [1,3], Stephen Gilliat [1], Alyssa Soles [1], Ferris J. Ghannoum [1], Kaelin Sbrocco [1], Hillary P. Handler [1,4], Orion Rainwater [4,5], Ryan Kang [1,6] and Marija Cvetanovic [1,4,*]

1 Cvetanovic Laboratory, Department of Neuroscience, College of Medicine, University of Minnesota, 2101 6th Street SE, Minneapolis, MN 55455, USA
2 Graduate Program for Neuroscience, Boston University, 700 Albany Street W516, Boston, MA 02118, USA
3 Baylor College of Medicine, 1 Baylor Plaza, Houston, TX 77030, USA
4 Institute for Translational Neuroscience, College of Medicine, University of Minnesota, 2101 6th Street SE, Minneapolis, MN 55455, USA
5 Department of Lab. Medicine and Pathology, College of Medicine, University of Minnesota, 2101 6th Street SE, Minneapolis, MN 55455, USA
6 Bowdoin College, Brunswick, ME 04011, USA
* Correspondence: mcvetano@umn.edu; Tel.: +1-612-626-491

Abstract: While astrocyte heterogeneity is an important feature of the healthy brain, less is understood about spatiotemporal heterogeneity of astrocytes in brain disease. Spinocerebellar ataxia type 1 (SCA1) is a progressive neurodegenerative disease caused by a CAG repeat expansion in the gene *Ataxin1* (*ATXN1*). We characterized astrocytes across disease progression in the four clinically relevant brain regions, cerebellum, brainstem, hippocampus, and motor cortex, of *Atxn1*$^{154Q/2Q}$ mice, a knock-in mouse model of SCA1. We found brain region-specific changes in astrocyte density and GFAP expression and area, early in the disease and prior to neuronal loss. Expression of astrocytic core homeostatic genes was also altered in a brain region-specific manner and correlated with neuronal activity, indicating that astrocytes may compensate or exacerbate neuronal dysfunction. Late in disease, expression of astrocytic homeostatic genes was reduced in all four brain regions, indicating loss of astrocyte functions. We observed no obvious correlation between spatiotemporal changes in microglia and spatiotemporal astrocyte alterations, indicating a complex orchestration of glial phenotypes in disease. These results support spatiotemporal diversity of glial phenotypes as an important feature of the brain disease that may contribute to SCA1 pathogenesis in a brain region and disease stage-specific manner.

Keywords: astrocytes; SCA1; neurodegeneration; microglia; morphology; gene expression; brain region

Citation: Rosa, J.-G.; Hamel, K.; Sheeler, C.; Borgenheimer, E.; Gilliat, S.; Soles, A.; Ghannoum, F.J.; Sbrocco, K.; Handler, H.P.; Rainwater, O.; et al. Spatial and Temporal Diversity of Astrocyte Phenotypes in Spinocerebellar Ataxia Type 1 Mice. *Cells* **2022**, *11*, 3323. https://doi.org/10.3390/cells11203323

Academic Editors: Michael Brenner and Vladimir Parpura

Received: 28 August 2022
Accepted: 17 October 2022
Published: 21 October 2022

1. Introduction

Astrocytes play critical roles in brain development and functioning of the adult brain [1–3]. One key way astrocytes regulate neuronal activity is by maintaining homeostasis of ions, neurotransmitters, and water in the brain [4–7]. Differences in the activity and metabolic needs of neurons across and within brain regions indicate the heterogeneous and region-specific activity of astrocytes [8–14]. While heterogeneity of astrocyte morphology was described by Ramon Y Cajal, recent advances in single-cell RNA sequencing (scRNA-seq), spatial transcriptomics, and single-cell mapping of astrocyte interactions have begun characterizing the molecular heterogeneity of astrocytes and its relevance for healthy brain function [15–17].

In brain diseases, astrocytes undergo a phenotypic change termed reactive gliosis, defined as morphological and gene expression changes that may result in altered astrocyte function [18]. Reactive astrocytes contribute to pathogenesis of neurodegenerative disease [19], and can either exacerbate or ameliorate neurodegeneration as has been shown in

Huntington's disease (HD) [20,21], Alzheimer's disease (AD) [22–24], and spinocerebellar ataxia type 1 (SCA1) [25]. It is still unknown how regional heterogeneity of astrocytes affects their changes in disease and whether it contributes to the selective regional neuronal vulnerability, which is one of the key features of neurodegenerative diseases [26,27].

SCA1 is an inherited neurodegenerative disorder caused by CAG repeat expansion in the *Ataxin-1 (ATXN1)* gene [28]. Repeat expansions of about 39 or more CAG repeats result in the expression of a toxic polyglutamine (polyQ) ATXN1 protein [29]. SCA1 is characterized by progressive degeneration that is most severe in the cerebellum and brainstem, but is also present in other brain regions, and manifests behaviorally as deficits in motor coordination (i.e., ataxia), swallowing, speaking, cognition, and mood [30–32]. While *Atxn1*-targeting antisense oligonucleotides (ASOs) show promise in pre-clinical trials [33], no disease-modifying treatments are currently available for SCA1.

Motor deficits and underlying cerebellar pathology have been the focus of the majority of SCA1 studies so far. However, *ATXN1* is expressed throughout the brain and likely affects brain regions beyond the cerebellum [34,35]. ATXN1 protein is thought to function as a transcriptional regulator as it interacts with several transcription factors, and gene expression changes are one of the earliest alterations in SCA1 mice. Pathology is described in the brainstem, motor cortex, and hippocampus of SCA1 patients, and is thought to contribute to SCA1 symptoms such as cognitive deficits, mood disorders, difficulties in respiration and swallowing, and premature lethality [36–41]. We have previously demonstrated that astrocytes actively contribute to disease pathogenesis in the cerebellum of SCA1 mice [25], but how astrocytes are altered in other SCA1-affected brain regions remains unknown.

To better understand brain-wide SCA1 astrocyte pathology, we have utilized the $Atxn1^{154Q/2Q}$ mouse model of SCA1 wherein 154 CAG repeats have been knocked-in into the endogenous mouse *Atxn1* gene [42]. Because the *Atxn1* gene is expressed under the endogenous promoter, $Atxn1^{154Q/2Q}$ mice express polyQ ATXN1 in cells throughout the central nervous system [https://www.proteinatlas.org/, accessed on March 2022] and in brain cells such as neurons, astrocytes, and microglia [http://www.brainrnaseq.org/, accessed on March 2022]. As a result, in addition to motor deficits, the $Atxn1^{154Q/2Q}$ mice exhibit premature lethality, failure to gain weight, and cognitive deficits [42,43]. Furthermore, there is evidence of hippocampal and brainstem changes in $Atxn1^{154Q/2Q}$ mice, although pathology in these brain regions has yet to be thoroughly characterized [42,44,45].

To investigate the region-specific and temporal changes in SCA1 astrocytes, we compared morphology and gene expression changes in astrocytes in the cerebellum, brainstem, hippocampus, and motor cortex of 12- and 20-week-old $Atxn1^{154Q/2Q}$ mice. We focused on these time points because they represent early- to mid- and late-disease stages. Early stages of disease are particularly of interest due to previous studies showing the significant potential of early therapeutic approaches [46,47].

We determined astrocytic density, area, and GFAP expression, and the expression of core astrocytic homeostatic genes including *Kcnj10, Aqp4, Slc1a2,* and *Gja1*. These astrocytic genes play a critical role in the maintenance of brain homeostasis by regulating local concentrations of potassium (*Kcnj10*), water (*Aqp4*), and glutamate (*Slc1a2*) [7,48–50].

We also investigated microglial and neuronal changes across these four brain regions. We examined changes in microglial density and area, and quantified neuronal density, activity, and loss of excitatory synapses in the examined regions.

2. Materials and Methods

2.1. Mice

The creation of the $Atxn1^{154Q/2Q}$ mice was previously described [42]. Because repeat length in trinucleotide repeat expansions is unstable and prone to further expansion [51], we routinely perform genetic sequencing of our mouse lines. We found that the number of repeats has recently expanded in our colony from 154 CAG to 166 CAG. Based on previous studies, this increase in CAG repeat number is expected to increase the severity of disease and lower the age of symptom onset. We used an approximately equal number of male and female mice in all experiments.

2.2. Immunofluorescent (IF) Staining

IF was performed on a minimum of six different floating 45-μm-thick brain slices from each mouse (six technical replicates per mouse per region or antibody of interest). Confocal images were acquired using a confocal microscope Olympus FV1000 (Leica, Deerfield, IL, USA) using a 20×, 40×, or 60× oil objective. Z-stacks consisting of twenty non-overlapping 1-μm-thick slices were taken of each stained brain slice per brain region (i.e., six z-stacks per mouse, each taken from a different brain slice). The laser power and detector gain were standardized and fixed between mice within a surgery cohort, and all images for mice within a cohort were acquired in a single imaging session to allow for quantitative comparison.

We used primary antibodies against Purkinje cell marker calbindin (mouse, Sigma-Aldrich, C9848, St. Louis, MO, USA), neuronal marker neuronal nuclei (NeuN) (rabbit, Ab104225, Abcam, Boston, MN, USA), astrocytic marker glial fibrillary acidic protein (GFAP) (chicken, AB5541, Millipore, Madison, WI, USA), microglial marker ionized calcium binding adaptor molecule 1 (Iba1) (rabbit, 019-19741, WAKO, Richmond, VA, USA), c-Fos (rabbit, ab190289, Abcam, Boston, MN, USA), PSD-95 (mouse, 810401, BioLegend, San Diego, CA, USA), and vesicular glutamate transporter 2 (VGLUT2) (guinea pig, AB2251-I, Millipore, Madison, WI, USA) as previously described [25,52]. We used specie-specific corresponding secondary antibodies conjugated with Alexa 488 and Alexa 594 (Jackson Immunoresearch Laboratories, West Grove, PA, USA). Quantitative analysis was performed using ImageJ (NIH, Bethesda, MA, USA). To quantify relative intensity of staining for GFAP and calbindin, we measured the average signal intensity in the region of interest and normalized it to that of the WT littermate controls. The density of neurons, astrocytes, microglia, and active neuronal cells was determined by normalizing the number of NeuN+, GFAP+, Iba1+, and c-Fos+ cells, respectively, with the area of the region of interest. We determined GFAP+ and Iba1+ percent area by creating a mask of GFAP and Iba1 staining, respectively, and recording the fraction of the region of interest covered by staining. To quantify atrophy of the cerebellar molecular layer, we took six measurements per image of the distance from the base of the Purkinje soma to the end of their dendrites, the average being the molecular layer width for that image. Numbers of VGLUT2 and PSD95 puncta, and their co-localizations, were quantified on 60× 4× zoomed images using Puncta Analyzer Plugin (Durham, NC, USA) in ImageJ. This plugin takes the advantage of the fact that presynaptic (VGLUT2) and postsynaptic (PSD95) proteins reside in separate cell compartments, but because of their close proximity they appear to co-localize at synapses. Puncta Analyzer Plugin is set to subtract background with rolling ball radius of 50 pixels. Threshold was selected for each image and channel to further remove any potential background noise, and particle size was set at 4 pixels. At least 3 optical sections per brain slice, and at least 2 brain slices per animal were analyzed, making a total of 16–27 image data sets per brain region for each genotype. The width of hippocampal neuronal layers (CA2 and CA3) was measured by dividing the area of each neuronal layer by the length of that layer as previously described [44].

2.3. RNA Extraction, Sequencing, and Analyses

Cerebellum, medulla, cerebral cortex, and hippocampal tissue was isolated from 26-week-old wild-type and *Atxn1*$^{154Q/2Q}$ mice and stored in RNAlater solution (Thermo Fisher Scientific, Coon Rapids, MN, USA). Total RNA was isolated using TRIzol reagent (Thermo Fisher Scientific, Coon Rapids, MN, USA) following the manufacturer's protocols. Tissue was homogenized using RNase-Free disposable pellet pestles in a motorized chuck. Purified RNA was sent to the University of Minnesota Genomics Center for quality control, including quantification using fluorimetry via RiboGreen assay kit (Thermo Fisher Scientific, Coon Rapids, MN, USA), and RNA integrity was assessed via capillary 34 electrophoresis using an Agilent BioAnalyzer 2100 to generate an RNA integrity number (RIN). RIN values for submitted RNA were above 8.0 for all samples except one medulla sample (RIN = 6.8). All submitted RNA samples had greater than 1 μg total mass. Library creation was completed using oligo-dT purification of polyadenylated RNA, which was reverse transcribed to create cDNA. cDNA was fragmented, blunt-ended, and ligated to barcode adaptors. Libraries were size selected to 320 bp ± 5% to produce average inserts of approximately 200 bp, and size distribution was validated using capillary electrophoresis and quantified using fluorimetry (PicoGreen, Thermo Fisher Scientific, Coon Rapids, MN, USA) and qPCR. Libraries were then normalized, pooled, and sequenced on an S4 flow cell by an Illumina NovaSeq 6000 using a 150-nucleotide, paired-end read strategy. The resulting FASTQ files were trimmed, aligned to the mouse reference genome (GRCm38), sorted, and counted using the Bulk RNAseq Analysis Pipeline from the Minnesota Supercomputing Institute's Collection of Hierarchical UMII/RIS Pipelines (v0.2.0). Genes less than 300 bp are too small to be accurately captured in standard RNAseq library preparations, so they were discarded from all downstream analyses.

Differential gene expression analysis was performed using the edgeR package (v3.30.3) in R (R Foundation for Statistical Computing v3.6.1). All four brain regions were analyzed independently. Genes with fewer than 10 counts across all samples in each region were excluded. Genes with FDR values less than or equal to 0.05 were considered significant.

2.4. Reverse Transcription and Quantitative Polymerase Chain Reaction (RT-qPCR)

Total RNA was extracted from dissected mouse cerebella, brainstem, cortex, and hippocampus using TRIzol (hermo Fisher Scientific, Coon Rapids, MN, USA), and RT-qPCR was performed as described previously [8]. We used IDT Primetime primers (IDT, Coralville, IW, USA). Relative mRNA levels were calculated using 18S RNA as a control and wild-type mice as a reference using $2^{-\Delta\Delta Ct}$ as previously described [8].

2.5. Statistics

Wherever possible, sample sizes were calculated using power analyses based on the standard deviations from our previous studies, significance level of 5%, and power of 90%. Statistical tests were performed with GraphPad Prism 7.0 (GraphPad Software, San Diego, CA, USA). Data was analyzed using two-way ANOVA (to assess the impact of genotype and treatment) followed by post-hoc two-tailed, unpaired Student's *t*-test (to test statistical significance of differences between groups of interest wherein only one variable was different (treatment or genotype), or one-way ANOVA followed by the Sidak's post-hoc test. Outliers were determined using GraphPad PRISM's Robust regression and Outlier removal (ROUT) with a Q = 1% for non-biased selection.

3. Results

3.1. Spatial Diversity of Astrocyte Morphology during Early Disease Stages in *Atxn1*$^{154Q/2q}$ Mice

Ataxin-1 is widely expressed throughout the brain in humans and in mice [44,53,54] yet how astrocytes are altered across relevant brain regions in SCA1 is unknown [35]. We first investigated morphological changes in astrocytes in the cerebellum, brainstem, hippocampus, and motor cortex of *Atxn1*$^{154Q/2Q}$ mice during early disease stage. These regions have presumptive roles in motor deficits, premature lethality, and cognitive dysfunction in SCA1. Previously described neuronal degeneration and gliosis in the brainstem of patients with SCA1 are thought to contribute to the loss of ability to protect airways and, subsequently, aspiration pneumonia and death [36,55].

To investigate morphological changes in SCA1 astrocytes, we quantified an increase in area occupied by glial fibrillary acidic protein (GFAP) (percentage of area that is GFAP+), which is a morphological characteristic of hypertrophy in reactive astrocytes, an increase in GFAP immunoreactivity, which is often used as a measure of reactive astrogliosis, and cell density where possible [56,57]. In the cerebellum, we examined Bergmann glia (BG), a subtype of cerebellar astrocytes that reside in the Purkinje and molecular layers and have a very intimate structural and functional relationship with Purkinje cells (PCs), neurons affected in SCA1 [58,59]. We detected a significant increase in GFAP+ percent area (Figure 1A), and in the intensity of GFAP staining [60], indicative of hypertrophy of SCA1 Bergmann glia. We have previously described SCA1 alterations in the dentate gyrus of the hippocampus in *Atxn1*$^{154Q/2Q}$ mice [45]. We examined astrocyte morphology and found an increase in intensity of GFAP expression and GFAP+ astrocyte percent area, indicative of astrocyte hypertrophy in the dentate gyrus of *Atxn1*$^{154Q/2Q}$ mice relative to wild-type controls (Figure 1B). In the brainstem, we characterized astrocyte morphology in the inferior olivary nucleus (ION) [41], because postmortem analysis of patient samples showed severe loss of volume in the ION [34], and due to the high-quality of GFAP staining relative to other brainstem nuclei (Figure 1C). Interestingly, we found that in the ION of *Atxn1*$^{154Q/2Q}$ mice, the intensity of GFAP staining and density of astrocytes were significantly reduced. Previous brain-wide investigation demonstrated neuronal loss in the primary motor cortex of SCA1 patients [34]. Because of the reported differences in gray and white matter astrocytes, we assessed astrocyte morphology in layer 6 of the motor cortex and the underlying corpus callosum of *Atxn1*$^{154Q/2Q}$ mice (Figure 1D). We detected a slight increase in GFAP intensity and percent area in the corpus callosum of *Atxn1*$^{154Q/2Q}$ mice that did not reach statistical significance and may indicate emerging hypertrophy of cortical astrocytes (Figure 1D). There were no changes in astrocyte density, GFAP intensity, or area in the layer 6 of motor cortex.

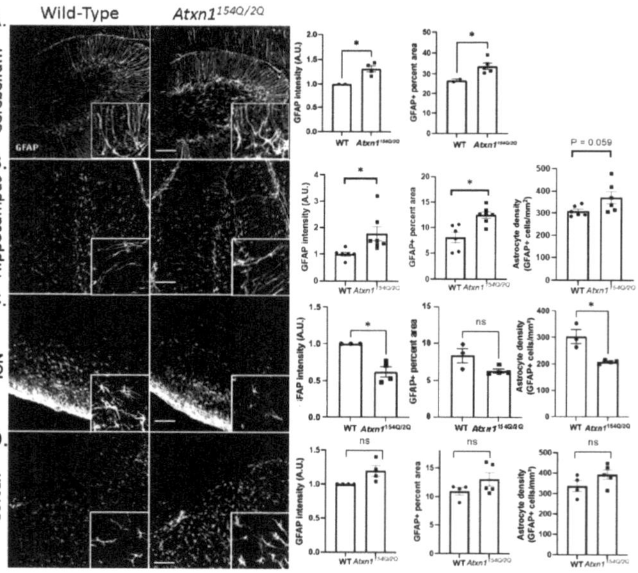

Figure 1. Spatial diversity of astrocytic morphology in *Atxn1*$^{154Q/2Q}$ mice early in disease. Brain sections from 12 weeks old *Atxn1*$^{154Q/2Q}$ and wild-type littermate controls were stained for GFAP. Confocal images of cerebellum ((**A**), lobule X), hippocampus ((**B**), dentate gyrus), brainstem ((**C**), ION (inferior olivary nucleus), and motor cortex ((**D**), layer 6 and corpus callosum) were used to quantify GFAP intensity, density of astrocytes, and GFAP area. Images were acquired at 20× (636.16 μm × 636.16 μm). Scale bars show 100 μm. Insets in the lower right corners of each image show zoomed images (79.5 μm × 79.5 μm) for qualitative assessment of astrocyte morphology. Data are presented as mean ± SEM with average values for each mouse represented by a dot. N = 3–7 mice of each genotype (WT and *Atxn1*$^{154Q/2Q}$) as noted by dots representing individual mice in each bar graph. ns: not significant, * $p < 0.05$ Students *t*-test.

These results indicate diverse morphological changes in SCA1 astrocytes across brain regions with astrocytic loss and atrophy in the brainstem, astrocyte hypertrophy in the cerebellum, and in the hippocampus and trending, but not significant astrocytic hypertrophy in the corpus callosum.

3.2. Spatial Diversity of Astrocytic Gene Expression Changes in SCA1

We next investigated molecular changes in SCA1 astrocytes across brain regions. Among the astrocytic roles critical for neuronal function is maintaining the homeostasis of the extracellular environment. Astrocytes accomplish this by removing excess extracellular glutamate via glutamate transporters encoded by the *Slc1a3* and *Slc1a2* genes, potassium ions via the potassium channel encoded by *Kcnj10*, and water through the expression of Aquaporin-4, encoded by *Aqp4* [61–64]. Both *Slc1a2* and *Slc1a3* are present in astrocytes throughout the brain, but *Slc1a3* is preferentially expressed in the cerebellum and *Slc1a2* in the hippocampus. Astrocyte intercommunication in a network allowing for calcium wave propagation, and also potassium and glutamate buffering, is facilitated by Connexin43 encoded by *Gja1*. We compared the molecular changes in SCA1 astrocytes across the cerebellum, brainstem (medulla), hippocampus, and cortex by quantifying the expression of these critical astrocytic core genes using RT-qPCR.

In the cerebellum of $Atxn1^{154Q/2Q}$ mice, we have found a large variability in the expression of *Aqp4*, *Gja1*, *Kcnj10*, and *Slc1a3*, and no significant change compared to wild-type controls (Figure 2A). We found a significant reduction in the expression of *Aqp4*, *Gja1*, *Slc1a2*, and *Kcnj10*, in the hippocampus, and with the exception of *Slc1a2*, also in the brainstem (Figure 2B,C) of $Atxn1^{154Q/2Q}$ mice at 12 weeks. In contrast, expression of these genes was increased in the cortex (Figure 2D). Given the importance of these homeostatic astrocytic genes for neuronal function, our results may indicate a novel mechanism by which astrocytes exacerbate neuronal dysfunction in the brainstem and hippocampus and compensate for, or ameliorate dysfunction in, the cortex of $Atxn1^{154Q/2Q}$ mice.

The large variability in the expression of *Aqp4*, *Gja1*, *Kcnj10*, and *Slc1a3*, in the cerebellum may reflect large variability in the reactive state of cerebellar astrocytes (as determined by the *Gfap* expression). Indeed, we found a positive correlation between the expression of *Aqp4*, *Gja1*, and *Slc1a3*, and the expression of *Gfap* (Supplementary Figure S1), and a trending correlation between *Gfap* and *Kcnj10* (data not shown). These results indicate that early in disease, cerebellar astrocytes with a higher *Gfap* expression (denoting that they are likely to be more reactive) also have a higher expression of homeostatic genes.

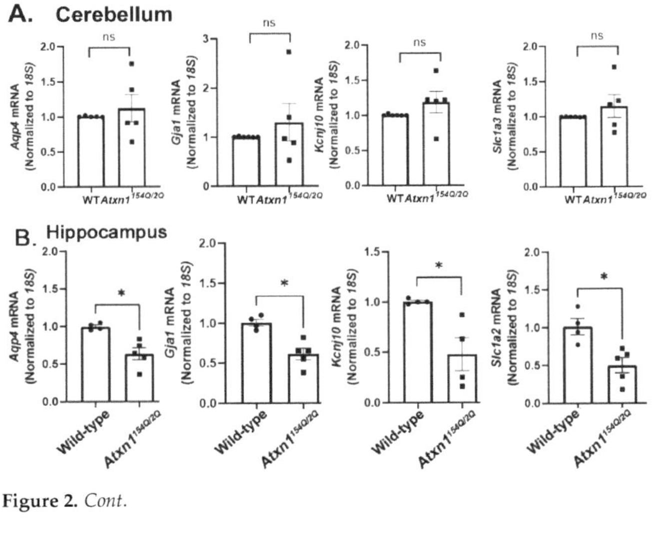

Figure 2. *Cont.*

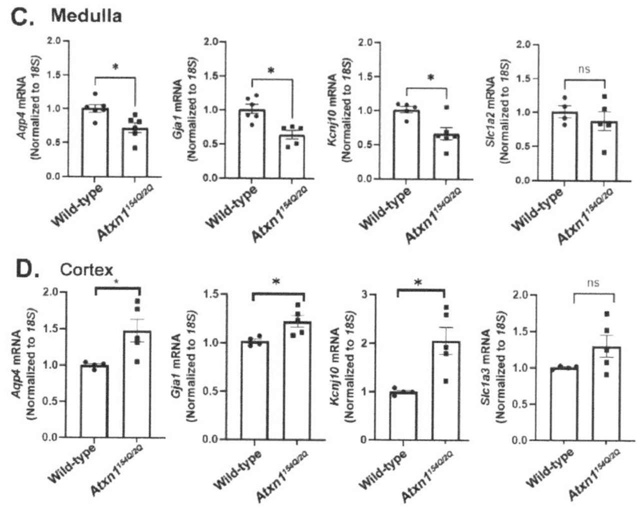

Figure 2. Spatial diversity of molecular changes in astrocytes during early SCA1. mRNA was extracted from the cerebellum (**A**), hippocampus (**B**), medulla (**C**), and cortex (**D**) of 12 weeks old *Atxn1*$^{154Q/2Q}$ mice and wild-type littermate controls (N = 4–6) and RTqPCR was used to evaluate expression of astrocyte specific genes. Data are presented as mean ± SEM with average values for each mouse. N = 4–6 mice of each genotype (WT and *Atxn1*$^{154Q/2Q}$) with individual mice represented by a dot in bar graphs. ns: not significant, * $p < 0.05$ Students *t*-test.

3.3. *Regional Diversity of Microglial SCA1 Changes*

We have previously shown that microglia are reactive in SCA1 cerebellar cortex and contribute to SCA1 pathogenesis in different SCA1 mouse models [65,66]. We investigated changes in microglial density and morphology in the cerebellum and other brain regions using IHC staining for the microglial marker ionized calcium-binding adaptor protein (Iba1). We quantified microglial density and the percent area covered by microglial Iba1+ processes (Figure 3). In the cerebellum, we found a significant increase in microglia density and Iba1+ percent area in 12-week-old *Atxn1*$^{154Q/2Q}$ mice (Figure 3A). Similarly, in the ION (Figure 3C) and cortex (Figure 3D), we found an increase in microglial density. Microglial area, indicative of hypertrophy, was significantly increased only in the cortex. We found no significant changes in microglial density or percent area in the hippocampus (Figure 3B).

Reactive changes in glia can be caused by neuronal loss. To determine whether observed glial changes may be caused by neuronal loss, we used neuronal marker NeuN to quantify neuronal numbers. We did not find evidence of neuronal loss in any of the examined brain regions at this early disease stage in *Atxn1*$^{154Q/2Q}$ mice (Supplementary Figure S2). While this result does not exclude neuronal dysfunction, our results suggest that spatially diverse changes in glial phenotypes precede neuronal loss.

To investigate whether observed changes in the glial phenotypes may be associated with more subtle neuronal changes, we quantified neuronal activity in the hippocampus and cortex, two regions with reduced and increased expression of core astrocyte genes, respectively. Intriguingly, the ratio of cFos+ neurons to total NeuN+ neurons was decreased in the hippocampus while it was not changed in the motor cortex of *Atxn1*$^{154Q/2Q}$ mice (Supplementary Figure S3A,B).

Together, these results suggest that microglia—similar to astrocytes—exhibit spatial diversity of SCA1 changes. Intriguingly, while early in disease astrocytes undergo both loss and hypertrophy, microglia mostly undergo hypertrophy, indicating a complex orchestration of astrocyte and microglial phenotypes early in SCA1. Moreover, decreased astrocytic support may contribute to reduced neuronal activity in the hippocampus.

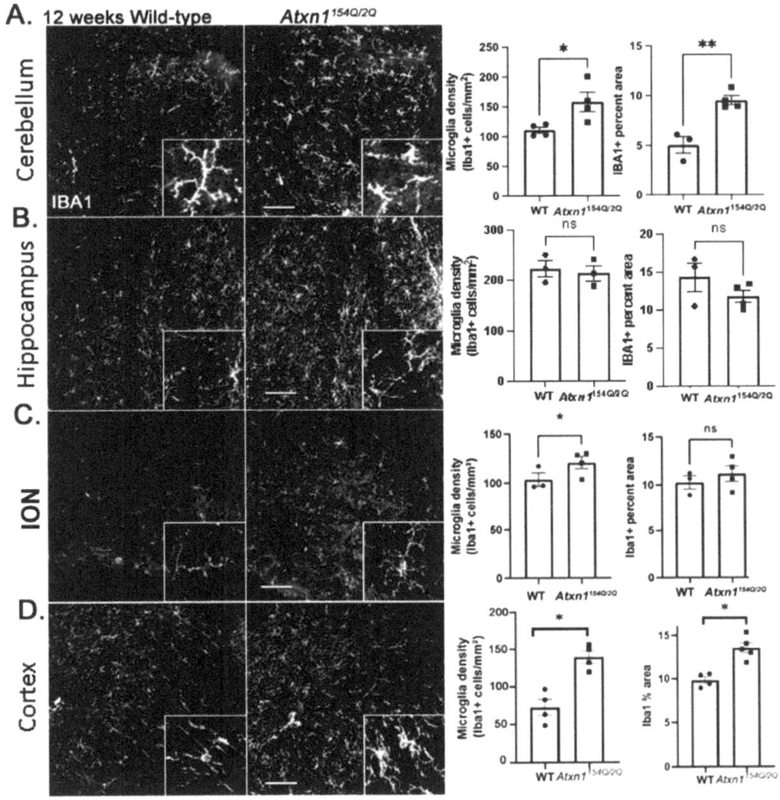

Figure 3. Early SCA1 microglial morphology in *Atxn1*$^{154Q/2Q}$ mice. Brain sections from 12 weeks old Atxn1154Q/2Q and wild-type littermate controls (N = 3–5) were stained for Iba1. Confocal images of cerebellum ((**A**), lobule X), hippocampus ((**B**), dentate gyrus), brainstem ((**C**), ION), and motor cortex ((**D**), layer 6 and underlying corpus callosum) were used to quantify density of Iba1 positive microglia and Iba1 area. Images were acquired at 20× (636.16 µm × 636.16 µm). Scale bars show 100 µm. Insets in the lower right corners of each image show zoomed images (79.5 × µm 79.5 µm) for qualitative assessment of microglia morphology. Data are presented as mean ± SEM with average values for each mouse. N = 3–5 mice of each genotype (WT and *Atxn1*$^{154Q/2Q}$) with individual mice represented by a dot in bar graphs. ns: not significant, * $p < 0.05$, ** $p < 0.01$. Student's *t*-test.

3.4. Temporal Diversity of Astrocyte Phenotypes in SCA1s

We next investigated how glial phenotypes change with disease progression. At late stage of disease (20 weeks), astrocytes in the cerebellum, hippocampus, and motor cortex showed signs of hypertrophy (Figure 4A,B,D). In sharp contrast, in the inferior olivary nucleus (ION) both astrocyte density and area occupied markedly declined (Figure 4C).

Using RNA sequencing, we compared the expression of astrocytic genes in these regions of *Atxn1*$^{154Q/2Q}$ mice and their wild-type littermate controls. We found a significant decrease in the expression of astrocytic homeostatic genes *Kcnj10*, *Aqp4*, *Gja1*, and *Slc1a2* in all four regions in the late stage *Atxn1*$^{154Q/2Q}$ mice (Figure 5). These results indicate temporal diversity of astrocyte phenotypes in SCA1 and suggest that brain-wide loss of astrocyte core gene expression characterizes late stages of disease in *Atxn1*$^{154Q/2Q}$ mice.

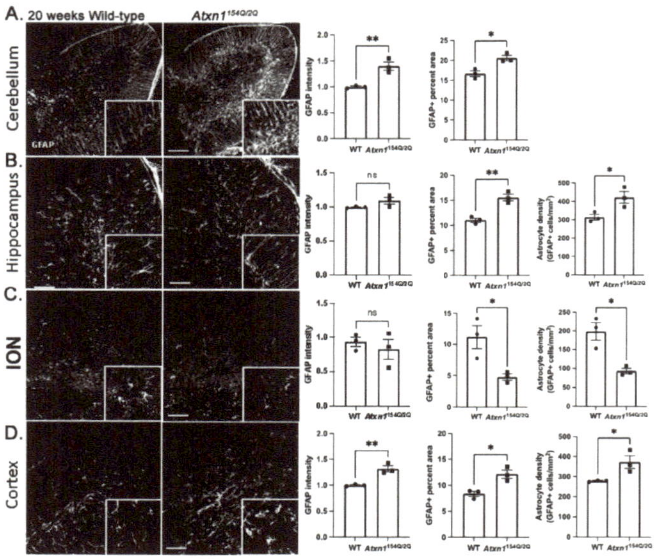

Figure 4. Spatial diversity of astrocytic morphology in Atxn1154Q/2Q mice at late disease stage. Brain sections from 20 weeks old Atxn1154Q/2Q and wild-type littermate controls were stained for GFAP. Confocal images of cerebellum ((**A**), lobule X), hippocampus ((**B**), dentate gyrus), brainstem ((**C**), ION), and motor cortex ((**D**), layer 6 and corpus callosum) were used to quantify GFAP intensity, density of astrocytes and GFAP area. Images were acquired at 20× (636.16 μm × 636.16 μm). Scale bars show 100 μm. Insets. in the lower right corners of each image show zoomed images (79.5 μm × 79.5 μm) for qualitative assessment of astrocyte morphology. Data are presented as mean ± SEM with average values for individual mice represented by a dot in bar graphs. N = 3–5 mice of each genotype (WT and *Atxn1*$^{154Q/2Q}$). ns: not significant, * $p < 0.05$, ** $p < 0.01$, Students *t*-test.

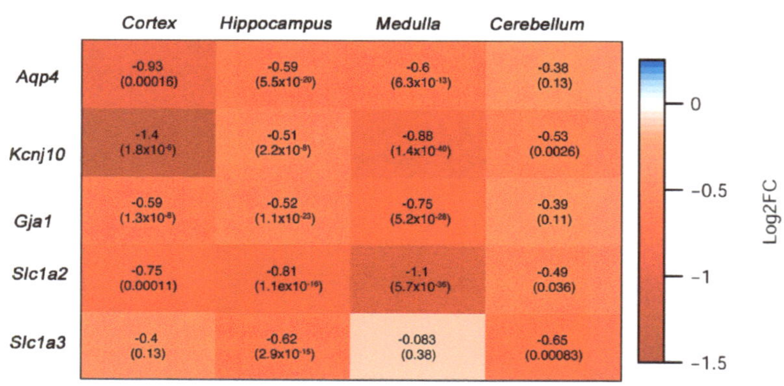

Figure 5. Late SCA1 stage astrocyte molecular analysis indicates brain-wide loss of core homeostatic functions. Cerebellum, brainstem (medulla), hippocampus, and cortex were isolated from 26 weeks old *Atxn1*$^{154Q/2Q}$ and wild-type control mice (N = 4 of each genotype) and isolated RNA was sent for RNA-sequencing (RNA-seq). Expression of astrocyte genes was determined using differential gene expression analysis with numbers in each square representing log2 fold change (log2FC) in *Atxn1*$^{154Q/2Q}$ compared to wild-type controls and numbers in brackets presenting False Discovery Rate adjusted *p* values.

Intriguingly, late in disease both the ION and hippocampus demonstrated a decrease in microglial density and in microglia area (although decrease in the microglia area was only trending in the hippocampus) (Figure 6). While density and area of cerebellar microglia was further increased with disease progression, we did not detect any change in microglial density or area in the cortex.

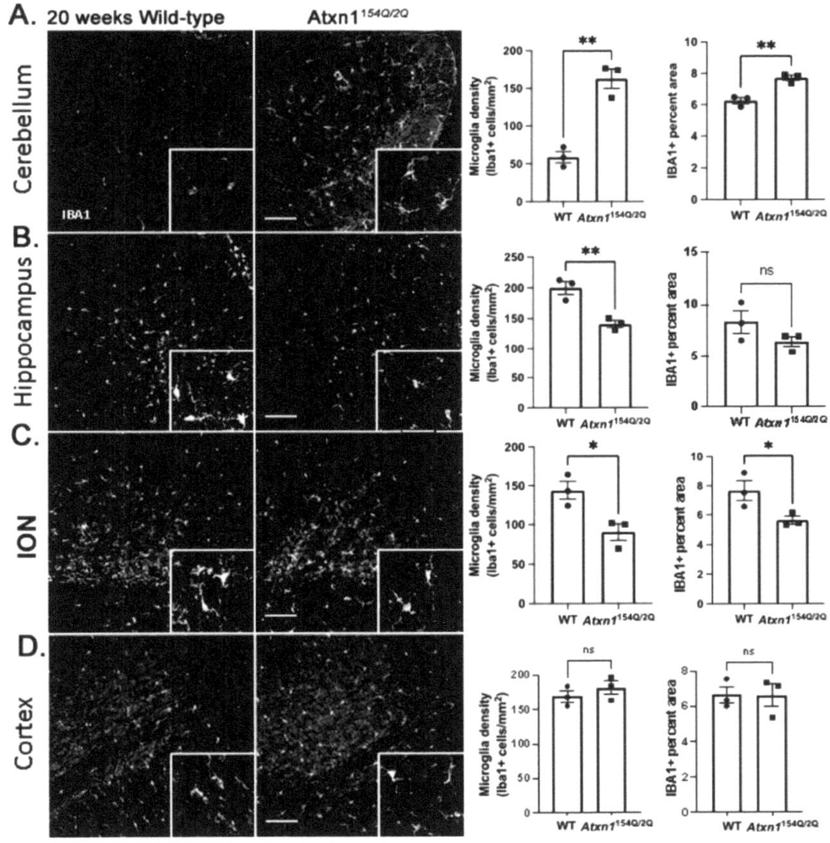

Figure 6. Diversity of late SCA1 microglial morphology phenotypes in *Atxn1*$^{154Q/2Q}$ mice. Brain sections from 20 weeks old *Atxn1*$^{154Q/2Q}$ and wild-type littermate controls (N = 3–5) were stained for Iba1. Confocal images of cerebellum ((**A**), lobule X), hippocampus ((**B**), dentate gyrus), brainstem ((**C**), ION), and motor cortex ((**D**), layer 6 and corpus callosum) were used to quantify density of Iba1-positive microglia and Iba1 area. Images were acquired at 20× (636.16 µm × 636.16 µm). Scale bars show 100 µm. Insets in the lower right corners of each image show zoomed images (79.5 µm × 79.5 µm) for qualitative assessment of microglia morphology. Data are presented as mean ± SEM with average values for each mouse represented by a dot. N = 3–5 mice of each genotype (WT and *Atxn1*$^{154Q/2Q}$). ns: not significant, * $p < 0.05$, ** $p < 0.01$. Students *t*-test.

Microglia have been suggested to lead to loss of synapses in disease. To investigate whether changes in microglia density are associated with loss of synapses, we quantified the number of excitatory synapses in brain regions with different changes in microglia: the cerebellum that exhibits microglial activation, and the cortex and hippocampus that respectively show reduction, or no change, in microglial density. We found significant reduction of VGLUT2/PSD95 synapses in the cerebellum, while there was no change in the VGLUT2/PSD95 synaptic quanta in the hippocampus and in the motor cortex (Supplementary Figure S4).

These results suggest that with disease progression, some brain regions (such as olives) may suffer from loss of both astrocytes and microglia, while other regions, such as the cerebellum, may suffer from reactive phenotypes in both cell types.

4. Discussion

Here, we report on the spectrum of reactive astrocyte phenotypes, across several clinically relevant brain regions of *Atxn1*$^{154Q/2Q}$ mice and at different stages of disease progression summarized in Table 1. Recent studies indicated heterogeneity of glial cells across healthy brain regions [16] and across different disease conditions [56,67]. Our results indicate spatial diversity of morphological and molecular changes in astrocytes during the early stages of disease in *Atxn1*$^{154Q/2Q}$ mice. This spectrum of reactive astrocytic phenotypes may indicate brain region-specific astrocytic dysfunctions and consequently their brain region-specific contributions to SCA1 disease pathogenesis. Given that these changes precede neuronal loss, it is crucial to examine glial heterogeneity as a means to distinguish protective versus deleterious astrocytic phenotypes and identify opportunities for early therapeutic intervention. For this reason, we focused on the astrocytic neuroprotective genes *Slc1a2*, *Kcnj10*, *Gja1*, and *Aqp4*, and found that their expression was reduced in the medulla and hippocampus, but was increased in the cortex of *Atxn1*$^{154Q/2Q}$ mice at 12 weeks of age and before notable neuronal loss [6]. *Kcnj10* encodes a potassium rectifier, Kir4.1, that is involved in maintaining potassium homeostasis [68] and has been implicated in several diseases, including Huntington's disease, ALS, and depression [68,69]. *Gja1* encodes the astrocyte gap-junction protein connexin 43 (Gja1), which is critical for astrocyte–astrocyte communication [50]. *Aquaporin* (*Aqp4*) is critical for both interstitial fluid drainage in the glymphatic system and water homeostasis [70]. Astrocytic *Slc1a* (2 and 3) genes encode glutamate transporters that are responsible for removing glutamate from synaptic space. These astrocytic genes are critical for neuronal function and have been implicated in the pathogenesis of several neurodegenerative diseases [71,72]. We propose that reduced expression of astrocyte neuro-supportive genes—as observed in *Atxn1*$^{154Q/2Q}$ mice—may exacerbate neuronal dysfunction and SCA1 pathogenesis in the hippocampus and medulla. Conversely, increased expression of these neuro-supportive astrocyte genes in the cortex may indicate compensatory roles of astrocytes that delay neuronal dysfunction in the SCA1-affected cortex. This is supported by our results showing preserved neuronal activity in the cortex and reduced neuronal activity in the hippocampus.

Early in disease, we found no evidence of increased GFAP expression, a hallmark of reactive astrogliosis, in the medulla. Likewise, we found only trending indications of increased GFAP expression in the cortex. Thus, our results may indicate that increased GFAP expression may not negatively correlate to changes in astrocyte homeostatic gene expression in disease conditions. This is consistent with a previous study which demonstrated that reduced expression of the astrocyte homeostatic gene *Kcnj10* in the striatum of Huntington's mice is not dependent on astrogliosis as defined by increased GFAP expression [68]. Moreover, in our study we have found a strong positive correlation of *Gfap* expression with the expression of *Kcnj10*, *Slc1a2*, and *Aqp4* in the cerebellum, indicating that reactive astrocytes (indicated by increased *Gfap* expression) may initially increase expression of core homeostatic genes as a neuro-supportive measure in disease.

Table 1. Summary of glial changes in SCA1 across brain regions and disease stages.

		Cerebellum	Hippocampus	Brainstem	Cortex
Astrocyte reactivity	Early	Increased ↑ GFAP intensity ↑ GFAP %area	Increased ↑ GFAP intensity ↑ GFAP %area - Astrocyte density	Decreased ↓ GFAP intensity - GFAP %area ↓ Astrocyte density	Not significant - GFAP intensity - GFAP %area - Astrocyte density
	Late	Increased ↑ GFAP intensity ↑ GFAP %area	Increased - GFAP intensity ↑ GFAP %area ↑ Astrocyte density	Decreased - GFAP intensity ↓ GFAP %area ↓ Astrocyte density	Increased ↑ GFAP intensity ↑ GFAP %area ↑ Astrocyte density
Astrocyte homeostasis	Early	Not significant - *Aqp4* - *Gja1* - *Kcnj10* - *Slc1a2*	Decreased ↓ *Aqp4* ↓ *Gja1* ↓ *Kcnj10* ↓ *Slc1a2*	Decreased ↓ *Aqp4* ↓ *Gja1* ↓ *Kcnj10* - *Slc1a2*	Increased ↑ *Aqp4* ↑ *Gja1* ↑ *Kcnj10* - *Slc1a3*
	Late	Decreased ↓ *Aqp4* ↓ *Gja1* ↓ *Kcnj10* ↓ *Slc1a2*	Decreased ↓ *Aqp4* ↓ *Gja1* ↓ *Kcnj10* ↓ *Slc1a2*	Decreased ↓ *Aqp4* ↓ *Gja1* ↓ *Kcnj10* ↓ *Slc1a2*	Decreased ↓ *Aqp4* ↓ *Gja1* ↓ *Kcnj10* ↓ *Slc1a2*
Microglia reactivity	Early	Increased ↑ IBA1 %area ↑ Microglia density	Not significant - IBA1 %area - Microglia density	Increased - IBA1 %area ↑ Microglia density	Increased ↑ IBA1 %area ↑ Microglia density
	Late	Increased ↑ IBA1 %area ↑ Microglia density	Decreased - IBA1 %area ↓ Microglia density	Decreased ↓ IBA1 %area ↓ Microglia density	Not significant - IBA1 %area - Microglia density

We also found a loss of astrocytes in the ION of 12-week-old *Atxn1*$^{154Q/2Q}$ mice, preceding neuronal loss. This is consistent with previous work showing lower levels of myo-inositol (Ins), an astrocytic marker, in the brainstem of 18- and 28-week-old *Atxn1*$^{154Q/2Q}$ mice [33]. Moreover, treatment with Atxn1-targeting antisense oligonucleotides (ASO), which are also taken up by astrocytes [73], was shown to rescue lower Ins levels in the brainstem and prolong the survival of *Atxn1*$^{154Q/2Q}$ mice [33]. Based on these results, we speculate that reduced expression of homeostatic genes and decreased astrocyte viability may contribute to brainstem dysfunction and perhaps even to premature lethality in SCA1 mice. However, while neuronal numbers were not altered in the ION of 12-week-old *Atxn1*$^{154Q/2Q}$ mice, we cannot exclude neuronal dysfunction. Future studies using conditional SCA1 mice to selectively express or delete mutant ATXN1 in astrocytes will provide insight into the cell autonomous effect of mutant ATXN1 on astrocyte pathology in the ION and other brain regions in SCA1.

Finally, we report that disease progression eliminates diversity of astrocyte gene expression changes across brain regions. On the other hand, morphological changes in astrocytes became more pronounced with disease. As disease progressed, the hippocampus and brainstem exhibited a decrease in microglial density. Together, these results indicate that profound loss of glial functions characterizes late disease stage in *Atxn1*$^{154Q/2Q}$ mice and is likely to exacerbate neuronal dysfunction and loss at this stage of disease (Figure 7). While future studies will investigate mechanisms underlying the loss of astrocytes and microglia, our results indicate that preserving glial functionality may provide therapeutic benefits in SCA1.

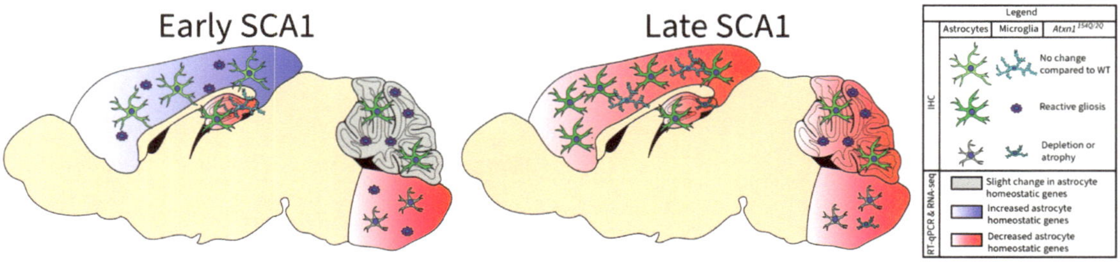

Figure 7. Simplified schematics of astrocytic changes in SCA1. Simplified schematic of astrocytic and microglia changes in SCA1. In early stages of SCA1 (left), brain regions exhibit an increase (cortex) or slight change (cerebellum) in the expression of astrocyte homeostatic genes, whereas regions in red exhibit loss of astrocyte homeostatic genes (hippocampus and medulla). Areas differ in patterns of reactive astrocytes, showing astrocyte atrophy (inferior olivary nucleus (ION)), hypertrophy (cerebellum), or no significant change relative to WT controls (hippocampus and cortex). Most brain regions (cerebellum, ION, cortex) show reactive microglia during early stages of SCA1. During late stages of SCA1 (right), all brain regions exhibit reduced expression of core genes necessary for astrocyte homeostatic function (cortex, cerebellum, hippocampus, and medulla). Most brain regions (cerebellum, hippocampus, and cortex) show reactive astrocytes during late stage of SCA1, although the ION continues to show significant reduction or atrophy of astrocytes. Reactive microglia persist in the cerebellum but are no longer present in the cortex in late SCA1. Some brain regions also show significantly fewer microglia (hippocampus and ION) relative to WT in late SCA1.

Supplementary Materials: The following supporting information can be downloaded at: https://www.mdpi.com/article/10.3390/cells11203323/s1, Figure S1: Pearson's correlation between *Gfap* and homeostatic astrocyte genes in the cerebellum; Figure S2: Neuronal density is not altered early in SCA1; Figure S3: Neuronal activity is altered in the hippocampus but preserved in motor cortex early in SCA1; Figure S4: Synaptic quantification in the cerebellum, hippocampus, and cortex.

Author Contributions: M.C. conceptualized the study. J.-G.R., C.S., K.H., E.B., A.S., H.P.H., F.J.G., R.K., K.S., O.R. and S.G. performed experiments. J.-G.R., M.C., H.P.H., F.J.G., K.S., R.K., S.G. and C.S. analyzed data. M.C. and J.-G.R., wrote the manuscript with input from O.R., C.S., K.H., A.S., E.B., H.P.H., R.K., K.S. and S.G. All authors have read and agreed to the published version of the manuscript.

Funding: This work was supported by National Institute of Health NINDS awards (R01 NS197387 and R01 NS109077 to M.C.).

Institutional Review Board Statement: Animal experimentation was approved by the Institutional Animal Care and Use Committee (IACUC) of University of Minnesota (protocol number 2109-39465A, date of approval Nov 11 2021); and was conducted in accordance with the National Institutes of Health's (NIH) Principles of Laboratory Animal Care (86–23, revised 1985), and the American Physiological Society's Guiding Principles in the Use of Animals.

Data Availability Statement: Data available on request from the authors.

Acknowledgments: We acknowledge all the members of the Orr and Cvetanovic laboratories for their thoughtful discussions and feedback on the study. Work in this study was aided by the Mouse Behavioral Core and Genomics Core at the University of Minnesota. We thank Laura Berg for editing.

Conflicts of Interest: The authors declare no conflict of interest.

References

1. Sheeler, C.; Rosa, J.; Ferro, A.; Mcadams, B.; Borgenheimer, E.; Cvetanovic, M. Glia in Neurodegeneration: The Housekeeper, the Defender and the Perpetrator. *Int. J. Mol. Sci.* **2020**, *21*, 9188. [CrossRef]
2. Barres, B.A. The Mystery and Magic of Glia: A Perspective on Their Roles in Health and Disease. *Neuron* **2008**, *60*, 430–440. [CrossRef] [PubMed]

3. Oliveira, J.F.; Sardinha, V.M.; Guerra-Gomes, S.; Araque, A.; Sousa, N. Do stars govern our actions? Astrocyte involvement in rodent behavior. *Trends Neurosci.* **2015**, *38*, 535–549. [CrossRef] [PubMed]
4. Bay, V.; Butt, A.M. Relationship between glial potassium regulation and axon excitability: A role for glial Kir4.1 channels. *Glia* **2012**, *60*, 651–660. [CrossRef] [PubMed]
5. Kucheryavykh, Y.V.; Kucheryavykh, L.Y.; Nichols, C.G.; Maldonado, H.M.; Baksi, K.; Reichenbach, A.; Skatchkov, S.N.; Eaton Misty, J. Downregulation of Kir4.1 inward rectifying potassium channel subunits by RNAi impairs potassium transfer and glutamate uptake by cultured cortical as-trocytes. *Glia* **2007**, *55*, 274–281. [CrossRef] [PubMed]
6. Miyazaki, T.; Yamasaki, M.; Hashimoto, K.; Kohda, K.; Yuzaki, M.; Shimamoto, K.; Tanaka, K.; Kano, M.; Watanabe, M. Glutamate transporter GLAST controls synaptic wrapping by Bergmann glia and ensures proper wiring of Purkinje cells. *Proc. Natl. Acad. Sci. USA* **2017**, *114*, 7438–7443. [CrossRef] [PubMed]
7. Nagelhus, E.A.; Ottersen, O.P. Physiological Roles of Aquaporin-4 in Brain. *Physiol. Rev.* **2013**, *93*, 1543–1562. [CrossRef]
8. Baroni, F.; Mazzoni, A. Heterogeneity of heterogeneities in neuronal networks. Front. Comput. *Neurosci.* **2014**, *8*, 161.
9. Herrero-Navarro, Á.; Puche-Aroca, L.; Moreno-Juan, V.; Sempere-Ferràndez, A.; Espinosa, A.; Susín, R.; Torres-Masjoan, L.; Leyva-Díaz, E.; Karow, M.; Figueres-Oñate, M.; et al. Astrocytes and neurons share region-specific transcriptional signatures that confer regional identity to neuronal reprogramming. *Sci. Adv.* **2021**, *7*, eabe8978. [CrossRef]
10. Martín, R.; Bajo-Grañeras, R.; Moratalla, R.; Perea, G.; Araque, A. Glial Cell Signaling. Circuit-specific signaling in astro-cyte-neuron networks in basal ganglia pathways. *Science* **2015**, *349*, 730–734. [CrossRef]
11. Zeisel, A.; Muñoz-Manchado, A.B.; Codeluppi, S.; Lönnerberg, P.; La Manno, G.; Juréus, A.; Marques, S.; Munguba, H.; He, L.; Betsholtz, C.; et al. Cell types in the mouse cortex and hippocampus revealed by single-cell RNA-seq. *Science* **2015**, *347*, 1138–1142. [CrossRef] [PubMed]
12. Saunders, A.; Macosko, E.Z.; Wysoker, A.; Goldman, M.; Krienen, F.M.; de Rivera, H.; Bien, E.; Baum, M.; Bortolin, L.; Wang, S.; et al. Molecular Diversity and Specializations among the Cells of the Adult Mouse Brain. *Cell* **2018**, *174*, 1015–1030.e16. [CrossRef] [PubMed]
13. Bayraktar, O.A.; Bartels, T.; Holmqvist, S.; Kleshchevnikov, V.; Martirosyan, A.; Polioudakis, D.; Ben Haim, L.; Young, A.M.H.; Batiuk, M.Y.; Prakash, K.; et al. Astrocyte layers in the mammalian cerebral cortex revealed by a single-cell in situ transcriptomic map. *Nat. Neurosci.* **2020**, *23*, 500–509. [CrossRef] [PubMed]
14. Clark, I.C.; Gutiérrez-Vázquez, C.; Wheeler, M.A.; Li, Z.; Rothhammer, V.; Linnerbauer, M.; Sanmarco, L.M.; Guo, L.; Blain, M.; Zandee, S.E.J.; et al. Barcoded viral tracing of single-cell interactions in central nervous system inflammation. *Science* **2021**, *372*, eabf1230. [CrossRef]
15. Pestana, F.; Edwards-Faret, G.; Belgard, T.G.; Martirosyan, A.; Holt, M.G. No Longer Underappreciated: The Emerging Concept of Astrocyte Heterogeneity in Neuroscience. *Brain Sci.* **2020**, *10*, 168. [CrossRef]
16. Soreq, L.; Rose, J.; Soreq, E.; Hardy, J.; Trabzuni, D.; Cookson, M.R.; Smith, C.; Ryten, M.; Patani, R.; Ule, J. Major Shifts in Glial Regional Identity Are a Transcriptional Hallmark of Human Brain Aging. *Cell Rep.* **2017**, *18*, 557–570. [CrossRef]
17. Ben Haim, L.; Rowitch, D.H. Functional diversity of astrocytes in neural circuit regulation. *Nat. Rev. Neurosci.* **2016**, *18*, 31–41. [CrossRef]
18. Pekny, M.; Wilhelmsson, U.; Pekna, M. The dual role of astrocyte activation and reactive gliosis. *Neurosci. Lett.* **2014**, *565*, 30–38. [CrossRef]
19. Lee, H.-G.; Wheeler, M.A.; Quintana, F.J. Function and therapeutic value of astrocytes in neurological diseases. *Nat. Rev. Drug Discov.* **2022**, *21*, 339–358. [CrossRef]
20. Diaz-Castro, B.; Gangwani, M.R.; Yu, X.; Coppola, G.; Khakh, B.S. Astrocyte molecular signatures in Huntington's disease. *Sci. Transl. Med.* **2019**, *11*, eaaw8546. [CrossRef]
21. Tong, X.; Ao, Y.; Faas, G.C.; Nwaobi, S.E.; Xu, J.; Haustein, M.D.; Anderson, M.A.; Mody, I.; Olsen, M.L.; Sofroniew, M.V.; et al. Astrocyte Kir4.1 ion channel deficits contribute to neuronal dys-function in Huntington's disease model mice. *Nat. Neurosci.* **2014**, *17*, 694–703. [CrossRef] [PubMed]
22. Habib, N.; McCabe, C.; Medina, S.; Varshavsky, M.; Kitsberg, D.; Dvir-Szternfeld, R.; Green, G.; Dionne, D.; Nguyen, L.; Marshall, J.L.; et al. Disease-associated astrocytes in Alzheimer's disease and aging. *Nat. Neurosci.* **2020**, *23*, 701–706. [CrossRef] [PubMed]
23. Mathys, H.; Davila-Velderrain, J.; Peng, Z.; Gao, F.; Mohammadi, S.; Young, J.Z.; Menon, M.; He, L.; Abdurrob, F.; Jiang, X.; et al. Single-cell transcriptomic analysis of Alzheimer's disease. *Nature* **2019**, *570*, 332–337. [CrossRef] [PubMed]
24. Matias, I.; Morgado, J.; Carvalho, F.; Gomes, A. Astrocyte Heterogeneity: Impact to Brain Aging and Disease. *Front. Aging Neurosci.* **2019**, *11*, 59. [CrossRef]
25. Kim, J.H.; Lukowicz, A.; Qu, W.; Johnson, A.; Cvetanovic, M. Astroglia contribute to the pathogenesis of spinocerebellar ataxia Type 1 (SCA1) in a biphasic, stage-of-disease specific manner. *Glia* **2018**, *66*, 1972–1987. [CrossRef]
26. Saxena, S.; Caroni, P. Selective Neuronal Vulnerability in Neurodegenerative Diseases: From Stressor Thresholds to Degeneration. *Neuron* **2011**, *71*, 35–48. [CrossRef]
27. Cragnolini, A.B.; Lampitella, G.; Virtuoso, A.; Viscovo, I.; Panetsos, F.; Papa, M.; Cirillo, G. Regional brain susceptibility to neurodegeneration: What is the role of glial cells? *Neural Regen. Res.* **2020**, *15*, 838–842.

28. Orr, H.; Chung, M.-Y.; Banfi, S.; Kwiatkowski, T.J.; Servadio, A.; Beaudet, A.L.; McCall, A.E.; Duvick, L.A.; Ranum, L.P.W.; Zoghbi, H. Expansion of an unstable trinucleotide CAG repeat in spinocerebellar ataxia type *Nat. Genet.* **1993**, *4*, 221–226. [CrossRef]
29. Zoghbi, H.; Orr, H. Pathogenic Mechanisms of a Polyglutamine-mediated Neurodegenerative Disease, Spinocerebellar Ataxia Type 1. *J. Biol. Chem.* **2009**, *284*, 7425–7429. [CrossRef]
30. Klinke, I.; Minnerop, M.; Schmitz-Hübsch, T.; Hendriks, M.; Klockgether, T.; Wüllner, U.; Helmstaedter, C. Neuropsychological features of patients with spinocerebellar ataxia (SCA) types 1, 2, 3, and 6. *Cerebellum* **2010**, *9*, 433–442. [CrossRef]
31. Zoghbi, H.; Pollack, M.S.; Bs, L.A.L.; Ferrell, R.E.; Daiger, S.P.; Beaudet, A.L. Spinocerebellar ataxia: Variable age of onset and linkage to human leukocyte antigen in a large kindred. *Ann. Neurol.* **1988**, *23*, 580–584. [CrossRef] [PubMed]
32. Koeppen, A.H. The pathogenesis of spinocerebellar ataxia. *Cerebellum* **2005**, *4*, 62–73. [CrossRef] [PubMed]
33. Friedrich, J.; Kordasiewicz, H.B.; O'Callaghan, B.L.; Handler, H.P.; Wagener, C.; Duvick, L.; Swayze, E.; Rainwater, O.; Hofstra, B.; Benneyworth, M.; et al. Antisense oligonucleotide–mediated ataxin-1 reduction prolongs survival in SCA1 mice and reveals disease-associated transcriptome profiles. *JCI Insight* **2018**, *3*, e123193. [CrossRef] [PubMed]
34. Rüb, U.; Bürk, K.; Timmann, D.; den Dunnen, W.; Seidel, K.; Farrag, K.; Brunt, E.; Heinsen, H.; Egensperger, R.; Bornemann, A.; et al. Spinocerebellar ataxia type 1 (SCA1): New pathoana-tomical and clinico-pathological insights. *Neuropathol. Appl. Neurobiol.* **2012**, *38*, 665–680. [CrossRef] [PubMed]
35. Seidel, K.; Siswanto, S.; Brunt, E.R.P.; Dunnen, W.D.; Korf, H.; Rueb, U. Brain pathology of spinocerebellar ataxias. *Acta Neuropathol.* **2012**, *124*, 1–21. [CrossRef] [PubMed]
36. Diallo, A.; Jacobi, H.; Cook, A.; Giunti, P.; Parkinson, M.H.; Labrum, R.; Durr, A.; Brice, A.; Charles, P.; Marelli, C.; et al. Prediction of Survival With Long-Term Disease Progression in Most Common Spinocerebellar Ataxia. *Mov. Disord.* **2019**, *34*, 1220–1227. [CrossRef]
37. Moriarty, A.; Cook, A.; Hunt, H.; Adams, M.E.; Cipolotti, L.; Giunti, P. A longitudinal investigation into cognition and disease pro-gression in spinocerebellar ataxia types 1, 2, 3, 6, and 7. *Orphanet J. Rare Dis.* **2016**, *11*, 82. [CrossRef]
38. Jacobi, H.; Reetz, K.; du Montcel, S.T.; Bauer, P.; Mariotti, C.; Nanetti, L.; Rakowicz, M.; Sulek, A.; Durr, A.; Charles, P.; et al. Biological and clinical characteristics of individuals at risk for spinocerebellar ataxia types 1, 2, 3, and 6 in the longitudinal RISCA study: Analysis of baseline data. *Lancet Neurol.* **2013**, *12*, 650–658. [CrossRef]
39. Koscik, T.R.; Sloat, L.; van der Plas, E.; Joers, J.M.; Deelchand, D.K.; Lenglet, C.; Öz, G.; Nopoulos, P.C. Brainstem and striatal volume changes are detectable in under 1 year and predict motor decline in spinocerebellar ataxia type 1. *Brain Commun.* **2020**, *2*, fcaa184. [CrossRef]
40. Genis, D.; Matilla, T.; Volpini, V.; Rosell, J.; Dávalos, A.; Ferrer, I.; Molins, A.; Estivill, X. Clinical, neuropathologic, and genetic studies of a large spinocerebellar ataxia type 1 (SCA1) kindred: (CAG)n expansion and early premonitory signs and symptoms. *Neurology* **1995**, *45*, 24–30. [CrossRef]
41. Rüb, U.; Schöls, L.; Paulson, H.; Auburger, G.; Kermer, P.; Jen, J.C.; Seidel, K.; Korf, H.W.; Deller, T. Clinical features, neurogenetics and neu-ropathology of the polyglutamine spinocerebellar ataxias type 1, 2, 3, 6 and 7. *Prog. Neurobiol.* **2013**, *104*, 38–66. [CrossRef]
42. Watase, K.; Weeber, E.J.; Xu, B.; Antalffy, B.; Yuva-Paylor, L.; Hashimoto, K.; Kano, M.; Atkinson, R.; Sun, Y.; Armstrong, D.L.; et al. A Long CAG Repeat in the Mouse Sca1 Locus Replicates SCA1 Features and Reveals the Impact of Protein Solubility on Selective Neurodegeneration. *Neuron* **2002**, *34*, 905–919. [CrossRef]
43. Asher, M.; Rosa, J.-G.; Rainwater, O.; Duvick, L.; Bennyworth, M.; Lai, R.-Y.; Kuo, S.-H.; Cvetanovic, M.; Sca, C. Cerebellar contribution to the cognitive alterations in SCA1: Evidence from mouse models. *Hum. Mol. Genet.* **2019**, *29*, 117–131. [CrossRef] [PubMed]
44. Suh, J.; Romano, D.M.; Nitschke, L.; Herrick, S.P.; DiMarzio, B.A.; Dzhala, V.; Bae, J.-S.; Oram, M.K.; Zheng, Y.; Hooli, B.; et al. Loss of Ataxin-1 Potentiates Alzheimer's Pathogenesis by Elevating Cerebral BACE1 Transcription. *Cell* **2019**, *178*, 1159–1175.e17. [CrossRef] [PubMed]
45. Cvetanovic, M.; Hu, Y.S.; Opal, P. Mutant Ataxin-1 Inhibits Neural Progenitor Cell Proliferation in SCA1. *Cerebellum* **2016**, *16*, 340–347. [CrossRef] [PubMed]
46. Cvetanovic, M.; Patel, J.; Marti, H.; Kini, A.R.; Opal, P. Vascular endothelial growth factor ameliorates the ataxic phenotype in a mouse model of spinocerebellar ataxia type 1. *Nat. Med.* **2011**, *17*, 1445–1447. [CrossRef]
47. Zu, T.; Duvick, L.A.; Kaytor, M.D.; Berlinger, M.S.; Zoghbi, H.Y.; Clark, H.B.; Orr, H.T. Recovery from polyglutamine-induced neurodegen-eration in conditional SCA1 transgenic mice. *J. Neurosci.* **2004**, *24*, 8853–8861. [CrossRef]
48. Nwaobi, S.E.; Cuddapah, V.; Patterson, K.; Randolph, A.C.; Olsen, M.L. The role of glial-specific Kir4.1 in normal and pathological states of the CNS. *Acta Neuropathol.* **2016**, *132*, 1–21. [CrossRef]
49. Rothstein, J.D.; Dykes-Hoberg, M.; Pardo, C.A.; Bristol, L.A.; Jin, L.; Kuncl, R.W.; Kanai, Y.; Hediger, M.A.; Wang, Y.; Schielke, J.P.; et al. Antisense knockout of glutamate transporters reveals a predominant role for astroglial glutamate transport in excitotoxicity and clearance of extracellular glutamate. *Neuron* **1996**, *16*, 675–686. [CrossRef]
50. Clasadonte, J.; Scemes, E.; Wang, Z.; Boison, D.; Haydon, P.G. Connexin 43-Mediated Astroglial Metabolic Networks Contribute to the Regulation of the Sleep-Connexin 43-Mediated Astroglial Metabolic Networks Contribute to the Regulation of the Sleep-Wake Cycle. *Neuron* **2017**, *95*, 1365–1380. [CrossRef]
51. Gatchel, J.R.; Zoghbi, H.Y. Diseases of unstable repeat expansion: Mechanisms and common principles. *Nat. Rev. Genet.* **2005**, *6*, 743–755. [CrossRef] [PubMed]

52. Cvetanovic, M. Decreased expression of glutamate transporter GLAST in Bergmann glia is associated with the loss of Purkinje neurons in the spinocerebellar ataxia type 1. *Cerebellum* **2015**, *14*, 8–11. [CrossRef] [PubMed]
53. Matilla, A.; Roberson, E.D.; Banfi, S.; Morales, J.; Armstrong, D.L.; Burright, E.N.; Orr, H.T.; Sweatt, J.D.; Zoghbi, H.Y.; Matzuk, M.M. Cloning and developmental expression analysis of the murine homolog of the spinocerebellar ataxia type 1 gene (Sca1). *Hum. Mol. Genet.* **1996**, *5*, 33–40.
54. Asher, M.; Johnson, A.; Zecevic, B.; Pease, D.; Cvetanovic, M. Ataxin-1 regulates proliferation of hippocampal neural precursors. *Neuroscience* **2016**, *322*, 54–65. [CrossRef] [PubMed]
55. Diallo, A.; Jacobi, H.; Tezenas du Montcel, S.; Klockgether, T. Natural history of most common spinocerebellar ataxia: A systematic review and meta-analysis. *J. Neurol.* **2021**, *268*, 2749–2756. [CrossRef]
56. Escartin, C.; Galea, E.; Lakatos, A.; O'Callaghan, J.P.; Petzold, G.C.; Serrano-Pozo, A.; Steinhäuser, C.; Volterra, A.; Carmignoto, G.; Agarwal, A.; et al. Reactive astrocyte nomenclature, definitions, and future directions. *Nat. Neurosci.* **2021**, *24*, 312–325. [CrossRef]
57. Pekny, M.; Pekna, M.; Messing, A.; Steinhäuser, C.; Lee, J.-M.; Parpura, V.; Hol, E.M.; Sofroniew, M.V.; Verkhratsky, A. Astrocytes: A central element in neurological diseases. *Acta Neuropathol.* **2015**, *131*, 323–345. [CrossRef]
58. Lippman Bell, J.J.; Lordkipanidze, T.; Cobb, N.; Dunaevsky, A. Bergmann glial ensheathment of dendritic spines regulates synapse number without affecting spine motility. *Neuron Glia Biol.* **2010**, *6*, 193–200. [CrossRef]
59. Farmer, W.T.; Abrahamsson, T.; Chierzi, S.; Lui, C.; Zaelzer, C.; Jones, E.V.; Bally, B.P.; Chen, G.G.; Théroux, J.-F.; Peng, J.; et al. Neurons diversify astrocytes in the adult brain through sonic hedgehog signaling. *Science* **2016**, *351*, 849–854. [CrossRef]
60. Cvetanovic, M.; Ingram, M.; Orr, H.; Opal, P. Early activation of microglia and astrocytes in mouse models of spinocerebellar ataxia type 1. *Neuroscience* **2015**, *289*, 289–299. [CrossRef]
61. Verkhratsky, A.; Nedergaard, M. Physiology of Astroglia. *Physiol. Rev.* **2018**, *98*, 239–389. [CrossRef] [PubMed]
62. Perea, G.; Sur, M.; Araque, A. Neuron-glia networks: Integral gear of brain function. *Front. Cell Neurosci.* **2014**, *8*, 378. [CrossRef] [PubMed]
63. Kofuji, P.; Newman, E.A. Potassium buffering in the central nervous system. *Neuroscience* **2004**, *129*, 1045–1056. [CrossRef] [PubMed]
64. Djukic, B.; Casper, K.B.; Philpot, B.D.; Chin, L.-S.; McCarthy, K.D. Conditional Knock-Out of Kir4.1 Leads to Glial Membrane Depolarization, Inhibition of Potassium and Glutamate Uptake, and Enhanced Short-Term Synaptic Potentiation. *J. Neurosci.* **2007**, *27*, 11354–11365. [CrossRef]
65. Ferro, A.; Qu, W.; Lukowicz, A.; Svedberg, D.; Johnson, A.; Cvetanovic, M. Inhibition of NF-κB signaling in IKKβ F/F;LysM Cre mice causes motor deficits but does not alter pathogenesis of Spinocerebellar ataxia type 1. *PLoS ONE* **2018**, *1*, e0200013.
66. Qu, W.; Johnson, A.; Kim, J.H.; Lukowicz, A.; Svedberg, D.; Cvetanovic, M. Inhibition of colony-stimulating factor 1 receptor early in disease ameliorates motor deficits in SCA1 mice. *J. Neuroinflamm.* **2017**, *14*, 107. [CrossRef]
67. Yu, X.; Nagai, J.; Marti-Solano, M.; Soto, J.S.; Coppola, G.; Babu, M.M.; Khakh, B.S. Context-Specific Striatal Astrocyte Molecular Responses Are Phenotypically Exploitable. *Neuron* **2020**, *108*, 1146–1162.e10. [CrossRef]
68. Cui, Y.; Yang, Y.; Ni, Z.; Dong, Y.; Cai, G.; Foncelle, A.; Ma, S.; Sang, K.; Tang, S.; Li, Y.; et al. Astroglial Kir4.1 in the lateral habenula drives neuronal bursts in depression. *Nature* **2018**, *554*, 323–327. [CrossRef]
69. Kelley, K.W.; Ben Haim, L.; Schirmer, L.; Tyzack, G.E.; Tolman, M.; Miller, J.G.; Tsai, H.-H.; Chang, S.M.; Molofsky, A.V.; Yang, Y.; et al. Kir4.1-Dependent Astrocyte-Fast Motor Neuron Interactions Are Required for Peak Strength. *Neuron* **2018**, *98*, 306–319.e7. [CrossRef]
70. Liu, E.; Peng, X.; Ma, H.; Zhang, Y.; Yang, X.; Zhang, Y.; Sun, L.; Yan, J. The Involvement of Aquaporin-4 in the Interstitial Fluid Drainage Impairment Following Subarachnoid Hemorrhage. *Front. Aging Neurosci.* **2021**, *12*, 611494. [CrossRef] [PubMed]
71. Matute, C.; Domercq, M.; Gomez, M.V.S. Glutamate-mediated glial injury: Mechanisms and clinical importance. *Glia* **2005**, *53*, 212–224. [CrossRef] [PubMed]
72. Pardo, A.; Wong, V.; Benson, L.M.; Dykes, M.; Tanaka, K.; Rothstein, J.D.; Maragakis, N.J. Loss of the astrocyte glutamate transporter GLT1 modifies disease in SOD1G93A mice. *Exp. Neurol.* **2006**, *201*, 120–130. [CrossRef] [PubMed]
73. McLoughlin, H.S.; Moore, L.R.; Chopra, R.; Komlo, R.; McKenzie, M.; Blumenstein, K.G.; Zhao, H.; Kordasiewicz, H.B.; Shakkottai, V.G.; Paulson, H.L. Oligonucleotide therapy mitigates disease in spinocerebellar ataxia type 3 mice. *Ann. Neurol.* **2018**, *84*, 64–77. [CrossRef]

cells

MDPI

Review

Role of Astrocytes in the Pathophysiology of Lafora Disease and Other Glycogen Storage Disorders

Jordi Duran [1,2,3]

[1] Institut Químic de Sarrià (IQS), Universitat Ramon Llull (URL), 08017 Barcelona, Spain; jordi.duran@iqs.url.edu
[2] Institute for Bioengineering of Catalonia (IBEC), The Barcelona Institute of Science and Technology, 08028 Barcelona, Spain
[3] Institute for Research in Biomedicine (IRB Barcelona), The Barcelona Institute of Science and Technology, 08028 Barcelona, Spain

Abstract: Lafora disease is a rare disorder caused by loss of function mutations in either the EPM2A or NHLRC1 gene. The initial symptoms of this condition are most commonly epileptic seizures, but the disease progresses rapidly with dementia, neuropsychiatric symptoms, and cognitive deterioration and has a fatal outcome within 5–10 years after onset. The hallmark of the disease is the accumulation of poorly branched glycogen in the form of aggregates known as Lafora bodies in the brain and other tissues. Several reports have demonstrated that the accumulation of this abnormal glycogen underlies all the pathologic traits of the disease. For decades, Lafora bodies were thought to accumulate exclusively in neurons. However, it was recently identified that most of these glycogen aggregates are present in astrocytes. Importantly, astrocytic Lafora bodies have been shown to contribute to pathology in Lafora disease. These results identify a primary role of astrocytes in the pathophysiology of Lafora disease and have important implications for other conditions in which glycogen abnormally accumulates in astrocytes, such as Adult Polyglucosan Body disease and the buildup of *Corpora amylacea* in aged brains.

Keywords: glycogen; aggregation; Lafora disease; neuroinflammation; neurodegeneration; epilepsy

Citation: Duran, J. Role of Astrocytes in the Pathophysiology of Lafora Disease and Other Glycogen Storage Disorders. *Cells* **2023**, *12*, 722. https://doi.org/10.3390/cells12050722

Academic Editors: Michael Brenner and Vladimir Parpura

Received: 14 December 2022
Revised: 5 February 2023
Accepted: 22 February 2023
Published: 24 February 2023

1. Brain Glycogen

Cells store glucose in the form of glycogen—a branched polymer of glucose—to minimize the increase in osmotic pressure that would be associated with the accumulation of free glucose. Glycogen molecules can store up to 55,000 glucose units in a water-soluble form that can be rapidly degraded when energy is required. This polysaccharide is synthesized through the coordinated action of glycogen synthase (GS), which joins glucose units by α-1,4-glycosidic linkages, and glycogen branching enzyme (GBE), which introduces branching points via α-1,6-glycosidic linkages [1]. Similarly, glycogen is broken down by the coordinated action of glycogen phosphorylase and glycogen debranching enzyme, which digest α-1,4- and α-1,6-glycosidic linkages, respectively. In normal glycogen, branches are introduced at even intervals and this branched structure is important for its function and solubility, generating a spherical molecule that can be rapidly degraded when the cell needs glucose. In contrast, in poorly branched glycogen, long α-1,4 linked glucose chains can form single or double helices [2–4] that exclude water, thereby impeding the degradation of glycogen and decreasing its solubility.

Some tissues, such as skeletal muscle and particularly the liver, accumulate high amounts of glycogen (up to 2% and 8% of wet weight, respectively) [5]. Muscle glycogen provides energy for contraction during intense exercise, while liver glycogen is mobilized during fasting periods to produce glucose, which is then released into the blood to be used by the rest of the body. All other tissues and cell types contain glycogen, although at lower concentrations than those found in the liver and skeletal muscle. Within the brain, glycogen concentration is relatively low—about 0.1% of wet weight—and it is present mostly in astrocytes [6–10]. However, neurons also have an active glycogen metabolism that plays an essential role in the function of these cells [11,12].

Given the low concentration of glycogen in the brain, its role in this organ was traditionally overlooked and widely presumed to serve as an emergency reservoir for pathophysiological conditions like hypoglycemia or ischemia [13]. However, it is now clear that glycogen plays key roles in the normal functioning of the brain under physiological conditions [14,15]. Brain glycogen is mobilized as supplemental fuel when energy needs increase due to neuronal activity [16,17], and substantial deficits in learning and memory arise when its use is blocked [12,18–21]. To support neuronal function, and according to the astrocyte-neuron lactate shuttle hypothesis, astrocytes degrade glycogen to generate lactate, which they then release and which is taken up and oxidized by neurons [22]. Another simpler hypothesis is that, in situations of high-energy demand, astrocytes use glucose obtained from their own glycogen store, thereby sparing interstitial glucose for neurons [23].

2. Lafora Disease

Despite the key physiological role of glycogen, this polysaccharide can also participate in brain pathology. In some conditions, glycogen abnormally accumulates in the nervous tissue. Lafora disease (LD, OMIM 254780) is probably the most striking example of the consequences of the abnormal buildup of glycogen in the brain. In LD, poorly branched glycogen accumulates in the brain and other tissues in the form of aggregates known as Lafora bodies (LBs). Glycogen in LBs is not accessible to glycogen phosphorylase and glycogen debranching enzyme. Thus, it cannot be degraded and progressively accumulates. Glycogen in LBs also contains high levels of covalently bound phosphate, which had been hypothesized to participate in glycogen insolubility in LD [24]. However, it seems now clear that abnormal glycogen branching, not hyperphosphorylation, underlies glycogen insolubility in LD [25,26]. In addition to poorly branched, hyperphosphorylated glycogen, LBs contain several proteins, including GS, ubiquitin, and the autophagy adaptor p62 [27–30].

The onset of LD occurs in adolescence, in previously healthy children, normally in the form of epileptic seizures that are difficult to distinguish from idiopathic generalized epilepsies. Seizures escalate over time, together with a rapid decline in cognitive function. The patient develops severe dementia and eventually enters a vegetative state with continuous seizures. Death invariably comes 5 to 10 years after the onset as a consequence of *status epilepticus* or complications derived from neurodegeneration [31–33]. LD is a rare disease with an estimated prevalence of ~4 cases per million individuals in the world [31]. However, the number of undiagnosed cases might be higher, particularly in developing countries. Current treatment remains palliative, with limited success in the modulation of symptoms.

LD is an autosomal recessive disease caused by mutations in two genes: EPM2A, encoding laforin, a dual phosphatase that contains a carbohydrate-binding domain [34,35], and EPM2B/NHLRC1, encoding malin, an E3-ubiquitin ligase [36]. Mutations in either of these genes cause an indistinguishable disease. The exact roles of malin and laforin in glycogen metabolism are not yet fully understood, but it is widely accepted that malin uses laforin as a scaffold to bind to glycogen and ubiquitinate proteins involved in glycogen metabolism [37–39]. The accumulation of poorly branched glycogen in LD suggests that malin and laforin form this functional complex to regulate glycogen synthesis and prevent glycogen insolubility [40]. To minimize the toxic consequences of the accumulation of poorly branched glycogen, proteins like the autophagy adaptor p62 promote its compaction in the form of LBs [29] (Figure 1). This protective mechanism is reminiscent of the condensation of ubiquitinated, misfolded proteins into larger structures to be degraded by autophagy [41–43].

To study LD, several mouse models of the disease lacking laforin or malin have been generated [44–47]. These animals present similar pathophysiological phenotypes that recapitulate the human disease; i.e., they accumulate LBs in several tissues, including the brain, show a progressive neuronal loss, behavioral impairments, neuroinflammation with reactive astrocytes and microglia, altered autophagy, and increased susceptibility to epileptogenic drugs such as kainate and pentylenetetrazole [27,29,44–48]. In these LD models, the accumulation of LBs increases with age and the pathological phenotypes also worsen progressively as the animals age [27,28,44,49,50].

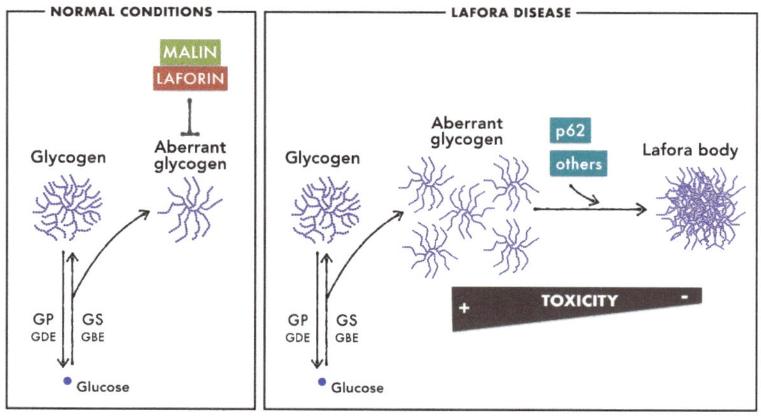

Figure 1. Brain glycogen metabolism in normal conditions and in LD. In normal conditions, malin and laforin prevent the accumulation of poorly branched glycogen that is generated as a side product of glycogen metabolism. In LD, due to the absence of malin or laforin, poorly branched glycogen accumulates in astrocytes and neurons. Proteins like the autophagy adaptor p62 promote its aggregation in the form of LBs to minimize the toxic consequences of its accumulation.

The role of LBs in the pathophysiology of LD has been unclear for many years. For instance, it was hypothesized that the primary cause of LD was an impairment in autophagy and that the accumulation of LBs was a consequence of this defect [48,51–53]. However, several groups, including ours, took advantage of mouse models of LD to demonstrate that excess glycogen underlies the pathology of this disease. Indeed, impeding or reducing glycogen synthesis in malin- or laforin-deficient mice prevents LB formation and prevents all the pathologic traits of the disease [27,45,54,55]. These models also showed that autophagy impairment is secondary to LB accumulation since autophagy markers are also normalized when glycogen accumulation is prevented [27,50]. Furthermore, we also used *Drosophila* and mouse models to demonstrate that forced accumulation of glycogen in neurons induces their death by apoptosis [56]. All these findings identified excessive glycogen accumulation as an inducer of neurodegeneration, and glycogen synthesis therefore became a putative target for the treatment of LD. Therapies based on the inhibition of glycogen synthesis are the focus of current research efforts [28,57–60].

3. Lafora Bodies in Neurons and Astrocytes

In the first description of LD in 1911, Dr. Gonzalo Rodriguez-Lafora reported the presence of LBs in neurons [61]. Until recently, it was widely believed that LBs accumulated exclusively in this cell type, and thus, all the pathologic traits of the disease were attributed to the toxic effects of neuronal LBs [31,62–64]. However, the premise that LBs are present exclusively in this cell population was inconsistent with the fact that, as mentioned before, brain glycogen is present mainly in astrocytes in normal conditions. It is now clear that LBs also accumulate in astrocytes. In 2011, we first reported the presence of LBs in these cells in a malin-deficient mouse model [46], but the significance of this discovery was underestimated. Several years later, we [65] and others [66] demonstrated that most LBs are present in astrocytes, particularly in regions like the hippocampus. We classified these bodies into neuronal (nLBs), and *Corpora amylacea*-like (CAL), the latter present in astrocytes, which were named this way because of their resemblance to *Corpora amylacea*, which are glycogen aggregates that accumulate in aged brains [67] (see below). Interestingly, CAL and nLBs differ not only in the cell type in which they accumulate but also in their shape and subcellular localization. CAL are polymorphic and present predominantly in astrocytic processes, and they show a patchy distribution, each patch corresponding to an individual astrocyte. In contrast, nLBs are normally present in the form of a single spherical aggregate close to the neuronal nucleus, and they resemble inclusion bodies formed by protein aggregates such as Lewy bodies [68]. The progressive accumulation of CAL and nLBs takes place in parallel, and both types of LB are already present at early stages in mouse models of LD [28,66].

Astrocytes have been shown to have phagocytic activity and can engulf apoptotic cells [69]. Thus, LBs present in astrocytes might not have originated in astrocytes themselves, but instead, they may have a neuronal origin; i.e., proceeding from the phagocytosis of an apoptotic body derived from a dead neuron. To decipher the origin of astrocytic LBs, as well as to understand their contribution to the pathology of LD, we generated a malin-deficient mouse in which GS was specifically deleted from astrocytes (malin^{KO} + GS$^{Gfap\text{-}KO}$ mice), thus preventing the synthesis of glycogen specifically in this cell type. The brains of these animals contained nLBs but were devoid of CAL, thereby unequivocally demonstrating that the latter originate in astrocytes [50] (Figure 2).

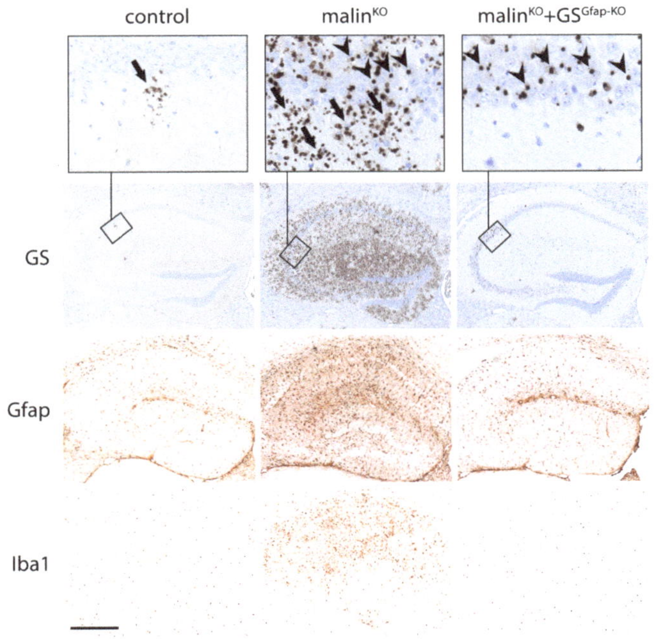

Figure 2. Accumulation of glycogen aggregates in neurons and astrocytes and neuroinflammation in mouse models of LD. GS, glial fibrillary acidic protein (Gfap) and ionized calcium-binding adapter molecule 1 (Iba1) immunostainings of mouse hippocampi are shown. GS immunostaining reveals the presence of few CAL aggregates in control mice. In contrast, the hippocampi of malin^{KO} mice show a conspicuous accumulation of CAL and nLBs, while the hippocampi of malin^{KO} + GS$^{Gfap\text{-}KO}$ mice only show nLBs (although not illustrated in this summary figure, the cell types containing the aggregates have been identified in several studies by using co-staining with cell-specific markers; e.g., see [46,65–67]). Gfap and Iba1 immunostainings, markers of astroglia and microglia, respectively, show a prominent neuroinflammation in the brains of malin^{KO} mice that is not present in malin^{KO} + GS$^{Gfap\text{-}KO}$ brains. Arrows: CAL, arrowheads: nLBs. Scale bar: 500 μm.

4. Role of Astrocytic LBs in the Pathophysiology of LD

The demonstration that LBs are also present in astrocytes opened up the possibility that these astrocytic LBs contribute to the pathology of LD. In fact, the quantification of CAL and nLBs showed that in brain regions like the hippocampus, CAL are clearly predominant over nLBs [65,66]. The analysis of malin^{KO}+GS$^{Gfap\text{-}KO}$ brains confirmed this quantification since the hippocampi of these mice are largely free of LBs [50] (Figure 2). Furthermore, RNA-Seq studies indicated that most of the upregulated genes in the brains of malin- and laforin-deficient mice encode pro-inflammatory mediators and that reactive glia, including astrocytes, are responsible for the expression of these inflammatory genes [70].

To understand the contribution of astrocytic LBs to the pathology of LD, the characteristic pathologic traits of the disease were analyzed in malin^{KO} + GS$^{Gfap\text{-}KO}$ mice. Neuroinflammation is considered one of the initial determinants of LD [71]. The brains of LD mouse models present clear astrogliosis, microgliosis, and increased expression of inflammatory genes [50,70]. In contrast, the analysis of malin^{KO} + GS$^{Gfap\text{-}KO}$ brains revealed normal levels of all these markers of neuroinflammation, thereby indicating that astrocytic LBs underlie neuroinflammation in LD [50] (Figure 2). The link between the excessive accumulation of astrocytic glycogen and neuroinflammation was further confirmed with another mouse model in which a constitutively active form of GS was expressed specifically in astrocytes. These mice, which accumulate high amounts of glycogen in astrocytes, present profound astrogliosis and microgliosis, as well as a marked increase in the expression of inflammatory genes [50]. These results confirmed that the excessive accumulation of glycogen in astrocytes induces neuroinflammation. However, the mechanism that links excessive astrocytic glycogen accumulation with neuroinflammation is currently unknown and is the focus of current research efforts.

As mentioned before, the initial symptom of LD is most commonly the presence of epileptic seizures, which worsen progressively with age. Animal models of LD reproduce this pathologic trait of the disease in the form of increased susceptibility to epileptogenic drugs like kainic acid. Astrocytes play essential roles in brain function, including the regulation of extracellular potassium and glutamate homeostasis, thus making them crucial actors in epilepsy [72]. In this regard, astrocytic glycogen has been shown to fuel potassium uptake into astrocytes, since the astrocytic sodium/potassium pump uses ATP obtained from glucose 6-phosphate originating from glycogen breakdown [73,74]. The non-clearance of extracellular potassium would result in neuronal hypersynchronization and burst firing, which would result in seizure generation and propagation. Thus, it has been suggested that alterations in glycogen metabolism contribute to the imbalance of glutamatergic and GABAergic neurotransmission associated with epileptic seizures [75]. Accordingly, it was reasonable to hypothesize that the impairment of astrocytic glycogen metabolism in LD compromises potassium uptake, which would increase excitability and thus be responsible for the epileptic phenotype of the disease. In line with this hypothesis, the presence of glycogen aggregates in astrocytes has also been described in patients with temporal lobe epilepsy [76]. Surprisingly, malin^{KO} + GS$^{Gfap\text{-}KO}$ mice do not show a significant amelioration of susceptibility to epilepsy [50], thereby indicating that astrocytic glycogen accumulation is not the main factor responsible for the epileptic phenotype of LD. Importantly, the deletion of GS specifically in astrocytes does not increase susceptibility to epilepsy per se [19]. Thus, the epileptic phenotype of LD might be attributable to neuronal LBs, most likely to those present in GABAergic interneurons, which would impair their function and generate an imbalance of glutamatergic and GABAergic transmission. In line with this notion, we described that parvalbumin interneurons of the hippocampus accumulate LBs [46] and this buildup is accompanied by damage to GABAergic neurons in mouse models of LD [77].

In summary, astrocytic glycogen accumulation drives the neuroinflammatory phenotype of LD but not the increased susceptibility to epilepsy, which might be attributable to neuronal LBs.

5. *Corpora amylacea*

The accumulation of glycogen in the nervous tissue is not exclusive to LD. As mentioned before, the presence of glycogen aggregates known as *Corpora amylacea* ("starch bodies" in Latin, due to their resemblance to starch) has also been observed in aged human brains [78]. Interestingly, these aggregates accumulate to a greater extent in neurodegenerative conditions like Alzheimer's, Parkinson's, Huntington's, and Pick's diseases, as well as in patients with temporal lobe epilepsy [67,79]. Although the cellular localization of *Corpora amylacea* has been a source of debate, several articles have described their presence in astrocytes [80–84]. Similar aggregates progressively accumulate with age in the astrocytes of control mice [30,65,67] (Figure 2). The composition of *Corpora amylacea* greatly resembles that of LBs, consisting of insoluble, poorly branched glycogen and a minor content of protein, including GS, ubiquitin, and p62 [67]. Interestingly, these glycogen aggregates are not found in the brains of aged GS knockout mice [30]. This observation thus indicates that, as for LBs, glycogen synthesis is a prerequisite for the formation of *Corpora amylacea*. In contrast, the formation of these aggregates is enhanced in models of accelerated aging, such as the Senescence Accelerated Mouse-Prone 8 (SAMP8) mouse [65]. The overexpression of protein targeting to glycogen (PTG), an activator of GS, also resulted in an increase in the formation of these glycogen aggregates [65]. This observation suggests that an imbalance between GS activity and GBE activity favors the formation of poorly branched glycogen, which would accumulate in the form of *Corpora amylacea*-like structures (see adult polyglucosan body disease below). Strikingly, the brains of control, SAMP8, and PTG-overexpressing animals show the presence of CAL but not nLBs [65]. Therefore, the latter seem to be exclusive to LD models.

Collectively, all of points explained above suggest that the progressive accumulation of *Corpora amylacea* in the nervous system contributes to the neurological decline associated with aging [79,85]. Mutations in malin and laforin would drastically increase the rate of this process; i.e., LD could be considered an accelerated aging process with respect to the consequences of glycogen accumulation in the brain. Furthermore, the increased presence of *Corpora amylacea* in neurodegenerative conditions like Alzheimer's and Parkinson's disease opens up the possibility that the toxicity induced by glycogen accumulation also participates in the pathology of these disorders. Alternatively, the presence of waste elements in *Corpora amylacea* has led some authors to hypothesize that these structures are waste containers in which deleterious or residual products are isolated for later removal by the natural immune system [78,86,87].

6. Adult Polyglucosan Body Disease

Another rare genetic condition in which glycogen accumulates abnormally in the nervous tissue is adult polyglucosan body disease (APBD, OMIM 263570). This is an autosomal recessive neurodegenerative disorder with onset normally in the 5th or 6th decade of life and slow progression, affecting the central and peripheral nervous system with severe leukodystrophy, atrophy of the spine and medulla, and cognitive impairment [88–90]. The disease is caused by mutations in GBE that result in the formation of glycogen with low solubility due to the lack of branching. Consequently, there is a progressive intracellular accumulation of glycogen aggregates (the so-called polyglucosan bodies), which are similar to LBs. In fact, the term "polyglucosan body" is also used to refer to LBs and *Corpora amylacea* [78,91] APBD is caused by mutations that generate a partial loss of GBE activity (with 5–20% residual activity), the most common of which is p.Y329S, found in patients of Ashkenazi Jewish descent [92]. Other mutations in GBE cause a clinically heterogeneous disorder collectively known as glycogen storage disease IV (OMIM 232500), with hepatic and neuromuscular presentations [93]. A mouse model with the GBE mutation p.Y329S presents a phenotype that is reminiscent of APBD in humans, with the accumulation of polyglucosan bodies and neurological dysfunction [94]. This model has allowed researchers to demonstrate that, like LD, APBD can be rescued by inhibiting GS [95], again evidencing the role of glycogen accumulation in the etiopathogeny of neurodegeneration.

In this regard, strategies targeting glycogen synthesis have proven effective in mouse models of both LD and APBD [59,60].

Similarly to LD, polyglucosan bodies have been reported to accumulate in astrocytes and neurons in APBD [96,97]. The exact contribution of astrocytic polyglucosan bodies to the pathophysiology of this disease is not clear. To unequivocally dissect this contribution, GS should be ablated specifically in astrocytes in the APBD mouse model, in a similar fashion as malin^{KO} + GS$^{Gfap\text{-}KO}$ mice for LD [50]. However, after all of the considerations above, it is reasonable to hypothesize that astrocytic polyglucosan bodies participate in the pathophysiology of APBD, probably by inducing neuroinflammation in a similar fashion as LBs in LD. In fact, the presence of polyglucosan bodies in astrocytes is sufficient to cause APBD [98].

7. RBCK1 Deficiency

RANBP2-Type and C3HC4-Type Zinc Finger Containing 1 protein (RBCK1, also known as HOIL1) is another E3 ubiquitin ligase that is related to glycogen metabolism. Mutations in the RBCK1 gene result in polyglucosan body myopathy with or without immunodeficiency (OMIM 615895). This disease affects children, and courses with progressive proximal muscle weakness and dilated cardiomyopathy, accompanied in some cases by severe immunodeficiency and a hyperinflammatory state [99,100]. The muscle and the heart of these patients show extensive polyglucosan body accumulation [100]. Although the disease is primarily a skeletal and cardiac myopathy, a mouse model of this condition also shows the presence of profuse aggregates of poorly branched glycogen in the nervous tissue, especially in the hippocampus, cerebellum, and spinal cord [101]. Interestingly, these polyglucosan bodies are localized mainly in astrocytes, and this accumulation is accompanied by astrogliosis and microgliosis, again linking neuroinflammation to the excessive accumulation of glycogen in astrocytes [101]. These observations indicate that the accumulation of polyglucosan bodies in the nervous tissue might also play a role in the pathophysiology of human RBCK1 deficiency. As with LD and APBD, the inhibition of glycogen synthesis rescues the pathological traits of the mouse model of RBCK1 deficiency, once again demonstrating that glycogen synthesis is a prerequisite for the formation of the polyglucosan bodies that underlie the disease [101]. Of note, a recent report has shown that the substrate of RBCK1 ubiquitination is not a protein but glycogen itself. More specifically, RBCK1 targets unbranched glucosaccharides, participating in a mechanism aimed at preventing polyglucosan body accumulation [102]. Thus, different E3 ubiquitin ligases (malin and RBCK1) are involved in preventing abnormal glycogen accumulation. Interestingly, ubiquitin is present in glycogen aggregates both in the absence of malin and of RBCK1 [48,103]. The genuine substrate of malin is still not clear, but the restoration of malin expression in a malin-deficient mouse model results in the degradation of the accumulated GS and laforin [104], thereby indicating that these two proteins are targets of malin ubiquitination. These experiments also showed that, once LBs have accumulated in the CNS, malin restoration is not able to promote their removal. This observation thus indicates that the role of malin is related to preventing the accumulation of abnormal glycogen (Figure 2) rather than eliminating it after its accumulation.

8. Concluding Remarks

Comparison of the pathology of LD, APBD, and RBCK1 deficiency opens up a number of questions. Why do LD and RBCK1 deficiency affect children while APBD affects adults? Why are the neurological presentations so different between the three diseases? Do LBs, *Corpora amylacea* and the polyglucosan bodies that accumulate in APBD have any distinguishing features amongst them that would offer insights into the differences among glycogen storage diseases? The observation that APBD patients do not present epilepsy reinforces the idea that the epileptic phenotype of LD is due to the accumulation of LBs in GABAergic interneurons. But then, if the malin-laforin complex, GBE, and RBCK1 are all important to prevent the accumulation of abnormal glycogen, why does the lack of one or the other result in the accumulation of polyglucosan bodies with different cell type-specificity?

Over recent years, the role of astrocytic dysfunction in neurodegenerative diseases previously thought to have an exclusively neuronal origin is becoming apparent. LD is not only an example of such diseases but also one in which the pathology is due in part to a defect originated primarily in astrocytes.

In conclusion, the study of LD has allowed the identification of the toxic consequences of the excessive accumulation of glycogen in astrocytes, a process that plays a key role in the pathophysiology of LD. This pathologic mechanism might have important implications for other conditions in which glycogen abnormally accumulates in astrocytes, such as in other rare conditions like APBD and RBCK1 deficiency, and more common neurodegenerative conditions like Alzheimer's, Parkinson's, Huntington's, and Pick's diseases, or even during normal aging. Further research is needed to understand the molecular mechanisms that link excess glycogen in astrocytes with neuroinflammation.

Funding: JD's laboratory is funded by grants from the Spanish Ministerio de Ciencia e Innovación (MCIU/FEDER/AEI) (PID2020-118699GB-I00) and the National Institutes of Health (NIH-NINDS) (NS097197-05S1P01). IRB Barcelona anb IBEC are the recipients of a Severo Ochoa Award of Excellence from the Spanish Ministry of Economy (MINECO).

Acknowledgments: We thank Tanya Yates for correcting the manuscript.

Conflicts of Interest: The author declares no conflict of interest.

References

1. Adeva-Andany, M.M.; González-Lucán, M.; Donapetry-García, C.; Fernández-Fernández, C.; Ameneiros-Rodríguez, E. Glycogen Metabolism in Humans. *BBA Clin.* **2016**, *5*, 85–100. [CrossRef]
2. Gessler, K.; Usón, I.; Takaha, T.; Krauss, N.; Smith, S.M.; Okada, S.; Sheldrick, G.M.; Saenger, W. V-Amylose at Atomic Resolution: X-ray Structure of a Cycloamylose with 26 Glucose Residues (Cyclomaltohexaicosaose). *Proc. Natl. Acad. Sci. USA* **1999**, *96*, 4246–4251. [CrossRef] [PubMed]
3. Meléndez, R.; Meléndez-Hevia, E.; Mas, F.; Mach, J.; Cascante, M. Physical Constraints in the Synthesis of Glycogen That Influence Its Structural Homogeneity: A Two-Dimensional Approach. *Biophys. J.* **1998**, *75*, 106–114. [CrossRef]
4. Meléndez-Hevia, E.; Waddell, T.G.; Shelton, E.D. Optimization of Molecular Design in the Evolution of Metabolism: The Glycogen Molecule. *Biochem. J.* **1993**, *295*, 477–483. [CrossRef] [PubMed]
5. Brown, A.M. Brain Glycogen Re-Awakened. *J. Neurochem.* **2004**, *89*, 537–552. [CrossRef]
6. Pfeiffer-Guglielmi, B.; Bröer, S.; Bröer, A.; Hamprecht, B. Isozyme Pattern of Glycogen Phosphorylase in the Rat Nervous System and Rat Astroglia-Rich Primary Cultures: Electrophoretic and Polymerase Chain Reaction Studies. *Neurochem. Res.* **2000**, *25*, 1485–1491. [CrossRef]
7. Oe, Y.; Baba, O.; Ashida, H.; Nakamura, K.C.; Hirase, H. Glycogen Distribution in the Microwave-Fixed Mouse Brain Reveals Heterogeneous Astrocytic Patterns. *Glia* **2016**, *64*, 1532–1545. [CrossRef] [PubMed]
8. Richter, K.; Hamprecht, B.; Scheich, H. Ultrastructural Localization of Glycogen Phosphorylase Predominantly in Astrocytes of the Gerbil Brain. *Glia* **1996**, *17*, 263–273. [CrossRef]
9. Cataldo, A.M.; Broadwell, R.D. Cytochemical Identification of Cerebral Glycogen and Glucose-6-Phosphatase Activity under Normal and Experimental Conditions. II. Choroid Plexus and Ependymal Epithelia, Endothelia and Pericytes. *J. Neurocytol.* **1986**, *15*, 511–524. [CrossRef]
10. Sagar, S.M.; Sharp, F.R.; Swanson, R.A. The Regional Distribution of Glycogen in Rat Brain Fixed by Microwave Irradiation. *Brain Res.* **1987**, *417*, 172–174. [CrossRef]
11. Saez, I.; Duran, J.; Sinadinos, C.; Beltran, A.; Yanes, O.; Tevy, M.F.; Martínez-Pons, C.; Milán, M.; Guinovart, J.J. Neurons Have an Active Glycogen Metabolism That Contributes to Tolerance to Hypoxia. *J. Cereb. Blood Flow Metab.* **2014**, *34*, 945–955. [CrossRef] [PubMed]
12. Duran, J.; Gruart, A.; Varea, O.; López-Soldado, I.; Delgado-García, J.M.; Guinovart, J.J. Lack of Neuronal Glycogen Impairs Memory Formation and Learning-Dependent Synaptic Plasticity in Mice. *Front. Cell. Neurosci.* **2019**, *13*, 374. [CrossRef]
13. Swanson, R.A.; Choi, D.W. Glial Glycogen Stores Affect Neuronal Survival during Glucose Deprivation In Vitro. *J. Cereb. Blood Flow Metab.* **1993**, *13*, 162–169. [CrossRef]
14. Dienel, G.A.; Carlson, G.M. Major Advances in Brain Glycogen Research: Understanding of the Roles of Glycogen Have Evolved from Emergency Fuel Reserve to Dynamic, Regulated Participant in Diverse Brain Functions. In *Brain Glycogen Metabolism*; Advances in Neurobiology; Springer: Berlin/Heidelberg, Germany, 2019; Volume 23, pp. 1–16. [CrossRef]
15. Dienel, G.A.; Cruz, N.F. Contributions of Glycogen to Astrocytic Energetics during Brain Activation. *Metab. Brain Dis.* **2015**, *30*, 281–298. [CrossRef]

16. Dienel, G.A. Brain Glucose Metabolism: Integration of Energetics with Function. *Physiol. Rev.* **2019**, *99*, 949–1045. [CrossRef]
17. Swanson, R.A.; Morton, M.M.; Sagar, S.M.; Sharp, F.R. Sensory Stimulation Induces Local Cerebral Glycogenolysis: Demonstration by Autoradiography. *Neuroscience* **1992**, *51*, 451–461. [CrossRef] [PubMed]
18. Duran, J.; Saez, I.; Gruart, A.; Guinovart, J.J.; Delgado-García, J.M. Impairment in Long-Term Memory Formation and Learning-Dependent Synaptic Plasticity in Mice Lacking Glycogen Synthase in the Brain. *J. Cereb. Blood Flow Metab.* **2013**, *33*, 550–556. [CrossRef]
19. Duran, J.; Brewer, M.K.; Hervera, A.; Gruart, A.; Del Rio, J.A.; Delgado-García, J.M.; Guinovart, J.J. Lack of Astrocytic Glycogen Alters Synaptic Plasticity but Not Seizure Susceptibility. *Mol. Neurobiol.* **2020**, *57*, 4657–4666. [CrossRef] [PubMed]
20. Gibbs, M.E.; Anderson, D.G.; Hertz, L. Inhibition of Glycogenolysis in Astrocytes Interrupts Memory Consolidation in Young Chickens. *Glia* **2006**, *54*, 214–222. [CrossRef]
21. Suzuki, A.; Stern, S.A.; Bozdagi, O.; Huntley, G.W.; Walker, R.H.; Magistretti, P.J.; Alberini, C.M. Astrocyte-Neuron Lactate Transport Is Required for Long-Term Memory Formation. *Cell* **2011**, *144*, 810–823. [CrossRef]
22. Pellerin, L.; Magistretti, P.J. Glutamate Uptake into Astrocytes Stimulates Aerobic Glycolysis: A Mechanism Coupling Neuronal Activity to Glucose Utilization. *Proc. Natl. Acad. Sci. USA* **1994**, *91*, 10625–10629. [CrossRef] [PubMed]
23. DiNuzzo, M.; Mangia, S.; Maraviglia, B.; Giove, F. Glycogenolysis in Astrocytes Supports Blood-Borne Glucose Channeling Not Glycogen-Derived Lactate Shuttling to Neurons: Evidence from Mathematical Modeling. *J. Cereb. Blood Flow Metab.* **2010**, *30*, 1895–1904. [CrossRef]
24. Roach, P.J. Glycogen Phosphorylation and Lafora Disease. *Mol. Aspects Med.* **2015**, *46*, 78–84. [CrossRef] [PubMed]
25. Gayarre, J.; Duran-Trío, L.; Criado Garcia, O.; Aguado, C.; Juana-López, L.; Crespo, I.; Knecht, E.; Bovolenta, P.; de Córdoba, S.R. The Phosphatase Activity of Laforin Is Dispensable to Rescue *Epm2a*$^{-/-}$ Mice from Lafora Disease. *Brain* **2014**, *137*, 806–818. [CrossRef] [PubMed]
26. Nitschke, F.; Sullivan, M.A.; Wang, P.; Zhao, X.; Chown, E.E.; Perri, A.M.; Israelian, L.; Juana-López, L.; Bovolenta, P.; de Córdoba, S.R.; et al. Abnormal Glycogen Chain Length Pattern, Not Hyperphosphorylation, Is Critical in Lafora Disease. *EMBO Mol. Med.* **2017**, *9*, 906–917. [CrossRef]
27. Duran, J.; Gruart, A.; García-Rocha, M.; Delgado-García, J.M.; Guinovart, J.J. Glycogen Accumulation Underlies Neurodegeneration and Autophagy Impairment in Lafora Disease. *Hum. Mol. Genet.* **2014**, *23*, 3147–3156. [CrossRef] [PubMed]
28. Varea, O.; Duran, J.; Aguilera, M.; Prats, N.; Guinovart, J.J. Suppression of Glycogen Synthesis as a Treatment for Lafora Disease: Establishing the Window of Opportunity. *Neurobiol. Dis.* **2021**, *147*, 105173. [CrossRef]
29. Pellegrini, P.; Hervera, A.; Varea, O.; Brewer, M.K.; López-Soldado, I.; Guitart, A.; Aguilera, M.; Prats, N.; Del Río, J.A.; Guinovart, J.J.; et al. Lack of p62 Impairs Glycogen Aggregation and Exacerbates Pathology in a Mouse Model of Myoclonic Epilepsy of Lafora. *Mol. Neurobiol.* **2022**, *59*, 1214–1229. [CrossRef] [PubMed]
30. Sinadinos, C.; Valles-Ortega, J.; Boulan, L.; Solsona, E.; Tevy, M.F.; Marquez, M.; Duran, J.; Lopez-Iglesias, C.; Calbó, J.; Blasco, E.; et al. Neuronal Glycogen Synthesis Contributes to Physiological Aging. *Aging Cell* **2014**, *13*, 935–945. [CrossRef]
31. Turnbull, J.; Tiberia, E.; Striano, P.; Genton, P.; Carpenter, S.; Ackerley, C.A.; Minassian, B.A. Lafora Disease. *Epileptic Disord.* **2016**, *18*, 38–62. [CrossRef]
32. Verhalen, B.; Arnold, S.; Minassian, B.A. Lafora Disease: A Review of Molecular Mechanisms and Pathology. *Neuropediatrics* **2018**, *49*, 357–362. [CrossRef]
33. Pondrelli, F.; Muccioli, L.; Licchetta, L.; Mostacci, B.; Zenesini, C.; Tinuper, P.; Vignatelli, L.; Bisulli, F. Natural History of Lafora Disease: A Prognostic Systematic Review and Individual Participant Data Meta-Analysis. *Orphanet J. Rare Dis.* **2021**, *16*, 362. [CrossRef]
34. Minassian, B.A.; Lee, J.R.; Herbrick, J.A.; Huizenga, J.; Soder, S.; Mungall, A.J.; Dunham, I.; Gardner, R.; Fong, C.Y.; Carpenter, S.; et al. Mutations in a Gene Encoding a Novel Protein Tyrosine Phosphatase Cause Progressive Myoclonus Epilepsy. *Nat. Genet.* **1998**, *20*, 171–174. [CrossRef]
35. Serratosa, J.M.; Gómez-Garre, P.; Gallardo, M.E.; Anta, B.; de Bernabé, D.B.; Lindhout, D.; Augustijn, P.B.; Tassinari, C.A.; Malafosse, R.M.; Topcu, M.; et al. A Novel Protein Tyrosine Phosphatase Gene Is Mutated in Progressive Myoclonus Epilepsy of the Lafora Type (EPM2). *Hum. Mol. Genet.* **1999**, *8*, 345–352. [CrossRef]
36. Chan, E.M.; Young, E.J.; Ianzano, L.; Munteanu, I.; Zhao, X.; Christopoulos, C.C.; Avanzini, G.; Elia, M.; Ackerley, C.A.; Jovic, N.J.; et al. Mutations in *NHLRC1* Cause Progressive Myoclonus Epilepsy. *Nat. Genet.* **2003**, *35*, 125–127. [CrossRef]
37. Cheng, A.; Zhang, M.; Gentry, M.S.; Worby, C.A.; Dixon, J.E.; Saltiel, A.R. A Role for AGL Ubiquitination in the Glycogen Storage Disorders of Lafora and Cori's Disease. *Genes Dev.* **2007**, *21*, 2399–2409. [CrossRef]
38. Worby, C.A.; Gentry, M.S.; Dixon, J.E. Malin Decreases Glycogen Accumulation by Promoting the Degradation of Protein Targeting to Glycogen (PTG). *J. Biol. Chem.* **2008**, *283*, 4069–4076. [CrossRef]
39. Solaz-Fuster, M.C.; Gimeno-Alcañiz, J.V.; Ros, S.; Fernandez-Sanchez, M.E.; Garcia-Fojeda, B.; Garcia, O.C.; Vilchez, D.; Dominguez, J.; Garcia-Rocha, M.; Sanchez-Piris, M.; et al. Regulation of Glycogen Synthesis by the Laforin–Malin Complex Is Modulated by the AMP-Activated Protein Kinase Pathway. *Hum. Mol. Genet.* **2008**, *17*, 667–678. [CrossRef]
40. Sullivan, M.A.; Nitschke, S.; Steup, M.; Minassian, B.A.; Nitschke, F. Pathogenesis of Lafora Disease: Transition of Soluble Glycogen to Insoluble Polyglucosan. *Int. J. Mol. Sci.* **2017**, *18*, 1743. [CrossRef]
41. Turco, E.; Savova, A.; Gere, F.; Ferrari, L.; Romanov, J.; Schuschnig, M.; Martens, S. Reconstitution Defines the Roles of p62, NBR1 and TAX1BP1 in Ubiquitin Condensate Formation and Autophagy Initiation. *Nat. Commun.* **2021**, *12*, 5212. [CrossRef]

42. Moscat, J.; Diaz-Meco, M.T. To Aggregate or Not to Aggregate? A New Role for p62. *EMBO Rep.* **2009**, *10*, 804. [CrossRef]
43. Bjørkøy, G.; Lamark, T.; Brech, A.; Outzen, H.; Perander, M.; Overvatn, A.; Stenmark, H.; Johansen, T. p62/SQSTM1 Forms Protein Aggregates Degraded by Autophagy and Has a Protective Effect on Huntingtin-Induced Cell Death. *J. Cell Biol.* **2005**, *171*, 603–614. [CrossRef]
44. Ganesh, S.; Delgado-Escueta, A.V.; Sakamoto, T.; Avila, M.R.; Machado-Salas, J.; Hoshii, Y.; Akagi, T.; Gomi, H.; Suzuki, T.; Amano, K.; et al. Targeted Disruption of the *Epm2a* Gene Causes Formation of Lafora Inclusion Bodies, Neurodegeneration, Ataxia, Myoclonus Epilepsy and Impaired Behavioral Response in Mice. *Hum. Mol. Genet.* **2002**, *11*, 1251–1262. [CrossRef]
45. Turnbull, J.; DePaoli-Roach, A.A.; Zhao, X.; Cortez, M.A.; Pencea, N.; Tiberia, E.; Piliguian, M.; Roach, P.J.; Wang, P.; Ackerley, C.A.; et al. PTG Depletion Removes Lafora Bodies and Rescues the Fatal Epilepsy of Lafora Disease. *PLoS Genet.* **2011**, *7*, e1002037. [CrossRef]
46. Valles-Ortega, J.; Duran, J.; Garcia-Rocha, M.; Bosch, C.; Saez, I.; Pujadas, L.; Serafin, A.; Cañas, X.; Soriano, E.; Delgado-García, J.M.; et al. Neurodegeneration and Functional Impairments Associated with Glycogen Synthase Accumulation in a Mouse Model of Lafora Disease. *EMBO Mol. Med.* **2011**, *3*, 667–681. [CrossRef]
47. DePaoli-Roach, A.A.; Tagliabracci, V.S.; Segvich, D.M.; Meyer, C.M.; Irimia, J.M.; Roach, P.J. Genetic Depletion of the Malin E3 Ubiquitin Ligase in Mice Leads to Lafora Bodies and the Accumulation of Insoluble Laforin. *J. Biol. Chem.* **2010**, *285*, 25372–25381. [CrossRef]
48. Criado, O.; Aguado, C.; Gayarre, J.; Duran-Trio, L.; Garcia-Cabrero, A.M.; Vernia, S.; San Millán, B.; Heredia, M.; Romá-Mateo, C.; Mouron, S.; et al. Lafora Bodies and Neurological Defects in Malin-Deficient Mice Correlate with Impaired Autophagy. *Hum. Mol. Genet.* **2012**, *21*, 1521–1533. [CrossRef]
49. García-Cabrero, A.M.; Marinas, A.; Guerrero, R.; de Córdoba, S.R.; Serratosa, J.M.; Sánchez, M.P. Laforin and Malin Deletions in Mice Produce Similar Neurologic Impairments. *J. Neuropathol. Exp. Neurol.* **2012**, *71*, 413–421. [CrossRef]
50. Duran, J.; Hervera, A.; Markussen, K.H.; Varea, O.; López-Soldado, I.; Sun, R.C.; Del Río, J.A.; Gentry, M.S.; Guinovart, J.J. Astrocytic Glycogen Accumulation Drives the Pathophysiology of Neurodegeneration in Lafora Disease. *Brain* **2021**, *144*, 2349–2360. [CrossRef]
51. Puri, R.; Suzuki, T.; Yamakawa, K.; Ganesh, S. Dysfunctions in Endosomal-Lysosomal and Autophagy Pathways Underlie Neuropathology in a Mouse Model for Lafora Disease. *Hum. Mol. Genet.* **2012**, *21*, 175–184. [CrossRef]
52. Aguado, C.; Sarkar, S.; Korolchuk, V.I.; Criado, O.; Vernia, S.; Boya, P.; Sanz, P.; de Córdoba, S.R.; Knecht, E.; Rubinsztein, D.C. Laforin, the Most Common Protein Mutated in Lafora Disease, Regulates Autophagy. *Hum. Mol. Genet.* **2010**, *19*, 2867–2876. [CrossRef]
53. Knecht, E.; Criado-García, O.; Aguado, C.; Gayarre, J.; Duran-Trio, L.; Garcia-Cabrero, A.M.; Vernia, S.; San Millán, B.; Heredia, M.; Romá-Mateo, C.; et al. Malin Knockout Mice Support a Primary Role of Autophagy in the Pathogenesis of Lafora Disease. *Autophagy* **2012**, *8*, 701–703. [CrossRef]
54. Turnbull, J.; Epp, J.R.; Goldsmith, D.; Zhao, X.; Pencea, N.; Wang, P.; Frankland, P.W.; Ackerley, C.A.; Minassian, B.A. PTG Protein Depletion Rescues Malin-Deficient Lafora Disease in Mouse. *Ann. Neurol.* **2014**, *75*, 442–446. [CrossRef]
55. Pederson, B.A.; Turnbull, J.; Epp, J.R.; Weaver, S.A.; Zhao, X.; Pencea, N.; Roach, P.J.; Frankland, P.W.; Ackerley, C.A.; Minassian, B.A. Inhibiting Glycogen Synthesis Prevents Lafora Disease in a Mouse Model. *Ann. Neurol.* **2013**, *74*, 297–300. [CrossRef]
56. Duran, J.; Tevy, M.F.; Garcia-Rocha, M.; Calbó, J.; Milán, M.; Guinovart, J.J. Deleterious Effects of Neuronal Accumulation of Glycogen in Flies and Mice. *EMBO Mol. Med.* **2012**, *4*, 719–729. [CrossRef]
57. Ahonen, S.; Nitschke, S.; Grossman, T.R.; Kordasiewicz, H.; Wang, P.; Zhao, X.; Guisso, D.R.; Kasiri, S.; Nitschke, F.; Minassian, B.A. *Gys1* Antisense Therapy Rescues Neuropathological Bases of Murine Lafora Disease. *Brain* **2021**, *144*, 2985–2993. [CrossRef]
58. Nitschke, S.; Chown, E.E.; Zhao, X.; Gabrielian, S.; Petković, S.; Guisso, D.R.; Perri, A.M.; Wang, P.; Ahonen, S.; Nitschke, F.; et al. An Inducible Glycogen Synthase-1 Knockout Halts but Does Not Reverse Lafora Disease Progression in Mice. *J. Biol. Chem.* **2021**, *296*, 100150. [CrossRef]
59. Gumusgoz, E.; Kasiri, S.; Guisso, D.R.; Wu, J.; Dear, M.; Verhalen, B.; Minassian, B.A. AAV-Mediated Artificial MiRNA Reduces Pathogenic Polyglucosan Bodies and Neuroinflammation in Adult Polyglucosan Body and Lafora Disease Mouse Models. *Neurotherapeutics* **2022**, *19*, 982–993. [CrossRef]
60. Gumusgoz, E.; Guisso, D.R.; Kasiri, S.; Wu, J.; Dear, M.; Verhalen, B.; Nitschke, S.; Mitra, S.; Nitschke, F.; Minassian, B.A. Targeting *Gys1* with AAV-SaCas9 Decreases Pathogenic Polyglucosan Bodies and Neuroinflammation in Adult Polyglucosan Body and Lafora Disease Mouse Models. *Neurotherapeutics* **2021**, *18*, 1414–1425. [CrossRef]
61. Lafora, G.R. Über das Vorkommen amyloider Körperchen im Innern der Ganglienzellen. *Virchows Arch. Pathol. Anat. Physiol. Klin. Med.* **1911**, *205*, 295–303. [CrossRef]
62. Machado-Salas, J.; Avila-Costa, M.R.; Guevara, P.; Guevara, J.; Durón, R.M.; Bai, D.; Tanaka, M.; Yamakawa, K.; Delgado-Escueta, A.V. Ontogeny of Lafora Bodies and Neurocytoskeleton Changes in Laforin-Deficient Mice. *Exp. Neurol.* **2012**, *236*, 131–140. [CrossRef]
63. Gentry, M.S.; Guinovart, J.J.; Minassian, B.A.; Roach, P.J.; Serratosa, J.M. Lafora Disease Offers a Unique Window into Neuronal Glycogen Metabolism. *J. Biol. Chem.* **2018**, *293*, 7117–7125. [CrossRef]
64. Minassian, B.A. Lafora's Disease: Towards a Clinical, Pathologic, and Molecular Synthesis. *Pediatr. Neurol* **2001**, *25*, 21–29. [CrossRef]
65. Augé, E.; Pelegrí, C.; Manich, G.; Cabezón, I.; Guinovart, J.J.; Duran, J.; Vilaplana, J. Astrocytes and Neurons Produce Distinct Types of Polyglucosan Bodies in Lafora Disease. *Glia* **2018**, *66*, 2094–2107. [CrossRef]
66. Rubio-Villena, C.; Viana, R.; Bonet, J.; Garcia-Gimeno, M.A.; Casado, M.; Heredia, M.; Sanz, P. Astrocytes: New Players in Progressive Myoclonus Epilepsy of Lafora Type. *Hum. Mol. Genet.* **2018**, *27*, 1290–1300. [CrossRef]

67. Augé, E.; Duran, J.; Guinovart, J.J.; Pelegrí, C.; Vilaplana, J. Exploring the Elusive Composition of Corpora Amylacea of Human Brain. *Sci. Rep.* **2018**, *8*, 13525. [CrossRef]
68. Kalia, L.V.; Lang, A.E. Parkinson's Disease. *Lancet* **2015**, *386*, 896–912. [CrossRef]
69. Konishi, H.; Koizumi, S.; Kiyama, H. Phagocytic Astrocytes: Emerging from the Shadows of Microglia. *Glia* **2022**, *70*, 1009–1026. [CrossRef]
70. Lahuerta, M.; Gonzalez, D.; Aguado, C.; Fathinajafabadi, A.; García-Giménez, J.L.; Moreno-Estellés, M.; Romá-Mateo, C.; Knecht, E.; Pallardó, F.V.; Sanz, P. Reactive Glia-Derived Neuroinflammation: A Novel Hallmark in Lafora Progressive Myoclonus Epilepsy That Progresses with Age. *Mol. Neurobiol.* **2020**, *57*, 1607–1621. [CrossRef]
71. Mollá, B.; Heredia, M.; Sanz, P. Modulators of Neuroinflammation Have a Beneficial Effect in a Lafora Disease Mouse Model. *Mol. Neurobiol.* **2021**, *58*, 2508–2522. [CrossRef]
72. Coulter, D.A.; Steinhäuser, C. Role of Astrocytes in Epilepsy. *Cold Spring Harb. Perspect. Med.* **2015**, *5*, a022434. [CrossRef]
73. Xu, J.; Song, D.; Xue, Z.; Gu, L.; Hertz, L.; Peng, L. Requirement of Glycogenolysis for Uptake of Increased Extracellular K^+ in Astrocytes: Potential Implications for K^+ Homeostasis and Glycogen Usage in Brain. *Neurochem. Res.* **2013**, *38*, 472–485. [CrossRef] [PubMed]
74. Obel, L.F.; Müller, M.S.; Walls, A.B.; Sickmann, H.M.; Bak, L.K.; Waagepetersen, H.S.; Schousboe, A. Brain Glycogen-New Perspectives on Its Metabolic Function and Regulation at the Subcellular Level. *Front. Neuroenergetics* **2012**, *4*, 3. [CrossRef]
75. DiNuzzo, M.; Mangia, S.; Maraviglia, B.; Giove, F. Does Abnormal Glycogen Structure Contribute to Increased Susceptibility to Seizures in Epilepsy? *Metab. Brain Dis.* **2015**, *30*, 307–316. [CrossRef] [PubMed]
76. Loiseau, H.; Marchal, C.; Vital, A.; Vital, C.; Rougier, A.; Loiseau, P. Occurrence of Polyglucosan Bodies in Temporal Lobe Epilepsy. *J. Neurol. Neurosurg. Psychiatry* **1992**, *55*, 1092–1093. [CrossRef]
77. Ortolano, S.; Vieitez, I.; Agis-Balboa, R.C.; Spuch, C. Loss of GABAergic Cortical Neurons Underlies the Neuropathology of Lafora Disease. *Mol. Brain* **2014**, *7*, 7. [CrossRef] [PubMed]
78. Cavanagh, J.B. Corpora-Amylacea and the Family of Polyglucosan Diseases. *Brain Res. Rev.* **1999**, *29*, 265–295. [CrossRef]
79. Rohn, T.T. Corpora Amylacea in Neurodegenerative Diseases: Cause or Effect? *Int. J. Neurol. Neurother.* **2015**, *2*, 031. [CrossRef]
80. Ramsey, H.J. Ultrastructure of Corpora Amylacea. *J. Neuropathol. Exp. Neurol.* **1965**, *24*, 25–39. [CrossRef]
81. Sbarbati, A.; Carner, M.; Colletti, V.; Osculati, F. Extrusion of Corpora Amylacea from the Marginal Gila at the Vestibular Root Entry Zone. *J. Neuropathol. Exp. Neurol.* **1996**, *55*, 196–201. [CrossRef]
82. Schipper, H.M.; Cissé, S. Mitochondrial Constituents of Corpora Amylacea and Autofluorescent Astrocytic Inclusions in Senescent Human Brain. *Glia* **1995**, *14*, 55–64. [CrossRef] [PubMed]
83. Wander, C.M.; Tseng, J.-H.; Song, S.; Al Housseiny, H.A.; Tart, D.S.; Ajit, A.; Ian Shih, Y.-Y.; Lobrovich, R.; Song, J.; Meeker, R.B.; et al. The Accumulation of Tau-Immunoreactive Hippocampal Granules and Corpora Amylacea Implicates Reactive Glia in Tau Pathogenesis during Aging. *iScience* **2020**, *23*, 101255. [CrossRef]
84. Augé, E.; Bechmann, I.; Llor, N.; Vilaplana, J.; Krueger, M.; Pelegrí, C. Corpora Amylacea in Human Hippocampal Brain Tissue Are Intracellular Bodies That Exhibit a Homogeneous Distribution of Neo-Epitopes. *Sci. Rep.* **2019**, *9*, 2063. [CrossRef] [PubMed]
85. Singhrao, S.K.; Neal, J.W.; Newman, G.R. Corpora Amylacea Could Be an Indicator of Neurodegeneration. *Neuropathol. Appl. Neurobiol.* **1993**, *19*, 269–276. [CrossRef]
86. Augé, E.; Cabezón, I.; Pelegrí, C.; Vilaplana, J. New Perspectives on *Corpora amylacea* in the Human Brain. *Sci. Rep.* **2017**, *7*, 41807. [CrossRef] [PubMed]
87. Riba, M.; Del Valle, J.; Augé, E.; Vilaplana, J.; Pelegrí, C. From *Corpora amylacea* to Wasteosomes: History and Perspectives. *Ageing Res. Rev.* **2021**, *72*, 101484. [CrossRef]
88. Bruno, C.; Servidei, S.; Shanske, S.; Karpati, G.; Carpenter, S.; McKee, D.; Barohn, R.J.; Hirano, M.; Rifai, Z.; DiMauro, S. Glycogen Branching Enzyme Deficiency in Adult Polyglucosan Body Disease. *Ann. Neurol.* **1993**, *33*, 88–93. [CrossRef] [PubMed]
89. Bigio, E.H.; Weiner, M.F.; Bonte, F.J.; White, C.L. Familial Dementia Due to Adult Polyglucosan Body Disease. *Clin. Neuropathol.* **1997**, *16*, 227–234. [PubMed]
90. Berkhoff, M.; Weis, J.; Schroth, G.; Sturzenegger, M. Extensive White-Matter Changes in Case of Adult Polyglucosan Body Disease. *Neuroradiology* **2001**, *43*, 234–236. [CrossRef]
91. Brewer, M.K.; Putaux, J.-L.; Rondon, A.; Uittenbogaard, A.; Sullivan, M.A.; Gentry, M.S. Polyglucosan Body Structure in Lafora Disease. *Carbohydr. Polym.* **2020**, *240*, 116260. [CrossRef]
92. Mochel, F.; Schiffmann, R.; Steenweg, M.E.; Akman, H.O.; Wallace, M.; Sedel, F.; Laforêt, P.; Levy, R.; Powers, J.M.; Demeret, S.; et al. Adult Polyglucosan Body Disease: Natural History and Key Magnetic Resonance Imaging Findings. *Ann. Neurol.* **2012**, *72*, 433–441. [CrossRef]
93. Bruno, C.; van Diggelen, O.P.; Cassandrini, D.; Gimpelev, M.; Giuffrè, B.; Donati, M.A.; Introvini, P.; Alegria, A.; Assereto, S.; Morandi, L.; et al. Clinical and Genetic Heterogeneity of Branching Enzyme Deficiency (Glycogenosis Type IV). *Neurology* **2004**, *63*, 1053–1058. [CrossRef]
94. Orhan Akman, H.; Emmanuele, V.; Kurt, Y.G.; Kurt, B.; Sheiko, T.; DiMauro, S.; Craigen, W.J. A Novel Mouse Model That Recapitulates Adult-Onset Glycogenosis Type 4. *Hum. Mol. Genet.* **2015**, *24*, 6801–6810. [CrossRef] [PubMed]
95. Chown, E.E.; Wang, P.; Zhao, X.; Crowder, J.J.; Strober, J.W.; Sullivan, M.A.; Xue, Y.; Bennett, C.S.; Perri, A.M.; Evers, B.M.; et al. GYS1 or PPP1R3C Deficiency Rescues Murine Adult Polyglucosan Body Disease. *Ann. Clin. Transl. Neurol.* **2020**, *7*, 2186–2198. [CrossRef]

96. Robitaille, Y.; Carpenter, S.; Karpati, G.; DiMauro, S.D. A Distinct Form of Adult Polyglucosan Body Disease with Massive Involvement of Central and Peripheral Neuronal Processes and Astrocytes: A Report of Four Cases and a Review of the Occurrence of Polyglucosan Bodies in Other Conditions Such as Lafora's Disease and Normal Ageing. *Brain* **1980**, *103*, 315–336. [CrossRef]
97. Gray, F.; Gherardi, R.; Marshall, A.; Janota, I.; Poirier, J. Adult Polyglucosan Body Disease (APBD). *J. Neuropathol. Exp. Neurol.* **1988**, *47*, 459–474. [CrossRef]
98. Dainese, L.; Monin, M.-L.; Demeret, S.; Brochier, G.; Froissart, R.; Spraul, A.; Schiffmann, R.; Seilhean, D.; Mochel, F. Abnormal Glycogen in Astrocytes Is Sufficient to Cause Adult Polyglucosan Body Disease. *Gene* **2013**, *515*, 376–379. [CrossRef]
99. Boisson, B.; Laplantine, E.; Prando, C.; Giliani, S.; Israelsson, E.; Xu, Z.; Abhyankar, A.; Israël, L.; Trevejo-Nunez, G.; Bogunovic, D.; et al. Immunodeficiency, Autoinflammation and Amylopectinosis in Humans with Inherited HOIL-1 and LUBAC Deficiency. *Nat. Immunol.* **2012**, *13*, 1178–1186. [CrossRef]
100. Nilsson, J.; Schoser, B.; Laforet, P.; Kalev, O.; Lindberg, C.; Romero, N.B.; Dávila López, M.; Akman, H.O.; Wahbi, K.; Iglseder, S.; et al. Polyglucosan Body Myopathy Caused by Defective Ubiquitin Ligase RBCK1. *Ann. Neurol.* **2013**, *74*, 914–919. [CrossRef]
101. Nitschke, S.; Sullivan, M.A.; Mitra, S.; Marchioni, C.R.; Lee, J.P.Y.; Smith, B.H.; Ahonen, S.; Wu, J.; Chown, E.E.; Wang, P.; et al. Glycogen Synthase Downregulation Rescues the Amylopectinosis of Murine RBCK1 Deficiency. *Brain* **2022**, *145*, 2361–2377. [CrossRef] [PubMed]
102. Kelsall, I.R.; McCrory, E.H.; Xu, Y.; Scudamore, C.L.; Nanda, S.K.; Mancebo-Gamella, P.; Wood, N.T.; Knebel, A.; Matthews, S.J.; Cohen, P. HOIL-1 Ubiquitin Ligase Activity Targets Unbranched Glucosaccharides and Is Required to Prevent Polyglucosan Accumulation. *EMBO J.* **2022**, *41*, e109700. [CrossRef]
103. Thomsen, C.; Malfatti, E.; Jovanovic, A.; Roberts, M.; Kalev, O.; Lindberg, C.; Oldfors, A. Proteomic Characterisation of Polyglucosan Bodies in Skeletal Muscle in RBCK1 Deficiency. *Neuropathol. Appl. Neurobiol.* **2022**, *48*, e12761. [CrossRef]
104. Varea, O.; Guinovart, J.J.; Duran, J. Malin Restoration as Proof of Concept for Gene Therapy for Lafora Disease. *Brain Commun.* **2022**, *4*, fcac168. [CrossRef]

cells

MDPI

Review

Role of Astrocytes in Parkinson's Disease Associated with Genetic Mutations and Neurotoxicants

Sanghoon Kim [1,†], Edward Pajarillo [1,†], Ivan Nyarko-Danquah [1], Michael Aschner [2] and Eunsook Lee [1,*]

[1] Department of Pharmaceutical Science, College of Pharmacy and Pharmaceutical Sciences, Florida A&M University, Tallahassee, FL 32307, USA
[2] Department of Molecular Pharmacology, Albert Einstein College of Medicine, Bronx, New York, NY 10461, USA
* Correspondence: eunsook.lee@famu.edu
† These authors contributed equally to this work.

Abstract: Parkinson's disease (PD) is a neurodegenerative disorder characterized by the loss of dopaminergic neurons and the aggregation of Lewy bodies in the basal ganglia, resulting in movement impairment referred to as parkinsonism. However, the etiology of PD is not well known, with genetic factors accounting only for 10–15% of all PD cases. The pathogenetic mechanism of PD is not completely understood, although several mechanisms, such as oxidative stress and inflammation, have been suggested. Understanding the mechanisms of PD pathogenesis is critical for developing highly efficacious therapeutics. In the PD brain, dopaminergic neurons degenerate mainly in the basal ganglia, but recently emerging evidence has shown that astrocytes also significantly contribute to dopaminergic neuronal death. In this review, we discuss the role of astrocytes in PD pathogenesis due to mutations in α-synuclein (PARK1), DJ-1 (PARK7), parkin (PARK2), leucine-rich repeat kinase 2 (LRRK2, PARK8), and PTEN-induced kinase 1 (PINK1, PARK6). We also discuss PD experimental models using neurotoxins, such as paraquat, rotenone, 6-hydroxydopamine, and MPTP/MPP+. A more precise and comprehensive understanding of astrocytes' modulatory roles in dopaminergic neurodegeneration in PD will help develop novel strategies for effective PD therapeutics.

Keywords: astrocytes; Parkinson's disease; α-synuclein; DJ-1; parkin; LRRK2; MPTP; paraquat; rotenone; 6-hydroxydopamine

Citation: Kim, S.; Pajarillo, E.; Nyarko-Danquah, I.; Aschner, M.; Lee, E. Role of Astrocytes in Parkinson's Disease Associated with Genetic Mutations and Neurotoxicants. *Cells* **2023**, *12*, 622. https://doi.org/10.3390/cells12040622

Academic Editors: Michael Brenner and Vladimir Parpura

Received: 16 December 2022
Revised: 2 February 2023
Accepted: 9 February 2023
Published: 15 February 2023

1. Introduction

Parkinson's disease (PD) is a neurodegenerative disease characterized by the loss of dopaminergic neurons in the nigrostriatal pathway and the formation of α-synuclein aggregates (Lewy bodies) in neurons. At the clinical level, the disease is characterized by bradykinesia, gait impairment, tremor, and loss of involuntary movement [1]. PD is the second most prevalent neurodegenerative disease after Alzheimer's disease (AD), affecting over a million Americans and causing a total economic burden in the U.S. of $51.9 billion in 2017 [2]. The etiology of PD is not well understood, but several factors appear to play a role, including gene dysfunctions and mutations. It has been shown that mutations in genes, such as leucine-rich repeat kinase 2 (LRRK2), α-synuclein, PTEN-induced kinase 1 (PINK1), and parkin, are linked to familial PD cases, but these account for only 10–15% of total PD cases (Table 1) [3]. Several cases of idiopathic PD have implicated that environmental factors, such as the pesticide rotenone and herbicide paraquat, contribute to PD development [4].

Numerous studies have investigated the molecular and cellular mechanisms underlying PD pathogenesis. Oxidative stress, inflammation, mitochondrial impairment, and dysregulation of autophagy and mitophagy have been reported in both in vitro and in vivo PD models [18–22]. Although dopaminergic neurodegeneration in the basal ganglia is the main feature of PD, other neural cell types, including astrocytes, have been shown to play a role in PD development [23].

Table 1. Astrocytes contribute to PD pathology in a variety of PD mutants.

Gene	Protein	Function	Mode of Inheritance	Time of Onset *	References
SNCA	α-synuclein	Glutamate transport, inflammatory response	Autosomal dominant	Early	[5–7]
PARK2	Parkin	Inflammatory response, mitochondrial function	Autosomal recessive	Early	[8,9]
PARK6	PINK1	Mitochondrial function	Autosomal recessive	Early	[10]
PARK7	DJ-1	Glutamate uptake, inflammatory response, mitochondrial function	Autosomal recessive	Early	[11–13]
PARK8	LRRK2	Autophagy, Lysosome function	Autosomal dominant	Late	[14,15]
PARK9	ATP13A2	Inflammatory response, Lysosome function	Autosomal recessive	Early	[16]
GBA	GCase (Glucocerebrosidase)	Autophagy, Lysosome function	Autosomal recessive	Late	[17]

* Early onset, ages 21–50; late-onset, ages older than 50.

Astrocytes were first described in the 19th century by Rudolf Virchow [24] and initially thought to function only in the maintenance and structural support for neurons, but recent studies have shown that astrocytes play a critical role in neurotransmitter homeostasis and synapse formation during development, potassium buffering, and maintenance of pH and the vascular lumen, just to name a few of their functions [25]. Accordingly, it is not surprising that they are implicated in multiple neurological disorders, including PD pathogenesis and progression [26,27]. Hence, studying the functional role of astrocytes is important for understanding the full spectrum of PD pathogenesis and for the development of novel therapeutic strategies to treat PD. In this review, we discuss the role of astrocytes in both genetic and neurotoxins-induced PD pathology, as well as potential neuroprotective strategies targeting astrocytes in PD.

2. Role of Astrocytes in PD-Related Gene Mutations-Induced Pathology

2.1. DJ-1

DJ-1 is encoded by the PARK7 gene, and its mutations are one of the causes of autosomal-recessive PD [28–30]. DJ-1 is expressed abundantly in astrocytes compared to neurons, and it is further upregulated in reactive astrocytes in chronic neurodegenerative disorders, including PD [31,32]. DJ-1 is associated with lipid rafts in astrocytic plasma membranes [11]. Lipid rafts are involved in many cellular signaling processes, such as membrane receptor trafficking, endocytosis, and signal transduction [33]. Deficient DJ-1 activity due to PARK7 gene mutation increases the degradation of the lipid raft proteins and, thus, disrupts lipid raft assembly [12]. DJ-1-deficient astrocytes display impaired glutamate uptake [12] since astrocytic glutamate transporter excitatory amino acid transporter 2 (EAAT2) protein is assembled in lipid rafts [34]. In turn, EAAT2 dysfunction leads to excess levels of synaptic glutamate and excitotoxic neuronal injury, which is related to a variety of neurodegenerative diseases [35], including PD. Lipid rafts are also involved in astrocytic immune responses to inflammatory stimulation, and thus, the disruption of lipid raft assembly in DJ-1 deficient astrocytes could increase inflammatory cytokine production in response to lipopolysaccharide (LPS) [11,36] and increase neuronal toxicity. The role of astrocytic DJ-1 in PD is supported by experimental PD models. Studies have shown that neurotoxic effects induced by PD toxicant models, such as rotenone and 6-hydroxydopamine (6-OHDA), caused greater cytotoxicity to neurons when co-cultured with DJ-1 deficient-astrocytes than those with wild-type (WT) astrocytes [37,38]. Moreover, overexpression of DJ-1 in astrocytes affords neuroprotection against rotenone toxicity [37].

In addition to its role in lipid raft signaling and glutamate uptake in the plasma membrane, astrocytic DJ-1 is critically involved in maintaining normal mitochondrial function as well as slowing PD progression by modulating mitochondrial motility and fission [39]. Studies have shown that mitochondrial dysfunction in astrocytes is closely associated with aberrant glutamate metabolism and excitotoxicity [40–42], suggesting that DJ-1 function in mitochondria may be inter-regulated with glutamate transporter in astrocytes. Moreover, the effect of DJ-1 dysfunction on mitochondria caused intracellular oxidative stress, activating the antioxidant transcription factor nuclear factor erythroid 2-related factor (Nrf2) and Nrf2's target genes [41,43,44]. Additionally, DJ-1 deficiency induces inflammatory activation in astrocytes [45]. It also impairs monocyte infiltration into the damaged brain owing to a decrease in the astrocyte-specific chemokine, chemokine (C-C motif) ligand 2 (CCL2) levels in mice [46], resulting in further inflammation and the delay of neuronal repair.

2.2. α-Synuclein

The α-Synuclein is encoded by the SNCA gene, and the aggregates of this protein caused by its mutations are a pathological hallmark of PD. The role of α-synuclein aggregates in neurons has been extensively studied [47], but their spreading to adjacent astrocytes and, thus, consequential toxic effects via astrocytes remain largely unexplored. The α-synuclein is predominantly expressed in presynaptic nerve terminals, playing an important role as a molecular chaperone in providing an adequate supply of synaptic vesicles in presynaptic terminals, assisting in the folding and refolding of synaptic SNARE proteins, the release of neurotransmitters, such as dopamine, and synaptic integrity [48], to name a few of its functions.

Astrocytes are essential for fatty acid metabolism in the brain [49]. Although the abundance of α-synuclein in astrocytes is low in comparison to neurons [50], it has been shown to play a role in fatty acid metabolism, which is important in energy homeostasis, membrane maintenance, and cell signaling. These suggest that α-synuclein may regulate fatty acid metabolism in astrocytes. Studies have shown that α-synuclein deficiency decreased the rate of arachidonic acid and palmitic acid metabolism in astrocytes, suggesting that α-synuclein and lipid metabolism in astrocytes may be implicated in PD pathology [51–53].

That astrocytes regulate exogenous α-synuclein is supported by the finding that α-synuclein-containing inclusions were found in astrocytes of postmortem PD brains [54]. Astrocytes take up α-synuclein secreted from neurons [55] via a toll-like receptor 4 (TLR4)-independent endocytosis pathway [5] and localize this abnormal protein into the lysosomes [56], suggesting that astrocytes have a role in its removal and degradation. Moreover, high levels of extracellular α-synuclein have been shown to induce a TLR4-dependent inflammatory response in primary astrocyte cultures [5], indicating a possible role of astrocytes in α-synuclein-induced inflammatory PD pathology. This accumulated astrocytic α-synuclein may dysregulate critical astrocytic functions, such as glutamate uptake, by decreasing the expression of the astrocytic glutamate transporters, glutamate-aspartate transporter (GLAST, EAAT1 in human), and glutamate transporter 1 (GLT-1, EAAT2 in human), as well as blood–brain barrier (BBB) integrity by disrupting the localization of the water channel Aquaporin-4 (AQP4). Mice, in which mutant α-synuclein was specifically overexpressed in astrocytes, developed astrogliosis prior to the onset of symptoms with dopaminergic neurodegeneration in substantia nigra pars compacta [6].

Since the α-synuclein aggregates are injurious pathologic factors in PD progression, several therapeutic interventions using astrocytes have been developed to effectively eliminate them. Studies on neuron-astrocyte co-cultures have shown that astrocytes from the midbrain inhibited neuronal α-synuclein aggregation and its transmission to other cells in a paracrine mode, resulting in attenuation of the α-synuclein-induced toxicities. Similar observations were made in an in vivo PD mouse model in which transplantation of healthy midbrain astrocytes into the midbrain of the α-synuclein expressing PD mice attenuated α-synuclein pathology and dopaminergic toxicity [57].

The astrocytic autophagy-lysosomal function could also be targeted to eliminate α-synuclein since this astrocytic function is important in the degradation of α-synuclein taken up by astrocytes. Moreover, the efficiency of this pathway is dependent on the astrocytic function of ATP13A2 [58], mutations of which are linked to autosomal recessive PD pathology, supporting the role of ATP13A2 in astrocytic lysosomal degradation of α-synuclein aggregates. The functional role of ATP13A2 in astrocytes is discussed in the next section.

2.3. ATP13A2

ATP13A2, encoded by PARK9, is a transmembrane lysosomal ATPase. It is highly expressed in the brain, particularly in the substantia nigra pars compacta [59], consistent with involvement in the pathophysiology of PD. Mutations in the PARK9 gene impair lysosomal functions and are linked to autosomal recessive early-onset parkinsonism [59,60]. The ATP13A2 mutations lead to abnormal α-synuclein aggregation and oxidative stress in mitochondria [61].

ATP13A2 regulates the astrocytic uptake and degradation of α-synuclein released by neurons. Loss-of-function mutations in ATP13A2 impair α-synuclein clearance by astrocytes and lead to the activation of nod-like receptor protein 3 (NLRP3) inflammasomes in astrocytes [61]. Given the protective role of astrocytes in decreasing the neuronal accumulation of α-synuclein, these protective functions are decreased in astrocytes carrying ATP13A2 mutations, resulting in increased accumulation and propagation of α-synuclein, as was demonstrated when induced pluripotent stem cell (iPSC)-derived dopaminergic neurons were cocultured with astrocytes from either healthy subjects, or patients carrying mutations in lysosomal *ATP13A2* [58].

2.4. LRRK2

LRRK2, encoded by the PARK8 gene, is a large 286 kDa protein with dual kinase and GTPase activity [62,63]. It regulates multiple functions, such as trafficking, stress response, autophagy, and inflammation [64–66]. Mutations in LRRK2, such as G2019S, R1441C/G/H, and I2020T, are the most common genetic cause of late-onset PD [67]. They are autosomal dominant [62,68] and induce hyper LRRK2 kinase activity. Activation of the LRRK2 kinase activity, even independent of its mutations, is also known to contribute to the pathogenesis of idiopathic PD [69], implicating LRRK2 in both familial and sporadic PD cases. LRRK2 is expressed in neurons, astrocytes, and microglia in the human brain [70], with astrocytic LRRK2 being localized primarily in lysosomes to regulate lysosome size, number, and function, as well as the autophagy-lysosomal pathway for the degradation and recycling of cellular components [14,15,66]. In physiological conditions, autophagy proceeds with the lipidation of the LC3-I protein to the LC3-II protein, which is then transported to the membrane of autophagic vesicles for the normal autophagy process. However, increasing LRRK2 kinase activity decreases the lipidation of LC3-I in primary mouse astrocytes [14], leading to dysregulation of autophagosome formation.

G2019S, R1441C, and Y1699C LRRK2 mutations induce enlarged lysosomes in astrocytes, impair lysosomal degradation of long-lived and damaged proteins, and reduce lysosomal pH due to increased LRRK2 kinase activity [15].

Interestingly, while "normal" astrocytes express low levels of α-synuclein, cultured iPSC-derived astrocytes carrying the PD LRRK2 G2019S mutation exhibit progressive endogenous α-synuclein accumulation, dysfunctional chaperone-mediated autophagy, and impaired macroautophagy, resulting in the decreased astrocytic ability to clear neuronal-derived extracellular α-synuclein aggregates when co-cultured with WT dopaminergic neurons [71]. Moreover, astrocytes with the LRRK2 G2019S mutation transferred α-synuclein to neighboring neurons, leading to α-synuclein accumulation and neurodegeneration, suggesting that LRRK2 mutated astrocytes can mediate toxicity in the course of PD pathogenesis [71].

Moreover, astrocytes with LRRK2 G2019S mutation show dysregulation of the actin-binding protein annexin A2, which is also a novel player in α-synuclein internalization by astrocytes, resulting in impairment of the astrocytic endo-lysosomal pathway to clear extracellular fibrillar α-synuclein [72].

2.5. PINK1 and Parkin

2.5.1. PINK1

PINK1, encoded by the PARK6 gene, is a protein involved in mitophagy, a process that selectively degrades damaged mitochondria following mitochondrial stress [73]. PINK1 loss-of-function mutations are linked to recessive PD cases [74]. The role of astrocytic PINK1 is underscored by the findings that activation of PINK1 by mitochondrial injury occurs predominantly in astrocytes among all neural cell types in the CNS [68], while it is almost absent in neurons [75]. This suggests that astrocytes represent a key neural cell type affected by PINK1 deficiency in familial PD cases [68]. Conditioned media from LPS/IFN-γ-treated PINK1-deficient astroglia induced a higher apoptotic rate in neurons compared to media from treated WT astroglia, suggesting that astrocytic PINK1 is important to protecting neurons against apoptotic neuronal injury [76]. Additionally, PINK1 expression in the brain is increased during embryonic development and has an important role in the development of astrocytes [77] and the cyto-protection against oxidative stress-induced apoptosis [76].

2.5.2. Parkin

Parkin, encoded by PARK2, is a cytoplasmic E3 ubiquitin ligase that plays a crucial role in mitophagy and clearance of reactive oxygen species (ROS) [78]. Parkin mutations are well known to be linked to the autosomal recessive PD. Constitutive Parkin expression is high in neurons as compared with astrocytes, but unfolded protein stress elicits a selective increase in astrocytic Parkin expression and a change in its distribution, while neuronal Parkin remains largely unchanged. Moreover, in stress conditions that cause unfolded protein production, protein levels of Parkin increase in astrocytes, but not in neurons, leading to neuroprotective effects and thus decreasing neuronal injury [79], suggesting the astrocytic Parkin's role in neuroprotection against stress condition, and pathogenic role of astrocytes in PD carrying a Parkin mutation. In mouse primary midbrain astrocyte cultures, Parkin deficiency led to the presence of fewer astrocytes by reducing their proliferation and increasing proapoptotic protein expression compared to WT controls. In addition, astrocytic Parkin deficiency caused structural mitochondrial defects [8], and exacerbated α-synuclein-induced impairment of mitochondrial respiration [80].

2.5.3. PINK1/Parkin Mitophagy

PINK1 is well known to play a critical role in mitophagy, in collaboration with Parkin in the PINK1-Parkin mitophagy pathway, which removes damaged mitochondria [81]. PINK1 activity induces Parkin translocation from the cytosol into mitochondria to bind depolarized mitochondria [81]. Transcellular mitophagy (transmitophagy), which is a mitochondrial transfer from neurons to astrocytes in the context of neuronal injury, occurs primarily in neighboring astrocytes to degrade axonal mitochondria [82]. Trans-neuronal mitophagy occurs in in vivo rat and mouse PD models [83], and astrocytes are the primary cell type responsible for the clearance of damaged mitochondria from neurons. These findings suggest that boosting PINK1/Parkin-mediated mitophagy in striatal astrocytes may help with the removal of damaged mitochondria in dopaminergic neurons.

Moreover, optimal mitophagy function can eliminate dysfunctional mitochondria to maintain mitochondrial homeostasis and protect against neuroinflammatory activation induced by ROS [84], indicating that impaired mitophagy induced by mutations in PINK1 and Parkin can provoke inflammatory activation in astrocytes [40]. There is a possible correlation between mitochondria perturbations and astrocyte inflammatory activation. Impaired mitophagy from PINK1 or Parkin mutations in neurons and glia may promote mitochondrial fragmentation and enhance inflammatory responses in astrocytes, leading to the potentiation of inflammatory activation and dysfunction of neighboring cells. Thus, the removal of damaged mitochondria from astrocytes may be broadly neuroprotective and beneficial against undesired chronic neuroinflammation in PD [85].

Impaired PINK1/Parkin-mediated mitophagy can trigger NLRP3 and other inflammatory pathways [85] by the mitochondria-specific lipid cardiolipin. Mitochondrial depolarization results in the translocation of cardiolipin from the inner mitochondrial membrane to the outer mitochondrial membrane, where it associates with NLRP3 [86]. An attractive hypothesis is that in PINK1 and Parkin mutant astrocytes, elevated levels of mitochondrial ROS and cardiolipin result in the sustained activity of the NLRP3 inflammasome. Notably, there are reports indicating that in vitro PINK1/Parkin mitophagy can be inhibited upon inflammasome activation since Parkin can be cleaved by caspase-1 and caspase-8 [87,88]. Collectively, these studies demonstrate the crucial role that mitochondria play in the activation of NLRP3 and underline the PINK1/Parkin pathway of mitophagy as a key mechanism limiting excessive inflammation and preserving CNS homeostasis.

Impaired mitophagy in Parkin- and PINK1-deficient astrocytes might further lead to a decrease in astrocyte proliferation and decreased GFAP-positive astrocytes in the substantia nigra, potentially contributing to the development of PD due to the delay of astrocyte-mediated repair of the brain microenvironment [10,77].

2.6. Glucocerebrosidase

Glucocerebrosidase (GCase), encoded by the *GBA1* gene, is a lysosomal enzyme that is involved in glycolipid metabolism. Heterozygous carriers of specific mutations in this gene are at an elevated risk of developing PD [89]. Human and murine astrocytes express high levels of this protein, suggesting its critical role in this cell type [90].

GCase is involved in lysosomal function, playing a role in autophagy. The iPSC-derived astrocytes carrying mutations in the *GBA1* gene exhibited astrogliosis with impaired lysosomal cathepsin activity, resulting in the accumulation of α-synuclein aggregates [91]. These results are supported by other studies reporting that GBA-deficient primary astrocytes showed defective autophagic and proteasomal machinery [92]. GBA-deficient mice also showed impairment of their autophagic pathway, as evidenced by reductions in both LC3-I and LC3-II fragmentation [93]. These studies demonstrate that astrocytic GCase plays a role in autophagy, and its functional deficiency due to mutations can result in the accumulation of toxic proteins in astrocytes, thus indirectly causing neurotoxicity. On the other hand, mutation of GBA1 in mouse primary astrocytes did not affect the degradation of α-synuclein, despite resulting in a decrease in lysosomal protease activity [94]. This indicates that the effects of astrocytic GBA1 deficiency could differ depending on the experimental setting, although the astrocytic role of GCase function in autophagy is consistent.

GCase-deficient astrocytes due to either GBA1 mutations or deletion are closely associated with astrogliosis and neurodegeneration in a mouse model [95]. Moreover, GBA1-deleted mice exhibited astroglial activation within the nigrostriatal pathways, accompanied by the accumulation of α-synuclein aggregates [96]. Abnormal α-synuclein accumulation was also detected clinically in post-mortem brain tissue from patients carrying a mutant GBA1 [89,97]. These findings suggest that GBA1-mutated astrocytes play a pathogenic role in astrocytic inflammatory responses, possibly contributing to neuronal injury and an increased risk of PD. Moreover, GBA1 mutations in primary astrocytes decreased immune responses by reducing several pro-inflammatory cytokines when exposed to LPS [94]. Although the role of *GBA1 mutation*-induced decrease in the production of cytokines in PD is not well understood at present, it may be similar to the decreased cytokine responsiveness observed in aging, which is the greatest risk for PD. GBA1 mutation-induced immune response dysregulation in astrocytes was normalized by inhibiting LRRK2 kinase activity, indicating functional intracellular crosstalk between GCase and LRRK2 in astrocytes [94].

GCase also plays a role in maintaining mitochondrial functions, such as mitophagy in astrocytes [92]. GBA1-deficient mouse primary astrocytes exhibited mitochondrial dysfunction characterized by reduced ATP generation, decreased membrane potential, and mitochondrial fragmentation, possibly due to impaired mitophagy caused by lysosomal defect [92,93].

2.7. Prospectives

Astrocyte reprogramming studies have shown some promising strategies by which astrocytes in the CNS can be transformed into functional neurons. These strategies are applicable to many gene mutations related to PD, including those in the α-synuclein gene (reviewed in [98]). Reprogramming methods, targeting specific transcription factors and mRNAs by modulating astrocytic microRNA (miRNA) mechanisms in the mouse brain, have been able to reprogram astrocytes into functional dopaminergic neurons (reviewed in [99]).

3. The Dysregulations of Astrocytic Cellular and Molecular Mechanisms in Experimental PD Models

The protective role of astrocytes for dopaminergic neurons in PD is well established by several studies. For example, 6-OHDA-induced dopaminergic neurodegeneration was exacerbated in the absence of astrocytic function, which was induced by chronic infusion of fluorocitrate into the substantia nigra of the rat brain [100]. Moreover, without astrocytic support, the substantia nigra displayed a higher microglial activation and reduced survival capability of the dopaminergic system in the 6-OHDA rat model [100]. This indicates that dopaminergic neurons in the substantia nigra pars compacta are more vulnerable to PD toxicants in the absence of astrocytic support. Extensive studies have been conducted to determine the astrocytic factors that modulate PD development. These include oxidative stress, inflammation, synaptic dysfunction, endoplasmic reticulum (ER) stress, trophic factor reduction, and apoptosis in astrocytes. PD neurotoxicants, such as paraquat, 1-methyl-4-phenyl-1,2,3,6-tetrahydropyridine (MPTP)/1-methyl-4-phenylpyridinium (MPP+), 6-OHDA and rotenone, have been extensively studied in astrocytes, in addition to neurons, to understand the role of astrocytes in the pathogenesis of idiopathic PD [37,38,101,102]. The following sections will focus on how these neurotoxins contribute to PD pathogenesis by affecting astrocytic cellular and molecular mechanisms.

3.1. The Role of Astrocytic Oxidative Stress in PD Development

Oxidative stress is a well-known important contributor to the progression of neurodegenerative disorders [103]. ROS are mainly produced in the mitochondria via oxidative phosphorylation and electron transfer reactions. Abnormal ROS production causes excessively abnormal peroxidation of lipids and proteins, resulting in DNA oxidation and strand breakage, leading to cellular injury. It has been shown that PD patients' blood and cerebrospinal fluid (CSF) have higher levels of oxidative stress compared to healthy subjects [103]. Causal factors of PD-related oxidative stress may include (1) impairment of mitochondrial function and degradation through genetic mutations and various toxicities and (2) dysregulation of antioxidant defense capabilities in the brain.

ROS produced by astrocytes are implicated in oxidative stress in various PD models, including primary neurons and astrocytes [103–106], leading to neuronal injury and PD-like neurological deficits in animals. Most PD-like neurotoxicants, such as rotenone, MPTP/MPP+, and 6-OHDA, trigger ROS production in astrocytes [13,38,107–111], and the ROS production is likely derived from the impaired mitochondrial function and dysregulated astrocytic antioxidant defense mechanisms.

Dysregulation of the astrocytic antioxidant system is well known to augment neuronal stress responses in primary astrocyte and mouse PD models [112]. Studies have shown that the expression of the antioxidant transcription factor Nrf2 and its target genes in astrocytes is reduced in an MPTP-induced PD mouse model [113], while overexpression of astrocytic Nrf2 protects neurons against MPTP-induced dopaminergic neurotoxicity in mice [113]. Activation of astrocytic Nrf2 also attenuated 6-OHDA-induced neurotoxicity in rats and Drosophila [114]. Moreover, the mutation of DJ-1, which causes an autosomal recessive PD, is closely associated with Nrf2 dysfunction. This is supported by the findings that DJ-1-deficient or -mutated astrocytes caused Nrf2 instability, reducing its transcriptional activities [115], and abolished astrocyte-mediated neuroprotection against PD [13,38]. However, DJ-1 overexpression in astrocytes protected dopaminergic neurons against rotenone-induced neurotoxicity in the substantia nigra of rat brain [107], suggesting that astrocytic oxidative stress plays a critical role in PD, at least in part by dysregulating Nrf2 and DJ-1. In addition, the co-culturing of primary rat astrocytes with neurons increased neuronal glutathione (GSH), preventing paraquat-induced damage to neurons [112]. Moreover, 6-OHDA-induced neuronal death was exacerbated in mesencephalic cultures pre-cultured with GSH-depleted astrocytes, but not complex I-inhibited astrocytes, compared to co-culturing with normal astrocytes. MPTP-induced neurotoxicity in mice was exacerbated by a deficiency in astrocytic SMP30, an essential component of the antioxidant vitamin C production [116]. These findings suggest that excessive astrocytic ROS may increase neuron vulnerability [117] and that enhancing astrocyte antioxidant defense systems could be an effective strategy for reducing oxidative stress in PD.

3.2. The Role of Astrocytic Inflammation in PD Development

Inflammation was first observed in the substantia nigra pars compacta of PD patients in the 1980s [118]. Later, more studies demonstrated that post-mortem brains, CSF, and serum of PD patients showed higher levels of proinflammatory cytokines, such as interleukin (IL)-1β, tumor necrosis factor (TNF)-α, IL-2, and IL-6, compared to those in healthy controls [119]. Although microglia are well-known inflammatory cytokine producers, astrocytes also release cytokines and inflammatory mediators under toxic conditions, leading to inflammation and cell death in dopaminergic neurons [120–122]. Numerous studies have reported the involvement of astrocytes in the MPTP neurotoxicity model, as MPTP triggered proinflammatory astrocytic activation and subsequent neurotoxicity [121]. This was supported by the findings that the inhibition of astrocytic activation in mice with iptakalim attenuated MPP^+-induced astrocytic inflammation as well as dopaminergic toxicity in the substantia nigra [121]. By using ALDH1L1 bacTRAP mice to identify genes being expressed in astrocytes, MPTP was observed to induce the production of the cytokines IL-1β, TNF-α, and CCL4 [123]. In addition to producing cytokines, evidence has shown that inflammatory reactive astrocytes amplify microglial activation under MPTP exposure [120], suggesting that astrocyte-microglia crosstalk contributes to neuroinflammation. Arundic acid, which exclusively suppresses astrocyte activation [124], attenuated MPTP-induced reactive astrocytes and dopaminergic neurotoxicity in mice [101,125]. Another PD toxin, 6-OHDA, was also shown to increase inflammatory astrocyte reactivity and expression of proinflammatory molecules, including inducible nitric oxide synthase (iNOS), nitric oxide (NO), cyclooxygenase 2 (COX-2), prostaglandin E2, and TNF-α in primary astrocytes [126]. A role for iNOS is indicated as its inhibition attenuated 6-OHDA-induced mitochondrial impairment and apoptosis in C6 astrocytes [127]. These findings indicate that astrocytes play a substantial role in mediating neurodegeneration via inflammation [102].

Studies have shown that astrocytic aquaporins and secretogranins are involved in the production of bioactive compounds from the cells in PD models. Paraquat at high exposure concentrations increased the expression of both secretogranin II and IL-6 and their colocalization in U118 astrocytes [47], while inhibition of secretogranin II decreased IL-6 protein levels in astrocytes, suggesting that astrocytic secretogranin could influence inflammation by regulating IL-6 production and release. Moreover, AQP4, a water-selective membrane channel regulating BBB permeability and immune response, is involved in MPTP-induced production of IL-1β and TNF-α in primary astrocyte cultures [120]. This is supported by the results that deletion of AQP4 exacerbated inflammatory astrocyte reactivity and levels of these cytokines, along with enhanced NF-κB activation in MPP^+-treated astrocytes [120] and amplified microglial activation in astrocyte-microglia co-cultures. These data indicate that astrocytic aquaporins play a critical role in the attenuation of astrocytic inflammation and microglial activation, serving as a potential target for treating MPTP neurotoxicity.

3.3. Dysregulation of Astrocytic Degradation of Damaged Proteins in PD Development

Eliminating protein aggregates, such as the Lewy bodies that can be formed by misfolded α-synuclein, is essential for the cell's health and physiological functions [128]. Several biological processes degrade damaged proteins and toxic substances, including endo-lysosomes, autophagy, and the ubiquitin-proteasomal pathway [129]. Given that α-synuclein could spread from neurons to astrocytes [56,130], astrocytes play a crucial role in removing α-synuclein via the activation of lysosomes and lysosomal enzymes [131].

Accumulating evidence suggests that astrocytes play a major role in removing accumulated debris caused by the degeneration of synapses and axons of dopaminergic neurons in a PD animal model [132]. Astrocytes can take up extracellular debris and damaged components via phagocytosis and endocytosis [56,133] and degrade these materials in lysosomes [134]. PD-related mutations in LRRK2, such as G2019S, R1441C, and Y1699C, cause lysosomal defects and abnormal lysosomal morphology that reduce degradative capacity in astrocytes [15], leading to the accumulation of cellular debris and cell death in astrocytes. A similar reduction in autophagic capacity in astrocytes is caused by neurotoxicants that produce PD-like symptoms, such as rotenone and paraquat. High concentrations of paraquat initially activate autophagy in astrocytes but subsequently reduce autophagosome formation and autophagic flux, leading to the accumulation of toxic substances in astrocytes [135]. These toxic substances can spread to neighboring cells, including neurons, by several mechanisms, such as exosomal propagation [136], passive transport [133], direct contact, and tunneling [137].

As described in earlier sections, loss-of-function mutations in ATP13A2 (PARK9) are associated with PD (for a review, see [138]). In physiological conditions, ATP13A2 regulates the formation and secretion of intraluminal vesicles and exosomes [138]. However, ATP13A2 mutations cause abnormal α-synuclein accumulation in neurons [139,140]. Moreover, ATP13A2 mutations reduce the ability of astrocytes to eliminate α-synuclein, resulting in α-synuclein accumulation in astrocytes and interneuronal transfer of α-synuclein in iPSC-derived dopaminergic neurons, damaging neurons in the astrocyte-neuron co-culture [58].

Rotenone induces inflammatory reactive astrocytes, and phosphorylated α-synuclein aggregates in dopaminergic neurons prior to dopaminergic neuronal loss in the substantia nigra [141], indicating that activation of astrocytes and aggregation of phosphorylated α-synuclein precede neuronal loss in rotenone neurotoxicity.

These findings indicate that targeting astrocytic degradation and clearance of protein aggregates could be an important therapeutic strategy for preventing the accumulation of toxic substances in PD.

3.4. The Role of Astrocytes in Inducing Excitotoxicity in PD Development

Glutamate is the main excitatory neurotransmitter in the CNS, inducing excitatory synaptic transmission, followed by its uptake into the adjacent astrocytes [142]. Studies have shown that excessive glutamate neurotransmission may be involved in PD pathogenesis, as the N-methyl-D-aspartate (NMDA) receptor antagonist, Ro 25–6981, attenuated Parkinsonian motor symptoms in the MPTP-induced PD mouse model [143].

Astrocytes maintain proper homeostasis of glutamate neurotransmission via glutamate-glutamine metabolism and glutamate uptake from the synaptic cleft via glutamate transporters, such as GLAST (EAAT1) and GLT-1 (EAAT2). Accordingly, the dysregulation of astrocytic glutamate transporters may lead to excessive glutamate accumulation in the extracellular synaptic cleft, leading to overstimulation of postsynaptic glutamate receptors and, thus, excitotoxic neuronal cell death. Although glutamate uptake cannot be measured in the brain of PD patients while alive, studies have shown that PD patients had a 48% ($p < 0.0001$) lower glutamate uptake in platelets compared to healthy controls [144], which correlated with increased severity of parkinsonian symptoms ($p < 0.05$). Although there is no report linking platelet glutamate uptake to astrocytic glutamate uptake, this finding implicates the potential dysfunction of glutamate uptake in PD.

Studies have shown that PD-related gene mutations, such as DJ-1 and LRRK2, are associated with impaired glutamate uptake in astrocytes resulting from impaired GLT-1 trafficking and localization and transcriptional repression [12,145]. The LRRK2 G2019S mutation decreased EAAT2 expression in PD human brains [145]. LRRK2 G2019S also reduced GLT-1 in the striatum of mice and primary astrocytes at the transcriptional and posttranslational levels, leading to a reduction in glutamate uptake [145]. Moreover, LRRK2 G2019S caused GLT-1 sequestration in Rab4-positive vesicles, preventing GLT-1 localization in the plasma membrane of primary astrocytes [145], leading to abnormal GLT-1 trafficking and increased degradation [12]. Nedd-4-mediated ubiquitination of GLT-1 decreased GLT-1 expression in an MPTP-PD mouse model, while its knockdown reversed the MPTP-induced decrease in GLT-1 expression and reduced both motor deficits and dopaminergic cell loss [146]. A 6-OHDA decreased striatal GLT-1 levels along with an increase in PD-like deficits [147]. Paraquat and MPTP also reduced astrocytic GLT-1 expression as well as glutamate uptake and led to dopaminergic neurotoxicity in mice [142]. In addition, an upregulation of EAAT2 increased glutamate uptake in human U251 astrocytes and afforded more protection against 6-OHDA toxicity in the neuroblastoma SH-SY5Y cell line [148].

3.5. Dysregulation of Astrocytic Growth Factor Synthesis in PD Development

Neurotrophic factors, such as brain-derived neurotrophic factor (BDNF), ciliary neurotrophic factor (CNTF), epidermal growth factor (EGF), acidic fibroblast growth factor (FGF), insulin-like growth factor (IGF), transforming growth factor-alpha (TGF-α), and glial cell line-derived neurotrophic factor (GDNF), have been shown to promote neuronal health and survival (for review, see [149]). Studies have shown that BDNF and nerve growth factor concentrations were decreased in the substantia nigra in PD patients [150,151], suggesting that growth factors were decreased in the PD brain.

Enhancing astrocytic levels of BDNF and GDNF ameliorated dopamine neurodegeneration induced by MPP^+ in rat-derived primary cultures of astrocytes and neurons [152], underscoring the role of astrocytic growth factors in neuroprotection. In the 6-OHDA PD animal model, astrocytic trophic factors were decreased, which contributed to neuronal injury. This is supported by the result that 6-OHDA inhibited ATP-dependent Ca^{2+} signaling, which is critical for trophic factor production in astrocytes [153]. It is well established that astrocytes afford a protective role in PD by supplying neurotrophic factors, such as nerve growth factor and BDNF [154], and that these protective effects are decreased in advanced PD [155].

Astrocytic growth factors induce neuroprotection by multiple cellular mechanisms. Healthy astrocytes elicit protective effects against MPTP-induced inflammation via IGF-1, IGF-1 receptor, and G protein-coupled estrogen receptor (GPER) in mice [156]. Astrocytic IGF-1 has been shown to attenuate MPP^+-induced upregulation of COX-2 and iNOS protein levels in primary astrocytes by binding to the IGF-1R and activating the GPER/PI3K/MAPK signaling pathway [156]. Astrocytic GDNF and BDNF have also been reported to protect neurons against 6-OHDA toxicities [157,158], although the efficacies of these trophic factors have not been conclusive [159,160]. The mesencephalic astrocyte-derived neurotrophic factor (MANF) [161] has been shown to offer neuroprotection in PD 6-OHDA-treated neuronal cell lines, exerting antioxidative stress, anti-inflammation, and antiapoptotic properties [122,162–164], as well as increasing antioxidant proteins Nrf2/HO-1 and Wnt signaling [122]. These findings imply that a reduction in astrocytic trophic factors contributes to adjacent neuronal injury, leading to exacerbation of nigrostriatal dopaminergic neurodegeneration in PD [117,153].

3.6. The Role of Astrocytic ER Stress in PD Development

In healthy cells, the ER regulates protein folding and trafficking, and thus, acute perturbations to ER homeostasis can alter the folding process, causing ER stress and inducing the unfolded protein response (UPR) to restore ER homeostasis and function [165]. However, chronic ER stress could lead to the accumulation of unfolded proteins, such as α-synuclein, and prolonged UPR activity can result in apoptotic cell death. PD brain tissues have shown accumulation of α-synuclein in the ER [166,167], suggesting that chronic ER stress is associated with α-synuclein misfolding and accumulation.

It has been shown that astrocytic ER stress triggered by excessive α-synuclein uptake leads to Golgi apparatus fragmentation and apoptosis in primary rat astrocytes [168]. Excessive α-synuclein impairs the normal degradation process in astrocytes, leading to higher α-synuclein accumulation in the trans-Golgi network and ER stress [137]. Moreover, LRRK2 G2019S mutation caused ER stress and apoptosis in astrocytes [169] by impairing the sarco/ER Ca^{2+}-ATPase, resulting in reduced ER Ca^{2+} levels. It also increased ER-mitochondria interaction, resulting in mitochondrial dysfunction [169]. This suggests that LRRK2 mutations increase the susceptibility of astrocytes to ER stress and mitochondrial dysfunction. Importantly, there was no significant effect on ER stress in LRRK2 G2019S neurons, but wild-type neurons co-cultured with LRRK2 G2019S astrocytes sustained more neuronal injury after α-synuclein treatment [169]. These results indicate that ER stress in astrocytes and neurons is differently affected by LRRK2 G2019S mutation, highlighting the role of astrocytic LRRK2 G2019S mutation and ER stress contributing to PD development. Moreover, astrocytic ER stress increases the production of proinflammatory cytokines and chemokines and also activates inflammatory microglia via paracrine signaling [169], indicating that astrocytic ER stress induces inflammatory astrocytic dysfunction and triggers adjacent microglial activation, which may lead to neuronal injury.

3.7. The Role of Astrocytic Apoptosis in PD Development

Astrocytic apoptosis could cause harmful damage to the surrounding neurons by releasing toxic astrocytic components and proinflammatory factors [170]. Astrocytic apoptosis can also impair astrocyte-neuron networks and destabilize synaptic neurotransmission, thus contributing to neuronal cell death. Prolonged astrocytic dysfunction and degeneration accelerated neuronal injury in the substantia nigra in a 6-OHDA PD rat model [100], indicating that astrocytic apoptosis could be an important contributor to PD pathology. Astrocytic apoptosis is caused by various cellular dysregulations, including oxidative stress, mitochondrial dysfunction, inflammation, disrupted Ca^{2+} homeostasis, membrane instability, and ER stress [168,171,172]. Rotenone causes astrocytic apoptosis, at least in part, by reducing astrocytic connexin 43 and increasing membrane and gap junction permeability [173]. Given that dysregulated gap junctions play an important role in neurological disorders, such as AD, HD, ischemia, and PD [173–175], and that connexins are critical in maintaining gap junctions to allow direct intercellular communication between adjacent cells [176], rotenone-induced dysfunctional gap junctions in astrocytes may cause the leakage of toxic substances to adjacent neural cells, including dopaminergic neurons. Astrocytic activation can lead to astrocyte apoptosis and stimulate microglial activation via cytokine and chemokine communication, indicating that the activation of astrocytes and microglia are closely connected to each other [177]. Major histocompatibility complex II levels were increased and mainly found in midbrain astrocytes and microglia rather than in neurons of the PD MPTP mouse model [178]. In addition, astrocytic apoptosis diminishes astrocyte-induced neuroprotection and antioxidant support to neurons, increasing neuronal susceptibility to toxic insults. These indicate that inhibiting astrocytic apoptosis might be an important therapeutic target to provide neuroprotection against PD insults.

4. Conclusions

In summary, astrocytes are essential neural cell types that protect neurons against PD-inducing insults. Healthy astrocytes exert their neuroprotective effects by releasing neurotrophic factors, producing antioxidants, reducing proinflammatory cytokines, and removing toxic aggregates such as α-synuclein and damaged mitochondria. However, astrocytic dysfunctions, such as oxidative stress, inflammation, autophagy impairment, and apoptosis, are commonly observed in pathological conditions in gene mutations associated with PD, such as DJ-1, α-synuclein, LRRK2, PINK1, and parkin, as well as experimental PD models using toxicants, such as paraquat, rotenone, MPTP/MPP+, and 6-OHDA (Figure 1). Understanding astrocytes' cellular and molecular responses in these PD models will greatly contribute to the development of PD therapeutics targeting astrocytes. In addition, the findings suggest that complex pathways are involved in PD pathology, and thus, a combination of approaches may be required for therapeutic intervention.

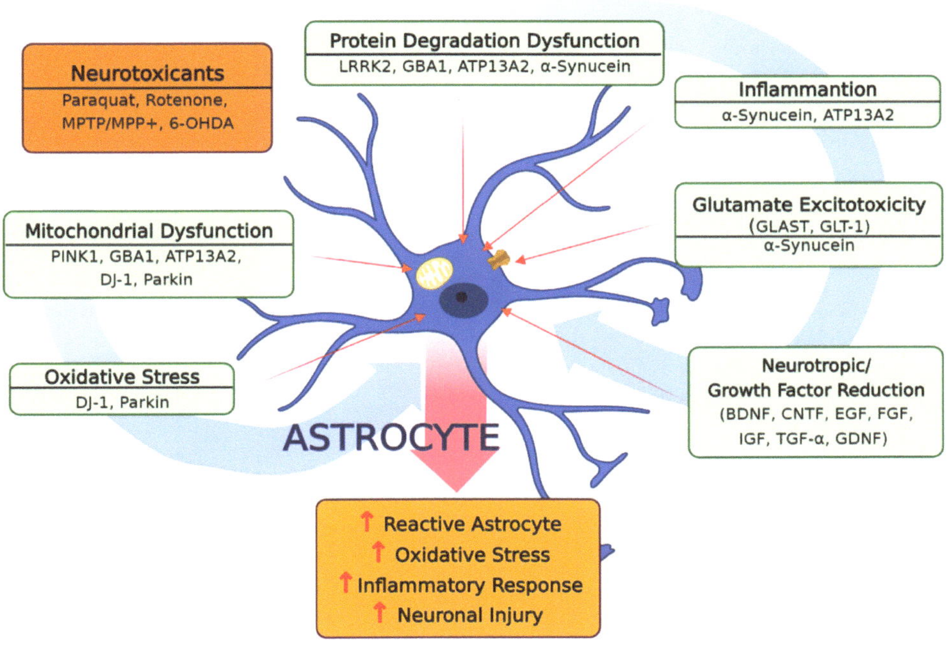

Figure 1. A summary of the astrocytic dysregulations contributing to dopaminergic neurodegeneration in PD gene mutations and PD neurotoxicants. Dysregulations in astrocytic mechanisms, such as oxidative stress, impairment of autophagy/mitophagy, inflammation, neurotrophic/growth factor reduction, and excitotoxicity, contribute to dopaminergic neuronal injury in PD. Several PD-mutated genes, including DJ-1, α-synuclein, LRRK2, PINK1, parkin, and PD-inducing experimental neurotoxicants, such as rotenone, MPTP/MPP$^+$, paraquat, and 6-OHDA, have been shown to modulate these molecular mechanisms in astrocytes, which could exacerbate PD pathology.

Author Contributions: Conceptualization—S.K., E.P. and E.L.; funding acquisition—E.L. and M.A.; project administration—E.L.; resources—E.L.; supervision—E.L.; validation—S.K., E.P., I.N.-D. and E.L.; visualization—S.K. and E.P.; writing—original draft—S.K., E.P. and I.N.-D.; writing—review and editing—S.K., E.P., I.N.-D., M.A. and E.L. All authors have read and agreed to the published version of the manuscript.

Funding: This work was supported by National Institutes of Health Grants, R01 ES031282 (to E.L.), R01 ES010563 (to M.A.), and NIMHD U54 MD007582.

Conflicts of Interest: The authors declare that this research was conducted in the absence of any commercial or financial relationships that could be construed as a potential conflict of interest.

References

1. Lieberman, A.; Dziatolowski, M.; Gopinathan, G.; Kupersmith, M.; Neophytides, A.; Korein, J. Evaluation of Parkinson's disease. *Adv. Biochem. Psychopharmacol.* **1980**, *23*, 277–286.
2. Yang, W.; Hamilton, J.L.; Kopil, C.; Beck, J.C.; Tanner, C.M.; Albin, R.L.; Dorsey, E.R.; Dahodwala, N.; Cintina, I.; Hogan, P.; et al. Current and projected future economic burden of Parkinson's disease in the U.S. *NPJ Park. Dis.* **2020**, *6*, 15. [CrossRef]
3. Shadrina, M.; Slominsky, P. Modeling Parkinson's Disease: Not Only Rodents? *Front. Aging Neurosci.* **2021**, *13*, 695718. [CrossRef]
4. Bove, J.; Prou, D.; Perier, C.; Przedborski, S. Toxin-induced models of Parkinson's disease. *NeuroRx* **2005**, *2*, 484–494. [CrossRef]
5. Rannikko, E.H.; Weber, S.S.; Kahle, P.J. Exogenous alpha-synuclein induces toll-like receptor 4 dependent inflammatory responses in astrocytes. *BMC Neurosci.* **2015**, *16*, 57. [CrossRef]
6. Gu, X.L.; Long, C.X.; Sun, L.; Xie, C.; Lin, X.; Cai, H. Astrocytic expression of Parkinson's disease-related A53T alpha-synuclein causes neurodegeneration in mice. *Mol. Brain.* **2010**, *3*, 12. [CrossRef]
7. Diniz, L.P.; Araujo, A.P.B.; Matias, I.; Garcia, M.N.; Barros-Aragão, F.G.; Reis, R.A.D.M.; Foguel, D.; Braga, C.; Figueiredo, C.P.; Romão, L.; et al. Astrocyte glutamate transporters are increased in an early sporadic model of synucleinopathy. *Neurochem. Int.* **2020**, *138*, 104758. [CrossRef]
8. Schmidt, S.; Linnartz, B.; Mendritzki, S.; Sczepan, T.; Lubbert, M.; Stichel, C.C.; Lubbert, H. Genetic mouse models for Parkinson's disease display severe pathology in glial cell mitochondria. *Hum. Mol. Genet.* **2011**, *20*, 1197–1211. [CrossRef]
9. Khasnavis, S.; Pahan, K. Cinnamon Treatment Upregulates Neuroprotective Proteins Parkin and DJ-1 and Protects Dopaminergic Neurons in a Mouse Model of Parkinson's Disease. *J. Neuroimmune Pharmacol.* **2014**, *9*, 569–581. [CrossRef]
10. Choi, I.; Kim, J.; Jeong, H.K.; Kim, B.; Jou, I.; Park, S.M.; Chen, L.; Kang, U.J.; Zhuang, X.; Joe, E.H. PINK1 deficiency attenuates astrocyte proliferation through mitochondrial dysfunction, reduced AKT and increased p38 MAPK activation, and downregulation of EGFR. *Glia* **2013**, *61*, 800–812. [CrossRef]
11. Kim, K.S.; Kim, J.S.; Park, J.Y.; Suh, Y.H.; Jou, I.; Joe, E.H.; Park, S.M. DJ-1 associates with lipid rafts by palmitoylation and regulates lipid rafts-dependent endocytosis in astrocytes. *Hum. Mol. Genet.* **2013**, *22*, 4805–4817. [CrossRef]
12. Kim, J.M.; Cha, S.H.; Choi, Y.R.; Jou, I.; Joe, E.H.; Park, S.M. DJ-1 deficiency impairs glutamate uptake into astrocytes via the regulation of flotillin-1 and caveolin-1 expression. *Sci. Rep.* **2016**, *6*, 28823. [CrossRef]
13. Mullett, S.J.; Hinkle, D.A. DJ-1 deficiency in astrocytes selectively enhances mitochondrial Complex I inhibitor-induced neurotoxicity. *J. Neurochem.* **2011**, *117*, 375–387. [CrossRef]
14. Manzoni, C.; Mamais, A.; Dihanich, S.; Abeti, R.; Soutar, M.P.M.; Plun-Favreau, H.; Giunti, P.; Tooze, S.A.; Bandopadhyay, R.; Lewis, P.A. Inhibition of LRRK2 kinase activity stimulates macroautophagy. *Biochim. Biophys. Acta* **2013**, *1833*, 2900–2910. [CrossRef]
15. Henry, A.G.; Aghamohammadzadeh, S.; Samaroo, H.; Chen, Y.; Mou, K.; Needle, E.; Hirst, W.D. Pathogenic LRRK2 mutations, through increased kinase activity, produce enlarged lysosomes with reduced degradative capacity and increase ATP13A2 expression. *Hum. Mol. Genet.* **2015**, *24*, 6013–6028. [CrossRef]
16. Qiao, C.; Yin, N.; Gu, H.-Y.; Zhu, J.-L.; Ding, J.-H.; Lu, M.; Hu, G. *Atp13a2* Deficiency Aggravates Astrocyte-Mediated Neuroinflammation via NLRP3 Inflammasome Activation. *CNS Neurosci. Ther.* **2016**, *22*, 451–460. [CrossRef]
17. Ran, C. Strong association between glucocerebrosidase mutations and Parkinson's disease in Sweden. *Neurobiol. Aging* **2016**, *45*, 212.e5–212.e11. [CrossRef]
18. Canet-Aviles, R.M.; Wilson, M.A.; Miller, D.W.; Ahmad, R.; McLendon, C.; Bandyopadhyay, S.; Baptista, M.J.; Ringe, D.; Petsko, G.A.; Cookson, M.R. The Parkinson's disease protein DJ-1 is neuroprotective due to cysteine-sulfinic acid-driven mitochondrial localization. *Proc. Natl. Acad. Sci. USA* **2004**, *101*, 9103–9108. [CrossRef]
19. Ahn, E.H.; Lei, K.; Kang, S.S.; Wang, Z.H.; Liu, X.; Hong, W.; Wang, Y.T.; Edgington-Mitchell, L.E.; Jin, L.; Ye, K. Mitochondrial dysfunction triggers the pathogenesis of Parkinson's disease in neuronal C/EBPbeta transgenic mice. *Mol. Psychiatry* **2021**, *26*, 7838–7850. [CrossRef]
20. Pajares, M.; Rojo, A.I.; Manda, G.; Bosca, L.; Cuadrado, A. Inflammation in Parkinson's Disease: Mechanisms and Therapeutic Implications. *Cells* **2020**, *9*, 1687. [CrossRef]
21. Wang, B.; Abraham, N.; Gao, G.; Yang, Q. Dysregulation of autophagy and mitochondrial function in Parkinson's disease. *Transl. Neurodegener.* **2016**, *5*, 19. [CrossRef]
22. Liu, J.; Liu, W.; Li, R.; Yang, H. Mitophagy in Parkinson's Disease: From Pathogenesis to Treatment. *Cells* **2019**, *8*, 712. [CrossRef]
23. Alexander, G.E. Biology of Parkinson's disease: Pathogenesis and pathophysiology of a multisystem neurodegenerative disorder. *Dialogues Clin. Neurosci.* **2004**, *6*, 259–280. [CrossRef]
24. Kettenmann, H.; Verkhratsky, A. Neuroglia: The 150 years after. *Trends Neurosci.* **2008**, *31*, 653–659. [CrossRef]
25. Jakel, S.; Dimou, L. Glial Cells and Their Function in the Adult Brain: A Journey through the History of Their Ablation. *Front. Cell. Neurosci.* **2017**, *11*, 24. [CrossRef]
26. Halliday, G.M.; Stevens, C.H. Glia: Initiators and progressors of pathology in Parkinson's disease. *Mov. Disord.* **2011**, *26*, 6–17. [CrossRef]
27. Verkhratsky, A.; Nedergaard, M.; Hertz, L. Why are astrocytes important? *Neurochem Res.* **2015**, *40*, 389–401. [CrossRef]
28. Bonifati, V.; Rizzu, P.; van Baren, M.J.; Schaap, O.; Breedveld, G.J.; Krieger, E.; Dekker, M.C.; Squitieri, F.; Ibanez, P.; Joosse, M.; et al. Mutations in the DJ-1 gene associated with autosomal recessive early-onset parkinsonism. *Science* **2003**, *299*, 256–259. [CrossRef]

29. Honbou, K.; Suzuki, N.N.; Horiuchi, M.; Niki, T.; Taira, T.; Ariga, H.; Inagaki, F. The crystal structure of DJ-1, a protein related to male fertility and Parkinson's disease. *J. Biol. Chem.* **2003**, *278*, 31380–31384. [CrossRef]
30. Ariga, H.; Takahashi-Niki, K.; Kato, I.; Maita, H.; Niki, T.; Iguchi-Ariga, S.M. Neuroprotective function of DJ-1 in Parkinson's disease. *Oxid. Med. Cell. Longev.* **2013**, *2013*, 683920. [CrossRef]
31. Bandopadhyay, R.; Kingsbury, A.E.; Cookson, M.R.; Reid, A.R.; Evans, I.M.; Hope, A.D.; Pittman, A.M.; Lashley, T.; Canet-Aviles, R.; Miller, D.W.; et al. The expression of DJ-1 (PARK7) in normal human CNS and idiopathic Parkinson's disease. *Brain* **2004**, *127*, 420–430. [CrossRef]
32. Mullett, S.J.; Hamilton, R.L.; Hinkle, D.A. DJ-1 immunoreactivity in human brain astrocytes is dependent on infarct presence and infarct age. *Neuropathology* **2009**, *29*, 125–131. [CrossRef]
33. Simons, K.; Ehehalt, R. Cholesterol, lipid rafts, and disease. *J. Clin. Invest.* **2002**, *110*, 597–603. [CrossRef]
34. Butchbach, M.E.; Tian, G.; Guo, H.; Lin, C.L. Association of excitatory amino acid transporters, especially EAAT2, with cholesterol-rich lipid raft microdomains: Importance for excitatory amino acid transporter localization and function. *J. Biol. Chem.* **2004**, *279*, 34388–34396. [CrossRef]
35. Roberts, P.J.; Davies, S.W. Excitatory receptors and their role in excitotoxicity. *Biochem. Soc. Trans.* **1987**, *15*, 218–219. [CrossRef] [PubMed]
36. Ashley, A.K.; Hinds, A.I.; Hanneman, W.H.; Tjalkens, R.B.; Legare, M.E. DJ-1 mutation decreases astroglial release of inflammatory mediators. *Neurotoxicology* **2016**, *52*, 198–203. [CrossRef]
37. Mullett, S.J.; Hinkle, D.A. DJ-1 knock-down in astrocytes impairs astrocyte-mediated neuroprotection against rotenone. *Neurobiol. Dis.* **2009**, *33*, 28–36. [CrossRef] [PubMed]
38. Lev, N.; Barhum, Y.; Ben-Zur, T.; Melamed, E.; Steiner, I.; Offen, D. Knocking out DJ-1 attenuates astrocytes neuroprotection against 6-hydroxydopamine toxicity. *J. Mol. Neurosci.* **2013**, *50*, 542–550. [CrossRef]
39. Larsen, N.J.; Ambrosi, G.; Mullett, S.J.; Berman, S.B.; Hinkle, D.A. DJ-1 knock-down impairs astrocyte mitochondrial function. *Neuroscience* **2011**, *196*, 251–264. [CrossRef]
40. Bantle, C.M.; Hirst, W.D.; Weihofen, A.; Shlevkov, E. Mitochondrial Dysfunction in Astrocytes: A Role in Parkinson's Disease? *Front. Cell. Dev. Biol.* **2020**, *8*, 608026. [CrossRef]
41. Peng, L.; Zhao, Y.; Li, Y.; Zhou, Y.; Li, L.; Lei, S.; Yu, S.; Zhao, Y. Effect of DJ-1 on the neuroprotection of astrocytes subjected to cerebral ischemia/reperfusion injury. *J. Mol. Med.* **2019**, *97*, 189–199. [CrossRef] [PubMed]
42. Thomas, K.J.; McCoy, M.K.; Blackinton, J.; Beilina, A.; van der Brug, M.; Sandebring, A.; Miller, D.; Maric, D.; Cedazo-Minguez, A.; Cookson, M.R. DJ-1 acts in parallel to the PINK1/parkin pathway to control mitochondrial function and autophagy. *Hum. Mol. Genet.* **2011**, *20*, 40–50. [CrossRef] [PubMed]
43. Dolgacheva, L.P.; Berezhnov, A.V.; Fedotova, E.I.; Zinchenko, V.P.; Abramov, A.Y. Role of DJ-1 in the mechanism of pathogenesis of Parkinson's disease. *J. Bioenerg. Biomembr.* **2019**, *51*, 175–188. [CrossRef] [PubMed]
44. Bjorkblom, B.; Adilbayeva, A.; Maple-Grodem, J.; Piston, D.; Okvist, M.; Xu, X.M.; Brede, C.; Larsen, J.P.; Moller, S.G. Parkinson disease protein DJ-1 binds metals and protects against metal-induced cytotoxicity. *J. Biol. Chem.* **2013**, *288*, 22809–22820. [CrossRef] [PubMed]
45. Waak, J.; Weber, S.S.; Waldenmaier, A.; Gorner, K.; Alunni-Fabbroni, M.; Schell, H.; Vogt-Weisenhorn, D.; Pham, T.T.; Reumers, V.; Baekelandt, V.; et al. Regulation of astrocyte inflammatory responses by the Parkinson's disease-associated gene DJ-1. *FASEB J.* **2009**, *23*, 2478–2489. [CrossRef]
46. Choi, D.J.; Yang, H.; Gaire, S.; Lee, K.A.; An, J.; Kim, B.G.; Jou, I.; Park, S.M.; Joe, E.H. Critical roles of astrocytic-CCL2-dependent monocyte infiltration in a DJ-1 knockout mouse model of delayed brain repair. *Glia* **2020**, *68*, 2086–2101. [CrossRef]
47. Zhang, G.; Xia, Y.; Wan, F.; Ma, K.; Guo, X.; Kou, L.; Yin, S.; Han, C.; Liu, L.; Huang, J.; et al. New Perspectives on Roles of Alpha-Synuclein in Parkinson's Disease. *Front. Aging Neurosci.* **2018**, *10*, 370. [CrossRef]
48. Bonini, N.M.; Giasson, B.I. Snaring the function of alpha-synuclein. *Cell* **2005**, *123*, 359–361. [CrossRef]
49. Moore, S.A.; Yoder, E.; Murphy, S.; Dutton, G.R.; Spector, A.A. Astrocytes, not neurons, produce docosahexaenoic acid (22:6 omega-3) and arachidonic acid (20:4 omega-6). *J. Neurochem.* **1991**, *56*, 518–524. [CrossRef]
50. Solano, S.M.; Miller, D.W.; Augood, S.J.; Young, A.B.; Penney, J.B., Jr. Expression of alpha-synuclein, parkin, and ubiquitin carboxy-terminal hydrolase L1 mRNA in human brain: Genes associated with familial Parkinson's disease. *Ann. Neurol.* **2000**, *47*, 201–210. [CrossRef]
51. Alecu, I.; Bennett, S.A.L. Dysregulated Lipid Metabolism and Its Role in alpha-Synucleinopathy in Parkinson's Disease. *Front. Neurosci.* **2019**, *13*, 328. [CrossRef] [PubMed]
52. Booth, H.D.E.; Hirst, W.D.; Wade-Martins, R. The Role of Astrocyte Dysfunction in Parkinson's Disease Pathogenesis. *Trends Neurosci.* **2017**, *40*, 358–370. [CrossRef] [PubMed]
53. Castagnet, P.I.; Golovko, M.Y.; Barcelo-Coblijn, G.C.; Nussbaum, R.L.; Murphy, E.J. Fatty acid incorporation is decreased in astrocytes cultured from alpha-synuclein gene-ablated mice. *J. Neurochem.* **2005**, *94*, 839–849. [CrossRef]
54. Braak, H.; Sastre, M.; Del Tredici, K. Development of alpha-synuclein immunoreactive astrocytes in the forebrain parallels stages of intraneuronal pathology in sporadic Parkinson's disease. *Acta Neuropathol.* **2007**, *114*, 231–241. [CrossRef]
55. Braidy, N.; Gai, W.P.; Xu, Y.H.; Sachdev, P.; Guillemin, G.J.; Jiang, X.M.; Ballard, J.W.; Horan, M.P.; Fang, Z.M.; Chong, B.H.; et al. Uptake and mitochondrial dysfunction of alpha-synuclein in human astrocytes, cortical neurons and fibroblasts. *Transl. Neurodegener.* **2013**, *2*, 20. [CrossRef] [PubMed]

56. Lee, H.J.; Suk, J.E.; Patrick, C.; Bae, E.J.; Cho, J.H.; Rho, S.; Hwang, D.; Masliah, E.; Lee, S.J. Direct transfer of alpha-synuclein from neuron to astroglia causes inflammatory responses in synucleinopathies. *J. Biol. Chem.* **2010**, *285*, 9262–9272. [CrossRef]
57. Yang, Q.; Wang, Y.; Zhao, C.; Pang, S.; Lu, J.; Chan, P. Alpha-Synuclein aggregation causes muscle atrophy through neuromuscular junction degeneration. *J. Cachexia Sarcopenia Muscle* **2022**, *14*, 226–242. [CrossRef]
58. Tsunemi, T.; Ishiguro, Y.; Yoroisaka, A.; Valdez, C.; Miyamoto, K.; Ishikawa, K.; Saiki, S.; Akamatsu, W.; Hattori, N.; Krainc, D. Astrocytes Protect Human Dopaminergic Neurons from alpha-Synuclein Accumulation and Propagation. *J. Neurosci.* **2020**, *40*, 8618–8628. [CrossRef]
59. Ramirez, A.; Heimbach, A.; Grundemann, J.; Stiller, B.; Hampshire, D.; Cid, L.P.; Goebel, I.; Mubaidin, A.F.; Wriekat, A.L.; Roeper, J.; et al. Hereditary parkinsonism with dementia is caused by mutations in ATP13A2, encoding a lysosomal type 5 P-type ATPase. *Nat. Genet.* **2006**, *38*, 1184–1191. [CrossRef]
60. Park, J.S.; Blair, N.F.; Sue, C.M. The role of ATP13A2 in Parkinson's disease: Clinical phenotypes and molecular mechanisms. *Mov. Disord.* **2015**, *30*, 770–779. [CrossRef]
61. Dang, T.; Cao, W.J.; Zhao, R.; Lu, M.; Hu, G.; Qiao, C. ATP13A2 protects dopaminergic neurons in Parkinson's disease: From biology to pathology. *J. Biomed. Res.* **2022**, *36*, 98–108. [CrossRef]
62. Zimprich, A.; Biskup, S.; Leitner, P.; Lichtner, P.; Farrer, M.; Lincoln, S.; Kachergus, J.; Hulihan, M.; Uitti, R.J.; Calne, D.B.; et al. Mutations in LRRK2 cause autosomal-dominant parkinsonism with pleomorphic pathology. *Neuron* **2004**, *44*, 601–607. [CrossRef] [PubMed]
63. Paisan-Ruiz, C.; Jain, S.; Evans, E.W.; Gilks, W.P.; Simon, J.; van der Brug, M.; Lopez de Munain, A.; Aparicio, S.; Gil, A.M.; Khan, N.; et al. Cloning of the gene containing mutations that cause PARK8-linked Parkinson's disease. *Neuron* **2004**, *44*, 595–600. [CrossRef] [PubMed]
64. Herbst, S.; Gutierrez, M.G. LRRK2 in Infection: Friend or Foe? *ACS Infect. Dis.* **2019**, *5*, 809–815. [CrossRef] [PubMed]
65. Wallings, R.L.; Tansey, M.G. LRRK2 regulation of immune-pathways and inflammatory disease. *Biochem. Soc. Trans.* **2019**, *47*, 1581–1595. [CrossRef]
66. Madureira, M.; Connor-Robson, N.; Wade-Martins, R. LRRK2: Autophagy and Lysosomal Activity. *Front. Neurosci.* **2020**, *14*, 498. [CrossRef]
67. Giasson, B.I.; Van Deerlin, V.M. Mutations in LRRK2 as a cause of Parkinson's disease. *Neurosignals* **2008**, *16*, 99–105. [CrossRef]
68. Greggio, E.; Jain, S.; Kingsbury, A.; Bandopadhyay, R.; Lewis, P.; Kaganovich, A.; van der Brug, M.P.; Beilina, A.; Blackinton, J.; Thomas, K.J.; et al. Kinase activity is required for the toxic effects of mutant LRRK2/dardarin. *Neurobiol. Dis.* **2006**, *23*, 329–341. [CrossRef]
69. Di Maio, R.; Hoffman, E.K.; Rocha, E.M.; Keeney, M.T.; Sanders, L.H.; De Miranda, B.R.; Zharikov, A.; Van Laar, A.; Stepan, A.F.; Lanz, T.A.; et al. LRRK2 activation in idiopathic Parkinson's disease. *Sci. Trans. Med.* **2018**, *10*, eaar5429. [CrossRef]
70. Miklossy, J.; Arai, T.; Guo, J.P.; Klegeris, A.; Yu, S.; McGeer, E.G.; McGeer, P.L. LRRK2 expression in normal and pathologic human brain and in human cell lines. *J. Neuropathol. Exp. Neurol.* **2006**, *65*, 953–963. [CrossRef]
71. di Domenico, A.; Carola, G.; Calatayud, C.; Pons-Espinal, M.; Munoz, J.P.; Richaud-Patin, Y.; Fernandez-Carasa, I.; Gut, M.; Faella, A.; Parameswaran, J.; et al. Patient-Specific iPSC-Derived Astrocytes Contribute to Non-Cell-Autonomous Neurodegeneration in Parkinson's Disease. *Stem Cell Rep.* **2019**, *12*, 213–229. [CrossRef] [PubMed]
72. Streubel-Gallasch, L.; Giusti, V.; Sandre, M.; Tessari, I.; Plotegher, N.; Giusto, E.; Masato, A.; Iovino, L.; Battisti, I.; Arrigoni, G.; et al. Parkinson's Disease-Associated LRRK2 Interferes with Astrocyte-Mediated Alpha-Synuclein Clearance. *Mol. Neurobiol.* **2021**, *58*, 3119–3140. [CrossRef] [PubMed]
73. Pickles, S.; Vigie, P.; Youle, R.J. Mitophagy and Quality Control Mechanisms in Mitochondrial Maintenance. *Curr. Biol.* **2018**, *28*, R170–R185. [CrossRef] [PubMed]
74. Kawajiri, S.; Saiki, S.; Sato, S.; Hattori, N. Genetic mutations and functions of PINK1. *Trends Pharmacol. Sci.* **2011**, *32*, 573–580. [CrossRef] [PubMed]
75. Barodia, S.K.; McMeekin, L.J.; Creed, R.B.; Quinones, E.K.; Cowell, R.M.; Goldberg, M.S. PINK1 phosphorylates ubiquitin predominantly in astrocytes. *NPJ Park. Dis.* **2019**, *5*, 29. [CrossRef]
76. Sun, L.; Shen, R.; Agnihotri, S.K.; Chen, Y.; Huang, Z.; Bueler, H. Lack of PINK1 alters glia innate immune responses and enhances inflammation-induced, nitric oxide-mediated neuron death. *Sci. Rep.* **2018**, *8*, 383. [CrossRef]
77. Choi, I.; Choi, D.J.; Yang, H.; Woo, J.H.; Chang, M.Y.; Kim, J.Y.; Sun, W.; Park, S.M.; Jou, I.; Lee, S.H.; et al. PINK1 expression increases during brain development and stem cell differentiation, and affects the development of GFAP-positive astrocytes. *Mol. Brain* **2016**, *9*, 5. [CrossRef]
78. Olszewska, D.A.; Lynch, T. Will crystal parkin help in understanding the future of Parkinson's disease? *Front. Neurol.* **2015**, *6*, 35. [CrossRef]
79. Ledesma, M.D.; Galvan, C.; Hellias, B.; Dotti, C.; Jensen, P.H. Astrocytic but not neuronal increased expression and redistribution of parkin during unfolded protein stress. *J. Neurochem.* **2002**, *83*, 1431–1440. [CrossRef]
80. Russ, K.; Teku, G.; Bousset, L.; Redeker, V.; Piel, S.; Savchenko, E.; Pomeshchik, Y.; Savistchenko, J.; Stummann, T.C.; Azevedo, C.; et al. TNF-alpha and alpha-synuclein fibrils differently regulate human astrocyte immune reactivity and impair mitochondrial respiration. *Cell Rep.* **2021**, *34*, 108895. [CrossRef]
81. Pickrell, A.M.; Youle, R.J. The roles of PINK1, parkin, and mitochondrial fidelity in Parkinson's disease. *Neuron* **2015**, *85*, 257–273. [CrossRef] [PubMed]

82. Davis, C.H.; Kim, K.Y.; Bushong, E.A.; Mills, E.A.; Boassa, D.; Shih, T.; Kinebuchi, M.; Phan, S.; Zhou, Y.; Bihlmeyer, N.A.; et al. Transcellular degradation of axonal mitochondria. *Proc. Natl. Acad. Sci. USA* **2014**, *111*, 9633–9638. [CrossRef] [PubMed]
83. Morales, I.; Sanchez, A.; Puertas-Avendano, R.; Rodriguez-Sabate, C.; Perez-Barreto, A.; Rodriguez, M. Neuroglial transmitophagy and Parkinson's disease. *Glia* **2020**, *68*, 2277–2299. [CrossRef] [PubMed]
84. Youle, R.J. Mitochondria-Striking a balance between host and endosymbiont. *Science* **2019**, *365*, eaaw9855. [CrossRef] [PubMed]
85. Gkikas, I.; Palikaras, K.; Tavernarakis, N. The Role of Mitophagy in Innate Immunity. *Front. Immunol.* **2018**, *9*, 1283. [CrossRef]
86. Iyer, S.S.; He, Q.; Janczy, J.R.; Elliott, E.I.; Zhong, Z.; Olivier, A.K.; Sadler, J.J.; Knepper-Adrian, V.; Han, R.; Qiao, L.; et al. Mitochondrial cardiolipin is required for Nlrp3 inflammasome activation. *Immunity* **2013**, *39*, 311–323. [CrossRef]
87. Kahns, S.; Kalai, M.; Jakobsen, L.D.; Clark, B.F.; Vandenabeele, P.; Jensen, P.H. Caspase-1 and caspase-8 cleave and inactivate cellular parkin. *J. Biol. Chem.* **2003**, *278*, 23376–23380. [CrossRef]
88. Yu, M.; Zhang, K.; Qi, W.; Huang, Z.; Ye, J.; Ma, Y.; Liao, M.; Ning, Z. Expression pattern of NLRP3 and its related cytokines in the lung and brain of avian influenza virus H9N2 infected BALB/c mice. *Virol. J.* **2014**, *11*, 229. [CrossRef]
89. Neumann, J.; Bras, J.; Deas, E.; O'Sullivan, S.S.; Parkkinen, L.; Lachmann, R.H.; Li, A.; Holton, J.; Guerreiro, R.; Paudel, R.; et al. Glucocerebrosidase mutations in clinical and pathologically proven Parkinson's disease. *Brain* **2009**, *132*, 1783–1794. [CrossRef]
90. Cahoy, J.D.; Emery, B.; Kaushal, A.; Foo, L.C.; Zamanian, J.L.; Christopherson, K.S.; Xing, Y.; Lubischer, J.L.; Krieg, P.A.; Krupenko, S.A.; et al. A transcriptome database for astrocytes, neurons, and oligodendrocytes: A new resource for understanding brain development and function. *J. Neurosci.* **2008**, *28*, 264–278. [CrossRef]
91. Aflaki, E.; Stubblefield, B.K.; McGlinchey, R.P.; McMahon, B.; Ory, D.S.; Sidransky, E. A characterization of Gaucher iPS-derived astrocytes: Potential implications for Parkinson's disease. *Neurobiol. Dis.* **2020**, *134*, 104647. [CrossRef] [PubMed]
92. Osellame, L.D.; Rahim, A.A.; Hargreaves, I.P.; Gegg, M.E.; Richard-Londt, A.; Brandner, S.; Waddington, S.N.; Schapira, A.H.V.; Duchen, M.R. Mitochondria and quality control defects in a mouse model of Gaucher disease–links to Parkinson's disease. *Cell Metab.* **2013**, *17*, 941–953. [CrossRef] [PubMed]
93. Osellame, L.D.; Duchen, M.R. Defective quality control mechanisms and accumulation of damaged mitochondria link Gaucher and Parkinson diseases. *Autophagy* **2013**, *9*, 1633–1635. [CrossRef] [PubMed]
94. Sanyal, A.; DeAndrade, M.P.; Novis, H.S.; Lin, S.; Chang, J.; Lengacher, N.; Tomlinson, J.J.; Tansey, M.G.; LaVoie, M.J. Lysosome and Inflammatory Defects in GBA1-Mutant Astrocytes Are Normalized by LRRK2 Inhibition. *Mov. Disord.* **2020**, *35*, 760–773. [CrossRef]
95. Farfel-Becker, T.; Vitner, E.B.; Kelly, S.L.; Bame, J.R.; Duan, J.; Shinder, V.; Merrill, A.H., Jr.; Dobrenis, K.; Futerman, A.H. Neuronal accumulation of glucosylceramide in a mouse model of neuronopathic Gaucher disease leads to neurodegeneration. *Hum. Mol. Genet.* **2014**, *23*, 843–854. [CrossRef]
96. Ginns, E.I.; Mak, S.K.; Ko, N.; Karlgren, J.; Akbarian, S.; Chou, V.P.; Guo, Y.; Lim, A.; Samuelsson, S.; LaMarca, M.L.; et al. Neuroinflammation and alpha-synuclein accumulation in response to glucocerebrosidase deficiency are accompanied by synaptic dysfunction. *Mol. Genet. Metab.* **2014**, *111*, 152–162. [CrossRef]
97. Goker-Alpan, O.; Giasson, B.I.; Eblan, M.J.; Nguyen, J.; Hurtig, H.I.; Lee, V.M.; Trojanowski, J.Q.; Sidransky, E. Glucocerebrosidase mutations are an important risk factor for Lewy body disorders. *Neurology* **2006**, *67*, 908–910. [CrossRef]
98. Talifu, Z.; Liu, J.Y.; Pan, Y.Z.; Ke, H.; Zhang, C.J.; Xu, X.; Gao, F.; Yu, Y.; Du, L.J.; Li, J.J. In vivo astrocyte-to-neuron reprogramming for central nervous system regeneration: A narrative review. *Neural. Regen. Res.* **2023**, *18*, 750–755. [CrossRef]
99. Wei, Z.D.; Shetty, A.K. Treating Parkinson's disease by astrocyte reprogramming: Progress and challenges. *Sci. Adv.* **2021**, *7*, eabg3198. [CrossRef]
100. Kuter, K.; Olech, L.; Glowacka, U. Prolonged Dysfunction of Astrocytes and Activation of Microglia Accelerate Degeneration of Dopaminergic Neurons in the Rat Substantia Nigra and Block Compensation of Early Motor Dysfunction Induced by 6-OHDA. *Mol. Neurobiol.* **2018**, *55*, 3049–3066. [CrossRef]
101. Kato, H.; Araki, T.; Imai, Y.; Takahashi, A.; Itoyama, Y. Protection of dopaminergic neurons with a novel astrocyte modulating agent (R)-(−)-2-propyloctanoic acid (ONO-2506) in an MPTP-mouse model of Parkinson's disease. *J. Neurol. Sci.* **2003**, *208*, 9–15. [CrossRef]
102. Himeda, T.; Kadoguchi, N.; Kamiyama, Y.; Kato, H.; Maegawa, H.; Araki, T. Neuroprotective effect of arundic acid, an astrocyte-modulating agent, in mouse brain against MPTP (1-methyl-4-phenyl-1,2,3,6-tetrahydropyridine) neurotoxicity. *Neuropharmacology* **2006**, *50*, 329–344. [CrossRef] [PubMed]
103. Wei, Z.; Li, X.; Li, X.; Liu, Q.; Cheng, Y. Oxidative Stress in Parkinson's Disease: A Systematic Review and Meta-Analysis. *Front. Mol. Neurosci.* **2018**, *11*, 236. [CrossRef] [PubMed]
104. Asanuma, M.; Okumura-Torigoe, N.; Miyazaki, I.; Murakami, S.; Kitamura, Y.; Sendo, T. Region-Specific Neuroprotective Features of Astrocytes against Oxidative Stress Induced by 6-Hydroxydopamine. *Int. J. Mol. Sci.* **2019**, *20*, 598. [CrossRef]
105. Bhatia, T.N.; Pant, D.B.; Eckhoff, E.A.; Gongaware, R.N.; Do, T.; Hutchison, D.F.; Gleixner, A.M.; Leak, R.K. Astrocytes Do Not Forfeit Their Neuroprotective Roles After Surviving Intense Oxidative Stress. *Front. Mol. Neurosci.* **2019**, *12*, 87. [CrossRef]
106. Mullett, S.J.; Di Maio, R.; Greenamyre, J.T.; Hinkle, D.A. DJ-1 expression modulates astrocyte-mediated protection against neuronal oxidative stress. *J. Mol. Neurosci.* **2013**, *49*, 507–511. [CrossRef] [PubMed]
107. De Miranda, B.R.; Rocha, E.M.; Bai, Q.; El Ayadi, A.; Hinkle, D.; Burton, E.A.; Timothy Greenamyre, J. Astrocyte-specific DJ-1 overexpression protects against rotenone-induced neurotoxicity in a rat model of Parkinson's disease. *Neurobiol. Dis.* **2018**, *115*, 101–114. [CrossRef] [PubMed]

108. Olesen, B.T.; Clausen, J.; Vang, O. Characterization of the transcriptional profile in primary astrocytes after oxidative stress induced by Paraquat. *Neurotoxicology* **2008**, *29*, 13–21. [CrossRef] [PubMed]
109. Pistollato, F.; Canovas-Jorda, D.; Zagoura, D.; Bal-Price, A. Nrf2 pathway activation upon rotenone treatment in human iPSC-derived neural stem cells undergoing differentiation towards neurons and astrocytes. *Neurochem. Int.* **2017**, *108*, 457–471. [CrossRef]
110. Sundar Boyalla, S.; Barbara Victor, M.; Roemgens, A.; Beyer, C.; Arnold, S. Sex- and brain region-specific role of cytochrome c oxidase in 1-methyl-4-phenylpyridinium-mediated astrocyte vulnerability. *J. Neurosci. Res.* **2011**, *89*, 2068–2082. [CrossRef]
111. Swarnkar, S.; Singh, S.; Goswami, P.; Mathur, R.; Patro, I.K.; Nath, C. Astrocyte activation: A key step in rotenone induced cytotoxicity and DNA damage. *Neurochem. Res.* **2012**, *37*, 2178–2189. [CrossRef]
112. Rathinam, M.L.; Watts, L.T.; Narasimhan, M.; Riar, A.K.; Mahimainathan, L.; Henderson, G.I. Astrocyte mediated protection of fetal cerebral cortical neurons from rotenone and paraquat. *Environ. Toxicol. Pharmacol.* **2012**, *33*, 353–360. [CrossRef]
113. Chen, P.C.; Vargas, M.R.; Pani, A.K.; Smeyne, R.J.; Johnson, D.A.; Kan, Y.W.; Johnson, J.A. Nrf2-mediated neuroprotection in the MPTP mouse model of Parkinson's disease: Critical role for the astrocyte. *Proc. Natl. Acad. Sci. USA* **2009**, *106*, 2933–2938. [CrossRef]
114. Guo, Q.; Wang, B.; Wang, X.; Smith, W.W.; Zhu, Y.; Liu, Z. Activation of Nrf2 in Astrocytes Suppressed PD-Like Phenotypes via Antioxidant and Autophagy Pathways in Rat and Drosophila Models. *Cells* **2021**, *10*, 1850. [CrossRef]
115. Clements, C.M.; McNally, R.S.; Conti, B.J.; Mak, T.W.; Ting, J.P. DJ-1, a cancer- and Parkinson's disease-associated protein, stabilizes the antioxidant transcriptional master regulator Nrf2. *Proc. Natl. Acad. Sci. USA* **2006**, *103*, 15091–15096. [CrossRef]
116. Kim, H.S.; Son, T.G.; Park, H.R.; Lee, Y.; Jung, Y.; Ishigami, A.; Lee, J. Senescence marker protein 30 deficiency increases Parkinson's pathology by impairing astrocyte activation. *Neurobiol. Aging* **2013**, *34*, 1177–1183. [CrossRef]
117. McNaught, K.S.; Jenner, P. Altered glial function causes neuronal death and increases neuronal susceptibility to 1-methyl-4-phenylpyridinium- and 6-hydroxydopamine-induced toxicity in astrocytic/ventral mesencephalic co-cultures. *J. Neurochem.* **1999**, *73*, 2469–2476. [CrossRef]
118. McGeer, P.L.; Itagaki, S.; Boyes, B.E.; McGeer, E.G. Reactive microglia are positive for HLA-DR in the substantia nigra of Parkinson's and Alzheimer's disease brains. *Neurology* **1988**, *38*, 1285–1291. [CrossRef]
119. Long-Smith, C.M.; Sullivan, A.M.; Nolan, Y.M. The influence of microglia on the pathogenesis of Parkinson's disease. *Prog. Neurobiol.* **2009**, *89*, 277–287. [CrossRef]
120. Sun, H.; Liang, R.; Yang, B.; Zhou, Y.; Liu, M.; Fang, F.; Ding, J.; Fan, Y.; Hu, G. Aquaporin-4 mediates communication between astrocyte and microglia: Implications of neuroinflammation in experimental Parkinson's disease. *Neuroscience* **2016**, *317*, 65–75. [CrossRef]
121. Yang, Y.J.; Zhang, S.; Ding, J.H.; Zhou, F.; Hu, G. Iptakalim protects against MPP+-induced degeneration of dopaminergic neurons in association with astrocyte activation. *Int. J. Neuropsychopharmacol.* **2009**, *12*, 317–327. [CrossRef] [PubMed]
122. Zhang, J.X.; Zhou, K.G.; Yin, Y.X.; Jin, L.J.; Tong, W.F.; Guo, J.; Yu, L.H.; Ye, X.C.; Jiang, M. Mesencephalic astrocyte-derived neurotrophic factor (MANF) prevents the neuroinflammation induced dopaminergic neurodegeneration. *Exp. Gerontol.* **2022**, *171*, 112037. [CrossRef] [PubMed]
123. Michalovicz, L.T.; Kelly, K.A.; Vashishtha, S.; Ben-Hamo, R.; Efroni, S.; Miller, J.V.; Locker, A.R.; Sullivan, K.; Broderick, G.; Miller, D.B.; et al. Astrocyte-specific transcriptome analysis using the ALDH1L1 bacTRAP mouse reveals novel biomarkers of astrogliosis in response to neurotoxicity. *J. Neurochem.* **2019**, *150*, 420–440. [CrossRef] [PubMed]
124. Mori, T.; Town, T.; Tan, J.; Tateishi, N.; Asano, T. Modulation of Astrocytic Activation by Arundic Acid (ONO-2506) Mitigates Detrimental Effects of the Apolipoprotein E4 Isoform after Permanent Focal Ischemia in Apolipoprotein E Knock-in Mice. *J. Cereb. Blood Flow Metab.* **2005**, *25*, 748–762. [CrossRef]
125. Kato, H.; Kurosaki, R.; Oki, C.; Araki, T. Arundic acid, an astrocyte-modulating agent, protects dopaminergic neurons against MPTP neurotoxicity in mice. *Brain Res.* **2004**, *1030*, 66–73. [CrossRef]
126. Wang, H.M.; Zhang, T.; Li, Q.; Huang, J.K.; Chen, R.F.; Sun, X.J. Inhibition of glycogen synthase kinase-3beta by lithium chloride suppresses 6-hydroxydopamine-induced inflammatory response in primary cultured astrocytes. *Neurochem. Int.* **2013**, *63*, 345–353. [CrossRef]
127. Gupta, S.; Goswami, P.; Biswas, J.; Joshi, N.; Sharma, S.; Nath, C.; Singh, S. 6-Hydroxydopamine and lipopolysaccharides induced DNA damage in astrocytes: Involvement of nitric oxide and mitochondria. *Mutat. Res. Genet. Toxicol. Environ. Mutagen.* **2015**, *778*, 22–36. [CrossRef]
128. Zhang, J.; Culp, M.L.; Craver, J.G.; Darley-Usmar, V. Mitochondrial function and autophagy: Integrating proteotoxic, redox, and metabolic stress in Parkinson's disease. *J. Neurochem.* **2018**, *144*, 691–709. [CrossRef]
129. Mebratu, Y.A.; Negasi, Z.H.; Dutta, S.; Rojas-Quintero, J.; Tesfaigzi, Y. Adaptation of Proteasomes and Lysosomes to Cellular Environments. *Cells* **2020**, *9*, 2221. [CrossRef]
130. Lee, H.J.; Patel, S.; Lee, S.J. Intravesicular localization and exocytosis of alpha-synuclein and its aggregates. *J. Neurosci.* **2005**, *25*, 6016–6024. [CrossRef]
131. Deleidi, M.; Maetzler, W. Protein clearance mechanisms of alpha-synuclein and amyloid-Beta in lewy body disorders. *Int. J. Alzheimers Dis.* **2012**, *2012*, 391438. [CrossRef]
132. Morales, I.; Sanchez, A.; Rodriguez-Sabate, C.; Rodriguez, M. Striatal astrocytes engulf dopaminergic debris in Parkinson's disease: A study in an animal model. *PLoS ONE* **2017**, *12*, e0185989. [CrossRef] [PubMed]

133. Cavaliere, F.; Cerf, L.; Dehay, B.; Ramos-Gonzalez, P.; De Giorgi, F.; Bourdenx, M.; Bessede, A.; Obeso, J.A.; Matute, C.; Ichas, F.; et al. In vitro alpha-synuclein neurotoxicity and spreading among neurons and astrocytes using Lewy body extracts from Parkinson disease brains. *Neurobiol. Dis.* **2017**, *103*, 101–112. [CrossRef]
134. Morales, I.; Puertas-Avendano, R.; Sanchez, A.; Perez-Barreto, A.; Rodriguez-Sabate, C.; Rodriguez, M. Astrocytes and retrograde degeneration of nigrostriatal dopaminergic neurons in Parkinson's disease: Removing axonal debris. *Transl. Neurodegener.* **2021**, *10*, 43. [CrossRef] [PubMed]
135. Janda, E.; Lascala, A.; Carresi, C.; Parafati, M.; Aprigliano, S.; Russo, V.; Savoia, C.; Ziviani, E.; Musolino, V.; Morani, F.; et al. Parkinsonian toxin-induced oxidative stress inhibits basal autophagy in astrocytes via NQO2/quinone oxidoreductase 2: Implications for neuroprotection. *Autophagy* **2015**, *11*, 1063–1080. [CrossRef] [PubMed]
136. Zhao, Z.H.; Chen, Z.T.; Zhou, R.L.; Zhang, X.; Ye, Q.Y.; Wang, Y.Z. Increased DJ-1 and alpha-Synuclein in Plasma Neural-Derived Exosomes as Potential Markers for Parkinson's Disease. *Front. Aging Neurosci.* **2018**, *10*, 438. [CrossRef]
137. Rostami, J.; Holmqvist, S.; Lindstrom, V.; Sigvardson, J.; Westermark, G.T.; Ingelsson, M.; Bergstrom, J.; Roybon, L.; Erlandsson, A. Human Astrocytes Transfer Aggregated Alpha-Synuclein via Tunneling Nanotubes. *J. Neurosci.* **2017**, *37*, 11835–11853. [CrossRef]
138. van Veen, S.; Sorensen, D.M.; Holemans, T.; Holen, H.W.; Palmgren, M.G.; Vangheluwe, P. Cellular function and pathological role of ATP13A2 and related P-type transport ATPases in Parkinson's disease and other neurological disorders. *Front. Mol. Neurosci.* **2014**, *7*, 48. [CrossRef]
139. Tsunemi, T.; Krainc, D. Zn(2)(+) dyshomeostasis caused by loss of ATP13A2/PARK9 leads to lysosomal dysfunction and alpha-synuclein accumulation. *Hum. Mol. Genet.* **2014**, *23*, 2791–2801. [CrossRef]
140. Tsunemi, T.; Perez-Rosello, T.; Ishiguro, Y.; Yoroisaka, A.; Jeon, S.; Hamada, K.; Rammonhan, M.; Wong, Y.C.; Xie, Z.; Akamatsu, W.; et al. Increased Lysosomal Exocytosis Induced by Lysosomal Ca(2+) Channel Agonists Protects Human Dopaminergic Neurons from alpha-Synuclein Toxicity. *J. Neurosci.* **2019**, *39*, 5760–5772. [CrossRef]
141. Rocha, S.M.; Bantle, C.M.; Aboellail, T.; Chatterjee, D.; Smeyne, R.J.; Tjalkens, R.B. Rotenone induces regionally distinct alpha-synuclein protein aggregation and activation of glia prior to loss of dopaminergic neurons in C57Bl/6 mice. *Neurobiol. Dis.* **2022**, *167*, 105685. [CrossRef] [PubMed]
142. Wang, R.; Zhao, X.; Xu, J.; Wen, Y.; Li, A.; Lu, M.; Zhou, J. Astrocytic JWA deletion exacerbates dopaminergic neurodegeneration by decreasing glutamate transporters in mice. *Cell Death Dis.* **2018**, *9*, 352. [CrossRef] [PubMed]
143. Loschmann, P.A.; De Groote, C.; Smith, L.; Wullner, U.; Fischer, G.; Kemp, J.A.; Jenner, P.; Klockgether, T. Antiparkinsonian activity of Ro 25-6981, a NR2B subunit specific NMDA receptor antagonist, in animal models of Parkinson's disease. *Exp. Neurol.* **2004**, *187*, 86–93. [CrossRef]
144. Ferrarese, C.; Tremolizzo, L.; Rigoldi, M.; Sala, G.; Begni, B.; Brighina, L.; Ricci, G.; Albizzati, M.G.; Piolti, R.; Crosti, F.; et al. Decreased platelet glutamate uptake and genetic risk factors in patients with Parkinson's disease. *Neurol. Sci.* **2001**, *22*, 65–66. [CrossRef] [PubMed]
145. Iovino, L.; Giusti, V.; Pischedda, F.; Giusto, E.; Plotegher, N.; Marte, A.; Battisti, I.; Di Iacovo, A.; Marku, A.; Piccoli, G.; et al. Trafficking of the glutamate transporter is impaired in LRRK2-related Parkinson's disease. *Acta Neuropathol.* **2022**, *144*, 81–106. [CrossRef]
146. Zhang, Y.; He, X.; Meng, X.; Wu, X.; Tong, H.; Zhang, X.; Qu, S. Regulation of glutamate transporter trafficking by Nedd4-2 in a Parkinson's disease model. *Cell Death Dis.* **2017**, *8*, e2574. [CrossRef]
147. Chung, E.K.; Chen, L.W.; Chan, Y.S.; Yung, K.K. Downregulation of glial glutamate transporters after dopamine denervation in the striatum of 6-hydroxydopamine-lesioned rats. *J. Comp. Neurol.* **2008**, *511*, 421–437. [CrossRef]
148. Wei, L.; Chen, C.; Ding, L.; Mo, M.; Zou, J.; Lu, Z.; Li, H.; Wu, H.; Dai, Y.; Xu, P.; et al. Wnt1 Promotes EAAT2 Expression and Mediates the Protective Effects of Astrocytes on Dopaminergic Cells in Parkinson's Disease. *Neural. Plast.* **2019**, *2019*, 1247276. [CrossRef]
149. Xiao, N.; Le, Q.T. Neurotrophic Factors and Their Potential Applications in Tissue Regeneration. *Arch. Immunol. Et Ther. Exp.* **2016**, *64*, 89–99. [CrossRef]
150. Mogi, M.; Togari, A.; Kondo, T.; Mizuno, Y.; Komure, O.; Kuno, S.; Ichinose, H.; Nagatsu, T. Brain-derived growth factor and nerve growth factor concentrations are decreased in the substantia nigra in Parkinson's disease. *Neurosci. Lett.* **1999**, *270*, 45–48. [CrossRef]
151. Howells, D.W. Reduced BDNF mRNA expression in the Parkinson's disease substantia nigra. *Exp. Neurol.* **2000**, *166*, 127–135. [CrossRef]
152. Du, F.; Li, R.; Huang, Y.; Li, X.; Le, W. Dopamine D3 receptor-preferring agonists induce neurotrophic effects on mesencephalic dopamine neurons. *Eur. J. Neurosci.* **2005**, *22*, 2422–2430. [CrossRef]
153. Streifel, K.M.; Gonzales, A.L.; De Miranda, B.; Mouneimne, R.; Earley, S.; Tjalkens, R. Dopaminergic neurotoxicants cause biphasic inhibition of purinergic calcium signaling in astrocytes. *PLoS ONE* **2014**, *9*, e110996. [CrossRef]
154. Miyazaki, I.; Asanuma, M. Neuron-Astrocyte Interactions in Parkinson's Disease. *Cells* **2020**, *9*, 2623. [CrossRef]
155. Rangasamy, S.B.; Soderstrom, K.; Bakay, R.A.; Kordower, J.H. Neurotrophic factor therapy for Parkinson's disease. *Prog. Brain Res.* **2010**, *184*, 237–264. [CrossRef]
156. Yuan, L.J.; Zhang, M.; Chen, S.; Chen, W.F. Anti-inflammatory effect of IGF-1 is mediated by IGF-1R cross talk with GPER in MPTP/MPP(+)-induced astrocyte activation. *Mol. Cell Endocrinol.* **2021**, *519*, 111053. [CrossRef]

157. Datta, I.; Ganapathy, K.; Razdan, R.; Bhonde, R. Location and Number of Astrocytes Determine Dopaminergic Neuron Survival and Function Under 6-OHDA Stress Mediated Through Differential BDNF Release. *Mol. Neurobiol.* **2018**, *55*, 5505–5525. [CrossRef] [PubMed]
158. Sandhu, J.K.; Gardaneh, M.; Iwasiow, R.; Lanthier, P.; Gangaraju, S.; Ribecco-Lutkiewicz, M.; Tremblay, R.; Kiuchi, K.; Sikorska, M. Astrocyte-secreted GDNF and glutathione antioxidant system protect neurons against 6OHDA cytotoxicity. *Neurobiol. Dis.* **2009**, *33*, 405–414. [CrossRef]
159. Kordower, J.H.; Palfi, S.; Chen, E.Y.; Ma, S.Y.; Sendera, T.; Cochran, E.J.; Cochran, E.J.; Mufson, E.J.; Penn, R.; Goetz, C.G.; et al. Clinicopathological findings following intraventricular glial-derived neurotrophic factor treatment in a patient with Parkinson's disease. *Ann. Neurol.* **1999**, *46*, 419–424. [CrossRef] [PubMed]
160. Lang, A.E.; Gill, S.; Patel, N.K.; Lozano, A.; Nutt, J.G.; Penn, R.; Brooks, D.J.; Hotton, G.; Moro, E.; Heywood, P.; et al. Randomized controlled trial of intraputamenal glial cell line-derived neurotrophic factor infusion in Parkinson disease. *Ann. Neurol.* **2006**, *59*, 459–466. [CrossRef]
161. Petrova, P.; Raibekas, A.; Pevsner, J.; Vigo, N.; Anafi, M.; Moore, M.K.; Peaire, A.E.; Shridhar, V.; Smith, D.I.; Kelly, J.; et al. MANF: A new mesencephalic, astrocyte-derived neurotrophic factor with selectivity for dopaminergic neurons. *J. Mol. Neurosci.* **2003**, *20*, 173–188. [CrossRef] [PubMed]
162. Zhang, J.; Tong, W.; Sun, H.; Jiang, M.; Shen, Y.; Liu, Y.; Gu, H.; Guo, J.; Fang, J.; Jin, L. Nrf2-mediated neuroprotection by MANF against 6-OHDA-induced cell damage via PI3K/AKT/GSK3beta pathway. *Exp. Gerontol.* **2017**, *100*, 77–86. [CrossRef]
163. Huang, J.; Chen, C.; Gu, H.; Li, C.; Fu, X.; Jiang, M.; Sun, H.; Xu, J.; Fang, J.; Jin, L. Mesencephalic astrocyte-derived neurotrophic factor reduces cell apoptosis via upregulating GRP78 in SH-SY5Y cells. *Cell. Biol. Int.* **2016**, *40*, 803–811. [CrossRef]
164. Voutilainen, M.H.; Back, S.; Porsti, E.; Toppinen, L.; Lindgren, L.; Lindholm, P.; Peranen, J.; Saarma, M.; Tuominen, R.K. Mesencephalic astrocyte-derived neurotrophic factor is neurorestorative in rat model of Parkinson's disease. *J. Neurosci.* **2009**, *29*, 9651–9659. [CrossRef] [PubMed]
165. Estebanez, B.; de Paz, J.A.; Cuevas, M.J.; Gonzalez-Gallego, J. Endoplasmic Reticulum Unfolded Protein Response, Aging and Exercise: An Update. *Front. Physiol.* **2018**, *9*, 1744. [CrossRef]
166. Colla, E.; Coune, P.; Liu, Y.; Pletnikova, O.; Troncoso, J.C.; Iwatsubo, T.; Schneider, B.L.; Lee, M.K. Endoplasmic reticulum stress is important for the manifestations of alpha-synucleinopathy in vivo. *J. Neurosci.* **2012**, *32*, 3306–3320. [CrossRef]
167. Colla, E.; Jensen, P.H.; Pletnikova, O.; Troncoso, J.C.; Glabe, C.; Lee, M.K. Accumulation of toxic alpha-synuclein oligomer within endoplasmic reticulum occurs in alpha-synucleinopathy in vivo. *J. Neurosci.* **2012**, *32*, 3301–3305. [CrossRef]
168. Liu, M.; Qin, L.; Wang, L.; Tan, J.; Zhang, H.; Tang, J.; Shen, X.; Tan, L.; Wang, C. Alpha-synuclein induces apoptosis of astrocytes by causing dysfunction of the endoplasmic reticulum-Golgi compartment. *Mol. Med. Rep.* **2018**, *18*, 322–332. [CrossRef] [PubMed]
169. Lee, J.H.; Han, J.H.; Kim, H.; Park, S.M.; Joe, E.H.; Jou, I. Parkinson's disease-associated LRRK2-G2019S mutant acts through regulation of SERCA activity to control ER stress in astrocytes. *Acta Neuropathol. Commun.* **2019**, *7*, 68. [CrossRef] [PubMed]
170. Takuma, K.; Baba, A.; Matsuda, T. Astrocyte apoptosis: Implications for neuroprotection. *Prog. Neurobiol.* **2004**, *72*, 111–127. [CrossRef]
171. Liu, Y.; Zeng, X.; Hui, Y.; Zhu, C.; Wu, J.; Taylor, D.H.; Ji, J.; Fan, W.; Huang, Z.; Hu, J. Activation of alpha7 nicotinic acetylcholine receptors protects astrocytes against oxidative stress-induced apoptosis: Implications for Parkinson's disease. *Neuropharmacology* **2015**, *91*, 87–96. [CrossRef] [PubMed]
172. Paquet, M.; Ribeiro, F.M.; Guadagno, J.; Esseltine, J.L.; Ferguson, S.S.; Cregan, S.P. Role of metabotropic glutamate receptor 5 signaling and homer in oxygen glucose deprivation-mediated astrocyte apoptosis. *Mol. Brain* **2013**, *6*, 9. [CrossRef]
173. Zhang, S.; Liang, R.; Zhou, F.; Huang, X.; Ding, J.H.; Hu, G. Reversal of rotenone-induced dysfunction of astrocytic connexin43 by opening mitochondrial ATP-sensitive potassium channels. *Cell Mol. Neurobiol.* **2011**, *31*, 111–117. [CrossRef] [PubMed]
174. Nakase, T.; Naus, C.C. Gap junctions and neurological disorders of the central nervous system. *Biochim. Biophys. Acta* **2004**, *1662*, 149–158. [CrossRef]
175. Kielian, T. Glial connexins and gap junctions in CNS inflammation and disease. *J. Neurochem.* **2008**, *106*, 1000–1016. [CrossRef]
176. Dermietzel, R.; Gao, Y.; Scemes, E.; Vieira, D.; Urban, M.; Kremer, M.; Bennett, M.V.; Spray, D.C. Connexin43 null mice reveal that astrocytes express multiple connexins. *Brain Res. Brain Res. Rev.* **2000**, *32*, 45–56. [CrossRef]
177. Norden, D.M.; Trojanowski, P.J.; Villanueva, E.; Navarro, E.; Godbout, J.P. Sequential activation of microglia and astrocyte cytokine expression precedes increased Iba-1 or GFAP immunoreactivity following systemic immune challenge. *Glia* **2016**, *64*, 300–316. [CrossRef]
178. Martin, H.L.; Santoro, M.; Mustafa, S.; Riedel, G.; Forrester, J.V.; Teismann, P. Evidence for a role of adaptive immune response in the disease pathogenesis of the MPTP mouse model of Parkinson's disease. *Glia* **2016**, *64*, 386–395. [CrossRef]

cells

MDPI

Review

Astrocyte Signature in Alzheimer's Disease Continuum through a Multi-PET Tracer Imaging Perspective

Igor C. Fontana [1,†], Miriam Scarpa [1,†], Mona-Lisa Malarte [1], Filipa M. Rocha [1,2,3], Sira Auselle-Bosch [1,4], Marina Bluma [1], Marco Bucci [1], Konstantinos Chiotis [1], Amit Kumar [1] and Agneta Nordberg [1,5,*]

1 Division of Clinical Geriatrics, Center for Alzheimer Research, Department of Neurobiology, Care Sciences and Society, Karolinska Institutet, 141 52 Stockholm, Sweden
2 Instituto de Ciência Biomédicas Abel Salazar da Universidade do Porto, 4050-313 Porto, Portugal
3 Faculdade de Engenharia, Universidade do Porto, 4200-465 Porto, Portugal
4 Faculty of Health and Life Sciences, Pompeu Fabra University, 08003 Barcelona, Spain
5 Theme Inflammation and Aging, Karolinska University Hospital, 141 57 Stockholm, Sweden
* Correspondence: agneta.k.nordberg@ki.se
† These authors contributed equally to this work.

Abstract: Reactive astrogliosis is an early event in the continuum of Alzheimer's disease (AD). Current advances in positron emission tomography (PET) imaging provide ways of assessing reactive astrogliosis in the living brain. In this review, we revisit clinical PET imaging and in vitro findings using the multi-tracer approach, and point out that reactive astrogliosis precedes the deposition of Aβ plaques, tau pathology, and neurodegeneration in AD. Furthermore, considering the current view of reactive astrogliosis heterogeneity—more than one subtype of astrocyte involved—in AD, we discuss how astrocytic body fluid biomarkers might fit into trajectories different from that of astrocytic PET imaging. Future research focusing on the development of innovative astrocytic PET radiotracers and fluid biomarkers may provide further insights into the heterogeneity of reactive astrogliosis and improve the detection of AD in its early stages.

Keywords: reactive astrogliosis; positron emission tomography; Alzheimer's disease; L-deprenyl; BU99008; SMBT-1; GFAP

Citation: Fontana, I.C.; Scarpa, M.; Malarte, M.-L.; Rocha, F.M.; Auselle-Bosch, S.; Bluma, M.; Bucci, M.; Chiotis, K.; Kumar, A.; Nordberg, A. Astrocyte Signature in Alzheimer's Disease Continuum through a Multi-PET Tracer Imaging Perspective. *Cells* **2023**, *12*, 1469. https://doi.org/10.3390/cells12111469

Academic Editors: Michael Brenner and Vladimir Parpura

Received: 1 February 2023
Revised: 2 May 2023
Accepted: 19 May 2023
Published: 24 May 2023

1. Introduction

Based on the latest consensus, the definition of Alzheimer's disease (AD) comprises clinical phenotypes and a biological construct composed of amyloid-β (Aβ) deposition, pathological tau, and neurodegeneration—the AT(N) research framework [1,2]. In an attempt to improve AD detection, the spotlight has now turned to the early stages of the disease (i.e., prior to the deposition of Aβ plaques) [3]. In this regard, the supplementation of a new category of the AT(N) framework has been suggested—the ATX(N)—in which "X" represents potential innovative candidate biomarkers (body fluids or imaging) that could reveal early pathophysiological changes in the AD continuum [3]. A promising target for the early detection of AD, which for many years has been neglected due to neurocentric approaches, is the astrocyte [4]. Astrocytes are the most abundant glial cells in the human brain, playing crucial roles in brain energetic metabolism, neurotransmitter recycling and release, and redox balance, among other functions (reviewed in Semyanov and Verkhratsky, 2021 [5]). In response to pathological insults, astrocytes become reactive, a phenomenon termed reactive astrogliosis [6]. According to a recent consensus, reactive astrogliosis is defined as *"the process whereby, in response to pathology, astrocytes engage in molecularly defined programs involving changes in transcriptional regulation, as well as biochemical, morphological, metabolic, and physiological remodeling, which ultimately result in gain of new function(s) or loss or upregulation of homeostatic ones."* [6].

Biological changes in AD, including reactive astrogliosis, can be detected in vivo by measuring fluid biomarkers from the cerebrospinal fluid (CSF) and plasma, or using brain imaging tools. Among the imaging modalities, only positron emission tomography (PET) allows for a spatiotemporal investigation of pathophysiological processes in vivo [7] ^{11}C-Deuterium-L-deprenyl (^{11}C-DED) (often synthesized without deuterium as ^{11}C-L-deprenyl) is a well-established PET radiotracer used in AD research that targets monoamine oxidase B (MAO-B) for the imaging of astrocytes [4,8–21]. Recently, ^{18}F-SMBT-1 and ^{11}C-BU99008, which bind to MAO-B and I2-imidazoline binding site (I_2BS), respectively, have been developed as alternative tools to assess reactive astrogliosis in AD brains [9,22,23] (Table 1). The characteristics of each of these three PET radiotracers are discussed below.

Table 1. PET radiotracers for imaging reactive astrocytes in Alzheimer's disease.

PET Radiotracer	Target *	Ligand Characteristics
^{11}C-Deuterium-L-deprenyl (^{11}C-DED)	MAO-B	The deuterium-substituted form of L-deprenyl, radiolabeled with carbon-11, provides high selectivity and sensitivity for imaging MAO-B with a lower trapping rate, mitigating its irreversible binding nature [24].
^{18}F-SMBT-1	MAO-B	A highly selective MAO-B tracer, with low nonspecific binding, high entry into the brain, and reversible kinetics [25,26].
^{11}C-BU99008	I_2BS	Reversible binding properties, with good entry into the brain and highly specific and selective binding to I_2BS [27].

* MAO-B and I_2BS are overexpressed in the outer mitochondrial membrane of reactive astrocytes.

To define how reactive astrogliosis fits in the ATX(N) framework, it is crucial to investigate the relationship between this phenomenon and Aβ deposition, tau pathology, and neurodegeneration in the AD continuum. In this context, the concept of multi-tracer studies has been designed to explore the associations between different AD pathophysiological hallmarks. In vitro autoradiography studies in postmortem human brain are a valuable complement since they provide a platform to validate the multi-tracer approach, and to investigate in detail the binding properties and associations among different PET radiotracers. In this review, we discuss how the translational multi-tracer approach, in vivo and in vitro, with the different astrocytic PET radiotracers DED, SMBT-1, and BU99008, provide important new information to understand the complex signature of reactive astrogliosis, including heterogeneity and possible associations with AT(N)-biomarkers in AD. Our main goal is to highlight the significance of reactive astrogliosis as a target for the early, presymptomatic detection of AD and its progression. Furthermore, we provide initial hints on how molecular/functional changes assessed by astrocytic PET tracers could have a trajectory different from fluid glial fibrillary acidic protein (GFAP) levels in the early stages of AD.

2. Reactive Astrogliosis, We Can See You!

For decades, an increased level of GFAP, evaluated by immunohistochemistry, has been pointed out as a universal marker of reactive astrogliosis [28]. However, recent transcriptomic analyses indicate that sole reliance on GFAP is unlikely to define the broad range of reactive astrogliosis in AD [29]. Therefore, different tools to detect reactive astrogliosis are required to depict both the heterogeneity of astrocytes and their role in AD brains [29,30]. A wide range of proteins is differentially expressed in reactive astrogliosis, including receptors, transporters, and enzymes, with great potential to become surrogate biomarkers for developing PET radiotracers to detect heterogenous populations of astrocytes in the brain in vivo [20,31]. Among these, MAO-B, an enzyme present in the outer membrane of mitochondria, was initially shown to localize in GFAP-positive astrocytes [32]. Remarkably, additional studies indicated that the overexpression of MAO-B in response to Aβ pathology had no or limited correlation with increased GFAP levels, and instead GFAP levels increase at later stages [33]. These observations lead to the possibility that increases in MAO-B and GFAP potentially reflect two mechanisms of reactive astrogliosis and perhaps different populations of astrocytes. Thus, targeting MAO-B and GFAP overexpression may offer the possibility of exploring the heterogeneity of reactive astrogliosis in AD. These findings fostered the development of ^{11}C-DED, which is the gold-standard PET radiotracer for imaging reactive astrogliosis in the living brain. Clinical PET imaging with ^{11}C-DED showed reactive astrogliosis in different brain diseases such as epilepsy, Creutzfeldt–Jakob disease, amyotrophic lateral sclerosis, and AD [9,34–36]. In the context of AD, ^{11}C-DED PET has provided insightful information on early pathological changes in the continuum of the disease, which we describe in detail in the next section.

Recently, ^{18}F-SMBT-1, a fluorine-18 radiolabeled PET radiotracer selective for MAO-B, was developed [37]. The human evaluation of ^{18}F-SMBT-1 highlighted its reversible binding properties with high blood–brain barrier (BBB) permeability and low non-specific binding. In addition, ^{18}F-SMBT-1 binding was most pronounced in basal ganglia and cortical regions, with lower detection in the cerebellum and white matter, in accordance with the regional expression of MAO-B in the human brain. Therefore, ^{18}F-SMBT-1 stands out as an additional tool for detecting reactive astrogliosis in AD brains [25]. The PET radiotracer ^{11}C-BU99008 was developed to be a non-MAO-B selective tool for imaging reactive astrogliosis [23,27]. This PET radiotracer binds with high affinity to I_2BS in the outer mitochondrial membrane [38]. I_2BS is expressed in both neurons and astrocytes [39], but similar to MAO-B, the overexpression of I_2BS is mostly observed in astrocytes (assessed by co-localization with GFAP immunoreactivity) and associated with pathology [40]. ^{11}C-BU99008 is characterized by its good BBB penetration, reversible kinetics, great specificity and selectivity, and regional distribution, according to the expression of I_2BS in the human brain [27,38].

Recently, we have provided a very detailed comparison between DED, BU99008, and SMBT-1 binding in vitro using postmortem AD and cognitively normal (CN) brains. Initially, we demonstrated that ^{3}H-DED and ^{3}H-BU99008 have similar regional distribution with higher binding in the hippocampus and frontal cortex of AD brains compared to CN ones [41]. However, ^{3}H-DED and ^{3}H-BU99008 differ in their number of binding sites in CN and AD brains [41]. In contrast to ^{3}H-DED, which detects a single binding site, ^{3}H-BU99008 has multiple binding sites with a wide range of affinities [41]. In the same context, SMBT-1 appears to have a binding pattern similar to ^{3}H-DED; however, it also displaces ^{3}H-BU99008 at multiple binding sites [21]. This could be explained by the presence of I_2BS in the substrate entrance channel of MAO-B [42], whereby the binding of SMBT-1 to MAO-B potentially causes steric hinderance affecting the binding of ^{3}H-BU99008 to the I_2BS binding site. Therefore, even though PET radiotracers such as BU99008 are developed to selectively target I_2BS, it might not be possible to discriminate between changes in MAO-B and I_2BS in AD brains.

3. Amyloid-β Pathology—A Consequence or Trigger of Reactive Astrogliosis?

The Aβ peptide (ranging from 36 to 43 amino acids) is a product of the consecutive cleavage of a transmembrane protein termed amyloid precursor protein. Once released in the form of monomers, Aβ tends to aggregate into soluble oligomers, protofibrils, and fibrils and then forms deposits of insoluble Aβ plaques [43]. The deposition of insoluble Aβ can be detected with ^{11}C-Pittsburgh compound B (PiB) [44], ^{18}F-NAV4694 [45], and the FDA-approved ^{18}F-Florbetaben [46], ^{18}F-Florbetapir [47], and ^{18}F-Flutemetamol [48]. Clinical multi-tracer PET studies using amyloid- and astrocyte-PET radiotracers have established higher ^{11}C-DED binding in the frontal and parietal cortices—regions enriched in astrocytes—of mildly cognitive-impaired (MCI, also referred to as prodromal AD [2]) and sporadic AD patients (sAD) compared to cognitive normal (CN) individuals [8,9,12]. Interestingly, increased ^{11}C-DED binding was more evident in MCI patients who were ^{11}C-PiB-positive (i.e., have a considerable amount of Aβ plaque deposits or are amyloid positive, A+), indicating that reactive astrogliosis is associated with Aβ pathology in MCI A+ individuals (prodromal AD) [9]. These findings were crucial to confirm previous immunohistochemical reports indicating a link between reactive astrogliosis and amyloidosis in AD [28]. However, these initial studies could not clarify whether reactive astrogliosis is a response to, or an instigator of, Aβ pathology in the early stages of AD. In this context, clinical multi-PET tracer studies in autosomal dominant AD (ADAD) cases offer the possibility of investigating whether reactive astrogliosis precedes Aβ deposition. The carriers of these genetic mutations develop AD symptoms onset at a mutation-specific age, allowing the preclinical trajectories of different biomarkers to be tracked at an individual level [12,14]. Remarkably, increased ^{11}C-DED binding was observed in the presymptomatic stages of ADAD approximately 10 years before the expected onset of clinical symptoms, followed by a decline across the disease continuum (Figure 1) [12]. Conversely, Aβ deposition progressively increased from early to advanced stages of the disease, until it reached a plateau [12].

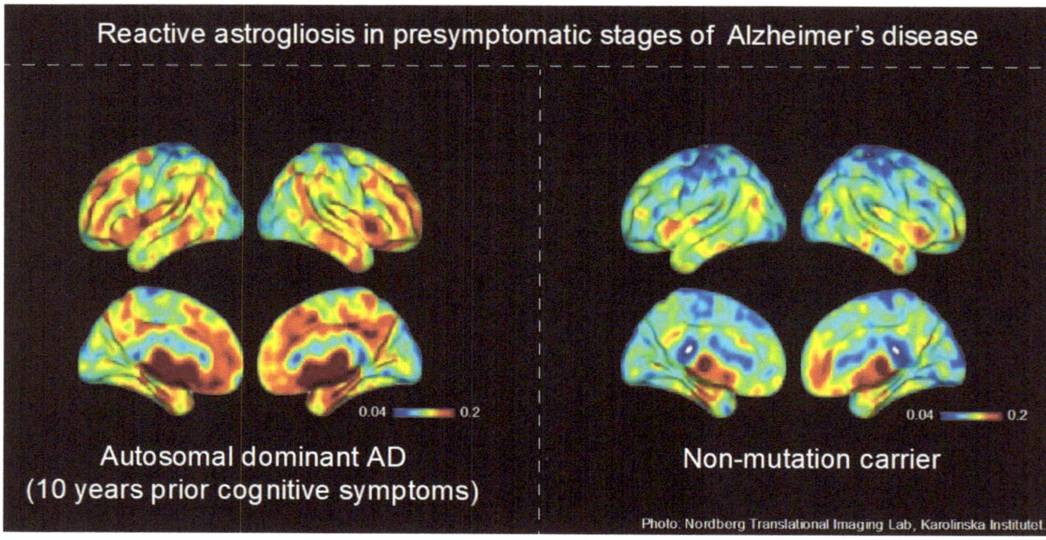

Figure 1. ^{11}C-L-Deprenyl binding in a presymptomatic autosomal dominant Alzheimer's disease (ADAD) carrier. In ADAD mutation carriers, increased ^{11}C-L-Deprenyl binding compared to mutation non-carriers is evident approximately 10 years before the presymptomatic stages. Figure credit: Nordberg Translational Imaging Lab, Karolinska Institutet.

The relationship between Aβ pathology and reactive astrogliosis has also been assessed with ^{18}F-SMBT-1 and ^{11}C-BU99008. Cross-sectional studies using ^{18}F-SMBT-1 demonstrated higher binding in multiple regions in AD brains compared to CN brains, including in the posterior cingulate gyrus, supramarginal gyrus, and lateral occipital lobe, and, to a lesser extent, in the hippocampus and globus pallidus [22]. In a multi-tracer study comparing CN, MCI, and sAD cases, reactive astrogliosis, assessed by ^{11}C-BU99008 binding, was evident in frontal, parietal, and occipital areas of Aβ-positive patients, where it positively correlated with Aβ plaque load measured by ^{18}F-Florbetaben PET [49]. Interestingly, ^{11}C-BU99008 binding was more pronounced in the cingulate, frontal, and temporal cortices of MCI compared to AD individuals [49].

As noted above, ^{11}CL-deprenyl binding is most prominent in the early stages of AD, prior to the deposition of Aβ plaques. Despite some limitations (including total number of AD subjects and mixing Aβ-positive and Aβ-negative individuals), findings on ^{11}C-BU99008 and ^{18}F-SMBT-1 provided initial hints that these PET-radiotracers follow similar patterns and could serve as surrogate markers of reactive astrogliosis. We suggest that pre-plaque soluble Aβ forms could be the main culprits that trigger reactive astrogliosis, which, in turn, may lead to the formation and spreading of Aβ in AD [50]. In a recent review article, we put forward a hypothesis in which we reported that the overexpression of α7 nicotinic acetylcholine receptors in reactive astrocytes may promote the formation of astrocytic Aβ plaques [51].

In vitro autoradiography studies using sAD and CN postmortem brain tissue support clinical research findings, pointing towards region-specific associations between reactive astrogliosis and Aβ pathology [10,18]. Comparative binding studies in sAD and CN tissues found that ^{3}H-DED had higher binding throughout the whole sAD hippocampus, whereas ^{3}H-PiB had low and uniform binding in this sAD brain region [10,18]. Conversely, in the frontal cortex, ^{3}H-PiB binding was high in all layers, while ^{3}H-DED showed a different pattern, with more pronounced binding in the superficial laminar sections [10,18]. Remarkably, GFAP levels determined by immunohistochemistry did not follow the same pattern as those of ^{3}H-DED binding in cortical regions. Instead, GFAP-positive cells were found uniformly distributed across the superficial cortical layers as well as in the cortical layer bordering the white matter and were concentrated around Aβ plaques in deeper layers [10]. The differences in the overexpression of GFAP and MAO-B (the latter demonstrated by ^{3}H-DED) observed in these results, together with studies in mouse models of disease [33], led to the hypothesis that increases in GFAP or MAO-B in AD reflect distinct subtypes of astrocytes displaying different responses to AD pathology, in a region-specific manner [10,13].

4. Pathological Tau and Reactive Astrogliosis

Tau is the main microtubule-associated protein in neuronal cells. The accumulation of hyperphosphorylated tau leads to the formation of neurofibrillary tangles, a key pathological hallmark of AD [52]. PET radiotracers that image tau pathology can be divided into first (^{18}F-AV-1451 (Flortaucipir), ^{18}F-THK5317, ^{18}F-THK5117, and ^{11}C-PBB3) [53] and second generation (^{18}F-MK6240, ^{18}F-PI2620, ^{18}F-RO948, ^{18}F-PM-PBB3, ^{18}F-APN-1607, ^{18}F-GTP1, and ^{18}F-JNJ31) [54]. To date, only ^{18}F-Flortaucipir has been FDA approved to target pathological tau for clinical diagnosis [55], but other tau tracers have been used in clinical research settings (following the required ethical approval). To our knowledge, only one clinical in vivo PET study has been conducted assessing astrocytic and tau PET radiotracers in the same set of individuals [22]. The authors demonstrated that ^{18}F-SMBT-1 binding positively correlates with ^{18}F-MK6240 in the temporoparietal cortex, and the supramarginal and posterior cingulate of patients across the AD continuum. This points towards a brain-region-specific association between reactive astrogliosis and tau pathology, although the degree of correlation is less than that of ^{18}F-SMBT-1 and amyloid-PET assessed by ^{18}F-NAV4694 in the same brain regions.

When our group assessed ^{3}H-THK5117 and ^{3}H-L-deprenyl binding in postmortem AD brains at advanced disease stages, we observed similar laminar distribution patterns [15]. Both ^{3}H-THK5117 and ^{3}H-L-deprenyl showed high binding in the superficial and deep layers of temporal cortices and diffuse binding throughout the middle frontal gyrus [15]. These observations are comparable with the immunostaining pattern of the GFAP antibody and AT8 antibody, which is specific for the phosphorylated form of tau [15].

In ADAD carriers of the ΔE9 mutation in the presenilin 1 gene (PSEN1ΔE9) or the Arctic mutation in the amyloid-beta precursor protein (APParc), ^{3}H-THK5117 and ^{3}H-DED displayed a similar laminar distribution (bilayer pattern) throughout the cortex, which was corroborated by GFAP and AT8 immunostaining [19]. Interestingly, a significant positive correlation between ^{3}H-DED and ^{3}H-THK5117 binding could only be observed in APParc brains, indicating there might be a closer relationship between tau and reactive astrocytosis in APParc brains than in PSEN1ΔE9 brains [19]. The APParc mutation is mostly characterized by fewer dense-core Aβ plaque deposits and more soluble Aβ forms than the PSEN1DE9 mutation. In this context, the stronger relationship between reactive astrogliosis and tau deposits in the APParc brains, compared to sAD and PSEN1ΔE9 brains [19], could be accounted for by higher exposure to soluble Aβ oligomers—the suggested top culprits of toxicity in AD pathophysiology—and by the fact that Aβ plaque pathology is not a requirement to trigger tau pathology and reactive astrogliosis.

5. Neurodegeneration and Reactive Astrogliosis: Two Separate Phenomena?

The fluorine-18 radiolabeled analogue of glucose—^{18}F-2-fluoro-2-deoxy-d-glucose (^{18}F-FDG)—PET radiotracer is a gold-standard biomarker to assess neurodegeneration in vivo. ^{18}F-FDG PET measures the rate of transport and trapping of the tracer, which has traditionally been attributed to neuronal glucose uptake, thus reflecting neuronal activity [56]. The first multi-tracer study that explored a possible link between neurodegeneration (evaluated with ^{18}F-FDG) and reactive astrogliosis (assessed with ^{11}C-DED) demonstrated no relationship between these two phenomena in either MCI or sAD individuals [9]. In a cross-sectional assessment of ADAD (presymptomatic and symptomatic), MCI, and sAD subjects, ^{18}F-FDG hypometabolism was mostly evident in the symptomatic/later stages, opposite to the trend of ^{11}C-DED and ^{11}C-PiB binding (higher since earlier stages) [12]. A subsequent study using a longitudinal approach demonstrated that ^{11}C-DED binding had significant positive associations with ^{18}F-FDG uptake in presymptomatic ADAD individuals across all brain regions of interest except in the hippocampus and frontal cortex—regions commonly affected by Aβ pathology—where no significant correlations were found [16]. These findings corroborate the view of astrocytes being contributors to the ^{18}F-FDG signal. In regions less affected by AD pathology, astrocytes may act in a compensatory way to maintain brain energetic homeostasis [57,58].

An exploratory analysis of ^{11}C-BU99008 in CN, MCI, and AD individuals demonstrated, on a voxel-wise basis, a correlation between reduced ^{18}F-FDG uptake and reduced ^{11}C-BU99008 binding in the parietal, frontal, and temporal lobes of ^{18}F-Florbetaben-positive subjects [49]. These findings were not statistically confirmed with a regional-based correlation analysis, thus blurring any conclusions regarding associations between reactive astrogliosis, assessed by ^{11}C-BU99008, and neurodegeneration [49]. Notably, no studies correlating SMBT-1 with neurodegeneration have been conducted so far.

Since neuronal and astrocytic ^{18}F-FDG uptake cannot be separately measured [59], alternative tools are required to investigate neurodegeneration in vivo. Moreover, FDG-PET is unlikely to reflect changes at the synaptic level (density, function, and structure). The recently developed novel PET radiotracer UCB-J, which targets the synaptic vesicle 2A (SV2A) as an index of synaptic density [60], might help define the potential link between reactive astrogliosis and neurodegeneration in AD [61]. Ongoing studies from our group have also shown that ^{3}H-UCB-J can target SV2A with high specificity in AD and CN brains [62]. Alternatively, the relationship between reactive astrogliosis and neurodegeneration can be explored using other imaging modalities such as magnetic resonance imaging (MRI). We have demonstrated a positive correlation between cortical microstructure and ^{11}C-DED PET binding in ADAD carriers [17], indicating that microstructural MRI changes may reflect reactive astrogliosis in the brain of these individuals. Furthermore, we have also reported that in the parahippocampus of MCI Aβ+ individuals, increased ^{11}C-DED binding correlated with grey matter (GM) density loss—a surrogate marker of brain atrophy assessed by MRI [11]. In this context, reactive astrogliosis might play a role in the structural brain changes in sAD. Overall, even though associations between ^{11}C-DED and ^{18}F-FDG uptake could only be observed in presymptomatic ADAD individuals, one cannot rule out that reactive astrogliosis may impair brain metabolism in the early stages of the disease, triggering downstream signaling pathways that might be associated with neurodegeneration in the AD dementia phase.

6. Relationship between Imaging and Fluid Astrocytic Biomarkers in AD

Apart from PET imaging, several CSF and blood biomarkers have been developed to track reactive astrogliosis heterogeneity. A few examples are GFAP, S100 calcium-binding protein B, aquaporin-4, and chitinase-3-like protein 1 (YKL-40) [20]. In fact, CSF levels of YKL-40 are associated with tau-PET [63], while GFAP (mainly plasma) correlated better with Aβ-PET [63–65]. This could indicate that (1) increases in CSF YKL-40 and plasma GFAP reflect distinct populations of astrocytes with unique responses to different AD-related pathologies, and (2) plasma GFAP could be a marker of Aβ pathology. In a cohort of Karolinska memory clinic patients, plasma GFAP levels correlated significantly with PET-detected amyloid deposits, but not with measure of pTau in CSF, whereas plasma pTau isoforms were associated with both measures of amyloid and tau pathology [66]. Nevertheless, the biological interpretation of these findings is still blurred and leaves a few questions. For instance, do higher GFAP levels in plasma reflect astrocytic responses originating in the central nervous system only? Or does peripheral GFAP also contribute to these changes (do chondrocytes, in a subpopulation of quiescent liver stellate cells, myoepithelial cells, and fibroblasts also express GFAP? [67])? Finally, most importantly, does plasma/CSF GFAP provide the same information as ^{11}C-DED PET?

When tracking reactive astrogliosis in ADAD and sAD individuals compared to healthy controls, we have observed a negative correlation between plasma GFAP levels and ^{11}C-L-deprenyl PET binding in vivo [68]. Furthermore, in a recent cross-sectional study on ADAD mutation carriers, plasma GFAP elevations emerge a decade before symptoms onset, prior to neurodegeneration, but after Aβ deposition commences [69,70]. In keeping with this, plasma GFAP seems to follow a different trajectory to ^{11}C-DED PET, which increases before the appearance of Aβ plaques. Hence, it might be possible that astrocytic PET imaging and plasma GFAP concentrations reflect independent aspects of reactive astrogliosis in AD, depicting different stages or subtypes of astrocytes. We hypothesize that ^{11}C-DED PET may illustrate a scenario in which 'Pre-plaque stage' soluble Aβ species trigger early molecular/functional changes in reactive astrocytes (characterized by MAO-B overexpression) and, with disease progression, Aβ aggregates into insoluble plaques (Aβ-plaque stage) and reactive astrocytes remodel their cytoskeleton (i.e., increase GFAP expression) and become hypertrophic. Why GFAP "leaks" into the blood is a question which needs further investigations.

7. Concluding Remarks

PET imaging plays a fundamental role not only in defining the biological construct of AD, but also in exploring the interplay between pathological hallmarks in AD progression. In this review, we revisited clinical biomarker findings and in vitro autoradiography studies providing spatiotemporal associations between reactive astrogliosis and Aβ, tau pathology, and neurodegeneration in order to evaluate the possibility that astrocytes contribute to the X in the ATX(N) research framework.

The spatiotemporal associations between reactive astrogliosis and other AD biomarkers seem to be complex. To shed light on this intricate relationship, we recently proposed a 'Two wave model of reactive astrogliosis' in the AD continuum, based on our multi-tracer clinical/translational studies ranging from in vitro to in vivo PET brain imaging [4]. In the early stages of AD, reactive astrogliosis is characterized by the increased expression of MAO-B (first wave of astrogliosis), which has a positive correlation with early Aβ deposition. The interplay between astrocytes and Aβ pathology at this stage is multifaceted. Since reactive astrogliosis was shown to precede significant build-up of Aβ plaque [12,23], we suggest that astrocytes could promote plaque formation, and/or that pre-plaque soluble Aβ (not detected with the current PET radiotracers) may trigger astrocyte dysfunction, for example, the abnormal production of reactive oxygen species, abnormal glutamate release, and hyperactivation of extrasynaptic NMDA receptors [50]. Across the continuum of the disease, as the prodromal stage progresses, in brain regions of advanced Aβ pathology (i.e., high amyloid-PET), astrocytes may become dystrophic, marking the end of the first wave of astrogliosis (i.e., ^{11}C-DED PET labeling declines) and the beginning of tau pathology detection (Figure 2).

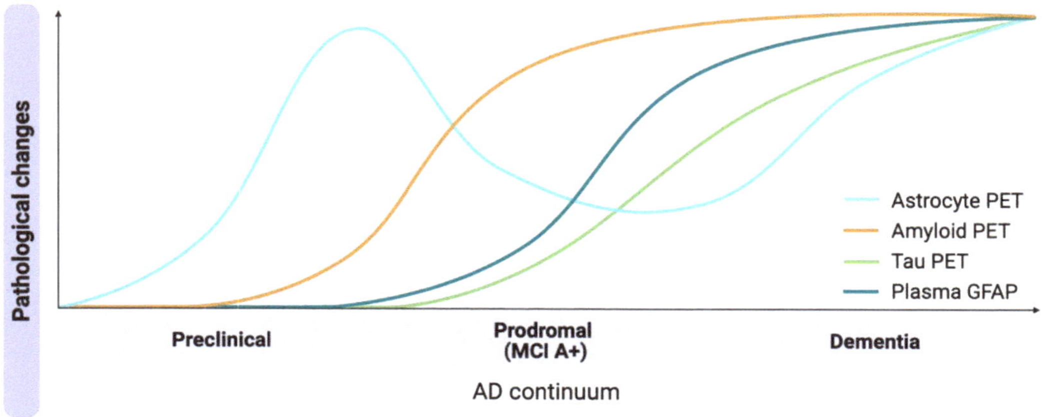

Figure 2. Reactive astrogliosis in the Alzheimer's disease (AD) continuum. The light blue curve illustrates increased astrocyte PET signal in preclinical AD assessed by ^{11}C-DED (or the version synthesized without deuterium, L-Deprenyl), a selective, irreversible monoamine oxidase type B, MAO-B, inhibitor. This is the first sign of reactive astrogliosis (i.e., the first wave). Following Aβ plaque deposition (orange curve), the astrocyte PET curve declines. As a consequence of Aβ deposition, GFAP levels increase in plasma (teal curve); however, further studies are required to define whether this phenomenon reflects reactive astrogliosis (a different state than that characterized by MAO-B overexpression as per ^{11}C-L-Deprenyl PET signal) or changes in peripheral cells that also express GFAP. Subsequently, tau pathology (green curve) and a second wave of reactive astrogliosis assessed by ^{11}C-L-Deprenyl can be detected, characterizing the later stages of AD (i.e., AD dementia). MCI A+, mild cognitive impairment with positive amyloid.

Although the mechanisms linking these three pathological hallmarks remain to be elucidated, it is possible that tau pathology could be exacerbated in the crosstalk between reactive astrocytes and activated microglial in response to soluble Aβ oligomers [71,72]. Positive associations between reactive astrogliosis and tau pathology, as well as with amyloid load at advanced stages of AD, can also be explained by the presence of resilient astrocytes (second wave of astrogliosis), which are resistant to pathological insults and cell death but functionally dormant. The involvement of astrocytes in neurodegeneration in later stages of AD remains unclear. Nevertheless, the positive correlation between ^{18}F-FDG levels and ^{11}C-DED binding in the presymptomatic phase of ADAD suggests that hyperactivity in brain energetic metabolism associated with reactive astrogliosis could be an initial trigger for neurodegeneration [57,58].

As the field of biomarker research moves at a fast pace, the combination of innovative astrocytic PET radiotracers and fluid biomarkers will provide further insights into the complexity of mechanisms associated with reactive astrogliosis across the continuum of AD, potentially shedding light on new tools for detecting the disease in the early stages.

Author Contributions: Conceptualization, I.C.F., M.S., A.K. and A.N.; writing—original draft preparation, I.C.F., M.S., M.-L.M., F.M.R., M.B. (Marina Bluma), M.B. (Marco Bucci), K.C., A.K.; writing—review and editing, I.C.F., M.S., A.K. and A.N.; visualization, S.A.-B. and I.C.F.; supervision, A.K. and A.N. All authors have read and agreed to the published version of the manuscript.

Funding: The financial support by the Swedish Research Council (projects 05817, 02965, 06086, 2017-02965, 2017-06086, 2020-01990), the Swedish Foundation for Strategic Research (SSF), the Stockholm County Council-Karolinska Institutet regional agreement on medical training and clinical research (ALF grant), the Swedish Brain Foundation, the Center for Innovative Medicine (CIMED) at Karolinska Institutet Stockholm Region, Sweden, Magnus Bergvall's Foundation, Tore Nilsons Foundation for Medical Research, Åhlens Foundation, The Recherche sur Alzheimer Foundation

(Paris, France) the Swedish Alzheimer's Foundation, the Foundation for Geriatric Diseases, the Loo and Hans Osterman Foundation for Medical Research at Karolinska Institutet, the Swedish Dementia Foundation, Alzheimer's Association (AARF-21-848395), and the Michael J Fox foundation are highly acknowledged, the Olle Engkvist Foundation (207-0584), the Foundation for Old Servants, Gun and Bertil Stohne's Foundation.

Data Availability Statement: Not applicable.

Acknowledgments: I.C.F. is employed full time by the Alzheimer's Association.

Conflicts of Interest: The authors declare no conflict of interest.

References

1. Jack, C.R., Jr.; Bennett, D.A.; Blennow, K.; Carrillo, M.C.; Dunn, B.; Haeberlein, S.B.; Holtzman, D.M.; Jagust, W.; Jessen, F.; Karlawish, J.; et al. NIA-AA Research Framework: Toward a biological definition of Alzheimer's disease. *Alzheimer's Dement.* **2018**, *14*, 535–562. [CrossRef]
2. Dubois, B.; Villain, N.; Frisoni, G.B.; Rabinovici, G.D.; Sabbagh, M.; Cappa, S.; Bejanin, A.; Bombois, S.; Epelbaum, S.; Teichmann, M.; et al. Clinical diagnosis of Alzheimer's disease: Recommendations of the International Working Group. *Lancet Neurol.* **2021**, *20*, 484–496. [CrossRef]
3. Hampel, H.; Cummings, J.; Blennow, K.; Gao, P.; Jack, C.R., Jr.; Vergallo, A. Developing the ATX(N) classification for use across the Alzheimer disease continuum. *Nat. Rev. Neurol.* **2021**, *17*, 580–589. [CrossRef] [PubMed]
4. Kumar, A.; Fontana, I.C.; Nordberg, A. Reactive astrogliosis: A friend or foe in the pathogenesis of Alzheimer's disease. *J. Neurochem.* **2021**, *164*, 309–324. [CrossRef] [PubMed]
5. Semyanov, A.; Verkhratsky, A. Astrocytic processes: From tripartite synapses to the active milieu. *Trends Neurosci.* **2021**, *44*, 781–792. [CrossRef]
6. Escartin, C.; Galea, E.; Lakatos, A.; O'Callaghan, J.P.; Petzold, G.C.; Serrano-Pozo, A.; Steinhauser, C.; Volterra, A.; Carmignoto, G.; Agarwal, A.; et al. Reactive astrocyte nomenclature, definitions, and future directions. *Nat. Neurosci.* **2021**, *24*, 312–325. [CrossRef] [PubMed]
7. Shegani, A.; Kealey, S.; Luzi, F.; Basagni, F.; Machado, J.D.M.; Ekici, S.D.; Ferocino, A.; Gee, A.D.; Bongarzone, S. Radiosynthesis, Preclinical, and Clinical Positron Emission Tomography Studies of Carbon-11 Labeled Endogenous and Natural Exogenous Compounds. *Chem. Rev.* **2022**, *123*, 105–229. [CrossRef]
8. Kadir, A.; Marutle, A.; Gonzalez, D.; Scholl, M.; Almkvist, O.; Mousavi, M.; Mustafiz, T.; Darreh-Shori, T.; Nennesmo, I.; Nordberg, A. Positron emission tomography imaging and clinical progression in relation to molecular pathology in the first Pittsburgh Compound B positron emission tomography patient with Alzheimer's disease. *Brain* **2011**, *134*, 301–317. [CrossRef]
9. Carter, S.F.; Scholl, M.; Almkvist, O.; Wall, A.; Engler, H.; Langstrom, B.; Nordberg, A. Evidence for astrocytosis in prodromal Alzheimer disease provided by 11C-deuterium-L-deprenyl: A multitracer PET paradigm combining 11C-Pittsburgh compound B and 18F-FDG. *J. Nucl. Med.* **2012**, *53*, 37–46. [CrossRef]
10. Marutle, A.; Gillberg, P.G.; Bergfors, A.; Yu, W.; Ni, R.; Nennesmo, I.; Voytenko, L.; Nordberg, A. (3)H-deprenyl and (3)H-PIB autoradiography show different laminar distributions of astroglia and fibrillar beta-amyloid in Alzheimer brain. *J. Neuroinflammation* **2013**, *10*, 90. [CrossRef]
11. Choo, I.L.; Carter, S.F.; Scholl, M.L.; Nordberg, A. Astrocytosis measured by (1)(1)C-deprenyl PET correlates with decrease in gray matter density in the parahippocampus of prodromal Alzheimer's patients. *Eur. J. Nucl. Med. Mol. Imaging* **2014**, *41*, 2120–2126. [CrossRef]
12. Scholl, M.; Carter, S.F.; Westman, E.; Rodriguez-Vieitez, E.; Almkvist, O.; Thordardottir, S.; Wall, A.; Graff, C.; Langstrom, B.; Nordberg, A. Early astrocytosis in autosomal dominant Alzheimer's disease measured in vivo by multi-tracer positron emission tomography. *Sci. Rep.* **2015**, *5*, 16404. [CrossRef] [PubMed]
13. Rodriguez-Vieitez, E.; Ni, R.; Gulyas, B.; Toth, M.; Haggkvist, J.; Halldin, C.; Voytenko, L.; Marutle, A.; Nordberg, A. Astrocytosis precedes amyloid plaque deposition in Alzheimer APPswe transgenic mouse brain: A correlative positron emission tomography and in vitro imaging study. *Eur. J. Nucl. Med. Mol. Imaging* **2015**, *42*, 1119–1132. [CrossRef]
14. Rodriguez-Vieitez, E.; Saint-Aubert, L.; Carter, S.F.; Almkvist, O.; Farid, K.; Scholl, M.; Chiotis, K.; Thordardottir, S.; Graff, C.; Wall, A.; et al. Diverging longitudinal changes in astrocytosis and amyloid PET in autosomal dominant Alzheimer's disease. *Brain* **2016**, *139*, 922–936. [CrossRef] [PubMed]
15. Lemoine, L.; Gillberg, P.G.; Svedberg, M.; Stepanov, V.; Jia, Z.; Huang, J.; Nag, S.; Tian, H.; Ghetti, B.; Okamura, N.; et al. Comparative binding properties of the tau PET tracers THK5117, THK5351, PBB3, and T807 in postmortem Alzheimer brains. *Alzheimer's Res. Ther.* **2017**, *9*, 96. [CrossRef]
16. Carter, S.F.; Chiotis, K.; Nordberg, A.; Rodriguez-Vieitez, E. Longitudinal association between astrocyte function and glucose metabolism in autosomal dominant Alzheimer's disease. *Eur. J. Nucl. Med. Mol. Imaging* **2019**, *46*, 348–356. [CrossRef]
17. Vilaplana, E.; Rodriguez-Vieitez, E.; Ferreira, D.; Montal, V.; Almkvist, O.; Wall, A.; Lleo, A.; Westman, E.; Graff, C.; Fortea, J.; et al. Cortical microstructural correlates of astrocytosis in autosomal-dominant Alzheimer disease. *Neurology* **2020**, *94*, e2026–e2036. [CrossRef]

18. Ni, R.; Rojdner, J.; Voytenko, L.; Dyrks, T.; Thiele, A.; Marutle, A.; Nordberg, A. In vitro Characterization of the Regional Binding Distribution of Amyloid PET Tracer Florbetaben and the Glia Tracers Deprenyl and PK11195 in Autopsy Alzheimer's Brain Tissue. *J. Alzheimer's Dis.* **2021**, *80*, 1723–1737. [CrossRef]
19. Lemoine, L.; Gillberg, P.G.; Bogdanovic, N.; Nennesmo, I.; Saint-Aubert, L.; Viitanen, M.; Graff, C.; Ingelsson, M.; Nordberg, A. Amyloid, tau, and astrocyte pathology in autosomal-dominant Alzheimer's disease variants: AbetaPParc and PSEN1DE9 *Mol. Psychiatry* **2021**, *26*, 5609–5619. [CrossRef]
20. Bellaver, B.; Ferrari-Souza, J.P.; Uglione da Ros, L.; Carter, S.F.; Rodriguez-Vieitez, E.; Nordberg, A.; Pellerin, L.; Rosa-Neto, P.; Leffa, D.T.; Zimmer, E.R. Astrocyte Biomarkers in Alzheimer Disease: A Systematic Review and Meta-analysis. *Neurology* **2021**, *96*, e2944–e2955. [CrossRef] [PubMed]
21. Fontana, I.C.; Kumar, A.; Okamura, N.; Nordberg, A.K. PET tracer SMBT-1 discriminates between BU99008 and Deprenyl binding sites on reactive astrocytes in Alzheimer's disease brains. *Alzheimer's Dement.* **2022**, *18*, e066484. [CrossRef]
22. Villemagne, V.L.; Harada, R.; Dore, V.; Furumoto, S.; Mulligan, R.; Kudo, Y.; Burnham, S.; Krishnadas, N.; Bourgeat, P.; Xia, Y.; et al. Assessing reactive astrogliosis with (18)F-SMBT-1 across the Alzheimer's disease spectrum. *J. Nucl. Med.* **2022**, *63*, 1560–1569. [CrossRef] [PubMed]
23. Calsolaro, V.; Matthews, P.M.; Donat, C.K.; Livingston, N.R.; Femminella, G.D.; Guedes, S.S.; Myers, J.; Fan, Z.; Tyacke, R.J.; Venkataraman, A.V.; et al. Astrocyte reactivity with late-onset cognitive impairment assessed in vivo using (11)C-BU99008 PET and its relationship with amyloid load. *Mol. Psychiatry* **2021**, *26*, 5848–5855. [CrossRef] [PubMed]
24. Fowler, J.S.; MacGregor, R.R.; Wolf, A.P.; Arnett, C.D.; Dewey, S.L.; Schlyer, D.; Christman, D.; Logan, J.; Smith, M.; Sachs, H.; et al. Mapping human brain monoamine oxidase A and B with 11C-labeled suicide inactivators and PET. *Science* **1987**, *235*, 481–485. [CrossRef]
25. Villemagne, V.L.; Harada, R.; Dore, V.; Furumoto, S.; Mulligan, R.; Kudo, Y.; Burnham, S.; Krishnadas, N.; Bozinovski, S.; Huang, K.; et al. First-in-human evaluation of (18)F-SMBT-1, a novel (18)F-labeled MAO-B PET tracer for imaging reactive astrogliosis. *J. Nucl. Med.* **2022**, *63*, 1551–1559. [CrossRef]
26. Harada, R.; Shimizu, Y.; Du, Y.; Ishikawa, Y.; Iwata, R.; Kudo, Y.; Yanai, K.; Okamura, N.; Furumoto, S. The Role of Chirality of [(18)F]SMBT-1 in Imaging of Monoamine Oxidase-B. *ACS Chem. Neurosci.* **2022**, *13*, 322–329. [CrossRef]
27. Tyacke, R.J.; Myers, J.F.M.; Venkataraman, A.; Mick, I.; Turton, S.; Passchier, J.; Husbands, S.M.; Rabiner, E.A.; Gunn, R.N.; Murphy, P.S.; et al. Evaluation of (11)C-BU99008, a PET Ligand for the Imidazoline2 Binding Site in Human Brain. *J. Nucl. Med.* **2018**, *59*, 1597–1602. [CrossRef]
28. Serrano-Pozo, A.; Frosch, M.P.; Masliah, E.; Hyman, B.T. Neuropathological alterations in Alzheimer disease. *Cold Spring Harb. Perspect. Med.* **2011**, *1*, a006189. [CrossRef]
29. Sadick, J.S.; O'Dea, M.R.; Hasel, P.; Dykstra, T.; Faustin, A.; Liddelow, S.A. Astrocytes and oligodendrocytes undergo subtype-specific transcriptional changes in Alzheimer's disease. *Neuron* **2022**, *110*, 1788–1805.e10. [CrossRef]
30. Burda, J.E.; O'Shea, T.M.; Ao, Y.; Suresh, K.B.; Wang, S.; Bernstein, A.M.; Chandra, A.; Deverasetty, S.; Kawaguchi, R.; Kim, J.H.; et al. Divergent transcriptional regulation of astrocyte reactivity across disorders. *Nature* **2022**, *606*, 557–564. [CrossRef]
31. Carter, S.F.; Herholz, K.; Rosa-Neto, P.; Pellerin, L.; Nordberg, A.; Zimmer, E.R. Astrocyte Biomarkers in Alzheimer's Disease. *Trends Mol. Med.* **2019**, *25*, 77–95. [CrossRef]
32. Ekblom, J.; Jossan, S.S.; Bergstrom, M.; Oreland, L.; Walum, E.; Aquilonius, S.M. Monoamine oxidase-B in astrocytes. *Glia* **1993**, *8*, 122–132. [CrossRef] [PubMed]
33. Olsen, M.; Aguilar, X.; Sehlin, D.; Fang, X.T.; Antoni, G.; Erlandsson, A.; Syvanen, S. Astroglial Responses to Amyloid-Beta Progression in a Mouse Model of Alzheimer's Disease. *Mol. Imaging Biol.* **2018**, *20*, 605–614. [CrossRef] [PubMed]
34. Engler, H.; Lundberg, P.O.; Ekbom, K.; Nennesmo, I.; Nilsson, A.; Bergstrom, M.; Tsukada, H.; Hartvig, P.; Langstrom, B. Multitracer study with positron emission tomography in Creutzfeldt-Jakob disease. *Eur. J. Nucl. Med. Mol. Imaging* **2003**, *30*, 85–95. [CrossRef] [PubMed]
35. Johansson, A.; Engler, H.; Blomquist, G.; Scott, B.; Wall, A.; Aquilonius, S.M.; Langstrom, B.; Askmark, H. Evidence for astrocytosis in ALS demonstrated by [11C](L)-deprenyl-D2 PET. *J. Neurol. Sci.* **2007**, *255*, 17–22. [CrossRef]
36. Kumlien, E.; Nilsson, A.; Hagberg, G.; Langstrom, B.; Bergstrom, M. PET with 11C-deuterium-deprenyl and 18F-FDG in focal epilepsy. *Acta Neurol. Scand.* **2001**, *103*, 360–366. [CrossRef]
37. Harada, R.; Hayakawa, Y.; Ezura, M.; Lerdsirisuk, P.; Du, Y.; Ishikawa, Y.; Iwata, R.; Shidahara, M.; Ishiki, A.; Kikuchi, A.; et al. (18)F-SMBT-1: A Selective and Reversible PET Tracer for Monoamine Oxidase-B Imaging. *J. Nucl. Med.* **2021**, *62*, 253–258. [CrossRef] [PubMed]
38. Tyacke, R.J.; Fisher, A.; Robinson, E.S.; Grundt, P.; Turner, E.M.; Husbands, S.M.; Hudson, A.L.; Parker, C.A.; Nutt, D.J. Evaluation and initial in vitro and ex vivo characterization of the potential positron emission tomography ligand, BU99008 (2-(4,5-dihydro-1H-imidazol-2-yl)-1- methyl-1H-indole), for the imidazoline(2) binding site. *Synapse* **2012**, *66*, 542–551. [CrossRef]
39. Keller, B.; Garcia-Sevilla, J.A. Immunodetection and subcellular distribution of imidazoline receptor proteins with three antibodies in mouse and human brains: Effects of treatments with I1- and I2-imidazoline drugs. *J. Psychopharmacol.* **2015**, *29*, 996–1012. [CrossRef]
40. Garcia-Sevilla, J.A.; Escriba, P.V.; Walzer, C.; Bouras, C.; Guimon, J. Imidazoline receptor proteins in brains of patients with Alzheimer's disease. *Neurosci. Lett.* **1998**, *247*, 95–98. [CrossRef]

41. Kumar, A.; Koistinen, N.A.; Malarte, M.L.; Nennesmo, I.; Ingelsson, M.; Ghetti, B.; Lemoine, L.; Nordberg, A. Astroglial tracer BU99008 detects multiple binding sites in Alzheimer's disease brain. *Mol. Psychiatry* **2021**, *26*, 5833–5847. [CrossRef] [PubMed]
42. McDonald, G.R.; Olivieri, A.; Ramsay, R.R.; Holt, A. On the formation and nature of the imidazoline I2 binding site on human monoamine oxidase-B. *Pharmacol. Res.* **2010**, *62*, 475–488. [CrossRef] [PubMed]
43. Hardy, J.A.; Higgins, G.A. Alzheimer's disease: The amyloid cascade hypothesis. *Science* **1992**, *256*, 184–185. [CrossRef]
44. Klunk, W.E.; Engler, H.; Nordberg, A.; Wang, Y.; Blomqvist, G.; Holt, D.P.; Bergstrom, M.; Savitcheva, I.; Huang, G.F.; Estrada, S.; et al. Imaging brain amyloid in Alzheimer's disease with Pittsburgh Compound-B. *Ann. Neurol.* **2004**, *55*, 306–319. [CrossRef]
45. Rowe, C.C.; Jones, G.; Dore, V.; Pejoska, S.; Margison, L.; Mulligan, R.S.; Chan, J.G.; Young, K.; Villemagne, V.L. Standardized Expression of 18F-NAV4694 and 11C-PiB beta-Amyloid PET Results with the Centiloid Scale. *J. Nucl. Med.* **2016**, *57*, 1233–1237. [CrossRef] [PubMed]
46. Martinez, G.; Vernooij, R.W.; Fuentes Padilla, P.; Zamora, J.; Flicker, L.; Bonfill Cosp, X. 18F PET with florbetaben for the early diagnosis of Alzheimer's disease dementia and other dementias in people with mild cognitive impairment (MCI). *Cochrane Database Syst. Rev.* **2017**, *11*, CD012883. [CrossRef]
47. Yang, L.; Rieves, D.; Ganley, C. Brain amyloid imaging—FDA approval of florbetapir F18 injection. *N. Engl. J. Med.* **2012**, *367*, 885–887. [CrossRef]
48. Rowe, C.C.; Ackerman, U.; Browne, W.; Mulligan, R.; Pike, K.L.; O'Keefe, G.; Tochon-Danguy, H.; Chan, G.; Berlangieri, S.U.; Jones, G.; et al. Imaging of amyloid beta in Alzheimer's disease with 18F-BAY94-9172, a novel PET tracer: Proof of mechanism. *Lancet Neurol.* **2008**, *7*, 129–135. [CrossRef]
49. Livingston, N.R.; Calsolaro, V.; Hinz, R.; Nowell, J.; Raza, S.; Gentleman, S.; Tyacke, R.J.; Myers, J.; Venkataraman, A.V.; Perneczky, R.; et al. Relationship between astrocyte reactivity, using novel (11)C-BU99008 PET, and glucose metabolism, grey matter volume and amyloid load in cognitively impaired individuals. *Mol. Psychiatry* **2022**, *27*, 2019–2029. [CrossRef]
50. Fontana, I.C.; Zimmer, A.R.; Rocha, A.S.; Gosmann, G.; Souza, D.O.; Lourenco, M.V.; Ferreira, S.T.; Zimmer, E.R. Amyloid-beta oligomers in cellular models of Alzheimer's disease. *J. Neurochem.* **2020**, *155*, 348–369. [CrossRef] [PubMed]
51. Fontana, I.C.; Kumar, A.; Nordberg, A. The role of astrocytic alpha7 nicotinic acetylcholine receptors in Alzheimer disease. *Nat. Rev. Neurol.* **2023**, *19*, 278–288. [CrossRef]
52. Iqbal, K.; Liu, F.; Gong, C.X.; Grundke-Iqbal, I. Tau in Alzheimer disease and related tauopathies. *Curr. Alzheimer Res.* **2010**, *7*, 656–664. [CrossRef]
53. Saint-Aubert, L.; Lemoine, L.; Chiotis, K.; Leuzy, A.; Rodriguez-Vieitez, E.; Nordberg, A. Tau PET imaging: Present and future directions. *Mol. Neurodegener.* **2017**, *12*, 19. [CrossRef]
54. Leuzy, A.; Chiotis, K.; Lemoine, L.; Gillberg, P.G.; Almkvist, O.; Rodriguez-Vieitez, E.; Nordberg, A. Tau PET imaging in neurodegenerative tauopathies-still a challenge. *Mol. Psychiatry* **2019**, *24*, 1112–1134. [CrossRef]
55. Tian, M.; Civelek, A.C.; Carrio, I.; Watanabe, Y.; Kang, K.W.; Murakami, K.; Garibotto, V.; Prior, J.O.; Barthel, H.; Zhou, R.; et al. International consensus on the use of tau PET imaging agent (18)F-flortaucipir in Alzheimer's disease. *Eur. J. Nucl. Med. Mol. Imaging* **2022**, *49*, 895–904. [CrossRef] [PubMed]
56. Sokoloff, L.; Reivich, M.; Kennedy, C.; Des Rosiers, M.H.; Patlak, C.S.; Pettigrew, K.D.; Sakurada, O.; Shinohara, M. The [14C]deoxyglucose method for the measurement of local cerebral glucose utilization: Theory, procedure, and normal values in the conscious and anesthetized albino rat. *J. Neurochem.* **1977**, *28*, 897–916. [CrossRef]
57. Zimmer, E.R.; Parent, M.J.; Souza, D.G.; Leuzy, A.; Lecrux, C.; Kim, H.I.; Gauthier, S.; Pellerin, L.; Hamel, E.; Rosa-Neto, P. [(18)F]FDG PET signal is driven by astroglial glutamate transport. *Nat. Neurosci.* **2017**, *20*, 393–395. [CrossRef] [PubMed]
58. Rocha, A.; Bellaver, B.; Souza, D.G.; Schu, G.; Fontana, I.C.; Venturin, G.T.; Greggio, S.; Fontella, F.U.; Schiavenin, M.L.; Machado, L.S.; et al. Clozapine induces astrocyte-dependent FDG-PET hypometabolism. *Eur. J. Nucl. Med. Mol. Imaging* **2022**, *49*, 2251–2264. [CrossRef]
59. Dienel, G.A.; Behar, K.L.; Rothman, D.L. Cellular Origin of [(18)F]FDG-PET Imaging Signals During Ceftriaxone-Stimulated Glutamate Uptake: Astrocytes and Neurons. *Neuroscientist* **2018**, *24*, 316–328. [CrossRef] [PubMed]
60. Nabulsi, N.B.; Mercier, J.; Holden, D.; Carre, S.; Najafzadeh, S.; Vandergeten, M.C.; Lin, S.F.; Deo, A.; Price, N.; Wood, M.; et al. Synthesis and Preclinical Evaluation of 11C-UCB-J as a PET Tracer for Imaging the Synaptic Vesicle Glycoprotein 2A in the Brain. *J. Nucl. Med.* **2016**, *57*, 777–784. [CrossRef]
61. Mecca, A.P.; O'Dell, R.S.; Sharp, E.S.; Banks, E.R.; Bartlett, H.H.; Zhao, W.; Lipior, S.; Diepenbrock, N.G.; Chen, M.K.; Naganawa, M.; et al. Synaptic density and cognitive performance in Alzheimer's disease: A PET imaging study with [(11) C]UCB-J. *Alzheimer's Dement.* **2022**, *18*, 2527–2536. [CrossRef]
62. Kumar, A.; Nordberg, A.K. Assessing synaptic integrity in Alzheimer's brain with UCB-J. *Alzheimer's Dement.* **2022**, *18*, e067047. [CrossRef]
63. Ferrari-Souza, J.P.; Ferreira, P.C.L.; Bellaver, B.; Tissot, C.; Wang, Y.T.; Leffa, D.T.; Brum, W.S.; Benedet, A.L.; Ashton, N.J.; De Bastiani, M.A.; et al. Astrocyte biomarker signatures of amyloid-beta and tau pathologies in Alzheimer's disease. *Mol. Psychiatry* **2022**, *27*, 4781–4789. [CrossRef]
64. Benedet, A.L.; Mila-Aloma, M.; Vrillon, A.; Ashton, N.J.; Pascoal, T.A.; Lussier, F.; Karikari, T.K.; Hourregue, C.; Cognat, E.; Dumurgier, J.; et al. Differences Between Plasma and Cerebrospinal Fluid Glial Fibrillary Acidic Protein Levels Across the Alzheimer Disease Continuum. *JAMA Neurol.* **2021**, *78*, 1471–1483. [CrossRef] [PubMed]

65. Pereira, J.B.; Janelidze, S.; Smith, R.; Mattsson-Carlgren, N.; Palmqvist, S.; Teunissen, C.E.; Zetterberg, H.; Stomrud, E.; Ashton, N.J.; Blennow, K.; et al. Plasma GFAP is an early marker of amyloid-beta but not tau pathology in Alzheimer's disease. *Brain* **2021**, *144*, 3505–3516. [CrossRef] [PubMed]
66. Bluma, M. Association between amyloid PET, CSF pTau, and plasma biomarkers in memory clinical patients. In Proceedings of the HAI (Human Amyloid Imaging) Meeting, Miami, FL, USA, 11–13 January 2023.
67. Middeldorp, J.; Hol, E.M. GFAP in health and disease. *Prog. Neurobiol.* **2011**, *93*, 421–443. [CrossRef]
68. Chiotis, K.; Johansson, C.; Rodriguez-Vieitez, E.; Pannee, J.; Ashton, N.J.; Blennow, K.; Zetterberg, H.; Graff, C.; Nordberg, A.K. Tracking reactive astrocytes in autosomal dominant Alzheimer disease with plasma GFAP and multi-modal PET imaging. In Proceedings of the Alzheimer's Association International Conference, Virtual, 31 July–4 August 2022.
69. Chatterjee, P.; Vermunt, L.; Gordon, B.A.; Pedrini, S.; Boonkamp, L.; Armstrong, N.J.; Xiong, C.; Singh, A.K.; Li, Y.; Sohrabi, H.R.; et al. Plasma glial fibrillary acidic protein in autosomal dominant Alzheimer's disease: Associations with Abeta-PET, neurodegeneration, and cognition. *Alzheimer's Dement.* **2022**, *92*, 1–15. [CrossRef] [PubMed]
70. Johansson, C.; Thordardottir, S.; Laffita-Mesa, J.; Rodriguez-Vieitez, E.; Zetterberg, H.; Blennow, K.; Graff, C. Plasma biomarker profiles in autosomal dominant Alzheimer's disease. *Brain* **2023**, *146*, 1132–1140. [CrossRef]
71. Liddelow, S.A.; Barres, B.A. Reactive Astrocytes: Production, Function, and Therapeutic Potential. *Immunity* **2017**, *46*, 957–967. [CrossRef]
72. Pascoal, T.A.; Benedet, A.L.; Ashton, N.J.; Kang, M.S.; Therriault, J.; Chamoun, M.; Savard, M.; Lussier, F.Z.; Tissot, C.; Karikari, T.K.; et al. Microglial activation and tau propagate jointly across Braak stages. *Nat. Med.* **2021**, *27*, 1592–1599. [CrossRef]

cells

MDPI

Review

Astrocytes: Dissecting Their Diverse Roles in Amyotrophic Lateral Sclerosis and Frontotemporal Dementia

Chiara F. Valori [1,2,*], Claudia Sulmona [3], Liliana Brambilla [3] and Daniela Rossi [3,*]

1 Molecular Neuropathology of Neurodegenerative Diseases, German Centre for Neurodegenerative Diseases (DZNE), 72072 Tübingen, Germany
2 Department of Neuropathology, University of Tübingen, 72076 Tübingen, Germany
3 Laboratory for Research on Neurodegenerative Disorders, Istituti Clinici Scientifici Maugeri IRCCS, 27100 Pavia, Italy
* Correspondence: chiara.valori@dzne.de (C.F.V.); daniela.rossi@icsmaugeri.it (D.R.); Tel.: +49-7071-9254-122 (C.F.V.); +39-0382-592064 (D.R.)

Abstract: Amyotrophic lateral sclerosis (ALS) and frontotemporal dementia (FTD) are fatal neurodegenerative disorders often co-occurring in the same patient, a feature that suggests a common origin of the two diseases. Consistently, pathological inclusions of the same proteins as well as mutations in the same genes can be identified in both ALS/FTD. Although many studies have described several disrupted pathways within neurons, glial cells are also regarded as crucial pathogenetic contributors in ALS/FTD. Here, we focus our attention on astrocytes, a heterogenous population of glial cells that perform several functions for optimal central nervous system homeostasis. Firstly, we discuss how post-mortem material from ALS/FTD patients supports astrocyte dysfunction around three pillars: neuroinflammation, abnormal protein aggregation, and atrophy/degeneration. Furthermore, we summarize current attempts at monitoring astrocyte functions in living patients using either novel imaging strategies or soluble biomarkers. We then address how astrocyte pathology is recapitulated in animal and cellular models of ALS/FTD and how we used these models both to understand the molecular mechanisms driving glial dysfunction and as platforms for pre-clinical testing of therapeutics. Finally, we present the current clinical trials for ALS/FTD, restricting our discussion to treatments that modulate astrocyte functions, directly or indirectly.

Keywords: astrocytes; ALS; FTD; animal models; cell models; rehabilitation; personalized medicine

Citation: Valori, C.F.; Sulmona, C.; Brambilla, L.; Rossi, D. Astrocytes: Dissecting Their Diverse Roles in Amyotrophic Lateral Sclerosis and Frontotemporal Dementia. *Cells* **2023**, *12*, 1450. https://doi.org/10.3390/cells12111450

Academic Editors: Michael Brenner and Vladimir Parpura

Received: 6 February 2023
Revised: 4 May 2023
Accepted: 19 May 2023
Published: 23 May 2023

1. Introduction

Amyotrophic Lateral Sclerosis (ALS) is the most common form of adult-onset motoneuron disease, with an incidence of 1–2 cases/year/100,000 individuals. The disease presents with a wide variety of symptoms arising from motoneuron degeneration. The site of symptom onset normally provides a first criterion to stratify patients. More specifically, bulbar onset is declared when the disease appears with difficulties in swallowing (dysphagia) and articulating speech (dysarthria), while fasciculation, tremors, and muscular tone loss are hallmarks of spinal-onset ALS. In addition to these motor disabilities, about 15% of patients develop cognitive and behavioral impairments that call for an additional diagnosis of Frontotemporal Dementia (FTD). Patients can be further stratified on the basis of the etiological origin of the disorder. In 90–95% of occurrences, it appears sporadically (sALS), with many environmental and behavioral agents implicated in its pathogenesis, whereas it is inherited in the remaining 5–10% of instances, mainly as an autosomal dominant character (fALS). Mutations in several genes have been associated with the disease. From a neuropathological standpoint, post-mortem analysis reveals the loss of the motor cortex and spinal cord motor neurons as well as reactive gliosis. In addition, the cytoplasm of surviving neurons and glial cells shows abnormal proteinaceous aggregates, known as inclusions, whose defining element is the protein Transactive Response DNA-binding Protein 43 (TDP-43), which is present in 95% of cases (reviewed in [1]).

FTD is a clinical umbrella term that defines the second most prevalent form of dementia, grouping patients with different manifestations of behavioral, cognitive, and linguistic impairments. The underlying neuropathological condition is the degeneration of neurons in the frontal and temporal lobes (frontotemporal lobe degeneration, FTLD). Hereafter, we will, therefore, refer to FTD when discussing findings relative to the clinical condition, and FTLD will be reserved to evidence gathered at the biochemical, pathological, and cellular levels. Importantly, about 20% of FTD individuals develop motor dysfunction, meeting the criteria for an additional ALS diagnosis. The majority of cases cannot be ascribed to a known cause, while about 30% of occurrences are of a genetic origin, as mutations in different genes, including some linked to fALS, have been associated with the disease [2]. Intriguingly, the neuropathological assessment of FTLD cortical specimens also revealed the presence of inclusions either consisting of TDP-43 in about 45% of cases or comprising the cytoskeletal protein tau in another 45% of patients [2].

Altogether, this amount of evidence not only highlights the overlap between ALS and FTD/FTLD at the clinical, genetic, and biochemical levels (Figure 1), but also raises important questions, the answers to which will likely provide the basis for a deeper understanding of mechanisms driving those pathological conditions. This, in turn, may help establish suitable therapeutic interventions. Examples of key questions are: what are the mechanisms leading different gene mutations (e.g., *C9orf72* and *GRN*) to cause the same neuropathological lesion (TDP-43 inclusions), though with a different clinical manifestation (ALS and/or FTD for *C9orf72*, purely FTD for *GRN*); and can we extend mechanisms uncovered using models of a specific familial form of the disease (i.e., ALS-*SOD1*) to other genetic forms and sporadic cases?

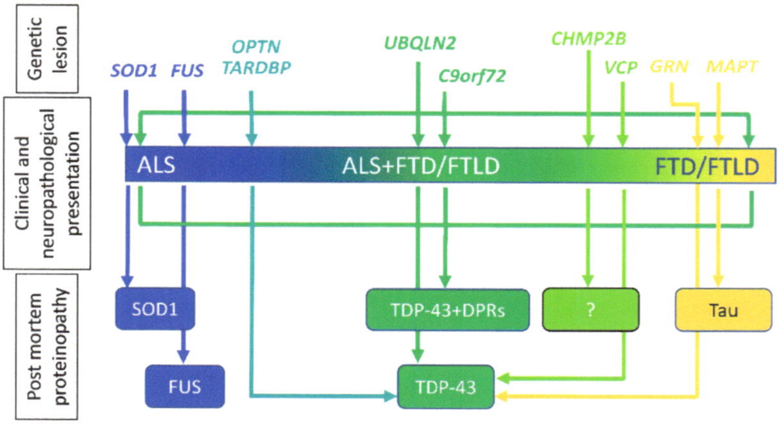

Figure 1. Familial ALS and FTD/FTLD: from genes to pathology. Several genes have been implicated in both ALS and FTD/FTLD. Curiously, some are equally associated with both disorders (i.e., *C9orf72*), while others (i.e., *SOD1* for ALS or *MAPT* for FTD/FTLD) predominantly cause one disease with a distinct neuropathological signature. Intriguingly, other genes (i.e., *TARDBP* for ALS or *GRN* for FTD/FTLD) are predominantly associated with different diseases, although they share the same pathological lesion (TDP-43 accumulation). *SOD1, superoxide dismutase 1; FUS, fused in sarcoma; OPTN, optineurin; TARDBP, TAR (transactive response) DNA Binding Protein; UBQLN2, Ubiquilin2; C9orf72, Chromosome 9 open reading frame 72; CHMP2B, Charged Multivesicular Body Protein 2B; VCP, Valosin Containing Protein; GRN, progranulin; MAPT, Microtubule Associated Protein Tau;* TDP-43, Transactive response DNA binding protein 43 kDa, the protein expressed from *TARDBP*. Blue: genes and proteins associated with a pure ALS phenotype. Yellow: genes and protein associated with FTD/FTLD. Green: genes and protein associated with the whole disease spectrum.

On a mechanistic standpoint, the disruption of several pathways has been associated with ALS or FTLD, such as oxidative stress, excitotoxicity, aberrant RNA metabolism, and impaired protein homeostasis [3]. Remarkably, cellular dysfunction may not be limited to neurons, but could affect also glial cells. The non-cell autonomous contribution of the latter has been a particularly hot topic in the context of ALS research in the last few years (reviewed in [4]) and is now gaining momentum as a potentially contributing factor also in FTD (reviewed in [5]). Importantly, a recent analysis of the transcriptional profile of the frontal and the motor cortices of ALS cases suggested that patients could be classified into three molecular subtypes, respectively, characterized by (i) glial activation, (ii) oxidative stress and synaptic dysregulation, or (iii) transcriptional dysregulation [6]. Remarkably, the subset of patients showing a pronounced glial involvement displayed the faster progression [7].

Here, we focus our attention on astrocytes, one of the most abundant glial cell populations in the central nervous system (CNS). After briefly discussing elements of astrocyte physiology and their response to insults, we will recapitulate recent evidence showing how their functions are perturbed in ALS/FTD patients as well as in cellular and animal models of the disease. Finally, we will address how setting the focus on astrocytes could lead to the validation of informative biomarkers as well as provide valuable therapeutic options.

2. Astrocytes: From Physiology to Pathology

The recognition of astrocytes as an array of highly heterogenous cells was achieved by neuropathological analyses in the 19th century (reviewed in [8]). More recently, this hallmark was further corroborated, at the molecular level, by several single-cell or single-nuclei RNA sequencing studies (sc/snRNA-seq) as well as by advanced imaging techniques supporting the existence of functionally distinct astrocyte subpopulations [9]. A recent ground-breaking investigation not only correlated the expression of gene modules to distinct morphologies, but also mechanistically demonstrated that reducing the expression of key genes in astrocytes could both alter their morphology as well as affect neural circuit function in mice [10]. In the same study, it was discovered that genes belonging to modules that were crucial for astrocyte morphology were strictly associated with gene datasets related to several neurological conditions, including ALS [10].

Several important implications stem from this evidence. Firstly, it explains the extraordinary versatility of astrocytes, which are implicated in a wide array of homeostatic activities in the CNS, ranging from providing metabolic and trophic support to neurons and regulating the extracellular concentration of neurotransmitters and K^+ ions to contributing to the protective functions of the blood–brain barrier (BBB) and the glymphatic system [11,12]. Secondly, it urges for an in-depth re-interpretation of the results obtained from animal models, in light of the growing number of differences acknowledged between rodent and human astrocytes [13,14]. Thirdly, it suggests that any findings observed in human inducible pluripotent stem cell (iPSC)-derived astrocytes should be regarded with keen awareness of the protocol used for differentiation, as there are now several methods available to generate region-specific cells [15]. On top of this, it is yet to be determined how faithfully those protocols allow for the recapitulation of the phenotype of distinct astrocyte subpopulations, by comparing their transcriptional profile with scRNA-seq studies.

Given such remarkable functional heterogeneity under physiological conditions, it is not surprising that astrocytes undergo substantial remodeling upon CNS insult, a phenomenon that has been known for decades as "reactive gliosis", an umbrella term used to encompass complex morphological and transcriptional changes sustaining a phenotypic switch that promotes an inflammatory milieu. Such remodeling is highly regulated and leads to several transcriptional changes, some of which are shared by different insults, while others are tailor-made [16,17]. An early study of the astrocyte transcriptome proposed that the nature of the insult determines whether astrocytes can either be converted into neurotoxic cells or develop a neuroprotective phenotype in a dichotomic way [18,19]. However, this hypothesis has been recently rejected in favor of a more articulated view of the complexity of the astrocyte response (reviewed in [20]). More specifically, in the context of neurodegenerative disorders, it is of outstanding importance to discern astrocytes that retain homeostatic activities and/or boost a neuroprotective phenotype from cells that exacerbate a toxic environment. Recent evidence supports the co-existence of these glial populations in ALS tissues [7], information that appears crucial to develop a suitably targeted therapeutic strategy.

Furthermore, given the intimate interplay between astrocytes and the cells in their immediate neighborhood (neurons and other glial populations), we envision that our understanding of the true nature and extent of cell-to-cell astrocytic signaling may massively improve in the near future with the refinement of visual transcriptomics techniques, thereby allowing unbiased in situ investigations of gene expression.

3. Astrocytes as Active Players in the Pathogenesis of the ALS/FTLD Spectrum

3.1. Evidence from Patients

Neuropathological analyses revealed three facets of astrocyte abnormalities in ALS/FTLD, namely, inflammation, the presence of protein inclusions, and atrophy/degeneration (Figure 2).

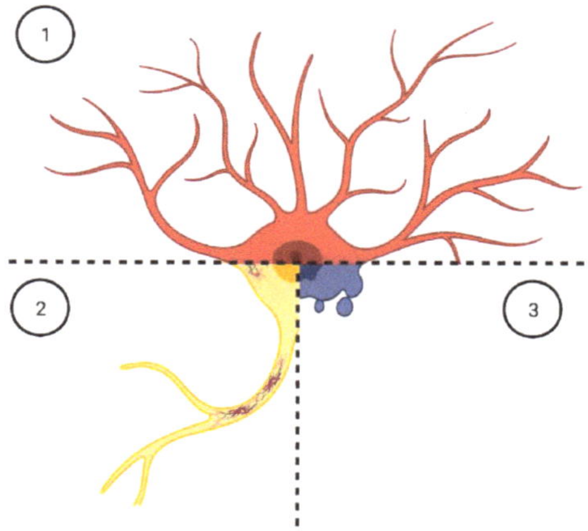

Figure 2. Facets of astrocyte pathology in ALS/FTLD. Neuropathological inspection of post-mortem material from ALS/FTLD patients revealed different aspects of astrocyte involvement, namely, inflammation (1), inclusion body formation (2), and atrophy/degeneration (3). Whether these phenotypes correspond to different moments along a continuous timeline or whether they rather affect distinct subpopulations of astrocytes deserve further investigations. Figure generated with BioRender.com.

As recapitulated in the previous section, reactive gliosis is a well-recognized aspect of neuroinflammation that can be found in most neurodegenerative conditions, including ALS (reviewed in [21]) and FTLD [22]. In the spinal cord of ALS patients, Schiffer et al. [23] reported that reactive astrocytes display distinct morphologies and an increased density at the entering points of the corticospinal tract in the grey matter. At these points, no pathological changes affecting the entering fibers were observed; however, more prominent gliosis was found to occur in the lateral funiculus in proximity to a high number of dystrophic neurites. This finding was interpreted by the authors as indicative of the fact that corticospinal fibers follow a "dying back" pattern. Further morphological abnormalities of the astrocytes include reduced association of astrocytic processes with motor neuron somata [24] and with the mural basement membrane of the blood vessels [25].

From a neuropathological standpoint, the investigation on ALS tissues revealed that astrocytes bear hyaline [26], ubiquitin-positive [27–30] or, rarely, TDP-43-positive inclusions [31]. Furthermore, inclusions of the microtubule associated protein tau are abundant in astrocytes in various subtypes of frontotemporal lobar degeneration with tau protein pathology (FTLD-tau) (reviewed in [32]), a group of pathologies that does not present ALS co-morbidity, however (Figure 1). Nonetheless, phosphorylated TDP-43 inclusions were reported in astrocytes from a small cohort of patients affected by atypical FTLD-tau [33]. Finally, atypical astrocyte and neuronal tauopathy has been detected with starkly increased frequency in patients harboring the A/A major risk genotype at the single nucleotide polymorphism (SNP) rs1990622 on *TMEM106b* [34], a well-established genetic disease modifier in the ALS/FTD spectrum (reviewed in [35]).

The last aspect of astrocyte dysfunction to be described in ALS/FTLD is atrophy/degeneration, an event that has been localized both in the spinal cord of sALS patients [28] and in different cortical layers as well as at the interface between the grey and the white matter of FTLD patients, where astrocyte apoptosis was found to correlate with disease severity [36–38].

These neuropathological findings characterized astrocyte damage at a macroscopic level but did not provide insights as to how the pathological process affects the molecular signature of these cells. To acquire this type of information, recently developed techniques, such as snRNA-seq and spatial transcriptomics, have been applied to post-mortem tissues. In particular, a recent study, aiming at investigating the transcriptome of non-neuronal cells in FTD patients carrying *GRN* mutations (FTLD-*GRN)*, revealed the appearance of a disease-associated astrocyte subpopulation characterized by the expression of WDR49 (WD repeat domain 49) and the perturbation of several grey matter astrocyte pathways, including signaling to neurons, other parenchymal cells, and endothelial cells [39]. Intriguingly, another group recently reported astrocyte dysfunction in FTLD-*GRN*, with the upregulation of *GJA1* (coding for Gap junction alpha-1 protein, also known as connexin 43) and *AQP4* (coding for the water channel protein Aquaporin-4), thus suggesting aberrant coupling within the glial syncytium as well as with the blood vessels [40]. Complementary, a proteomic study, followed by expression-weighted cell type enrichment (EWCE), uncovered altered cellular metabolism in astrocytes in the grey matter of the frontal and temporal lobes of FTLD-*GRN* patients [41]. Although the overexpression of astrocyte-specific proteins, such as the glial fibrillary acidic protein (GFAP), was detected also in patients with frontotemporal lobar degeneration with *MAPT* mutation(s) (FTLD-*MAPT*), the EWCE analysis pointed towards functionally defective oligodendrocytes [41]. The involvement of distinct astrocyte subpopulations in FTLD-TDP vs. FTLD-tau was further corroborated by a study combining a post-mortem magnetic resonance imaging (MRI) investigation of the whole brain hemisphere with histological analysis [42]. Specifically, FTLD-tau patients displayed iron accumulation in astrocytes and microglia of the deep grey matter layer and in the white matter, while, in FTLD-TDP patients, iron-positive astrocytes were detected in the proximity of the blood vessels of the superficial cortical layers [42]. Although no peer-reviewed articles have been published to date (28 April 2023) on human post-mortem specimens from ALS patients, several publications describing snRNA-seq analyses have been recently deposited in a preprint repository [43–47]. Thus, we envision that we will soon gain new stimulating insights into astrocyte dysfunction in ALS/FTLD. In the meantime, spatial transcriptomics performed in a mouse model of the disease revealed the existence of gene modules associated with astrocytes displaying a distinct behavior along phenotype progression, thus further supporting the existence of different subpopulations [48]. Intriguingly, Maniatis et al. identified a correspondence between some gene expression modules in mice and those in the spinal cord of ALS patients [48]. In a small-scale study, spatial transcriptomics allowed for the identification of the dysregulation of *GRM3* (coding for Glutamate metabotropic receptor 3) and *USP47* (coding for Ubiquitin Specific Peptidase 47) genes in the cerebellum of an ALS-*C9orf72* patient, a finding which was corroborated in different brain areas of a larger cohort of patients, including ALS-*C9orf72*, ALS-*SOD1*, and sALS cases, using an in situ hybridization technique [49]. However, one issue that should be carefully considered is the insufficient resolution of published spatial transcriptomics studies, as they combine the RNA extracted from 100 μm-diameter areas into a single spot, thereby precluding the possibility of pinpointing the exact cell type displaying differential gene expression. This is a critical aspect since *GRM3* is expressed by a wide variety of cells in addition to astrocytes. We envision that the refinement of spatial transcriptomics to achieve single-cell resolution will clarify such ambiguities in the near future. Taken together, this amount of evidence implies that the astrocytic response against the neurodegenerative process occurring in FTLD is two-fold: on the one hand, it presents multifaceted features, combining neuroprotective with detrimental aspects; on the other hand, it is tailored to the specific type of genetic or biochemical lesion that triggers the disease.

Previous evidence supported the hypothesis that astrocytes play a complex role in the pathogenesis of ALS and FTLD. A notable limitation of these findings is that they were recorded in post-mortem material, thereby providing a snapshot of the final stage of astrocyte dysfunction. To overcome this drawback and gain information as to how glial pathology evolves, there is an urgent need to monitor astrocytes during disease progression. The development of astrocyte-specific ligands, such as [11C](L)-deprenyl-D2, which enables the measurement of monoamine oxidase B (MAO-B) levels, has indeed served as a powerful tool to track astrocyte activation by positron emission tomography (PET) in ALS [50]. Complementary, tools to monitor the activity of distinct brain structures relying on astrocyte function, such as the glymphatic system or the BBB, could provide information as to whether astrocytes undergo a specific functional deficit along disease progression. To this end, the development of a novel neuroimaging method, known as DTI-ALPS (diffusion tensor image analysis along the perivascular space), has enabled the ability to monitor the efficiency of the glymphatic system [51]. This method has been rapidly implemented in neuroimaging facilities, thereby providing insights into an astrocyte-specific dysfunction in living patients affected by several neurological conditions, including Alzheimer's disease [52] and Parkinson's disease [53]. We anticipate that similar studies will be soon performed on cohorts of ALS and FTD patients as well. The activity of the glymphatic system could also be inferred by administering the gadolinium-based MRI-contrast agent gadobutrol, by allowing its penetration in precisely selected areas of the brain parenchyma by magnetic resonance–guided focused ultrasound surgery (MRgFUS), and, finally, by monitoring its dynamics by contrast-fluid-attenuated inversion recovery [54]. Notably, a case-report study presented the results obtained from an ALS patient using MRgFUS, thereby supporting the feasibility of this technique in the context of motor neuron disorders [54].

A complementary approach to monitor astrocyte impairment along disease progression would be to identify and validate soluble biomarkers that can report such dysfunction. For example, assessing the BBB integrity may serve as a proxy to monitor astrocyte activity since these cells play a crucial role in the formation and maintenance of the barrier [55]. Interestingly, dysfunction of the BBB has been long recognized in human ALS cases [56,57] and, more recently, its increased permeability has been associated with a worse prognosis and seems to be more pronounced in patients affected by ALS/FTD than in those with ALS only [58]. In agreement with this observation, another recent study confirmed the association between enhanced BBB impairment and increased risk of death in ALS patients with spinal onset [59]. However, in a larger study aiming at evaluating BBB dysfunction in patients with different forms of dementia, no association could be found with FTD patients [60].

These conflicting results urged for the identification and validation of alternative astrocyte-specific biomarkers, as BBB dysfunction may also arise from impaired functionality of other cellular partners contributing to its homeostasis. To reach this goal, several studies were performed to investigate the levels of proteins, typically expressed by activated astrocytes, in the cerebrospinal fluid of patients affected by different neurodegenerative diseases. Interestingly, chitinase-3-like protein 1 (CHI3L1, or YKL-40) expression was reported to discriminate patients who received a double ALS/FTD diagnosis from those with FTD with no impairment of motor capabilities [61]. Moreover, significantly elevated levels of CHI3L1 have been demonstrated in other studies [62,63], thus suggesting that monitoring astrocyte dysfunction could be a useful tool not only to monitor disease progression but possibly also for differential diagnosis. Interestingly, CHI3L1 appears to be specifically expressed by a subpopulation of white matter astrocytes [62], thereby further reinforcing the hypothesis that different astrocyte populations provide distinct contributions to disease pathogenesis.

Another glial protein, which is currently under investigation as a potential biomarker, is the cytoskeletal protein GFAP. In serum, its concentration is higher in ALS patients with cognitive impairment [64] and negatively correlates with cognitive scores [65]. Furthermore, the abundance of GFAP in the plasma of FTD patients, alone or in combination with the neurofilament light chain protein, revealed that the simultaneous assessment of these two proteins could serve as a differential diagnosis biomarker assay to discriminate between FTD and Alzheimer disease patients [66] as well as between molecularly distinct FTLD subtypes [67]. Furthermore, plasma GFAP could act as a progression biomarker as its level correlates with a higher burden of cognitive impairment during FTD disease progression [66]. In the future, we envision that other molecules of astrocyte origin might be explored as potential biomarkers, such as specific microRNAs [68].

These biomarker studies point towards the intriguing possibility that the dysfunction of astrocyte subpopulations could be the ultimate disease modifier, aggravating neurodegeneration in different CNS regions and, in turn, leading to the development of more complex disease manifestations. Further indirect evidence supports this hypothesis: the composition of the gut microbiota, which has an important tuning role on astrocytes homeostasis [69–72], also modulates disease progression in ALS patients (reviewed in [73]). In addition, a recent study demonstrated that gut dysbiosis worsens disease course in patients with spinal-onset, while oral dysbiosis negatively impacts patients with bulbar onset of ALS [74].

Although no published studies are currently available that explore the gut microbiota in FTD cases, we can speculate that similar findings may be extended to those patients in light of the recent finding showing that the ALS/FTD-associated *C9orf72* gene plays a role in subduing a pathological inflammatory response triggered by gut bacteria [75].

3.2. Evidence from Animal Models

The observations presented in the previous section indisputably support the notion that several intertwined mechanisms are likely to contribute to the pathogenesis of the ALS/FTLD spectrum. It is, therefore, imperative to develop animal and cellular models that allow for the dissection of the different mechanistic aspects and, thus, support the development of suitable therapies. The ideal animal model should recapitulate every aspect of the disease: the phenotype should be similar to clinical symptoms and should correlate with analogous molecular and neuropathological abnormalities, including astrocyte dysfunction. A comprehensive discussion on all the available animal models of ALS/FTLD goes beyond the scope of this review, and can be appreciated elsewhere [76]. Here, we summarize evidence supporting the role of astrocytes in ALS/FTLD animal models.

In 1993, the groundbreaking discovery of *Superoxide Dismutase 1* (*SOD1*), as the first gene associated with fALS [77], led to the generation of several transgenic mouse strains expressing various mutant versions of its translated product, the Cu/Zn-SOD (SOD1) enzyme. Importantly, the phenotype of these animals closely resembles human ALS, as they develop tremors and muscular weakness, rapidly progressing into paralysis and, ultimately, death. From the histopathological standpoint, they also recapitulate many aspects of the human condition, including severe motoneuron loss, ubiquitin-positive inclusions, and microgliosis. In terms of astrocyte dysfunction, reactive gliosis, inclusions, and cell demise have been described. Intriguingly, in one such transgenic mouse line, the earliest detected abnormality was the presence of SOD1- and ubiquitin-positive inclusions in astrocytes, thus suggesting that these cells are directly impaired by mutant SOD1 expression, and, consequently, motoneuron loss could be a consequence of glial dysfunction [27]. Selective transgene-driven mutant SOD1 expression in astrocytes allowed for a test of this hypothesis. An early report demonstrated that $SOD1^{G86R}$ (the murine orthologue of the fALS-associated $hSOD1^{G85R}$) expression in GFAP-positive astrocytes is sufficient to trigger reactive gliosis but fails to trigger motoneuron demise [78]. Yet, several subsequent studies depicted a different scenario. Firstly, wild-type motoneurons surrounded by mutant SOD1-expressing glial cells displayed ubiquitin-positive inclusions in chimeric mice [79]. Furthermore, transplanting entopic mouse [80] or human astrocytes [81] that express mutant SOD1 into wild-type rodents instigated motor dysfunction and motoneuron degeneration. These studies consistently support the hypothesis that astrocytes expressing ALS-linked SOD1 mutants develop a neurotoxic phenotype. One may, therefore, argue that this feature is a hallmark of a distinct fALS subtype. In addition, transgenic rats expressing $hTDP\text{-}43^{M337V}$ under the control of an inducible, astrocyte-specific promoter rapidly developed motor impairment, motoneuron loss, and ubiquitin-positive inclusions in astrocytes as well as astrogliosis and microgliosis [82]. Similarly, the expression of another fALS-associated mutant hTDP-43 in glial cells in the fruit flies resulted in neuromuscular junction impairment [83]. From a mechanistic standpoint, it is believed that inclusion body formation leads to loss of TDP-43 physiological functions, which, in turn, contribute to the pathogenesis of ALS/FTLD. In keeping with this, when TDP-43 expression was reduced predominantly in astrocytes by RNA interference, transgenic mice showed motor impairment progressing into paralysis, aberrant electromyography parameters, and motoneuron loss [84]. Analogously, astrocytes from transgenic mice having TDP-43 depleted from GFAP-positive cells display altered morphology and acquire a reactive transcriptome pattern [85]. This finding was intriguingly echoed by the recent discovery of the association between astrocyte morphology-defining genes and genes associated with ALS [10]. Furthermore, astrocyte-specific TDP-43 depletion was reported to trigger motor impairment in the absence of motoneuron loss in transgenic mice [85]. Finally, targeted overexpression of mutant FUS in spinal cord astrocytes by viral infection was sufficient to induce a neuroinflammatory phenotype, motoneuron loss, and motor impairment in mice [86].

Taken together, this amount of evidence consistently supports the notion that ALS/FTLD-associated proteins can influence the homeostasis of astrocytes, often turning their phenotype into a neurotoxic one. This conclusion has profound repercussions in terms of therapy development, as it hints that targeting astrocytes might be a valuable option to slow down or even halt neurodegeneration. To support the validity of this hypothesis, one should first ask whether healthy astrocytes preserve their ability to support their surrounding neurons during disease unfolding. In line with this, transplanting wild-type rat glial-restricted precursors [87], human iPSCs [88], or embryonic stem-cell-derived [89] astrocyte-committed glial precursors were shown to ameliorate the phenotype of ALS animal models. Furthermore, more recent studies demonstrated that astrocytes retained their neuroprotective abilities when co-cultured with motoneurons previously exposed to TDP-43 aggregates [90]. Finally, healthy astrocytes do not acquire a neurotoxic phenotype when exposed to the dipeptide repeat protein polyGA [91], a by-product of the hexanucleotide repeat expansion of the *C9orf72* gene [92].

A complementary approach may be based on reversing potentially aberrant pathways taking place cell-autonomously in astrocytes or affecting their communication with surrounding neuronal and glial partners. In this specific case, the mandatory prerequisite is to identify the de-regulated mechanisms. The next section reviews several studies that have significantly contributed to shed light into this issue.

4. Cell-Autonomous Mechanisms of Astrocyte Dysfunction in ALS/FTLD

As discussed in the previous sections, neuropathological assessment of astrocyte response to the ALS/FTLD spectrum can be summarized into three major events: inflammation, protein aggregation, and atrophy/cell death. As inflammation implicates the release of soluble mediators, and the latter can act in a paracrine way, this process cannot be included among cell-autonomous occurrences. Thus, we will address it in the next section, in the context of the aberrant interactions between astrocytes and other cells. Another important point to bear in mind is that these three phenomena might occur in parallel, or in different subpopulations of astrocytes, or they might rather capture different stages of a continuous detrimental process. In support of the latter possibility, we analyzed the spinal cord of hSOD1^{G93A} transgenic mice and found out that ubiquitin-positive morphologically aberrant astrocytes are degenerating cells, as they express the apoptotic marker active caspase-3 [28,93]. Further analyses of the same transgenic model revealed that activated astrocytes overexpress the BH3-interacting domain death agonist and other members of the BH3 protein family, which are well-known for their role in the signaling cascade driving mitochondrial stress into apoptosis. However, the absence of other markers of cell demise suggests that these studies have captured cells in an early stage of the detrimental process. Related to this point, it is important to remark that increased astrocyte vulnerability is not restricted to ALS-*SOD1* models but was confirmed also in differentiated patient-derived iPSCs carrying mutations in the *TARDBP* [94] or *VCP* [95] genes. Intriguingly, human astrocytes derived from fibroblasts of patients carrying *C9orf72* hexanucleotide repeat expansions displayed increased vulnerability to metabolic stressors [96,97]. However, astrocytes differentiated from a healthy donor and exposed to TDP-43 high molecular weight species neither underwent apoptosis nor mounted a pro-inflammatory phenotype [90], thus suggesting that the presence of aberrant TDP-43 in astrocytes was not sufficient to trigger the development of a detrimental phenotype. More recently, evidence of abnormal cell morphology and impaired proteostasis has been independently detected in iPSCs-derived astrocytes carrying mutations in the *MAPT* [98,99]. Moreover, defective autophagy, leading to endoplasmic reticulum stress in astrocytes, has been recognized as an early event in organoid slice cultures generated from patients harboring hexanucleotide repeat expansions in the *C9orf72* gene, which is also implicated in the pathogenesis of ALS and FTD [100]. Curiously, another group investigated astrocytes derived from a patient carrying a mutation in the gene *CHAMP2B* and uncovered enhanced autophagy [101]. This discrepancy could be explained by several differences between the two studies, such as the use of different models (i.e., 2D vs. 3D cultures), the difference in the genetics (i.e., *CHAMP2B* vs. *C9orf72*), and the biochemical lesions (i.e., yet unknown ubiquitinated protein vs. TDP-43). However, a more intriguing possibility is that astrocytes rely on a tightly controlled autophagy flux to maintain their homeostasis and any disturbance, whether it goes towards hyper- or hypo-activation, is detrimental. Thus, this calls for further investigations aiming at comprehending the homeostasis of specific proteins in astrocytes.

Parallel to atrophic astrocytes, a subpopulation of hyperproliferating cells was identified in both mouse [102,103] and rat [104] models of ALS-*SOD1*. However, the relevance of these cells to disease progression was likely limited, as their depletion was shown not to prolong the lifespan of transgenic ALS mice [103]. On the contrary, astrocyte apoptosis is likely to be a significant contributor to ALS pathogenesis, as the number of degenerating astrocytes ramps up during disease progression [28,93], and their accumulation correlates with earlier death in experimental mice [105]. Furthermore, pharmacological manipulations halting astrocyte apoptosis prolong survival in transgenic mice [28,93].

Based on these findings, it is of outstanding importance to understand the molecular mechanisms leading to astrocyte cell death. To this end, our group speculated that hSOD1^{G93A} astrocytes might develop enhanced vulnerability to glutamate toxicity since these cells are contiguous to glutamatergic synapses [28,93]. We established that not only are ALS astrocytes prone to succumb when challenged with this neurotransmitter, but, also, the deadly cascade is triggered by the activation of metabotropic glutamate receptor 5 (mGluR5) and progresses through aberrant Ca^{2+} signaling [28,93]. Importantly, impaired Ca^{2+} homeostasis has been independently identified by several groups and attributed either to the aberrant release of Ca^{2+} from the intracellular endoplasmic reticulum stores [28,106–110] or to excessive import of this ion from the extracellular environment [111]. It has also been recently demonstrated that aberrant Ca^{2+} signaling leads to impaired proteostasis and mitochondrial dysfunction (reviewed in [112]), thus potentially fueling detrimental vicious cycles within astrocytes. In agreement with this, iPSC-derived astrocytes modeling FTLD-tau displayed increased vulnerability to the mitochondrial poison rotenone, a well-known molecule inducing oxidative stress [98]. Moreover, human astrocytes carrying *CHAMP2B* mutations displayed aberrant mitochondrial morphologies and functions, thus leading to enhanced oxidative stress [101].

Taken together, these findings suggest that the phenotype of astrocytes undergoes substantial changes in the context of ALS and FTLD neurodegeneration, therefore, encouraging a comprehensive understanding of the pathways involved. Many groups independently applied unbiased analyses of the transcriptome of mouse astrocytes along disease progression [113–116] as well as of human astrocytes in monolayer cultures [117,118], human iPSC-derived cerebral organoids [119,120] or human post-mortem material [39]. Interestingly, a meta-analysis of these gene expression studies has been recently released [121] and showed that fALS astrocytes undertake a pro-inflammatory phenotype that shares similarities with the neurotoxic phenotype induced by a cytokine cocktail and originally named "A1" [19]. The relevance for disease progression of such a harmful phenotypic switch has been demonstrated in an ALS mouse model, where genetic manipulations were introduced to abrogate the expression of the pro-inflammatory factors driving the transition towards the "A1" status. Noteworthy, this study revealed that abolishing the key pro-inflammatory elements prolonged mouse survival [122]. However, the "A1" transcriptome fingerprint does not comprehensively describe the phenotype of fALS astrocytes and shares little overlap with the transcriptome changes displayed by sALS patients [121]. Further evidence that the "A1" status does not fully recapitulate the features of activated astrocytes even in fALS is provided by the proteomic analysis of conditioned medium harvested from astrocytes expressing hSOD1^{G93A} [123]. Finally, astrocytes carrying mutations in different genes associated with fALS (i.e., *VCP* and *SOD1*) mount a transcriptional response when they are treated with the "A1" inducing pro-inflammatory cytokine cocktail. This response not only is distinct among the different genotypes, but it also contains elements of both "A1" and "A2" statuses [118].

Altogether, these studies reinforce the idea that others have already emphasized, i.e., the requirement for a more precise description of the reactive astrocyte response [20]. In this regard, a recent elegant functional genomic study combined CRISPR-mediated gene ablation with scRNA-seq and computational analysis. The authors discovered that human iPSC-derived astrocytes responded to the "A1" induction protocol primarily by activating the transcription factor NF-kB. Then, they could diverge towards two distinct inflammatory states: the first one was driven by the transcription factor STAT3 and fueled by IL-6 signaling; the second one was regulated by interferon signaling [124]. Although not yet investigated in the context of ALS/FTLD research, these findings were particularly intriguing since STAT3 may also drive a neuroprotective phenotype when triggered by ephrin B1 signaling, a pathway activated by injured neurons, but was impaired in an ALS mouse model [125]. Furthermore, STAT3 signaling has been recently linked to astrocyte-mediated BBB dysfunction in different neurological conditions [126]. Finally, in cerebral organoids modeling FTLD-*MAPT*, scRNA-seq revealed increased IL-6 signaling in astrocytes [119]. Another example of the complexity of astrocyte reactivity comes from the analysis of the dynamics of the protein megalencephalic leukoencephalopathy with subcortical cysts 1. Under physiological conditions, it is considered a marker of perivascular astrocytes [127], but it is consistently overexpressed in astrocytic cell models of ALS [121] as well as in ALS patients [128]. Interestingly, this protein might play a role in self-limiting the inflammatory response [129] and might also counteract connexin 43 impairment in ALS [107] since it stabilizes connexin 43-containing gap junctions [130].

Complementary, a spatial transcriptomics study performed in the spinal cord of the hSOD1^{G93A} mouse model of fALS uncovered the existence of several astrocyte-related expression modules with distinct expression trajectories along disease progression [48]. Overall, these studies gave us valuable insights into the exquisite complexity of astrocyte inflammatory response to CNS insult in the context of ALS/FTD neurodegeneration.

Another relevant facet of astrocyte response in the ALS/FTLD disease spectrum emerging from several unbiased studies is the downregulation of various branches of their metabolism [97,101,131,132]. When looking into specific metabolites, deficient lactate synthesis and release was consistently reported in several astrocyte models [96,133,134]. Furthermore, hSOD1^{G93A} mouse astrocytes were shown to display deficient cholesterol biosynthesis [113], a finding that was recently extended to female hTDP-43^{A315T} mice [135] and, more importantly, to human cerebral organoids cultivated from FTLD-tau patients [120].

5. Non-Cell Autonomous Mechanisms of Astrocyte Dysfunction in ALS/FTLD

The ever-growing body of evidence addressed in the previous section suggests that the pathological milieu occurring in ALS/FTLD directly disrupts the homeostasis of astrocytes. Consequently, they may interact with their surrounding cell partners in new and aberrant ways. Here, we will specifically discuss how their interplay with neurons, other glial cells, and the BBB becomes dysfunctional in diseased conditions, specifically in that of ALS and FTD.

5.1. Neurons

As previously described, the expression of ALS/FTLD proteins in astrocytes is sufficient to trigger neurodegeneration in several animal models [79–82,84]. These findings strongly argue that astrocytes can become inherently toxic to neurons, a hypothesis that has been further corroborated by several independent co-culture studies of mouse [40,133,136–142] and human cells [40,142–150]. Several observations indicate that astrocyte toxicity is mediated by soluble factors [123,139,150–154], although no consensus has been achieved as to which molecule(s) is (are) the culprit(s). Early studies pointed at excessive extracellular glutamate due to reduced clearance by astrocytes, oxidative stress species, growth factors, and neuroinflammatory mediators as major neurodamaging agents (reviewed in [21] and further corroborated in more recent studies, such as [142,155,156]). However, in sharp contrast with the aforementioned evidence, a recent study investigating glutamate metabolism in astrocytes expressing FTD-associated *CHMP2B* variants unexpectedly uncovered enhanced uptake and metabolism of the neurotransmitter [157]. The authors suggested that the increased metabolism was due to an augmented expression in astrocytes of glutamine synthetase, the rate limiting enzyme converting glutamate to glutamine. Interestingly, they did not clarify whether this was specific to *CHMP2B* astrocytes, or if it may also be present in other FTD-associated mutant astrocytes. They further proposed that enhanced glutamate uptake represented an attempt of the astrocytes to remove glutamate that was released in excess by neurons, which failed to incorporate it in the tricarboxylic acid cycle due to reduced glutamate dehydrogenase 2 (GDH2) levels [157]. Whether the reduction in GDH2 expression is specific for neurons carrying *CHMP2B* mutations or is a general feature of this cell type in all forms of FTD deserves further investigations.

The discovery that ALS/FTD astrocytes display metabolic impairment led to the hypothesis that motor neuron toxicity could arise from a lack of adequate sustenance. In agreement with this vision, two strategies proved to be informative. As mouse ALS astrocytes display defective lactate release, a supplementation of mouse ALS astrocyte-motor neuron co-culture medium with this metabolite was first shown to rescue neuronal demise [133]. A similar result was achieved also by pre-treating human *C9orf72* hexanucleotide repeat expansions (HRE) astrocytes with inosine, which boosts their metabolism and, in turn, increases the synthesis and release of lactate [96]. Consistent with this finding, a recent study revealed that the PET tracer (SP-4-2)-[[2,2′-(1,2-dimethyl-1,2-ethanediylidene)bis[N-methylhydrazinecarbothioamidato-κN^2,κS]](2-)]-copper (CuATSM) reduced elevated basal and ATP-linked mitochondrial respiration and increased glycolysis, leading to enhanced lactate release in human astrocytes derived from sALS patients or fALS cases carrying different genetic lesions. However, the molecular mechanisms underlying these events have not been fully elucidated and require further attention [158]. Interestingly, both studies involving human cells uncovered that, in spite of being consistently toxic to motoneurons, astrocytes displayed an unexpected patient-specific heterogeneous response to the treatment, thereby advocating for urgency of developing personalized medicine for the cure of ALS [96,158].

More recently, microRNAs were identified as another class of potential mediators of astrocyte-mediated neurotoxicity. In particular, mouse astrocytes expressing hSOD1^{G93A} downregulate miRNA146a. Restoring its expression in astrocytes reverted their neurotoxic phenotype [110,159]. Intriguingly, human astrocytes transdifferentiated from sALS, and fALS fibroblasts also displayed dysregulated miRNA146a expression and secretion. However, depending on the patient, the expression of miRNA146a could be either reduced or enhanced. Restoration of miRNA146a levels in depleted ALS astrocytes counteracted their reactive/inflammatory phenotype [160]. As re-establishing miRNA146a expression in astrocytes protected motoneuron-like cells in co-cultures, it is reasonable to postulate that any study addressing the therapeutic value of agents increasing its levels should be focused on patients displaying such deregulation [160]. Noteworthy, some astrocyte lines were found to exhibit miRNA146a upregulation and to trigger neurodegeneration in co-culture systems. However, the impact on motor neuron survival of decreasing its levels was not elucidated. Intriguingly, miRNA146a modulates the expression of a number of genes, such as the TNFR-associated factor 6 and the IL-1R-associated kinase [159], both inducers of the NF-kB inflammatory pathway [161].

Another astrocyte-derived microRNA of interest is miR-494-3p. This small RNA is released by astrocytes in the exosomes, and its expression is reduced in cells differentiated from patients carrying *C9orf72* HRE as well as in the spinal cord of sALS cases [162]. Restoring miR-494-3p levels induced the expression of its target, Semaphorin 3A, a critical gene for the maintenance of motor neuronal axons, and, importantly, rescued motor neuron demise in co-culture systems [162]. Furthermore, it is possible that other mechanisms might contribute to the neuroprotective effects of miR-494-3p restoration. For instance, miR-494-3p administration might also reduce the expression of its target gene protein phosphatase and tensin homolog (*PTEN*) [163], an event that sustains ALS-motoneuron survival [164] and mitigates NF-kB-mediated inflammation [165]. It is, therefore, tempting to speculate that these non-coding RNAs play a pivotal role in tuning inflammation in the context of ALS/FTLD.

Taken together, these studies suggest that the toxic interplay between astrocytes and neurons in ALS/ FTLD is likely mediated by a combination of several different aberrant cellular pathways in both cell types. This consideration implies that efforts should be made to prioritize the most robust candidates in the quest to identify the toxic element of the deadly astrocyte-neuron interplay in ALS. To this end, Mishra et al. [123] took an interesting approach by first performing a proteomic analysis of the conditioned medium from hSOD1^{G93A} expressing astrocytes. The results were then submitted to a novel bioinformatics analysis along with pre-existing data describing the transcriptome of motoneurons in order to identify and prioritize noxious ligand-receptor interactions. This strategy led to the discovery of aberrant signaling originating from the release of the amyloid precursor protein (APP) from astrocytes and the activation of the death receptor-6 (DR6) on motoneurons [123]. Importantly, the relevance of this finding was demonstrated by showing that the motoneuron demise, upon exposure to the ALS-astrocyte conditioned medium, can be prevented by either blocking APP release from astrocytes prior to producing the conditioned medium, or ablating the expression of DR6 on motoneurons [123].

5.2. Other Glial Cells

The cross-talk between different populations of glial cells has received less attention than the cell–cell communication signaling occurring between astrocytes and neurons. The characterization of transgenic animals with selective TDP-43 depletion from astrocytes, however, led to the discovery that astrocytes themselves acquire a neuroinflammatory phenotype, and microglia increase their reactivity markers [85]. These mice did not undergo motoneuron loss, but displayed a selective reduction in the number of mature oligodendrocytes, while the abundance and proliferation of oligodendrocyte precursors remained unaffected, a finding that may explain the motor deficits that these mice experience [85]. Intriguingly, the administration of an inflammatory astrocyte-conditioned medium to mature oligodendrocytes induced cell death in vitro, thus suggesting that mature oligodendrocyte degeneration could be a general response to an inflammatory microenvironment [166]. This observation supports the hypothesis that also oligodendrocytes become dysfunctional in ALS/FTLD, a topic which has recently gained more and more attention [167–169]. Finally, astrocyte-mediated microglial activation was also independently observed in a cell model of ALS-*SOD1*. More specifically, a conditioned medium from ALS-astrocytes was reported to induce a robust inflammatory response in a microglia-like cell line [110]. This microglial activation did not occur when miRNA-146a expression was restored in ALS astrocytes, thereby suggesting that it was linked to reduced expression of this microRNA [110]. This evidence, together with the aforementioned observation that restoring miRNA-146a in ALS astrocytes reverts their neurotoxic phenotype, strongly argues that miRNA-146a regulates a variety of pivotal targets to sustain astrocyte homeostatic properties.

5.3. BBB

The BBB and the blood–spinal cord barrier are the structures that tightly regulate the exchange of substances between the bloodstream and the CNS parenchyma. Perivascular astrocytes locate their endfeet around the capillaries and establish functional interactions with components of the BBB. They regulate its functions by exchanging signals with the other cell constituents, namely, endothelial cells and pericytes [127]. Notably, BBB leakage has been long recognized in ALS patients [56,57,170] as well as in several cellular and animal models of the disease [57,171–177]. It has been linked to endothelial cells, where reduced expression of proteins regulating their junctions might impair the tightness of their coupling [57,176]. In addition, endothelial cells in ALS patients [178,179] as well as in different experimental models of ALS [179] upregulate transporter proteins, such as P-glycoprotein, thereby suggesting an incorrect balancing between influx and extrusion of substances from the brain and spinal cord parenchyma. Intriguingly, such endothelial dysfunction is secondary to signaling arising from astrocytes, and several mechanisms have been proposed to drive this event. These include excessive release of glutamate [179], reactive oxidative species, and the pro-inflammatory cytokine TNFα [126,180]. BBB dysfunction has been comparatively less investigated in the context of FTLD. However, a recent snRNA-seq study performed on several cortical areas of FTD patients with *GRN* mutations suggested that diseased astrocytes display enhanced chemokine (C-X-C motif) ligand 1 signaling towards vascular cells, a cascade known to be associated with augmented BBB permeability [39]. Finally, additional strain on the BBB tightness could come from the overexpression [40,181–185] and/or mislocalization of the protein AQP4, a molecule of outstanding importance for the regulation of brain water homeostasis [186].

6. Astrocytes as Therapeutic Targets in ALS/FTLD

Several lines of evidence indicate that astrocyte dysfunction is not only a consistent feature of the ALS/FTLD spectrum but also contributes to the establishment of a neurotoxic environment through aberrant interactions with neurons and other neural cell types. Thus, astrocytes may represent an attractive target for therapeutic intervention in ALS/FTD. The previously mentioned transplantation experiments in animal models [87–89] already provide proof-of-principle evidence that restoring astrocyte function could be beneficial in ALS. Notably, the neuroprotective potential of transplanted glial precursor cells could be further boosted by engineering them to overexpress the glial cell line-derived neurotrophic factor (GDNF, [187]), a pro-survival protein that astrocytes tend to upregulate during the late-stages of disease progression in a mouse model of ALS and sALS patients [188]. Notably, blocking the signaling cascade that mediates GDNF upregulation in astrocytes was reported to lead to faster motor neuron loss and reduction in the lifespan of hSOD1^{G93A} mice [188].

Neural progenitor cells engineered to differentiate to astrocytes and to secrete GDNF (CNS10-NPC-GDNF) were produced under good manufacturing practice for a phase I/IIa clinical trial (NCT02943850) to test the safety of unilateral spinal cord injection in a small cohort of ALS patients [189]. The study met the primary end-point for safety but showed only a small delay in motor function loss in the treated side, most likely due to an unanticipated minimal migration of the injected cells from the dorsal towards the lumbar spinal cord [189]. These overall encouraging results will be hopefully reinforced by the results of an analogous phase I/IIa trial from an independent ALS clinic (NCT03482050) and by an ongoing study designed to evaluate the injection of CNS10-NPC-GDNF in the motor cortex of ALS patients (NCT05306457), thus possibly corroborating the therapeutic neuroprotective power of healthy astrocytes.

A complementary approach would be to identify a key molecule/pathway within astrocytes in order to design a suitable therapeutic strategy to be validated in animal models In keeping with this strategy, an evident candidate was recognized in mutant SOD1s Consistently, early genetic manipulations [190,191] or viral-mediated gene therapy [192,193] demonstrated that reducing the expression of this harmful protein in astrocytes slowed motor decline and prolonged survival in mouse models of ALS-*SOD1*. The success of these and several other studies aimed at silencing *SOD1* in various CNS populations (reviewed in [194]) led to testing the therapeutic potential of SOD1 ablation in ALS-*SOD1* patients Regrettably, although a phase I/IIa clinical trial showed that the administration of an antisense oligonucleotide to lower *SOD1* expression in ALS patients was safe [195], a subsequent study very recently failed to demonstrate that the intervention slowed disease progression [196]. Notably, at the moment of publication of this latter study, most of the patients enrolled in this trial were still alive and participating in a long-term follow-up investigation, thereby leaving hope that the treatment might still be effective in the long term and supporting its recent accelerated approval for fALS treatment by the Food and Drug Administration (25 April 2023. It would be of outstanding importance to determine whether the treatment actually ablated mutant *SOD1* expression in astrocytes or any other cell type.

Among the different mechanisms mediating astrocyte toxicity towards neurons, functional impairment of the astrocyte plasmalemmal glutamate transporter EAAT2 has gained considerable interest as a potential therapeutic target from the milestone discovery that the antibiotic ceftriaxone enhances its expression and ameliorates the phenotype in mouse models of ALS [197] and FTLD-tau [156]. In spite of such promising pre-clinical results, ceftriaxone administration neither delayed the disease progression nor extended the survival of ALS patients in a phase III clinical trial [198]. Notably, a recent review of the literature discussed the extensive pre-clinical testing of ceftriaxone in the context of several neurological conditions and proposed that the drug should be administered prior to symptom onset to fulfil its therapeutic potential [199]. An alternative explanation for the failure of the clinical trial could be that ceftriaxone should be formulated to restrict its delivery to astrocytes to achieve the optimal effect on EAAT2 expression. Finally, the recent conflicting evidence that astrocytes expressing *CHMP2B* variants rather increase glutamate uptake and metabolism [157] suggests that not every patient might profit from EAAT2 overexpression, further highlighting the need for patient stratification and personalized medicine.

Another signaling pathway that raised considerable interest as a potential therapeutic target within astrocytes is the antioxidant response mediated by the transcription factor nuclear factor erythroid 2-related factor 2 (NRF2) [200]. The rationale for this choice was based on the overwhelming amount of evidence implicating defective NRF2 signaling in ALS patients [201,202] as well as in a wide variety of cell and animal models of the ALS/FTD spectrum [142,180,203–211]. Furthermore, pre-clinical studies convincingly showed that the selective rescue of functional deficit of astrocyte NRF2 is neuroprotective both in vitro [142,212–214] and in animal models [215], thus setting a strong rationale for astrocyte-targeted therapeutics. However, in spite of major progress in preclinical settings, specific targeting of astrocytes is yet to be implemented in the clinical practice [216].

Consistently, currently ongoing clinical trials are not investigating formulations to deliver NRF2-pathway-stimulating compounds specifically to astrocytes but rather exploring the therapeutic potential of nutritional supplementations with such biological action to the entire CNS. More specifically, studies have been registered aiming at evaluating whether supplementation with curcumin (found in turmeric), a polyphenol that stimulates the NRF2 signaling pathway [217], could delay progression in ALS patients (NCT04499963; NCT04654689). Although there are no current studies aiming at recruiting patients with FTD, it would be of interest to test the impact of curcumin also on individuals affected by this type of dementia. We envision that the subgroup of patients with FTLD-tau might gain the highest profit from supplementation with curcumin since this phenolic compound also binds to tau fibrils [218]. We can, therefore, speculate that it might interfere with inclusion body formation as well as enhance the anti-oxidant defense response.

In the previous section, we discussed how restoring the metabolic balance in ALS/FTD astrocytes might halt neuronal death. Intriguingly, the results of a pilot clinical trial (NCT02288091) showing the safety of inosine supplementation to ALS patients have been recently published [219]. Although the rationale for such study was to elevate blood concentration of uric acid, a known antioxidant whose levels correlate with slower progression rate and extended survival of ALS patients [220], we know now that inosine can also prevent astrocyte-derived motoneuron demise [96]. Therefore, we can speculate that any potential therapeutic benefit arising from inosine supplementation might be mediated also by rescuing astrocyte metabolic dysfunction. Similar considerations could be drawn about the compound CuATSM. The exact mechanisms of action of this drug are not yet known. Regardless, there are several clinical trials aiming at investigating its potential therapeutic effect in ALS (NCT04082832, NCT03136809, NCT04313166); these trials are perhaps prompted by CuATSM efficacy in ALS mouse models [221–224].

7. Conclusions

In this review, we have extensively discussed considerable evidence demonstrating that astrocytes play a major role in the pathogenesis of ALS and FTLD. These observations suggest that any therapeutic regimen designed to halt neurodegeneration should include agents rescuing glial dysfunctions. Along with pharmacological interventions, rehabilitative training should be also considered to favor morphological remodeling and to improve the functional performance of astrocytes. Physical exercise [225–227] and environmental enrichment [225,228,229] were indeed reported to importantly reduce reactive astrocytosis and to alleviate the neuroinflammatory response in various animal models of injury and disease. In addition, several studies indicated that the expression of ALS- or FTD-associated mutant genes was sufficient to induce deleterious cell-autonomous changes in astrocytes as well as in other cell types of the CNS. During disease progression, neurodegeneration and microgliosis definitely increase the reactive status of astrocytes, thus intensifying several pathways, some of which can amplify the toxic environment, while others attempt to compensate by dampening neuronal injury (Figure 3). On the basis of these assumptions, we can conclude that a comprehensive understanding of the distinct responses of astrocytes requires urgent attention, with a particular focus on the mechanism(s) by which they integrate the different triggers of the disease (i.e., genetic or sporadic) as well as inter-individual genetic variability and personal habits. This information would hopefully guide the development of personalized treatments tailored to maximize the benefit for the different stages of the disease and patient individuality.

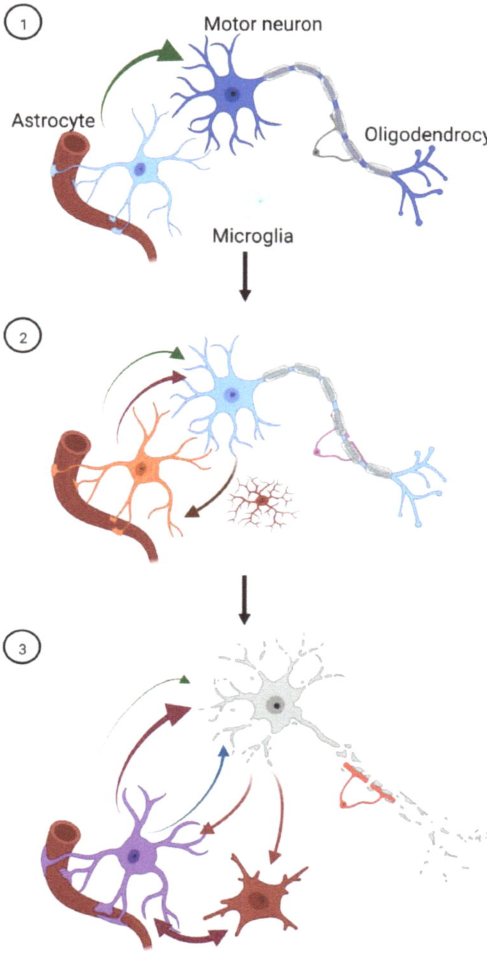

Figure 3. Hypothetical stage-dependent involvement of astrocytes in the pathogenesis of ALS and FTD/FTLD. Under physiological conditions (1), astrocytes engage in homeostatic signaling with neurons (thick green arrow) and other glial cell types to maintain the optimal milieu within the CNS. The expression of ALS/FTD-mutant proteins (2) or other environmental insults are sufficient to trigger an inflammatory response in astrocytes, diverting them from their housekeeping role (medium green arrow) towards a potentially harmful mutual interplay with neurons (red arrows). As neurodegeneration settles in (3), astrocytes are likely to acquire a full-blown neuroinflammatory phenotype that exerts further neurotoxic signaling, in cooperation with other glial cells (red arrows), compensatory neuroprotective messengers (blue arrow), and reduced trophic support (thin green arrow). Figure generated with Biorender.com.

Author Contributions: All of the authors (C.F.V., C.S., L.B., D.R.) contributed to the manuscript preparation. All authors have read and agreed to the published version of the manuscript.

Funding: Work in the laboratory of D.R. is supported by funds from AFM-Téléthon (#16177, #18221, #21565), JPND2022-DC4MND, and the "Ricerca Corrente" funding scheme of Italian Ministry of Health.

Institutional Review Board Statement: Not applicable.

Informed Consent Statement: Not applicable.

Data Availability Statement: Not applicable.

Conflicts of Interest: The authors declare no conflict of interest.

References

1. Feldman, E.L.; Goutman, S.A.; Petri, S.; Mazzini, L.; Savelieff, M.G.; Shaw, P.J.; Sobue, G. Amyotrophic lateral sclerosis. *Lancet* **2022**, *400*, 1363–1380. [CrossRef] [PubMed]
2. Kirola, L.; Mukherjee, A.; Mutsuddi, M. Recent Updates on the Genetics of Amyotrophic Lateral Sclerosis and Frontotemporal Dementia. *Mol. Neurobiol.* **2022**, *59*, 5673–5694. [CrossRef] [PubMed]
3. Hardiman, O.; Al-Chalabi, A.; Chio, A.; Corr, E.M.; Logroscino, G.; Robberecht, W.; Shaw, P.J.; Simmons, Z.; van den Berg, L.H. Amyotrophic lateral sclerosis. *Nat. Rev. Dis. Primers* **2017**, *3*, 17071. [CrossRef] [PubMed]
4. Vahsen, B.F.; Gray, E.; Thompson, A.G.; Ansorge, O.; Anthony, D.C.; Cowley, S.A.; Talbot, K.; Turner, M.R. Non-neuronal cells in amyotrophic lateral sclerosis—From pathogenesis to biomarkers. *Nat. Rev. Neurol.* **2021**, *17*, 333–348. [CrossRef]
5. Ghasemi, M.; Keyhanian, K.; Douthwright, C. Glial Cell Dysfunction in C9orf72-Related Amyotrophic Lateral Sclerosis and Frontotemporal Dementia. *Cells* **2021**, *10*, 249. [CrossRef]
6. Tam, O.H.; Rozhkov, N.V.; Shaw, R.; Kim, D.; Hubbard, I.; Fennessey, S.; Propp, N.; Consortium, N.A.; Fagegaltier, D.; Harris, B.T.; et al. Postmortem Cortex Samples Identify Distinct Molecular Subtypes of ALS: Retrotransposon Activation, Oxidative Stress, and Activated Glia. *Cell Rep.* **2019**, *29*, 1164–1177.e1165. [CrossRef]
7. Eshima, J.; O'Connor, S.A.; Marschall, E.; Consortium, N.A.; Bowser, R.; Plaisier, C.L.; Smith, B.S. Molecular subtypes of ALS are associated with differences in patient prognosis. *Nat. Commun.* **2023**, *14*, 95. [CrossRef] [PubMed]
8. Verkhratsky, A.; Zorec, R.; Parpura, V. Stratification of astrocytes in healthy and diseased brain. *Brain Pathol.* **2017**, *27*, 629–644. [CrossRef]
9. Khakh, B.S.; Deneen, B. The Emerging Nature of Astrocyte Diversity. *Annu. Rev. Neurosci.* **2019**, *42*, 187–207. [CrossRef]
10. Endo, F.; Kasai, A.; Soto, J.S.; Yu, X.; Qu, Z.; Hashimoto, H.; Gradinaru, V.; Kawaguchi, R.; Khakh, B.S. Molecular basis of astrocyte diversity and morphology across the CNS in health and disease. *Science* **2022**, *378*, eadc9020. [CrossRef]
11. Rossi, D. Astrocyte physiopathology: At the crossroads of intercellular networking, inflammation and cell death. *Prog. Neurobiol.* **2015**, *130*, 86–120. [CrossRef] [PubMed]
12. Verkhratsky, A.; Parpura, V.; Vardjan, N.; Zorec, R. Physiology of Astroglia. *Adv. Exp. Med. Biol.* **2019**, *1175*, 45–91. [CrossRef] [PubMed]
13. Zhang, Y.; Sloan, S.A.; Clarke, L.E.; Caneda, C.; Plaza, C.A.; Blumenthal, P.D.; Vogel, H.; Steinberg, G.K.; Edwards, M.S.; Li, G.; et al. Purification and Characterization of Progenitor and Mature Human Astrocytes Reveals Transcriptional and Functional Differences with Mouse. *Neuron* **2016**, *89*, 37–53. [CrossRef] [PubMed]
14. Li, J.; Pan, L.; Pembroke, W.G.; Rexach, J.E.; Godoy, M.I.; Condro, M.C.; Alvarado, A.G.; Harteni, M.; Chen, Y.-W.; Stiles, L.; et al. Conservation and divergence of vulnerability and responses to stressors between human and mouse astrocytes. *Nat. Commun.* **2021**, *12*, 3958. [CrossRef]
15. Franklin, H.; Clarke, B.E.; Patani, R. Astrocytes and microglia in neurodegenerative diseases: Lessons from human in vitro models. *Prog. Neurobiol.* **2021**, *200*, 101973. [CrossRef]
16. Yu, X.; Nagai, J.; Marti-Solano, M.; Soto, J.S.; Coppola, G.; Babu, M.M.; Khakh, B.S. Context-Specific Striatal Astrocyte Molecular Responses Are Phenotypically Exploitable. *Neuron* **2020**, *108*, 1146–1162.e1110. [CrossRef]
17. Burda, J.E.; O'Shea, T.M.; Ao, Y.; Suresh, K.B.; Wang, S.; Bernstein, A.M.; Chandra, A.; Deverasetty, S.; Kawaguchi, R.; Kim, J.H.; et al. Divergent transcriptional regulation of astrocyte reactivity across disorders. *Nature* **2022**, *606*, 557–564. [CrossRef]
18. Zamanian, J.L.; Xu, L.; Foo, L.C.; Nouri, N.; Zhou, L.; Giffard, R.G.; Barres, B.A. Genomic analysis of reactive astrogliosis. *J. Neurosci.* **2012**, *32*, 6391–6410. [CrossRef]
19. Liddelow, S.A.; Guttenplan, K.A.; Clarke, L.E.; Bennett, F.C.; Bohlen, C.J.; Schirmer, L.; Bennett, M.L.; Munch, A.E.; Chung, W.S.; Peterson, T.C.; et al. Neurotoxic reactive astrocytes are induced by activated microglia. *Nature* **2017**, *541*, 481–487. [CrossRef]
20. Escartin, C.; Galea, E.; Lakatos, A.; O'Callaghan, J.P.; Petzold, G.C.; Serrano-Pozo, A.; Steinhauser, C.; Volterra, A.; Carmignoto, G.; Agarwal, A.; et al. Reactive astrocyte nomenclature, definitions, and future directions. *Nat. Neurosci.* **2021**, *24*, 312–325. [CrossRef]
21. Valori, C.F.; Guidotti, G.; Brambilla, L.; Rossi, D. Astrocytes in Motor Neuron Diseases. *Adv. Exp. Med. Biol.* **2019**, *1175*, 227–272. [CrossRef]
22. Black, S.E. Focal Cortical Atrophy Syndromes. *Brain Cogn.* **1996**, *31*, 188–229. [CrossRef] [PubMed]
23. Schiffer, D.; Cordera, S.; Cavalla, P.; Migheli, A. Reactive astrogliosis of the spinal cord in amyotrophic lateral sclerosis. *J. Neurol. Sci.* **1996**, *139*, 27–33. [CrossRef] [PubMed]
24. O'Reilly, S.A.; Roedica, J.; Nagy, D.; Hallewell, R.A.; Alderson, K.; Marklund, S.L.; Kuby, J.; Kushner, P.D. Motor neuron-astrocyte interactions and levels of Cu,Zn superoxide dismutase in sporadic amyotrophic lateral sclerosis. *Exp. Neurol.* **1995**, *131*, 203–210. [CrossRef] [PubMed]

25. Manberg, A.; Skene, N.; Sanders, F.; Trusohamn, M.; Remnestal, J.; Szczepinska, A.; Aksoylu, I.S.; Lonnerberg, P.; Ebarasi, L.; Wouters, S.; et al. Altered perivascular fibroblast activity precedes ALS disease onset. *Nat. Med.* **2021**, *27*, 640–646. [CrossRef]
26. Kato, S.; Saito, M.; Hirano, A.; Ohama, E. Recent advances in research on neuropathological aspects of familial amyotrophic lateral sclerosis with superoxide dismutase 1 gene mutations: Neuronal Lewy body-like hyaline inclusions and astrocytic hyaline inclusions. *Histol. Histopathol.* **1999**, *14*, 973–989. [CrossRef]
27. Bruijn, L.I.; Becher, M.W.; Lee, M.K.; Anderson, K.L.; Jenkins, N.A.; Copeland, N.G.; Sisodia, S.S.; Rothstein, J.D.; Borchelt, D.R.; Price, D.L.; et al. ALS-linked SOD1 mutant G85R mediates damage to astrocytes and promotes rapidly progressive disease with SOD1-containing inclusions. *Neuron* **1997**, *18*, 327–338. [CrossRef]
28. Martorana, F.; Brambilla, L.; Valori, C.F.; Bergamaschi, C.; Roncoroni, C.; Aronica, E.; Volterra, A.; Bezzi, P.; Rossi, D. The BH4 domain of Bcl-X(L) rescues astrocyte degeneration in amyotrophic lateral sclerosis by modulating intracellular calcium signals. *Hum. Mol. Genet.* **2012**, *21*, 826–840. [CrossRef]
29. Mendonca, D.M.; Chimelli, L.; Martinez, A.M. Expression of ubiquitin and proteasome in motorneurons and astrocytes of spinal cords from patients with amyotrophic lateral sclerosis. *Neurosci. Lett.* **2006**, *404*, 315–319. [CrossRef]
30. Pasinelli, P.; Houseweart, M.K.; Brown, R.H., Jr.; Cleveland, D.W. Caspase-1 and -3 are sequentially activated in motor neuron death in Cu,Zn superoxide dismutase-mediated familial amyotrophic lateral sclerosis. *Proc. Natl. Acad. Sci. USA* **2000**, *97*, 13901–13906. [CrossRef]
31. Nishihira, Y.; Tan, C.F.; Onodera, O.; Toyoshima, Y.; Yamada, M.; Morita, T.; Nishizawa, M.; Kakita, A.; Takahashi, H. Sporadic amyotrophic lateral sclerosis: Two pathological patterns shown by analysis of distribution of TDP-43-immunoreactive neuronal and glial cytoplasmic inclusions. *Acta Neuropathol.* **2008**, *116*, 169–182. [CrossRef] [PubMed]
32. Reid, M.J.; Beltran-Lobo, P.; Johnson, L.; Perez-Nievas, B.G.; Noble, W. Astrocytes in Tauopathies. *Front. Neurol.* **2020**, *11*, 572850. [CrossRef] [PubMed]
33. Gelpi, E.; Aldecoa, I.; Lopez-Villegas, D.; Abellan-Vidal, M.T.; Mercadel-Fananas, P.; Fortea, J.; Ribosa, R.; Morenas, E.; Gomez-Anson, B.; Molina-Porcel, L.; et al. Atypical astroglial pTDP-43 pathology in astroglial predominant tauopathy. *Neuropathol. Appl. Neurobiol.* **2021**, *47*, 1109–1113. [CrossRef] [PubMed]
34. Llibre-Guerra, J.J.; Lee, S.E.; Suemoto, C.K.; Ehrenberg, A.J.; Kovacs, G.G.; Karydas, A.; Staffaroni, A.; Franca Resende, E.D.P.; Kim, E.-J.; Hwang, J.-H.; et al. A novel temporal-predominant neuro-astroglial tauopathy associated with TMEM106B gene polymorphism in FTLD/ALS-TDP. *Brain Pathol.* **2021**, *31*, 267–282. [CrossRef]
35. Feng, T.; Lacrampe, A.; Hu, F. Physiological and pathological functions of TMEM106B: A gene associated with brain aging and multiple brain disorders. *Acta Neuropathol.* **2021**, *141*, 327–339. [CrossRef]
36. Su, J.H.; Nichol, K.E.; Sitch, T.; Sheu, P.; Chubb, C.; Miller, B.L.; Tomaselli, K.J.; Kim, R.C.; Cotman, C.W. DNA damage and activated caspase-3 expression in neurons and astrocytes: Evidence for apoptosis in frontotemporal dementia. *Exp. Neurol.* **2000**, *163*, 9–19. [CrossRef]
37. Martinac, J.A.; Craft, D.K.; Su, J.H.; Kim, R.C.; Cotman, C.W. Astrocytes degenerate in frontotemporal dementia: Possible relation to hypoperfusion. *Neurobiol. Aging* **2001**, *22*, 195–207. [CrossRef]
38. Broe, M.; Kril, J.; Halliday, G.M. Astrocytic degeneration relates to the severity of disease in frontotemporal dementia. *Brain* **2004**, *127*, 2214–2220. [CrossRef]
39. Gerrits, E.; Giannini, L.A.A.; Brouwer, N.; Melhem, S.; Seilhean, D.; Le Ber, I.; Brainbank Neuro-CEB Neuropathology Network; Kamermans, A.; Kooij, G.; de Vries, H.E.; et al. Neurovascular dysfunction in GRN-associated frontotemporal dementia identified by single-nucleus RNA sequencing of human cerebral cortex. *Nat. Neurosci.* **2022**, *25*, 1034–1048. [CrossRef]
40. Marsan, E.; Velmeshev, D.; Ramsey, A.; Patel, R.K.; Zhang, J.; Koontz, M.; Andrews, M.G.; de Majo, M.; Mora, C.; Blumenfeld, J.; et al. Astroglial toxicity promotes synaptic degeneration in the thalamocortical circuit in frontotemporal dementia with GRN mutations. *J. Clin. Investig.* **2023**, *133*, e164919. [CrossRef]
41. Miedema, S.S.M.; Mol, M.O.; Koopmans, F.T.W.; Hondius, D.C.; van Nierop, P.; Menden, K.; de Veij Mestdagh, C.F.; van Rooij, J.; Ganz, A.B.; Paliukhovich, I.; et al. Distinct cell type-specific protein signatures in GRN and MAPT genetic subtypes of frontotemporal dementia. *Acta Neuropathol. Commun.* **2022**, *10*, 100. [CrossRef] [PubMed]
42. Tisdall, M.D.; Ohm, D.T.; Lobrovich, R.; Das, S.R.; Mizsei, G.; Prabhakaran, K.; Ittyerah, R.; Lim, S.; McMillan, C.T.; Wolk, D.A.; et al. Ex vivo MRI and histopathology detect novel iron-rich cortical inflammation in frontotemporal lobar degeneration with tau versus TDP-43 pathology. *Neuroimage Clin.* **2022**, *33*, 102913. [CrossRef] [PubMed]
43. Pineda, S.S.; Lee, H.; Fitzwalter, B.E.; Mohammadi, S.; Pregent, L.J.; Gardashli, M.E.; Mantero, J.; Engelberg-Cook, E.; DeJesus-Hernandez, M.; van Blitterswijk, M.; et al. Single-cell profiling of the human primary motor cortex in ALS and FTLD. *bioRxiv* **2021**. [CrossRef]
44. Limone, F.; Mordes, D.; Couto, A.; Pietiläinen, O.; Joseph, B.J.; Burberry, A.; Ghosh, S.D.; Meyer, D.; Goldman, M.; Bortolin, L.; et al. Single-nucleus sequencing reveals enriched expression of genetic risk factors sensitises Motor Neurons to degeneration in ALS. *bioRxiv* **2021**. [CrossRef]
45. Masrori, P.; Bijnens, B.; Davie, K.; Poovathingal, S.K.; Storm, A.; Hersmus, N.; Fumagalli, L.; Van Den Bosch, L.; Fiers, M.; Thal, D.R.; et al. Hexanucleotide repeat expansions in C9orf72 alter microglial responses and prevent a coordinated glial reaction in ALS. *bioRxiv* **2022**. [CrossRef]
46. Li, J.; Jaiswal, M.K.; Chien, J.-F.; Kozlenkov, A.; Zhou, P.; Gardashli, M.; Pregent, L.J.; Engelberg-Cook, E.; Dickson, D.W.; Belzil, V.V.; et al. Divergent impacts of C9orf72 repeat expansion on neurons and glia in ALS and FTD. *bioRxiv* **2022**. [CrossRef]

47. Wang, H.-L.V.; Veire, A.M.; Gendron, T.F.; Gearing, M.; Glass, J.D.; Jin, P.; Corces, V.G.; McEachin, Z.T. Single nucleus multiome analysis of the prefrontal cortex from C9orf72 ALS/FTD patients illuminates pathways affected during disease progression. *bioRxiv* **2023**. [CrossRef]
48. Maniatis, S.; Aijo, T.; Vickovic, S.; Braine, C.; Kang, K.; Mollbrink, A.; Fagegaltier, D.; Andrusivova, Z.; Saarenpaa, S.; Saiz-Castro, G.; et al. Spatiotemporal dynamics of molecular pathology in amyotrophic lateral sclerosis. *Science* **2019**, *364*, 89–93. [CrossRef]
49. Gregory, J.M.; McDade, K.; Livesey, M.R.; Croy, I.; Marion de Proce, S.; Aitman, T.; Chandran, S.; Smith, C. Spatial transcriptomics identifies spatially dysregulated expression of GRM3 and USP47 in amyotrophic lateral sclerosis. *Neuropathol. Appl. Neurobiol.* **2020**, *46*, 441–457. [CrossRef]
50. Johansson, A.; Engler, H.; Blomquist, G.; Scott, B.; Wall, A.; Aquilonius, S.M.; Langstrom, B.; Askmark, H. Evidence for astrocytosis in ALS demonstrated by [11C](L)-deprenyl-D2 PET. *J. Neurol. Sci.* **2007**, *255*, 17–22. [CrossRef]
51. Taoka, T.; Masutani, Y.; Kawai, H.; Nakane, T.; Matsuoka, K.; Yasuno, F.; Kishimoto, T.; Naganawa, S. Evaluation of glymphatic system activity with the diffusion MR technique: Diffusion tensor image analysis along the perivascular space (DTI-ALPS) in Alzheimer's disease cases. *Jpn. J. Radiol.* **2017**, *35*, 172–178. [CrossRef] [PubMed]
52. Hsu, J.L.; Wei, Y.C.; Toh, C.H.; Hsiao, I.T.; Lin, K.J.; Yen, T.C.; Liao, M.F.; Ro, L.S. Magnetic Resonance Images Implicate That Glymphatic Alterations Mediate Cognitive Dysfunction in Alzheimer Disease. *Ann. Neurol.* **2022**, *93*, 164–174. [CrossRef] [PubMed]
53. Cai, X.; Chen, Z.; He, C.; Zhang, P.; Nie, K.; Qiu, Y.; Wang, L.; Wang, L.; Jing, P.; Zhang, Y. Diffusion along perivascular spaces provides evidence interlinking compromised glymphatic function with aging in Parkinson's disease. *CNS Neurosci. Ther.* **2022**, *29*, 111–121. [CrossRef]
54. Meng, Y.; Abrahao, A.; Heyn, C.C.; Bethune, A.J.; Huang, Y.; Pople, C.B.; Aubert, I.; Hamani, C.; Zinman, L.; Hynynen, K.; et al. Glymphatics Visualization after Focused Ultrasound-Induced Blood-Brain Barrier Opening in Humans. *Ann. Neurol.* **2019**, *86*, 975–980. [CrossRef] [PubMed]
55. Langen, U.H.; Ayloo, S.; Gu, C. Development and Cell Biology of the Blood-Brain Barrier. *Annu. Rev. Cell Dev. Biol.* **2019**, *35*, 591–613. [CrossRef]
56. Apostolski, S.; Nikolić, J.; Bugarski-Prokopljević, C.; Miletić, V.; Pavlović, S.; Filipović, S. Serum and CSF immunological findings in ALS. *Acta Neurol. Scand.* **1991**, *83*, 96–98. [CrossRef] [PubMed]
57. Winkler, E.A.; Sengillo, J.D.; Sullivan, J.S.; Henkel, J.S.; Appel, S.H.; Zlokovic, B.V. Blood–spinal cord barrier breakdown and pericyte reductions in amyotrophic lateral sclerosis. *Acta Neuropathol.* **2013**, *125*, 111–120. [CrossRef]
58. Li, J.-Y.; Cai, Z.-Y.; Sun, X.-H.; Shen, D.-c.; Yang, X.-Z.; Liu, M.-S.; Cui, L.-Y. Blood-brain barrier dysfunction and myelin basic protein in survival of amyotrophic lateral sclerosis with or without frontotemporal dementia. *Neurol. Sci.* **2022**, *43*, 3201–3210. [CrossRef]
59. Assialioui, A.; Domínguez, R.; Ferrer, I.; Andrés-Benito, P.; Povedano, M. Elevated Cerebrospinal Fluid Proteins and Albumin Determine a Poor Prognosis for Spinal Amyotrophic Lateral Sclerosis. *Int. J. Mol. Sci.* **2022**, *23*, 11063. [CrossRef]
60. Skillback, T.; Delsing, L.; Synnergren, J.; Mattsson, N.; Janelidze, S.; Nagga, K.; Kilander, L.; Hicks, R.; Wimo, A.; Winblad, B.; et al. CSF/serum albumin ratio in dementias: A cross-sectional study on 1861 patients. *Neurobiol. Aging* **2017**, *59*, 1–9. [CrossRef]
61. Abu-Rumeileh, S.; Steinacker, P.; Polischi, B.; Mammana, A.; Bartoletti-Stella, A.; Oeckl, P.; Baiardi, S.; Zenesini, C.; Huss, A.; Cortelli, P.; et al. CSF biomarkers of neuroinflammation in distinct forms and subtypes of neurodegenerative dementia. *Alzheimer's Res. Ther.* **2019**, *12*, 2. [CrossRef] [PubMed]
62. Vu, L.; An, J.; Kovalik, T.; Gendron, T.; Petrucelli, L.; Bowser, R. Cross-sectional and longitudinal measures of chitinase proteins in amyotrophic lateral sclerosis and expression of CHI3L1 in activated astrocytes. *J. Neurol. Neurosurg. Psychiatry* **2020**, *91*, 350–358. [CrossRef] [PubMed]
63. Woollacott, I.O.C.; Nicholas, J.M.; Heller, C.; Foiani, M.S.; Moore, K.M.; Russell, L.L.; Paterson, R.W.; Keshavan, A.; Schott, J.M.; Warren, J.D.; et al. Cerebrospinal Fluid YKL-40 and Chitotriosidase Levels in Frontotemporal Dementia Vary by Clinical, Genetic and Pathological Subtype. *Dement. Geriatr. Cogn. Disord.* **2020**, *49*, 56–76. [CrossRef]
64. Falzone, Y.M.; Domi, T.; Mandelli, A.; Pozzi, L.; Schito, P.; Russo, T.; Barbieri, A.; Fazio, R.; Volonte, M.A.; Magnani, G.; et al. Integrated evaluation of a panel of neurochemical biomarkers to optimize diagnosis and prognosis in amyotrophic lateral sclerosis. *Eur. J. Neurol.* **2022**, *29*, 1930–1939. [CrossRef] [PubMed]
65. Verde, F.; Milone, I.; Maranzano, A.; Colombo, E.; Torre, S.; Solca, F.; Doretti, A.; Gentile, F.; Manini, A.; Bonetti, R.; et al. Serum levels of glial fibrillary acidic protein in patients with amyotrophic lateral sclerosis. *Ann. Clin. Transl. Neurol.* **2023**, *10*, 118–129. [CrossRef] [PubMed]
66. Zhu, N.; Santos-Santos, M.; Illán-Gala, I.; Montal, V.; Estellés, T.; Barroeta, I.; Altuna, M.; Arranz, J.; Muñoz, L.; Belbin, O.; et al. Plasma glial fibrillary acidic protein and neurofilament light chain for the diagnostic and prognostic evaluation of frontotemporal dementia. *Transl. Neurodegener.* **2021**, *10*, 50. [CrossRef] [PubMed]
67. Cousins, K.A.Q.; Shaw, L.M.; Chen-Plotkin, A.; Wolk, D.A.; Van Deerlin, V.M.; Lee, E.B.; McMillan, C.T.; Grossman, M.; Irwin, D.J. Distinguishing Frontotemporal Lobar Degeneration Tau From TDP-43 Using Plasma Biomarkers. *JAMA Neurol.* **2022**, *79*, 1155–1164. [CrossRef]
68. Ramirez, A.E.; Gil-Jaramillo, N.; Tapias, M.A.; Gonzalez-Giraldo, Y.; Pinzon, A.; Puentes-Rozo, P.J.; Aristizabal-Pachon, A.F.; Gonzalez, J. MicroRNA: A Linking between Astrocyte Dysfunction, Mild Cognitive Impairment, and Neurodegenerative Diseases. *Life* **2022**, *12*, 1439. [CrossRef]

69. Rothhammer, V.; Mascanfroni, I.D.; Bunse, L.; Takenaka, M.C.; Kenison, J.E.; Mayo, L.; Chao, C.C.; Patel, B.; Yan, R.; Blain, M.; et al. Type I interferons and microbial metabolites of tryptophan modulate astrocyte activity and central nervous system inflammation via the aryl hydrocarbon receptor. *Nat. Med.* **2016**, *22*, 586–597. [CrossRef]
70. Sanmarco, L.M.; Wheeler, M.A.; Gutierrez-Vazquez, C.; Polonio, C.M.; Linnerbauer, M.; Pinho-Ribeiro, F.A.; Li, Z.; Giovannoni, F.; Batterman, K.V.; Scalisi, G.; et al. Gut-licensed IFNgamma(+) NK cells drive LAMP1(+)TRAIL(+) anti-inflammatory astrocytes. *Nature* **2021**, *590*, 473–479. [CrossRef]
71. Margineanu, M.B.; Sherwin, E.; Golubeva, A.; Peterson, V.; Hoban, A.; Fiumelli, H.; Rea, K.; Cryan, J.F.; Magistretti, P.J. Gut microbiota modulates expression of genes involved in the astrocyte-neuron lactate shuttle in the hippocampus. *Eur. Neuropsychopharmacol.* **2020**, *41*, 152–159. [CrossRef] [PubMed]
72. Rothhammer, V.; Borucki, D.M.; Tjon, E.C.; Takenaka, M.C.; Chao, C.C.; Ardura-Fabregat, A.; de Lima, K.A.; Gutierrez-Vazquez, C.; Hewson, P.; Staszewski, O.; et al. Microglial control of astrocytes in response to microbial metabolites. *Nature* **2018**, *557*, 724–728. [CrossRef] [PubMed]
73. Martin, S.; Battistini, C.; Sun, J. A Gut Feeling in Amyotrophic Lateral Sclerosis: Microbiome of Mice and Men. *Front. Cell Infect. Microbiol.* **2022**, *12*, 839526. [CrossRef] [PubMed]
74. Kim, H.S.; Son, J.; Lee, D.; Tsai, J.; Wang, D.; Chocron, E.S.; Jeong, S.; Kittrell, P.; Murchison, C.F.; Kennedy, R.E.; et al. Gut- and oral-dysbiosis differentially impact spinal- and bulbar-onset ALS, predicting ALS severity and potentially determining the location of disease onset. *BMC Neurol.* **2022**, *22*, 62. [CrossRef]
75. Burberry, A.; Wells, M.F.; Limone, F.; Couto, A.; Smith, K.S.; Keaney, J.; Gillet, G.; van Gastel, N.; Wang, J.-Y.; Pietilainen, O.; et al. C9orf72 suppresses systemic and neural inflammation induced by gut bacteria. *Nature* **2020**, *582*, 89–94. [CrossRef]
76. Todd, T.W.; Petrucelli, L. Modelling amyotrophic lateral sclerosis in rodents. *Nat. Rev. Neurosci.* **2022**, *23*, 231–251. [CrossRef]
77. Rosen, D.R.; Siddique, T.; Patterson, D.; Figlewicz, D.A.; Sapp, P.; Hentati, A.; Donaldson, D.; Goto, J.; O'Regan, J.P.; Deng, H.X.; et al. Mutations in Cu/Zn superoxide dismutase gene are associated with familial amyotrophic lateral sclerosis. *Nature* **1993**, *362*, 59–62. [CrossRef]
78. Gong, Y.H.; Parsadanian, A.S.; Andreeva, A.; Snider, W.D.; Elliott, J.L. Restricted expression of G86R Cu/Zn superoxide dismutase in astrocytes results in astrocytosis but does not cause motoneuron degeneration. *J. Neurosci.* **2000**, *20*, 660–665. [CrossRef]
79. Clement, A.M.; Nguyen, M.D.; Roberts, E.A.; Garcia, M.L.; Boillee, S.; Rule, M.; McMahon, A.P.; Doucette, W.; Siwek, D.; Ferrante, R.J.; et al. Wild-type nonneuronal cells extend survival of SOD1 mutant motor neurons in ALS mice. *Science* **2003**, *302*, 113–117. [CrossRef]
80. Papadeas, S.T.; Kraig, S.E.; O'Banion, C.; Lepore, A.C.; Maragakis, N.J. Astrocytes carrying the superoxide dismutase 1 (SOD1G93A) mutation induce wild-type motor neuron degeneration in vivo. *Proc. Natl. Acad. Sci. USA* **2011**, *108*, 17803–17808. [CrossRef]
81. Chen, H.; Qian, K.; Chen, W.; Hu, B.; Blackbourn, L.W., 4th; Du, Z.; Ma, L.; Liu, H.; Knobel, K.M.; Ayala, M.; et al. Human-derived neural progenitors functionally replace astrocytes in adult mice. *J. Clin. Investig.* **2015**, *125*, 1033–1042. [CrossRef] [PubMed]
82. Tong, J.; Huang, C.; Bi, F.; Wu, Q.; Huang, B.; Liu, X.; Li, F.; Zhou, H.; Xia, X.G. Expression of ALS-linked TDP-43 mutant in astrocytes causes non-cell-autonomous motor neuron death in rats. *EMBO J.* **2013**, *32*, 1917–1926. [CrossRef] [PubMed]
83. Estes, P.S.; Daniel, S.G.; McCallum, A.P.; Boehringer, A.V.; Sukhina, A.S.; Zwick, R.A.; Zarnescu, D.C. Motor neurons and glia exhibit specific individualized responses to TDP-43 expression in a Drosophila model of amyotrophic lateral sclerosis. *Dis. Model. Mech.* **2013**, *6*, 721–733. [CrossRef]
84. Yang, C.; Wang, H.; Qiao, T.; Yang, B.; Aliaga, L.; Qiu, L.; Tan, W.; Salameh, J.; McKenna-Yasek, D.M.; Smith, T.; et al. Partial loss of TDP-43 function causes phenotypes of amyotrophic lateral sclerosis. *Proc. Natl. Acad. Sci. USA* **2014**, *111*, E1121–E1129. [CrossRef]
85. Peng, A.Y.T.; Agrawal, I.; Ho, W.Y.; Yen, Y.-C.; Pinter, A.J.; Liu, J.; Phua, Q.X.C.; Koh, K.B.; Chang, J.-C.; Sanford, E.; et al. Loss of TDP-43 in astrocytes leads to motor deficits by triggering A1-like reactive phenotype and triglial dysfunction. *Proc. Natl. Acad. Sci. USA* **2020**, *117*, 29101–29112. [CrossRef] [PubMed]
86. Jensen, B.K.; McAvoy, K.J.; Heinsinger, N.M.; Lepore, A.C.; Ilieva, H.; Haeusler, A.R.; Trotti, D.; Pasinelli, P. Targeting TNFalpha produced by astrocytes expressing amyotrophic lateral sclerosis-linked mutant fused in sarcoma prevents neurodegeneration and motor dysfunction in mice. *Glia* **2022**, *70*, 1426–1449. [CrossRef]
87. Lepore, A.C.; Rauck, B.; Dejea, C.; Pardo, A.C.; Rao, M.S.; Rothstein, J.D.; Maragakis, N.J. Focal transplantation-based astrocyte replacement is neuroprotective in a model of motor neuron disease. *Nat. Neurosci.* **2008**, *11*, 1294–1301. [CrossRef]
88. Kondo, T.; Funayama, M.; Tsukita, K.; Hotta, A.; Yasuda, A.; Nori, S.; Kaneko, S.; Nakamura, M.; Takahashi, R.; Okano, H.; et al. Focal transplantation of human iPSC-derived glial-rich neural progenitors improves lifespan of ALS mice. *Stem Cell Rep.* **2014**, *3*, 242–249. [CrossRef]
89. Izrael, M.; Slutsky, S.G.; Admoni, T.; Cohen, L.; Granit, A.; Hasson, A.; Itskovitz-Eldor, J.; Krush Paker, L.; Kuperstein, G.; Lavon, N.; et al. Safety and efficacy of human embryonic stem cell-derived astrocytes following intrathecal transplantation in SOD1(G93A) and NSG animal models. *Stem Cell Res. Ther.* **2018**, *9*, 152. [CrossRef]
90. Smethurst, P.; Risse, E.; Tyzack, G.E.; Mitchell, J.S.; Taha, D.M.; Chen, Y.R.; Newcombe, J.; Collinge, J.; Sidle, K.; Patani, R. Distinct responses of neurons and astrocytes to TDP-43 proteinopathy in amyotrophic lateral sclerosis. *Brain* **2020**, *143*, 430–440. [CrossRef]

91. Marchi, P.M.; Marrone, L.; Brasseur, L.; Coens, A.; Webster, C.P.; Bousset, L.; Destro, M.; Smith, E.F.; Walther, C.G.; Alfred, V.; et al. *C9ORF72*-derived poly-GA DPRs undergo endocytic uptake in iAstrocytes and spread to motor neurons. *Life Sci. Alliance* **2022**, *5*, e202101276. [CrossRef] [PubMed]
92. Schmitz, A.; Pinheiro Marques, J.; Oertig, I.; Maharjan, N.; Saxena, S. Emerging Perspectives on Dipeptide Repeat Proteins in C9ORF72 ALS/FTD. *Front. Cell. Neurosci.* **2021**, *15*, 637548. [CrossRef] [PubMed]
93. Rossi, D.; Brambilla, L.; Valori, C.F.; Roncoroni, C.; Crugnola, A.; Yokota, T.; Bredesen, D.E.; Volterra, A. Focal degeneration of astrocytes in amyotrophic lateral sclerosis. *Cell Death Differ.* **2008**, *15*, 1691–1700. [CrossRef] [PubMed]
94. Serio, A.; Bilican, B.; Barmada, S.J.; Ando, D.M.; Zhao, C.; Siller, R.; Burr, K.; Haghi, G.; Story, D.; Nishimura, A.L.; et al. Astrocyte pathology and the absence of non-cell autonomy in an induced pluripotent stem cell model of TDP-43 proteinopathy. *Proc. Natl. Acad. Sci. USA* **2013**, *110*, 4697–4702. [CrossRef] [PubMed]
95. Hall, C.E.; Yao, Z.; Choi, M.; Tyzack, G.E.; Serio, A.; Luisier, R.; Harley, J.; Preza, E.; Arber, C.; Crisp, S.J.; et al. Progressive Motor Neuron Pathology and the Role of Astrocytes in a Human Stem Cell Model of VCP-Related ALS. *Cell Rep.* **2017**, *19*, 1739–1749. [CrossRef]
96. Allen, S.P.; Hall, B.; Castelli, L.M.; Francis, L.; Woof, R.; Siskos, A.P.; Kouloura, E.; Gray, E.; Thompson, A.G.; Talbot, K.; et al. Astrocyte adenosine deaminase loss increases motor neuron toxicity in amyotrophic lateral sclerosis. *Brain* **2019**, *142*, 586–605. [CrossRef]
97. Allen, S.P.; Hall, B.; Woof, R.; Francis, L.; Gatto, N.; Shaw, A.C.; Myszczynska, M.; Hemingway, J.; Coldicott, I.; Willcock, A.; et al. C9orf72 expansion within astrocytes reduces metabolic flexibility in amyotrophic lateral sclerosis. *Brain* **2019**, *142*, 3771–3790. [CrossRef]
98. Hallmann, A.L.; Arauzo-Bravo, M.J.; Mavrommatis, L.; Ehrlich, M.; Ropke, A.; Brockhaus, J.; Missler, M.; Sterneckert, J.; Scholer, H.R.; Kuhlmann, T.; et al. Astrocyte pathology in a human neural stem cell model of frontotemporal dementia caused by mutant TAU protein. *Sci. Rep.* **2017**, *7*, 42991. [CrossRef]
99. Mahali, S.; Karch, C. Defective proteostasis in patient-derived iPSC-astrocytes and neurons carrying a MAPT IVS10+16 mutation. *Alzheimer's Dement.* **2021**, *17*, e058727. [CrossRef]
100. Szebenyi, K.; Wenger, L.M.D.; Sun, Y.; Dunn, A.W.E.; Limegrover, C.A.; Gibbons, G.M.; Conci, E.; Paulsen, O.; Mierau, S.B.; Balmus, G.; et al. Human ALS/FTD brain organoid slice cultures display distinct early astrocyte and targetable neuronal pathology. *Nat. Neurosci.* **2021**, *24*, 1542–1554. [CrossRef]
101. Chandrasekaran, A.; Dittlau, K.S.; Corsi, G.I.; Haukedal, H.; Doncheva, N.T.; Ramakrishna, S.; Ambardar, S.; Salcedo, C.; Schmidt, S.I.; Zhang, Y.; et al. Astrocytic reactivity triggered by defective autophagy and metabolic failure causes neurotoxicity in frontotemporal dementia type 3. *Stem Cell Rep.* **2021**, *16*, 2736–2751. [CrossRef] [PubMed]
102. Chen, Y.; Guan, Y.; Liu, H.; Wu, X.; Yu, L.; Wang, S.; Zhao, C.; Du, H.; Wang, X. Activation of the Wnt/beta-catenin signaling pathway is associated with glial proliferation in the adult spinal cord of ALS transgenic mice. *Biochem. Biophys. Res. Commun.* **2012**, *420*, 397–403. [CrossRef] [PubMed]
103. Lepore, A.C.; Dejea, C.; Carmen, J.; Rauck, B.; Kerr, D.A.; Sofroniew, M.V.; Maragakis, N.J. Selective ablation of proliferating astrocytes does not affect disease outcome in either acute or chronic models of motor neuron degeneration. *Exp. Neurol.* **2008**, *211*, 423–432. [CrossRef] [PubMed]
104. Diaz-Amarilla, P.; Olivera-Bravo, S.; Trias, E.; Cragnolini, A.; Martinez-Palma, L.; Cassina, P.; Beckman, J.; Barbeito, L. Phenotypically aberrant astrocytes that promote motoneuron damage in a model of inherited amyotrophic lateral sclerosis. *Proc. Natl. Acad. Sci. USA* **2011**, *108*, 18126–18131. [CrossRef] [PubMed]
105. Komine, O.; Yamashita, H.; Fujimori-Tonou, N.; Koike, M.; Jin, S.; Moriwaki, Y.; Endo, F.; Watanabe, S.; Uematsu, S.; Akira, S.; et al. Innate immune adaptor TRIF deficiency accelerates disease progression of ALS mice with accumulation of aberrantly activated astrocytes. *Cell Death Differ.* **2018**, *25*, 2130–2146. [CrossRef]
106. Vergouts, M.; Doyen, P.J.; Peeters, M.; Opsomer, R.; Hermans, E. Constitutive downregulation protein kinase C epsilon in hSOD1(G93A) astrocytes influences mGluR5 signaling and the regulation of glutamate uptake. *Glia* **2018**, *66*, 749–761. [CrossRef]
107. Almad, A.A.; Doreswamy, A.; Gross, S.K.; Richard, J.P.; Huo, Y.; Haughey, N.; Maragakis, N.J. Connexin 43 in astrocytes contributes to motor neuron toxicity in amyotrophic lateral sclerosis. *Glia* **2016**, *64*, 1154–1169. [CrossRef]
108. Kawamata, H.; Ng, S.K.; Diaz, N.; Burstein, S.; Morel, L.; Osgood, A.; Sider, B.; Higashimori, H.; Haydon, P.G.; Manfredi, G.; et al. Abnormal intracellular calcium signaling and SNARE-dependent exocytosis contributes to SOD1G93A astrocyte-mediated toxicity in amyotrophic lateral sclerosis. *J. Neurosci.* **2014**, *34*, 2331–2348. [CrossRef]
109. Norante, R.P.; Peggion, C.; Rossi, D.; Martorana, F.; De Mario, A.; Lia, A.; Massimino, M.L.; Bertoli, A. ALS-Associated SOD1(G93A) Decreases SERCA Pump Levels and Increases Store-Operated Ca^{2+} Entry in Primary Spinal Cord Astrocytes from a Transgenic Mouse Model. *Int. J. Mol. Sci.* **2019**, *20*, 5151. [CrossRef]
110. Barbosa, M.; Gomes, C.; Sequeira, C.; Gonçalves-Ribeiro, J.; Pina, C.C.; Carvalho, L.A.; Moreira, R.; Vaz, S.H.; Vaz, A.R.; Brites, D. Recovery of Depleted miR-146a in ALS Cortical Astrocytes Reverts Cell Aberrancies and Prevents Paracrine Pathogenicity on Microglia and Motor Neurons. *Front. Cell Dev. Biol.* **2021**, *9*. [CrossRef]
111. Milosevic, M.; Bataveljic, D.; Nikolic, L.; Bijelic, D.; Andjus, P. The effect of amyotrophic lateral sclerosis-linked exogenous SOD1-G93A on electrophysiological properties and intracellular calcium in cultured rat astrocytes. *Amyotroph Lateral Scler Front. Degener* **2016**, *17*, 443–451. [CrossRef]

112. Lim, D.; Dematteis, G.; Tapella, L.; Genazzani, A.A.; Cali, T.; Brini, M.; Verkhratsky, A. Ca(2+) handling at the mitochondria-ER contact sites in neurodegeneration. *Cell Calcium.* **2021**, *98*, 102453. [CrossRef] [PubMed]
113. Baker, D.J.; Blackburn, D.J.; Keatinge, M.; Sokhi, D.; Viskaitis, P.; Heath, P.R.; Ferraiuolo, L.; Kirby, J.; Shaw, P.J. Lysosomal and phagocytic activity is increased in astrocytes during disease progression in the SOD1 (G93A) mouse model of amyotrophic lateral sclerosis. *Front. Cell Neurosci.* **2015**, *9*, 410. [CrossRef]
114. Miller, S.J.; Glatzer, J.C.; Hsieh, Y.-C.; Rothstein, J.D. Cortical astroglia undergo transcriptomic dysregulation in the G93A SOD1 ALS mouse model. *J. Neurogenet.* **2018**, *32*, 322–335. [CrossRef] [PubMed]
115. Miller, S.J.; Zhang, P.-W.; Glatzer, J.; Rothstein, J.D. Astroglial transcriptome dysregulation in early disease of an ALS mutant SOD1 mouse model. *J. Neurogenet.* **2017**, *31*, 37–48. [CrossRef] [PubMed]
116. Liu, W.; Venugopal, S.; Majid, S.; Ahn, I.S.; Diamante, G.; Hong, J.; Yang, X.; Chandler, S.H. Single-cell RNA-seq analysis of the brainstem of mutant SOD1 mice reveals perturbed cell types and pathways of amyotrophic lateral sclerosis. *Neurobiol. Dis.* **2020**, *141*, 104877. [CrossRef] [PubMed]
117. Ferraiuolo, L.; Meyer, K.; Sherwood, T.W.; Vick, J.; Likhite, S.; Frakes, A.; Miranda, C.J.; Braun, L.; Heath, P.R.; Pineda, R.; et al. Oligodendrocytes contribute to motor neuron death in ALS via SOD1-dependent mechanism. *Proc. Natl. Acad. Sci. USA* **2016**, *113*, E6496–E6505. [CrossRef]
118. Taha, D.M.; Clarke, B.E.; Hall, C.E.; Tyzack, G.E.; Ziff, O.J.; Greensmith, L.; Kalmar, B.; Ahmed, M.; Alam, A.; Thelin, E.P.; et al. Astrocytes display cell autonomous and diverse early reactive states in familial amyotrophic lateral sclerosis. *Brain* **2022**, *145*, 481–489. [CrossRef]
119. Bowles, K.R.; Silva, M.C.; Whitney, K.; Bertucci, T.; Berlind, J.E.; Lai, J.D.; Garza, J.C.; Boles, N.C.; Mahali, S.; Strang, K.H.; et al. ELAVL4, splicing, and glutamatergic dysfunction precede neuron loss in MAPT mutation cerebral organoids. *Cell* **2021**, *184*, 4547–4563. [CrossRef]
120. Glasauer, S.M.K.; Goderie, S.K.; Rauch, J.N.; Guzman, E.; Audouard, M.; Bertucci, T.; Joy, S.; Rommelfanger, E.; Luna, G.; Keane-Rivera, E.; et al. Human tau mutations in cerebral organoids induce a progressive dyshomeostasis of cholesterol. *Stem Cell Rep.* **2022**, *17*, 2127–2140. [CrossRef]
121. Ziff, O.J.; Clarke, B.E.; Taha, D.M.; Crerar, H.; Luscombe, N.M.; Patani, R. Meta-analysis of human and mouse ALS astrocytes reveals multi-omic signatures of inflammatory reactive states. *Genome Res.* **2022**, *32*, 71–84. [CrossRef] [PubMed]
122. Guttenplan, K.A.; Weigel, M.K.; Adler, D.I.; Couthouis, J.; Liddelow, S.A.; Gitler, A.D.; Barres, B.A. Knockout of reactive astrocyte activating factors slows disease progression in an ALS mouse model. *Nat. Commun.* **2020**, *11*, 3753. [CrossRef]
123. Mishra, V.; Re, D.B.; Le Verche, V.; Alvarez, M.J.; Vasciaveo, A.; Jacquier, A.; Doulias, P.T.; Greco, T.M.; Nizzardo, M.; Papadimitriou, D.; et al. Systematic elucidation of neuron-astrocyte interaction in models of amyotrophic lateral sclerosis using multi-modal integrated bioinformatics workflow. *Nat. Commun.* **2020**, *11*, 5579. [CrossRef] [PubMed]
124. Leng, K.; Rose, I.V.L.; Kim, H.; Xia, W.; Romero-Fernandez, W.; Rooney, B.; Koontz, M.; Li, E.; Ao, Y.; Wang, S.; et al. CRISPRi screens in human iPSC-derived astrocytes elucidate regulators of distinct inflammatory reactive states. *Nat. Neurosci.* **2022**, *25*, 1528–1542. [CrossRef] [PubMed]
125. Tyzack, G.E.; Hall, C.E.; Sibley, C.R.; Cymes, T.; Forostyak, S.; Carlino, G.; Meyer, I.F.; Schiavo, G.; Zhang, S.C.; Gibbons, G.M.; et al. A neuroprotective astrocyte state is induced by neuronal signal EphB1 but fails in ALS models. *Nat. Commun.* **2017**, *8*, 1164. [CrossRef]
126. Kim, H.; Leng, K.; Park, J.; Sorets, A.G.; Kim, S.; Shostak, A.; Embalabala, R.J.; Mlouk, K.; Katdare, K.A.; Rose, I.V.L.; et al. Reactive astrocytes transduce inflammation in a blood-brain barrier model through a TNF-STAT3 signaling axis and secretion of alpha 1-antichymotrypsin. *Nat. Commun.* **2022**, *13*, 6581. [CrossRef]
127. Morales, J.E.; De, A.; Miller, A.A.; Chen, Z.; McCarty, J.H. Mlc1-Expressing Perivascular Astrocytes Promote Blood–Brain Barrier Integrity. *J. Neurosci.* **2022**, *42*, 1406–1416. [CrossRef]
128. Figueroa-Romero, C.; Hur, J.; Bender, D.E.; Delaney, C.E.; Cataldo, M.D.; Smith, A.L.; Yung, R.; Ruden, D.M.; Callaghan, B.C.; Feldman, E.L. Identification of Epigenetically Altered Genes in Sporadic Amyotrophic Lateral Sclerosis. *PLoS ONE* **2012**, *7*, e52672. [CrossRef]
129. Brignone, M.S.; Lanciotti, A.; Serafini, B.; Mallozzi, C.; Sbriccoli, M.; Veroni, C.; Molinari, P.; Elorza-Vidal, X.; Petrucci, T.C.; Estevez, R.; et al. Megalencephalic Leukoencephalopathy with Subcortical Cysts Protein-1 (MLC1) Counteracts Astrocyte Activation in Response to Inflammatory Signals. *Mol. Neurobiol.* **2019**, *56*, 8237–8254. [CrossRef]
130. Lanciotti, A.; Brignone, M.S.; Belfiore, M.; Columba-Cabezas, S.; Mallozzi, C.; Vincentini, O.; Molinari, P.; Petrucci, T.C.; Visentin, S.; Ambrosini, E. Megalencephalic Leukoencephalopathy with Subcortical Cysts Disease-Linked MLC1 Protein Favors Gap-Junction Intercellular Communication by Regulating Connexin 43 Trafficking in Astrocytes. *Cells* **2020**, *9*, 1425. [CrossRef]
131. Sun, S.; Sun, Y.; Ling, S.C.; Ferraiuolo, L.; McAlonis-Downes, M.; Zou, Y.; Drenner, K.; Wang, Y.; Ditsworth, D.; Tokunaga, S.; et al. Translational profiling identifies a cascade of damage initiated in motor neurons and spreading to glia in mutant SOD1-mediated ALS. *Proc. Natl. Acad. Sci. USA* **2015**, *112*, E6993–E7002. [CrossRef] [PubMed]
132. Stella, R.; Bonadio, R.S.; Cagnin, S.; Massimino, M.L.; Bertoli, A.; Peggion, C. Perturbations of the Proteome and of Secreted Metabolites in Primary Astrocytes from the hSOD1(G93A) ALS Mouse Model. *Int. J. Mol. Sci.* **2021**, *22*, 7028. [CrossRef] [PubMed]
133. Ferraiuolo, L.; Higginbottom, A.; Heath, P.R.; Barber, S.; Greenald, D.; Kirby, J.; Shaw, P.J. Dysregulation of astrocyte-motoneuron cross-talk in mutant superoxide dismutase 1-related amyotrophic lateral sclerosis. *Brain* **2011**, *134*, 2627–2641. [CrossRef] [PubMed]

134. Madji Hounoum, B.; Mavel, S.; Coque, E.; Patin, F.; Vourc'h, P.; Marouillat, S.; Nadal-Desbarats, L.; Emond, P.; Corcia, P.; Andres, C.R.; et al. Wildtype motoneurons, ALS-Linked SOD1 mutation and glutamate profoundly modify astrocyte metabolism and lactate shuttling. *Glia* **2017**, *65*, 592–605. [CrossRef]
135. Egawa, N.; Izumi, Y.; Suzuki, H.; Tsuge, I.; Fujita, K.; Shimano, H.; Izumikawa, K.; Takahashi, N.; Tsukita, K.; Enami, T.; et al. TDP-43 regulates cholesterol biosynthesis by inhibiting sterol regulatory element-binding protein 2. *Sci. Rep.* **2022**, *12*, 7988. [CrossRef]
136. Tortarolo, M.; Vallarola, A.; Lidonnici, D.; Battaglia, E.; Gensano, F.; Spaltro, G.; Fiordaliso, F.; Corbelli, A.; Garetto, S.; Martini, E.; et al. Lack of TNF-alpha receptor type 2 protects motor neurons in a cellular model of amyotrophic lateral sclerosis and in mutant SOD1 mice but does not affect disease progression. *J. Neurochem.* **2015**, *135*, 109–124. [CrossRef]
137. Phatnani, H.P.; Guarnieri, P.; Friedman, B.A.; Carrasco, M.A.; Muratet, M.; O'Keeffe, S.; Nwakeze, C.; Pauli-Behn, F.; Newberry, K.M.; Meadows, S.K.; et al. Intricate interplay between astrocytes and motor neurons in ALS. *Proc. Natl. Acad. Sci. USA* **2013**, *110*, E756–E765. [CrossRef]
138. Nagai, M.; Re, D.B.; Nagata, T.; Chalazonitis, A.; Jessell, T.M.; Wichterle, H.; Przedborski, S. Astrocytes expressing ALS-linked mutated SOD1 release factors selectively toxic to motor neurons. *Nat. Neurosci.* **2007**, *10*, 615–622. [CrossRef]
139. Kia, A.; McAvoy, K.; Krishnamurthy, K.; Trotti, D.; Pasinelli, P. Astrocytes expressing ALS-linked mutant FUS induce motor neuron death through release of tumor necrosis factor-alpha. *Glia* **2018**, *66*, 1016–1033. [CrossRef]
140. Di Giorgio, F.P.; Carrasco, M.A.; Siao, M.C.; Maniatis, T.; Eggan, K. Non-cell autonomous effect of glia on motor neurons in an embryonic stem cell-based ALS model. *Nat. Neurosci.* **2007**, *10*, 608–614. [CrossRef]
141. Bilsland, L.G.; Nirmalananthan, N.; Yip, J.; Greensmith, L.; Duchen, M.R. Expression of mutant SOD1 in astrocytes induces functional deficits in motoneuron mitochondria. *J. Neurochem.* **2008**, *107*, 1271–1283. [CrossRef] [PubMed]
142. Provenzano, F.; Nyberg, S.; Giunti, D.; Torazza, C.; Parodi, B.; Bonifacino, T.; Usai, C.; Kerlero de Rosbo, N.; Milanese, M.; Uccelli, A.; et al. Micro-RNAs Shuttled by Extracellular Vesicles Secreted from Mesenchymal Stem Cells Dampen Astrocyte Pathological Activation and Support Neuroprotection in In-Vitro Models of ALS. *Cells* **2022**, *11*, 3923. [CrossRef] [PubMed]
143. Re, D.B.; Le Verche, V.; Yu, C.; Amoroso, M.W.; Politi, K.A.; Phani, S.; Ikiz, B.; Hoffmann, L.; Koolen, M.; Nagata, T.; et al. Necroptosis drives motor neuron death in models of both sporadic and familial ALS. *Neuron* **2014**, *81*, 1001–1008. [CrossRef] [PubMed]
144. Meyer, K.; Ferraiuolo, L.; Miranda, C.J.; Likhite, S.; McElroy, S.; Renusch, S.; Ditsworth, D.; Lagier-Tourenne, C.; Smith, R.A.; Ravits, J.; et al. Direct conversion of patient fibroblasts demonstrates non-cell autonomous toxicity of astrocytes to motor neurons in familial and sporadic ALS. *Proc. Natl. Acad. Sci. USA* **2014**, *111*, 829–832. [CrossRef] [PubMed]
145. Marchetto, M.C.; Muotri, A.R.; Mu, Y.; Smith, A.M.; Cezar, G.G.; Gage, F.H. Non-cell-autonomous effect of human SOD1 G37R astrocytes on motor neurons derived from human embryonic stem cells. *Cell Stem Cell* **2008**, *3*, 649–657. [CrossRef]
146. Haidet-Phillips, A.M.; Hester, M.E.; Miranda, C.J.; Meyer, K.; Braun, L.; Frakes, A.; Song, S.; Likhite, S.; Murtha, M.J.; Foust, K.D.; et al. Astrocytes from familial and sporadic ALS patients are toxic to motor neurons. *Nat. Biotechnol.* **2011**, *29*, 824–828. [CrossRef]
147. Di Giorgio, F.P.; Boulting, G.L.; Bobrowicz, S.; Eggan, K.C. Human embryonic stem cell-derived motor neurons are sensitive to the toxic effect of glial cells carrying an ALS-causing mutation. *Cell Stem Cell* **2008**, *3*, 637–648. [CrossRef]
148. Song, S.; Miranda, C.J.; Braun, L.; Meyer, K.; Frakes, A.E.; Ferraiuolo, L.; Likhite, S.; Bevan, A.K.; Foust, K.D.; McConnell, M.J.; et al. Major histocompatibility complex class I molecules protect motor neurons from astrocyte-induced toxicity in amyotrophic lateral sclerosis. *Nat. Med.* **2016**, *22*, 397–403. [CrossRef]
149. Ikiz, B.; Alvarez, M.J.; Ré, D.B.; Le Verche, V.; Politi, K.; Lotti, F.; Phani, S.; Pradhan, R.; Yu, C.; Croft, G.F.; et al. The Regulatory Machinery of Neurodegeneration in In Vitro Models of Amyotrophic Lateral Sclerosis. *Cell Rep.* **2015**, *12*, 335–345. [CrossRef]
150. Arredondo, C.; Cefaliello, C.; Dyrda, A.; Jury, N.; Martinez, P.; Díaz, I.; Amaro, A.; Tran, H.; Morales, D.; Pertusa, M.; et al. Excessive release of inorganic polyphosphate by ALS/FTD astrocytes causes non-cell-autonomous toxicity to motoneurons. *Neuron* **2022**, *110*, 1656–1670. [CrossRef]
151. Fritz, E.; Izaurieta, P.; Weiss, A.; Mir, F.R.; Rojas, P.; Gonzalez, D.; Rojas, F.; Brown, R.H., Jr.; Madrid, R.; van Zundert, B. Mutant SOD1-expressing astrocytes release toxic factors that trigger motoneuron death by inducing hyperexcitability. *J. Neurophysiol.* **2013**, *109*, 2803–2814. [CrossRef] [PubMed]
152. Ramirez-Jarquin, U.N.; Rojas, F.; van Zundert, B.; Tapia, R. Chronic infusion of SOD1(G93A) astrocyte-secreted factors induces spinal motoneuron degeneration and neuromuscular dysfunction in healthy rats. *J. Cell Physiol.* **2017**, *232*, 2610–2615. [CrossRef]
153. Rojas, F.; Cortes, N.; Abarzua, S.; Dyrda, A.; van Zundert, B. Astrocytes expressing mutant SOD1 and TDP43 trigger motoneuron death that is mediated via sodium channels and nitroxidative stress. *Front. Cell Neurosci.* **2014**, *8*, 24. [CrossRef]
154. Rojas, F.; Gonzalez, D.; Cortes, N.; Ampuero, E.; Hernandez, D.E.; Fritz, E.; Abarzua, S.; Martinez, A.; Elorza, A.A.; Alvarez, A.; et al. Reactive oxygen species trigger motoneuron death in non-cell-autonomous models of ALS through activation of c-Abl signaling. *Front. Cell Neurosci.* **2015**, *9*, 203. [CrossRef] [PubMed]
155. Lee, S.; Kim, S.; Kang, H.-Y.; Lim, H.R.; Kwon, Y.; Jo, M.; Jeon, Y.-M.; Kim, S.R.; Kim, K.; Ha, C.M.; et al. The overexpression of TDP-43 in astrocytes causes neurodegeneration via a PTP1B-mediated inflammatory response. *J. Neuroinflamm.* **2020**, *17*, 299. [CrossRef]

156. Decker, J.M.; Krüger, L.; Sydow, A.; Dennissen, F.J.; Siskova, Z.; Mandelkow, E.; Mandelkow, E.-M. The Tau/A152T mutation, a risk factor for frontotemporal-spectrum disorders, leads to NR2B receptor-mediated excitotoxicity. *EMBO Rep.* **2016**, *17*, 552–569. [CrossRef] [PubMed]
157. Aldana, B.I.; Zhang, Y.; Jensen, P.; Chandrasekaran, A.; Christensen, S.K.; Nielsen, T.T.; Nielsen, J.E.; Hyttel, P.; Larsen, M.R.; Waagepetersen, H.S.; et al. Glutamate-glutamine homeostasis is perturbed in neurons and astrocytes derived from patient iPSC models of frontotemporal dementia. *Mol. Brain* **2020**, *13*, 125. [CrossRef] [PubMed]
158. Dennys, C.N.; Roussel, F.; Rodrigo, R.; Zhang, X.; Sierra Delgado, A.; Hartlaub, A.; Saelim-Ector, A.; Ray, W.; Heintzman, S.; Fox, A.; et al. CuATSM effectively ameliorates ALS patient astrocyte-mediated motor neuron toxicity in human in vitro models of amyotrophic lateral sclerosis. *Glia* **2022**, *71*, 350–365. [CrossRef]
159. Gomes, C.; Cunha, C.; Nascimento, F.; Ribeiro, J.A.; Vaz, A.R.; Brites, D. Cortical Neurotoxic Astrocytes with Early ALS Pathology and miR-146a Deficit Replicate Gliosis Markers of Symptomatic SOD1G93A Mouse Model. *Mol. Neurobiol.* **2019**, *56*, 2137–2158. [CrossRef]
160. Gomes, C.; Sequeira, C.; Likhite, S.; Dennys, C.N.; Kolb, S.J.; Shaw, P.J.; Vaz, A.R.; Kaspar, B.K.; Meyer, K.; Brites, D. Neurotoxic Astrocytes Directly Converted from Sporadic and Familial ALS Patient Fibroblasts Reveal Signature Diversities and miR-146a Theragnostic Potential in Specific Subtypes. *Cells* **2022**, *11*, 1186. [CrossRef]
161. Saba, R.; Sorensen, D.L.; Booth, S.A. MicroRNA-146a: A Dominant, Negative Regulator of the Innate Immune Response. *Front. Immunol.* **2014**, *5*, 578. [CrossRef] [PubMed]
162. Varcianna, A.; Myszczynska, M.A.; Castelli, L.M.; O'Neill, B.; Kim, Y.; Talbot, J.; Nyberg, S.; Nyamali, I.; Heath, P.R.; Stopford, M.J.; et al. Micro-RNAs secreted through astrocyte-derived extracellular vesicles cause neuronal network degeneration in C9orf72 ALS. *eBioMedicine* **2019**, *40*, 626–635. [CrossRef] [PubMed]
163. Li, X.-T.; Wang, H.-Z.; Wu, Z.-W.; Yang, T.-Q.; Zhao, Z.-H.; Chen, G.-L.; Xie, X.-S.; Li, B.; Wei, Y.-X.; Huang, Y.-l.; et al. miR-494-3p Regulates Cellular Proliferation, Invasion, Migration, and Apoptosis by PTEN/AKT Signaling in Human Glioblastoma Cells. *Cell. Mol. Neurobiol.* **2015**, *35*, 679–687. [CrossRef] [PubMed]
164. Yang, D.J.; Wang, X.L.; Ismail, A.; Ashman, C.J.; Valori, C.F.; Wang, G.; Gao, S.; Higginbottom, A.; Ince, P.G.; Azzouz, M.; et al. PTEN regulates AMPA receptor-mediated cell viability in iPS-derived motor neurons. *Cell Death Dis.* **2014**, *5*, e1096. [CrossRef] [PubMed]
165. Zhang, S.; He, K.; Zhou, W.; Cao, J.; Jin, Z. miR-494-3p regulates lipopolysaccharide-induced inflammatory responses in RAW264.7 cells by targeting PTEN. *Mol. Med. Rep.* **2019**, *19*, 4288–4296. [CrossRef]
166. Guttenplan, K.A.; Weigel, M.K.; Prakash, P.; Wijewardhane, P.R.; Hasel, P.; Rufen-Blanchette, U.; Münch, A.E.; Blum, J.A.; Fine, J.; Neal, M.C.; et al. Neurotoxic reactive astrocytes induce cell death via saturated lipids. *Nature* **2021**, *599*, 102–107. [CrossRef]
167. Raffaele, S.; Boccazzi, M.; Fumagalli, M. Oligodendrocyte Dysfunction in Amyotrophic Lateral Sclerosis: Mechanisms and Therapeutic Perspectives. *Cells* **2021**, *10*, 565. [CrossRef]
168. Traiffort, E.; Morisset-Lopez, S.; Moussaed, M.; Zahaf, A. Defective Oligodendroglial Lineage and Demyelination in Amyotrophic Lateral Sclerosis. *Int. J. Mol. Sci.* **2021**, *22*, 3426. [CrossRef]
169. Valori, C.F.; Neumann, M. Contribution of RNA/DNA Binding Protein Dysfunction in Oligodendrocytes in the Pathogenesis of the Amyotrophic Lateral Sclerosis/Frontotemporal Lobar Degeneration Spectrum Diseases. *Front Neurosci.* **2021**, *15*, 724891. [CrossRef]
170. Garbuzova-Davis, S.; Hernandez-Ontiveros, D.G.; Rodrigues, M.C.; Haller, E.; Frisina-Deyo, A.; Mirtyl, S.; Sallot, S.; Saporta, S.; Borlongan, C.V.; Sanberg, P.R. Impaired blood-brain/spinal cord barrier in ALS patients. *Brain Res.* **2012**, *1469*, 114–128. [CrossRef]
171. Garbuzova-Davis, S.; Saporta, S.; Haller, E.; Kolomey, I.; Bennett, S.P.; Potter, H.; Sanberg, P.R. Evidence of compromised blood-spinal cord barrier in early and late symptomatic SOD1 mice modeling ALS. *PLoS ONE* **2007**, *2*, e1205. [CrossRef] [PubMed]
172. Garbuzova-Davis, S.; Haller, E.; Saporta, S.; Kolomey, I.; Nicosia, S.V.; Sanberg, P.R. Ultrastructure of blood-brain barrier and blood-spinal cord barrier in SOD1 mice modeling ALS. *Brain Res.* **2007**, *1157*, 126–137. [CrossRef] [PubMed]
173. Milane, A.; Fernandez, C.; Dupuis, L.; Buyse, M.; Loeffler, J.P.; Farinotti, R.; Meininger, V.; Bensimon, G. P-glycoprotein expression and function are increased in an animal model of amyotrophic lateral sclerosis. *Neurosci. Lett.* **2010**, *472*, 166–170. [CrossRef]
174. Stamenkovic, S.; Pavicevic, A.; Mojovic, M.; Popovic-Bijelic, A.; Selakovic, V.; Andjus, P.; Bacic, G. In vivo EPR pharmacokinetic evaluation of the redox status and the blood brain barrier permeability in the SOD1(G93A) ALS rat model. *Free Radic. Biol. Med.* **2017**, *108*, 258–269. [CrossRef]
175. Zhong, Z.; Deane, R.; Ali, Z.; Parisi, M.; Shapovalov, Y.; O'Banion, M.K.; Stojanovic, K.; Sagare, A.; Boillee, S.; Cleveland, D.W.; et al. ALS-causing SOD1 mutants generate vascular changes prior to motor neuron degeneration. *Nat. Neurosci.* **2008**, *11*, 420–422. [CrossRef]
176. Nicaise, C.; Mitrecic, D.; Demetter, P.; De Decker, R.; Authelet, M.; Boom, A.; Pochet, R. Impaired blood-brain and blood-spinal cord barriers in mutant SOD1-linked ALS rat. *Brain Res.* **2009**, *1301*, 152–162. [CrossRef]
177. Zamudio, F.; Loon, A.R.; Smeltzer, S.; Benyamine, K.; Navalpur Shanmugam, N.K.; Stewart, N.J.F.; Lee, D.C.; Nash, K.; Selenica, M.-L.B. TDP-43 mediated blood-brain barrier permeability and leukocyte infiltration promote neurodegeneration in a low-grade systemic inflammation mouse model. *J. Neuroinflamm.* **2020**, *17*, 283. [CrossRef]

178. Jablonski, M.R.; Jacob, D.A.; Campos, C.; Miller, D.S.; Maragakis, N.J.; Pasinelli, P.; Trotti, D. Selective increase of two ABC drug efflux transporters at the blood-spinal cord barrier suggests induced pharmacoresistance in ALS. *Neurobiol. Dis.* **2012**, *47*, 194–200. [CrossRef] [PubMed]
179. Mohamed, L.A.; Markandaiah, S.S.; Bonanno, S.; Pasinelli, P.; Trotti, D. Excess glutamate secreted from astrocytes drives upregulation of P-glycoprotein in endothelial cells in amyotrophic lateral sclerosis. *Exp. Neurol.* **2019**, *316*, 27–38. [CrossRef]
180. Qosa, H.; Lichter, J.; Sarlo, M.; Markandaiah, S.S.; McAvoy, K.; Richard, J.-P.; Jablonski, M.R.; Maragakis, N.J.; Pasinelli, P.; Trotti, D. Astrocytes drive upregulation of the multidrug resistance transporter ABCB1 (P-Glycoprotein) in endothelial cells of the blood–brain barrier in mutant superoxide dismutase 1-linked amyotrophic lateral sclerosis. *Glia* **2016**, *64*, 1298–1313. [CrossRef]
181. Watanabe-Matsumoto, S.; Moriwaki, Y.; Okuda, T.; Ohara, S.; Yamanaka, K.; Abe, Y.; Yasui, M.; Misawa, H. Dissociation of blood-brain barrier disruption and disease manifestation in an aquaporin-4-deficient mouse model of amyotrophic lateral sclerosis. *Neurosci. Res.* **2018**, *133*, 48–57. [CrossRef]
182. Nicaise, C.; Soyfoo, M.S.; Authelet, M.; De Decker, R.; Bataveljic, D.; Delporte, C.; Pochet, R. Aquaporin-4 overexpression in rat ALS model. *Anat. Rec. (Hoboken)* **2009**, *292*, 207–213. [CrossRef] [PubMed]
183. Khalfallah, Y.; Kuta, R.; Grasmuck, C.; Prat, A.; Durham, H.D.; Vande Velde, C. TDP-43 regulation of stress granule dynamics in neurodegenerative disease-relevant cell types. *Sci. Rep.* **2018**, *8*, 7551. [CrossRef]
184. Dai, J.; Lin, W.; Zheng, M.; Liu, Q.; He, B.; Luo, C.; Lu, X.; Pei, Z.; Su, H.; Yao, X. Alterations in AQP4 expression and polarization in the course of motor neuron degeneration in SOD1G93A mice. *Mol. Med. Rep.* **2017**, *16*, 1739–1746. [CrossRef]
185. Bataveljic, D.; Nikolic, L.; Milosevic, M.; Todorovic, N.; Andjus, P.R. Changes in the astrocytic aquaporin-4 and inwardly rectifying potassium channel expression in the brain of the amyotrophic lateral sclerosis SOD1(G93A) rat model. *Glia* **2012**, *60*, 1991–2003. [CrossRef]
186. Mader, S.; Brimberg, L. Aquaporin-4 Water Channel in the Brain and Its Implication for Health and Disease. *Cells* **2019**, *8*, 90. [CrossRef]
187. Thomsen, G.M.; Avalos, P.; Ma, A.A.; Alkaslasi, M.; Cho, N.; Wyss, L.; Vit, J.P.; Godoy, M.; Suezaki, P.; Shelest, O.; et al. Transplantation of Neural Progenitor Cells Expressing Glial Cell Line-Derived Neurotrophic Factor into the Motor Cortex as a Strategy to Treat Amyotrophic Lateral Sclerosis. *Stem Cells* **2018**, *36*, 1122–1131. [CrossRef] [PubMed]
188. Brambilla, L.; Guidotti, G.; Martorana, F.; Iyer, A.M.; Aronica, E.; Valori, C.F.; Rossi, D. Disruption of the astrocytic TNFR1-GDNF axis accelerates motor neuron degeneration and disease progression in amyotrophic lateral sclerosis. *Hum. Mol. Genet.* **2016**, *25*, 3080–3095. [CrossRef] [PubMed]
189. Baloh, R.H.; Johnson, J.P.; Avalos, P.; Allred, P.; Svendsen, S.; Gowing, G.; Roxas, K.; Wu, A.; Donahue, B.; Osborne, S.; et al. Transplantation of human neural progenitor cells secreting GDNF into the spinal cord of patients with ALS: A phase 1/2a trial. *Nat. Med.* **2022**, *28*, 1813–1822. [CrossRef]
190. Wang, L.; Gutmann, D.H.; Roos, R.P. Astrocyte loss of mutant SOD1 delays ALS disease onset and progression in G85R transgenic mice. *Hum. Mol. Genet.* **2011**, *20*, 286–293. [CrossRef]
191. Yamanaka, K.; Chun, S.J.; Boillee, S.; Fujimori-Tonou, N.; Yamashita, H.; Gutmann, D.H.; Takahashi, R.; Misawa, H.; Cleveland, D.W. Astrocytes as determinants of disease progression in inherited amyotrophic lateral sclerosis. *Nat. Neurosci.* **2008**, *11*, 251–253. [CrossRef] [PubMed]
192. Foust, K.D.; Salazar, D.L.; Likhite, S.; Ferraiuolo, L.; Ditsworth, D.; Ilieva, H.; Meyer, K.; Schmelzer, L.; Braun, L.; Cleveland, D.W.; et al. Therapeutic AAV9-mediated suppression of mutant SOD1 slows disease progression and extends survival in models of inherited ALS. *Mol. Ther.* **2013**, *21*, 2148–2159. [CrossRef] [PubMed]
193. Rochat, C.; Bernard-Marissal, N.; Kallstig, E.; Pradervand, S.; Perrin, F.E.; Aebischer, P.; Raoul, C.; Schneider, B.L. Astrocyte-targeting RNA interference against mutated superoxide dismutase 1 induces motoneuron plasticity and protects fast-fatigable motor units in a mouse model of amyotrophic lateral sclerosis. *Glia* **2022**, *70*, 842–857. [CrossRef] [PubMed]
194. Fang, T.; Je, G.; Pacut, P.; Keyhanian, K.; Gao, J.; Ghasemi, M. Gene Therapy in Amyotrophic Lateral Sclerosis. *Cells* **2022**, *11*, 2066. [CrossRef] [PubMed]
195. Miller, T.; Cudkowicz, M.; Shaw, P.J.; Andersen, P.M.; Atassi, N.; Bucelli, R.C.; Genge, A.; Glass, J.; Ladha, S.; Ludolph, A.L.; et al. Phase 1–2 Trial of Antisense Oligonucleotide Tofersen for SOD1 ALS. *N. Engl. J. Med.* **2020**, *383*, 109–119. [CrossRef] [PubMed]
196. Miller, T.M.; Cudkowicz, M.E.; Genge, A.; Shaw, P.J.; Sobue, G.; Bucelli, R.C.; Chiò, A.; Van Damme, P.; Ludolph, A.C.; Glass, J.D.; et al. Trial of Antisense Oligonucleotide Tofersen for SOD1 ALS. *N. Engl. J. Med.* **2022**, *387*, 1099–1110. [CrossRef] [PubMed]
197. Rothstein, J.D.; Patel, S.; Regan, M.R.; Haenggeli, C.; Huang, Y.H.; Bergles, D.E.; Jin, L.; Dykes Hoberg, M.; Vidensky, S.; Chung, D.S.; et al. Beta-lactam antibiotics offer neuroprotection by increasing glutamate transporter expression. *Nature* **2005**, *433*, 73–77. [CrossRef]
198. Cudkowicz, M.E.; Titus, S.; Kearney, M.; Yu, H.; Sherman, A.; Schoenfeld, D.; Hayden, D.; Shui, A.; Brooks, B.; Conwit, R.; et al. Safety and efficacy of ceftriaxone for amyotrophic lateral sclerosis: A multi-stage, randomised, double-blind, placebo-controlled trial. *Lancet Neurol.* **2014**, *13*, 1083–1091. [CrossRef]
199. Abulseoud, O.A.; Alasmari, F.; Hussein, A.M.; Sari, Y. Ceftriaxone as a Novel Therapeutic Agent for Hyperglutamatergic States: Bridging the Gap Between Preclinical Results and Clinical Translation. *Front. Neurosci.* **2022**, *16*, 841036. [CrossRef]
200. Jimenez-Villegas, J.; Ferraiuolo, L.; Mead, R.J.; Shaw, P.J.; Cuadrado, A.; Rojo, A.I. NRF2 as a therapeutic opportunity to impact in the molecular roadmap of ALS. *Free Radic. Biol. Med.* **2021**, *173*, 125–141. [CrossRef]

201. Sarlette, A.; Krampfl, K.; Grothe, C.; Neuhoff, N.; Dengler, R.; Petri, S. Nuclear erythroid 2-related factor 2-antioxidative response element signaling pathway in motor cortex and spinal cord in amyotrophic lateral sclerosis. *J. Neuropathol. Exp. Neurol.* **2008**, *67*, 1055–1062. [CrossRef] [PubMed]
202. Bakkar, N.; Kousari, A.; Kovalik, T.; Li, Y.; Bowser, R. RBM45 Modulates the Antioxidant Response in Amyotrophic Lateral Sclerosis through Interactions with KEAP1. *Mol. Cell Biol.* **2015**, *35*, 2385–2399. [CrossRef] [PubMed]
203. Cassina, P.; Cassina, A.; Pehar, M.; Castellanos, R.; Gandelman, M.; de Leon, A.; Robinson, K.M.; Mason, R.P.; Beckman, J.S.; Barbeito, L.; et al. Mitochondrial dysfunction in SOD1G93A-bearing astrocytes promotes motor neuron degeneration: Prevention by mitochondrial-targeted antioxidants. *J. Neurosci.* **2008**, *28*, 4115–4122. [CrossRef] [PubMed]
204. Duan, W.; Li, X.; Shi, J.; Guo, Y.; Li, Z.; Li, C. Mutant TAR DNA-binding protein-43 induces oxidative injury in motor neuron-like cell. *Neuroscience* **2010**, *169*, 1621–1629. [CrossRef] [PubMed]
205. Goode, A.; Rea, S.; Sultana, M.; Shaw, B.; Searle, M.S.; Layfield, R. ALS-FTLD associated mutations of SQSTM1 impact on Keap1-Nrf2 signalling. *Mol Cell Neurosci.* **2016**, *76*, 52–58. [CrossRef]
206. Moujalled, D.; Grubman, A.; Acevedo, K.; Yang, S.; Ke, Y.D.; Moujalled, D.M.; Duncan, C.; Caragounis, A.; Perera, N.D.; Turner, B.J.; et al. TDP-43 mutations causing amyotrophic lateral sclerosis are associated with altered expression of RNA-binding protein hnRNP K and affect the Nrf2 antioxidant pathway. *Hum. Mol. Genet.* **2017**, *26*, 1732–1746. [CrossRef]
207. Mimoto, T.; Miyazaki, K.; Morimoto, N.; Kurata, T.; Satoh, K.; Ikeda, Y.; Abe, K. Impaired antioxydative Keap1/Nrf2 system and the downstream stress protein responses in the motor neuron of ALS model mice. *Brain Res.* **2012**, *1446*, 109–118. [CrossRef]
208. Pehar, M.; Vargas, M.R.; Robinson, K.M.; Cassina, P.; Diaz-Amarilla, P.J.; Hagen, T.M.; Radi, R.; Barbeito, L.; Beckman, J.S. Mitochondrial superoxide production and nuclear factor erythroid 2-related factor 2 activation in p75 neurotrophin receptor-induced motor neuron apoptosis. *J. Neurosci.* **2007**, *27*, 7777–7785. [CrossRef]
209. Tian, Y.P.; Che, F.Y.; Su, Q.P.; Lu, Y.C.; You, C.P.; Huang, L.M.; Wang, S.G.; Wang, L.; Yu, J.X. Effects of mutant TDP-43 on the Nrf2/ARE pathway and protein expression of MafK and JDP2 in NSC-34 cells. *Genet. Mol. Res.* **2017**, *16*. [CrossRef]
210. Wang, F.; Lu, Y.; Qi, F.; Su, Q.; Wang, L.; You, C.; Che, F.; Yu, J. Effect of the human SOD1-G93A gene on the Nrf2/ARE signaling pathway in NSC-34 cells. *Mol. Med. Rep.* **2014**, *9*, 2453–2458. [CrossRef]
211. Foster, A.D.; Downing, P.; Figredo, E.; Polain, N.; Stott, A.; Layfield, R.; Rea, S.L. ALS-associated TBK1 variant p.G175S is defective in phosphorylation of p62 and impacts TBK1-mediated signalling and TDP-43 autophagic degradation. *Mol. Cell Neurosci.* **2020**, *108*, 103539. [CrossRef] [PubMed]
212. Diaz-Amarilla, P.; Miquel, E.; Trostchansky, A.; Trias, E.; Ferreira, A.M.; Freeman, B.A.; Cassina, P.; Barbeito, L.; Vargas, M.R.; Rubbo, H. Electrophilic nitro-fatty acids prevent astrocyte-mediated toxicity to motor neurons in a cell model of familial amyotrophic lateral sclerosis via nuclear factor erythroid 2-related factor activation. *Free Radic. Biol. Med.* **2016**, *95*, 112–120. [CrossRef]
213. Vargas, M.R.; Pehar, M.; Cassina, P.; Beckman, J.S.; Barbeito, L. Increased glutathione biosynthesis by Nrf2 activation in astrocytes prevents p75NTR-dependent motor neuron apoptosis. *J. Neurochem.* **2006**, *97*, 687–696. [CrossRef] [PubMed]
214. Vargas, M.R.; Pehar, M.; Cassina, P.; Martinez-Palma, L.; Thompson, J.A.; Beckman, J.S.; Barbeito, L. Fibroblast growth factor-1 induces heme oxygenase-1 via nuclear factor erythroid 2-related factor 2 (Nrf2) in spinal cord astrocytes: Consequences for motor neuron survival. *J. Biol. Chem.* **2005**, *280*, 25571–25579. [CrossRef] [PubMed]
215. Vargas, M.R.; Johnson, D.A.; Sirkis, D.W.; Messing, A.; Johnson, J.A. Nrf2 activation in astrocytes protects against neurodegeneration in mouse models of familial amyotrophic lateral sclerosis. *J. Neurosci.* **2008**, *28*, 13574–13581. [CrossRef]
216. Valori, C.F.; Possenti, A.; Brambilla, L.; Rossi, D. Challenges and Opportunities of Targeting Astrocytes to Halt Neurodegenerative Disorders. *Cells* **2021**, *10*, 2019. [CrossRef]
217. Ashrafizadeh, M.; Ahmadi, Z.; Mohammadinejad, R.; Farkhondeh, T.; Samarghandian, S. Curcumin Activates the Nrf2 Pathway and Induces Cellular Protection Against Oxidative Injury. *Curr. Mol. Med.* **2020**, *20*, 116–133. [CrossRef]
218. Mohorko, N.; Repovš, G.; Popović, M.; Kovacs, G.G.; Bresjanac, M. Curcumin Labeling of Neuronal Fibrillar Tau Inclusions in Human Brain Samples. *J. Neuropathol. Exp. Neurol.* **2010**, *69*, 405–414. [CrossRef]
219. Nicholson, K.; Chan, J.; Macklin, E.A.; Levine-Weinberg, M.; Breen, C.; Bakshi, R.; Grasso, D.L.; Wills, A.-M.; Jahandideh, S.; Taylor, A.A.; et al. Pilot trial of inosine to elevate urate levels in amyotrophic lateral sclerosis. *Ann. Clin. Transl. Neurol.* **2018**, *5*, 1522–1533. [CrossRef]
220. Zhang, F.; Zhang, Q.; Ke, Y.; Hao, J.; Lu, L.; Lu, N.; Chen, X. Serum uric acid levels in patients with amyotrophic lateral sclerosis: A meta-analysis. *Sci. Rep.* **2018**, *8*, 1100. [CrossRef]
221. Williams, J.R.; Trias, E.; Beilby, P.R.; Lopez, N.I.; Labut, E.M.; Bradford, C.S.; Roberts, B.R.; McAllum, E.J.; Crouch, P.J.; Rhoads, T.W.; et al. Copper delivery to the CNS by CuATSM effectively treats motor neuron disease in SODG93A mice co-expressing the Copper-Chaperone-for-SOD. *Neurobiol. Dis.* **2016**, *89*, 1–9. [CrossRef] [PubMed]
222. Vieira, F.G.; Hatzipetros, T.; Thompson, K.; Moreno, A.J.; Kidd, J.D.; Tassinari, V.R.; Levine, B.; Perrin, S.; Gill, A. CuATSM efficacy is independently replicated in a SOD1 mouse model of ALS while unmetallated ATSM therapy fails to reveal benefits. *IBRO Rep.* **2017**, *2*, 47–53. [CrossRef] [PubMed]
223. Kuo, M.T.H.; Beckman, J.S.; Shaw, C.A. Neuroprotective effect of CuATSM on neurotoxin-induced motor neuron loss in an ALS mouse model. *Neurobiol. Dis.* **2019**, *130*, 104495. [CrossRef]

224. Lum, J.S.; Brown, M.L.; Farrawell, N.E.; McAlary, L.; Ly, D.; Chisholm, C.G.; Snow, J.; Vine, K.L.; Karl, T.; Kreilaus, F.; et al. CuATSM improves motor function and extends survival but is not tolerated at a high dose in SOD1G93A mice with a C57BL/6 background. *Sci. Rep.* **2021**, *11*, 19392. [CrossRef] [PubMed]
225. Rodriguez, J.J.; Terzieva, S.; Olabarria, M.; Lanza, R.G.; Verkhratsky, A. Enriched environment and physical activity reverse astrogliodegeneration in the hippocampus of AD transgenic mice. *Cell Death Dis.* **2013**, *4*, e678. [CrossRef]
226. Lundquist, A.J.; Parizher, J.; Petzinger, G.M.; Jakowec, M.W. Exercise induces region-specific remodeling of astrocyte morphology and reactive astrocyte gene expression patterns in male mice. *J. Neurosci. Res.* **2019**, *97*, 1081–1094. [CrossRef]
227. Yin, M.; Pu, T.; Wang, L.; Marshall, C.; Wu, T.; Xiao, M. Astroglial water channel aquaporin 4-mediated glymphatic clearance function: A determined factor for time-sensitive treatment of aerobic exercise in patients with Alzheimer's disease. *Med. Hypotheses* **2018**, *119*, 18–21. [CrossRef]
228. Sampedro-Piquero, P.; De Bartolo, P.; Petrosini, L.; Zancada-Menendez, C.; Arias, J.L.; Begega, A. Astrocytic plasticity as a possible mediator of the cognitive improvements after environmental enrichment in aged rats. *Neurobiol. Learn Mem.* **2014**, *114*, 16–25. [CrossRef]
229. Diniz, D.G.; de Oliveira, M.A.; de Lima, C.M.; Foro, C.A.; Sosthenes, M.C.; Bento-Torres, J.; da Costa Vasconcelos, P.F.; Anthony, D.C.; Diniz, C.W. Age, environment, object recognition and morphological diversity of GFAP-immunolabeled astrocytes. *Behav. Brain Funct.* **2016**, *12*, 28. [CrossRef]

cells

MDPI

Review

Roles of Astrocytic Endothelin ET_B Receptor in Traumatic Brain Injury

Shotaro Michinaga [1], Shigeru Hishinuma [1] and Yutaka Koyama [2,*]

[1] Department of Pharmacodynamics, Meiji Pharmaceutical University, 2-522-1 Noshio, Tokyo 204-8588, Japan
[2] Laboratory of Pharmacology, Kobe Pharmaceutical University, 4-19-1 Motoyama-Kita Higashinada, Kobe 668-8558, Japan
* Correspondence: koyama-y@kobepharma-u.ac.jp; Tel.: +81-78-441-7572

Abstract: Traumatic brain injury (TBI) is an intracranial injury caused by accidents, falls, or sports. The production of endothelins (ETs) is increased in the injured brain. ET receptors are classified into distinct types, including ET_A receptor (ET_A-R) and ET_B receptor (ET_B-R). ET_B-R is highly expressed in reactive astrocytes and upregulated by TBI. Activation of astrocytic ET_B-R promotes conversion to reactive astrocytes and the production of astrocyte-derived bioactive factors, including vascular permeability regulators and cytokines, which cause blood–brain barrier (BBB) disruption, brain edema, and neuroinflammation in the acute phase of TBI. ET_B-R antagonists alleviate BBB disruption and brain edema in animal models of TBI. The activation of astrocytic ET_B receptors also enhances the production of various neurotrophic factors. These astrocyte-derived neurotrophic factors promote the repair of the damaged nervous system in the recovery phase of patients with TBI. Thus, astrocytic ET_B-R is expected to be a promising drug target for TBI in both the acute and recovery phases. This article reviews recent observations on the role of astrocytic ET_B receptors in TBI.

Keywords: astrocyte; traumatic brain injury; endothelin; ET_B receptor; blood–brain barrier; neuroinflammation

Citation: Michinaga, S.; Hishinuma, S.; Koyama, Y. Roles of Astrocytic Endothelin ET_B Receptor in Traumatic Brain Injury. *Cells* **2023**, *12*, 719. https://doi.org/10.3390/cells12050719

Academic Editors: Michael Brenner and Vladimir Parpura

Received: 25 November 2022
Revised: 8 February 2023
Accepted: 22 February 2023
Published: 24 February 2023

1. Introduction

Traumatic brain injury (TBI) is critical damage to the brain caused by a sudden insult, such as traffic accidents, falls, collisions, and sporting activities. TBI is a major cause of death and disability worldwide. Even in surviving patients, TBI causes severe sequelae in motor, sensory, mental, and cognitive functions, resulting in decreased quality of life (QOL). Therefore, much effort has been devoted to realizing effective therapies for TBI, that is, treatments to protect the brain from damage in the acute phase and promote the recovery of neurological function in TBI patients with sequelae.

Brain damage caused by TBI is classified as primary or secondary [1,2]. Primary damage includes direct physical injury to the brain parenchyma, such as skull fractures, intracranial hemorrhage, compression/deformation of nerve tissue, diffuse axonal injury, and crushing of blood vessels. Biochemical, cellular, and physiological alterations induced by primary damage propagate from the impact core to the peripheral area and aggravate brain injury (secondary damage) [1,2]. Pathological events that induce secondary damage in TBI around the impact core include excitotoxicity, cerebral hypoperfusion, brain edema, and neuroinflammation. While primary damage is irreversible and difficult to reduce, secondary damage is partly reversible and remediable. Therefore, therapies for TBI in the acute phase focus on reducing secondary damage. Current treatments for the acute phase of TBI include decompressive craniotomy, hyperosmolar treatment, barbiturates, sedation, and hypothermia therapy. However, these treatments are insufficient and may have adverse effects in some cases. Several candidate drugs have shown beneficial effects in preclinical studies using experimental TBI animal models, but clinical trials have failed to show significant effects in patients with TBI [3–5]. In addition, for TBI patients with sequelae, treatments to promote the recovery of neurological functions impaired by TBI are required, which are currently performed by physical therapy. Although many studies have shown that physical therapy promotes synaptic regeneration in damaged nervous systems [6–8], no medication is clinically used to enhance its efficiency. Therefore, research and development of medicines applied in the acute and recovery phases of TBI have been extensively conducted.

Many studies have clarified the roles of astrocytes in nerve damage and recovery processes in several brain disorders, including TBI [9–11]. Based on these studies, the regulation of astrocytic functions has been proposed as a novel therapeutic strategy for TBI. Endothelin (ET) is one of the factors regulating the pathophysiological functions of astrocytes in damaged nerve tissues [12]. ET receptor signaling-mediated pathophysiological reactions include ischemia, neuropathic pain, and disruption of the blood–brain barrier (BBB) [12]. We previously observed that ET_B receptor (ET_B-R) but not ET_A receptor (ET_A-R) was highly distributed in astrocytes and that an ET_B-R antagonist but not an ET_A-R antagonist alleviated pathological conditions, including the proliferation of reactive astrocytes, BBB disruption, and brain edema in TBI model mice [13]. In this review, the roles of the ET_B receptor (ET_B-R) in astrocytic functions in TBI are reviewed. In addition, the possibility of astrocytic ET_B-R as a novel drug target for TBI is discussed.

2. Pathophysiological Responses of Astrocytes to TBI

In response to brain disorders, astrocytes change their phenotype to that of reactive astrocytes, which are characterized by increased glial fibrillary acidic protein (GFAP) expression and hypertrophy. Reactive astrocytes are involved in the progression of many brain pathologies and the regeneration of the injured nervous system. In patients with TBI, phenotypic conversion to reactive astrocytes is predominantly observed in damaged areas [14]. Similarly, reactive astrocytes were also increased in experimental TBI model animals [13–16]. Brain edema occurs during the acute phase of TBI. Increased intracranial pressure accompanied by brain edema causes impairment of the nervous system and often results in the death of patients with TBI. In addition, neuroinflammation in the acute phase exacerbates neuronal damage caused by TBI and causes various neurological dysfunctions in patients affecting motor, sensory, and cognitive activities. Disruption of the BBB underlies the development of brain edema and neuroinflammation caused by TBI. That is, the hyperpermeability of brain microvascular endothelial cells, which constitute the BBB, can allow the infiltration of inflammatory cells and serum proteins into the cerebral parenchyma damaged by TBI. Astrocytes support the integrity of the BBB, and their end feet surround a large part of the basolateral side of brain microvessels. The permeability of brain microvascular endothelial cells responsible for the BBB is regulated by their interaction with astrocytes. Functional alterations in astrocytes in TBI lead to excessive hyperpermeability of the BBB, which allows the entry of inflammatory cells and serum proteins (Figure 1).

Increased production of various cytokines and chemokines in the acute phase of TBI has been reported [17,18]. The production of astrocytic cytokines and chemokines is stimulated by several signaling molecules released from damaged cells [19–21]. In TBI, astrocytic IL-33 is increased in the human and mouse brain and promotes the accumulation of microglia/macrophages at the site of injury [22]. Xue et al. [23] showed that astrocytes produce C-C Motif Chemokine Ligand 7 (CCL7), which promotes microglia-mediated inflammation in a TBI rat model. Other astrocytic vascular permeability regulators, such as matrix metalloproteinase 9 (MMP9) and vascular endothelial growth factor-A (VEGF-A), are also upregulated by TBI and cause disruption of the BBB [13,24–27]. These results indicate that astrocytes have a detrimental effect on BBB function during the acute phase of TBI. However, some reports suggest that astrocytes have supporting roles in BBB function, by which brain edema and neuroinflammation in TBI are reduced. Hu et al. [28] showed that the ablation of astrocytes exacerbated the infiltration of monocytes into the cerebral parenchyma and neuronal loss in mice with brain stab injuries. Gao et al. [29] found that programmed cell death 1 (PD-L1) signaling in reactive astrocytes prevented excessive neuroimmune and neuronal damage in a controlled cortical impact-induced TBI mouse model. Astrocyte-derived exosomes also protect hippocampal neurons by suppressing mitochondrial oxidative stress and apoptosis in rats with TBI [30]. We also found that astrocyte-derived vascular protective factors, such as angiopoietin-1 (ANG-1) and sonic hedgehog (SHH), were increased in TBI model mice [31,32]. These findings suggest that astrocytes suppress TBI-induced neuroinflammation and BBB disruption and exert protective actions against neuronal damage in TBI. Additionally, expression levels of MMP-9 and VEGF-A increased at 6 h to 5 days after TBI, whereas expression levels of ANG-1 and SHH increased at 3 to 10 days after TBI [13,31,32]. This finding shows a possibility that conversion to reactive astrocytes enhances both detrimental and supportive actions on BBB function depending on the phase of TBI.

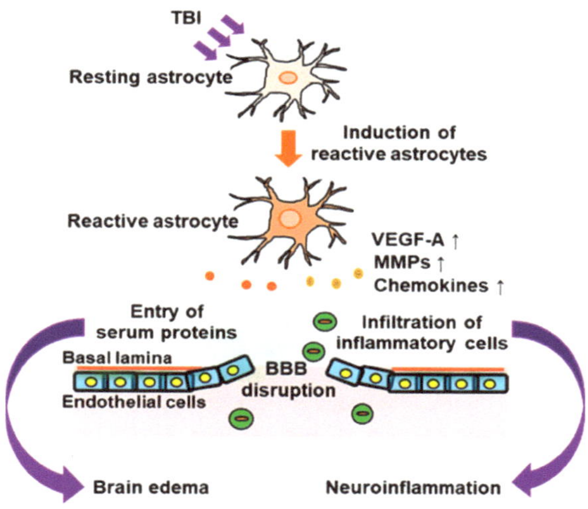

Figure 1. Involvement of astrocytes in BBB disruption in the acute phase of traumatic brain injury (TBI). In the acute phase of TBI, activation of astrocytes occurs around the damaged brain area. The astrocytic activation is accompanied by increased expressions of astrocyte-derived vascular permeability factors, such as vascular endothelial growth factor-A (VEGF-A), matrix metalloproteinase 9 (MMP9), and chemokines. This impairs the barrier function of the blood–brain barrier (BBB) and allows entry of inflammatory cells and serum proteins into the brain parenchyma, which leads to neuroinflammation and brain edema.

During the recovery phase of TBI, new synapses are formed in damaged areas, which are supported by neurogenesis from neural progenitors and axonal elongation [33]. This remodeling of the damaged nervous system is the mechanism that underlies the recovery of brain functions impaired by TBI. Some astrocyte-derived factors have been shown to promote the remodeling of the nervous system damaged by TBI. Astrocyte-produced apolipoprotein E [34] and S100b [35] promote neurogenesis and recovery of cognitive function impairments in TBI. Thrombospondins (TSPs) are astrocyte-secreted proteins that promote synaptogenesis [36,37]. Cheng et al. reported that TSP-1 was increased in TBI and that TSP-1 knockout mice exhibited significantly worse neurological deficits in motor and cognitive functions [38]. Production of neurotrophin family neurotrophic factors is upregulated by TBI and promotes neuroprotection in the acute phase, as well as regeneration in the recovery phase [39]. Astrocytes are the major source of nerve growth factor (NGF), which is also upregulated by TBI [40]. Administration of NGF to the rat brain reversed the decrease in cholinergic nerves induced by TBI and enhanced cognitive function [41]. Treatment to increase the production of brain-derived neurotrophic factor (BDNF) in astrocytes restored neuronal function impaired by TBI [42,43]. Hao et al. showed that exogenous neurotrophin-3 (NT-3) administration to TBI model rats promoted neural stem cell proliferation and synaptogenesis [44]. These findings indicate that the ability of astrocytes to produce neurotrophic factors is beneficial in promoting nerve regeneration during the recovery phase of TBI.

3. Endothelin in TBI

3.1. Endothelin Receptor Signal Pathways and Pathophysiological Reactions

The ET ligand family, ET-1, ET-2, and ET-3, was initially discovered as vasoconstrictor peptides produced by vascular endothelial cells. ET-1 is present in the brain and is one of the factors involved in pathophysiological responses of damaged nerve tissues [45,46]. ET-2 is largely limited to the gastrointestinal tract, sex organs, and pituitary gland, and ET-3 is abundantly expressed in the intestine, pituitary gland, and brain [12]. Receptors for ETs, which are seven-transmembrane G-protein-coupled receptors, are classified into two distinct types: ET_A-R and ET_B-R, which are encoded by the EDNRA and EDNRB genes, respectively (Figure 2). ET_A-R shows higher affinities for ET-1 and ET-2 than for ET-3, whereas ET_B-R has an equal affinity for all three ET ligands. Both ET_A-R and ET_B-R are linked to the Gq protein and increase intracellular Ca^{2+} by activating phospholipase C (PLC). However, ET_A-R and ET_B-R have different regulatory mechanisms for adenylate cyclase-mediated signals. ET_A-R is Gs-linked to increase cAMP, whereas ET_B-R is linked to Gi and suppresses the signal (Figure 2). Both ET_A-R and ET_B-R are also linked to the $G_{12/13}$ protein (Figure 2). The $G_{12/13}$ protein-mediated signal activates the Rho protein, a low molecular weight G-protein, and stimulates Rho-associated protein kinase (ROCK), which regulates cellular proliferation, Ca^{2+}, and cytoskeletal actin reorganization [12].

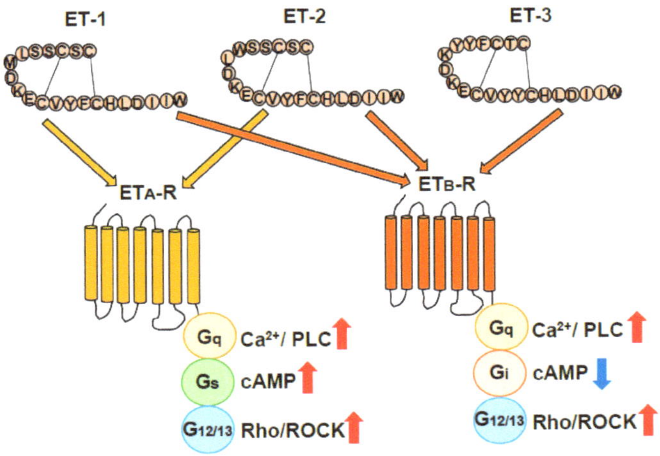

Figure 2. Classification of endothelin receptors. Endothelins (ETs) exert their physiological actions through two G-protein-coupled receptor subtypes, ET_A (ET_A-R) and ET_B receptors (ET_B-R). ET_A receptor signal pathways mediate Gq, Gs, and $G_{12/13}$ resulting in activation of Ca^{2+}/phospholipase C (PLC), cAMP, and $G_{12/13}$/Rho/Rock signal pathways. ET_B-R signal pathways mediate Gq, Gi, and $G_{12/13}$ resulting in activation of Ca^{2+}/PLC and Rho/Rock signal pathways and suppression of cAMP signaling pathways [12].

Several studies imply that ET-R-mediated calcium-dependent signaling contributes to neuroinflammation. ET-1 increases intracellular Ca^{2+} by the influx of calcium and release of intracellular calcium stores. ET_A-R-mediated calcium-dependent responses include activation and degranulation of neutrophils by calcium influx [47]. ET_B-R-mediated calcium-dependent responses include the chemotactic migration of neutrophils by the release of intracellular calcium [47]. In addition, astrocytic dysregulation of Ca^{2+} homeostasis promotes the release of inflammatory factors, which cause neuroinflammation in Alzheimer's disease [48]. Because ET_B-R is highly expressed in astrocytes and ET_B-R signaling promotes the production of astrocytic bioactive factors [12,13], ET_B-R signaling may be involved in neuroinflammation by astrocytic dysregulation of Ca^{2+} homeostasis. These observations suggest that activation of ET-Rs and neuroinflammation are correlated to calcium homeostasis dysregulation.

The pathophysiological roles of ETs in the cardiovascular system, such as arterial hypertension, myocardial infarction, preeclampsia, and coronary atherosclerosis, have been well investigated. In the clinical state, several ET-R antagonists, including bosentan (non-selective), ambrisentan (ET_A-R selective), and macitentan (non-selective), have been applied as therapeutic drugs for pulmonary arterial hypertension. ET antagonists are also expected to be therapeutic drugs for cardiovascular and other disorders [49]. In patients with several neurodegenerative disorders, cerebral ischemia, and subarachnoid hemorrhage, expression levels of ETs are increased [50–53]. Some studies showed that in addition to vascular endothelial cells, astrocytes are also a primary source of ET-1. The expression level of astrocytic ET-1 is increased by several factors, including cytokines, hypoxia, and ET-1 itself [54–56].

3.2. Relationships of Endothelin and TBI

Brain ET-1 production was found to be increased by TBI, whereas productions of ET-2 and ET-3 have not been investigated. Studies in patients with TBI and experimental animal models indicate that increases in brain ET-1 are closely related to the pathological conditions of TBI. Maier et al. [57] showed that ET-1 is increased in the plasma and cerebrospinal fluid of patients with TBI. Chen et al. [58] also found that ET-1 levels were significantly higher in the severe TBI group than in the mild/moderate TBI and control groups. Additionally, ET-1 is also increased in the cerebrospinal fluid and is associated with unfavorable outcomes in children after severe TBI [59]. In experimental TBI models, ET-1 levels are increased in the brain [14,27,56,60,61]. Histological observations have shown that reactive astrocytes produce ET-1 in TBI [56]. An examination using cultured cells showed that a direct physical effect on astrocytes is partially involved in TBI-induced astrocytic ET-1 production [31]. TBI causes cerebral circulation dysfunction, such as vasospasm and hyperpermeability of brain microvessels, which aggravates secondary damage [62]. Experimental animal models of cerebral ischemia and subarachnoid hemorrhage have shown that increases in brain ET-1 cause vasospasm through ET_A-R [63,64]. ET_A-R mediates vasospasm through protein kinase C, whereas ET_B-R mediates vasodilation through nitric oxide synthesis [64]. These experimental results suggest that antagonism of ET_A-R may also alleviate vasospasm in the brain damaged by TBI.

ET_B-Rs are expressed in astrocytes and brain vascular endothelial cells [56,65–68], whereas ET_A-Rs are highly expressed in brain blood vessels but not in astrocytes [13]. Expression of astrocytic ET_B-Rs is also upregulated in response to brain disorders, as well as astrocytic ET-1 [67–69]. We found that ET_B-R expression in reactive astrocytes was significantly increased in the injured cerebrum of TBI mice [56]. The concomitant increases in both astrocytic ET ligands and receptors suggest an autocrine/paracrine mechanism mediated by ET-1/ET_B-R signaling, which becomes more prominent in the regulation of astrocytic pathophysiological actions in TBI (Figure 3). Therefore, the role of ET_B-R-mediated signaling in astrocytic pathophysiological responses has been investigated.

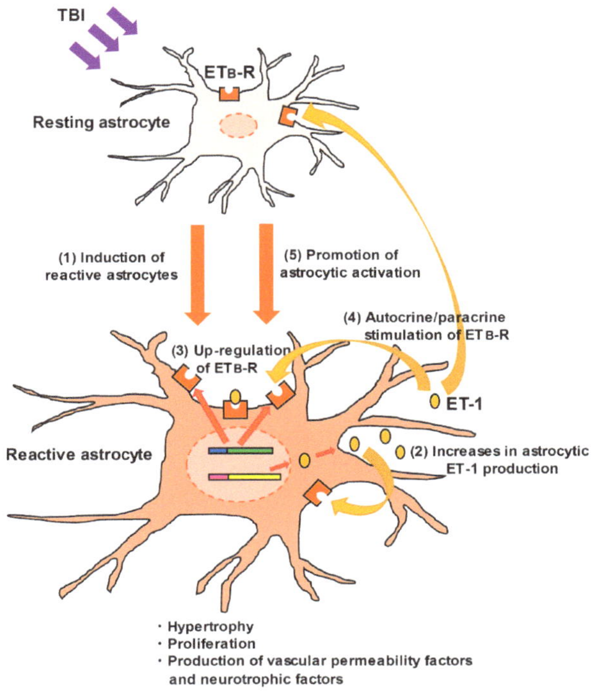

Figure 3. Promotion of astrocytic activation through ET-1/ET_B-R signaling. In TBI, ET-1 induces reactive astrocytes through the following mechanisms: (1) TBI promotes the conversion of resting astrocytes to reactive ones. (2) ET-1 is produced and released from reactive astrocytes. The physical impact of TBI directly stimulates ET-1 production, although the mechanism is unclear [27,31]. (3) As well as ET-1, expression of astrocytic ET_B-R is upregulated, accompanied by astrocytic activation. (4) Astrocyte-derived ET-1 stimulates ET_B-R through an autocrine/paracrine mechanism. (5) ET_B-R stimulation further promotes conversion to reactive astrocytes and enhances their functions.

4. Regulation of Astrocytic Functions by ET_B-R

4.1. ET_B-R-Mediated Conversion to Reactive Astrocytes

Conversion of resting astrocytes to reactive phenotypes in response to brain injury is characterized by increased GFAP expression and hypertrophy. Hyperplasia of reactive astrocytes causes glial scar formation in injured areas, which prevents axonal elongation during the recovery of damaged nerve tissues. ET-1 is one of the factors that induce conversion to reactive astrocytes in injured nerve tissues. Intracerebroventricular administration of $Ala^{1,3,11,15}$-ET-1, a selective ET_B-R agonist, promotes the reactive conversion of astrocytes in the rat brain [65]. In experimental brain injury models, inhibition of ET_B-R decreases the number of reactive astrocytes [69–72]. We also found that BQ788, a selective ET_B-R antagonist, reduced the induction of reactive astrocytes in TBI model mice [13].

4.2. ET_B-R-Mediated Proliferation Signal Pathways in Astrocytes

The proliferation of reactive astrocytes is regulated by multiple intracellular signaling pathways. Several studies have suggested that signal transducer and activator of transcription 3 (STAT3) is a key transcription factor for astrocytic proliferation in brain injury [73–75]. Selective nuclear localization of phosphorylated STAT3 in reactive astrocytes was observed in the cerebrum of TBI model rats [76]. We also found that phosphorylated STAT3 increased in the cerebrum of mice with TBI [77]. In cultured astrocytes, ET-1 promotes STAT3 activation through ET_B-R, and inhibition of STAT3 reduces cell proliferation [77], suggesting that

STAT3 is involved in ET-induced astrocytic proliferation. Transcriptional target molecules of STAT3 include cyclin D1 and S-phase kinase-associated protein 2 (Skp-2), cell cycle regulators that stimulate the G1/S transition [77,78]. Histological observations in brain injury models have shown that cyclin D1 is expressed in reactive astrocytes [79]. The increase in cyclin D1 expression mediates astrocytic proliferation triggered by ET_B-R/STAT3 signaling [77]. Other transcription factors than STAT3 are also involved in astrocytic cyclin D1 expression. We found that ET-1-activated specificity protein-1 (SP-1) promotes the expression of cyclin D1 and the proliferation of cultured astrocytes through ET_B-R [80]. ET-1-induced SP-1 activation and cyclin D1 expression were suppressed by inhibitors of mitogen-activated protein kinases (MAPKs) [80]. Gadea et al. [71] also showed that ET_B-R-mediated astrocyte proliferation and activation occur through the activation of ERK- and JNK-dependent pathways. ET-induced astrocytic proliferation is coordinately regulated by both cell adhesion-dependent and independent pathways under ET_B-R activation [81]. The expression of cyclin D1 through the ERK and JNK pathways is in a cell adhesion-independent pathway, whereas astrocytic cyclin D3, another G1/S cyclin, is increased by the stimulation of ET_B-R in a cell adhesion-dependent manner. In the adhesion-dependent mechanism, a small G-protein (Rho) and focal adhesion kinase (FAK) mediate the cell proliferation signal from ET_B-R to cyclin D3 [82]. This diversity of proliferation pathways characterizes the ET_B-R signal in hyperplasia of reactive astrocytes, leading to glial scar formation (Figure 4).

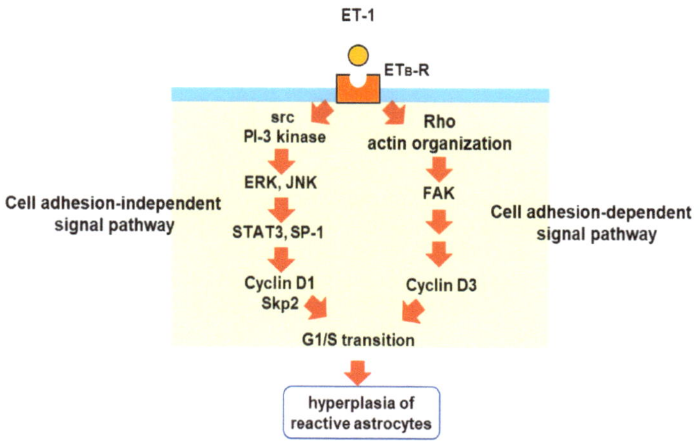

Figure 4. Proliferation signal pathways triggered by astrocytic ET_B-R. ET_B-R-mediated intracellular signals for astrocytic proliferation include cell adhesion-dependent and -independent pathways. In the cell adhesion-dependent pathway, stimulation of ET_B-R causes cytoskeletal actin reorganization through a small G-protein, rho. The formation of focal adhesions accompanied by cytoskeletal reorganization activates FAK and increases cyclinD3 expression. In the cell adhesion-independent signal pathway, stimulation of ET_B-R increases the activities of STAT3 and SP-1 through ERK and JNK. These transcription factors increase the expressions of cyclinD1 and S-phase kinase-associated protein 2 (Skp-2). The upregulation of these cell cycle regulatory proteins by ET-1 induces the proliferation of reactive astrocytes in brain disorders, including TBI.

4.3. ET_B-R- Mediated Production of Bioactive Factors in Astrocytes

In the damaged brain, reactive astrocytes produce multiple bioactive factors that increase the permeability of microvascular endothelial cells. The production of these factors disrupts the BBB, which leads to brain edema and infiltration of inflammatory cells. Activation of ET_B-R in cultured astrocytes and rat brains stimulated the production of MMP-9, MMP-2, VEGF-A, and stromelysin-1 [83–87], which increases vascular permeability. At brain injury sites, activated astrocytes produce monocyte chemoattractant protein-1 (MCP-1)/C-C motif chemokine 2 (CCL2) and cytokine-induced neutrophil chemoattractant-1 (CINC-1)/C-X-C motif chemokine ligand 1 (CXCL1) [88–90]. In addition to increasing the vascular permeability of brain microvessels, these chemokines serve as chemoattractants that lead to blood inflammatory cells in the brain. The production of astrocytic MCP-1/CCL2 and CINC-1/CXCL1 is increased by the stimulation of ET_B-R [21,90]. In contrast, stimulation of ET_B-R decreased the expression of astrocytic ANG-1 and SHH [31,32], which are factors that enhance the integrity of brain microvessels and protect the BBB from brain injury [31,32,86]. These actions of ET-1 on vascular permeability-regulating factors suggest that the activation of astrocytic ET_B-R signaling impairs BBB function, resulting in brain edema and neuroinflammation.

As described above, astrocytes also play a role in supporting neuroprotection and repair of damaged nerve systems. Stimulation of ET_B-R increases the production of factors promoting neuroprotection/neuronal regeneration, including glial cell line-derived neurotrophic factor (GDNF), BDNF, NT-3, NGF, and basic fibroblast growth factor (bFGF) [91–93]. Ephrin is a membrane-associated protein family that regulates adhesion-dependent cellular functions by interacting with Eph receptors [94]. Several ephrin subtypes are expressed in astrocytes. Increased expression of astrocytic ephrins in brain disorders inhibited axon elongation and synaptogenesis during the regeneration of damaged nerve tissues [95–100]. The expression of ephrin-A2, -A4, -B2, and -B3 in cultured astrocytes was decreased by stimulation of ET_B-R [101]. Both the increase in neurotrophic factors and the decrease in ephrins suggest that stimulation of astrocytic ET_B receptors plays a role in promoting neuroprotection/neuronal regeneration of damaged nerve tissues.

5. Roles of Astrocytic ET_B-R in the Acute Phase of TBI

Under normal conditions, the function of the BBB is maintained by a balance between astrocyte-derived factors that increase vascular permeability and enhance the integrity of brain microvascular endothelial cells. In the acute phase of TBI, the production of astrocytic factors that enhance vascular permeability causes BBB disruption in the injured areas, which results in brain edema and neuroinflammation. ET-1 is upregulated in the acute phase of TBI and regulates the production of these astrocytic factors through ET_B-R [13]. Therefore, we investigated the effects of ET_B-R antagonists on BBB disruption in fluid percussion injury (FPI)-induced TBI model mice [13]. In TBI model mice, BBB permeability was remarkably increased one to five days after FPI. Furthermore, an increased expression level of ET_B-R and proliferation of reactive astrocytes were also observed [13,56]. Thus, proliferation signal pathways of astrocytic ET_B-R are activated and promote BBB permeability in TBI. Intracerebroventricular administration of the ET_B-R antagonist BQ788 from 2 to 5 days after FPI significantly reduced TBI-induced BBB disruption and brain edema in the acute phase (three to five days after FPI), accompanied by a reduction in the number of GFAP-positive reactive astrocytes [13,31,32]. Thus, an ET_B-R antagonist inhibits excessive proliferation signal pathways of astrocytic ET_B-R, resulting in the reduction in BBB disruption and brain edema. On the other hand, administration of the ET_A-R antagonist FR139317 from 2 to 5 days after FPI had no significant effects on BBB disruption and increased reactive astrocytes [13]. Guo et al. [102] found that BQ788 reduced neutrophils and monocytes and decreased the expression levels of cytokines and inflammatory mediators in the spinal cord after contusion injury. Although no selective ET_B receptor antagonists, including BQ788, have been clinically applied, some ET antagonists have been used as therapeutics for pulmonary arterial hypertension [49]. We also found that the intravenous administration of bosentan, a non-selective ET antagonist, reduced FPI-induced BBB disruption and brain edema [56]. These results indicate that antagonism of ET_B-R during the acute phase of TBI is effective in protecting the BBB.

BBB protection by ET_B-R antagonists is mediated by the normalization of the expression of astrocyte-derived vascular permeability regulators. TBI increased the expression levels of MMP-9 and VEGF-A in reactive astrocytes [13]. Administration of BQ788 and bosentan significantly decreased the expression of MMP-9 and VEGF-A in FPI-induced TBI model mice [13,56]. On the other hand, expression levels of astrocytic ANG-1 and SHH, which were increased by TBI, were further increased by the administration of BQ788 and bosentan [31,32,56]. Supporting the results in a TBI animal model, we confirmed that the altered expression of these permeability regulators by ET-1 was inhibited by BQ788 in cultured astrocytes [31,32,56]. These results imply that antagonism of ET_B-R in the acute phase of TBI can alleviate BBB disruption by controlling the expression of astrocyte-derived bioactive factors (Figure 5A).

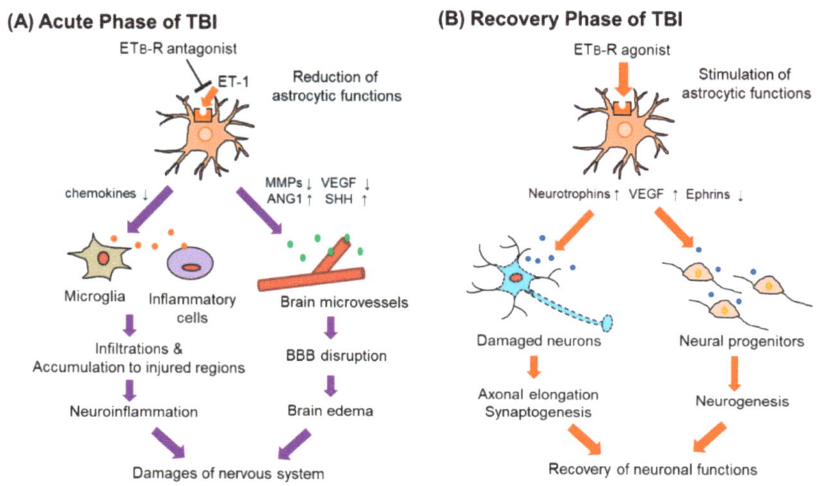

Figure 5. Astrocytic ET_B-R as a therapeutic target of TBI. Astrocytic ET_B-R may have different significance as a therapeutic target for patients with TBI in the acute and recovery phases. (**A**) In acute TBI, increased ET-1 increases the expression of MMPs and VEGF-A through the astrocytic ET_B-R. It reduces angiopoietin-1 (ANG-1) and sonic hedgehog (SHH) production. Altered production of these vascular permeability regulators causes BBB disruption that induces brain edema. Stimulation of ET_B-Rs also attracts inflammatory cells to the site of injury through increased production of chemokines. Administration of ET_B-R antagonists in the acute phase of TBI reduces these astrocytic functions and may protect the nervous system. (**B**) In the recovery phase of TBI, the administration of ET_B-R agonists is expected to be effective in recovering neuronal function impaired by TBI. Stimulation of astrocytic ET_B-Rs promotes the production of neurotrophic factors, which promotes synaptogenesis and neurogenesis. ET_B-R agonists also reduce the expression of ephrins. Then, regeneration of the damaged nervous system is promoted.

6. Roles of Astrocytic ET_B-R in the Recovery Phase of TBI

Neuroinflammation and cerebral ischemia in the acute phase of TBI induce neuronal degradation in injured areas, leading to impairment of motor, cognitive, and emotional functions in patients with TBI [1,2]. To improve these nerve dysfunctions, it is important to reduce damage in the acute phase of TBI and promote neurogenesis and synaptic regeneration in the recovery phase, which is conventionally performed by physical therapy. The direct application of several neurotrophic factors was initially trialed to recover nerve dysfunction in patients with brain insults and neurodegenerative diseases. However, owing to the difficulty of intracerebral delivery and the occurrence of side effects, none of them has been clinically used [5,103]. Recently, a clinical trial of genetically modified bone marrow-derived mesenchymal stem cells (SB623) in patients with TBI in the recovery phase was conducted to examine the effectiveness of recovery of impaired nerve functions [104]. However, no low molecular weight drugs used for peripheral administration have been clinically applied, and a new therapeutic strategy is desired.

Reactive astrocytes produce many factors that regulate the regeneration of the nervous system, including neurotrophic factors and cell adhesion molecules [105,106]. Regulating the production of astrocyte-derived factors is proposed as an effective approach to promote the regeneration of the damaged nervous system [107]. Some astrocyte-derived factors have been shown to play roles in nerve regeneration after TBI [108,109], suggesting that impaired nerve function can be recovered by regulating astrocytic function. ET_B-R signaling stimulates the production of neurotrophic factors, including NGF, bFGF, BDNF, NT-3, and VEGF, [83–85,91–93]. In addition, ET_B-R signaling decreased the expression of ephrins, which are negative regulators of neurite outgrowth and neurogenesis in the repair of damaged nerve systems [101]. The effects of ET_B-R signaling on the expression of trophic factors and ephrins suggest that stimulation of astrocytic ET_B-R promotes the regeneration of the nervous system. In TBI model mice, we found that increased expression of ET_B-R and reactive astrocytes were also observed not only in the acute phase but also in the recovery phase (seven days after FPI) [13,56]. Thus, the proliferation signal pathway of astrocytic ET_B-R is also activated in the recovery phase and enhances the production of astrocytic neurotrophic factors. In addition, ET-1 may affect neural progenitor cells to promote the regeneration of the nervous system through ET_B-R. During the development of the enteric nervous system, the proliferation and migration of neural progenitor cells are regulated by the ET-3/ET_B-R signal [110,111]. Nishikawa et al. observed in an ex vivo culture system of mouse cortical tissue that an ET_B-R agonist, IRL-1620 (sovateltide), enhanced interkinetic nuclear migration of cerebral neural progenitor cells, whereas BQ788, an ET_B-R antagonist, reduced it [112]. This report also showed that BQ788 decreased the number of newborn neurons. Although the role of ET_B-R in nerve regeneration after TBI has not been investigated, some studies in animal models have suggested that stimulation of ET_B-R promotes regeneration of the damaged nervous system. Leonard et al. [113,114] showed that intravenous treatment with IRL-1620 enhanced angiogenesis and neurogenesis and reduced neurological damage in a rat model of permanent cerebral ischemia. Briyal et al. [115] suggested that IRL-1620 prevents beta-amyloid-induced oxidative stress and cognitive impairment in rats. These findings raise the possibility that stimulation of ET_B-R during the recovery phase of TBI also has a beneficial effect on the improvement of impaired nerve function (Figure 5B).

7. Significance of Astrocytic ET_B-R as a Therapeutic Target of TBI

In this article, we have discussed the importance of ET_B-R signaling in regulating the pathophysiological functions of astrocytes, which show both detrimental and beneficial effects on the nervous system in response to brain disorders. Both astrocytic actions were promoted by stimulation with ET_B-R. This indicates that the selective use of ET_B-R antagonists/agonists according to the pathological course of TBI is necessary to obtain therapeutic benefits. In the acute phase of TBI, the increased production of ET-1 induces BBB disruption and brain edema through astrocytic ET_B-R. Therefore, ET_B-R antagonists may be effective in reducing secondary damage. Some drugs with ET_B-R antagonistic activity, such as bosentan and macitentan, have been clinically applied for pulmonary hypertension [49]. These ET antagonists are expected to expand their clinical application to protect against nerve damage during the acute phase of TBI. In contrast, stimulation of astrocytic ET_B-R increases the production of several neurotrophic factors and decreases the expression of ephrin subtypes. Because these ET_B-R-mediated functional alterations of astrocytes can promote regeneration of the impaired nervous system, the application of ET_B-R agonists may show beneficial effects in patients with TBI during the recovery phase. In clinical trials of selective ET_B agonists, IRL-1620 has been evaluated in patients with acute cerebral ischemic stroke [116].

Despite societal demands, no effective drug treatments are currently in clinical use for patients with TBI. In addition, targeting drugs for astrocytes have not been developed although multiple studies suggest that astrocytes play key roles in the pathogenesis of TBI. Because astrocytic ET_B-R signaling is closely involved in the pathogenesis of TBI and regulates astrocyte functions including activation, proliferation, and production of bioactive factors, targeting drugs for astrocytic ET_B-R may resolve two problems at the same time Although further preclinical verification is required regarding its efficacy, astrocytic ET_B-R is expected to be a novel target for drugs that aim to protect/improve brain functions impaired by TBI.

Author Contributions: Writing, review, and editing, Y.K., S.H. and S.M. All authors have read and agreed to the published version of the manuscript.

Funding: This research was funded by the Japan Society for the Promotion of Science (grant numbers: 20K16016 and 21K06609).

Institutional Review Board Statement: Not applicable.

Informed Consent Statement: Not applicable.

Data Availability Statement: Not applicable.

Conflicts of Interest: The authors declare no conflict of interest.

References

1. Jarrahi, A.; Braun, M.; Ahluwalia, M.; Gupta, R.V.; Wilson, M.; Munie, S.; Ahluwalia, P.; Vender, J.R.; Vale, F.L.; Dhandapani, K.M.; et al. Revisiting traumatic brain injury: From molecular mechanisms to therapeutic interventions. *Biomedicines* **2020**, *8*, 389. [CrossRef]
2. Thapa, K.; Khan, H.; Singh, T.G.; Kaur, A. Traumatic brain injury: Mechanistic insight on pathophysiology and potential therapeutic targets. *J. Mol. Neurosci.* **2021**, *9*, 1725–1742. [CrossRef]
3. Maas, A.I.; Roozenbeek, B.; Manley, G.T. Clinical trials in traumatic brain injury: Past experience and current developments. *Neurotherapeutics* **2010**, *7*, 115–126. [CrossRef] [PubMed]
4. Jha, R.M.; Kochanek, P.M.; Simard, J.M. Pathophysiology and treatment of cerebral edema in traumatic brain injury. *Neuropharmacology* **2019**, *145*, 230–246. [CrossRef]
5. Lerouet, D.; Marchand-Leroux, C.; Besson, V.C. Neuropharmacology in traumatic brain injury: From preclinical to clinical neuroprotection? *Fundam. Clin. Pharmacol.* **2021**, *35*, 524–538. [CrossRef]
6. Maugeri, G.; D'Agata, V.; Trovato, B.; Roggio, F.; Castorina, A.; Vecchio, M.; Di Rosa, M.; Musumeci, G. The role of exercise on peripheral nerve regeneration: From animal model to clinical application. *Heliyon* **2021**, *7*, e08281. [CrossRef] [PubMed]
7. Theisen, C.C.; Sachdeva, R.; Austin, S.; Kulich, D.; Kranz, V.; Houle, J.D. Exercise and peripheral nerve grafts as a strategy to promote regeneration after acute or chronic spinal cord injury. *J. Neurotrauma* **2017**, *34*, 1909–1914. [CrossRef] [PubMed]
8. Sachdeva, R.; Theisen, C.C.; Ninan, V.; Twiss, J.L.; Houlé, J.D. Exercise dependent increase in axon regeneration into peripheral nerve grafts by propriospinal but not sensory neurons after spinal cord injury is associated with modulation of regeneration-associated genes. *Exp. Neurol.* **2016**, *276*, 72–82. [CrossRef] [PubMed]
9. Lee, H.G.; Wheeler, M.A.; Quintana, F.J. Function and therapeutic value of astrocytes in neurological diseases. *Nat. Rev. Drug Discov.* **2022**, *21*, 339–358. [CrossRef] [PubMed]
10. Gorshkov, K.; Aguisanda, F.; Thorne, N.; Zheng, W. Astrocytes as targets for drug discovery. *Drug Discov. Today* **2018**, *23*, 673–680. [CrossRef]
11. Zhou, Y.; Shao, A.; Yao, Y.; Tu, S.; Deng, Y.; Zhang, J. Dual roles of astrocytes in plasticity and reconstruction after traumatic brain injury. *Cell Commun. Signal.* **2020**, *18*, 62. [CrossRef] [PubMed]
12. Koyama, Y. Endothelin ET_B receptor-mediated astrocytic activation: Pathological roles in brain disorders. *Int. J. Mol. Sci.* **2021**, *22*, 4333. [CrossRef]
13. Michinaga, S.; Kimura, A.; Hatanaka, S.; Minami, S.; Asano, A.; Ikushima, Y.; Matsui, S.; Toriyama, Y.; Fujii, M.; Koyama, Y. Delayed administration of BQ788, an ET_B antagonist, after experimental traumatic brain injury promotes recovery of blood-brain barrier function and a reduction of cerebral edema in mice. *J. Neurotrauma* **2018**, *35*, 1481–1494. [CrossRef] [PubMed]
14. Castejón, O.J. Morphological astrocytic changes in complicated human brain trauma. A light and electron microscopic study. *Brain Inj.* **1998**, *12*, 409–427. [CrossRef]
15. Dunn, C.; Sturdivant, N.; Venier, S.; Ali, S.; Wolchok, J.; Balachandran, K. Blood-brain barrier breakdown and astrocyte reactivity evident in the absence of behavioral changes after repeated traumatic brain injury. *Neurotrauma Rep.* **2021**, *2*, 399–410. [CrossRef] [PubMed]
16. Zhai, Y.; Ye, S.Y.; Wang, Q.S.; Xiong, R.P.; Fu, S.Y.; Du, H.; Xu, Y.W.; Peng, Y.; Huang, Z.Z.; Yang, N.; et al. Overexpressed ski efficiently promotes neurorestoration, increases neuronal regeneration, and reduces astrogliosis after traumatic brain injury. *Gene Ther.* **2022**. [CrossRef] [PubMed]
17. Prabhakar, N.K.; Khan, H.; Grewal, A.K.; Singh, T.G. Intervention of neuroinflammation in the traumatic brain injury trajectory: In vivo and clinical approaches. *Int. Immunopharmacol.* **2022**, *108*, 108902. [CrossRef]

18. Goodman, J.C.; Van, M.; Gopinath, S.P.; Robertson, C.S. Pro-inflammatory and proapoptotic elements of the neuroinflammatory response are activated in traumatic brain injury. *Acta Neurochir. Suppl.* **2008**, *102*, 437–439. [CrossRef]
19. Pedrazzi, M.; Patrone, M.; Passalacqua, M.; Ranzato, E.; Colamassaro, D.; Sparatore, B.; Pontremoli, S.; Melloni, E. Selective proinflammatory activation of astrocytes by high-mobility group box 1 protein signaling. *J. Immunol.* **2007**, *179*, 8525–8532. [CrossRef]
20. Gorina, R.; Font-Nieves, M.; Márquez-Kisinousky, L.; Santalucia, T.; Planas, A.M. Astrocyte TLR4 activation induces a proinflammatory environment through the interplay between MyD88-dependent NFκB signaling, MAPK, and Jak1/Stat1 pathways. *Glia* **2011**, *59*, 242–255. [CrossRef]
21. Koyama, Y.; Kotani, M.; Sawamura, T.; Kuribayashi, M.; Konishi, R.; Michinaga, S. Different actions of endothelin-1 on chemokine production in rat cultured astrocytes: Reduction of CX3CL1/fractalkine and an increase in CCL2/MCP-1 and CXCL1/CINC-1. *J. Neuroinflamm.* **2013**, *10*, 51. [CrossRef] [PubMed]
22. Wicher, G.; Wallenquist, U.; Lei, Y.; Enoksson, M.; Li, X.; Fuchs, B.; Abu Hamdeh, S.; Marklund, N.; Hillered, L.; Nilsson, G.; et al. Interleukin-33 promotes recruitment of microglia/macrophages in response to traumatic brain injury. *J. Neurotrauma* **2017**, *34*, 3173–3182. [CrossRef] [PubMed]
23. Xue, J.; Zhang, Y.; Zhang, J.; Zhu, Z.; Lv, Q.; Su, J. Astrocyte-derived CCL7 promotes microglia-mediated inflammation following traumatic brain injury. *Int. Immunopharmacol.* **2021**, *99*, 107975. [CrossRef] [PubMed]
24. Ding, J.Y.; Kreipke, C.W.; Schafer, P.; Schafer, S.; Speirs, S.L.; Rafols, J.A. Synapse loss regulated by matrix metalloproteinases in traumatic brain injury is associated with hypoxia inducible factor-1alpha expression. *Brain Res.* **2009**, *1268*, 125–1234. [CrossRef] [PubMed]
25. Suehiro, E.; Fujisawa, H.; Akimura, T.; Ishihara, H.; Kajiwara, K.; Kato, S.; Fujii, M.; Yamashita, S.; Maekawa, T.; Suzuki, M. Increased matrix metalloproteinase-9 in blood in association with activation of interleukin-6 after traumatic brain injury: Influence of hypothermic therapy. *J. Neurotrauma* **2004**, *21*, 1706–1711. [CrossRef] [PubMed]
26. Gao, W.; Zhao, Z.; Yu, G.; Zhou, Z.; Zhou, Y.; Hu, T.; Jiang, R.; Zhang, J. VEGI attenuates the inflammatory injury and disruption of blood-brain barrier partly by suppressing the TLR4/NF-κB signaling pathway in experimental traumatic brain injury. *Brain Res.* **2015**, *1622*, 230–239. [CrossRef]
27. Michinaga, S.; Onishi, K.; Shimizu, K.; Mizuguchi, H.; Hishinuma, S. Pharmacological inhibition of transient receptor potential vanilloid 4 reduces vasogenic edema after traumatic brain injury in mice. *Biol. Pharm. Bull.* **2021**, *44*, 1759–1766. [CrossRef]
28. Hu, X.; Li, S.; Shi, Z.; Lin, W.J.; Yang, Y.; Li, Y.; Li, H.; Xu, Y.; Zhou, M.; Tang, Y. Partial ablation of astrocytes exacerbates cerebral infiltration of monocytes and neuronal loss after brain stab injury in mice. *Cell Mol. Neurobio.* **2022**. [CrossRef]
29. Gao, X.; Li, W.; Syed, F.; Yuan, F.; Li, P.; Yu, Q. PD-L1 signaling in reactive astrocytes counteracts neuroinflammation and ameliorates neuronal damage after traumatic brain injury. *J. Neuroinflamm.* **2022**, *19*, 43. [CrossRef]
30. Zhang, W.; Hong, J.; Zhang, H.; Zheng, W.; Yang, Y. Astrocyte-derived exosomes protect hippocampal neurons after traumatic brain injury by suppressing mitochondrial oxidative stress and apoptosis. *Aging (Albany NY)* **2021**, *13*, 21642–21658. [CrossRef]
31. Michinaga, S.; Tanabe, A.; Nakaya, R.; Fukutome, C.; Inoue, A.; Iwane, A.; Minato, Y.; Tujiuchi, Y.; Miyake, D.; Mizuguchi, H.; et al. Angiopoietin-1/Tie-2 signal after focal traumatic brain injury is potentiated by BQ788, an ET_B receptor antagonist, in the mouse cerebrum: Involvement in recovery of blood-brain barrier function. *J. Neurochem.* **2020**, *154*, 330–348. [CrossRef]
32. Michinaga, S.; Inoue, A.; Sonoda, K.; Mizuguchi, H.; Koyama, Y. Down-regulation of astrocytic sonic hedgehog by activation of endothelin ET_B receptors: Involvement in traumatic brain injury-induced disruption of blood brain barrier in a mouse model. *Neurochem. Int.* **2021**, *146*, 105042. [CrossRef]
33. Jamjoom, A.A.B.; Rhodes, J.; Andrews, P.J.D.; Grant, S.G.N. The synapse in traumatic brain injury. *Brain* **2021**, *144*, 18–31. [CrossRef]
34. Yu, T.S.; Tensaouti, Y.; Stephanz, E.P.; Chintamen, S.; Rafikian, E.E.; Yang, M.; Kernie, S.G. Astrocytic ApoE underlies maturation of hippocampal neurons and cognitive recovery after traumatic brain injury in mice. *Commun. Biol.* **2021**, *4*, 1303. [CrossRef]
35. Baecker, J.; Wartchow, K.; Sehm, T.; Ghoochani, A.; Buchfelder, M.; Kleindienst, A. Treatment with the neurotrophic protein S100B increases synaptogenesis after traumatic brain injury. *J. Neurotrauma* **2020**, *37*, 1097–1107. [CrossRef] [PubMed]
36. Christopherson, K.S.; Ullian, E.M.; Stokes, C.C.; Mullowney, C.E.; Hell, J.W.; Agah, A.; Lawler, J.; Mosher, D.F.; Bornstein, P.; Barres, B.A. Thrombospondins are astrocyte-secreted proteins that promote CNS synaptogenesis. *Cell* **2005**, *120*, 421–433. [CrossRef] [PubMed]
37. Lu, Z.; Kipnis, J. Thrombospondin 1—A key astrocyte-derived neurogenic factor. *FASEB J.* **2010**, *24*, 1925–1934. [CrossRef] [PubMed]
38. Cheng, C.; Yu, Z.; Zhao, S.; Liao, Z.; Xing, C.; Jiang, Y.; Yang, Y.G.; Whalen, M.J.; Lo, E.H.; Sun, X.; et al. Thrombospondin-1 gene deficiency worsens the neurological outcomes of traumatic brain injury in mice. *Int. J. Med. Sci.* **2017**, *14*, 927–936. [CrossRef]
39. Lin, P.H.; Kuo, L.T.; Luh, H.T. The roles of neurotrophins in traumatic brain injury. *Life* **2021**, *12*, 26. [CrossRef]
40. Goss, J.R.; O'Malley, M.E.; Zou, L.; Styren, S.D.; Kochanek, P.M.; DeKosky, S.T. Astrocytes are the major source of nerve growth factor upregulation following traumatic brain injury in the rat. *Exp. Neurol.* **1998**, *149*, 301–309. [CrossRef]
41. Dixon, C.E.; Flinn, P.; Bao, J.; Venya, R.; Hayes, R.L. Nerve growth factor attenuates cholinergic deficits following traumatic brain injury in rats. *Exp. Neurol.* **1997**, *146*, 479–490. [CrossRef]

42. Furtado, A.B.V.; Gonçalves, D.F.; Hartmann, D.D.; Courtes, A.A.; Cassol, G.; Nunez-Figueredo, Y.; Argolo, D.S.; do Nascimento, R.P.; Costa, S.L.; da Silva, V.D.A.; et al. JM-20 treatment after mild traumatic brain injury reduces glial cell pro-inflammatory signaling and behavioral and cognitive deficits by increasing neurotrophin expression. *Mol. Neurobiol.* **2021**, *58*, 4615–4627. [CrossRef] [PubMed]
43. Zhao, J.; Qu, D.; Xi, Z.; Huan, Y.; Zhang, K.; Yu, C.; Yang, D.; Kang, J.; Lin, W.; Wu, S.; et al. Mitochondria transplantation protects traumatic brain injury via promoting neuronal survival and astrocytic BDNF. *Transl. Res.* **2021**, *235*, 102–114. [CrossRef]
44. Hao, P.; Duan, H.; Hao, F.; Chen, L.; Sun, M.; Fan, K.S.; Sun, Y.E.; Williams, D.; Yang, Z.; Li, X. Neural repair by NT3-chitosan via enhancement of endogenous neurogenesis after adult focal aspiration brain injury. *Biomaterials* **2017**, *140*, 88–102. [CrossRef] [PubMed]
45. Schinelli, S. Pharmacology and physiopathology of the brain endothelin system: An overview. *Curr. Med. Chem.* **2006**, *13*, 627–638. [CrossRef] [PubMed]
46. Dashwood, M.R.; Loesch, A. Endothelin-1 as a neuropeptide: Neurotransmitter or neurovascular effects? *J. Cell Commun. Signal.* **2010**, *4*, 51–62. [CrossRef] [PubMed]
47. Tykocki, N.R.; Watts, S.W. The interdependence of endothelin-1 and calcium: A review. *Clin. Sci.* **2010**, *119*, 361–372. [CrossRef]
48. Di Benedetto, G.; Burgaletto, C.; Bellanca, C.M.; Munafò, A.; Bernardini, R.; Cantarella, G. Role of Microglia and astrocytes in Alzheimer's disease: From neuroinflammation to Ca^{2+} homeostasis dysregulation. *Cells* **2022**, *11*, 728. [CrossRef]
49. Enevoldsen, F.C.; Sahana, J.; Wehland, M.; Grimm, D.; Infanger, M.; Krüger, M. Endothelin receptor antagonists: Status quo and future perspectives for targeted therapy. *J. Clin. Med.* **2020**, *9*, 824. [CrossRef]
50. Sapira, V.; Cojocaru, I.M.; Lilios, G.; Grigorian, M.; Cojocaru, M. Study of endothelin-1 in acute ischemic stroke. *Rom. J. Intern. Med.* **2010**, *48*, 329–332.
51. Palmer, J.C.; Barker, R.; Kehoe, P.G.; Love, S. Endothelin-1 is elevated in Alzheimer's disease and upregulated by amyloid-β. *J. Alzheimers* **2012**, *29*, 853–861. [CrossRef] [PubMed]
52. Yan, Y.; Fu, J. Plasma ApoA-1 and endothelin-1 levels changes in early Parkinson disease and its relationship with cognitive function and cerebral white matter structure change. *Pak. J. Pharm. Sci.* **2021**, *34*, 487–492.
53. Bellapart, J.; Nasrallah, F.; Winearls, J.; Lassig-Smith, M.; Stuart, J.; Boots, R.; Winter, C.; Flaws, D.; Bulmer, A.; Jones, L.; et al. Diagnostic sensitivity of plasma endothelin-1 for cerebral vasospasm after subarachnoid hemorrhage, a multicentre double-blind Sstudy. *J. Intensive Care Med.* **2022**, *37*, 769–775. [CrossRef] [PubMed]
54. Liu, X.; Deng, F.; Yu, Z.; Xie, Y.; Hu, C.; Chen, L. Inhibition of endothelin A receptor protects brain microvascular endothelial cells against hypoxia-induced injury. *Int. J. Mol. Med.* **2014**, *34*, 313–320. [CrossRef] [PubMed]
55. Hostenbach, S.; D'Haeseleer, M.; Kooijman, R.; De Keyser, J. Modulation of cytokine-induced astrocytic endothelin-1 production as a possible new approach to the treatment of multiple sclerosis. *Front. Pharmacol.* **2020**, *10*, 1491. [CrossRef]
56. Michinaga, S.; Inoue, A.; Yamamoto, H.; Ryu, R.; Inoue, A.; Mizuguchi, H.; Koyama, Y. Endothelin receptor antagonists alleviate blood-brain barrier disruption and cerebral edema in a mouse model of traumatic brain injury: A comparison between bosentan and ambrisentan. *Neuropharmacology* **2020**, *175*, 108182. [CrossRef] [PubMed]
57. Maier, B.; Lehnert, M.; Laurer, H.L.; Marzi, I. Biphasic elevation in cerebrospinal fluid and plasma concentrations of endothelin 1 after traumatic brain injury in human patients. *Shock* **2007**, *27*, 610–614. [CrossRef]
58. Chen, L.X.; Zhang, W.F.; Wang, M.; Jia, P.F. Relationship of calcitonin gene-related peptide with disease progression and prognosis of patients with severe traumatic brain injury. *Neural Regen. Res.* **2018**, *13*, 1782–1786. [CrossRef]
59. Salonia, R.; Empey, P.E.; Poloyac, S.M.; Wisniewski, S.R.; Klamerus, M.; Ozawa, H.; Wagner, A.K.; Ruppel, R.; Bell, M.J.; Feldman, K.; et al. Endothelin-1 is increased in cerebrospinal fluid and associated with unfavorable outcomes in children after severe traumatic brain injury. *J. Neurotrauma* **2010**, *27*, 1819–1825. [CrossRef]
60. Curvello, V.; Pastor, P.; Hekierski, H.; Armstead, W.M. Inhaled nitric oxide protects cerebral autoregulation and reduces hippocampal necrosis after traumatic brain injury through inhibition of ET-1, ERK MAPK and IL-6 upregulation in pigs. *Neurocrit. Care* **2019**, *30*, 467–477. [CrossRef]
61. Liu, T.; Liao, X.Z.; Zhou, M.T. Ulinastatin alleviates traumatic brain injury by reducing endothelin-1. *Transl. Neurosci.* **2021**, *12*, 1–8. [CrossRef]
62. Jullienne, A.; Obenaus, A.; Ichkova, A.; Savona-Baron, C.; Pearce, W.J.; Badaut, J. Chronic cerebrovascular dysfunction after traumatic brain injury. *J. Neurosci. Res.* **2016**, *94*, 609–622. [CrossRef]
63. Clozel, M.; Watanabe, H. BQ-123, a peptidic endothelin ET_A receptor antagonist, prevents the early cerebral vasospasm following subarachnoid hemorrhage after intracisternal but not intravenous injection. *Life Sci.* **1993**, *52*, 825–834. [CrossRef]
64. Yeung, P.K.; Shen, J.; Chung, S.S.; Chung, S.K. Targeted over-expression of endothelin-1 in astrocytes leads to more severe brain damage and vasospasm after subarachnoid hemorrhage. *BMC Neurosci.* **2013**, *14*, 131. [CrossRef]
65. Ishikawa, N.; Takemura, M.; Koyama, Y.; Shigenaga, Y.; Okada, T.; Baba, A. Endothelins promote the activation of astrocytes in rat neostriatum through ET_B receptors. *Eur. J. Neurosci.* **1997**, *9*, 895–901. [CrossRef]
66. Rogers, S.D.; Peters, C.M.; Pomonis, J.D.; Hagiwara, H.; Ghilardi, J.R.; Mantyh, P.W. Endothelin B receptors are expressed by astrocytes and regulate astrocyte hypertrophy in the normal and injured CNS. *Glia* **2003**, *41*, 180–190. [CrossRef] [PubMed]
67. Wilhelmsson, U.; Li, L.; Pekna, M.; Berthold, C.H.; Blom, S.; Eliasson, C.; Renner, O.; Bushong, E.; Ellisman, M.; Morgan, T.E.; et al. of glial fibrillary acidic protein and vimentin prevents hypertrophy of astrocytic processes and improves post-traumatic regeneration. *J. Neurosci.* **2004**, *24*, 5016–5021. [CrossRef] [PubMed]

68. Peters, C.M.; Rogers, S.D.; Pomonis, J.D.; Egnaczyk, G.F.; Keyser, C.P.; Schmidt, J.A.; Ghilardi, J.R.; Maggio, J.E.; Mantyh, P.W. Endothelin receptor expression in the normal and injured spinal cord: Potential involvement in injury-induced ischemia and gliosis. *Exp. Neurol.* **2003**, *180*, 1–13. [CrossRef] [PubMed]
69. LeComte, M.D.; Shimada, I.S.; Sherwin, C.; Spees, J.L. Notch1-STAT3-ETBR signaling axis controls reactive astrocyte proliferation after brain injury. *Proc. Natl. Acad. Sci. USA* **2015**, *112*, 8726–8731. [CrossRef]
70. Koyama, Y.; Takemura, M.; Fujiki, K.; Ishikawa, N.; Shigenaga, Y.; Baba, A. BQ788, an endothelin ET_B receptor antagonist, attenuates stab wound injury-induced reactive astrocytes in rat brain. *Glia* **1999**, *26*, 268–271. [CrossRef]
71. Gadea, A.; Schinelli, S.; Gallo, V. Endothelin-1 regulates astrocyte proliferation and reactive gliosis via a JNK/c-Jun signaling pathway. *J. Neurosci.* **2008**, *28*, 2394–2408. [CrossRef]
72. Yamasaki, R.; Fujii, T.; Wang, B.; Masaki, K.; Kido, M.A.; Yoshida, M.; Matsushita, T.; Kira, J.I. Allergic inflammation leads to neuropathic pain via glial cell activation. *J. Neurosci.* **2016**, *36*, 11929–11945. [CrossRef] [PubMed]
73. Herrmann, J.E.; Imura, T.; Song, B.; Qi, J.; Ao, Y.; Nguyen, T.K.; Korsak, R.A.; Takeda, K.; Akira, S.; Sofroniew, M.V. STAT3 is a critical regulator of astrogliosis and scar formation after spinal cord injury. *J. Neurosci.* **2008**, *28*, 7231–7243. [CrossRef] [PubMed]
74. Tsuda, M.; Kohro, Y.; Yano, T.; Tsujikawa, T.; Kitano, J.; Tozaki-Saitoh, H.; Koyanagi, S.; Ohdo, S.; Ji, R.R.; Salter, M.W.; et al. JAK-STAT3 pathway regulates spinal astrocyte proliferation and neuropathic pain maintenance in rats. *Brain* **2011**, *134*, 1127–1139. [CrossRef] [PubMed]
75. Wanner, I.B.; Anderson, M.A.; Song, B.; Levine, J.; Fernandez, A.; Gray-Thompson, Z.; Ao, Y.; Sofroniew, M.V. Glial scar borders are formed by newly proliferated, elongated astrocytes that interact to corral inflammatory and fibrotic cells via STAT3-dependent mechanisms after spinal cord injury. *J. Neurosci.* **2013**, *33*, 2870–12886. [CrossRef]
76. Oliva, A.A., Jr.; Kang, Y.; Sanchez-Molano, J.; Furones, C.; Atkins, C.M. STAT3 signaling after traumatic brain injury. *J. Neurochem.* **2012**, *120*, 710–720. [CrossRef]
77. Koyama, Y.; Sumie, S.; Nakano, Y.; Nagao, T.; Tokumaru, S.; Michinaga, S. Endothelin-1 stimulates expression of cyclin D1 and S-phase kinase-associated protein 2 by activating the transcription factor STAT3 in cultured rat astrocytes. *J. Biol. Chem.* **2019**, *294*, 3920–3933. [CrossRef]
78. Kitazawa, S.; Kitazawa, R.; Maeda, S. Transcriptional regulation of rat cyclin D1 gene by CpG methylation status in promoter region. *J. Biol. Chem.* **1999**, *274*, 28787–28793. [CrossRef]
79. Di Giovanni, S.; Movsesyan, V.; Ahmed, F.; Cernak, I.; Schinelli, S.; Stoica, B.; Faden, A.I. Cell cycle inhibition provides neuroprotection and reduces glial proliferation and scar formation after traumatic brain injury. *Proc. Natl. Acad. Sci. USA* **2005**, *102*, 8333–8338. [CrossRef] [PubMed]
80. Michinaga, S.; Ishida, A.; Takeuchi, R.; Koyama, Y. Endothelin-1 stimulates cyclin D1 expression in rat cultured astrocytes via activation of Sp1. *Neurochem. Int.* **2013**, *63*, 25–34. [CrossRef] [PubMed]
81. Cazaubon, S.; Chaverot, N.; Romero, I.A.; Girault, J.A.; Adamson, P.; Strosberg, A.D.; Couraud, P.O. Growth factor activity of endothelin-1 in primary astrocytes mediated by adhesion-dependent and -independent pathways. *J. Neurosci.* **1997**, *17*, 6203–6212. [CrossRef] [PubMed]
82. Koyama, Y.; Yoshioka, Y.; Shinde, M.; Matsuda, T.; Baba, A. Focal adhesion kinase mediates endothelin-induced cyclin D3 expression in rat cultured astrocytes. *J. Neurochem.* **2004**, *90*, 904–912. [CrossRef]
83. Koyama, Y.; Tanaka, K. Endothelins stimulate the production of stromelysin-1 in cultured rat astrocytes. *Biochem. Biophys. Res. Commun.* **2008**, *371*, 659–663. [CrossRef] [PubMed]
84. Koyama, Y.; Tanaka, K. Intracerebroventricular administration of an endothelin ET(B)-receptor agonist increases expression of matrix metalloproteinase-2 and -9 in rat brain. *J. Pharmacol. Sci.* **2010**, *114*, 433–443. [CrossRef]
85. Koyama, Y.; Maebara, Y.; Hayashi, M.; Nagae, R.; Tokuyama, S.; Michinaga, S. Endothelins reciprocally regulate VEGF-A and angiopoietin-1 production in cultured rat astrocytes: Implications on astrocytic proliferation. *Glia* **2012**, *60*, 1954–1963. [CrossRef] [PubMed]
86. Koyama, Y.; Hayashi, M.; Nagae, R.; Tokuyama, S.; Konishi, T. Endothelin-1 increases the expression of VEGF-R1/Flt-1 receptors in rat cultured astrocytes through ET_B receptors. *J. Neurochem.* **2014**, *130*, 759–769. [CrossRef]
87. Kim, J.Y.; Ko, A.R.; Hyun, H.W.; Kang, T.C. ET_B receptor-mediated MMP-9 activation induces vasogenic edema via ZO-1 protein degradation following status epilepticus. *Neuroscience* **2015**, *304*, 355–367. [CrossRef]
88. Gourmala, N.G.; Buttini, M.; Limonta, S.; Sauter, A.; Boddeke, H.W. Differential and time-dependent expression of monocyte chemoattractant protein-1 mRNA by astrocytes and macrophages in rat brain: Effects of ischemia and peripheral lipopolysaccharide administration. *J. Neuroimmunol.* **1997**, *74*, 35–44. [CrossRef]
89. Ishizuka, K.; Kimura, T.; Igata-yi, R.; Katsuragi, S.; Takamatsu, J.; Miyakawa, T. Identification of monocyte chemoattractant protein-1 in senile plaques and reactive microglia of Alzheimer's disease. *Psychiatry Clin. Neurosci.* **1997**, *51*, 135–138. [CrossRef]
90. Koyama, Y.; Baba, A.; Matsuda, T. Production of monocyte chemoattractant protein-1 and cytokine-induced neutrophil chemoattractant-1 in rat brain is stimulated by intracerebroventricular administration of an endothelin ET_B receptor agonist. *Neuroreport* **2007**, *18*, 1275–1279. [CrossRef]
91. Koyama, Y.; Tsujikawa, K.; Matsuda, T.; Baba, A. Endothelin-1 stimulates glial cell line-derived neurotrophic factor expression in cultured rat astrocytes. *Biochem. Biophys. Res. Commun.* **2003**, *303*, 1101–1105. [CrossRef]
92. Koyama, Y.; Tsujikawa, K.; Matsuda, T.; Baba, A. Intracerebroventricular administration of an endothelin ET_B receptor agonist increases expressions of GDNF and BDNF in rat brain. *Eur. J. Neurosci.* **2003**, *18*, 887–894. [CrossRef] [PubMed]

93. Koyama, Y.; Tsujikawa, K.; Matsuda, T.; Baba, A. Endothelin increases expression of exon III- and exon IV-containing brain-derived neurotrophic factor transcripts in cultured astrocytes and rat brain. *J. Neurosci. Res.* **2005**, *80*, 809–816. [CrossRef]
94. Goldshmit, Y.; McLenachan, S.; Turnley, A. Roles of Eph receptors and ephrins in the normal and damaged adult CNS. *Brain Res. Rev.* **2006**, *52*, 327–345. [CrossRef]
95. Jiao, J.W.; Feldheim, D.A.; Chen, D.F. Ephrins as negative regulators of adult neurogenesis in diverse regions of the central nervous system. *Proc. Natl. Acad. Sci. USA* **2008**, *105*, 8778–8783. [CrossRef]
96. Wu, L.Y.; Yu, X.L.; Feng, L.Y. Connexin 43 stabilizes astrocytes in a stroke-like milieu to facilitate neuronal recovery. *Acta Pharmacol. Sin.* **2015**, *36*, 928–938. [CrossRef] [PubMed]
97. Li, Y.; Chen, Y.; Tan, L.; Pan, J.Y.; Lin, W.W.; Wu, J.; Hu, W.; Chen, X.; Wang, X.D. RNAimediatedephrin-B2 silencing attenuates astroglial-fibrotic scar formation and improves spinal cord axon growth. *CNS Neurosci. Ther.* **2017**, *23*, 779–789. [CrossRef]
98. Ren, Z.; Chen, X.; Yang, J.; Kress, B.T.; Tong, J.; Liu, H.; Takano, T.; Zhao, Y.; Nedergaard, M. Improved axonal regeneration after spinal cord injury in mice with conditional deletion of ephrin B2 under the GFAP promoter. *Neuroscience* **2013**, *241*, 89–99. [CrossRef] [PubMed]
99. Dixon, K.J.; Mier, J.; Gajavelli, S.; Turbic, A.; Bullock, R.; Turnley, A.M.; Liebl, D.J. EphrinB3 restricts endogenous neural stem cell migration after traumatic brain injury. *Stem Cell Res.* **2016**, *17*, 504–513. [CrossRef] [PubMed]
100. Duffy, P.; Wang, X.; Siegel, C.S.; Tu, N.; Henkemeyer, M.; Cafferty, W.B.; Strittmatter, S.M. Myelin-derived ephrinB3 restricts axonal regeneration and recovery after adult CNS injury. *Proc. Natl. Acad. Sci. USA* **2012**, *109*, 5063–5068. [CrossRef]
101. Koyama, Y.; Tsuboi, S.; Mizogui, F. Endothelin-1 decreases the expression of Ephrin-A and B subtypes in cultured rat astrocytes through ETB receptors. *Neurosci. Lett.* **2021**, *741*, 135393. [CrossRef]
102. Guo, J.; Li, Y.; He, Z.; Zhang, B.; Li, Y.; Hu, J.; Han, M.; Xu, Y.; Li, Y.; Gu, J.; et al. Targeting endothelin receptors A and B attenuates the inflammatory response and improves locomotor function following spinal cord injury in mice. *Int. J. Mol. Med.* **2014**, *34*, 74–82. [CrossRef]
103. Mira, R.G.; Lira, M.; Cerpa, W. Traumatic brain injury: Mechanisms of glial response. *Front. Physiol.* **2021**, *12*, 740939. [CrossRef] [PubMed]
104. Kawabori, M.; Weintraub, A.H.; Imai, H.; Zinkevych, L.; McAllister, P.; Steinberg, G.K.; Frishberg, B.M.; Yasuhara, T.; Chen, J.W.; Cramer, S.C.; et al. Cell therapy for chronic TBI: Interim analysis of the randomized controlled STEMTRA Trial. *Neurology* **2021**, *96*, e1202–e1214. [CrossRef] [PubMed]
105. Chiareli, R.A.; Carvalho, G.A.; Marques, B.L.; Mota, L.S.; Oliveira-Lima, O.C.; Gomes, R.M.; Birbrair, A.; Gomez, R.S.; Simão, F.; Klempin, F.; et al. The Role of astrocytes in the neurorepair process. *Front. Cell Dev. Biol.* **2021**, *9*, 665795. [CrossRef]
106. Tan, C.X.; Eroglu, C. Cell adhesion molecules regulating astrocyte-neuron interactions. *Curr. Opin. Neurobiol.* **2021**, *69*, 170–177. [CrossRef]
107. Michinaga, S.; Koyama, Y. Pathophysiological responses and roles of astrocytes in traumatic brain injury. *Int. J. Mol. Sci.* **2021**, *22*, 6418. [CrossRef]
108. Wu, J.; Li, H.; He, J.; Tian, X.; Luo, S.; Li, J.; Li, W.; Zhong, J.; Zhang, H.; Huang, Z.; et al. Downregulation of microRNA-9-5p promotes synaptic remodeling in the chronic phase after traumatic brain injury. *Cell Death Dis.* **2021**, *12*, 9. [CrossRef]
109. Nikolakopoulou, A.M.; Koeppen, J.; Garcia, M.; Leish, J.; Obenaus, A.; Ethell, I.M. Astrocytic ephrin-B1 regulates synapse remodeling following traumatic brain injury. *ASN Neuro* **2016**, *8*, 1–18. [CrossRef] [PubMed]
110. Barlow, A.; De Graaff, E.; Pachnis, V. Enteric nervous system progenitors are coordinately controlled by the G protein-coupled receptor EDNRB and the receptor tyrosine kinase RET. *Neuron* **2003**, *40*, 905–916. [CrossRef]
111. Pawolski, V.; Schmidt, M.H.H. Neuron-glia interaction in the developing and adult enteric nervous system. *Cells* **2020**, *10*, 47. [CrossRef] [PubMed]
112. Nishikawa, K.; Ayukawa, K.; Hara, Y.; Wada, K.; Aoki, S. Endothelin/endothelin-B receptor signals regulate ventricle-directed interkinetic nuclear migration of cerebral cortical neural progenitors. *Neurochem. Int.* **2011**, *58*, 261–272. [CrossRef] [PubMed]
113. Leonard, M.G.; Briyal, S.; Gulati, A. Endothelin B receptor agonist, IRL-1620, reduces neurological damage following permanent middle cerebral artery occlusion in rats. *Brain Res.* **2011**, *1420*, 48–58. [CrossRef] [PubMed]
114. Leonard, M.G.; Gulati, A. Endothelin B receptor agonist, IRL-1620, enhances angiogenesis and neurogenesis following cerebral ischemia in rats. *Brain Res.* **2013**, *1528*, 28–41. [CrossRef]
115. Briyal, S.; Shepard, C.; Gulati, A. Endothelin Receptor Type B Agonist, IRL-1620, Prevents Beta Amyloid (Aβ) induced oxidative stress and cognitive impairment in normal and diabetic rats. *Pharmacol. Biochem. Behav.* **2014**, *120*, 65–72. [CrossRef]
116. Gulati, A.; Agrawal, N.; Vibha, D.; Misra, U.K.; Paul, B.; Jain, D.; Pandian, J.; Borgohain, R. Safety and efficacy of sovateltide (IRL-1620) in a multicenter randomized controlled clinical trial in patients with acute cerebral ischemic stroke. *CNS Drugs* **2021**, *35*, 5–104. [CrossRef] [PubMed]

cells

MDPI

Article

Chemically Functionalized Single-Walled Carbon Nanotubes Prevent the Reduction in Plasmalemmal Glutamate Transporter EAAT1 Expression in, and Increase the Release of Selected Cytokines from, Stretch-Injured Astrocytes in Vitro

Nika गržeta Krpan [1], Anja Harej Hrkać [1], Tamara Janković [1], Petra Dolenec [1], Elena Bekyarova [2], Vladimir Parpura [3] and Kristina Pilipović [1,*]

1 Department of Basic and Clinical Pharmacology and Toxicology, Faculty of Medicine, University of Rijeka, HR-51000 Rijeka, Croatia; nika.grzeta@uniri.hr (N.G.K.); aharej@uniri.hr (A.H.H.); tamara.jankovic@medri.uniri.hr (T.J.); petra.dolenec@medri.uniri.hr (P.D.)

2 Department of Chemistry, University of California, Riverside, CA 92521, USA; elenab@ucr.edu

3 International Translational Neuroscience Research Institute, Zhejiang Chinese Medical University, Hangzhou 310053, China; vlad@zcmu.edu.cn

* Correspondence: kristina.pilipovic@uniri.hr; Tel.: +385-(051)-651-242

Abstract: We tested the effects of water-soluble single-walled carbon nanotubes, chemically functionalized with polyethylene glycol (SWCNT-PEG), on primary mouse astrocytes exposed to a severe in vitro simulated traumatic brain injury (TBI). The application of SWCNT-PEG in the culture media of injured astrocytes did not affect cell damage levels, when compared to those obtained from injured, functionalization agent (PEG)-treated cells. Furthermore, SWCNT-PEG did not change the levels of oxidatively damaged proteins in astrocytes. However, this nanomaterial prevented the reduction in plasmalemmal glutamate transporter EAAT1 expression caused by the injury, rendering the level of EAAT1 on par with that of control, uninjured PEG-treated astrocytes; in parallel, there was no significant change in the levels of GFAP. Additionally, SWCNT-PEG increased the release of selected cytokines that are generally considered to be involved in recovery processes following injuries. As a loss of EAATs has been implicated as a culprit in the suffering of human patients from TBI, the application of SWCNT-PEG could have valuable effects at the injury site, by preventing the loss of astrocytic EAAT1 and consequently allowing for a much-needed uptake of glutamate from the extracellular space, the accumulation of which leads to unwanted excitotoxicity. Additional potential therapeutic benefits could be reaped from the fact that SWCNT-PEG stimulated the release of selected cytokines from injured astrocytes, which would promote recovery after injury and thus counteract the excess of proinflammatory cytokines present in TBI.

Keywords: astrocytes; brain injury; traumatic; cytokines mouse; nanotubes; carbon

Citation: Gržeta Krpan, N.; Harej Hrkać, A.; Janković, T.; Dolenec, P.; Bekyarova, E.; Parpura, V.; Pilipović, K. Chemically Functionalized Single-Walled Carbon Nanotubes Prevent the Reduction in Plasmalemmal Glutamate Transporter EAAT1 Expression in, and Increase the Release of Selected Cytokines from, Stretch-Injured Astrocytes in Vitro. *Cells* **2024**, *13*, 225. https://doi.org/10.3390/cells13030225

Academic Editor: Jessica Filosa

Received: 21 June 2023
Revised: 15 January 2024
Accepted: 17 January 2024
Published: 25 January 2024

1. Introduction

Traumatic brain injury (TBI) represents a significant public health problem due to the fact that it is one of the leading causes of death and disability worldwide [1]. Pathological changes that occur in the brain following trauma result from the primary, irreversible injury and the subsequent secondary injury mechanisms that develop over time. According to the Glasgow Coma Scale (GCS), a classification based on the neurological status of patients, there are three degrees of severity of TBI, mild, moderate, and severe. Severe TBI has a 30–40% mortality rate and it can cause significant physical, psychosocial, and social deficits in the majority of surviving patients [2]. Survivors, after moderate and severe TBI, have significant long-term deficits that develop due to the widespread cellular damage and the neurotoxic environment around the site of the injury, the latter in part due to excessive glutamate [3], as well as prolonged and intense neuroinflammation mediated by cytokines [4]. There are no effective treatments to help mediate the repair and regeneration of the brain tissue after TBI. The current therapeutic approach for TBI solely relies on acute stabilization of the patient and administration of supportive therapy [5].

It is known that TBI, as a part of the tissue response to injury, induces the activation of astrocytes, one of the most abundant glial cells in the brain. This process, known as reactive astrogliosis, is involved in tissue remodeling, repair of the blood–brain barrier (BBB), regulation of synaptic plasticity, formation of a glial scar, and changes in the extracellular matrix [6]. Reactive astrocytes are characterized by molecular, cellular, and functional changes that arise in the response to brain damage. A common feature of this type of reaction is cellular hypertrophy and upregulation of the glial fibrillary acidic protein (GFAP) expression [7]. Indeed, GFAP is one of the key markers for the activation of reactive astrogliosis following brain injury or other types of stress in the central nervous system (CNS).

Astrocytes are the major buffering cells for the excitatory amino acids (EAAs), in particular glutamate, that is abundantly released from the injured neuronal cells [8]. Removal of glutamate from the extracellular space via astrocytic plasmalemmal glutamate transporters, which belong to the excitatory amino acid transporter (EAAT) family, is crucial for glutamate homeostasis. Glutamate uptake by astrocytes prevents elevation of glutamate to excitation levels in the extracellular space, which is a critical factor for the survival of neurons [9,10]. Two major types of Na^+-dependent glutamate transporters in astrocytes of the CNS are EAAT1 (in rodents also referred to as GLAST) and EAAT2 (in rodents also referred to as GLT-1) [11]. Their expression is developmentally regulated, so in neonatal mice, which we used here, astrocytes express low levels of GLT-1/EAAT 2 when compared to GLAST/EAAT1 expression [12]. In both preclinical and clinical studies, significant dysregulation of these transporters following TBI was detected [13–15]. Of particular interest to our present study is the fact that there is a loss of EAAT1 and 2 expression in astrocytes in human patients suffering from TBI [15]. Taming this loss may provide a fertile ground for therapeutic intervention.

Carbon nanotubes (CNTs) may be employed in this quest. CNTs are promising nanomaterial candidates in treating brain disorders [16]. They have excellent physicochemical properties and the ability to interact with neurons, neuronal circuits, and astrocytes. Most of the previous studies have been focused on bio-imaging and drug delivery applications of CNTs, whereas their application as a therapeutic agent has not yet been sufficiently investigated [16]. Some difficulties regarding the widespread applications of CNTs in treating disorders of the CNS are related to their biocompatibility. Namely, especially with pristine, non-functionalized CNTs, it was detected that they can increase free radical production, the accumulation of which leads to oxidative stress; induce tissue inflammatory responses; and cause DNA damage [16]. However, these issues can be alleviated by changing the physicochemical properties of the CNTs, making them more prone to dispersion in the aqueous media and thus enhancing their biocompatibility.

SWCNTs chemically functionalized with polyethylene glycol (PEG) or poly-m-aminobenzene sulfonic acid are aqueous colloidal solutes. These nanomaterials, when applied on a short term in cell culture (3 days), promoted outgrowth of selected neurites [17] and induced changes in the morphology of astrocytes by enhancing the GFAP expression, as measured by immunostaining, and caused the appearance of a more mature phenotype of astrocytes [18]. Furthermore, SWCNT-PEG (in cell culture for 4 days), but not PEG itself, caused an increase in expression of GLAST/EAAT1 in astrocytes along with an increase in the astrocyte uptake of this excitatory transmitter [19].

It is still largely unknown how the traumatically injured astrocytes would react to exposure to water-soluble chemically functionalized SWCNTs. Until now, only one study reported the effects of the application of SWCNT-PEG (for a longer term of 28 days) in a model of the traumatic injury of the CNS in vivo [20]. The study showed that the injection of SWCNT-PEG at the injury site in rats with complete transection of the spinal cord, at the thoracic nine level, decreased the lesion volume and modestly improved recovery of hind limb locomotor function, and without the change in GFAP expression.

The aim of this study was to initially investigate the effects of SWCNT-PEG on the survival of astrocytes exposed to severe in vitro (cell culture) TBI within the first 24 h and then to subsequently study if SWCNT-PEG caused changes in the protein expression of GFAP and EAAT1 in the injured astrocytes. Furthermore, we also examined if there were significant changes in the secretome profile of these cells, which was analyzed by using the multiplex cytokine array, in the media collected from cultures of injured astrocytes. It should be noted that the stretch injury as a model of TBI used here does not produce the overt necrosis that may be present in whole-animal TBI. Nonetheless, our study provides for the profile with respect to the use of SWCNT-PEG as a potential curative agent acting upon injured astrocytes.

2. Materials and Methods

2.1. Experimental Animals

Mice used for this research were born and bred in the Laboratory for Mice Breeding and Engineering Rijeka (LAMRI). For the preparation of the in vitro culture of astrocytes, C57BL/6 mice of both sexes at 0–2 days of age were sacrificed. For each analysis, three to four groups of mice from different mothers were sacrificed on different days for the isolation of astrocytes to perform experiments in biological triplicates or quadruplicates. All the procedures on mice were performed at the Department of Basic and Clinical Pharmacology and Toxicology of the Faculty of Medicine of the University of Rijeka. The use of animals for experiments was approved by the Faculty's Animal Welfare Committee, as well as by the Veterinary Department of the Ministry of Agriculture, and the experiments were performed in accordance with the Croatian Animal Protection Act, which was matched with the current European Union legislation.

2.2. In Vitro Experiments

2.2.1. Astrocyte Culture

Astrocytes were isolated from mice neocortices with some modifications of a previously described procedure [21]. After removing mice brains, neocortices were isolated, cleaned of meninges, and put into ice-cold cell culture medium lacking phenol-red (Dulbecco's Modified Eagle Medium, DMEM; Cat. No. P04-01161, PAN-Biotech, Aidenbach, Germany). Neocortices were homogenized with an automatic pipettor and then passed through 230 μm and 140 μm pore-diameter sterile filters, respectively. The filtrate was pelleted by centrifugation, the supernatant was decanted, and the pellet was resuspended in 2.5 mL of completed medium containing DMEM with phenol-red (Cat. No. P04-04510) and 10% (v/v) fetal bovine serum (FBS) (both from PAN-Biotech, Germany) with added 1% v/v of antibiotic–antimycotic solution (Cat. No. A5955; Sigma-Aldrich, Saint Louis, MO, USA). The resulting cell suspension was passed through a 73 μm pore-diameter sterile filter. Cells in the filtrate were counted and seeded at a density of 2×10^5 cells/cm^2 into 75 cm^2 flasks and incubated at 37 °C in a 5% CO_2—95% air atmosphere incubator with relative humidity of 95% for 7 days. At that time, flasks were shaken twice on an orbital incubator–shaker (Unimax 1010 Orbital Platform Shaker, Heidolph, Schwabach, Germany), first for 30 min at 180 rpm, and then, after replacing the media, again for 6 h at 240 rpm. The supernatant was discarded, the fresh medium was added to the adherent cells, i.e., astrocytes, and flasks were then put back into the incubator for an additional 7–14 days until they were used in experiments. This procedure yields purified, ~99% astrocytes culture, as confirmed using immunocytochemistry on GFAP and EAAT1 (Supplementary Materials; Figure S1).

2.2.2. Cell Injury Model

Purified astrocytes were detached from flasks and after mechanical trituration seeded into wells bottomed with silastic membranes that were top-coated with collagen type I

(BioFlex® culture plates, Cat. No. BF-3001C; FlexCell® International Corporation, Burlington, NC, USA). We applied 150,000 cells in 2.5 mL growth medium per well. Upon reaching confluence (within three to five days), astrocytes were subjected to injury by the controlled delivery of pressurized nitrogen. Pressure was regulated by the Cell Injury Controller II (Virginia Commonwealth University, Richmond, VA, USA). We applied a 19–29 kPa (3.8–4.2 psi) peak pressure pulse, 50 ms in duration, perpendicular to the silastic membrane. This caused a membrane displacement of 7.5 mm and biaxial stretch of the cells attached to them. This injury in vitro corresponds to the severe traumatic brain injury (sTBI) in vivo [22]. After the injury, the cells were returned to the incubator.

2.2.3. Single-Walled Carbon Nanotubes Preparation

Purified single-walled carbon nanotubes were obtained from Carbon Solutions Inc., Riverside, CA, USA (Cat No. P3-SWNT; www.carbonsolution.com (accessed on 15 January 2024)) and functionalized with PEG, as described in detail in the Supplementary Materials. The attached PEG functionalities rendered the carbon nanotubes soluble in purified water upon ultrasonication for 1 h. The stock SWCNT-PEG aqueous colloidal solution was at 2 mg/mL. We used SWCNT-PEG due to their known biological effects on astrocytes, as we disclose in the introduction [17–20]. We characterized SWCNT-PEG with atomic force microscopy, thermogravimetric analysis, mid-infrared spectroscopy, and ultraviolet-visible–near-infrared spectroscopy (Supplementary Materials; Scheme S1 and Figures S2–S5) to determine their morphology and chemical composition. The SWCNT-PEG material contains 72.3 weight percent (wt%) SWCNTs, 22.6 wt% PEG, and 5.1 wt% metal nanoparticles (nickel and yttrium mixture, in the atomic ratio of 4:1). SWCNT-PEG exist as individual tubes as well as axially aligned small bundles. The individual SWCNT-PEG have diameters between 1.2 nm and 1.7 nm with an average diameter of 1.5 nm, while bundles have a diameter of 2 nm to 7 nm, and the length in the range of 0.3–1.5 μm.

2.2.4. Incubation of the Primary Astrocyte Culture with SWCNT-PEG or the Functionalization Agent PEG

SWCNT-PEG were added to the astrocyte cultures 1 h after the injury at a concentration of 5 μg/mL and kept for 23 h. This concentration was chosen according to the previous research using this very material, for 3 or 4 days on cultured astrocytes [18,19], showing its biocompatibility along with the effect on astrocytes, which included increases in GFAP and EAAT1 levels, respectively, as measured by immunostaining, and functionally enhanced uptake of glutamate by these glial cells [19]. Before adding SWCNT-PEG to astrocytes with sTBI (sTBI+CNT group), the stock solution was diluted in 150 μL of the cell culture medium and applied to the culture medium to reach the final concentration of 5 μg/mL. In uninjured astrocytes (control group) and sham-treated astrocytes with sTBI (sTBI group), instead of SWCNT-PEG stock solution, we used purified water containing the functionalization agent PEG, to mix in with 150 μL of the cell culture medium to reach the final concentration of 1 μg/mL of PEG, as previously described [23]. This concentration of PEG corresponds to the PEG weight load in SWCNT-PEG. The final ratio of purified water (in PEG or SWCNT-PEG solution) to cell culture medium was 1:400. Thus, we had three experimental groups: (i) control, uninjured astrocytes receiving PEG; severe TBI (sTBI), injured astrocytes receiving PEG; and sTBI+CNT, injured astrocytes receiving SWCNT-PEG.

2.3. *Determination of Severity of the Astrocyte Injury*

For the determination of the severity of the astrocyte injury, release of lactate dehydrogenase (LDH) into cell culture media was measured using the CytoTox96® Non-Radioactive Cytotoxicity assay (Promega Corporation, Madison, WI, USA). Aliquots of astrocyte culture media were collected at 15 min, 1 h, 3 h, 6 h, and 24 h after the injury, and transferred into a 96-well plate for sample analysis. Samples were treated with the reagents from the kit in accordance with the manufacturer's instruction manual and analyzed by the spectrophotometric method using the BioTek EL808 Microplate Reader (Agilent Technologies, Santa Clara, CA, USA).

2.4. Dot Blot

The dot blot method was used to detect the levels of oxidatively damaged proteins by measuring the protein carbonyl content. 2,4-dinitrophenylhydrazine (DNPH) reacts with the carbonyl group on the protein to form 2,4-dinitrophenylhydrazone (DNP hydrazone), which is detected immunochemically. Two aliquots of each sample were used, one of which was subjected to the derivatization reaction, while the other aliquot represented the negative control. Protein samples were derivatized with 10 mM DNPH in the presence of 12% sodium dodecyl sulfate (SDS) for 15 min at room temperature (20–25 °C) in the dark. A neutralization solution (2 M Tris, 30% glycerol, and 19% 2-mercaptoethanol) was then added to the samples, which were then spotted onto a polyvinylidene difluoride (PVDF) membrane using a dot blot device (Cleaver Scientific Ltd., Warwickshire, UK). The membrane was dried for 15 min and washed twice in succession with concentrated acetic acid and then water. The membrane was then rinsed with wash buffer (10 mM Tris-HCl, pH 7.5, 150 mM NaCl, 0.05% Tween; i.e., TBST), blocked by incubation with 5% bovine serum albumin (BSA) in TBST for one hour. After blocking, an anti-DNP antibody (OxyBlot™, EMD Millipore, Billerica, MA, USA), diluted 1:100 in TBST with 1% BSA, was added to the membranes and kept overnight at 4 °C. The next day, the membrane was washed 3 times for 5 min with TBST and then was incubated with biotinylated goat anti-rabbit secondary antibody (1:3000) (Table 1) for one hour at room temperature. After that, the membrane was washed with TBST again and incubated with streptavidin conjugated with horseradish peroxidase (HRP). A detection solution (chemiluminescent substrate SuperSignal™ West Pico PLUS; Thermo Fisher Scientific, Waltham, MA, USA) was added to the membrane and visualized on a Kodak Image Station 440CF device, analyzed in Kodak 1D Image Analysis Software, Version 3.6 (Eastman Kodak, Rochester, NY, USA).

Table 1. Information on used antibodies.

Antibody	Immunogen	Manufacturer; Catalog Number; RRID; Host Species and Antibody Type	Dilution
anti-GFAP	native GFAP purified from pig spinal cord	Cell Signaling Technology (Danvers, MA, USA); 3670; AB_561049; mouse monoclonal	1:1000
anti-EAAT1	synthetic peptide corresponding to residues surrounding Glu230 of human EAAT1	Cell Signaling Technology; 5685; AB_10694915; rabbit monoclonal	1:1000
anti-β-actin	β-actin (C4) raised against gizzard actin from chicken	Santa Cruz Biotechnology (Dallas, TX, USA); sc-47778; AB_2714189; mouse monoclonal	1:1000
biotinylated goat anti-rabbit	gamma immunoglobulins	Thermo Fisher Scientific; 65-6140; AB_2533969; goat polyclonal	1:3000
biotinylated goat anti-mouse	gamma immunoglobulins	Thermo Fisher Scientific; A16070; AB_2534743; goat polyclonal	1:3000

Abbreviations: RRID, research resource identifier.

2.5. Western Blotting

For the Western blotting, samples were prepared by lysing the cells at 24 h after the injury with 100 μL/well of ice-cold radioimmunoprecipitation assay (RIPA) buffer, complemented with protease and phosphatase inhibitors (cOmplete™, Cat. No. 11873580001, Roche, Basel, Switzerland). Cells were scraped from the silastic membranes and placed in a vessel on the orbital shaker on ice for 20 min. The resulting cell lysate was collected, and cell debris pelleted by centrifugation at 4 °C. The supernatants were collected and stored at

−80 °C. The concentration of proteins in supernatants was determined using the Bradford method [24].

Proteins were loaded onto 10% polyacrylamide gels at a concentration of ~1 μg/μL. The electrophoresis was run at 200 V in the Mini-PROTEAN® Tetra System (Bio-Rad Laboratories Inc., Hercules, CA, USA). The transfer of proteins from gels to nitrocellulose membranes was performed in the Trans-Blot® Turbo™ Transfer System (Bio-Rad Laboratories Inc., Hercules, CA, USA). Afterwards, membranes were treated for 1 h with 5% dry milk or 5% BSA in TBST to block non-specific binding sites. The primary antibodies used were mouse anti-GFAP, rabbit anti-EAAT1, and mouse anti-β-actin (all at 1:1000) (Table 1); β-actin stain was used as a loading control. After the overnight incubation at 4 °C, membranes were washed with Tris-buffered saline (TBS) containing 0.1% Tween (TBST2) for 30 min and incubated with secondary antibodies for 1 h. The secondary antibodies used were goat anti-rabbit IgG and goat anti-mouse IgG, both conjugated with biotin (Table 1). After the incubation, membranes were again washed with TBST2 and then incubated with the streptavidin–HRP conjugate for 30 min. Visualization of proteins, achieved by adding the chemiluminescent substrate SuperSignal™ West Pico PLUS (Thermo Fisher Scientific, Waltham, MA, USA), was performed using the Kodak Image Station 440; bands were analyzed using the Kodak 1D program (v.3.6.5, Kodak Scientific Imaging Systems, Kennewick, WA, USA).

2.6. Cytokine Assay

For the measurement of cytokines released by astrocytes, cell culture media aliquots were collected at 24 h after the injury. Media samples were screened for 62 or 40 cytokines by using the Mouse Cytokine Array C3 kit (for data in Figure 4 and Table S1) or Mouse Inflammation Array C1 kit (for data in Figure S7 and Table S2), respectively, and according to the manufacturer's instructions (Ray Biotech Inc., Norcross, GA, USA). In brief, membranes provided in the kit were sequentially incubated with equal volumes (200 μL) of the cell culture medium, primary biotinylated antibodies and HRP-conjugated streptavidin solution. Visualization of signals on membranes was performed on a Kodak Image Station 440 and their relative intensities were analyzed using the Kodak 1D program (v. 3.6.5, Kodak Scientific Imaging Systems, USA).

2.7. Statistical Analyses

All analyses were performed on at least three separate, independent experiments (biological replicates). The sample size required for an individual set of experiments was pre-assessed using power analysis (set at 80% and $\alpha = 0.05$), and the normality of data was established using the Shapiro–Wilk test for normality. For all the statistical analyses, Statistica software version 13.0 (StatSoft, Inc., Tulsa, OK, USA) was used. Comparisons between multiple groups were done using one-way analysis of variance (ANOVA) followed by the Duncan's post hoc multiple comparison test. Results are expressed as mean ± standard error of means (SEM). In all the comparisons, $p < 0.05$ was considered to indicate statistical significance. To remove the variations between the Western blotting sessions, the results of the densitometric analyses were corrected with the Factor Correction program version 16.1 [25].

3. Results

In our experiments, astrocytes were plated on silastic membranes and, upon reaching confluence, exposed to stretch injury simulating a severe TBI. Astrocytes were then treated with 5 μg/mL SWCNT-PEG (sTBI+CNTs group), or 1 μg/mL PEG (sTBI group), starting from 1 h post-injury for 23 h; see Materials and Methods for concertation considerations for SWCNT-PEG and PEG. Control astrocytes that were uninjured received 1 μg/mL PEG (control group) according to the above schedule. The addition of PEG to both sTBI and control groups provided assurance that the effects we observed in the sTBI+CNT group were the result of the SWCNT backbone action, rather than that of PEG, as a functionalization agent. We assessed the effects of SWCNT-PEG on the astrocyte viability and cell protein oxidative damage upon injury. We also investigated if exposure to SWCNT-PEG changes the expression of proteins GFAP and EAAT1, and if this nanomaterial affects the release of cytokines from the injured astrocytes.

3.1. SWCNT-PEG Do Not Affect Cell Viability of Astrocytes Exposed to Severe Stretch Injury

The therapeutic potential of CNTs is somewhat hampered by their possible toxicity, although it has been shown that their harmfulness is considerably lower in cases of purified and surface-functionalized nanotubes [26]. Specifically, SWCNT-PEG, the very material we use here, did not exert adverse effects on astrocytes when used at a concentration of 5 μg/mL for up to 4 days in culture [19]. The effects of SWCNT-PEG on the viability of astrocytes exposed to severe in vitro trauma have not been assessed until now.

We assessed the cytotoxicity by measuring the plasma membrane permeability to LDH by quantifying LDH activity in the cell culture medium, as this cytosolic enzyme diffuses (through a leaky membrane) into the extracellular space. Control, uninjured astrocytes show low levels of LDH at all time points recorded, prior to and after they received the PEG treatment (Figure 1). This is in a stark contrast to astrocytes that received a severe stretch injury. In order to establish that the level of cell injury between the experimental conditions was comparable, LDH levels were first measured at 15 min and 1 h after the in vitro TBI, i.e., before the application of either SWCNT-PEG or PEG at 1 h post in vitro TBI. LDH concentrations were additionally determined at 3 h, 6 h, 12 h, and 24 h post-injury and with SWCNT-PEG or PEG added to cells. At all the investigated time points, LDH levels were significantly higher in the culture media of the injured cells compared to the control conditions ($p < 0.05$). Exposure to either SWCNT-PEG or PEG did not affect the amount of LDH released from the injured astrocytes at any of the designated time points ($p > 0.05$).

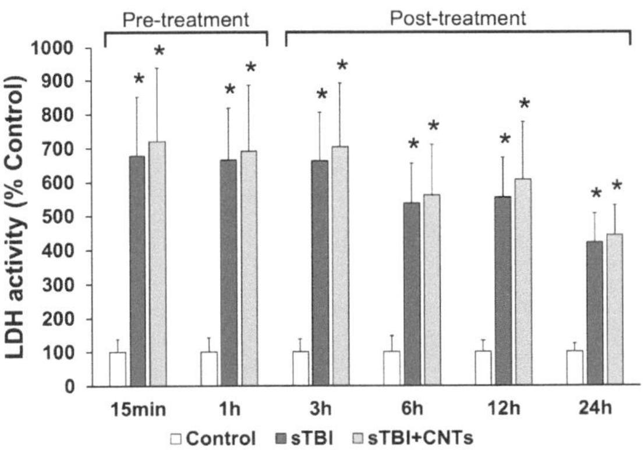

Figure 1. The lack of effect of chemically functionalized single-walled carbon nanotubes (CNTs) on the survival of the mouse primary astrocytes exposed to severe traumatic brain injury (sTBI) in vitro. Injured cells were treated with PEG (sTBI) or SWCNT-PEG (sTBI+CNTs) at 1 h post-stretch injury. Cell culture media were collected at different time points post-injury (15 min, 1 h, 3 h, 6 h, 12 h, and 24 h), from which lactate dehydrogenase (LDH) activity was quantified. Results are expressed as means ± SEM (N = 3). * $p < 0.05$, significantly different from the control (uninjured and PEG-treated) group (one-way ANOVA). There was not a significant change in LDH activity over time in the control group.

3.2. Severe Stretch Injury of Cultured Astrocytes and the Application of SWCNT-PEG Do Not Cause Changes in the Protein Oxidation Levels

Protein carbonyl levels were detected using the dot blot immunoassay to monitor if the oxidative damage of proteins is potentially caused by the in vitro severe TBI and/or the SWCNT-PEG treatment (Figure 2). Figure 2A shows representative dot blots of experimental group samples, and Figure 2B the associated densitometric analyses corrected for the protein load determined by Ponceau S staining and presented as a percentage of the results from the control group samples. It is evident from the results that there is no statistically significant change in the levels of carbonylated proteins following cell injury or the addition of the investigated nanomaterial (or functionalization agent), determined at 24 h after severe cell stretch injury.

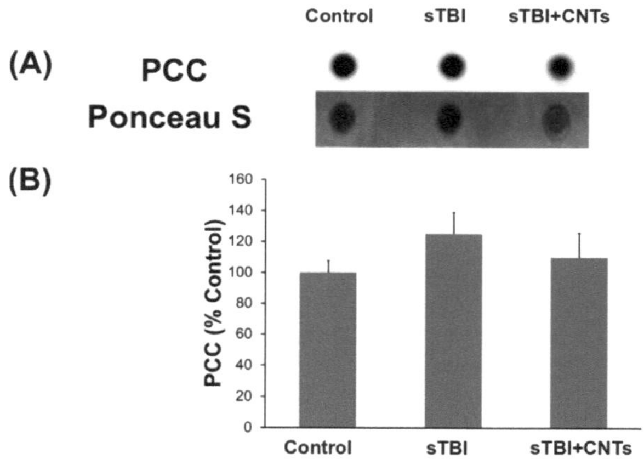

Figure 2. The lack of effect by chemically functionalized single-walled carbon nanotubes (CNTs) on the levels of oxidative protein damage, determined by the measurement of the protein carbonyl content (PCC), in mouse primary astrocytes exposed to severe traumatic brain injury (sTBI) in vitro. Shown are (**A**) representative dot blots and (**B**) the results of the densitometric analyses. The histogram shows the PCC levels in control (uninjured cells receiving PEG) conditions, as well as at 24 h after the injury and following the addition of PEG (sTBI) or SWCNT-PEG (sTBI+CNTs). Values are shown as the proportion (%) of PCC levels detected in the cells of the control group, corrected for the protein load determined by Ponceau S staining. Results are expressed as means ± SEM. Results of the analyses from four separate biological replicates are presented.

3.3. *Application of SWCNT-PEG to Astrocytes Rescues the Injury-Induced Loss of Plasmalemmal Glutamate Transporter EAAT1*

Changes in the expression of GFAP following brain injury are the most characteristic markers of astrocyte reactivity [7]. In our investigation, Western blot analyses of GFAP protein expression (Figure 3A) showed no significant change in GFAP levels at 24 h after severe stretch injury ($F(2;8) = 0.583$; $p = 0.580$). This is in line with the previously published research by Wanner et al. [27] in which no significant increase in GFAP synthesis, but mainly induction of relocalization of existing GFAP in astrocytes, was detected. In our experiment, the addition of nanotube colloid solution also did not affect the GFAP content, which is in agreement with the previously published work in spinal cord in vivo injury treated with SWCNT-PEG for 28 days [20], but at odds with an increase in the immunoreactivity of GFAP in cultured astrocytes exposed to SWCNT-PEG for 3 days [19].

In human patients suffering from TBI, there is a loss of EAAT1 expression in astrocytes [15]. We first confirmed that our stretch-injury model appropriately simulates such EAAT1 expression in vitro. We then assessed whether we can revert this loss by using SWCNT-PEG as a proof-of-principle demonstration of the translational use of this nanomaterial. Namely, it has been previously established that SWCNT-PEG cause an increase in the immunoreactivity of glutamate transporter EAAT1 on the plasma membrane of astrocytes [19].

We, therefore, evaluated if the application of the investigated nanomaterial to astrocytes exposed to severe stretch impacts the total expression level of EAAT1, and the results of this analysis are presented in Figure 3B. Analyses showed that the expression of EAAT1 was affected by severe stretch injury (F(2;9) = 11.157; $p = 0.004$). Post hoc comparison revealed that the in vitro trauma causes a decrease in the EAAT1 levels ($p = 0.002$) and that the addition of the SWCNT-PEG, but not PEG, significantly rescues the injury-induced loss of this plasmalemmal glutamate transporter ($p = 0.005$). This finding points to a possible translational use of SWCNT-PEG to treat injured astrocytes in the attempt to rescue their ability to uptake the excess of extracellular glutamate that occurs in TBI.

We separately conducted experiments in which we tested whether SWCNT-PEG treatment affects GFAP or EAAT1 expressions in astrocytes not exposed to the stretch paradigm, but no significant differences were detected in levels of both investigated proteins. between control, non-treated, and PEG- or SWCNT-PEG-treated astrocytes (Figure S6).

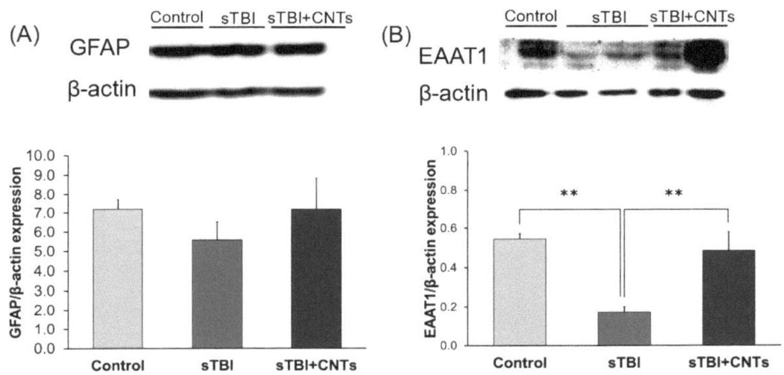

Figure 3. The effects of chemically functionalized single-walled carbon nanotubes (CNTs) on the protein expression of GFAP and EAAT1 in mouse primary astrocytes exposed to severe traumatic brain injury (sTBI) in vitro. Injured cells were treated with PEG (sTBI) or SWCNT-PEG (sTBI+CNTs) at 1 h post-stretch injury. Cells were lysed at 23 h after the treatment. Densitometric analyses of (**A**) GFAP and (**B**) EAAT1 expressions. There was no change in the expression of GFAP protein in injured cells, both treated with SWCNT-PEG or PEG, in comparison with the control cells. Injury significantly decreased levels of EAAT1 in cells treated with PEG in comparison to control cells, while the addition of SWCNT-PEG restored injury-induced loss of EAAT1 to a level similar to that seen in the uninjured, PEG-treated control. Results of the densitometric analyses were corrected for the β-actin contents. Data are expressed as mean ± SEM. ** $p < 0.001$ (one-way ANOVA).

3.4. Cytokine Array

Astrocytes are considered secretory cells of the CNS, and their ability to secrete cytokines can be associated with a variety of inflammatory roles and pathologies [28,29]. Thus, we assessed the secretion of cytokines from astrocytes in our three experimental groups.

The antibody array we used allowed us to assay simultaneously 62 mouse cytokines. Analyses of the media, conditioned by the control (uninjured, PEG-treated) astrocytes, sampled at 24 h after the onset of the experiment, detected the secretion of multiple compounds (Figure 4 and Supplementary Materials, Table S1; the latter table contains all the abbreviations used in the text below). In the media derived from the injured cells treated with PEG, no significant changes in the levels of the cytokines were found compared to the media collected from the control astrocytes. Conversely, results of the microarray analysis of the media conditioned by astrocytes that were subjected to the in vitro severe TBI and subsequently exposed to SWCNT-PEG revealed that the investigated treatment significantly affected the levels of the following compounds: axl, BLC, G-CSF, GM-CSF, IL-3-Rb, IL-4, IL-5, IL-6, IL-9, and IL-10. In the media sampled from the injured cells treated with nanotubes, levels of axl, BLC, GM-CSF, IL-4, IL-5, IL-6, and IL-10 were higher compared to both the concentrations determined in the media conditioned by the control cells and the injured, PEG-treated, astrocytes. Additionally, statistical analyses revealed that the G-CSF and IL-9 levels were increased in the samples from the CNT-treated, injured cells (sTBI+CNTs) only when compared to the results obtained from the PEG-treated, stretched astrocytes. Regarding the IL-3-Rb levels, they were significantly higher in the media sampled from the SWCN-PEG-treated, injured only compared to the media taken from the cells maintained in the control conditions.

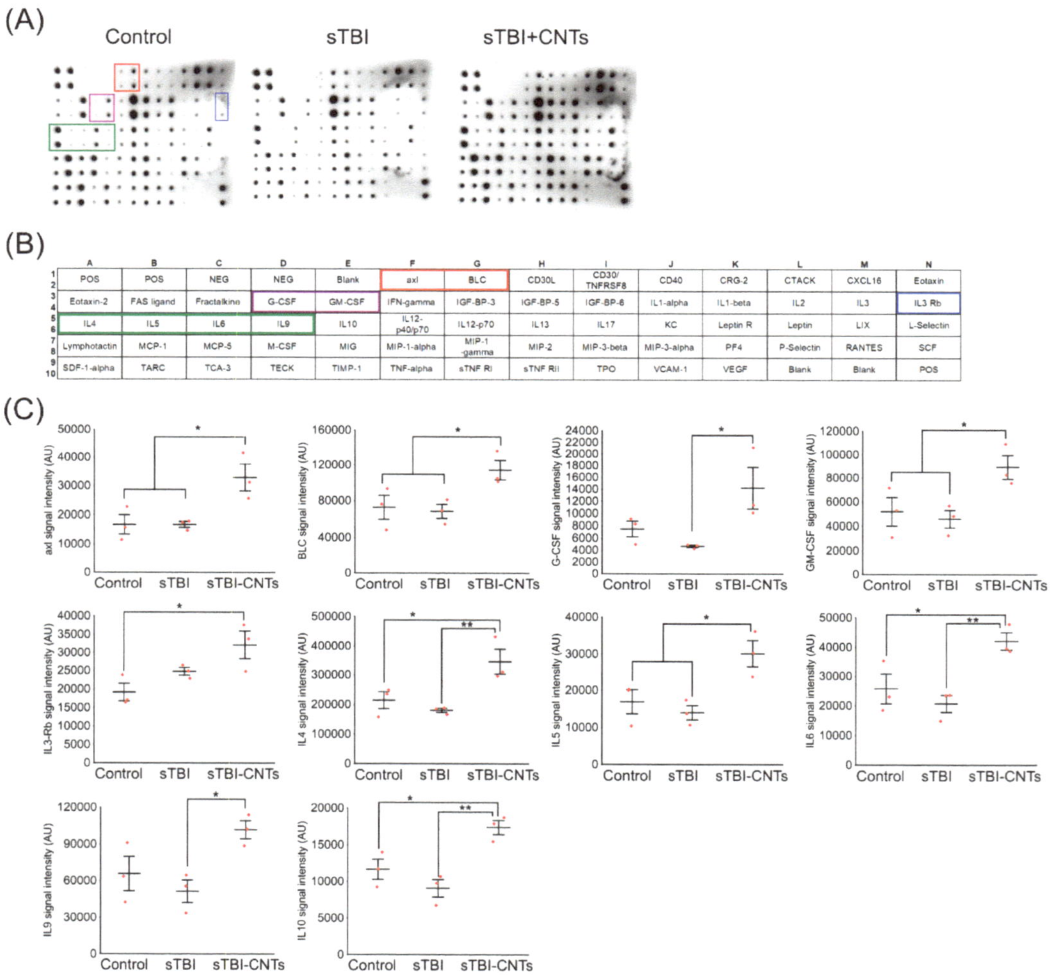

Figure 4. Chemically functionalized single-walled carbon nanotubes (CNTs) increase the secretion of selected cytokines from astrocytes exposed to severe traumatic brain injury (sTBI) in vitro. (**A**) Representative membranes showing the content/pattern of cytokines, released from the control group of astrocytes, i.e., uninjured and PEG-treated cells (control), astrocytes subjected to stretch injury and treated with PEG (sTBI) or treated with SWCNT-PEG (sTBI+CNTs). (**B**) Table with the list of cytokines available within the Mouse Cytokine Array C3 kit. Letters (A–N) and numbers (1–10) indicate coordinates (x,y) on the dot blot. Colored boxes indicate selected proteins with significant changes in their expression. (**C**) Summary graphs showing significant increases in signal intensity of selected (boxed in (**A**,**B**)) cytokines in the cell culture media samples from the SWCNT-PEG-treated injured astrocytes. Data are expressed as mean ± SEM. * $p < 0.05$, ** $p < 0.001$.

In separate experiments, we also tested if the SWCNT-PEG exposure has an effect on astrocytes not exposed to the stretch injury. Other than the decrease in the release of the single cytokine, MIP-1 gamma, from SWCNT-PEG-treated astrocytes compared to control, non-treated cells, no other differences in the cytokine levels were detected between the experimental groups (Figure S7; Table S2). It should be noted that the antibody array kit (see Materials and Methods and compare Tables S1 and S2) used in these experiments did not test the levels of all the affected cytokines in the above TBI experiments (Figure 4 and Table S1), so that axl and IL-5 levels where not tested. However, this technicality should not distract from the fact that SWCNT-PEG stimulated the release of selected cytokines from injured astrocytes, which would promote recovery after injury and thus counteract the excess of proinflammatory cytokines present in TBI.

4. Discussion

In the present study, we used an in vitro astrocyte stretch injury model to demonstrate the effects of SWNTC-PEGs on the cell survival, oxidative protein damage level, as well as on the expression of GFAP and EAAT1, and the secretion of cytokines from the injured astrocytes.

To the best of our knowledge, this is the first research in which the effects of the nanomaterial were assessed in an in vitro model of severe TBI. SWCNT-PEG were previously investigated in the spinal cord injury model in rats [20]. In that study, it was shown that the post-injury administration of SWCNT-PEG in the lesion site promotes axonal survival and repair, with a reduction in the lesion volume, and an increase in the number of neuronal fibers in the lesion epicenter. Additionally, a moderate recovery of motility in SWCNT-PEG-injected rats was accomplished.

4.1. SWCNT-PEG Do Not Affect Viability of Astrocytes Exposed to Severe Stretch Injury

With the emerging prospective applications of CNTs in biomedicine, questions regarding their potential toxicity have also been raised. Indeed, several studies have thus far exposed some toxic effects in cells following CNT exposure [30]. Firstly, it seems that the CNT toxicity is less likely to occur in cases where they were engineered and immobilized, for example on electrodes, but higher when the CNTs are utilized as free particles and not functionalized. This appears to be mainly due to the capacity of nanoparticles in general to enter the cells and disperse in the cytoplasm [31].

In our study, SWCNT-PEG did not affect the post-injury survival of astrocytes, within the first 24 h following stretch. Additionally, no overt oxidative damage to proteins extracted from the cultures astrocytes was detected, regardless of whether cells post-injury were vehicle- or SWCNT-PEG-treated.

Previously, one study observed the effects of SWCNTs on cell lines with different phagocytic properties and concluded that the higher the phagocytic activity a cell has, the higher the toxic effects will be, due to the higher uptake of the nanotubes [32]. The mechanism of toxicity is linked to the oxidative stress and formation of reactive oxygen species (ROS) [33]. It is possible that in the in vitro conditions of our research, astrocytes did not have pronounced phagocytic function, and it would be interesting to determine whether the toxicity changes in co-cultures with neurons and microglia. On the other side, in our study, we used SWCNTs functionalized with PEG that are proven to be less toxic when internalized compared to the non-functionalized SWCNTs [34]. It is important to point out that SWCNT-PEG show dose-dependent toxicity, and the concentration used in our research (5 μg/mL) has not been shown to cause a significant rise in ROS production [34].

The obtained results confirm that SWCNT-PEG are not toxic to primary astrocyte cells in culture, although they do not promote better survival of the same cells after the injury. These results also need to be critically evaluated because research methods for monitoring the increase or decrease in cell death need to be used for more than 24 h. Namely, previous studies have shown cytotoxicity of differently chemically functionalized SWCNTs after 48 h and up to 4 days. Therefore, in order to detect the true function of SWCNTs in in vitro astrocytes after injury, more cytotoxicity studies need to be conducted before switching to in vivo models.

4.2. Application of SWCNT-PEG to Astrocytes Exposed to Severe Stretch Injury Prevents the Decrease in the Expression of EAAT1 in Severely Stretched Astrocytes

One of the many functions of astrocytes is the regulation of extracellular glutamate via the EAAT transporters that these cells can express. This function is of key importance in brain trauma wherein large quantities of glutamate are released to induce excitotoxicity and promote cell death, and secondary brain injury [13].

The results of our study show that in vitro TBI causes a decrease in EAAT1 expression in astrocytes. This finding is in accordance with the results of previous studies, both in vitro and in vivo, which were conducted on different models of ischemic [35–39] and trauma-caused brain injury [14,15,40]. One of the possible reasons for this decrease could be related to changes in astrocyte morphology and subsequent downregulation of gene expression due to the cell stretching. Astrocytes are strongly adaptable cells that can quickly change their morphological and functional properties depending on their surroundings, for example in tissue damaged by TBI [41]. Several studies have confirmed that a change in the actin cytoskeleton of astrocytes is connected to the lowered expression of EAAT1 transporters [15,42–44].

Our results also show that the application of SWCNT-PEG to the injured astrocytes rescues stress-injury-evoked loss of EAAT1. Levels of this protein 24 h after the injury do not differ significantly from those detected in the uninjured cells, which points to the potentially protective effects of SWCNT-PEG for the brain tissue after the injury by means of the enhanced glutamate clearance by the astrocytes. So far, this effect of SWCNT-PEG has been studied only on the uninjured cells, which showed a similar effect on the EAAT1 upregulation [19]. The mechanism of this upregulation is still not clear, but a few studies have found that SWCNTs also have an effect on the cytoskeleton; more specifically, they initiate the grouping of actin into filaments and cause changes in cell surface function and the processes of endo/exocytosis [45]. It could be possible that with this remodeling, changes caused by injury, and a subsequent decrease in EAAT1 production, are mitigated or reversed.

Some of the studies found that in vivo TBI caused reactive astrogliosis, higher expression of GFAP, and a rise in EAAT1-positive astrocytes 24 h after the injury. This was followed by a prolonged decrease in signal [15,39], probably caused by the glutamate overload of astrocytes [15]. Since our results were obtained in in vitro conditions, and no changes in GFAP levels were found, they cannot be compared to these findings. However, it would be interesting to see whether the levels of EAAT1 would still be elevated in SWCNT-PEG-treated astrocytes in periods longer than 24 h. It would also be of interest to investigate if there are effects on other plasma membrane proteins to assess if the SWCNT-PEG actions shown here might be a general cell surface phenomenon.

In the previous work in vitro using SWCNT-PEG at the same concertation as used here [18,19,23], SWCNT-PEG, when applied for a prolonged time (3 or 4 days vs. 24 h here), caused an increase in GFAP and EAAT1 levels in uninjured (vs both uninjured and injured here) astrocytes grown on polyethylene-imine-coated glass strata (vs silastic membranes here). The fact that, in the present work, we did not see changes in GFAP and/or EAAT1 expression, thus, could be attributed to any individual or combined experimental differences.

4.3. SWCNT-PEG Modulate the Secretory Function of Astrocytes Exposed to Severe Stretch Injury

Neurons and glia, the primary cells of the CNS, communicate with each other via signals that include neurotransmitters, ions, and other extracellular molecules, and the emerging evidence points to astrocytes as secretory cells of the CNS [46]. In brain injury, astrocytes are the cells responsible for the secretion of various factors involved in signal transduction and the stimulation of the migration of other glial cells to the site of the lesion.

However, in certain instances, astrocyte-secreted proteins can have detrimental effects and promote neuroinflammation and neurodegeneration, in addition to their beneficial actions that include the promotion of tissue remodeling, axonal regeneration, glial scar formation, and neural circuit rewiring [47].

Our results indicate that the stretch injury itself had no significant effect on the levels of the selected cytokines/chemokines released from astrocytes. However, in the cell culture media conditioned by injured astrocytes exposed to SWCNT-PEG, higher concentrations of axl, BLC, G-CSF, GM-CSF, IL-3-Rb, IL-4, IL-5, IL-6, IL-9, and IL-10 were detected.

Axl is a receptor tyrosine kinase that, upon binding of the growth factor GAS6, regulates numerous physiological processes such as cell survival, proliferation, migration, and differentiation. Axl can be cleaved by proteases and released as a soluble version of the protein. We detected a low expression of soluble axl in the media released from the control and injured astrocytes, but a higher expression of axl was detected in culture media of stretched astrocytes treated with SWCNT-PEG. Usually, axl is upregulated in astrocytes after TBI, and it promotes a phenotype of astrocytes that can ingest damaged neurons and other cells after TBI [48]. We found a high concentration of released axl in culture media after treatment with SWCNT-PEG, so it can be argued that this astrocytic phenotype transformation might promote the recovery of neurological cells after TBI.

IL-6 is known as a cytokine with different roles in injured cells of the CNS. It is produced at high concentrations in the injured brain. Glial-derived IL-6 is a key factor in diverse neuroinflammatory cascades, so IL-6 gene-deficient astrocytes in vitro show an altered cytokine profile compared to control cell cultures [49]. Lau and Yu [50] demonstrated that astrocytes, after scratch injury, were positively immunostained with IL-1α, IL-6, and TNFα, and the release of these cytokines into culture media was almost double that in control astrocytes. We have detected a high concentration of released IL-6 in media of injured cells treated with CNTs compared to injured and control-condition cells. This should be beneficial for recovery from injury as IL-6 is necessary for the timely resolution of injury, e.g., cerebrovascular repair after an intracerebral hemorrhage [51].

It is known that IL-4 and IL-5 can increase the amount of nerve growth factor, which is secreted by astrocytes after neural trauma, independent of cell growth [52]. IL-4 has its own role in immunity, but it has also a role in higher functions of the normal brain such as learning and memory. It has many important functions in a variety of cellular events, such as proinflammatory and anti-inflammatory effects on astrocytes depending on the treatment [53]. There is evidence that pretreatment with IL-4 in primary mouse astrocytes can decrease the production of NO and inducible NO synthase protein, as well as modulate the secretion of TNFα after liposaccharide stimulation. As we recorded an increased release of IL-4 from injured astrocytes treated with SCWNT-PEGs, this result raises an issue of whether there could be paracrine activity of this cytokine in injury or in the recovery thereof.

In our study, cytokine IL-10 was elevated in culture media of injured astrocytes treated with CNTs. This cytokine has a known immunomodulatory role after TBI. Kamm et al. [54] found that, in the brains of adult male Sprague Dawley rats, there is an elevated expression of IL-10 mRNA immediately following TBI, with levels of protein IL-10 being initially stable, but beginning to rise rapidly after 2 h. Studies with IL-10 knock-out mice showed that IL-10 is useful after TBI, but the usefulness of IL-10 administered after TBI as a therapeutic agent depends on the administration route and experimental plan [55].

It is known that the IL-9 receptor is expressed in brain cells such as microglia, astrocytes, oligodendrocyte progenitor cells, and oligodendrocytes, and its expression is induced in the brain and spinal cord during encephalomyelitis [56]. In our case, the amount of IL-9 was higher in injured astrocytes treated with CNTs. The exact role of IL-9 is a subject of ongoing investigations and remains rather unclear, as there is evidence both for its protective role in some conditions, e.g., in experimental autoimmune encephalomyelitis [57], as well as its ameliorative effects of IL-9 neutralization, e.g., in experimental strokes [58].

Results of the microarray cytokine measurement also showed a significant increase in the release of G-CSF and GM-CSF, cytokines that stimulate the production and maturation of granulocytes and, thus, support the immune system, aiding in the recovery from certain medical conditions. However, these cytokines have effects beyond simply stimulating the proliferation of neutrophils and monocytes [59]. G-CSF was found to regulate diverse functions in the nervous system and has shown ameliorative effects in certain pathological conditions, as it exhibited neurotrophic effects on primary cholinergic neurons in vitro [60], and in vivo promoted neuronal survival after focal cerebral ischemia [61,62]. One TBI retrospective study showed that plasma G-CSF levels are rapidly elevated in patients and decline after 14 days in those with mild, but not severe, brain trauma, and that higher circulating G-CSF levels correlate with better outcomes at 6 months post-injury [63]. Administration of G-CSF has also been suggested as a potential therapeutic strategy for the prevention of long-term complications of TBI, including dementia [64]. It has also been suggested that G-CSF can enhance neurological function by promoting the sprouting of the corticospinal tract into the denervated side of the cervical spinal cord. Additionally, G-CSF may mitigate microglia degeneration and restore a proper balance in synaptic pruning within the hippocampus, leading to improved neurological outcomes [65]. Similarly, studies have shown that GM-CSF may also have significant protective properties in the chronic consequences of experimental TBI [66,67].

Regarding the other cytokines that were found to be affected by the stretch injury and the nanotubes application, specifically, IL-1Rb (IL-1 receptor type II) and BLC (also known as CXCL13), limited information is available on their role in TBI pathophysiology. IL-1Rb, which comes in a membrane-bound and soluble form, is a decoy receptor, meaning that it captures IL-1 but does not initiate signal transduction, and by doing that it regulates IL-1-mediated signaling [68]. Conversely, BLC has been mostly associated with proinflammatory effects, as it was found to be frequently elevated in cerebrospinal fluid in a variety of inflammatory CNS conditions [69,70], and it has been suggested as a danger signal that is upregulated to recruit peripheral immune cells to the CNS through the damaged BBB [71]. The exact implications of our findings on the levels of excreted IL-1Rb and BLC from the injured astrocytes are difficult to define, and further research is warranted.

Taken together, our results indicate that the SWCNT-PEG effect on cytokines released from injured astrocytes is aligned with promoting recovery after injury. Should SWCNT-PEG be used in the treatment of the TBI, this nanomaterial would not only encounter injured but also uninjured astrocytes. Consequently, we tested the effect that SWCNT-PEG may have on cytokine release from uninjured astrocytes. We found the change in the level of the sole cytokine, i.e., the decrease in the release of MIP-1 gamma (also referred to as CCL9, i.e., chemokine ligand 9, macrophage inflammatory protein-related protein-2 and CCF 18) from SWCNT-PEG-treated uninjured astrocytes. As this protein plays a role in the development of the TBI in a mouse model [72], its decrease in the surrounding uninjured astrocyte may point out to a potential and additional protective action of SWCNT-PEG in the parenchyma surrounding the TBI site. While out of scope of the present study, of additional interest would be to compare the cytokine release pattern from astrocytes exposed to SWCNT-PEG to that of astrocytes exposed to known astrocytic secretagogues, such as adenosine triphosphate. There is a three-pronged rationale for this endeavor: (i) this nucleotide is increased in the extracellular space in experimental TBI [73]; (ii) it can cause glutamate release from cultured uninjured astrocytes via regulated exocytosis, i.e., vesicular release [74,75]; and SCWNT-PEGs affect vesicular recycling in cultured uninjured astrocytes [76].

5. Conclusions

We studied the effects of SWCNT-PEG administration on cultured astrocytes that were subjected to in vitro stretch injury. Our results showed that SWCNT-PEG do not increase cell death or oxidative damage to proteins in injured cells. We also showed that SWCNT-PEG prevented the loss of the plasmalemmal transporter EAAT1 in injured cells, which could point to the enhanced protective role of astrocytes against glutamate toxicity in brain injury. Additionally, SWCN-PEGs altered the cytokine secretory function of astrocytes aligned with promoting recovery after injury, which points to the possible mechanism of their protective effects in CNS injury by affecting the brain tissue microenvironment. Future additional studies regarding the possible long-term toxicity of SWCNTs on different cell types of neuronal origin are necessary, as well as tests to determine if the results from in vitro investigations translate to the in vivo model of TBI.

Supplementary Materials: The following supporting information can be downloaded at: https://www.mdpi.com/article/10.3390/cells13030225/s1, Figure S1: Purity of primary astrocyte culture; Scheme S1: Schematic illustration of the SWCNT-PEG structure; Figure S2: AFM height (left) and amplitude (right) images of SWCNT-PEG; Figure S3: Mid-infrared spectra of SWCNT-PEG and the starting carbon nanotube materials for the functionalization SWCNT-COOH; Figure S4: Thermogravimetric curve showing weight loss of SWCNT-PEG as a function of temperature during heating at 5 °C/min in air; Figure S5: Ultraviolet-visible–near-infrared red spectra of SWCNT-PEG and the starting carbon nanotube materials for the functionalization of SWCNT-COOHs; Table S1: Results of the multiplex cytokine (listed in alphabetical order) array analysis of culture media samples from control (uninjured and PEG-treated) or injured astrocytes, the latter treated with the vehicle PEG (sTBI) or SWCNT-PEG (sTBI+CNTs); Figure S6: The effects of chemically functionalized single-walled carbon nanotubes (CNTs) on the protein expression of (A) GFAP and (B) EAAT1 in mouse primary astrocytes; Figure S7: The effects of SWCNT-PEG on the secretion of selected cytokines from primary mouse astrocytes; Table S2: Results of the multiplex cytokine (listed in alphabetical order) array analysis of culture media samples from control, non-treated astrocytes, and cells treated with vehicle (PEG) or SWCNT-PEG (CNTs).

Author Contributions: Conceptualization, K.P. and V.P.; methodology, K.P., N.G.K., P.D. and T.J.; formal analysis, N.G.K., A.H.H. and K.P.; investigation, N.G.K., A.H.H., P.D. and K.P.; resources, K.P. and E.B; writing—original draft preparation, N.G.K., A.H.H., K.P. and V.P.; writing—review and editing, K.P., E.B. and V.P; visualization, N.G.K., A.H.H. and K.P.; supervision, K.P.; project administration, K.P.; funding acquisition, K.P. All authors have read and agreed to the published version of the manuscript.

Funding: This research was fully supported by the Croatian Science Foundation, grant UIP-2017-05-9517 to K.P. VP's work is supported by the Cheung Kong (Yangtze River) Scholar Program, Ministry of Education, China and by the Kun Peng Action of Zhejiang Province, China.

Institutional Review Board Statement: The study was conducted according to the guidelines of the European Community Council Directive (86/609/EEC) and the Croatian laws and rules (NN 135/06; NN 37/13; NN 125/13; NN 39/17), approved by the Faculty Ethical Committee and the Croatian Ministry of Agriculture (approval issued on 5 June 2018 for project "Single-walled carbon nanotubes in experimental traumatic brain injury").

Informed Consent Statement: Not applicable.

Data Availability Statement: The data that support the findings of this study are available within the article and Supplementary Materials or upon request from the corresponding author.

Acknowledgments: We thank Marina Jakovac, Maja Rukavina, and Tanja Mešanović for their expert and valued technical assistance. We thank Michael Brenner for constructive comments on the first draft of this manuscript. K.P. thanks Gordana Župan for their encouragement to pursue this work.

Conflicts of Interest: The authors declare no conflicts of interest.

References

1. Meaney, D.F.; Morrison, B.; Dale Bass, C. The Mechanics of Traumatic Brain Injury: A Review of What We Know and What We Need to Know for Reducing Its Societal Burden. *J. Biomech. Eng.* **2014**, *136*, 021008. [CrossRef] [PubMed]
2. Khellaf, A.; Khan, D.Z.; Helmy, A. Recent Advances in Traumatic Brain Injury. *J. Neurol.* **2019**, *266*, 2878–2889. [CrossRef] [PubMed]
3. Gruenbaum, B.F.; Zlotnik, A.; Fleidervish, I.; Frenkel, A.; Boyko, M. Glutamate Neurotoxicity and Destruction of the Blood–Brain Barrier: Key Pathways for the Development of Neuropsychiatric Consequences of TBI and Their Potential Treatment Strategies. *Int. J. Mol. Sci.* **2022**, *23*, 9628. [CrossRef] [PubMed]

4. Cederberg, D.; Visse, E.; Marklund, N.; Siesjö, P. Prolonged and Intense Neuroinflammation after Severe Traumatic Brain Injury Assessed by Cerebral Microdialysis with 300 KDa Membranes. *J. Neuroimmunol.* **2023**, *377*, 578020. [CrossRef] [PubMed]
5. Betancur, M.I.; Mason, H.D.; Alvarado-Velez, M.; Holmes, P.V.; Bellamkonda, R.V.; Karumbaiah, L. Chondroitin Sulfate Glycosaminoglycan Matrices Promote Neural Stem Cell Maintenance and Neuroprotection Post-Traumatic Brain Injury. *ACS Biomater. Sci. Eng.* **2017**, *3*, 420–430. [CrossRef] [PubMed]
6. Zhou, Y.; Shao, A.; Yao, Y.; Tu, S.; Deng, Y.; Zhang, J. Dual Roles of Astrocytes in Plasticity and Reconstruction after Traumatic Brain Injury. *Cell Commun. Signal.* **2020**, *18*, 62. [CrossRef] [PubMed]
7. Escartin, C.; Galea, E.; Lakatos, A.; O'Callaghan, J.P.; Petzold, G.C.; Serrano-Pozo, A.; Steinhäuser, C.; Volterra, A.; Carmignoto, G.; Agarwal, A.; et al. Reactive Astrocyte Nomenclature, Definitions, and Future Directions. *Nat. Neurosci.* **2021**, *24*, 312–325. [CrossRef]
8. Lepore, A.C.; O'Donnell, J.; Kim, A.S.; Yang, E.J.; Tuteja, A.; Haidet-Phillips, A.; O'Banion, C.P.; Maragakis, N.J. Reduction in Expression of the Astrocyte Glutamate Transporter, GLT1, Worsens Functional and Histological Outcomes Following Traumatic Spinal Cord Injury. *Glia* **2011**, *59*, 1996–2005. [CrossRef]
9. Allaman, I.; Bélanger, M.; Magistretti, P.J. Astrocyte-Neuron Metabolic Relationships: For Better and for Worse. *Trends Neurosci.* **2011**, *34*, 76–87. [CrossRef]
10. Siracusa, R.; Fusco, R.; Cuzzocrea, S. Astrocytes: Role and Functions in Brain Pathologies. *Front. Pharmacol.* **2019**, *10*, 1114. [CrossRef]
11. Pajarillo, E.; Rizor, A.; Lee, J.; Aschner, M.; Lee, E. The Role of Astrocytic Glutamate Transporters GLT-1 and GLAST in Neurological Disorders: Potential Targets for Neurotherapeutics. *Neuropharmacology* **2019**, *161*, 107559. [CrossRef]
12. Furuta, A.; Rothstein, J.D.; Martin, L.J. Glutamate Transporter Protein Subtypes Are Expressed Differentially during Rat CNS Development. *J. Neurosci.* **1997**, *17*, 8363–8375. [CrossRef] [PubMed]
13. Beschorner, R.; Dietz, K.; Schauer, N.; Mittelbronn, M.; Schluesener, H.J.; Trautmann, K.; Meyermann, R.; Simon, P. Expression of EAAT1 Reflects a Possible Neuroprotective Function of Reactive Astrocytes and Activated Microglia Following Human Traumatic Brain Injury. *Histol. Histopathol.* **2007**, *22*, 515–526. [CrossRef] [PubMed]
14. Piao, C.-S.; Holloway, A.L.; Hong-Routson, S.; Wainwright, M.S. Depression Following Traumatic Brain Injury in Mice Is Associated with Down-Regulation of Hippocampal Astrocyte Glutamate Transporters by Thrombin. *J. Cereb. Blood Flow Metab.* **2019**, *39*, 58–73. [CrossRef] [PubMed]
15. van Landeghem, F.K.H.; Weiss, T.; Oehmichen, M.; von Deimling, A. Decreased Expression of Glutamate Transporters in Astrocytes after Human Traumatic Brain Injury. *J. Neurotrauma* **2006**, *23*, 1518–1528. [CrossRef] [PubMed]
16. Xiang, C.; Zhang, Y.; Guo, W.; Liang, X.-J. Biomimetic Carbon Nanotubes for Neurological Disease Therapeutics as Inherent Medication. *Acta Pharm. Sin. B* **2020**, *10*, 239–248. [CrossRef] [PubMed]
17. Ni, Y.; Hu, H.; Malarkey, E.B.; Zhao, B.; Montana, V.; Haddon, R.C.; Parpura, V. Chemically Functionalized Water Soluble Single-Walled Carbon Nanotubes Modulate Neurite Outgrowth. *J. Nanosci. Nanotechnol.* **2005**, *5*, 1707–1712. [CrossRef]
18. Gottipati, M.K.; Kalinina, I.; Bekyarova, E.; Haddon, R.C.; Parpura, V. Chemically Functionalized Water-Soluble Single-Walled Carbon Nanotubes Modulate Morpho-Functional Characteristics of Astrocytes. *Nano Lett.* **2012**, *12*, 4742–4747. [CrossRef]
19. Gottipati, M.K.; Bekyarova, E.; Haddon, R.C.; Parpura, V. Chemically Functionalized Single-Walled Carbon Nanotubes Enhance the Glutamate Uptake Characteristics of Mouse Cortical Astrocytes. *Amino Acids* **2015**, *47*, 1379–1388. [CrossRef]
20. Roman, J.A.; Niedzielko, T.L.; Haddon, R.C.; Parpura, V.; Floyd, C.L. Single-Walled Carbon Nanotubes Chemically Functionalized with Polyethylene Glycol Promote Tissue Repair in a Rat Model of Spinal Cord Injury. *J. Neurotrauma* **2011**, *28*, 2349–2362. [CrossRef]
21. Wanner, I.-B. An in Vitro Trauma Model to Study Rodent and Human Astrocyte Reactivity. *Methods Mol. Biol.* **2012**, *814*, 189–219. [CrossRef] [PubMed]
22. Ellis, E.F.; McKinney, J.S.; Willoughby, K.A.; Liang, S.; Povlishock, J.T. A New Model for Rapid Stretch-Induced Injury of Cells in Culture: Characterization of the Model Using Astrocytes. *J. Neurotrauma* **1995**, *12*, 325–339. [CrossRef] [PubMed]
23. Gottipati, M.K.; Verkhratsky, A.; Parpura, V. Probing Astroglia with Carbon Nanotubes: Modulation of Form and Function. *Philos. Trans. R. Soc. Lond. B Biol. Sci.* **2014**, *369*, 20130598. [CrossRef] [PubMed]
24. Bradford, M.M. A Rapid and Sensitive Method for the Quantitation of Microgram Quantities of Protein Utilizing the Principle of Protein-Dye Binding. *Anal. Biochem.* **1976**, *72*, 248–254. [CrossRef] [PubMed]
25. Ruijter, J.M.; Thygesen, H.H.; Schoneveld, O.J.L.M.; Das, A.T.; Berkhout, B.; Lamers, W.H. Factor Correction as a Tool to Eliminate Between-Session Variation in Replicate Experiments: Application to Molecular Biology and Retrovirology. *Retrovirology* **2006**, *3*, 2. [CrossRef] [PubMed]
26. Bardi, G.; Nunes, A.; Gherardini, L.; Bates, K.; Al-Jamal, K.T.; Gaillard, C.; Prato, M.; Bianco, A.; Pizzorusso, T.; Kostarelos, K. Functionalized Carbon Nanotubes in the Brain: Cellular Internalization and Neuroinflammatory Responses. *PLoS ONE* **2013**, *8*, e80964. [CrossRef] [PubMed]
27. Wanner, I.B.; Deik, A.; Torres, M.; Rosendahl, A.; Neary, J.T.; Lemmon, V.P.; Bixby, J.L. A New in Vitro Model of the Glial Scar Inhibits Axon Growth. *Glia* **2008**, *56*, 1691–1709. [CrossRef] [PubMed]
28. Greco, T.M.; Seeholzer, S.H.; Mak, A.; Spruce, L.; Ischiropoulos, H. Quantitative Mass Spectrometry-Based Proteomics Reveals the Dynamic Range of Primary Mouse Astrocyte Protein Secretion. *J. Proteome Res.* **2010**, *9*, 2764–2774. [CrossRef]

29. Verkhratsky, A.; Matteoli, M.; Parpura, V.; Mothet, J.-P.; Zorec, R. Astrocytes as Secretory Cells of the Central Nervous System: Idiosyncrasies of Vesicular Secretion. *EMBO J.* **2016**, *35*, 239–257. [CrossRef]
30. Yuan, X.; Zhang, X.; Sun, L.; Wei, Y.; Wei, X. Cellular Toxicity and Immunological Effects of Carbon-Based Nanomaterials. *Part. Fibre Toxicol.* **2019**, *16*, 18. [CrossRef]
31. Simon-Deckers, A.; Gouget, B.; Mayne-L'hermite, M.; Herlin-Boime, N.; Reynaud, C.; Carrière, M. In Vitro Investigation of Oxide Nanoparticle and Carbon Nanotube Toxicity and Intracellular Accumulation in A549 Human Pneumocytes. *Toxicology* **2008**, *253*, 137–146. [CrossRef] [PubMed]
32. Foldbjerg, R.; Irving, E.S.; Wang, J.; Thorsen, K.; Sutherland, D.S.; Autrup, H.; Beer, C. The Toxic Effects of Single-Walled Carbon Nanotubes Are Linked to the Phagocytic Ability of Cells. *Toxicol. Res.* **2014**, *3*, 228–241. [CrossRef]
33. Nel, A.; Xia, T.; Mädler, L.; Li, N. Toxic Potential of Materials at the Nanolevel. *Science* **2006**, *311*, 622–627. [CrossRef] [PubMed]
34. Zhang, Y.; Xu, Y.; Li, Z.; Chen, T.; Lantz, S.M.; Howard, P.C.; Paule, M.G.; Slikker, W.; Watanabe, F.; Mustafa, T.; et al. Mechanistic Toxicity Evaluation of Uncoated and PEGylated Single-Walled Carbon Nanotubes in Neuronal PC12 Cells. *ACS Nano* **2011**, *5*, 7020–7033. [CrossRef] [PubMed]
35. Fukamachi, S.; Furuta, A.; Ikeda, T.; Ikenoue, T.; Kaneoka, T.; Rothstein, J.D.; Iwaki, T. Altered Expressions of Glutamate Transporter Subtypes in Rat Model of Neonatal Cerebral Hypoxia-Ischemia. *Brain Res. Dev. Brain Res.* **2001**, *132*, 131–139. [CrossRef] [PubMed]
36. Moretto, M.B.; Arteni, N.S.; Lavinsky, D.; Netto, C.A.; Rocha, J.B.T.; Souza, D.O.; Wofchuk, S. Hypoxic-Ischemic Insult Decreases Glutamate Uptake by Hippocampal Slices from Neonatal Rats: Prevention by Guanosine. *Exp. Neurol.* **2005**, *195*, 400–406. [CrossRef] [PubMed]
37. Namura, S.; Maeno, H.; Takami, S.; Jiang, X.-F.; Kamichi, S.; Wada, K.; Nagata, I. Inhibition of Glial Glutamate Transporter GLT-1 Augments Brain Edema after Transient Focal Cerebral Ischemia in Mice. *Neurosci. Lett.* **2002**, *324*, 117–120. [CrossRef]
38. Rao, V.L.; Bowen, K.K.; Dempsey, R.J. Transient Focal Cerebral Ischemia Down-Regulates Glutamate Transporters GLT-1 and EAAC1 Expression in Rat Brain. *Neurochem. Res.* **2001**, *26*, 497–502. [CrossRef]
39. Shen, Y.; Lu, H.; Xu, R.; Tian, H.; Xia, X.; Zhou, F.H.; Wang, L.; Dong, J.; Sun, L. The Expression of GLAST and GLT1 in a Transient Cerebral Ischemia Mongolian Gerbil Model. *Neuropsychiatr. Dis. Treat.* **2020**, *16*, 789–800. [CrossRef]
40. Rao, V.L.; Başkaya, M.K.; Doğan, A.; Rothstein, J.D.; Dempsey, R.J. Traumatic Brain Injury Down-Regulates Glial Glutamate Transporter (GLT-1 and GLAST) Proteins in Rat Brain. *J. Neurochem.* **1998**, *70*, 2020–2027. [CrossRef]
41. Maragakis, N.J.; Rothstein, J.D. Mechanisms of Disease: Astrocytes in Neurodegenerative Disease. *Nat. Clin. Pract. Neurol.* **2006**, *2*, 679–689. [CrossRef]
42. Lau, C.L.; O'Shea, R.D.; Broberg, B.V.; Bischof, L.; Beart, P.M. The Rho Kinase Inhibitor Fasudil Up-Regulates Astrocytic Glutamate Transport Subsequent to Actin Remodelling in Murine Cultured Astrocytes. *Br. J. Pharmacol.* **2011**, *163*, 533–545. [CrossRef] [PubMed]
43. Sheean, R.K.; Lau, C.L.; Shin, Y.S.; O'Shea, R.D.; Beart, P.M. Links between L-Glutamate Transporters, Na+/K+-ATPase and Cytoskeleton in Astrocytes: Evidence Following Inhibition with Rottlerin. *Neuroscience* **2013**, *254*, 335–346. [CrossRef]
44. Zagami, C.J.; O'Shea, R.D.; Lau, C.L.; Cheema, S.S.; Beart, P.M. Regulation of Glutamate Transporters in Astrocytes: Evidence for a Relationship between Transporter Expression and Astrocytic Phenotype. *Neurotox. Res.* **2005**, *7*, 143–149. [CrossRef] [PubMed]
45. Holt, B.D.; Short, P.A.; Rape, A.D.; Wang, Y.; Islam, M.F.; Dahl, K.N. Carbon Nanotubes Reorganize Actin Structures in Cells and Ex Vivo. *ACS Nano* **2010**, *4*, 4872–4878. [CrossRef] [PubMed]
46. Sofroniew, M.V. Astrocyte Reactivity: Subtypes, States, and Functions in CNS Innate Immunity. *Trends Immunol.* **2020**, *41*, 758–770. [CrossRef] [PubMed]
47. Lazic, A.; Balint, V.; Stanisavljevic Ninkovic, D.; Peric, M.; Stevanovic, M. Reactive and Senescent Astroglial Phenotypes as Hallmarks of Brain Pathologies. *Int. J. Mol. Sci.* **2022**, *23*, 4995. [CrossRef]
48. Zhou, H.; Hu, L.; Li, J.; Ruan, W.; Cao, Y.; Zhuang, J.; Xu, H.; Peng, Y.; Zhang, Z.; Xu, C.; et al. AXL Kinase-Mediated Astrocytic Phagocytosis Modulates Outcomes of Traumatic Brain Injury. *J. Neuroinflamm.* **2021**, *18*, 154. [CrossRef]
49. Semple, B.D.; Frugier, T.; Morganti-Kossmann, M.C. CCL2 Modulates Cytokine Production in Cultured Mouse Astrocytes. *J. Neuroinflamm.* **2010**, *7*, 67. [CrossRef]
50. Lau, L.T.; Yu, A.C. Astrocytes Produce and Release Interleukin-1, Interleukin-6, Tumor Necrosis Factor Alpha and Interferon-Gamma Following Traumatic and Metabolic Injury. *J. Neurotrauma* **2001**, *18*, 351–359. [CrossRef]
51. Choi, B.-R.; Johnson, K.R.; Maric, D.; McGavern, D.B. Monocyte-Derived IL-6 Programs Microglia to Rebuild Damaged Brain Vasculature. *Nat. Immunol.* **2023**, *24*, 1110–1123. [CrossRef] [PubMed]
52. Awatsuji, H.; Furukawa, Y.; Hirota, M.; Murakami, Y.; Nii, S.; Furukawa, S.; Hayashi, K. Interleukin-4 and -5 as Modulators of Nerve Growth Factor Synthesis/Secretion in Astrocytes. *J. Neurosci. Res.* **1993**, *34*, 539–545. [CrossRef] [PubMed]
53. Gadani, S.P.; Cronk, J.C.; Norris, G.T.; Kipnis, J. IL-4 in the Brain: A Cytokine to Remember. *J. Immunol.* **2012**, *189*, 4213–4219. [CrossRef] [PubMed]
54. Kamm, K.; Vanderkolk, W.; Lawrence, C.; Jonker, M.; Davis, A.T. The Effect of Traumatic Brain Injury upon the Concentration and Expression of Interleukin-1beta and Interleukin-10 in the Rat. *J. Trauma* **2006**, *60*, 152–157. [CrossRef] [PubMed]
55. Garcia, J.M.; Stillings, S.A.; Leclerc, J.L.; Phillips, H.; Edwards, N.J.; Robicsek, S.A.; Hoh, B.L.; Blackburn, S.; Doré, S. Role of Interleukin-10 in Acute Brain Injuries. *Front. Neurol.* **2017**, *8*, 244. [CrossRef] [PubMed]

56. Ding, X.; Cao, F.; Cui, L.; Ciric, B.; Zhang, G.-X.; Rostami, A. IL-9 Signaling Affects Central Nervous System Resident Cells during Inflammatory Stimuli. *Exp. Mol. Pathol.* **2015**, *99*, 570–574. [CrossRef] [PubMed]
57. Yoshimura, S.; Thome, R.; Konno, S.; Mari, E.R.; Rasouli, J.; Hwang, D.; Boehm, A.; Li, Y.; Zhang, G.-X.; Ciric, B.; et al. IL-9 Controls Central Nervous System Autoimmunity by Suppressing GM-CSF Production. *J. Immunol.* **2020**, *204*, 531–539. [CrossRef] [PubMed]
58. Tan, S.; Shan, Y.; Lin, Y.; Liao, S.; Zhang, B.; Zeng, Q.; Wang, Y.; Deng, Z.; Chen, C.; Hu, X.; et al. Neutralization of Interleukin-9 Ameliorates Experimental Stroke by Repairing the Blood-Brain Barrier via down-Regulation of Astrocyte-Derived Vascular Endothelial Growth Factor-A. *FASEB J.* **2019**, *33*, 4376–4387. [CrossRef]
59. Kovacic, J.C.; Muller, D.W.M.; Graham, R.M. Actions and Therapeutic Potential of G-CSF and GM-CSF in Cardiovascular Disease. *J. Mol. Cell. Cardiol.* **2007**, *42*, 19–33. [CrossRef]
60. Konishi, Y.; Chui, D.H.; Hirose, H.; Kunishita, T.; Tabira, T. Trophic Effect of Erythropoietin and Other Hematopoietic Factors on Central Cholinergic Neurons in Vitro and in Vivo. *Brain Res.* **1993**, *609*, 29–35. [CrossRef]
61. Schäbitz, W.-R.; Kollmar, R.; Schwaninger, M.; Juettler, E.; Bardutzky, J.; Schölzke, M.N.; Sommer, C.; Schwab, S. Neuroprotective Effect of Granulocyte Colony-Stimulating Factor after Focal Cerebral Ischemia. *Stroke* **2003**, *34*, 745–751. [CrossRef] [PubMed]
62. Schneider, A.; Krüger, C.; Steigleder, T.; Weber, D.; Pitzer, C.; Laage, R.; Aronowski, J.; Maurer, M.H.; Gassler, N.; Mier, W.; et al. The Hematopoietic Factor G-CSF Is a Neuronal Ligand That Counteracts Programmed Cell Death and Drives Neurogenesis. *J. Clin. Investig.* **2005**, *115*, 2083–2098. [CrossRef] [PubMed]
63. Huang, X.; Dai, Y.; Ma, X.; Wang, S.; Xu, X.; Pei, X.; Li, R.; Wang, H. Different Changes in Granulocyte-Colony Stimulating Factor and Its Correlation with Inflammatory Biomarkers in Patients after Traumatic Brain Injury. *NeuroReport* **2020**, *31*, 293. [CrossRef] [PubMed]
64. Corrigan, F.; Arulsamy, A.; Teng, J.; Collins-Praino, L.E. Pumping the Brakes: Neurotrophic Factors for the Prevention of Cognitive Impairment and Dementia after Traumatic Brain Injury. *J. Neurotrauma* **2017**, *34*, 971–986. [CrossRef] [PubMed]
65. Qiu, X.; Ping, S.; Kyle, M.; Chin, L.; Zhao, L.-R. SCF + G-CSF Treatment in the Chronic Phase of Severe TBI Enhances Axonal Sprouting in the Spinal Cord and Synaptic Pruning in the Hippocampus. *Acta Neuropathol. Commun.* **2021**, *9*, 63. [CrossRef] [PubMed]
66. Nishihara, T.; Ochi, M.; Sugimoto, K.; Takahashi, H.; Yano, H.; Kumon, Y.; Ohnishi, T.; Tanaka, J. Subcutaneous Injection Containing IL-3 and GM-CSF Ameliorates Stab Wound-Induced Brain Injury in Rats. *Exp. Neurol.* **2011**, *229*, 507–516. [CrossRef] [PubMed]
67. Shultz, S.R.; Tan, X.L.; Wright, D.K.; Liu, S.J.; Semple, B.D.; Johnston, L.; Jones, N.C.; Cook, A.D.; Hamilton, J.A.; O'Brien, T.J. Granulocyte-Macrophage Colony-Stimulating Factor Is Neuroprotective in Experimental Traumatic Brain Injury. *J. Neurotrauma* **2014**, *31*, 976–983. [CrossRef]
68. Boraschi, D.; Italiani, P.; Weil, S.; Martin, M.U. The Family of the Interleukin-1 Receptors. *Immunol. Rev.* **2018**, *281*, 197–232. [CrossRef]
69. Harrer, C.; Otto, F.; Radlberger, R.F.; Moser, T.; Pilz, G.; Wipfler, P.; Harrer, A. The CXCL13/CXCR5 Immune Axis in Health and Disease—Implications for Intrathecal B Cell Activities in Neuroinflammation. *Cells* **2022**, *11*, 2649. [CrossRef]
70. Pilz, G.; Wipfler, P.; Otto, F.; Hitzl, W.; Afazel, S.; Haschke-Becher, E.; Trinka, E.; Harrer, A. Cerebrospinal Fluid CXLC13 Indicates Disease Course in Neuroinfection: An Observational Study. *J. Neuroinflamm.* **2019**, *16*, 13. [CrossRef]
71. Rayasam, A.; Kijak, J.A.; Kissel, L.; Choi, Y.H.; Kim, T.; Hsu, M.; Joshi, D.; Laaker, C.J.; Cismaru, P.; Lindstedt, A.; et al. CXCL13 Expressed on Inflamed Cerebral Blood Vessels Recruit IL-21 Producing TFH Cells to Damage Neurons Following Stroke. *J. Neuroinflamm.* **2022**, *19*, 125. [CrossRef] [PubMed]
72. Ciechanowska, A.; Popiolek-Barczyk, K.; Ciapała, K.; Pawlik, K.; Oggioni, M.; Mercurio, D.; de Simoni, M.G.; Mika, J. Traumatic brain injury in mice induces changes in the expression of the XCL1/XCR1 and XCL1/ITGA9 axes. *Pharmacol. Rep.* **2020**, *72*, 1579–1592. [CrossRef] [PubMed]
73. Moro, N.; Ghavim, S.S.; Sutton, R.L. Massive Efflux of Adenosine Triphosphate into the Extracellular Space Immediately after Experimental Traumatic Brain Injury. *Exp. Ther. Med.* **2021**, *21*, 575. [CrossRef]
74. Malarkey, E.B.; Parpura, V. Temporal Characteristics of Vesicular Fusion in Astrocytes: Examination of Synaptobrevin 2-Laden Vesicles at Single Vesicle Resolution. *J. Physiol.* **2011**, *589*, 4271–4300. [CrossRef] [PubMed]
75. Montana, V.; Ni, Y.; Sunjara, V.; Hua, X.; Parpura, V. Vesicular Glutamate Transporter-Dependent Glutamate Release from Astrocytes. *J. Neurosci.* **2004**, *24*, 2633–2642. [CrossRef]
76. Gottipati, M.K.; Bekyarova, E.; Haddon, R.C.; Parpura, V. Chemically Functionalized Water-Soluble Single-Walled Carbon Nanotubes Obstruct Vesicular/Plasmalemmal Recycling in Astrocytes Down-Stream of Calcium Ions. *Cells* **2020**, *9*, 1597. [CrossRef]

cells

MDPI

Review

The Role of Aquaporins in Spinal Cord Injury

Terese A. Garcia [1], Carrie R. Jonak [1] and Devin K. Binder [1,2,3,*]

1 Division of Biomedical Sciences, School of Medicine, University of California, Riverside, CA 92521, USA
2 Center for Glial-Neuronal Interactions, University of California, Riverside, CA 92521, USA
3 Neuroscience Graduate Program, University of California, Riverside, CA 92521, USA
* Correspondence: dbinder@ucr.edu; Tel.: +1-951-827-2224

Abstract: Edema formation following traumatic spinal cord injury (SCI) exacerbates secondary injury, and the severity of edema correlates with worse neurological outcome in human patients. To date, there are no effective treatments to directly resolve edema within the spinal cord. The aquaporin-4 (AQP4) water channel is found on plasma membranes of astrocytic endfeet in direct contact with blood vessels, the glia limitans in contact with the cerebrospinal fluid, and ependyma around the central canal. Local expression at these tissue–fluid interfaces allows AQP4 channels to play an important role in the bidirectional regulation of water homeostasis under normal conditions and following trauma. In this review, we consider the available evidence regarding the potential role of AQP4 in edema after SCI. Although more work remains to be carried out, the overall evidence indicates a critical role for AQP4 channels in edema formation and resolution following SCI and the therapeutic potential of AQP4 modulation in edema resolution and functional recovery. Further work to elucidate the expression and subcellular localization of AQP4 during specific phases after SCI will inform the therapeutic modulation of AQP4 for the optimization of histological and neurological outcomes.

Keywords: astrocytes; aquaporin; aquaporin-4; spinal cord injury; edema

Citation: Garcia, T.A.; Jonak, C.R.; Binder, D.K. The Role of Aquaporins in Spinal Cord Injury. *Cells* **2023**, *12*, 1701. https://doi.org/10.3390/cells12131701

Academic Editors: Michael Brenner and Vladimir Parpura

Received: 28 February 2023
Revised: 8 June 2023
Accepted: 21 June 2023
Published: 23 June 2023

1. Introduction

Traumatic spinal cord injury (SCI) is a debilitating and life-altering condition [1,2]. SCI can lead to loss of motor, sensory, and autonomic function [3–5]. These impairments can be additionally linked to comorbidities that introduce additional obstacles for SCI patients in their daily lives [6,7]. For example, psychological comorbidities of SCI that have recently received a great deal of attention include depression, anxiety, and post-traumatic stress disorder [8]. The initial mechanical insult to the spinal cord (termed primary injury) is followed by a cascade of events that may exacerbate the injury (termed secondary injury). Histopathological processes occurring in the injured spinal cord include breakdown of the blood–spinal cord barrier (BSCB), edema (water accumulation), hemorrhage, inflammation, excitotoxicity, cavitation, and glial scar formation. One major goal of SCI research is to identify potentially reversible causes of secondary injury.

Animal models of SCI include contusion, compression, and transection [9]. Contusion injuries involve impact to the exposed spinal cord with an impactor or a weight drop system [10–12]. The most common model involves creating an injury with a computer-controlled impactor (Infinite Horizons) which creates user-selectable force levels (30–300 kilodynes, i.e., 0.3–3 N) and allows the impact tip dwell time to be modulated. Real-time probe force and displacement graphs provide the operator with a precise detailed record of the kinematics of each impact. These parameters can be modified to generate mild, moderate, or severe SCI. Compression injuries are created by constricting the exposed cord with clips, forceps, or weight application [13–15]. Transection models involve durotomy (opening of the dura) and myelotomy (incision into the spinal cord parenchyma, such as complete transection or hemisection of the cord). In general, contusion models more accurately model typical human SCI [16–18].

2. Spinal Cord Edema

After acute SCI, edema develops at the injury site and spreads rostrally and caudally, leading to increased cord water content and cord tissue volume [19]. Edema is generally divided into *vasogenic* and *cytotoxic*, but there may be a combination of the two. *Vasogenic* edema results from breakdown of the BSCB, resulting in an accumulation of fluid in the interstitial (extracellular) space of the cord. *Cytotoxic* edema results from fluid flow into astrocytes resulting in cellular (hence "cytotoxic") swelling. Contusion SCI results in traumatic BSCB disruption, and hence is thought to be largely a model of vasogenic edema, in contrast, compression SCI causes ischemia to cord vessels and swelling of astrocytes, and hence is thought to be largely a model of cytotoxic edema. It is clear that injury severity correlates with the development of edema at the injury site [20], and the extent of edema impacts secondary damage and neurological outcome following injury [19,21–23].

What is the mechanism by which edema contributes to secondary injury? Studies in both rodent SCI models and in humans have now clearly indicated that spinal cord edema at the injury site leads to the swelling of the cord, compression of the cord against the surrounding dura within the spinal canal, an increase in intraparenchymal cord pressure, ischemia, and eventual infarction [24–26]. Thus, edema can essentially lead to or exacerbate a spinal cord "stroke". Vascular disruption and BSCB breakdown contribute to fluid, protein, and blood accumulation within the parenchyma of the cord, further exacerbating edema [27,28]. In a rat model of thoracic contusion SCI, spinal cord edema (as assessed by measurement of water content at the lesion site at multiple time points) was found to occur by 6 h and to peak at 72 h [29]. Interestingly, the BSCB is thought to be repaired by 2–3 weeks after trauma [27], suggesting that the acute period (at least through 72 h post-injury) may represent a "therapeutic window" for intervention to alleviate edema and mitigate secondary injury.

Recent human studies of SCI have extensively demonstrated the therapeutic importance of edema reduction in functional outcome. What is apparent from a review of the neurosurgical literature on SCI is that decompressive laminectomy, duraplasty, and even myelotomy do not fully relieve edema or local pressure in the spinal cord. The key studies in this regard have been conducted as part of the ISCoPE (Injured Spinal Cord Pressure Evaluation) trials. These are the first studies in humans to evaluate intraparenchymal spinal cord pressure. These studies find massive increases in intraspinal pressure at the injury site that falls off farther away, indicating that peak edema at the injury site is likely responsible for secondary injury [30–34]; indeed, intraspinal pressure and spinal cord perfusion pressure predict neurological outcome [33].

The above studies establish the importance of edema within the cord contributing to secondary injury. Ample clinical data over decades with the use of methylprednisolone [35] have indicated minimal salutary effects of intravenous methylprednisolone on reducing cord edema or improving outcome [36]. As more is being uncovered about the mechanisms underlying edema formation following injury to the central nervous system (CNS), interventions aimed at this issue are becoming more clinically important. To date, there is no effective treatment for edema following SCI. Thus, the discovery and characterization of molecular water channels (the aquaporins) and their examination in the context of SCI are of critical scientific and translational importance.

3. Aquaporins

Aquaporins (AQPs) are a family of membrane proteins that function as water channels in many cell types and tissues in which fluid transport is crucial [37–40]. AQPs are small hydrophobic integral membrane proteins (~30 kDa monomer) that facilitate bidirectional water transport in response to osmotic gradients [41]. Multiple mammalian aquaporins have been identified, including AQP0, AQP1, AQP2, AQP4, and AQP5, which transport water only (aquaporins), and AQP3, AQP7, and AQP9, which also transport glycerol (aquaglyceroporins) [39,41].

Aquaporin-4 (AQP4) [42–45] (Figure 1) is of particular interest in neuroscience as it is expressed in the brain and spinal cord by astrocytes and ependymal cells, especially at specialized membrane domains including astrocytic endfeet in contact with blood vessels and astrocyte membranes that ensheathe glutamatergic synapses [40,46–49]. AQP9 is also expressed in astrocytes at low levels, and AQP1 is expressed in choroid plexus epithelial cells (Figure 1C, right inset). AQP4 has two major isoforms (AQP4-M1 and AQP4-M23). AQP4-M23, but not AQP1-M1, forms square arrays in the astrocyte plasma membrane, known as orthogonal arrays of particles (OAPs) [50]. OAPs can be seen in freeze-fracture electron micrographs [51–55]. It has been demonstrated that activity-induced water fluxes in the neocortex may represent water movement via aquaporin channels in response to physiological activity [56,57].

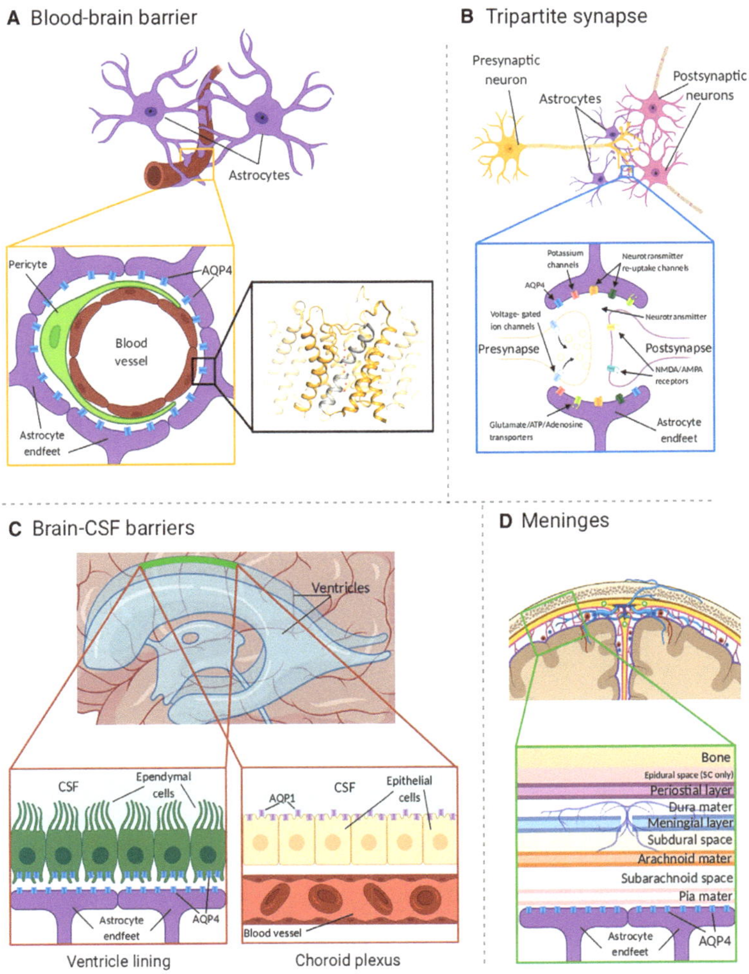

Figure 1. AQP4 localization in the CNS. (**A**) AQP4 (blue) is located within astrocyte endfeet processes surrounding blood vessels in both brain tissue and the BBB. The inset shows the crystal structure of human AQP4 (PDB code 3GD8). AQP4 assembles as a tetramer with each monomer comprising six

transmembrane helices and two half-helices (grey). The two half-helices harbor the aromatic-arginine (ar/R) motif that functions as a selectivity filter. Within the pore, water molecules (red spheres) align in a single file. (**B**) AQP4 is also localized at the astrocyte component of the tripartite synapse. During neurotransmission, neurons release mediators and neurotransmitters from synaptic nerve terminals (affecter cells) into the synaptic cleft to communicate with other neurons (effector cells). This synaptic activity induces an increase in intracellular Ca^{2+} concentration, which is accompanied by changed water and solute concentrations in astrocytes, leading to the release of glutamate and other gliotransmitters. This gliotransmission results in feedback to the presynaptic neurons to modulate neuro-transmission. AQP4 plays an essential role in maintaining water homeostasis during this process. (**C**) In ventricles, AQPs are present within ependymal cells lining the brain–CSF interfaces (left inset). AQP4 is localized to the basolateral membrane of ependymal cells and the endfeet of contacting astrocytes. AQP1 (purple) is localized to the apical membrane of the choroid plexus epithelium (right inset). (**D**) CSF within the subarachnoid and cisternal spaces flows into the brain specifically via periarterial spaces and then exchanges with brain interstitial fluid facilitated by AQP4 water channels that are positioned within perivascular astrocyte endfoot processes. Figure and figure legend adopted from [42] with permission.

In the spinal cord, AQP4 expression is found ubiquitously in both gray and white matter and is prominent not only in astrocytic endfeet but also at the glia limitans, where the spinal cord directly contacts cerebrospinal fluid [49] (Figure 2). This pattern of expression suggests that AQP4 may play a crucial role in spinal cord water homeostasis.

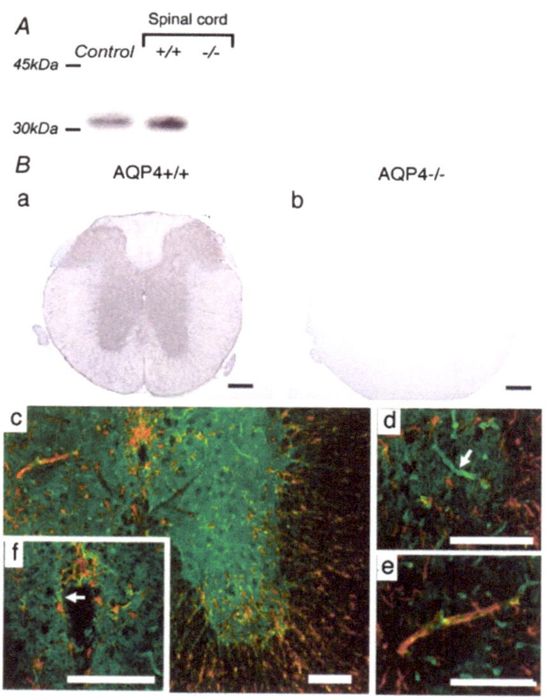

Figure 2. *Cont.*

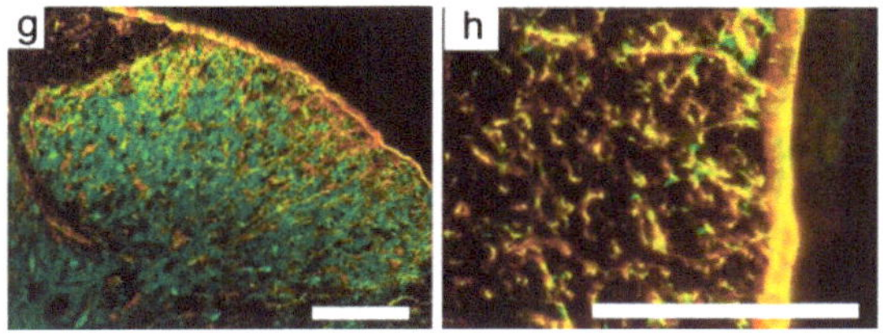

Figure 2. AQP4 expression in mouse spinal cord. (**A**) Western blot analysis demonstrates an approximately 32 kDa band in control cerebral cortex (Control) and wild-type spinal cord (+/+). No expression was detected in the spinal cords of AQP4-deficient mice ($^{-/-}$). (**B**) Immunohistochemistry for AQP4 reveals extensive expression in gray and white matter in the cervical spinal cords of wild-type mice (+/+) (**a**). No specific immunostaining was found in the spinal cords of AQP4 $^{-/-}$ mice ($^{-/-}$). (**b**). Dense AQP4 staining (green) appeared extensively in gray matter (**c**), especially in astrocytic endfeet surrounding capillaries ((**d**), white arrow). GFAP-immunopositive (red) glial processes extending to capillaries were observed (**e**). Faint AQP4 and GFAP staining was detected in ependymal cells lining the central canal ((**f**), white arrow). GFAP and AQP4 were co-localized in fibrous thin astrocytes in the superficial dorsal horn (**g**). In white matter, AQP4 was co-expressed prominently with GFAP-immunoreactive radial fibrous glial processes surrounding the blood vessels and the glia limitans (**h**). ((**c**–**h**); red: GFAP, green: AQP4). Black scale bar = 0.2 mm. White scale bar = 0.1 mm. Figure and figure legend adopted from [49] with permission.

4. Aquaporin 4 Expression after SCI

The first and still most comprehensive study of AQP4 changes in rodent SCI was published in 2006 [58]. The authors demonstrated the robust baseline expression of AQP4 in gray and white matter astrocytes, specifically in astrocytic processes surrounding neurons and blood vessels. This immunoreactivity was particularly strong in the glia limitans externa and interna. These findings in the baseline rat spinal cord were very similar to those published in 2004 in the mouse spinal cord [49]. Indeed, the authors commented that the spatial distribution of AQP4 suggests the critical role that astrocytes expressing AQP4 play in the transport of water from blood/cerebrospinal fluid to spinal cord parenchyma and vice versa. Following thoracic contusion SCI, biphasic changes in astrocytic AQP4 levels were observed, including early downregulation and subsequent persistent upregulation. This early downregulation occurred in the acute and subacute periods following injury (up to 14 days) and subsequently there was increased AQP4 expression in both gray and white matter of the spinal cord (day 21 to 11 months post-injury) [58].

Interestingly, this group also studied neurological outcome based on injury severity (mild, moderate, and severe) and correlation to AQP4 expression. At 3 weeks post-injury, animals showing more impairment on the Basso, Beattie, and Bresnahan locomotor scale [59] had lower levels of AQP4 protein. The authors suggest that changes in expression and/or localization of AQP4 (and its anchoring proteins dystrophin and α-syntrophin) may lead to impaired water clearance in the chronically injured cord [58,60].

In a human study, several spinal cord tissue samples were analyzed for the expression of AQP4. Tissue was obtained at autopsy from individuals that had been injured for a duration of 1–2 years (so of course no data were available from human tissue in the acute/subacute period). The authors found that human SCI tissue samples contained GFAP-positive astrocytes both with and without AQP4 labeling at the lesion epicenter [60]. In contrast to the relative lack of expression of AQP4 in astrocytes at the lesion epicenter, AQP4 labeling was increased in spared white matter compared to uninjured cord tissue [60].

Subsequently, other studies using different SCI models and species have also indicated chronic increases in AQP4 expression after SCI [61–63] or AQP4 modulation by pharmacological agents after SCI [25,64–74]. These studies have previously been reviewed [75,76]. Overall, the studies suggest that there is a long-term upregulation of

AQP4 in the chronically-injured spinal cord, but whether this is compensatory or a result of reactive astrocytosis is unclear.

5. SCI Phenotype in AQP4$^{-/-}$ Mice

The above studies of AQP4 regulation in rodent and human tissue strongly suggest that there is a functional role for AQP4 in spinal cord water homeostasis, and hence the extent of edema following SCI. However, gene expression data and pharmacological modulation data do not unambiguously determine the functional role of AQP4 in SCI Therefore, several studies have been conducted to test the role of AQP4 in SCI by using AQP4$^{-/-}$ mice.

AQP4$^{-/-}$ mice were originally generated by targeted gene disruption in 1997 [77], and recently an astroglial conditional deletion of AQP4 has been generated [78]. AQP4$^{-/-}$ mice and astroglial conditional AQP4 knockouts are grossly normal phenotypically, do not manifest overt neurological abnormalities, altered blood–brain barrier properties, abnormal baseline intracranial pressure, impaired osmoregulation, or obvious brain dysmorphology [79–81].

In vivo studies of these mice demonstrated a functional role for AQP4 in brain water transport. AQP4$^{-/-}$ mice have markedly decreased accumulation of brain water (cerebral edema) following water intoxication or focal cerebral ischemia [82] and impaired clearance of brain water in models of vasogenic edema [81]. Clearance of seizure-induced edema may also be AQP4-dependent [83]. Impaired water flux into (in the case of cytotoxic edema) and out of (in the case of vasogenic edema) the brain makes sense based on the bidirectional nature of water flux across the AQP4 plasma membrane channel at the blood–brain barrier The recently-generated astroglial-conditional AQP4 knockout mouse line demonstrates a 31% reduction in brain water uptake after systemic hypoosmotic stress [78]. Similarly, mice deficient in dystrophin or α-syntrophin, in which there is mislocalization of AQP4 protein [84–86], show attenuated cerebral edema in response to hypoosmotic stress [85,87]

Based on the brain studies above, one would expect similar results in the spinal cord of AQP4$^{-/-}$ mice; in other words, amelioration with AQP4 deletion in models of cytotoxic edema and exacerbation in models of vasogenic edema. That is exactly what has been observed.

In 2008, Saadoun et al. reported the phenotype of AQP4$^{-/-}$ mice subjected to thoracic compression injury [88]. This is thought to be an ischemic injury leading to cytotoxic edema. The authors found that 2 days after thoracic compression injury, AQP4$^{-/-}$ mice had improved neuronal survival and myelin sparing along with reduced tissue water content and intraparenchymal cord pressure compared to wild-type mice. In addition, AQP4$^{-/-}$ mice showed improved functional outcome in the Basso mouse locomotor scale (BMS) [89], inclined plane test, and footprint analysis. Lastly, AQP4$^{-/-}$ mice showed improved sensory outcome as measured by spinal-somatosensory-evoked responses. Thus, AQP4$^{-/-}$ mice have improved outcome in this model of cytotoxic spinal cord edema.

In 2010, Kimura et al. reported the phenotype of AQP4$^{-/-}$ mice subjected to thoracic contusion injury [90]. Contusion is thought to be an injury leading to the breakdown of the BSCB and vasogenic edema. The authors found that AQP4$^{-/-}$ mice showed greater levels of demyelination, cyst formation, and greater neuronal loss 42 days following thoracic contusion injury (Figure 3). In human and rat models of SCI, removal of necrotic tissue by phagocytic cells at the lesion epicenter typically leads to the formation of fluid-filled cavities/cysts in their place, regardless of the size of the species [91]. These results are interesting, as mouse models of SCI do not typically display cyst formation. In addition, Kimura et al. [89] showed that injured tissue from AQP4$^{-/-}$ mice contained increased water content compared to controls at 2 and 4 weeks following injury. Lastly, AQP4$^{-/-}$ mice showed reduced functional recovery in both the BMS and footprint analysis tasks at 42 days post-injury (Figure 4). In contrast to cytotoxic edema produced by compression injury, contusion injury results in vasogenic edema and disruption of the blood spinal cord barrier, not cytotoxic cell swelling. Thus, AQP4$^{-/-}$ mice have worsened outcome in this model of vasogenic spinal cord edema. From this, it appears that AQP4 expression is required for clearance of vasogenic edema and plays a protective role following contusion SCI.

In 2014, Wu et al. reported the phenotype of AQP4$^{-/-}$ mice after rubrospinal tract hemisection [92]. Transection is thought to be an injury leading to the breakdown of the BSCB and vasogenic edema. AQP4$^{-/-}$ mice showed increased spinal cord edema at 72 h after hemisection compared to wild-type littermates. In addition, there was reduced migration of astrocytes to the lesion (at week 1); greater lesion volume, glial scar formation, and cyst volume (at week 6); and increased retrograde axonal degeneration. This study is in agreement with the protective role of AQP4 following vasogenic edema [92].

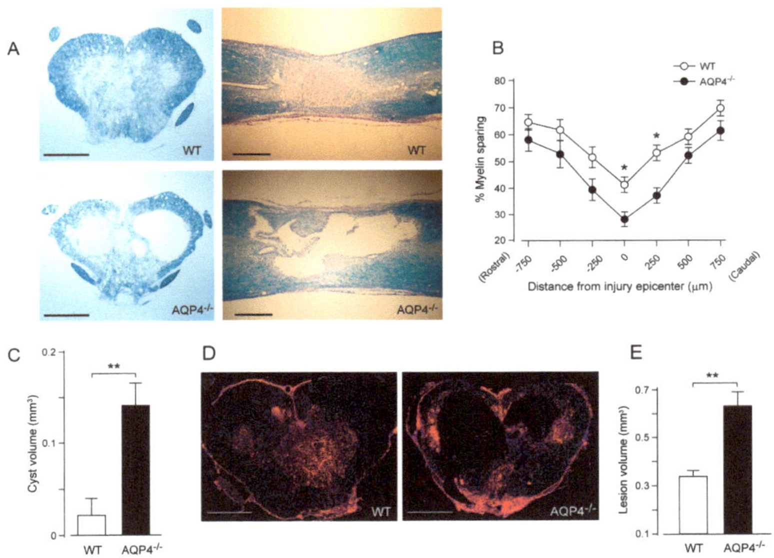

Figure 3. Greater tissue damage and prominent cyst formation in AQP4$^{-/-}$ mice following a thoracic contusion injury. (**A**) Representative images of Luxol Fast Blue (LFB) staining at 42 days post-injury (dpi) (upper left, WT cross section; upper right, WT longitudinal section; lower left, AQP4$^{-/-}$ cross section, lower right; AQP4$^{-/-}$ longitudinal section). Note the greater demyelination and prominent cyst formation in AQP4$^{-/-}$ mice. (**B**) Stereological quantification of LFB staining shows significantly increased myelin loss in AQP4$^{-/-}$ mice. Data are represented as mean ± SEM, $n = 7$ each group; statistical significance was evaluated using a two-way ANOVA with Bonferroni *post-hoc* test, * $p < 0.05$. (**C**) Quantification of cyst volume shows significantly greater cyst volume in AQP4$^{-/-}$ mice (black bar) compared with WT (white bar). Data are represented as mean ± SEM, $n = 7$ each group; statistical comparisons were made using a Student's t-test, ** $p < 0.01$. (**D**) Representative images of fibronectin staining at the injury epicenter (red, fibronectin; blue, DAPI; left, WT; right, AQP4$^{-/-}$). (**E**) Stereological quantification of lesion volume delineated by fibronectin shows significantly greater lesion volume in AQP4$^{-/-}$ mice (black bar) compared with WT mice (white bar). Data are represented as mean ± SEM, $n = 7$ each group; statistical comparisons were made using a Student's t-test, ** $p < 0.01$. Scale bars: A, D, 400 μm. Figure and figure legend adopted from [90] with permission.

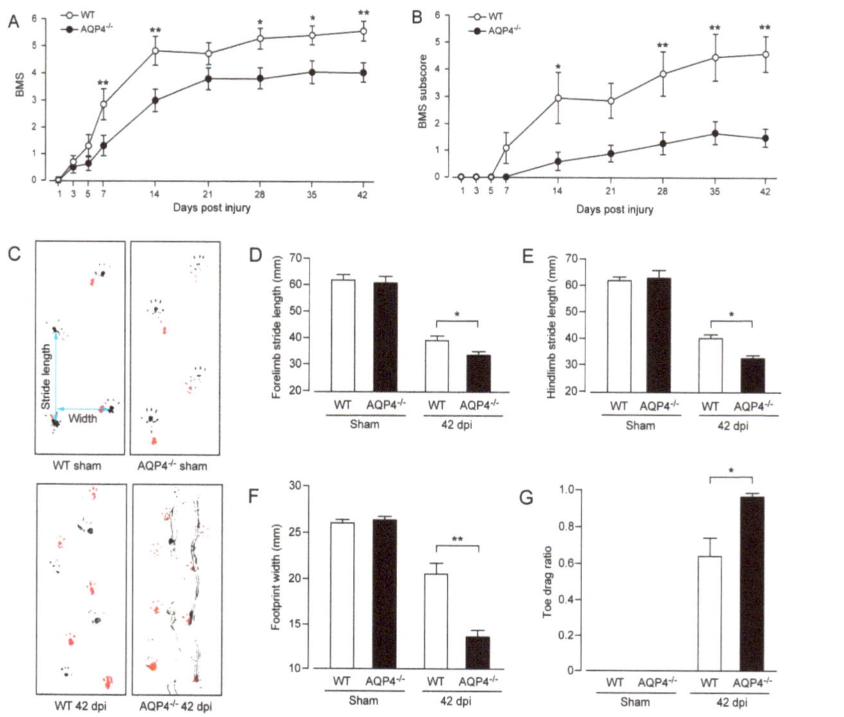

Figure 4. Impaired locomotor recovery following contusion SCI in AQP4$^{-/-}$ mice. AQP4$^{-/-}$ mice had a worse functional outcome as revealed by BMS (**A**) and BMS subscores (**B**) compared with WT mice (white circles, WT; black circles, AQP4$^{-/-}$). Data are represented as mean ± SEM, n = 10 each group. Statistical comparisons were made using repeated-measures ANOVA with Bonferroni *post-hoc* test, * $p < 0.05$, ** $p < 0.01$. (**C**) Representative images of footprint analysis (upper left, WT sham; lower left, WT 42 days post-injury, i.e., dpi; upper right, AQP4$^{-/-}$ sham; lower right, AQP4$^{-/-}$ 42 dpi). AQP4$^{-/-}$ mice showed significantly decreased stride length in both forelimbs (**D**) and hindlimbs (**E**), decreased width of hindlimb footprints (**F**), and increased toe drag ratio (**G**) (white bars, WT; black bars, AQP4$^{-/-}$). Data are represented as mean ± SEM, n = 10 each group. Statistical significance was evaluated using a Student's t-test or Mann–Whitney U test, * $p < 0.05$, ** $p < 0.01$. Figure and figure legend adopted from [90] with permission.

6. AQP4 Modulation in SCI

The studies in AQP4$^{-/-}$ mice together with the clinical data (summarized above) indicating the critical importance of edema in secondary injury after SCI strongly support the concept of modulation of AQP4 as a therapeutic strategy early after SCI. Studies of spinal cord water content changes in the rat model of thoracic SCI indicate that approximately between 6 h and 72 h (3 days) after acute injury there may be a "therapeutic window" to ameliorate edema and improve outcome. Unfortunately, despite intense effort, selective inhibitors or activators of AQP4 have been difficult to develop [93,94].

Nevertheless, several studies have attempted to target AQP4 and related molecules for therapeutic effect. In 2016, Zhang et al. found that, in mice, the antioxidant astaxanthin alleviated cerebral edema after controlled cortical impact TBI by reducing NKCC1, AQP4 mRNA, and protein levels [95]. In 2018, in a thoracic contusion model, Yan et al. found that pretreatment of rats with inhibitors of AQP4 or NKCC1 attenuated edema and tissue damage after SCI [96]. These studies, while interesting, did not selectively target AQP4 nor assess subcellular distribution of AQP4 or channel activities.

In 2020, a landmark paper was published in *Cell* targeting AQP4 subcellular localization to treat edema after SCI [97]. The authors demonstrated that (1) AQP4 cell-surface abundance increases in response to hypoxia-induced cell swelling in a calmodulin-dependent manner; (2) calmodulin directly binds the AQP4 carboxyl terminus, causing a conformational change and driving cell surface localization; and (3) inhibition of calmodulin in a rat SCI model with trifluoperazine inhibited AQP4 localization to the BSCB, abrogated CNS edema, and accelerated functional recovery compared with untreated animals. The SCI model used in this study was a dorsal column crush injury in 6–8-week-old rats. This would be expected to lead to cytotoxic edema (as a compression model). Hence, inhibition of AQP4 surface localization by the calmodulin inhibitor trifluoperazine was efficacious at limiting post-injury edema and improving neurological outcome.

These findings lead to the novel therapeutic strategy of modulation of AQP4 subcellular localization to treat CNS edema [42]. Given the difficulties in developing pore-blocking AQP4 inhibitors, targeting AQP4 subcellular localization opens up new treatment avenues for CNS edema. In particular, the inhibition of AQP4 membrane localization (as with trifluoperazine) is a novel treatment option for cytotoxic edema when it is desirable to limit AQP4 activity. This may have implications for the treatment of conditions other than CNS edema as well. For example, there is prominent AQP4 subcellular mislocalization in epilepsy [98], for which renormalization of its subcellular localization has also been suggested as a therapeutic strategy [99].

7. AQP4, Glial Scar Formation, and Neuroinflammation

Glial scar formation and neuroinflammation are other key processes after CNS injury. In the context of SCI, glial scar formation around the injury involves migration of reactive astrocytes to the site of injury. What role might AQP4 play in SCI-associated glial scar formation and associated neuroinflammatory processes?

Several studies have indicated the involvement of AQP4 in glial scar formation. Interestingly, in AQP4$^{-/-}$ mice, astrocytes demonstrate impaired migration and wound healing rate compared with wild-type mice [100]. AQP4 is polarized to the leading edge of the plasma membrane in migrating wild-type astroglia, suggesting that AQP4 facilitates water influx across the leading edge of a migrating cell [100]. Glial scar formation was impaired in AQP4$^{-/-}$ mice after cortical stab injury [100]. Impaired astrocyte migration toward a cervical spinal cord transection injury was also observed in AQP4$^{-/-}$ mice [92]. This was associated with more severe atrophy and loss of axotomized rubrospinal neurons [92]. These authors concluded that AQP4 not only promotes edema clearance (as shown by Kimura et al. [90]) but also glial scar formation after SCI [92]. Glial scar formation was also reduced after TBI in mice with AQP4 gene silencing by lentiviral vector-delivered small hairpin RNA [101] and in AQP4$^{-/-}$ mice [102].

In addition, there is some evidence that AQP4 deletion affects neuroinflammation [103,104]. For example, neuroinflammation was greatly reduced in AQP4$^{-/-}$ mice in various models (active immunization experimental allergic encephalomyelitis (EAE), adoptive-transfer EAE, and intracerebral lipopolysaccharide injection) [105]. Pharmacologic inhibition of AQP4 with TGN-020 has also modulated inflammation in various models [106,107]. However, the precise pathways by which AQP4 modulates neuroinflammatory cascades remain obscure.

8. Summary and Future Directions

The role and importance of AQP4 following SCI is still not completely understood or fully explored. In addition, there remains a lack of mechanistic explanation of AQP4 regulation following injury (mRNA expression, protein expression, and subcellular targeting/distribution). As discussed above, the lack of availability of selective aquaporin inhibitors and activators has also significantly limited research approaches. Overall, however, the evidence indicates that AQP4 water channels play critical roles in the formation and resolution of distinct forms of edema following SCI. In particular, AQP4 water channels, as in the brain, play a protective role in contusion or transection SCI (vasogenic edema), but a deleterious role in compression SCI (cytotoxic edema). These and other considerations lead to a variety of key future directions:

- Further study should be pursued to more fully characterize the extent of vasogenic and/or cytotoxic edema in different animal models. These results would help to elucidate conflicting findings between studies, both within and across injury models. In the end, understanding the precise role of AQP4 following a wide range of injuries will help to better inform the timing of treatments directed at AQP4 modulation for optimal therapeutic benefit and enhanced neurological outcome.
- In the same vein, a greater understanding of the mechanisms and timing of AQP4 regulation after SCI is needed. This would include regulation both at the mRNA level as well as the subcellular targeting of the AQP4 protein.
- Are there distinct roles of AQP4 isoforms M1-AQP4 and M23-AQP4 after injury? This is a largely unexplored question. It is known that the M23-AQP4 isoform is critical to forming membrane OAPs (orthogonal arrays of particles), and an M23-AQP4-null mouse has been generated demonstrating massive depletion of brain AQP4 [108,109]. However, the relative roles of M1-AQP4 and M23-AQP4 after SCI remain to be explored.
- Test AQP4 modulation approaches in more detail. A *Cell* paper made it clear that, rather than pharmacologically targeting the AQP4 membrane pore itself, therapeutic manipulation could be accomplished by modulating the subcellular targeting of AQP4 [97]. Thus, further research into understanding the subcellular targeting and regulation [110,111] of AQP4 will be helpful both to develop new inhibition as well as activation strategies depending on the edema and injury context. And of course, inhibition of AQP4 membrane localization may help to limit cytotoxic edema, but what would be the analogous approach to *improve* AQP4 expression and membrane targeting to limit vasogenic edema (by increasing AQP4-mediated vasogenic edema clearance)?
- Elucidate AQP4 molecular partners. AQP4 has been shown to be associated with other membrane channels, such as the inwardly-rectifying potassium channel Kir4.1 [47,112], the mechanosensitive cation channel TRPV4 [113], and the ABC protein/TRP channel complex SUR1-TRPM4 [114]. Further understanding of these interactions may also lead to novel mechanistic interventions. For example, TRPM4 knockout blocked astrocyte swelling in a mouse cerebellar cold injury model [114], and the SUR1-TRPM4 inhibitor glyburide inhibits cerebral edema [115]. Such interactions and mechanisms should be targeted specifically for development of novel SCI therapies.
- Even if the optimal AQP4 modulatory drug(s) are identified, drug delivery considerations should be considered. How will a given drug access and penetrate spinal cord tissue when it is injured, edematous, and ischemic? It is notable in this regard that in the *Cell* paper [97], the drug (trifluoperazine) was injected directly into the lesion site. Efficacy in preclinical trials must be balanced with assessment of drug concentration at the target tissue.
- Understand how therapeutic intervention to relieve edema impacts intraspinal pressure, spinal cord blood flow, and neurological outcome. The iSCoPE trials clearly indicated the key importance of ISP in neurological outcome [30–34]. This has now been shown in animal models as well and serves as a key target for therapeutic intervention [116,117]. But there has yet to be a study targeting edema to clearly demonstrate all of the links between relieving edema, improving intraspinal pressure, improving spinal cord perfusion, and improving neurological outcome.
- Osmotic removal of edema fluid. It has been demonstrated that, through establishing an external osmotic gradient, water can be removed from the brain in a controlled manner under normal and pathological brain swelling conditions. Such an "osmotic treatment device" (OTD) was able to reduce brain tissue water content and improve neurological outcome in mouse models of cytotoxic edema and traumatic brain injury

without causing histological damage [118–120]. These results established proof-of-principle for the concept of direct osmotherapy for the treatment of CNS edema. It has been hypothesized that a similar OTD placed on the dura mater of the spinal cord at the site of injury can withdraw excess water from the cord parenchyma and thus ameliorate SCI edema and improve vascular perfusion and neurological outcome [29]. Proof-of-principle for this concept could potentially be combined with local drug delivery (for example, AQP4 activators or inhibitors depending on the type of edema).

Ideally, an integrated strategy for the treatment of SCI-related edema should involve (1) microstructural considerations of tuning AQP4 expression and distribution for the restoration of local water transport/homeostasis at the lesion site; and (2) macrostructural considerations of macroscopic tissue edema and pressure. For example, one can envision a combination of AQP4 modulatory approaches together with macroscopic edema removal via OTD to optimally treat post-traumatic edema. Molecular "optimization" of AQP4 would allow for rapid osmotic equilibration within injured tissue, then the external OTD could gently remove excess edema fluid at the surface, thus decreasing intraspinal pressure and restoring blood flow and spinal cord homeostasis. Such approaches could marry creative control of astrocyte water channel physiology to bioengineering approaches. These considerations lead to entirely novel concepts for limiting edema and secondary damage following traumatic SCI.

Author Contributions: Original draft preparation, T.A.G. and C.R.J.; review, editing, and figure selection, D.K.B. All authors have read and agreed to the published version of the manuscript.

Funding: This research received no external funding.

Institutional Review Board Statement: Not applicable.

Informed Consent Statement: Not applicable.

Data Availability Statement: Not applicable.

Conflicts of Interest: The authors declare no conflict of interest.

References

1. New, P.W.; Rawicki, H.B.; Bailey, M.J. Nontraumatic spinal cord injury: Demographic characteristics and complications. *Arch. Phys. Med. Rehabil.* **2002**, *83*, 996–1001. [CrossRef] [PubMed]
2. David, G.; Mohammadi, S.; Martin, A.R.; Cohen-Adad, J.; Weiskopf, N.; Thompson, A.; Freund, P. Traumatic and nontraumatic spinal cord injury: Pathological insights from neuroimaging. *Nat. Rev. Neurol.* **2019**, *15*, 718–731. [CrossRef] [PubMed]
3. Jeffries, M.A.; Tom, V.J. Peripheral Immune Dysfunction: A Problem of Central Importance after Spinal Cord Injury. *Biology* **2021**, *10*, 928. [CrossRef] [PubMed]
4. Dudley-Javoroski, S.; Lee, J.; Shields, R.K. Cognitive function, quality of life, and aging: Relationships in individuals with and without spinal cord injury. *Physiother. Theory Pract.* **2022**, *38*, 36–45. [CrossRef]
5. Sun, X.; Jones, Z.; Chen, X.; Zhou, L.; So, K.; Ren, Y. Multiple organ dysfunction and systemic inflammation after spinal cord injury: A complex relationship. *J. Neuroinflamm.* **2016**, *13*, 260. [CrossRef]
6. Stein, D.M.; Menaker, J.; McQuillan, K.; Handley, C.; Aarabi, B.; Scalea, T.M. Risk factors for organ dysfunction and failure in patients with acute traumatic cervical spinal cord injury. *Neurocrit. Care* **2010**, *13*, 29–39. [CrossRef]
7. Yeh, T.S.; Huang, Y.P.; Wang, H.I.; Pan, S.L. Spinal cord injury and Parkinson's disease: A population-based, propensity score-matched, longitudinal follow-up study. *Spinal Cord* **2016**, *54*, 1215–1219. [CrossRef]
8. Usta Saglam, N.G.; Poyraz, C.A.; Dogan, D.; Erhan, B. Suicidal ideation, post-traumatic stress disorder, and depression in traumatic spinal cord injury: What resilience tells us. *J. Spinal Cord Med.* **2022**, *46*, 309–316. [CrossRef]
9. Zhang, N.; Fang, M.; Chen, H.; Gou, F.; Ding, M. Evaluation of spinal cord injury animal models. *Neural Regen. Res.* **2014**, *9*, 2008–2012.
10. Beattie, M.S. Anatomic and behavioral outcome after spinal cord injury produced by a displacement controlled impact device. *J. Neurotrauma* **1992**, *9*, 157–159; discussion 159–160. [CrossRef]
11. Behrmann, D.L.; Bresnahan, J.C.; Beattie, M.S.; Shah, B.R. Spinal cord injury produced by consistent mechanical displacement of the cord in rats: Behavioral and histologic analysis. *J. Neurotrauma* **1992**, *9*, 197–217. [CrossRef] [PubMed]
12. Gruner, J.A. A monitored contusion model of spinal cord injury in the rat. *J. Neurotrauma* **1992**, *9*, 123–126; discussion 126–128. [CrossRef] [PubMed]

13. Faulkner, J.R.; Herrmann, J.E.; Woo, M.J.; Tansey, K.E.; Doan, N.B.; Sofroniew, M.V. Reactive astrocytes protect tissue and preserve function after spinal cord injury. *J. Neurosci.* **2004**, *24*, 2143–2155. [CrossRef] [PubMed]
14. Huang, W.L.; George, K.J.; Ibba, V.; Liu, M.C.; Averill, S.; Quartu, M.; Hamlyn, P.J.; Priestley, J.V. The characteristics of neuronal injury in a static compression model of spinal cord injury in adult rats. *Eur. J. Neurosci.* **2007**, *25*, 362–372. [CrossRef]
15. Rivlin, A.S.; Tator, C.H. Effect of duration of acute spinal cord compression in a new acute cord injury model in the rat. *Surg. Neurol.* **1978**, *10*, 38–43.
16. Rosenzweig, E.S.; McDonald, J.W. Rodent models for treatment of spinal cord injury: Research trends and progress toward useful repair. *Curr. Opin. Neurol.* **2004**, *17*, 121–131. [CrossRef]
17. Kwon, B.K.; Oxland, T.R.; Tetzlaff, W. Animal models used in spinal cord regeneration research. *Spine* **2002**, *27*, 1504–1510. [CrossRef]
18. Stokes, B.T.; Jakeman, L.B. Experimental modelling of human spinal cord injury: A model that crosses the species barrier and mimics the spectrum of human cytopathology. *Spinal Cord* **2002**, *40*, 101–109. [CrossRef]
19. Tator, C.H.; Fehlings, M.G. Review of the secondary injury theory of acute spinal cord trauma with emphasis on vascular mechanisms. *J. Neurosurg.* **1991**, *75*, 15–26. [CrossRef]
20. Leypold, B.G.; Flanders, A.E.; Burns, A.S. The early evolution of spinal cord lesions on MR imaging following traumatic spinal cord injury. *AJNR Am. J. Neuroradiol.* **2008**, *29*, 1012–1016. [CrossRef]
21. Bozzo, A.; Marcoux, J.; Radhakrishna, M.; Pelletier, J.; Goulet, B. The role of magnetic resonance imaging in the management of acute spinal cord injury. *J. Neurotrauma* **2011**, *28*, 1401–1411. [CrossRef]
22. Miyanji, F.; Furlan, J.C.; Aarabi, B.; Arnold, P.M.; Fehlings, M.G. Acute cervical traumatic spinal cord injury: MR imaging findings correlated with neurologic outcome—Prospective study with 100 consecutive patients. *Radiology* **2007**, *243*, 820–827. [CrossRef] [PubMed]
23. Shepard, M.J.; Bracken, M.B. Magnetic resonance imaging and neurological recovery in acute spinal cord injury: Observations from the National Acute Spinal Cord Injury Study 3. *Spinal Cord* **1999**, *37*, 833–837. [CrossRef] [PubMed]
24. Borgens, R.B.; Liu-Snyder, P. Understanding secondary injury. *Q. Rev. Biol.* **2012**, *87*, 89–127. [CrossRef] [PubMed]
25. Hu, A.M.; Li, J.J.; Sun, W.; Yang, D.G.; Yang, M.L.; Du, L.J.; Gu, R.; Gao, F.; Li, J.; Chu, H.Y.; et al. Myelotomy reduces spinal cord edema and inhibits aquaporin-4 and aquaporin-9 expression in rats with spinal cord injury. *Spinal Cord* **2015**, *53*, 98–102. [CrossRef] [PubMed]
26. Saadoun, S.; Papadopoulos, M.C. Aquaporin-4 in brain and spinal cord oedema. *Neuroscience* **2010**, *168*, 1036–1046. [CrossRef]
27. Whetstone, W.D.; Hsu, J.Y.; Eisenberg, M.; Werb, Z.; Noble-Haeusslein, L.J. Blood-spinal cord barrier after spinal cord injury: Relation to revascularization and wound healing. *J. Neurosci. Res.* **2003**, *74*, 227–239. [CrossRef]
28. Sharma, H.S. Early microvascular reactions and blood-spinal cord barrier disruption are instrumental in pathophysiology of spinal cord injury and repair: Novel therapeutic strategies including nanowired drug delivery to enhance neuroprotection. *J. Neural Transm.* **2011**, *118*, 155–176. [CrossRef]
29. Hale, C.; Yonan, J.; Batarseh, R.; Chaar, R.; Jonak, C.R.; Ge, S.; Binder, D.; Rodgers, V.G.J. Implantable Osmotic Transport Device Can Reduce Edema After Severe Contusion Spinal Cord Injury. *Front. Bioeng. Biotechnol.* **2020**, *8*, 806. [CrossRef]
30. Hogg, F.R.A.; Gallagher, M.J.; Chen, S.; Zoumprouli, A.; Papadopoulos, M.C.; Saadoun, S. Predictors of Intraspinal Pressure and Optimal Cord Perfusion Pressure After Traumatic Spinal Cord Injury. *Neurocrit. Care* **2019**, *30*, 421–428. [CrossRef]
31. Hogg, F.R.A.; Kearney, S.; Gallagher, M.J.; Zoumprouli, A.; Papadopoulos, M.C.; Saadoun, S. Spinal Cord Perfusion Pressure Correlates with Anal Sphincter Function in a Cohort of Patients with Acute, Severe Traumatic Spinal Cord Injuries. *Neurocrit. Care* **2021**, *35*, 794–805. [CrossRef]
32. Hogg, F.R.A.; Kearney, S.; Zoumprouli, A.; Papadopoulos, M.C.; Saadoun, S. Acute Spinal Cord Injury: Correlations and Causal Relations Between Intraspinal Pressure, Spinal Cord Perfusion Pressure, Lactate-to-Pyruvate Ratio, and Limb Power. *Neurocrit. Care* **2021**, *34*, 121–129. [CrossRef] [PubMed]
33. Saadoun, S.; Chen, S.; Papadopoulos, M.C. Intraspinal pressure and spinal cord perfusion pressure predict neurological outcome after traumatic spinal cord injury. *J. Neurol. Neurosurg. Psychiatry* **2017**, *88*, 452–453. [CrossRef] [PubMed]
34. Saadoun, S.; Papadopoulos, M.C. Acute, Severe Traumatic Spinal Cord Injury: Monitoring from the Injury Site and Expansion Duraplasty. *Neurosurg. Clin. N. Am.* **2021**, *32*, 365–376. [CrossRef] [PubMed]
35. Bracken, M.B.; Shepard, M.J.; Collins, W.F.; Holford, T.R.; Young, W.; Baskin, D.S.; Eisenberg, H.M.; Flamm, E.; Leo-Summers, L.; Maroon, J.; et al. A randomized, controlled trial of methylprednisolone or naloxone in the treatment of acute spinal-cord injury. Results of the Second National Acute Spinal Cord Injury Study. *N. Engl. J. Med.* **1990**, *322*, 1405–1411. [CrossRef]
36. Fehlings, M.G.; Wilson, J.R.; Harrop, J.S.; Kwon, B.K.; Tetreault, L.A.; Arnold, P.M.; Singh, J.M.; Hawryluk, G.; Dettori, J.R. Efficacy and Safety of Methylprednisolone Sodium Succinate in Acute Spinal Cord Injury: A Systematic Review. *Global Spine J.* **2017**, *7* (Suppl. 3), 116S–137S. [CrossRef] [PubMed]
37. Verkman, A.S. More than just water channels: Unexpected cellular roles of aquaporins. *J. Cell. Sci.* **2005**, *118 Pt 15*, 3225–3232. [CrossRef] [PubMed]
38. Amiry-Moghaddam, M.; Ottersen, O.P. The molecular basis of water transport in the brain. *Nat. Rev. Neurosci.* **2003**, *4*, 991–1001. [CrossRef]

39. Agre, P.; King, L.S.; Yasui, M.; Guggino, W.B.; Ottersen, O.P.; Fujiyoshi, Y.; Engel, A.; Nielsen, S. Aquaporin water channels—From atomic structure to clinical medicine. *J. Physiol.* **2002**, *542 Pt 1*, 3–16. [CrossRef]
40. Jarius, S.; Paul, F.; Franciotta, D.; Waters, P.; Zipp, F.; Hohlfeld, R.; Vincent, A.; Wildemann, B. Mechanisms of disease: Aquaporin-4 antibodies in neuromyelitis optica. *Nat. Clin. Pract. Neurol.* **2008**, *4*, 202–214. [CrossRef]
41. Verkman, A.S. Physiological importance of aquaporin water channels. *Ann. Med.* **2002**, *34*, 192–200. [CrossRef] [PubMed]
42. Salman, M.M.; Kitchen, P.; Halsey, A.; Wang, M.X.; Tornroth-Horsefield, S.; Conner, A.C.; Badaut, J.; Iliff, J.J.; Bill, R.M. Emerging roles for dynamic aquaporin-4 subcellular relocalization in CNS water homeostasis. *Brain* **2022**, *145*, 64–75. [CrossRef]
43. Hasegawa, H.; Ma, T.; Skach, W.; Matthay, M.A.; Verkman, A.S. Molecular cloning of a mercurial-insensitive water channel expressed in selected water-transporting tissues. *J. Biol. Chem.* **1994**, *269*, 5497–5500. [CrossRef] [PubMed]
44. Yang, B.; Ma, T.; Verkman, A.S. cDNA cloning, gene organization, and chromosomal localization of a human mercurial insensitive water channel. Evidence for distinct transcriptional units. *J. Biol. Chem.* **1995**, *270*, 22907–22913. [CrossRef]
45. Jung, J.S.; Bhat, R.V.; Preston, G.M.; Guggino, W.B.; Baraban, J.M.; Agre, P. Molecular characterization of an aquaporin cDNA from brain: Candidate osmoreceptor and regulator of water balance. *Proc. Natl. Acad. Sci. USA* **1994**, *91*, 13052–13056. [CrossRef] [PubMed]
46. Nielsen, S.; Nagelhus, E.A.; Amiry-Moghaddam, M.; Bourque, C.; Agre, P.; Ottersen, O.P. Specialized membrane domains for water transport in glial cells: High-resolution immunogold cytochemistry of aquaporin-4 in rat brain. *J. Neurosci.* **1997**, *17*, 171–180. [CrossRef] [PubMed]
47. Nagelhus, E.A.; Mathiisen, T.M.; Ottersen, O.P. Aquaporin-4 in the central nervous system: Cellular and subcellular distribution and coexpression with Kir4.1. *Neuroscience* **2004**, *129*, 905–913. [CrossRef]
48. Rash, J.E.; Yasumura, T.; Hudson, C.S.; Agre, P.; Nielsen, S. Direct immunogold labeling of aquaporin-4 in square arrays of astrocyte and ependymocyte plasma membranes in rat brain and spinal cord. *Proc. Natl. Acad. Sci. USA* **1998**, *95*, 11981–11986. [CrossRef]
49. Oshio, K.; Binder, D.K.; Yang, B.; Schecter, S.; Verkman, A.S.; Manley, G.T. Expression of aquaporin water channels in mouse spinal cord. *Neuroscience* **2004**, *127*, 685–693. [CrossRef]
50. Furman, C.S.; Gorelick-Feldman, D.A.; Davidson, K.G.; Yasumura, T.; Neely, J.D.; Agre, P.; Rash, J.E. Aquaporin-4 square array assembly: Opposing actions of M1 and M23 isoforms. *Proc. Natl. Acad. Sci. USA* **2003**, *100*, 13609–13614. [CrossRef]
51. Verbavatz, J.M.; Ma, T.; Gobin, R.; Verkman, A.S. Absence of orthogonal arrays in kidney, brain and muscle from transgenic knockout mice lacking water channel aquaporin-4. *J. Cell Sci.* **1997**, *110 Pt 22*, 2855–2860. [CrossRef] [PubMed]
52. Yang, B.; Brown, D.; Verkman, A.S. The mercurial insensitive water channel (AQP-4) forms orthogonal arrays in stably transfected Chinese hamster ovary cells. *J. Biol. Chem.* **1996**, *271*, 4577–4580. [CrossRef] [PubMed]
53. Frigeri, A.; Gropper, M.A.; Umenishi, F.; Kawashima, M.; Brown, D.; Verkman, A.S. Localization of MIWC and GLIP water channel homologs in neuromuscular, epithelial and glandular tissues. *J. Cell Sci.* **1995**, *108 Pt 9*, 2993–3002. [CrossRef] [PubMed]
54. Wolburg, H.; Berg, K. Distribution of orthogonal arrays of particles in the Muller cell membrane of the mouse retina. *Glia* **1988**, *1*, 246–252. [CrossRef] [PubMed]
55. Wolburg, H.; Wolburg-Buchholz, K.; Fallier-Becker, P.; Noell, S.; Mack, A.F. Structure and functions of aquaporin-4-based orthogonal arrays of particles. *Int. Rev. Cell Mol. Biol.* **2011**, *287*, 1–41. [PubMed]
56. Niermann, H.; Amiry-Moghaddam, M.; Holthoff, K.; Witte, O.W.; Ottersen, O.P. A novel role of vasopressin in the brain: Modulation of activity-dependent water flux in the neocortex. *J. Neurosci.* **2001**, *21*, 3045–3051. [CrossRef]
57. Holthoff, K.; Witte, O.W. Directed spatial potassium redistribution in rat neocortex. *Glia* **2000**, *29*, 288–292. [CrossRef]
58. Nesic, O.; Lee, J.; Ye, Z.; Unabia, G.C.; Rafati, D.; Hulsebosch, C.E.; Perez-Polo, J.R. Acute and chronic changes in aquaporin 4 expression after spinal cord injury. *Neuroscience* **2006**, *143*, 779–792. [CrossRef]
59. Basso, D.M.; Beattie, M.S.; Bresnahan, J.C. A sensitive and reliable locomotor rating scale for open field testing in rats. *J. Neurotrauma* **1995**, *12*, 1–21. [CrossRef]
60. Nesic, O.; Guest, J.D.; Zivadinovic, D.; Narayana, P.A.; Herrera, J.J.; Grill, R.J.; Mokkapati, V.U.; Gelman, B.B.; Lee, J. Aquaporins in spinal cord injury: The janus face of aquaporin 4. *Neuroscience* **2010**, *168*, 1019–1035. [CrossRef]
61. Leonard, A.V.; Thornton, E.; Vink, R. Substance P as a mediator of neurogenic inflammation after balloon compression induced spinal cord injury. *J. Neurotrauma* **2013**, *30*, 1812–1823. [CrossRef]
62. Huang, Y.; Li, S.N.; Zhou, X.Y.; Zhang, L.X.; Chen, G.X.; Wang, T.H.; Xia, Q.J.; Liang, N.; Zhang, X. The Dual Role of AQP4 in Cytotoxic and Vasogenic Edema Following Spinal Cord Contusion and Its Possible Association with Energy Metabolism via COX5A. *Front. Neurosci.* **2019**, *13*, 584. [CrossRef]
63. Pan, Y.L.; Guo, Y.; Ma, Y.; Wang, L.; Zheng, S.Y.; Liu, M.M.; Huang, G.C. Aquaporin-4 expression dynamically varies after acute spinal cord injury-induced disruption of blood spinal cord barrier in rats. *Neuropathology* **2019**, *39*, 181–186. [CrossRef] [PubMed]
64. Vitellaro-Zuccarello, L.; Mazzetti, S.; Madaschi, L.; Bosisio, P.; Fontana, E.; Gorio, A.; De Biasi, S. Chronic erythropoietin-mediated effects on the expression of astrocyte markers in a rat model of contusive spinal cord injury. *Neuroscience* **2008**, *151*, 452–466. [CrossRef] [PubMed]
65. Guptarak, J.; Wiktorowicz, J.E.; Sadygov, R.G.; Zivadinovic, D.; Paulucci-Holthauzen, A.A.; Vergara, L.; Nesic, O. The cancer drug tamoxifen: A potential therapeutic treatment for spinal cord injury. *J. Neurotrauma* **2014**, *31*, 268–283. [CrossRef] [PubMed]

66. Wu, Q.; Jing, Y.; Yuan, X.; Zhang, X.; Li, B.; Liu, M.; Wang, B.; Li, H.; Liu, S.; Xiu, R. Melatonin treatment protects against acute spinal cord injury-induced disruption of blood spinal cord barrier in mice. *J. Mol. Neurosci.* **2014**, *54*, 714–722. [CrossRef] [PubMed]
67. Jing, Y.; Wu, Q.; Yuan, X.; Li, B.; Liu, M.; Zhang, X.; Liu, S.; Li, H.; Xiu, R. Microvascular protective role of pericytes in melatonin-treated spinal cord injury in the C57BL/6 mice. *Chin. Med. J.* **2014**, *127*, 2808–2813.
68. Zu, J.; Wang, Y.; Xu, G.; Zhuang, J.; Gong, H.; Yan, J. Curcumin improves the recovery of motor function and reduces spinal cord edema in a rat acute spinal cord injury model by inhibiting the JAK/STAT signaling pathway. *Acta Histochem.* **2014**, *116*, 1331–1336. [CrossRef]
69. Mao, L.; Wang, H.D.; Pan, H.; Qiao, L. Sulphoraphane enhances aquaporin-4 expression and decreases spinal cord oedema following spinal cord injury. *Brain Inj.* **2011**, *25*, 300–306. [CrossRef]
70. Fan, Z.K.; Wang, Y.F.; Cao, Y.; Zhang, M.C.; Zhang, Z.; Lv, G.; Lu, W.; Zhang, Y.Q. The effect of aminoguanidine on compression spinal cord injury in rats. *Brain Res.* **2010**, *1342*, 1–10. [CrossRef]
71. Wang, Y.F.; Fan, Z.K.; Cao, Y.; Yu, D.S.; Zhang, Y.Q.; Wang, Y.S. 2-Methoxyestradiol inhibits the up-regulation of AQP4 and AQP1 expression after spinal cord injury. *Brain Res.* **2011**, *1370*, 220–226. [CrossRef] [PubMed]
72. Ge, R.; Zhu, Y.; Diao, Y.; Tao, L.; Yuan, W.; Xiong, X.C. Anti-edema effect of epigallocatechin gallate on spinal cord injury in rats. *Brain Res.* **2013**, *1527*, 40–46. [CrossRef] [PubMed]
73. Liu, X.; Wang, Y.; Yang, J.; Liu, Y.; Zhou, D.; Hou, M.; Xiang, L. Anti-edema effect of melatonin on spinal cord injury in rats. *Biomed. Pap. Med. Fac. Univ. Palacky Olomouc Czech Repub.* **2015**, *159*, 220–226. [CrossRef] [PubMed]
74. Li, Y.; Gu, R.; Zhu, Q.; Liu, J. Changes of Spinal Edema and Expression of Aquaporin 4 in Methylprednisolone-treated Rats with Spinal Cord Injury. *Ann. Clin. Lab. Sci.* **2018**, *48*, 453–459.
75. Yonan, J.; Binder, D.K. Aquaporin-4 and spinal cord injury. *World J. Neurol.* **2016**, *6*, 1–13. [CrossRef]
76. Halsey, A.M.; Conner, A.C.; Bill, R.M.; Logan, A.; Ahmed, Z. Aquaporins and Their Regulation after Spinal Cord Injury. *Cells* **2018**, *7*, 174. [CrossRef]
77. Ma, T.; Yang, B.; Gillespie, A.; Carlson, E.J.; Epstein, C.J.; Verkman, A.S. Generation and phenotype of a transgenic knockout mouse lacking the mercurial-insensitive water channel aquaporin-4. *J. Clin. Investig.* **1997**, *100*, 957–962. [CrossRef]
78. Haj-Yasein, N.N.; Vindedal, G.F.; Eilert-Olsen, M.; Gundersen, G.A.; Skare, O.; Laake, P.; Klungland, A.; Thoren, A.E.; Burkhardt, J.M.; Ottersen, O.P.; et al. Glial-conditional deletion of aquaporin-4 (AQP4) reduces blood-brain water uptake and confers barrier function on perivascular astrocyte endfeet. *Proc. Natl. Acad. Sci. USA* **2011**, *108*, 17815–17820. [CrossRef]
79. Saadoun, S.; Tait, M.J.; Reza, A.; Davies, D.C.; Bell, B.A.; Verkman, A.S.; Papadopoulos, M.C. AQP4 gene deletion in mice does not alter blood-brain barrier integrity or brain morphology. *Neuroscience* **2009**, *161*, 764–772. [CrossRef]
80. Binder, D.K.; Oshio, K.; Ma, T.; Verkman, A.S.; Manley, G.T. Increased seizure threshold in mice lacking aquaporin-4 water channels. *Neuroreport* **2004**, *15*, 259–262. [CrossRef]
81. Papadopoulos, M.C.; Manley, G.T.; Krishna, S.; Verkman, A.S. Aquaporin-4 facilitates reabsorption of excess fluid in vasogenic brain edema. *FASEB J.* **2004**, *18*, 1291–1293. [CrossRef]
82. Manley, G.T.; Fujimura, M.; Ma, T.; Noshita, N.; Filiz, F.; Bollen, A.W.; Chan, P.; Verkman, A.S. Aquaporin-4 deletion in mice reduces brain edema after acute water intoxication and ischemic stroke. *Nat. Med.* **2000**, *6*, 159–163. [CrossRef] [PubMed]
83. Lee, D.J.; Amini, M.; Hamamura, M.J.; Hsu, M.S.; Seldin, M.M.; Nalcioglu, O.; Binder, D.K. Aquaporin-4-dependent edema clearance following status epilepticus. *Epilepsy Res.* **2011**, *98*, 264–268. [CrossRef] [PubMed]
84. Frigeri, A.; Nicchia, G.P.; Nico, B.; Quondamatteo, F.; Herken, R.; Roncali, L.; Svelto, M. Aquaporin-4 deficiency in skeletal muscle and brain of dystrophic mdx mice. *FASEB J.* **2001**, *15*, 90–98. [CrossRef] [PubMed]
85. Vajda, Z.; Pedersen, M.; Fuchtbauer, E.M.; Wertz, K.; Stodkilde-Jorgensen, H.; Sulyok, E.; Doczi, T.; Neely, J.D.; Agre, P.; Frokiaer, J.; et al. Delayed onset of brain edema and mislocalization of aquaporin-4 in dystrophin-null transgenic mice. *Proc. Natl. Acad. Sci. USA* **2002**, *99*, 13131–13136. [CrossRef] [PubMed]
86. Neely, J.D.; Amiry-Moghaddam, M.; Ottersen, O.P.; Froehner, S.C.; Agre, P.; Adams, M.E. Syntrophin-dependent expression and localization of aquaporin-4 water channel protein. *Proc. Natl. Acad. Sci. USA* **2001**, *98*, 14108–14113. [CrossRef]
87. Amiry-Moghaddam, M.; Otsuka, T.; Hurn, P.D.; Traystman, R.J.; Haug, F.M.; Froehner, S.C.; Adams, M.E.; Neely, J.D.; Agre, P.; Ottersen, O.P.; et al. An alpha-syntrophin-dependent pool of AQP4 in astroglial end-feet confers bidirectional water flow between blood and brain. *Proc. Natl. Acad. Sci. USA* **2003**, *100*, 2106–2111. [CrossRef]
88. Saadoun, S.; Bell, B.A.; Verkman, A.S.; Papadopoulos, M.C. Greatly improved neurological outcome after spinal cord compression injury in AQP4-deficient mice. *Brain* **2008**, *131 Pt 4*, 1087–1098. [CrossRef]
89. Basso, D.M.; Fisher, L.C.; Anderson, A.J.; Jakeman, L.B.; McTigue, D.M.; Popovich, P.G. Basso Mouse Scale for locomotion detects differences in recovery after spinal cord injury in five common mouse strains. *J. Neurotrauma* **2006**, *23*, 635–659. [CrossRef]
90. Kimura, A.; Hsu, M.; Seldin, M.; Verkman, A.S.; Scharfman, H.E.; Binder, D.K. Protective role of aquaporin-4 water channels after contusion spinal cord injury. *Ann. Neurol.* **2010**, *67*, 794–801. [CrossRef]
91. Inman, D.M.; Steward, O. Physical size does not determine the unique histopathological response seen in the injured mouse spinal cord. *J. Neurotrauma* **2003**, *20*, 33–42. [CrossRef] [PubMed]
92. Wu, Q.; Zhang, Y.J.; Gao, J.Y.; Li, X.M.; Kong, H.; Zhang, Y.P.; Xiao, M.; Shields, C.B.; Hu, G. Aquaporin-4 mitigates retrograde degeneration of rubrospinal neurons by facilitating edema clearance and glial scar formation after spinal cord injury in mice. *Mol. Neurobiol.* **2014**, *49*, 1327–1337. [CrossRef] [PubMed]

93. Verkman, A.S.; Anderson, M.O.; Papadopoulos, M.C. Aquaporins: Important but elusive drug targets. *Nat. Rev. Drug Discov.* **2014**, *13*, 259–277. [CrossRef] [PubMed]
94. Abir-Awan, M.; Kitchen, P.; Salman, M.M.; Conner, M.T.; Conner, A.C.; Bill, R.M. Inhibitors of Mammalian Aquaporin Water Channels. *Int. J. Mol. Sci.* **2019**, *20*, 1589. [CrossRef]
95. Zhang, M.; Cui, Z.; Cui, H.; Cao, Y.; Zhong, C. Wang. Astaxanthin alleviates cerebral edema by modulating NKCC1 and AQP4 expression after traumatic brain injury in mice. *BMC Neurosci.* **2016**, *17*, 60. [CrossRef]
96. Yan, X.; Liu, J.; Wang, X.; Li, W.; Chen, J.; Sun, H. Pretreatment with AQP4 and NKCC1 Inhibitors Concurrently Attenuated Spinal Cord Edema and Tissue Damage after Spinal Cord Injury in Rats. *Front. Physiol.* **2018**, *9*, 6. [CrossRef]
97. Kitchen, P.; Salman, M.M.; Halsey, A.M.; Clarke-Bland, C.; MacDonald, J.A.; Ishida, H.; Vogel, H.J.; Almutiri, S.; Logan, A.; Kreida, S. Targeting Aquaporin-4 Subcellular Localization to Treat Central Nervous System Edema. *Cell* **2020**, *181*, 784–799.e19. [CrossRef]
98. Binder, D.K.; Nagelhus, E.A.; Ottersen, O.P. Aquaporin-4 and epilepsy. *Glia* **2012**, *60*, 1203–1214. [CrossRef]
99. Szu, J.I.; Binder, D.K. Mechanisms Underlying Aquaporin-4 Subcellular Mislocalization in Epilepsy. *Front. Cell. Neurosci.* **2022**, *16*, 900588. [CrossRef]
100. Saadoun, S.; Papadopoulos, M.C.; Watanabe, H.; Yan, D.; Manley, G.T.; Verkman, A.S. Involvement of aquaporin-4 in astroglial cell migration and glial scar formation. *J. Cell. Sci.* **2005**, *118*, 5691–5698. [CrossRef]
101. Yong, Y.X.; Li, Y.M.; Lian, J.; Luo, C.M.; Zhong, D.X.; Han, K. Inhibitory role of lentivirus-mediated aquaporin-4 gene silencing in the formation of glial scar in a rat model of traumatic brain injury. *J. Cell. Biochem.* **2019**, *120*, 368–379. [CrossRef] [PubMed]
102. Lu, D.C.; Zador, Z.; Yao, J.; Fazlollahi, F.; Manley, G.T. Aquaporin-4 Reduces Post-Traumatic Seizure Susceptibility by Promoting Astrocytic Glial Scar Formation in Mice. *J. Neurotrauma.* **2021**, *38*, 1193–1201. [CrossRef] [PubMed]
103. Fukuda, A.M.; Badaut, J. Aquaporin 4: A player in cerebral edema and neuroinflammation. *J. Neuroinflamm.* **2012**, *9*, 279. [CrossRef]
104. Lan, Y.L.; Fang, D.Y.; Zhao, J.; Ma, T.H.; Li, S. A research update on the potential roles of aquaporin 4 in neuroinflammation. *Acta. Neurol. Belg.* **2016**, *116*, 127–134. [CrossRef] [PubMed]
105. Li, L.; Zhang, H.; Varrin-Doyer, M.; Zamvil, S.S.; Verkman, A.S. Proinflammatory role of aquaporin-4 in autoimmune neuroinflammation. *FASEB J.* **2011**, *25*, 1556–1566. [CrossRef] [PubMed]
106. Li, J.; Jia, Z.; Zhang, Q.; Dai, J.; Kong, J.; Fan, Z.; Li, G. Inhibition of ERK1/2 phosphorylation attenuates spinal cord injury induced astrocyte activation and inflammation through negatively regulating aquaporin-4 in rats. *Brain Res. Bull.* **2021**, *170*, 162–173. [CrossRef]
107. Li, Y.; Lu, H.; Lv, X.; Tang, Q.; Li, W.; Zhu, H.; Long, Y. Blockade of Aquaporin 4 Inhibits Irradiation-Induced Pulmonary Inflammation and Modulates Macrophage Polarization in Mice. *Inflammation* **2018**, *41*, 2196–2205. [CrossRef]
108. de Bellis, M.; Cibelli, A.; Mola, M.G.; Pisani, F.; Barile, B.; Mastrodonato, M.; Banitalebi, S.; Amiry-Moghaddam, M.; Abbrescia, P.; Frigeri, A.; et al. Orthogonal arrays of particle assembly are essential for normal aquaporin-4 expression level in the brain. *Glia* **2021**, *69*, 473–488. [CrossRef]
109. Pisani, F.; Simone, L.; Mola, M.G.; De Bellis, M.; Frigeri, A.; Nicchia, G.P.; Svelto, M. Regulation of aquaporin-4 expression in the central nervous system investigated using M23-AQP4 null mouse. *Glia* **2021**, *69*, 2235–2251. [CrossRef]
110. Nesverova, V.; Tornroth-Horsefield, S. Phosphorylation-Dependent Regulation of Mammalian Aquaporins. *Cells* **2019**, *8*, 82. [CrossRef]
111. McCoy, E.S.; Haas, B.R.; Sontheimer, H. Water permeability through aquaporin-4 is regulated by protein kinase C and becomes rate-limiting for glioma invasion. *Neuroscience* **2010**, *168*, 971–981. [CrossRef] [PubMed]
112. Connors, N.C.; Kofuji, P. Potassium channel Kir4.1 macromolecular complex in retinal glial cells. *Glia* **2006**, *53*, 124–131. [CrossRef] [PubMed]
113. Benfenati, V.; Caprini, M.; Dovizio, M.; Mylonakou, M.N.; Ferroni, S.; Ottersen, O.P.; Amiry-Moghaddam, M. An aquaporin-4/transient receptor potential vanilloid 4 (AQP4/TRPV4) complex is essential for cell-volume control in astrocytes. *Proc. Natl. Acad. Sci. USA* **2011**, *108*, 2563–2568. [CrossRef] [PubMed]
114. Stokum, J.A.; Kwon, M.S.; Woo, S.K.; Tsymbalyuk, O.; Vennekens, R.; Gerzanich, V.; Simard, J.M. SUR1-TRPM4 and AQP4 form a heteromultimeric complex that amplifies ion/water osmotic coupling and drives astrocyte swelling. *Glia* **2018**, *66*, 108–125. [CrossRef] [PubMed]
115. King, Z.A.; Sheth, K.N.; Kimberly, W.T.; Simard, J.M. Profile of intravenous glyburide for the prevention of cerebral edema following large hemispheric infarction: Evidence to date. *Drug Des. Dev. Ther.* **2018**, *12*, 2539–2552. [CrossRef] [PubMed]
116. Yang, C.; He, T.; Wang, Q.; Wang, G.; Ma, J.; Chen, Z.; Li, Q.; Wang, L.; Quan, Z. Elevated intraspinal pressure drives edema progression after acute compression spinal cord injury in rabbits. *Exp. Neurol.* **2022**, *357*, 114206. [CrossRef]
117. Yang, C.H.; Quan, Z.X.; Wang, G.J.; He, T.; Chen, Z.Y.; Li, Q.C.; Yang, J.; Wang, Q. Elevated intraspinal pressure in traumatic spinal cord injury is a promising therapeutic target. *Neural Regen. Res.* **2022**, *17*, 1703–1710. [CrossRef]
118. McBride, D.W.; Donovan, V.; Hsu, M.S.; Obenaus, A.; Rodgers, V.G.; Binder, D.K. Reduction of Cerebral Edema via an Osmotic Transport Device Improves Functional Outcome after Traumatic Brain Injury in Mice. *Acta Neurochir. Suppl.* **2016**, *121*, 285–289.

119. McBride, D.W.; Szu, J.I.; Hale, C.; Hsu, M.S.; Rodgers, V.G.; Binder, D.K. Reduction of cerebral edema after traumatic brain injury using an osmotic transport device. *J. Neurotrauma* **2014**, *31*, 1948–1954. [CrossRef]
120. McBride, D.W.; Hsu, M.S.; Rodgers, V.G.; Binder, D.K. Improved survival following cerebral edema using a novel hollow fiber-hydrogel device. *J. Neurosurg.* **2012**, *116*, 1389–1394. [CrossRef]

cells MDPI

Article

The CaMKII/MLC1 Axis Confers Ca^{2+}-Dependence to Volume-Regulated Anion Channels (VRAC) in Astrocytes

Maria Stefania Brignone [1], Angela Lanciotti [1,†], Antonio Michelucci [2,†], Cinzia Mallozzi [1], Serena Camerini [3], Luigi Catacuzzeno [2], Luigi Sforna [2], Martino Caramia [2], Maria Cristina D'Adamo [4], Marina Ceccarini [5], Paola Molinari [6], Pompeo Macioce [1], Gianfranco Macchia [3], Tamara Corinna Petrucci [1], Mauro Pessia [7,8], Sergio Visentin [6] and Elena Ambrosini [1,*]

1 Department of Neuroscience, Istituto Superiore di Sanità, 00169 Rome, Italy
2 Department of Chemistry, Biology and Biotechnology, University of Perugia, 06123 Perugia, Italy
3 Core Facilities (FAST), Istituto Superiore di Sanità, 00169 Rome, Italy
4 Department of Medicine and Surgery, LUM Giuseppe Degennaro University, 70010 Bari, Italy
5 National Centre for Rare Diseases, Istituto Superiore di Sanità, 00169 Rome, Italy
6 National Centre for Drug Research and Evaluation (FARVA), Istituto Superiore di Sanità, 00169 Rome, Italy
7 Department of Physiology and Biochemistry, Faculty of Medicine and Surgery, University of Malta, MSD2080 Msida, Malta
8 Department of Physiology, College of Medicine and Health Sciences, United Arab Emirates University, Al Ain P.O. Box 17666, United Arab Emirates
* Correspondence: elena.ambrosini@iss.it; Tel.: +39-06-4990-2037
† These authors contributed equally to this work.

Citation: Brignone, M.S.; Lanciotti, A.; Michelucci, A.; Mallozzi, C.; Camerini, S.; Catacuzzeno, L.; Sforna, L.; Caramia, M.; D'Adamo, M.C.; Ceccarini, M.; et al. The CaMKII/MLC1 Axis Confers Ca^{2+}-Dependence to Volume-Regulated Anion Channels (VRAC) in Astrocytes. *Cells* **2022**, *11*, 2656. https://doi.org/10.3390/cells11172656

Academic Editors: Michael Brenner and Vladimir Parpura

Received: 7 July 2022
Accepted: 24 August 2022
Published: 26 August 2022

Publisher's Note: MDPI stays neutral with regard to jurisdictional claims in published maps and institutional affiliations.

Abstract: Astrocytes, the main glial cells of the central nervous system, play a key role in brain volume control due to their intimate contacts with cerebral blood vessels and the expression of a distinctive equipment of proteins involved in solute/water transport. Among these is MLC1, a protein highly expressed in perivascular astrocytes and whose mutations cause megalencephalic leukoencephalopathy with subcortical cysts (MLC), an incurable leukodystrophy characterized by macrocephaly, chronic brain edema, cysts, myelin vacuolation, and astrocyte swelling. Although, in astrocytes, MLC1 mutations are known to affect the swelling-activated chloride currents ($ICl_{,swell}$) mediated by the volume-regulated anion channel (VRAC), and the regulatory volume decrease, MLC1's proper function is still unknown. By combining molecular, biochemical, proteomic, electrophysiological, and imaging techniques, we here show that MLC1 is a Ca^{2+}/Calmodulin-dependent protein kinase II (CaMKII) target protein, whose phosphorylation, occurring in response to intracellular Ca^{2+} release, potentiates VRAC-mediated $ICl_{,swell}$. Overall, these findings reveal that MLC1 is a Ca^{2+}-regulated protein, linking volume regulation to Ca^{2+} signaling in astrocytes. This knowledge provides new insight into the MLC1 protein function and into the mechanisms controlling ion/water exchanges in the brain, which may help identify possible molecular targets for the treatment of MLC and other pathological conditions caused by astrocyte swelling and brain edema.

Keywords: brain edema; astrocytes; Ca^{2+} release; $ICl_{,swell}$; VRAC; CaMKII; leukodystrophy; rare diseases

1. Introduction

In the central nervous system (CNS), ion and water movements are strictly controlled by astrocytes, the glial cell population mainly responsible for the maintenance of tissue homeostasis and ensuring correct brain functionality [1–3]. Alterations of these regulative processes can lead to serious complications, due to brain water accumulation and edema formation, as observed in several CNS injuries such as ischemia, trauma, infection, tumors, and inflammation [3–5]. The astrocyte's favorable morphological structure and anatomical distribution, with end-foot processes contacting blood vessels, allow these cells to connect parenchymal tissue to the blood circulation, thus favoring the uptake and removal of accumulating solutes and water [2,6]. To accomplish this task, astrocytes also express a large number of transporters and ion channel proteins [1,5–7]. Understanding how these molecules interplay would greatly help identify potential targets to treat the devastating consequences of brain edema [8]. Interesting advances in the comprehension of these mechanisms emerged from the study of genetic diseases distinguished by chronic brain edema and caused by mutations in specific astrocytic proteins, such as MLC1 in the megalencephalic leukoencephalopathy with subcortical cysts (MLC) disease [9]. This rare leukodystrophy, which causes severe motor and cognitive impairments and epilepsy [10,11], is characterized by diffuse white matter edema and subcortical cysts, observed by magnetic resonance imaging, and by myelin vacuolation and astrocyte swelling, as revealed by histological analysis of MLC brain tissue. In 80% of MLC patients, the disease is linked to mutations in the *MLC1* gene encoding the MLC1 protein [12], a 377 amino acid protein containing eight predicted transmembrane domains and short cytoplasmic amino and carboxylic tails. In the human and mouse brain MLC1 is almost exclusively expressed by astrocyte end-feet contacting blood vessels and meninges, as well as by the Bergmann glia of the cerebellum [13,14]. Although, to date, MLC1′s proper function has remained elusive, both cellular and animal models of MLC have proven that a lack of functional MLC1 affects the activation of two different chloride currents: an outward-rectifying anion current, mainly due to volume-regulated anion channels (VRAC); and a strong inward-rectifying chloride current mediated by the voltage-gated ClC-2 channels [15–18]. Both currents are affected by knock-out (KO) or mutations of the *MLC1* gene, but they may differently contribute to MLC pathogenesis. Indeed, while the ClC-2-activated current is likely involved in extracellular potassium ion (K^+) homeostasis during neuronal activation [17], VRAC, the main component of the swelling-activated Cl^-current ($ICl_{,swell}$), is mainly responsible for regulatory volume decrease (RVD) in response to osmotic changes [18]. Interestingly, MLC1 was described to interact with a number of effector proteins directly involved in osmotic homeostasis, such as AQP4, Na, K-ATPase, TRPV4, and Kir4.1 [19–22]. All these experimental and clinical observations suggest that MLC1 mutation-induced ion/water exchange defects are responsible for homeostatic disorders such as edema or myelin vacuolation, but how this protein regulates cell volume and the underlying anion currents is unknown. To clarify this issue, we reasoned about the possible connection between MLC1 and intracellular Ca^{2+} signaling. This explorative working hypothesis stems from our previous findings indicating MLC1 expression in caveolar rafts [21–23], plasma membrane (PM) microdomains hosting several proteins involved in Ca^{2+} signaling [24], and MLC1′s potentiation of Ca^{2+} influx through TRPV4 channels in response to osmotic stress [20]. In addition, defects in intracellular Ca^{2+} homeostasis were observed in different MLC pathological models [25,26]. Notably, in astrocytes, Ca^{2+} regulates all MLC1-modulated biological processes, such as $ICl_{,swell}$ and RVD [27]. By integrating complementary experimental approaches, encompassing biochemical, molecular biology, imaging, proteomic, and electrophysiological techniques being applied to astrocyte-based cellular models of MLC, we revealed new Ca^{2+}-dependent functional properties of the MLC1 protein that can unravel the molecular mechanisms controlling volume changes in astrocytes and whose dysfunction accounts for brain edema in MLC patients.

2. Materials and Methods

2.1. Cell Cultures and Treatments

Astrocyte-enriched cultures were generated from newborn CD1 Swiss mice and maintained in culture as previously described [28]. Generation/maintenance of U251 cell lines expressing wild type (MLC1-WT) or mutated MLC1 (MLC1-S280L) has been previously described [20]. Cell treatments were performed in serum-free (SF) medium, as follows: 5 min with adenosine triphosphate (ATP, 100 μM, Sigma-Aldrich, Saint Louis, MO, USA), 30 min with isotonic solution (140 mM NaCl, 5 mM KCl, 2 mM $MgCl_2$, 2 mM $CaCl_2$, 20 mM D-glucose, 5 mM HEPES, pH 7.4) or hypotonic solution prepared by adding 30% distilled water to the isotonic solution (Hypo) [29]. For high K^+ stimulation, cells were treated for

30 min with the same isotonic buffer, with the KCl concentration increased up to 60 mM (HK), as previously described [30]. For CaMKII inhibition U251 MLC1-WT cells and mouse astrocytes were treated for 15 min with 10 μM of KN93 (10 μM, Sigma-Aldrich, Saint Louis, MO, USA) in isotonic solution or co-stimulated for 15 min with 10 μM KN93 in hypotonic solution. Cycloheximide (CHX) treatment (100 μg/mL, Sigma-Aldrich) was maintained for 1, 3, 4, or 5 h.

2.2. Generation of U251 Cell Lines Expressing MLC1-T17A and T17D Mutants

Mutagenesis of the amino acid (aa) Threonine (T) into Alanine (A) or aspartic acid (D) (T17A and T17D, respectively) was obtained by using oligonucleotides containing the mutations required. pQCXIN-MLC1-WT plasmid construct [20] was digested with MfeI and NotI restriction enzymes (NEB, New England Biolabs Ltd., Hitchin, UK). The DNA fragment obtained was run on agarose gel and extracted/purified with a Gel Extraction Kit (Qiagen, Hilden, Germany). gBlocks™ Gene oligonucleotides (IDT, Integrated DNA Technologies, Inc., Coralville, IA, USA) containing a T17A or T17D mutation and MfeI/NotI sequence at their ends (18 ng for each fragment) were incubated for 1 h at 50 °C with 100 ng of the MfeI/NotI digested plasmid in 1X Gibson Assembly® Master Mix (NEB, New England Biolabs). Then, 2.0 μL of the reaction mixture were used to transform DH5α competent cells. Positive colonies grown on the selective agar medium were screened by plasmid DNA sequencing (Eurofins, Genomics, Ebersberg, Germany). Plasmids containing the desired mutations were used for stable U251 line generation, as previously described [20].

2.3. Immunofluorescence Staining (IF)

IF staining was performed as described in [26,30], using the following primary antibodies (Abs): anti-MLC1 polyclonal Ab (pAb) (1:50, Atlas AB, AlbaNova University Center, Stockholm, Sweden) and anti-Xpress monoclonal Ab (mAb) (1:50, ThermoFischer Scientific, Rockford, IL, USA) to detect the Xpress tag present at the NH_2 terminal of the recombinant MLC1 and anti-phospho (p)CaMKII pAb (1:50, Santa Cruz Biotechnology, Dallas, TX, USA). As secondary Abs, a biotinyated goat anti-rabbit IgG H+L Ab (4.3 μg/mL; Jackson Immunoresearch Laboratories, West Grove, PA, USA), followed by streptavidin-TRITC (2 μg/mL; Jackson) and Alexa Fluor 488 goat anti-mouse IgG Ab (1:300, Invitrogen, Thermo Fisher Scientific, Rockford, IL, USA) were used. Coverslips were sealed in Fluoroshield with DAPI (Sigma-Aldrich) and analyzed with a fluorescence microscope (Carl Zeiss, Jena, Germany, ZEISS Axiocam 512); images were acquired using ZEN 3.1 blu edition software. The distribution of immunofluorescence pixel intensity along a freely defined line (indicated by a dotted arrow in the IF images) spanning the whole cell was evaluated using the profile analysis tool of the NIH ImageJ software in 50–60 cells/condition, as described in [31]. The colocalization signal was evaluated by Pearson's correlation coefficient (PCC). Statistics analysis was performed using one way ANOVA and an unpaired two-tailed Student's *t*-test.

2.4. Pull-Down Assay and Immunoprecipitation

Pull-down assay was performed using rat brain extracts and sepharose-immobilized GST-fused MLC1 (C-t and N-t domains) and with His-fused MLC1 full-length protein (FL), all obtained as previously described [20]. Rat brain was extracted with ice-cold binding buffer (25 mM Tris-HCl, pH 7.4, 150 mM NaCl, 1.5 mM $MgCl_2$, 1% Triton X-100, 0.5% sodium deoxycolate), containing a protease inhibitor mixture (Roche Molecular Biochemicals, Mannheim, Germany). Brain extracts (2 mg/mL) were incubated with the indicated fusion proteins (20 μL of beads suspension) followed by centrifugation and WB analysis. For immunoprecipitation of U251 cells and primary mouse astrocytes, cells were lysed in ice-cold RIPA buffer. The supernatants were incubated with 50% (*w/v*) protein G plus agarose beads (Santa Cruz Biotechnology) for 2 h at 4 °C and left overnight (ON) at 4 °C in a rotating wheel with the anti-Xpress mAb or the anti-pCaMKII pAb, respectively, previously cross-linked with dimethyl pimelimidatedihydrochloride (DMP, Sigma Aldrich). The immunoprecipitated eluates were processed for western blot analysis (WB).

2.5. Protein Extract Preparation and Western Blotting (WB)

Total protein extracts from astrocytoma cell lines, quantification, and WB analysis were performed as previously described [20,21]. Nitrocellulose membranes were blotted ON at 4 °C using anti-MLC1 pAb (1:1500, in-house generated [22]), anti-Actin mAb (1:2000), anti-pCaMKII pAb (1:200), anti-PP1 pAb (1:1000), anti-PP2A pAb (1:1000), anti-phosphothreonine mAb (1:500) (all from Santa Cruz Biotechnology), anti β-DG mAb (1:250; NCL-43 DAG, Novocastra Lab, Ltd., Newcastle upon Tyne, UK), and anti-Xpress mAb (1:2000, Thermo Fisher Scientific), in PBS + 3% BSA. Densitometric analyses of WB experiments were performed using NIH ImageJ software and Bio-Rad ChemiDoc XRS system.

2.6. In Vitro Kinase Assay for CaMKII

The phosphorylated active enzyme CaMKII was immunoprecipitated from U251 cells using a specific anti-pCaMKII (Thr286) pAb (2μg, Santa Cruz Biotechnology). The enzyme was then used to phosphorylate in vitro the agarose-bound-GST N-terminal peptide and the His-fused MLC1 FL protein obtained as described above. The in vitro kinase assay was performed by incubating the immunoprecipitated pCaMKII or the commercial recombinant purified pCaMKII (0.5 μg, Invitrogen, Thermo Fisher Scientific) with the recombinant proteins in kinase buffer (25 mM Hepes, pH 7.4, 50 mM KCl, 10 mM $MgCl_2$, 1 mM DTT, 2 mM $CaCl_2$, 0.1 mM PMSF, phosphatase inhibitor cocktail) containing 2 μM CaM and 1–2 μCi [γ^{32}P] ATP (>3000 Ci/mmol) for 10 min at 30 °C under gentle stirring. When indicated, the CaMKII inhibitor KN93 (10 μM, Sigma-Aldrich) was used in the reaction mixture. The kinase reaction was stopped by adding 4X SDS loading buffer and the samples were resolved on 12% SDS-PAGE. The dried gels were exposed to X-ray film for autoradiography. When indicated, myelin basic protein (MBP) was added to the reaction mixture.

2.7. Mass Spectrometry Analysis

For MS analysis, the phosphorylation reaction was carried out in kinase buffer containing 2 μM CaM and 0.3 mM ATP, using commercial recombinant purified pCaMKII (see above). Proteins eluted from the beads were separated on a 1D-gel NuPAGE 4–12% (Novex, Invitrogen, Carlsbad, CA, USA). The bands were stained with Coomassie blue (Colloidal Blue Staining kit, Invitrogen), cut, and digested with trypsin (Promega, Madison, WI, USA), Asp-N-protease (Promega), or Glu-C endoproteinase (Sigma-Aldrich). To identify phosphorylated sites, the peptide mixture was analyzed using LS-MS/MS as described [32]: MS/MS and MS/MS/MS were acquired in order to search for phosphopeptides carrying mass increases and neutral losses, respectively, and all the data were analyzed using the SEQUEST algorithm [33]. Data are available via ProteomeXchange with identifier PXD030549.

2.8. Electrophysiology

The whole cell dialyzed configuration, the external standard, and Ca^{2+}-free (0 Ca^{2+}) solutions, as well as the internal solution, were used for electrophysiological recordings of the $ICl_{,swell}$, as previously described [34]. Briefly, the external standard solution contained: NaCl 140 mM, KCl 5 mM, $CaCl_2$ 2 mM, $MgCl_2$ 2 mM, MOPS 5 mM, and glucose 10 mM (pH 7.40). The external Ca^{2+}-free solution contained NaCl 140 mM, KCl 5 mM, $MgCl_2$ 4 mM, MOPS 5 mM, and glucose 10 mM (pH 7.40). The internal solution contained KCl 155 mM, EGTA-K 1 mM, MOPS 5 mM, and $MgCl_2$ 1 mM (pH 7.20). The external hypotonic solution was obtained by adding 30% distilled water to the external solution. VRAC-mediated $ICl_{,swell}$ were blocked by DCPIB (10 μM), while KN93 (10 μM, Sigma-Aldrich) was used for CaMKII inhibition. Both compounds were dissolved in DMSO (Sigma-Aldrich), with the highest DMSO concentration in the recording solutions being 0.1%. Both the degree and kinetic of inactivation of the hypotonic-induced $ICl_{,swell}$ were assessed in current traces elicited at +100 mV. The degree of inactivation was obtained by dividing the current amplitude taken at the steady-state ($I_{steady\text{-}state}$) and the current amplitude taken at the peak (instantaneous current or $I_{istantaneous}$), as follows: 1-($I_{steady\text{-}state}/I_{istantaneous}$). The kinetic of inactivation was calculated by fitting the current trace with a monoexponential decay function, which provided the time constant (τ) of current inactivation. To verify the involvement of any Ca^{2+} possibly released from intracellular stores, we assessed the maximal $ICl_{,swell}$ activated using hypotonic solution with 0 mM Ca^{2+} (to avoid contamination from external Ca^{2+} influx) in U251 (MLC1-WT) cells pretreated or not with 1 μM thapsigargin (TG, Sigma-Aldrich), a SERCA pump irreversible inhibitor. Before the activation of the $ICl_{,swell}$ with the Ca^{2+} free hypotonic solution, cells were incubated with 1 μM TG in an isotonic Ca^{2+}-free solution for 5 min, a time period sufficient to completely deplete the intracellular Ca^{2+} stores in non-excitable cells. To verify the involvement of IP3 receptor (IP3R)-mediated Ca^{2+} release from internal stores, we assessed the maximal $ICl_{,swell}$ activated by the re-addition of Ca^{2+} to the hypotonic solution in WT cells in the presence and absence of 1 mM 2-Aminoethoxydiphenyl borate (2-APB, Sigma-Aldrich), an IP3R inhibitor, added in the internal pipette solution. We used this protocol to reach a high concentration of 2-APB in the cell cytoplasm and to avoid the direct inhibition of VRAC occurring in U251 cells in response to the extracellular Ca^{2+} re-addition in the constant presence of 100 μM 2-APB, as previously described [35].

2.9. Fura-2-Based Ca^{2+} Imaging

Cytoplasmic Ca^{2+} was measured using the fluorescence video-imaging technique with the Ca^{2+} indicator Fura-2-AM (Invitrogen, Thermo Fisher Scientific), as previously described [20]. U251 cells seeded on poly-l-lysine treated glass coverslips were loaded with 2 μM Fura-2-AM dissolved in recording buffer, with the following composition (mM): 140 NaCl, 5 KCl, 1 $MgCl_2$, 2.5 $CaCl_2$, 5.5 D-glucose, and 10 HEPES/NaOH at RT (pH 7.4). In Ca^{2+}-free solutions, Ca^{2+} was replaced with $MgCl_2$ and 0.5 mM EGTA was added (0-Ca^{2+} solution). A custom-made local solution exchange system allowed the solution bathing the cells to be rapidly switched between control and test solutions. Ca^{2+} concentrations were expressed as the ratio of emissions at 340 nm to 380 nm.

2.10. Statistical Analysis

All the statistical analyses were performed using GraphPad prism software (Version 9.3, GraphPad, San Diego, CA 92108, USA). Results were expressed as mean ± standard error of mean (SEM). Data were first subjected to a normality test (D'Agostino and Pearson Omnibus Normality test); when data followed a normal distribution, a Student's *t*-test was applied; otherwise, non-parametric tests, such as Wilcoxon test or Kruskal-Wallis test, followed by Dunn's Multiple Comparison post hoc test, when necessary, were performed. Statistically significant *p* values are * $p < 0.05$, ** $p < 0.01$ and *** $p < 0.001$, while not significant differences are indicated with the "ns" abbreviation.

3. Results

3.1. Ca^{2+}-Dependent Protein Kinases II (CaMKII) Binds MLC1 Protein in Astrocytes

To explore the link between MLC1 and Ca^{2+} signaling in astrocytes, we focused on the possible involvement of the Ca^{2+}-dependent CaMKII enzyme. The recognized role of the Ca^{2+}-dependent kinases (PKC and CaMKII) in astrocyte volume regulation [3] and the observation of the presence of a potential CaMKII phosphorylation target site (RXXS/T) in a highly conserved amino acid (aa) sequence of the MLC1 protein (Figure 1A) underpinned this choice. CaMKII is a multifunctional serine/threonine kinase regulated by the Ca^{2+}/Calmodulin (CaM) complex and recognized as a crucial mediator of many physiological effects caused by elevations of intracellular Ca^{2+} [36]. To study the relationship between MLC1 and CaMKII, we used the already characterized U251 human astrocytoma cell lines stably overexpressing the human recombinant MLC1 wild-type (MLC1-WT), [20,23,26]. We first verified a possible interaction between the phosphorylated (p), activated CaMKII, and MLC1 protein in these cells, by performing immunofluorescence (IF) staining and co-immunoprecipitation (IP) experiments. These analyses showed that pCaMKII/MLC1 colocalization was poorly detectable in the cytosol of U251 cells in unstimulated conditions (Figure 1B), but it increased in perinuclear/ER areas after cell treatment with hyposmotic and high K^+ solutions (Figure 1C–E, asterisks in panel C and arrowheads in panel D, respectively), with both conditions favoring intracellular Ca^{2+} increase and MLC1 functional activation [15,17]. Co-IP experiments in the same cells confirmed an MLC1/CaMKII interaction at the biochemical level, by showing that pCaMKII co-immunoprecipitated with MLC1 (Figure 1F, arrowhead and arrow, respectively). Accordingly, pCaMKII and MLC1 co-IP were observed when mouse astrocyte protein extracts were used, revealing that the endogenous MLC1 protein expressed in astrocytes from primary cultures also interacted with the pCaMKII (Figure S1).

To better clarify the molecular details of MLC1/CaMKII binding, we also performed a pull-down assay of rat brain protein extracts using recombinant GST-MLC1-NH_2-t (GST-MLC1-N-t) and GST-MLC1-COOH-t (GST-MLC1-C-t) domains or the full length MLC1 recombinant protein (aa 2-377, FL) containing a Histidine (His) tag, which were immobilized, respectively, on glutathione-agarose or Histidine-binding resin, as previously described [19,22]. pCaMKII was observed among the pulled-down proteins when GST-MLC1-C-t and the FL protein were used, but not when the assay was performed with the recombinant GST-MLC1-N-t peptide (Figure 1G). The MLC1 GST-MLC1-C-t and FL protein also bound the serine/threonine protein phosphatase 1 (PP1) and protein phosphatase 2A (PP2A), (Figure 1G) with both enzymes involved in the de-phosphorylation of the CaMKII and of their target proteins [37,38]. In the same experiment, no binding occurred with beta dystroglycan (β-DG), a protein already found not to interact with either the MLC1 N- or C-terminal [19], and thus used as negative control.

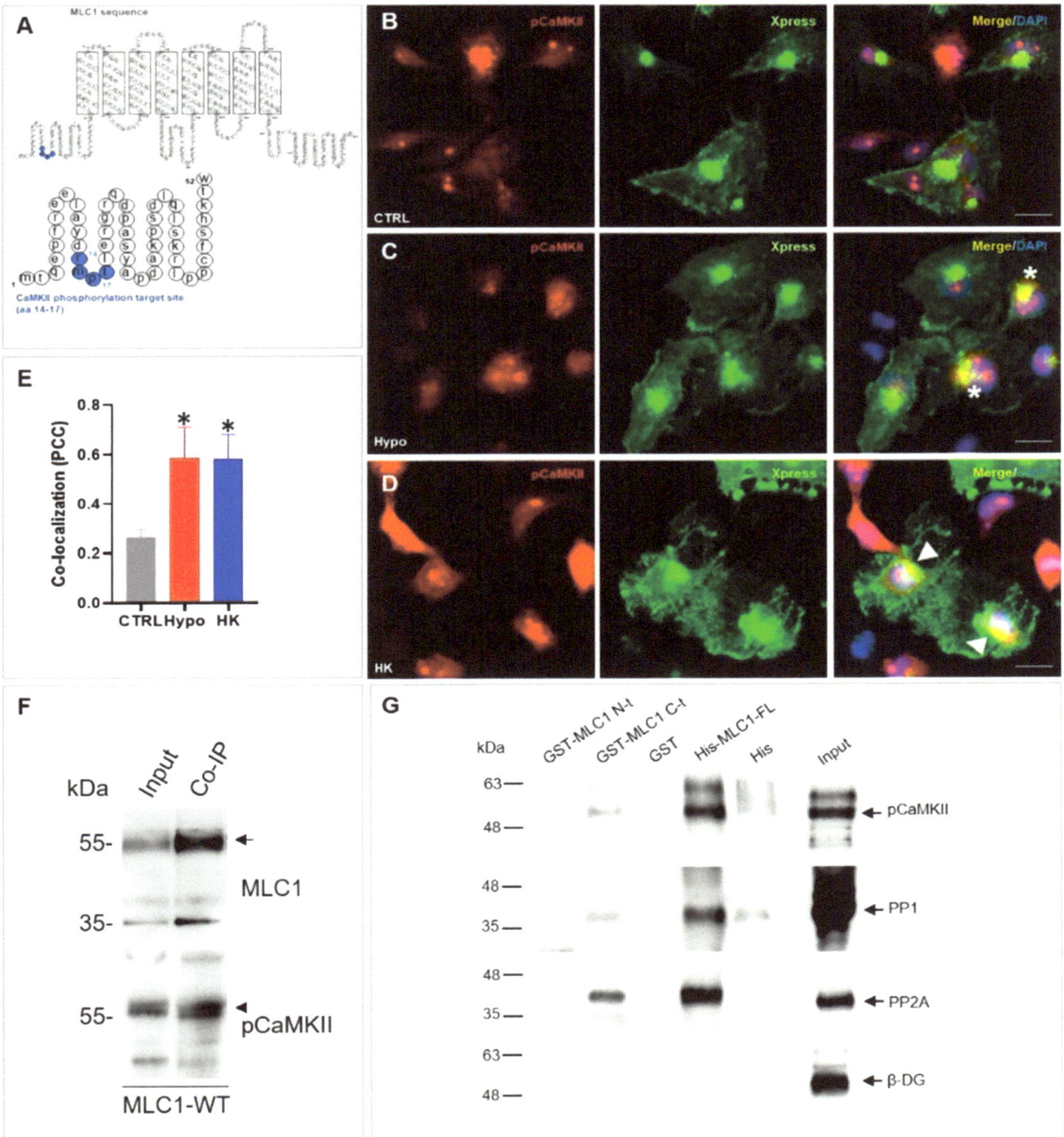

Figure 1. CaMKII binds the MLC1 protein in U251 cells. (**A**) Schematic representation of MLC1 protein sequence showing a potential CaMKII phosphorylation target site in the MLC1 NH_2 terminal (aa 14–17). (**B**–**D**) IF staining of MLC1-WT U251 cells with anti-phospho (p)CaMKII pAb (red) in combination with anti-Xpress mAb (green) to detect the Xpress tag localized in the MLC1 NH_2 terminal shows low levels of pCaMKII expression in the cytosol of U251 cells in control conditions (CTRL, (**B**)) and an increase of its expression and a partial colocalization with MLC1 in the perinuclear areas after a 30-min treatment with hyposmotic (Hypo) and high K^+ (HK) solutions (asterisks and arrowheads in (**C**,**D**), respectively). Scale bars: 20 µm. Panel (**E**) shows the quantification of the colocalization signal (yellow) evaluated using ImageJ software. The degree of

colocalization was estimated using the Pearson correlation coefficient (PCC). Data are shown as means ± SEM of 50–60 cells for each condition (* $p < 0.001$ vs. CTR, unpaired two-tailed Student's t-test). (**F**) Immunoprecipitation (IP) of MLC1 protein from MLC1-WT U251 cells using the anti-Xpress mAb. Immunoblotting performed with the anti-MLC1 pAb (arrow) and anti-pCaMKII pAb (arrowhead) showed that MLC1/pCaMKII co-immunoprecipate. (**G**) A pull-down assay of rat brain extract was carried out with sepharose-immobilized GST-MLC1-NH_2 and -COOH terminal domains (GST-MLC1-N-t and GST-MLC1-C-t, respectively), and with GST alone as control. Agarose-immobilized Histidine (His)-fused MLC1 full-length protein (His-MLC1-FL) and Histidines (His) alone, as a control, were also used. WB of protein eluate reveals the binding of pCaMKII, and of the phosphatases PP1 and PP2A to His-MLC1-FL and GST-MLC1-C-t, but not to GST-MLC1-N-t. Beta dystroglycan (β-DG) was used as negative control. The starting material derived from rat brain extract is indicated as Input. One representative experiment out of the three performed is shown.

3.2. *CaMKII Phosphorylates MLC1 Protein at the Amino Acid Residue Threonine 17*

The above results identified MLC1 as a CaMKII-binding protein. To verify whether MLC1 was also phosphorylated by CaMKII, we performed an in vitro kinase assay. The assay was carried out in the presence of [γ^{32}P] ATP, using the full-length MLC1 protein (FL) and the GST-MLC1 NH_2-t (where the putative CaMKII phosphorylation motif had been observed) as substrates for the CaMKII enzyme precipitated from U251 cells. As shown in Figure 2A, the immunoprecipitated phosphorylated enzyme (pCaMKII) was fully functional, being able to phosphorylate the myelin basic protein (MBP), a control substrate used to monitor CaMKII activity. The CaMKII enzyme also phosphorylated the FL-MLC1 protein (Figure 2B) and the MLC1-N-t peptide (Figure 2C). The specificity of this phosphorylation was confirmed by introducing to the reaction mixture the CaMKII inhibitor KN93, which strongly reduced MLC1 protein phosphorylation in both cases (Figure 2B,C). To identify the specific aa residue targeted by the CaMKII enzyme, a kinase assay was performed with the use of the CaMKII recombinant enzyme and the GST-MLC1 NH_2-t peptide and the reaction mixture was subjected to LC-MS/MS analysis. These experiments allowed the recognition of a phosphorylation site at Threonine (T), in position 17 of the MLC1-N-t peptide, within the supposed CaMKII consensus sequence RXXS/T (aa 14–17, Figure 2D).

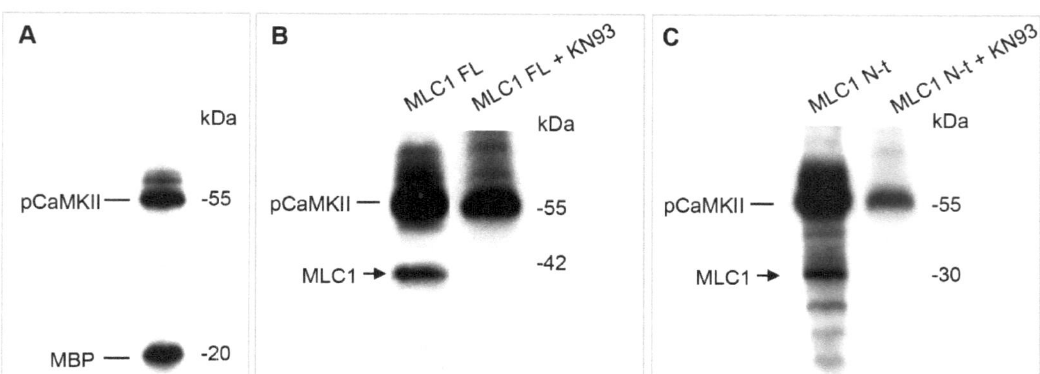

Figure 2. *Cont.*

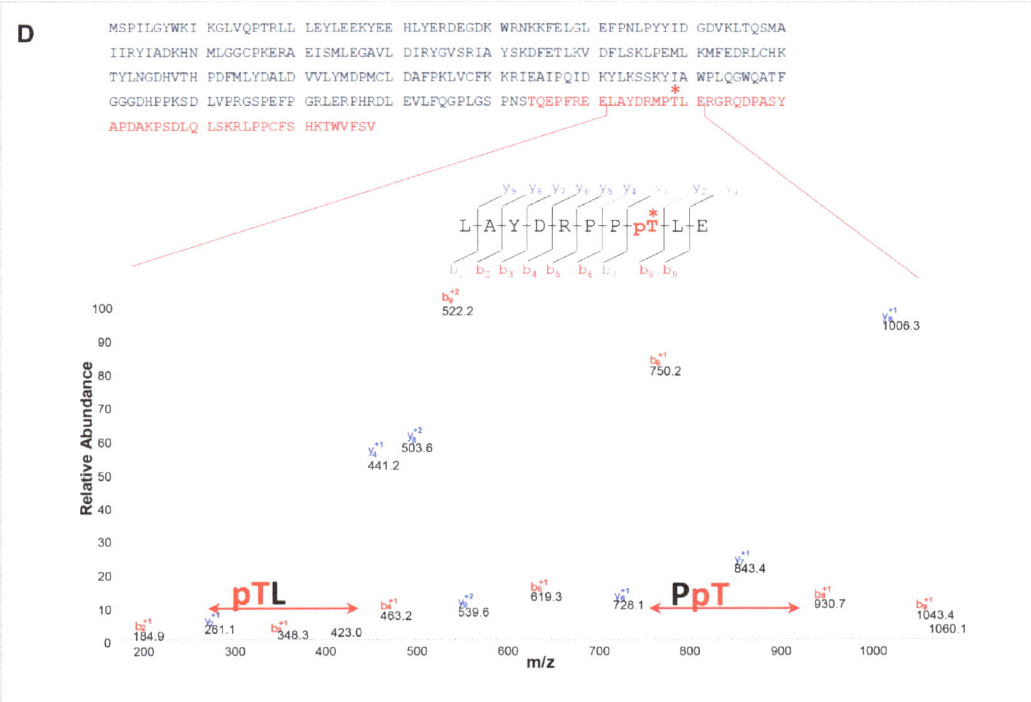

Figure 2. CaMKII phosphorylates threonine 17 residue in the MLC1 NH_2 terminal domain. (**A–C**) In vitro kinase assay using CaMKII enzyme immunoprecipitated from U251 cells. The not-specific substrate MBP (**A**), the His-MLC1 full-length protein (FL, (**B**)), or the sepharose-immobilized GST-MLC1 NH_2-terminal domain (MLC1 N-t, (**C**)) were incubated with the CaMKII enzyme, with or without the CaMKII inhibitor KN93, and the reaction was carried out in the presence of [γ^{32}P] ATP. The [32P]-labeled proteins revealed by autoradiography indicated that MLC-FL and MLC1 N-t were specifically phosphorylated by CaMKII. (**D**) Mass spectrometry analysis of the phosphorylated MLC1 NH_2 terminal peptide after GluC digestion. In the upper part of the figure, the sequence of GST-MLC1 NH_2-terminal region is shown, with the MLC1-NH_2-terminus highlighted in red. The lower panel shows the MS3 spectrum of the double-charged phosphorylated peptide LAYDRMPTLE obtained after phosphoric acid loss caused by CID fragmentation. This neutral loss confirmed the presence of a phosphorylated site. B and Y series ions (reported in red and blue, respectively) in the MS3 spectrum localize the phosphorylation at T17. Phospho-T (pT) in the peptide sequence indicates a phosphorylated T, while the asterisk marks the localization of the pT in the sequence of MLC1-N-terminal.

3.3. CaMKII-Mediated Phosphorylation of MLC1 in Response to Intracellular Ca^{2+} Increase Affects MLC1 Protein Assembly and Stability in U251 Cells

The activation of CaMKII and the consequent CaMKII-mediated phosphorylation can strongly modulate the target protein structural organization, intracellular localization, and functionality [38,39]. To clarify the effects of CaMKII-mediated phosphorylation on MLC1 structural and functional properties, we generated two U251 cell lines, where the CaMKII phosphorylation site T17 was mutated into (i) alanine (T17A), a non-phosphorylatable amino acid, or (ii) aspartic acid (T17D), an amino acid mimicking protein constitutive phosphorylation. Characterization of these cell lines by WB and IF revealed that T17D substitution favored dimer formation (Figure 3A, arrow) and a general increase of MLC1 protein distribution, either in the cytoplasm or at cell PM, when compared to MLC1-WT expressing cells (Figure 3B, arrowheads and panel C). No strong alterations of MLC1 intracellular distribution were observed in cells expressing the non-phosphorylatable MLC1-T17A mutant by IF staining (Figure 3B–C), suggesting that CaMKII phosphorylation did not affect MLC1 protein trafficking. In the U251 cell lines that we generated, MLC1 regulation at transcriptional level could not occur, since the MLC1 cDNA is cloned under the control of a constitutive promoter (the cytomegalovirus promoter, CMV). Therefore, we tested whether CaMKII phosphorylation could increase MLC1-T17D protein levels by affecting protein stability. To this aim, we treated cells for different durations with cycloheximide (CHX), an inhibitor of protein biosynthesis, and verified by WB the degradation kinetics of the MLC1-T17D and T17A mutants in comparison with the MLC1-WT protein. CHX treatment showed that T17D mutation improved MLC1 stability, this latter protein still being detectable between 3 and 4 h of CHX treatment, a time when the MLC1-WT was already degraded (Figure 3D). In parallel, we also observed that the MLC1-T17A mutant had a slightly reduced stability when compared to MLC1-WT (Figure S2). These combined analyses encompassing WB, IF, and CHX treatment provided evidence that CaMKII-dependent phosphorylation favors MLC1 dimer formation and protein stabilization.

Consistent with these results, we also observed that in both U251 cells expressing the recombinant MLC1-WT and in primary mouse astrocytes expressing the endogenous MLC1 protein, cell treatment with the CaMKII inhibitor KN93 reduced the formation of the MLC1 dimer in response to hyposmotic stimulation (Figure S3).

To have a more direct confirmation of the CaMKII-mediated phosphorylation of MLC1 at the T17 aa residue in an in vivo system, we monitored whether phosphorylation occurred in the MLC1-WT protein, and not in the non-phosphorylatable mutant T17A, in response to stimuli inducing an increase in intracellular Ca^{2+} levels and CaMKII activation. To this aim, MLC1-WT and MLC1-T17A U251 cells were treated with ATP, one of the most efficient physiological activators of intracellular Ca^{2+} release in astrocytes and astrocytoma cells [40], and MLC1 protein was immunoprecipitated from control and stimulated cells. To monitor the presence of threonine (Thr) phosphorylation on MLC1 protein, we performed WB analysis of the immunoprecipitated protein eluates using a specific anti-phospho(p)Thr mAb, as previously described [41]. These experiments showed increased levels of p-Thr at the molecular weight expected for the MLC1 protein in the eluate derived from cells expressing the WT MLC1, but not from cells expressing the T17A mutant, in response to ATP stimulation (Figure 3E, arrow), thus providing evidence that CaMKII phosphorylation of MLC1 occurs at T17 in U251 cells and not only in in vitro systems.

Considering the assembled evidence of the above studies, we can conclude that, in astrocytes, MLC1 stabilization/dimerization is favored by the phosphorylation of the T17 aa residue induced by the CaMKII enzyme activated by intracellular Ca^{2+} release.

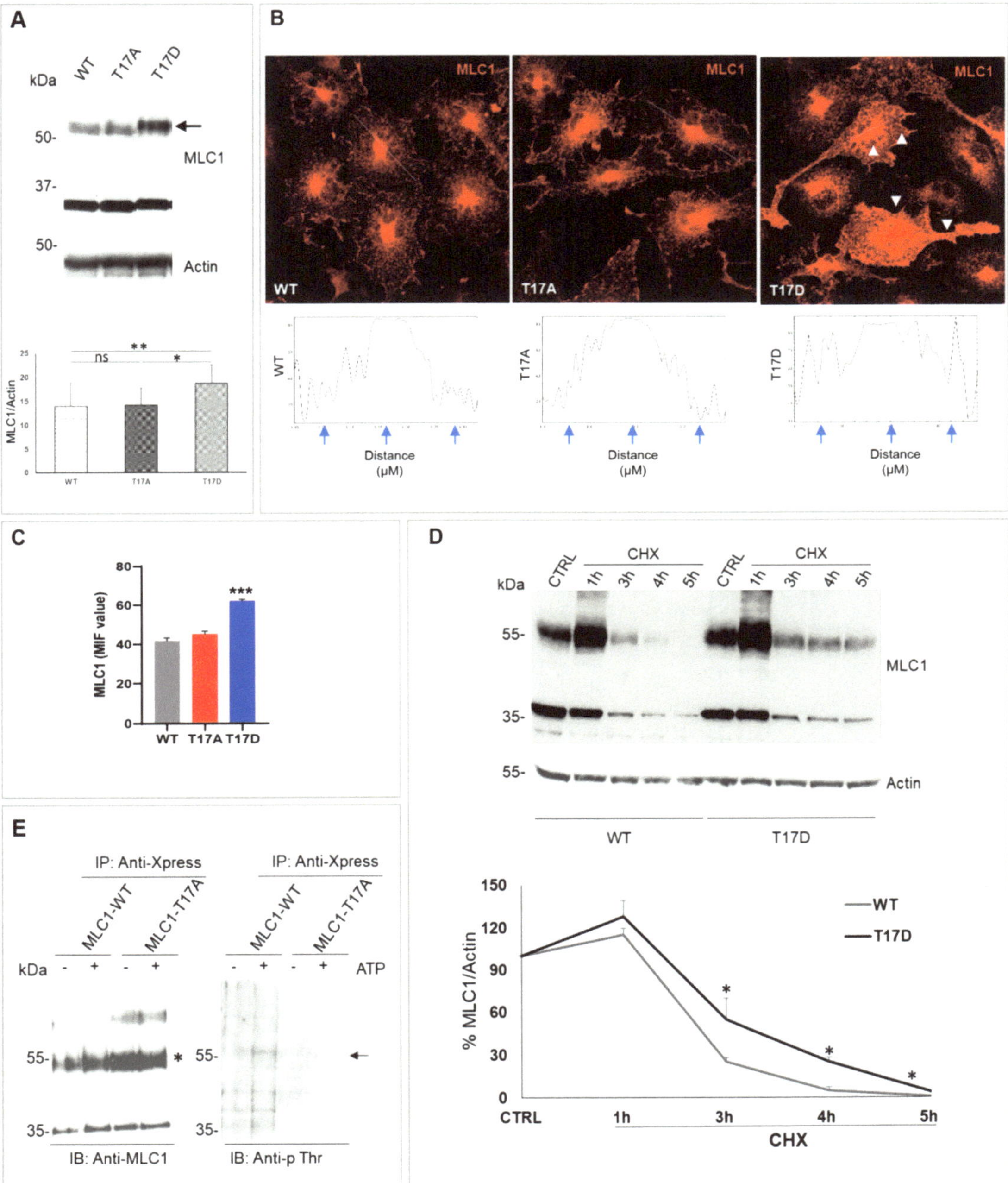

Figure 3. CaMKII-mediated phosphorylation of the MLC1 T17 residue favors MLC1 protein dimerization and stabilization. (**A**) WB analysis of U251 mutant cell lines shows that T17D substitution favors MLC1 dimer formation (arrow) when compared to MLC1-WT and T17A expressing cells. The bar graph below the WB represents the densitometry analysis of the MLC1 protein bands normalized

with the amount of actin (means ± SEM of 3 independent experiments; * $p < 0.05$, ** $p < 0.001$ calculated using non-parametric test). Panel (**B**) shows IF staining of cells expressing MLC1-WT, T17A, and T17D mutants with anti-MLC1 pAb (red). An increase of the MLC1-T17D mutant localization at PM and in intracellular compartments (arrowheads in B) when compared to WT and T17A MLC1 is observed. Scale bars: 20 μm. Below each IF panel, the distribution of IF pixel intensity along a freely defined line (representatively indicated in each IF images) spanning the whole cell confirms a general increase of the MLC1-T17D protein fluorescence intensity. Fluorescence intensity peaks are marked by arrows. One representative intensity plot is shown for each IF panel. (**C**) Mean fluorescence intensity (MIF) of 50–60 cells/conditions from 3 independent experiments was calculated (means ± SEM values of mean, *** $p < 0.0001$ calculated using one-way ANOVA, unpaired two-tailed Student's *t*-test). (**D**) WB analysis of U251 cells expressing MLC1-WT and the T17D mutant, untreated (CTRL) or treated with cycloheximide (CHX, 100 μg/mL) for 1, 3, 4, and 5 h revealed an increase of the MLC1-T17D protein half-life (stability) when compared to MLC1-WT. A graph indicating the densitometry analysis of MLC1 protein bands normalized with the amount of actin is shown below. Data are expressed as the percentage of the value measured in control untreated cells (means ± SEM of 3 replicates for each type of experiments; * $p < 0.05$ calculated using non-parametric test). (**E**) WB of protein eluates derived from immunoprecipitation of MLC1-WT and MLC1-T17A expressing cells with anti-Xpress mAb in control conditions or after ATP stimulation (5 min, 100 μM). Immunoblotting was performed with the anti-MLC1 pAb (asterisk), as positive control of IP procedures, and with anti-phosphothreonine (anti-p Thr) mAb to assess Thr phosphorylation levels. As indicated by the arrow, Thr phosphorylation signal increases after ATP stimulation in MLC1-WT, corresponding with the MLC1 protein molecular weight, and not in MLC1-T17A expressing cells.

3.4. MLC1 Potentiates VRAC-Mediated $ICl_{,swell}$ in U251 Cells

As MLC1 was involved in the activation of Cl-currents (e.g., $ICl_{,swell}$) and cell volume regulation in several MLC disease models [15–17], we investigated the role of CaMKII phosphorylation on MLC1 functionality by performing electrophysiological experiments on U251 cells. We first assessed whether in our U251-based cell model MLC1 was also involved in $ICl_{,swell}$ potentiation. To this aim, patch-clamp recordings were performed on U251 cells expressing MLC1-WT and, as controls, on U251 cells transfected with an empty vector (Ø) or expressing the already characterized pathological mutant MLC1-S280L [20,26], (Figure 4). Step potentials from −100 to +100 mV were imposed before (left) and during (center) perfusion with hyposmotic solution; the current obtained by subtracting the current elicited under hypotonic (Hypo) solution and that in isotonic (Iso) solution represents the hyposmosis-induced current ($ICl_{,swell}$, right), (Figure 4A). The quantitative analysis showed that the hyposmosis-induced $ICl_{,swell}$ density was significantly larger in MLC1-WT than in Ø and MLC1-S280L cells (Figure 4B,C). Notably, the expression of MLC1-WT also affected the $ICl_{,swell}$ inactivation (Figure 4 and Figure S4). Specifically, the inactivation kinetic, calculated as a time constant (τ), was significantly faster in MLC1-WT (112.1 ± 8.0 ms) compared to MLC1-Ø (154.3 ± 14.5 ms) cells, and the degree of inactivation, calculated as the ratio of the current amplitudes taken at the steady state and at the peak ($1-(I_{steady\ state}/I_{peak})$), was significantly larger in MLC1-WT (0.86 ± 0.03) compared to MLC1-Ø (0.71 ± 0.02) cells. Overall, these findings further support the ability of MLC1 to also act as a $ICl_{,swell}$ modulator when expressed in U251 cells.

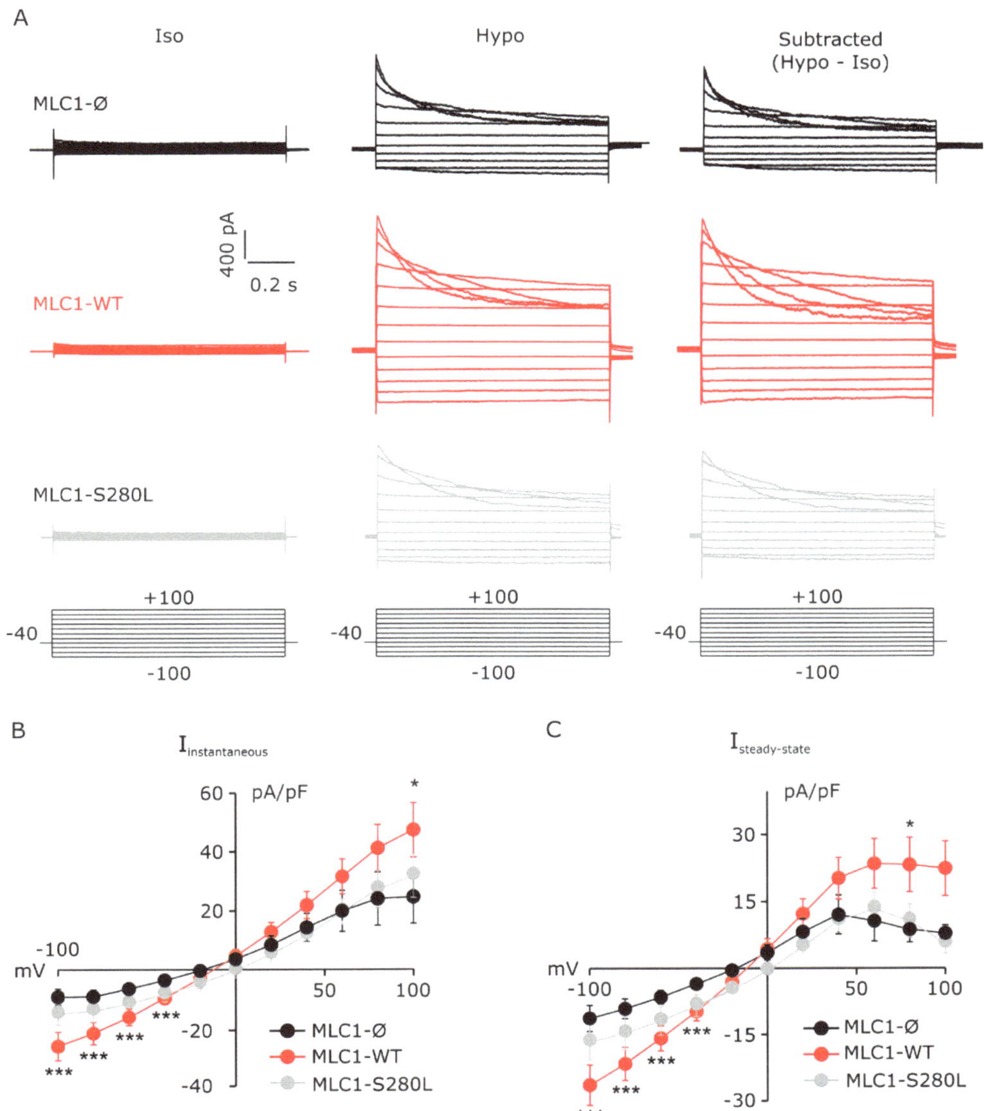

Figure 4. MLC1 upregulates $ICl_{,swell}$ currents in U251 astrocytoma cells. (**A**) Representative I–V relationships recorded from −100 to +100 mV pulses (Delta 20 mV, holding potential −40 mV, duration 1 s) in U251 cells infected with an empty vector (Ø) (**top**), expressing MLC1-WT (**middle**) or MLC1-S280L mutant (**bottom**) in control condition (**left**), following application of hyposmotic solution (**center**) and after digital subtraction (Hypo-Iso, **right**). (**B**,**C**) Mean I–V relationships of instantaneous and steady state current obtained in MLC1-Ø (black; $n = 6$), MLC1-WT (red; $n = 9$), and MLC1-S280L cells (grey; $n = 6$) after digital subtraction (Hypo-Iso) (mean ± SEM; * $p < 0.05$; *** $p < 0.001$).

3.5. MLC1 Potentiates VRAC-Mediated $ICl_{,swell}$ in U251 Cells through a CaMKII-Dependent Mechanism

Once it was determined that the expression of functional MLC1 in U251 cells was required for the enhancement of the $ICl_{,swell}$, we investigated if the CaMKII-dependent phosphorylation of MLC1 represented a key step for its activation and the modulation of $ICl_{,swell}$ by MLC1. To this aim, we assessed the $ICl_{,swell}$ density in U251 expressing MLC1-WT or carrying the T17A or T17D mutations. Voltage ramps from −100 to +100 mV from a holding potential of −40 mV were applied. In Figure 5A, we illustrate the protocol used and the time course of $ICl_{,swell}$ activation in U251 cells with the hypotonic solution, together with representative current ramps (Inset: left) under hypotonic conditions and after application of 10 μM DCPIB, which selectively inhibits VRAC. The inset on the right of Figure 5A shows the typical $ICl_{,swell}$ induced by step potentials during hyposmotic solution perfusion in the absence or presence of 10 μM DCPIB, which almost fully blocked the hyposmosis-activated currents. The results from these experiments revealed that the hyposmotic solution induced an inward current at −80 mV ($ICl_{,swell}$) in all cell models, as shown in the bar plot of Figure 5B. As described above, MLC1-WT cells exhibited a greater current density (~54 pA/pF) than cells transfected with the empty vector (MLC1-Ø, not containing MLC1) (~32 pA/pF). By analyzing $ICl_{,swell}$ from T17D and T17A mutants we observed that the T17D cells, exhibited an $ICl_{,swell}$ density similar to WT cells (~56 pA/pF), while the $ICl_{,swell}$ density observed in T17A cells (~32 pA/pF) was similar to MLC1-Ø cells (Figure 5B). These data suggest that CaMKII phosphorylation of MLC1 is required for upregulating $ICl_{,swell}$. In support of this view, we showed that 10 μM of the CaMKII inhibitor KN93, acutely added to the Hypo solution, was able to reduce $ICl_{,swell}$ by ~40% in MLC1-WT cells (Figure 5C,D). Notably, this was about the same difference between $ICl_{,swell}$ recorded in MLC1-Ø and MLC1-WT or between T17D and T17A cells. Interestingly, in MLC1-Ø, as well as in cells expressing MLC1 proteins that cannot be modulated by the CaMKII-mediated phosphorylation (MLC1-T17D and MLC1-T17A), KN93 was not able to reduce the hyposmosis-induced $ICl_{,swell}$.

3.6. Hyposmotic Challenge Induces Ca^{2+} Influx and Ca^{2+} Release in U251 Cells

It is known that CaMKII is activated by the Ca^{2+} binding protein Calmodulin upon increase of cytosolic Ca^{2+}. Increased cytosolic Ca^{2+} due to cell swelling has also been shown to activate $ICl_{,swell}$ in astrocytes [42,43]. These results identify Ca^{2+} as a key player in the activation pathways of both MLC1 and $ICl_{,swell}$. Since specific Ca^{2+} sources are known to be linked to specific functions, we aimed to distinguish the roles of external Ca^{2+} influx and Ca^{2+} release from internal stores in the CaMKII/MLC1-dependent enhancement of the $ICl_{,swell}$. First, we carried out Fura-2-based Ca^{2+} recording experiments, to discriminate between the contribution of external Ca^{2+} influx and intracellular Ca^{2+} release in the hyposmosis-induced Ca^{2+} movements in U251. A hyposmotic challenge lasting 15 min was imposed on Fura-2-loaded MLC1-WT cells in the presence or absence of external Ca^{2+} (Figure 6A–C). When control solution containing 2.5 mM Ca^{2+} was used, most cells responded with a fast and transient Ca^{2+} rise, very variable in amplitude. A slow rise of Ca^{2+} followed the initial peak, varying in both speed of onset and amplitude. On the other hand, in Ca^{2+}-free solution, the initial peak was absent, and only a slowly raising phase, with a similar onset rate and amplitude, was observed. We can conclude that the first Ca^{2+} peak was mainly due to Ca^{2+} entry, with the possible minor participation of Ca^{2+} release from internal stores induced by Ca^{2+} entry (so called CICR), which, instead, was represented by the asynchronous Ca^{2+} transients predominant in the later phase. Finally, the slowly rising phase recorded in 0-Ca^{2+} depicts a homogeneous response among cells, due to release from unidentified intracellular stores, and is not attributable to a CICR mechanism. To offer a quantitative evaluation of the very variable response to hyposmotic challenge observed in the presence of Ca^{2+}, we calculated the integral (i.e., area) of the Ca^{2+} signal at different time intervals, in both the presence and absence of extracellular Ca^{2+}. The resulting graph (Figure 6D,E) confirms the initial Ca^{2+} raise, followed by a larger Ca^{2+}

response, which decays at later times in the presence of Ca^{2+}, and a slight and monophasic rise in the absence of Ca^{2+}.

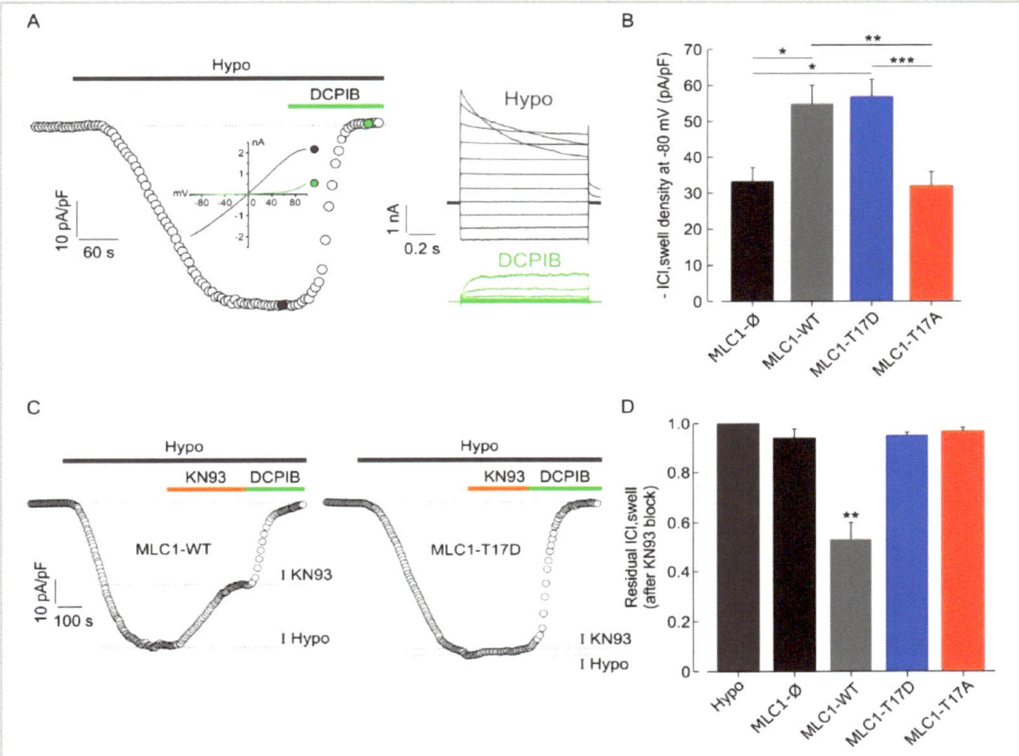

Figure 5. Phosphorylation of MLC1-T17 potentiates $ICl_{,swell}$ in U251 cells. (**A**) Representative time course of ICl, swell density (pA/pF) measured at −80 mV (the equilibrium potential for K+ under our recording conditions), taken from current ramps (from −100 to 100 mV, 600 ms duration, holding potential −40 mV) applied during the application of hypotonic solution (grey trace) and upon addition of 10 μM DCPIB (green trace). Right: families of current traces evoked by applying 1 s voltage steps from −100 to 100 mV, in steps of 20 mV (HP −40 mV) in the presence of a hypotonic solution (grey traces), and upon addition of 10 μM DCPIB (green traces). (**B**) Bar plot showing the average current density measured at −80 mV during exposure to 30% hypotonic solution, in U251 cells transfected with the empty vector (MLC1-Ø, n = 10), with MLC1-WT protein (n = 14), and with MLC1-T17D (phosphorylation mimicking mutant, n = 9) or MLC1-T17A (not phosphorylatable mutant, n = 14). (**C**) Representative time courses of $ICl_{,swell}$ density (pA/pF) measured at −80 mV from current ramps, during application of hypotonic solution (grey bar) and upon addition of 10 μM KN93 and 10 μM DCPIB, in MLC1-WT (**left**) and MLC1-T17D (**right**) U251 cells. The dashed lines in the time courses indicate the levels where the peaks of currents were measured. (**D**) Bar plot showing the mean residual $ICl_{,swell}$ in the presence of 10 μM KN93, as normalized to the $ICl_{,swell}$ elicited during exposure to 30% of hypotonic solution in the absence of the KN93 (Hypo), in the different cell lines (Mean ± SEM n = 3–5; * $p < 0.05$; ** $p < 0.01$, *** $p < 0.001$).

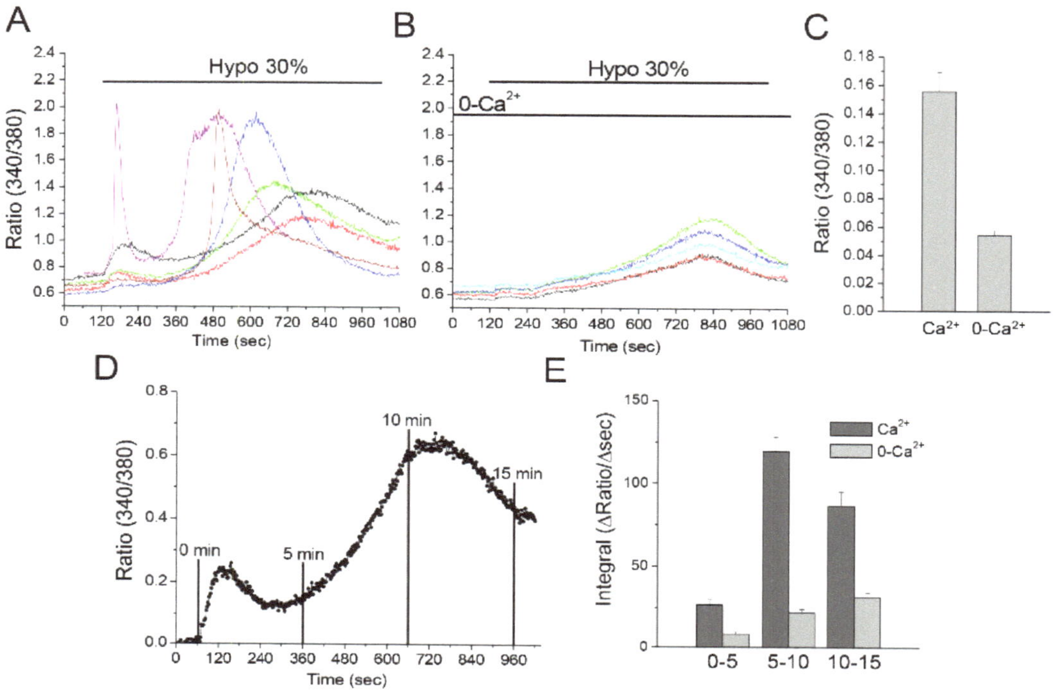

Figure 6. Ca^{2+} movements in U251 cells in response to hyposmotic challenge. Fura-2-loaded MLC1-WT expressing U251 cells were exposed to hyposmotic solution for 15 min in the presence of Ca^{2+} (2.5 mM), or in a 0-Ca^{2+} solution (Ca^{2+} replaced by Mg^{2+} and with EGTA 0.5 mM). Exemplificative traces recorded in the presence (**A**) or absence of Ca^{2+} (**B**) are shown. The amplitude of the Ca^{2+} peak was calculated as the difference between the maximum value reached within the first 120 s of application of hyposmotic solution and the baseline value. Peak amplitudes are plotted in panel (**C**) (mean ± SEM of 90 and 65 cells, from 3 recordings for each condition). Areas underneath Ca^{2+} transients at defined time intervals (0–5, 5–10 and 10–15 min) were calculated after baseline subtraction, as shown in panel (**D**), to represent actual Ca^{2+} movements. Mean ± SEM of ΔRatio/Δtime at different time intervals are shown in panel (**E**) (n = 72 and 65 cells, in Ca^{2+} and 0-Ca^{2+}, respectively).

3.7. CaMKII/MLC1-Dependent Enhancement of $ICl_{,swell}$ Requires Release of Ca^{2+} from Intracellular Stores Induced by Extracellular Ca^{2+} Influx

Once it was ascertained that the hypotonic stimulus triggered both Ca^{2+} influx and Ca^{2+} release from intracellular stores, we asked if changes in cytosolic Ca^{2+} levels were important for the CaMKII/MLC1-dependent enhancement of $ICl_{,swell}$. We first carried out patch-clamp experiments aimed at distinguishing between the specific contributions of the two Ca^{2+} sources. To study the role of Ca^{2+} influx, after a stabilization period in 0-Ca^{2+} isotonic solution, cells were exposed to hyposmotic solution, in the absence of external Ca^{2+}, to assess the amount of $ICl_{,swell}$ activated by cell swelling without the contribution of Ca^{2+} influx. After a stable $IICl_{,swell}$ activation was reached, cells were probed with a hypotonic solution containing 2 mM Ca^{2+}, to evaluate a possible further increase of the current due to the entry of external Ca^{2+} (Figure 7A). The results obtained with the different cell models showed that only in cells expressing MLC1-WT was the $ICl_{,swell}$ activated by the Ca^{2+}-free hypotonic solution further increased (by ~35%) upon addition of external Ca^{2+}, while in all the other cell types (MLC1-Ø, MLC1-T17D, and MLC1-T17A) the current density remained unchanged (Figure 7B and Figure S5). These results were confirmed by similar tests carried out with 0-Ca^{2+} and with 2 mM Ca^{2+} in the activating hyposmotic solution, but this time done in separate MLC1-WT cell groups. Moreover, in this case, we found that the hyposmosis-activated $ICl_{,swell}$ density in the presence of extracellular Ca^{2+} (~54 pA/pF) was markedly greater than that elicited in the cell group probed in the absence of extracellular Ca^{2+} (~32 pA/pF) (Figure 7C). The above results suggest that additional activation of $ICl_{,swell}$ by external Ca^{2+} influx occurred only where MLC1 activity could be modulated by Ca^{2+}-dependent CaMKII, which was the major feature distinguishing MLC1-WT from all the other cell lines analyzed here. Overall, these experiments indicated that Ca^{2+} influx is needed for MLC1-dependent $ICl_{,swell}$ potentiation in U251 cells. Then, we wanted to test whether Ca^{2+} release was also involved. To this end, we assessed the maximal $ICl_{,swell}$ activated by hypotonic solution with 0-Ca^{2+} (to avoid contamination from external Ca^{2+} influx and ensuing activation of the CICR process) in MLC1-WT cells and in the same cells pretreated for 5 min with 1 μM Thapsigargin (TG), a non-competitive inhibitor of the ER Ca^{2+} ATPase (SERCA), known to cause a massive Ca^{2+} release from the ER (i.e., with internal stores fully depleted). A representative experiment carried out in WT cells pretreated with TG, with the hypotonic solution with 0-Ca^{2+} activated $ICl_{,swell}$, is shown in Figure 7D. The same protocol was applied to WT cells not pretreated with TG, to be used as control. As shown in Figure 7E, in the absence of Ca^{2+} influx (0-Ca^{2+} condition) the TG pretreatment did not significantly alter the $ICl_{,swell}$ density. As the influx of Ca^{2+} from extracellular space is able to further activate the $ICl_{,swell}$ in WT cells, we then asked whether Ca^{2+} entering into the cell could directly enhance the $ICl_{,swell}$ via the CaMKII/MLC1 pathway, or indirectly, by stimulating the release of Ca^{2+} from the intracellular stores via CICR. To this end, we assessed the $ICl_{,swell}$ density activated by hypotonic solution with 0-Ca^{2+}, in MLC1-WT cells pretreated or not with TG, and then verified if $ICl_{,swell}$ density could be further increased upon switching to the hypotonic solution with 2 mM Ca^{2+} (Figure 7F; i.e., we followed the same protocol as the experiment shown in Figure 5). The analysis of these results revealed that only in cells not treated with TG, was the $ICl_{,swell}$ activated by the Ca^{2+}-free hypotonic solution further increased (by ~35%) upon addition of external Ca^{2+}, while in TG-treated cells, the current density remained unchanged (Figure 7F). Similar results were obtained by including in the pipette solution the IP3 receptor (IP3R) inhibitor 2-APB (see Materials and Methods). These observations revealed that when the internal stores are fully depleted of Ca^{2+} or when the release of Ca^{2+} from IP3R is blocked, and no CICR process can be activated, the external Ca^{2+} influx per se is ineffective in increasing $ICl_{,swell}$ Notably, Ca^{2+} influx potentiates $ICl_{,swell}$ by activating the CICR mechanism only in cells expressing the CaMKII phosphorylatable MLC1 protein.

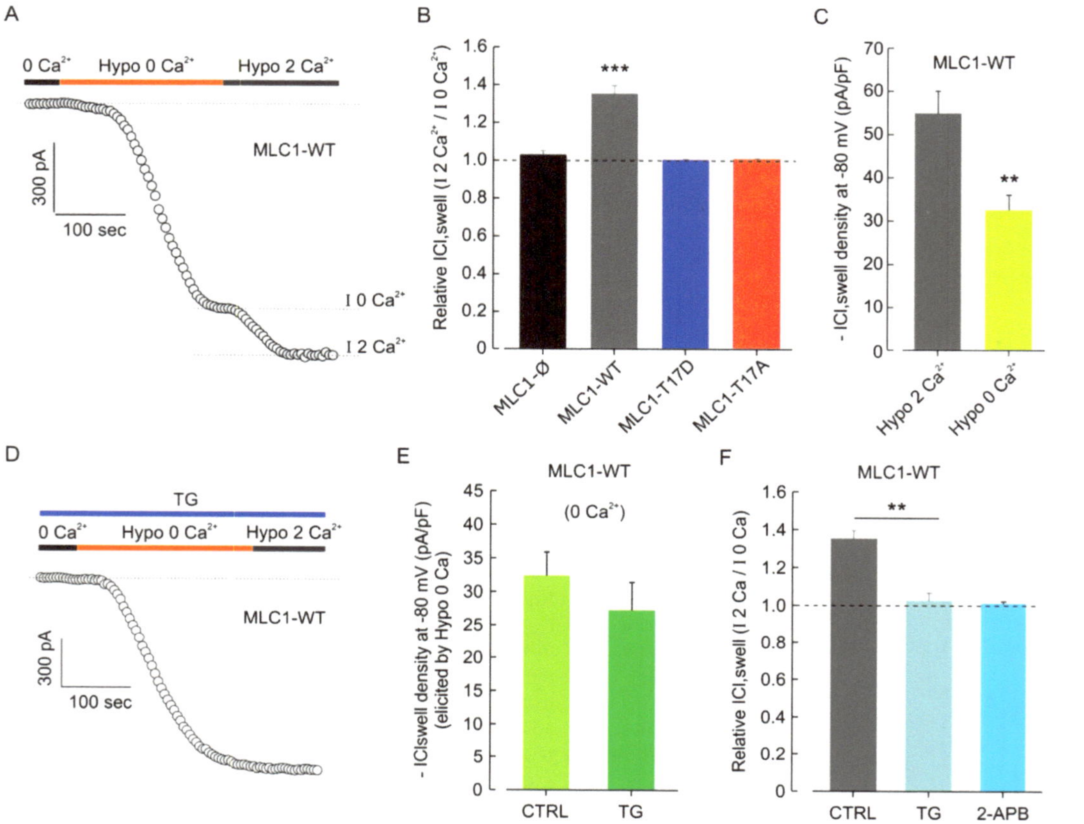

Figure 7. Extracellular and endoplasmic reticulum Ca^{2+}-dependence of $ICl_{,swell}$ in U251 cell lines carrying WT and mutated MLC1. (**A**) Representative time course of $ICl_{,swell}$ during application of a Ca^{2+}-free hypotonic solution (Hypo 0 Ca^{2+}) and a hypotonic solution containing 2 mM Ca^{2+} (Hypo 2 Ca^{2+}). The dashed lines indicate the levels where the maximal hypotonic-activated current in the absence (I 0 Ca^{2+}) and presence (I 2 Ca^{2+}) of extracellular Ca^{2+} were measured. (**B**) Bar plot showing the mean $ICl_{,swell}$ current elicited by the addition of 2 mM external Ca^{2+} (dashed lines), normalized to the current activated by Hypo 0 Ca^{2+}, in MLC1-Ø (n = 5), MLC1-WT (n = 8), MLC1-T17D (n = 6), and MLC1-T17A (n = 5) U251 cell lines (mean ± SEM, *** $p < 0.01$, using one-sample t-test). (**C**) Bar plot showing the average current density during exposure to hypotonic solution, in the continuous presence (grey bar, n = 14) or absence (light green bar, n = 9) of external Ca^{2+}, assessed in MLC1-WT cells (mean ± SEM, ** $p < 0.01$, using two-sample t-test). (**D**) Representative time course of $ICl_{,swell}$ during application of a Ca^{2+}-free hypotonic solution (Hypo 0 Ca^{2+}) followed by a hypotonic solution containing 2 mM Ca2+ (Hypo 2 Ca^{2+}), after 1 μM of thapsigargin (TG) pretreatment (5 min to deplete ER Ca^{2+}stores). (**E**) Bar plot showing the average current density during exposure to hypotonic 0 Ca^{2+} solution, in MLC1-WT cells in the absence (CTRL, n = 9) and presence (TG, n = 9) of TG pre-treatment (mean ± SEM, p = 0.36). (**F**) Bar plot showing the mean fractional $ICl_{,swell}$ elicited by Hypo 2 Ca^{2+}, normalized to the current activated by Hypo 0 Ca^{2+}, in MLC1-WT cells either in the absence (n = 14) or presence (n = 9) of TG or of the IP3R inhibitor 2-APB (n = 7) (mean± SEM, ** $p < 0.01$).

4. Discussion

Pathological astrocyte swelling often occurs in response to CNS damage, leading to cellular (cytotoxic) edema. Cytotoxic edema represents a first step in the formation of tissue edema, as it generates the driving force for the vasogenic edema formation, causing tissue swelling [7]. This pathological cascade may reflect MLC disease pathogenesis, where an increased water content in brain is due to mutations in MLC1, a protein favoring astrocyte regulatory volume decrease after hyposmosis-induced swelling [13,44]. Understanding MLC1 function and the molecular mechanisms regulating volume changes in astrocytes may pave the way for the identification of possible molecular target(s) to treat brain edema in MLC. Following this aim, in the present study we revealed that MLC1 functional activation, assessed by MLC1-mediated VRAC current potentiation, is favored by CAMKII phosphorylation of MLC1 in response to stimuli inducing intracellular Ca^{2+} increase in astrocytes. Since Ca^{2+} release is strongly linked to astrocyte activation in response to several pathological stimulations, these findings may have interesting implications for the comprehension of MLC1′s proper function and its central role in the control of volume changes and astrocyte activation processes in different brain pathological conditions, as discussed below.

4.1. CamKII Binds and Phosphorylates MLC1

We, here, have provided evidence that CaMKII binds MLC1 at its COOH terminal, similarly to what has been observed for the Ca^{2+} channel TRPV4 [43]. In addition, IF staining showed that treatments inducing Ca^{2+} release in astrocytes (hyposmosis, high K^+) favored MLC1/CaMKII interaction, suggesting a possible functional outcome of this interaction. The identification by proteomic analysis of the T17 aa as the CaMKII target site and the characterization of the U251 cell lines expressing the MLC1 mutant insensitive to CaMKII phosphorylation (T17A) or mimicking constitutive phosphorylation (T17D), proved that CaMKII favors MLC1 protein assembly (dimer formation), an indication of MLC1 functional activation [19–21,23,25,26], and increases protein half-life (stabilization), but does not alter MLC1 trafficking. The occurrence of phosphorylation in response to ATP stimulation, only in MLC1-WT, and not in MLC1-T17A mutant, confirmed that CaMKII also phosphorylates MLC1 in intact cells and not only in in vitro assays, in response to intracellular Ca^{2+} release. The functions of many proteins involved in volume regulation, such as AQP4 [39] and calcium (TRPV4) [45] and chloride channels (Best-1, Ano-1, VRAC, ClC3) [46–48], are modulated by CaMKII interaction and phosphorylation. Interestingly, the coexisting binding of the PP1 and PP2A phosphatases at the MLC1 C-terminal suggests that MLC1 protein may be subjected to dynamic phosphorylation/de-phosphorylation events, likely involved in the modulation of its functionality.

4.2. CaMKII-Mediated Phosphorylation of MLC1 Confers Ca^{2+}-Dependence to VRAC Activation

VRAC is the channel mainly responsible for both $ICl_{,swell}$ and volume regulation in several types of cell. Once activated, VRAC is permeable to osmolytes such as taurine, glutamate, and Cl^- anions, whose efflux results in cell water outflow and restoration of the original cell volume [49,50]. Lack of functional MLC1 reduces VRAC currents [13,15]; conversely, MLC1 overexpression increases VRAC currents [51]. We here demonstrate that, in U251 cells, the overexpression of MLC1-WT, but not of the pathological mutant form (MLC1-S280L), affects several biophysical features of the current, such as its amplitude, inactivation degree, and kinetics. The biochemical and electrophysiological characterization of MLC1-T17A and T17D mutants, demonstrated that this specific post-translational modification is essential for MLC1 to modulate the biophysical and functional properties of VRAC and to confer Ca^{2+} dependence to this channel, which would be otherwise insensitive to Ca^{2+} ions in glioma cells [52]. The observation that, in presence of external Ca^{2+}, $ICl_{,swell}$ is increased in MLC1-WT, but not in T17A or T17D mutants, is in accordance with the ability of Ca^{2+} influx to promote CaMKII-dependent MLC1 phosphorylation only in MLC1-WT expressing cells. Moreover, the abolition of a Ca^{2+} influx-mediated increase of $ICl_{,swell}$ after pretreatment with TG and 2-APB, confirmed the contribution of internal Ca^{2+} stores in this process, possibly via activation of the CICR mechanism. As summarized in Figure 8, our results are consistent with the following cascade of events: (1) induction of an initial

Ca^{2+} influx by hypotonic cell swelling, most likely due to opening of stretch-activated Ca^{2+} channels (such as TRPV4); (2) triggering of Ca^{2+} release from intracellular stores through the CIRC mechanism; (3) Ca^{2+}-induced CaMKII activation; (4) CaMKII-mediated MLC1 phosphorylation and consequent dimer stabilization of the newly activated protein and/or of that already present at the PM; (5) MLC1 potentiation of VRAC current. Besides the MLC1-mediated VRAC potentiation, the molecular mechanisms of VRAC activation is still an open question. The initial hypothesis involving changes in ion strength is now paralleled by other mechanisms, among which are membrane tension and signaling pathways. The heterogeneity of the activation mechanism may be attributed to cell-specific properties of VRAC and different interactions of the channel with Ca^{2+}-sensitive effectors. In astrocytes, VRAC is also stimulated by neurotransmitters and neuromodulators such as ATP, through an increase of intracellular Ca^{2+}, PKC, and CaM/CaMKII pathways activation [53]. Although these studies leave unresolved the mechanism by which PKC and CaM/CaMKII activate VRAC, they are in keeping with our data on U251 cells expressing MLC1 and the mechanism suggested above (Figure 8). We here propose MLC1 as the essential component of the Ca^{2+}-dependent modulation of VRAC in astrocytes. This notion is further supported by the fact that in astrocytoma/glioblastoma cells, where MLC1 is virtually absent (or minimally expressed), changes in intracellular Ca^{2+} do not affect $ICl_{,swell}$ [52]. On the contrary, in primary astrocytes where MLC1 is highly expressed, the Ca^{2+}-modulation of the $ICl_{,swell}$ was reported [42,43]. It would also be plausible to assume that other similar proteins may replace MLC1 in providing Ca^{2+}-sensitivity to VRAC in other cell types where the Ca^{2+} dependence of $ICl_{,swell}$ has been demonstrated. To date, no direct interaction between MLC1 and the LRRC8A protein, the most generally recognized molecular component of the VRAC channels [54,55], has been found in astrocytes and brain tissue [56]. In addition, we did not detect LRRC8A among the molecular interactors previously co-purified with MLC1 with the use of the Histidine-tagged protein affinity methods, or in the co-IP here presented (not shown), confirming the previously published evidence. However, MLC1 may act by indirectly influencing the post-translational modifications of some LRRC8 accessory subunits, forming the fully functional channel, in a cell specific way, as we previously observed for the LRRC8C subunits [56]. In light of the recent findings indicating a possible role for the Tweety-homologs proteins (Ttyh1-3) in the activation of the $ICl_{,swell}$ in astrocytes [57], it will also be extremely interesting to study the relationship between MLC1 and this class of proteins in the future.

4.3. *Significance of Astrocyte Ca^{2+} Signaling in MLC Pathogenesis and Beyond*

Overall, the data here presented reveal that, in astrocytes, MLC1 protein function is regulated by Ca^{2+} changes occurring in response to osmotic stress. Interestingly, in different in vitro and in vivo models, osmotic stress has been found to elicit Ca^{2+} responses in astrocyte end-feet, where MLC1 is preferentially expressed [14,22]. We can hypothesize that the osmotic- or other stress signals inducing CICR in perivascular astrocyte end-feet lead to MLC1 activation and VRAC potentiation aimed at stimulating an efficient RVD. Alterations to these processes may cause abnormal astrocyte end-feet swelling, resulting in the dysfunction of both the blood–brain barrier (BBB) and of the neuro-vascular unit (NVU), the functional entity composed of neurons, astrocytes, pericytes, endothelial cells, and basal lamina that couples neuronal activity to cerebral blood flow in response to energy demand [58]. In addition, since adjacent astrocytes are interconnected via Ca^{2+}-permeable gap-junctions, the CICR mechanism generates Ca^{2+} waves propagating through the astrocyte syncytium [59]. Ca^{2+} wave spreading is known to modulate several fundamental astrocyte activities (i.e., release/uptake of ions and gliotransmitters, support to BBB and NVU functionality) [58]. By potentiating VRAC channels, known to favor intercellular Ca^{2+} wave propagation via extracellular ATP release [60], and by favoring the function of the gap-junction protein Cx43 that connects adjacent astrocytes [61], MLC1 might be itself involved in the control of Ca^{2+} wave spreading. It is worth noting that alterations of Ca^{2+} diffusion at the perivascular end-feet can cause serious consequences for NVU and neuron functionality, since a Ca^{2+} increase in this compartment adjusts cerebral blood flow during synaptic activity [62]. Indeed, electrical field stimulation in brain slices and in vivo induces an increase in intracellular Ca^{2+} in perivascular astrocyte end-feet, followed by vascular smooth muscle Ca^{2+} oscillations and arteriolar dilation [63]. Thus, we cannot exclude that alterations of Ca^{2+} dynamics in astrocyte end-feet due to MLC1 mutations might influence arteriolar contractility and BBB permeability, causing edema and NVU dysfunctions. In accordance with this hypothesis, a recent study provided evidence of alterations of vascular smooth muscle contractility and NVU coupling in MLC1 KO mice [64]. The emerging role of MLC1 in the maintenance of BBB/NVU integrity has also been recently highlighted in an elegant in vivo study by Morales and collaborators [65], who generated and characterized transgenic mice where inducible and selective ablation of perivascular astrocytes expressing MLC1 (but not of other astrocyte populations) could be obtained.

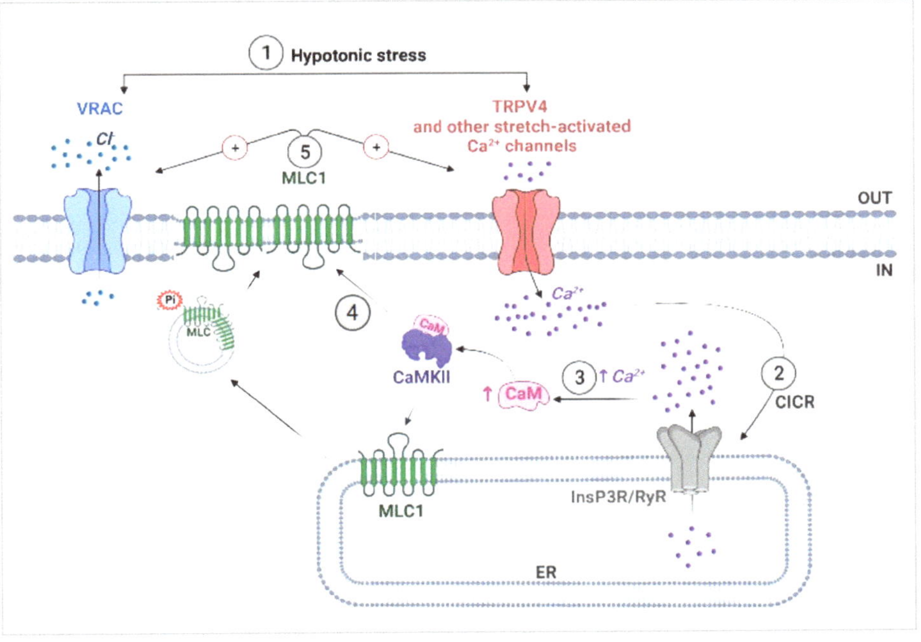

Figure 8. Scheme showing the proposed molecular mechanisms of the MLC1-induced Ca^{2+} regulation of VRAC. (1) Exposure to a hypotonic stress induces astrocyte (or astrocyte end-feet) swelling, which triggers the opening of TRPV4 and other Ca^{2+} channels (voltage-dependent and mechanosensitive Ca^{2+} channels). (2) The entry of extracellular Ca^{2+} promotes Ca^{2+} release from the ER, likely mediated by the InsP3 (or/and Ryanodine) receptors, through the CICR mechanism. (3) The consequent elevation of intracellular Ca^{2+} is sensed by the Ca^{2+}-binding protein calmodulin (CaM), which binds and activates CaMKII. (4) By binding MLC1 and phosphorylating the threonine aa at position 17 (T17), CaMKII promotes MLC1 dimerization, stabilization, and functional activation. (5) Phosphorylated MLC1 potentiates TRPV4 [20] and VRAC channels, favoring $ICl_{,swell}$ and regulatory volume decrease.

Dysfunction of NVU and of cerebral blood flow are also implicated in the pathogenesis of cerebral disorders, such as traumatic brain injury, Alzheimer's (AD) and Parkinson's (PD) diseases, multiple sclerosis (MS), and glioblastoma malignancy [58,66]. Interestingly, increased expression of MLC1 at the perivascular astrocyte end-feet has been observed in the brain of patients affected by AD, MS, and also prion disease [31], which might indicate an attempt to ameliorate these pathological processes by increased MLC1 expression in activated astrocytes. In this scenario, it is noteworthy that Ca^{2+} release from intracellular stores occurs in astrocytes in response to different damaging stimuli, leading to astrocyte swelling and activation [67].

5. Conclusions

In the present study, we revealed for the first time that MLC1 is a Ca^{2+}-sensitive protein linking VRAC-dependent volume regulation to Ca^{2+} dynamics in astrocytes. Although further analyses are needed to fully understand MLC1′s function and the details of MLC/VRAC cooperation, the results here presented open new perspectives for comprehension of the astrocyte-mediated molecular events governing brain water/ion exchanges in normal and pathological conditions. Considering the potential of astrocytes as pharmacological targets, this knowledge may be of interest for the identification of astrocyte-based therapeutic approaches for MLC and other brain diseases where astrocyte swelling and brain edema have a central role in the pathological process.

Supplementary Materials: The following supporting information can be downloaded at: https://www.mdpi.com/article/10.3390/cells11172656/s1, Figure S1: Phospho (p)CaMKII interacts with MLC1 protein in primary mouse astrocytes; Figure S2: MLC1-T17A mutant protein shows a reduced stability when compared to the MLC1-WT protein. Figure S3: CaMKII inhibition reduces MLC1 protein dimerization in U251 cells and primary mouse astrocytes; Figure S4: Effects of MLC1 WT on the degree and kinetics of inactivation of the hypotonic-induced $ICl_{,swell}$; Figure S5. $ICl_{,swell}$ density measured in the presence and absence of extracellular Ca^{2+} in U251 cell populations.

Author Contributions: Study conception and design: M.S.B., A.L., A.M., C.M., S.C., L.C., L.S., M.C. (Martino Caramia), M.C.D., M.C. (Marina Ceccarini), P.M. (Paola Molinari), P.M. (Pompeo Macioce), G.M., M.P., S.V. and E.A. Data curation: M.S.B., A.L., A.M., L.C. and C.M. Acquisition of data: M.S.B. and A.L. Investigation: M.S.B., A.L., A.M., C.M., S.C., L.S., M.C. (Martino Caramia), M.C.D., M.C. (Marina Ceccarini), P.M. (Paola Molinari), P.M. (Pompeo Macioce) and G.M. Supervision: T.C.P., S.V. and E.A. Writing—original draft: A.M., S.V. and E.A. Writing—review and editing: M.S.B., A.L., A.M., L.C., M.P., S.V. and E.A. All authors helped with critical reading of the manuscript and contributed to the ideas presented therein. All authors have read and agreed to the published version of the manuscript.

Funding: This work was supported by Italian Ministry of Health, Ricerca Finalizzata, (Grant N. GR9-2013-02355882 to A.L.); European Leukodystrophies Association (ELA) (Grant N. ELA 2016-002F3 to E.A. and M.S.B) and 5x1000 project of the Istituto Superiore di Sanità (Project code: ISS5x1000_21-949432e8c9be to A.L.).

Institutional Review Board Statement: CD1 Swiss mice were purchased from Harlan Laboratories (San Pietro Al Natisone, Udine, Italy). The animal study protocol was approved by the National Center for Animal Research and Welfare review board of the Istituto Superiore di Sanità and by the Italian Ministry of Health (Authorization 271/SSA/2010). The procedure for obtaining rat brain extracts was approved by the Ministry of Health review board (authorization number: 152/2016-PR) based on the European Communities Council Directive N. 2010/63/EU and the Italian Law Decree n° 26/2014.

Data Availability Statement: The data that support the findings of this study are available upon request from the authors.

Acknowledgments: We thank Antonella Calogero for providing the U251 astrocytoma cell lines and Antonietta Bernardo for fluorescence image quantification analysis.

Conflicts of Interest: The authors declare no conflict of interest.

References

1. Verkhratsky, A.; Parpura, V.; Li, B.; Scuderi, C. Astrocytes: The Housekeepers and Guardians of the CNS. In *Astrocytes in Psychiatric Disorders*; Advances in Neurobiology; Li, B., Parpura, V., Verkhratsky, A., Scuderi, C., Eds.; Springer International Publishing: Cham, Switzerland, 2021; Volume 26, pp. 21–53, ISBN 978-3-030-77374-8.
2. MacAulay, N. Molecular Mechanisms of Brain Water Transport. *Nat. Rev. Neurosci.* **2021**, *22*, 326–344. [CrossRef]
3. Wilson, C.S.; Mongin, A.A. Cell Volume Control in Healthy Brain and Neuropathologies. *Curr. Top. Membr.* **2018**, *81*, 385–455. [CrossRef] [PubMed]

4. Benga, O.; Huber, V.J. Brain Water Channel Proteins in Health and Disease. *Mol. Asp. Med.* **2012**, *33*, 562–578. [CrossRef]
5. Reed, M.M.; Blazer-Yost, B. Channels and Transporters in Astrocyte Volume Regulation in Health and Disease. *Cell Physiol. Biochem.* **2022**, *56*, 12–30. [CrossRef] [PubMed]
6. Lafrenaye, A.; Simard, J. Bursting at the Seams: Molecular Mechanisms Mediating Astrocyte Swelling. *IJMS* **2019**, *20*, 330. [CrossRef] [PubMed]
7. Stokum, J.A.; Kurland, D.B.; Gerzanich, V.; Simard, J.M. Mechanisms of Astrocyte-Mediated Cerebral Edema. *Neurochem. Res.* **2015**, *40*, 317–328. [CrossRef]
8. Jha, R.M.; Raikwar, S.P.; Mihaljevic, S.; Casabella, A.M.; Catapano, J.S.; Rani, A.; Desai, S.; Gerzanich, V.; Simard, J.M. Emerging Therapeutic Targets for Cerebral Edema. *Expert Opin. Ther. Targets* **2021**, *25*, 917–938. [CrossRef]
9. Min, R.; van der Knaap, M.S. Genetic Defects Disrupting Glial Ion and Water Homeostasis in the Brain. *Brain Pathol.* **2018**, *28*, 372–387. [CrossRef]
10. Hamilton, E.M.C.; Tekturk, P.; Cialdella, F.; van Rappard, D.F.; Wolf, N.I.; Yalcinkaya, C.; Çetinçelik, Ü.; Rajaee, A.; Kariminejad, A.; Paprocka, J.; et al. Megalencephalic Leukoencephalopathy with Subcortical Cysts: Characterization of Disease Variants. *Neurology* **2018**, *90*, e1395–e1403. [CrossRef]
11. van der Knaap, M.S.; Barth, P.G.; Vrensen, G.F.J.M.; Valk, J. Histopathology of an Infantile-Onset Spongiform Leukoencephalopathy with a Discrepantly Mild Clinical Course. *Acta Neuropathol.* **1996**, *92*, 206–212. [CrossRef]
12. Leegwater, P.A.J.; Yuan, B.Q.; van der Steen, J.; Mulders, J.; Könst, A.A.M.; Boor, P.K.I.; Mejaski-Bosnjak, V.; van der Maarel, S.M.; Frants, R.R.; Oudejans, C.B.M.; et al. Mutations of MLC1 (KIAA0027), Encoding a Putative Membrane Protein, Cause Megalencephalic Leukoencephalopathy with Subcortical Cysts. *Am. J. Hum. Genet.* **2001**, *68*, 831–838. [CrossRef] [PubMed]
13. Bosch, A.; Estévez, R. Megalencephalic Leukoencephalopathy: Insights Into Pathophysiology and Perspectives for Therapy. *Front. Cell. Neurosci.* **2020**, *14*, 627887. [CrossRef] [PubMed]
14. Boor, P.K.I.; de Groot, K.; Waisfisz, Q.; Kamphorst, W.; Oudejans, C.B.M.; Powers, J.M.; Pronk, J.C.; Scheper, G.C.; van der Knaap, M.S. MLC1: A Novel Protein in Distal Astroglial Processes. *J. Neuropathol. Exp. Neurol.* **2005**, *64*, 412–419. [CrossRef] [PubMed]
15. Ridder, M.C.; Boor, I.; Lodder, J.C.; Postma, N.L.; Capdevila-Nortes, X.; Duarri, A.; Brussaard, A.B.; Estévez, R.; Scheper, G.C.; Mansvelder, H.D.; et al. MegalencephalicLeucoencephalopathy with Cysts: Defect in Chloride Currents and Cell Volume Regulation. *Brain* **2011**, *134*, 3342–3354. [CrossRef] [PubMed]
16. Hoegg-Beiler, M.B.; Sirisi, S.; Orozco, I.J.; Ferrer, I.; Hohensee, S.; Auberson, M.; Gödde, K.; Vilches, C.; de Heredia, M.L.; Nunes, V.; et al. Disrupting MLC1 and GlialCAM and ClC-2 Interactions in Leukodystrophy Entails Glial Chloride Channel Dysfunction. *Nat. Commun.* **2014**, *5*, 3475. [CrossRef] [PubMed]
17. Sirisi, S.; Elorza-Vidal, X.; Arnedo, T.; Armand-Ugón, M.; Callejo, G.; Capdevila-Nortes, X.; López-Hernández, T.; Schulte, U.; Barrallo-Gimeno, A.; Nunes, V.; et al. Depolarization Causes the Formation of a Ternary Complex between GlialCAM, MLC1 and ClC-2 in Astrocytes: Implications in Megalencephalic Leukoencephalopathy. *Hum. Mol. Genet.* **2017**, *26*, 2436–2450. [CrossRef]
18. Formaggio, F.; Saracino, E.; Mola, M.G.; Rao, S.B.; Amiry-Moghaddam, M.; Muccini, M.; Zamboni, R.; Nicchia, G.P.; Caprini, M.; Benfenati, V. LRRC8A Is Essential for Swelling-activated Chloride Current and for Regulatory Volume Decrease in Astrocytes. *FASEB J.* **2019**, *33*, 101–113. [CrossRef]
19. Brignone, M.S.; Lanciotti, A.; Macioce, P.; Macchia, G.; Gaetani, M.; Aloisi, F.; Petrucci, T.C.; Ambrosini, E. The B1 Subunit of the Na,K-ATPase Pump Interacts with Megalencephalic Leucoencephalopathy with Subcortical Cysts Protein 1 (MLC1) in Brain Astrocytes: New Insights into MLC Pathogenesis. *Hum. Mol. Genet.* **2011**, *20*, 90–103. [CrossRef]
20. Lanciotti, A.; Brignone, M.S.; Molinari, P.; Visentin, S.; De Nuccio, C.; Macchia, G.; Aiello, C.; Bertini, E.; Aloisi, F.; Petrucci, T.C.; et al. Megalencephalic Leukoencephalopathy with Subcortical Cysts Protein 1 Functionally Cooperates with the TRPV4 Cation Channel to Activate the Response of Astrocytes to Osmotic Stress: Dysregulation by Pathological Mutations. *Hum. Mol. Genet.* **2012**, *21*, 2166–2180. [CrossRef] [PubMed]
21. Lanciotti, A.; Brignone, M.S.; Camerini, S.; Serafini, B.; Macchia, G.; Raggi, C.; Molinari, P.; Crescenzi, M.; Musumeci, M.; Sargiacomo, M.; et al. MLC1 Trafficking and Membrane Expression in Astrocytes: Role of Caveolin-1 and Phosphorylation. *Neurobiol. Dis.* **2010**, *37*, 581–595. [CrossRef]
22. Ambrosini, E.; Serafini, B.; Lanciotti, A.; Tosini, F.; Scialpi, F.; Psaila, R.; Raggi, C.; Di Girolamo, F.; Petrucci, T.C.; Aloisi, F. Biochemical Characterization of MLC1 Protein in Astrocytes and Its Association with the Dystrophin-Glycoprotein Complex. *Mol. Cell. Neurosci.* **2008**, *37*, 480–493. [CrossRef] [PubMed]
23. Brignone, M.S.; Lanciotti, A.; Camerini, S.; De Nuccio, C.; Petrucci, T.C.; Visentin, S.; Ambrosini, E. MLC1 Protein: A Likely Link between Leukodystrophies and Brain Channelopathies. *Front. Cell. Neurosci.* **2015**, *9*, 66. [CrossRef] [PubMed]
24. Weerth, S.H.; Holtzclaw, L.A.; Russell, J.T. Signaling Proteins in Raft-like Microdomains Are Essential for Ca2+ Wave Propagation in Glial Cells. *Cell Calcium.* **2007**, *41*, 155–167. [CrossRef]
25. Petrini, S.; Minnone, G.; Coccetti, M.; Frank, C.; Aiello, C.; Cutarelli, A.; Ambrosini, E.; Lanciotti, A.; Brignone, M.S.; D'Oria, V.; et al. Monocytes and Macrophages as Biomarkers for the Diagnosis of Megalencephalic Leukoencephalopathy with Subcortical Cysts. *Mol. Cell. Neurosci.* **2013**, *56*, 307–321. [CrossRef] [PubMed]
26. Lanciotti, A.; Brignone, M.S.; Visentin, S.; De Nuccio, C.; Catacuzzeno, L.; Mallozzi, C.; Petrini, S.; Caramia, M.; Veroni, C.; Minnone, G.; et al. Megalencephalic Leukoencephalopathy with Subcortical Cysts Protein-1 Regulates Epidermal Growth Factor Receptor Signaling in Astrocytes. *Hum. Mol. Genet.* **2016**, *25*, 1543–1558. [CrossRef]

27. Cardin, V.; Lezama, R.; Torres-Márquez, M.E.; Pasantes-Morales, H. Potentiation of the Osmosensitive Taurine Release and Cell Volume Regulation by Cytosolic Ca2+ Rise in Cultured Cerebellar Astrocytes. *Glia* **2003**, *44*, 119–128. [CrossRef]
28. Eleuteri, C.; Olla, S.; Veroni, C.; Umeton, R.; Mechelli, R.; Romano, S.; Buscarinu, M.; Ferrari, F.; Calò, G.; Ristori, G.; et al. A Staged Screening of Registered Drugs Highlights Remyelinating Drug Candidates for Clinical Trials. *Sci. Rep.* **2017**, *7*, 45780. [CrossRef]
29. Sforna, L.; Cenciarini, M.; Belia, S.; Michelucci, A.; Pessia, M.; Franciolini, F.; Catacuzzeno, L. Hypoxia Modulates the Swelling-Activated Cl Current in Human Glioblastoma Cells: Role in Volume Regulation and Cell Survival: Hypoxia-Activated Cl Current in Glioblastoma. *J. Cell. Physiol.* **2017**, *232*, 91–100. [CrossRef]
30. Arnedo, T.; López-Hernández, T.; Jeworutzki, E.; Capdevila-Nortes, X.; Sirisi, S.; Pusch, M.; Estévez, R. Functional Analyses of Mutations in HEPACAM Causing Megalencephalic Leukoencephalopathy. *Hum. Mutat.* **2014**, *35*, 1175–1178. [CrossRef]
31. Brignone, M.S.; Lanciotti, A.; Serafini, B.; Mallozzi, C.; Sbriccoli, M.; Veroni, C.; Molinari, P.; Elorza-Vidal, X.; Petrucci, T.C.; Estévez, R.; et al. Megalencephalic Leukoencephalopathy with Subcortical Cysts Protein-1 (MLC1) Counteracts Astrocyte Activation in Response to Inflammatory Signals. *Mol. Neurobiol.* **2019**, *56*, 8237–8254. [CrossRef]
32. Cherubini, F.; Serio, D.; Guccini, I.; Fortuni, S.; Arcuri, G.; Condò, I.; Rufini, A.; Moiz, S.; Camerini, S.; Crescenzi, M.; et al. Src Inhibitors Modulate Frataxin Protein Levels. *Hum. Mol. Genet.* **2015**, *24*, 4296–4305. [CrossRef] [PubMed]
33. Yates, J.R.; Eng, J.K.; McCormack, A.L.; Schieltz, D. Method to Correlate Tandem Mass Spectra of Modified Peptides to Amino Acid Sequences in the Protein Database. *Anal. Chem.* **1995**, *67*, 1426–1436. [CrossRef] [PubMed]
34. Sforna, L.; Michelucci, A.; Morena, F.; Argentati, C.; Franciolini, F.; Vassalli, M.; Martino, S.; Catacuzzeno, L. Piezo1 Controls Cell Volume and Migration by Modulating Swelling-Activated Chloride Current through Ca^{2+} Influx. *J. Cell. Physiol.* **2022**, *237*, 1857–1870. [CrossRef] [PubMed]
35. Lemonnier, L.; Prevarskaya, N.; Mazurier, J.; Shuba, Y.; Skryma, R. 2-APB inhibits volume-regulated anion channels independently from intracellular calcium signaling modulation. *FEBS Lett.* **2004**, *556*, 121–126. [CrossRef]
36. Swulius, M.T.; Waxham, M.N. Ca^{2+}/Calmodulin-Dependent Protein Kinases. *Cell. Mol. Life Sci.* **2008**, *65*, 2637–2657. [CrossRef]
37. Leslie, S.N.; Nairn, A.C. CAMP Regulation of Protein Phosphatases PP1 and PP2A in Brain. *Biochim. Biophys. Acta BBA Mol. Cell Res.* **2019**, *1866*, 64–73. [CrossRef]
38. Kim, K.-R.; Jeong, H.-J.; Kim, Y.; Lee, S.Y.; Kim, Y.; Kim, H.-J.; Lee, S.-H.; Cho, H.; Kang, J.-S.; Ho, W.-K. Calbindin Regulates Kv4.1 Trafficking and Excitability in Dentate Granule Cells via CaMKII-Dependent Phosphorylation. *Exp. Mol. Med.* **2021**, *53*, 1134–1147. [CrossRef]
39. Li, S.; Jiang, B.; Lu, H.; Huang, S.; Fan, B.; Chen, Z.; Li, Y. Inhibition of CAMK II Modulates Water Permeability by Reducing AQP4 Expression in Astrocytes after Oxygen-Glucose Deprivation. *Neuroreport* **2021**, *32*, 1058–1064. [CrossRef]
40. Hartmann, J.; Verkhratsky, A. Relations between Intracellular Ca2+ Stores and Store-Operated Ca2+ Entry in Primary Cultured Human Glioblastoma Cells. *J. Physiol.* **1998**, *513 Pt 2*, 411–424. [CrossRef]
41. Li, H.; Liu, S.; Sun, X.; Yang, J.; Yang, Z.; Shen, H.; Li, X.; Liu, Y.; Chen, G. Critical Role for Annexin A7 in Secondary Brain Injury Mediated by Its Phosphorylation after Experimental Intracerebral Hemorrhage in Rats. *Neurobiol. Dis.* **2018**, *110*, 82–92. [CrossRef]
42. Mola, M.G.; Saracino, E.; Formaggio, F.; Amerotti, A.G.; Barile, B.; Posati, T.; Cibelli, A.; Frigeri, A.; Palazzo, C.; Zamboni, R.; et al. Cell Volume Regulation Mechanisms in Differentiated Astrocytes. *Cell. Physiol. Biochem.* **2021**, *55*, 196–212. [CrossRef] [PubMed]
43. O'Connor, E.; Kimelberg, H. Role of Calcium in Astrocyte Volume Regulation and in the Release of Ions and Amino Acids. *J. Neurosci.* **1993**, *13*, 2638–2650. [CrossRef] [PubMed]
44. van der Knaap, M.S.; Boor, I.; Estévez, R. Megalencephalic Leukoencephalopathy with Subcortical Cysts: Chronic White Matter Oedema Due to a Defect in Brain Ion and Water Homoeostasis. *Lancet Neurol.* **2012**, *11*, 973–985. [CrossRef]
45. Goswami, C.; Kuhn, J.; Heppenstall, P.A.; Hucho, T. Importance of Non-Selective Cation Channel TRPV4 Interaction with Cytoskeleton and Their Reciprocal Regulations in Cultured Cells. *PLoS ONE* **2010**, *5*, e11654. [CrossRef] [PubMed]
46. Saberbaghi, T.; Wong, R.; Rutka, J.T.; Wang, G.-L.; Feng, Z.-P.; Sun, H.-S. Role of Cl- Channels in Primary Brain Tumour. *Cell Calcium* **2019**, *81*, 1–11. [CrossRef] [PubMed]
47. Duran, C.; Chien, L.-T.; Hartzell, H.C. Drosophila Bestrophin-1 Currents Are Regulated by Phosphorylation via a CaMKII Dependent Mechanism. *PLoS ONE* **2013**, *8*, e58875. [CrossRef]
48. Cuddapah, V.A.; Sontheimer, H. Molecular Interaction and Functional Regulation of ClC-3 by Ca2+/Calmodulin-Dependent Protein Kinase II (CaMKII) in Human Malignant Glioma. *J. Biol. Chem.* **2010**, *285*, 11188–11196. [CrossRef]
49. Jentsch, T.J.; Lutter, D.; Planells-Cases, R.; Ullrich, F.; Voss, F.K. VRAC: Molecular Identification as LRRC8 Heteromers with Differential Functions. *Pflügers Arch. Eur. J. Physiol.* **2016**, *468*, 385–393. [CrossRef]
50. Bertelli, S.; Remigante, A.; Zuccolini, P.; Barbieri, R.; Ferrera, L.; Picco, C.; Gavazzo, P.; Pusch, M. Mechanisms of Activation of LRRC8 Volume Regulated Anion Channels. *Cell. Physiol. Biochem.* **2021**, *55*, 41–56. [CrossRef]
51. Capdevila-Nortes, X.; López-Hernández, T.; Apaja, P.M.; López de Heredia, M.; Sirisi, S.; Callejo, G.; Arnedo, T.; Nunes, V.; Lukacs, G.L.; Gasull, X.; et al. Insights into MLC Pathogenesis: GlialCAM Is an MLC1 Chaperone Required for Proper Activation of Volume-Regulated Anion Currents. *Hum. Mol. Genet.* **2013**, *22*, 4405–4416. [CrossRef]
52. Catacuzzeno, L.; Michelucci, A.; Sforna, L.; Aiello, F.; Sciaccaluga, M.; Fioretti, B.; Castigli, E.; Franciolini, F. Identification of Key Signaling Molecules Involved in the Activation of the Swelling-Activated Chloride Current in Human Glioblastoma Cells. *J. Membr. Biol.* **2014**, *247*, 45–55. [CrossRef] [PubMed]

53. Mongin, A.A.; Kimelberg, H.K. ATP Regulates Anion Channel-Mediated Organic Osmolyte Release from Cultured Rat Astrocytes via Multiple Ca $^{2+}$ -Sensitive Mechanisms. *Am. J. Physiol.-Cell Physiol.* **2005**, *288*, C204–C213. [CrossRef] [PubMed]
54. Voss, F.K.; Ullrich, F.; Münch, J.; Lazarow, K.; Lutter, D.; Mah, N.; Andrade-Navarro, M.A.; von Kries, J.P.; Stauber, T.; Jentsch, T.J. Identification of LRRC8 heteromers as an essential component of the volume-regulated anion channel VRAC. *Science* **2014**, *344*, 634–638. [CrossRef]
55. Qiu, Z.; Dubin, A.E.; Mathur, J.; Tu, B.; Reddy, K.; Miraglia, L.J.; Reinhardt, J.; Orth, A.P.; Patapoutian, A. SWELL1, a plasma membrane protein, is an essential component of volume-regulated anion channel. *Cell* **2014**, *157*, 447–458. [CrossRef] [PubMed]
56. Elorza-Vidal, X.; Sirisi, S.; Gaitán-Peñas, H.; Pérez-Rius, C.; Alonso-Gardón, M.; Armand-Ugón, M.; Lanciotti, A.; Brignone, M.S.; Prat, E.; Nunes, V.; et al. GlialCAM/MLC1 modulates LRRC8/VRAC currents in an indirect manner: Implications for megalencephalic leukoencephalopathy. *Neurobiol. Dis.* **2018**, *119*, 88–99. [CrossRef] [PubMed]
57. Han, Y.E.; Kwon, J.; Won, J.; An, H.; Jang, M.W.; Woo, J.; Lee, J.S.; Park, M.G.; Yoon, B.E.; Lee, S.E.; et al. Tweety-homolog (Ttyh) Family Encodes the Pore-forming Subunits of the Swelling-dependent Volume-regulated Anion Channel (VRACswell) in the Brain. *Exp. Neurobiol.* **2019**, *28*, 183–215. [CrossRef]
58. Schaeffer, S.; Iadecola, C. Revisiting the Neurovascular Unit. *Nat. Neurosci.* **2021**, *24*, 1198–1209. [CrossRef]
59. Semyanov, A. Spatiotemporal Pattern of Calcium Activity in Astrocytic Network. *Cell Calcium* **2019**, *78*, 15–25. [CrossRef]
60. Fujii, Y.; Maekawa, S.; Morita, M. Astrocyte Calcium Waves Propagate Proximally by Gap Junction and Distally by Extracellular Diffusion of ATP Released from Volume-Regulated Anion Channels. *Sci. Rep.* **2017**, *7*, 13115. [CrossRef]
61. Lanciotti, A.; Brignone, M.S.; Belfiore, M.; Columba-Cabezas, S.; Mallozzi, C.; Vincentini, O.; Molinari, P.; Petrucci, T.C.; Visentin, S.; Ambrosini, E. Megalencephalic Leukoencephalopathy with Subcortical Cysts Disease-Linked MLC1 Protein Favors Gap-Junction Intercellular Communication by Regulating Connexin 43 Trafficking in Astrocytes. *Cells* **2020**, *9*, 1425. [CrossRef]
62. Lind, B.L.; Jessen, S.B.; Lønstrup, M.; Joséphine, C.; Bonvento, G.; Lauritzen, M. Fast Ca $^{2+}$ Responses in Astrocyte End-Feet and Neurovascular Coupling in Mice. *Glia* **2018**, *66*, 348–358. [CrossRef] [PubMed]
63. Eilert-Olsen, M.; Hjukse, J.B.; Thoren, A.E.; Tang, W.; Enger, R.; Jensen, V.; Pettersen, K.H.; Nagelhus, E.A. Astroglial Endfeet Exhibit Distinct Ca $^{2+}$ Signals during Hypoosmotic Conditions. *Glia* **2019**, *67*, 2399–2409. [CrossRef] [PubMed]
64. Gilbert, A.; Elorza-Vidal, X.; Rancillac, A.; Chagnot, A.; Yetim, M.; Hingot, V.; Deffieux, T.; Boulay, A.-C.; Alvear-Perez, R.; Cisternino, S.; et al. Megalencephalic Leukoencephalopathy with Subcortical Cysts Is a Developmental Disorder of the Gliovascular Unit. *eLife* **2021**, *10*, e71379. [CrossRef]
65. Morales, J.E.; De, A.; Miller, A.A.; Chen, Z.; McCarty, J.H. Mlc1-Expressing Perivascular Astrocytes Promote Blood-Brain Barrier Integrity. *J. Neurosci.* **2022**, *42*, 1406–1416. [CrossRef] [PubMed]
66. Kane, J.R. The Role of Brain Vasculature in Glioblastoma. *Mol. Neurobiol.* **2019**, *56*, 6645–6653. [CrossRef]
67. Verkhratsky, A. Astroglial Calcium Signaling in Aging and Alzheimer's Disease. *Cold Spring Harb. Perspect. Biol.* **2019**, *11*, a035188. [CrossRef]

cells

Article

Inward Operation of Sodium-Bicarbonate Cotransporter 1 Promotes Astrocytic Na$^+$ Loading and Loss of ATP in Mouse Neocortex during Brief Chemical Ischemia

Katharina Everaerts [1,#], Pawan Thapaliya [2,#], Nils Pape [1], Simone Durry [1], Sara Eitelmann [1], Eleni Roussa [3], Ghanim Ullah [2] and Christine R. Rose [1,*]

1 Institute of Neurobiology, Heinrich Heine University Düsseldorf, Universitätsstraße 1, D-40225 Düsseldorf, Germany; katharina.everaerts@hhu.de (K.E.); nils.pape@hhu.de (N.P.); simone.durry@hhu.de (S.D.); sara.eitelmann@hhu.de (S.E.)
2 Department of Physics, University of South Florida, Tampa, FL 33620, USA; pkthapaliya@usf.edu (P.T.); gullah@usf.edu (G.U.)
3 Institute of Anatomy and Cell Biology, Department of Molecular Embryology, Faculty of Medicine, Albert-Ludwigs-Universität Freiburg, Albertstrasse 17, D-79104 Freiburg, Germany; eleni.roussa@anat.uni-freiburg.de
* Correspondence: rose@hhu.de
\# These authors contributed equally to this work.

Abstract: Ischemic conditions cause an increase in the sodium concentration of astrocytes, driving the breakdown of ionic homeostasis and exacerbating cellular damage. Astrocytes express high levels of the electrogenic sodium-bicarbonate cotransporter1 (NBCe1), which couples intracellular Na$^+$ homeostasis to regulation of pH and operates close to its reversal potential under physiological conditions. Here, we analyzed its mode of operation during transient energy deprivation via imaging astrocytic pH, Na$^+$, and ATP in organotypic slice cultures of the mouse neocortex, complemented with patch-clamp and ion-selective microelectrode recordings and computational modeling. We found that a 2 min period of metabolic failure resulted in a transient acidosis accompanied by a Na$^+$ increase in astrocytes. Inhibition of NBCe1 increased the acidosis while decreasing the Na$^+$ load. Similar results were obtained when comparing ion changes in wild-type and *Nbce1*-deficient mice. Mathematical modeling replicated these findings and further predicted that NBCe1 activation contributes to the loss of cellular ATP under ischemic conditions, a result confirmed experimentally using FRET-based imaging of ATP. Altogether, our data demonstrate that transient energy failure stimulates the inward operation of NBCe1 in astrocytes. This causes a significant amelioration of ischemia-induced astrocytic acidification, albeit at the expense of increased Na$^+$ influx and a decline in cellular ATP.

Keywords: astrocyte; sodium; pH; neocortex; ischemia; ATP; imaging; modeling; excitotoxicity

Citation: Everaerts, K.; Thapaliya, P.; Pape, N.; Durry, S.; Eitelmann, S.; Roussa, E.; Ullah, G.; Rose, C.R. Inward Operation of Sodium-Bicarbonate Cotransporter 1 Promotes Astrocytic Na$^+$ Loading and Loss of ATP in Mouse Neocortex during Brief Chemical Ischemia. *Cells* **2023**, *12*, 2675. https://doi.org/10.3390/cells12232675

Academic Editors: Vladimir Parpura and Michael Brenner

Received: 6 July 2023
Revised: 14 November 2023
Accepted: 16 November 2023
Published: 21 November 2023

1. Introduction

Astrocytes are central to brain function. Long-established roles of astrocytes include the uptake of K$^+$ from the extracellular space (ECS) and the regulation of extracellular K$^+$ homeostasis [1]. Uptake of K$^+$ is mainly mediated by the astrocytic Na$^+$/K$^+$-ATPase (NKA), a major consumer of cellular ATP [2,3]. In addition to its role in extracellular K$^+$ homeostasis, the NKA is responsible for the export of Na$^+$ and the maintenance of a low intracellular Na$^+$ concentration ([Na$^+$]$_i$) [4]. Astrocytes, moreover, play a key role in the uptake and regulation of the neurotransmitter glutamate, which is mediated by the Na$^+$-dependent high-affinity transporters GLAST and GLT-1 (EAAT1 and EAAT2, respectively) [5,6]. Both processes rely on an intact energy metabolism and ATP production to enable NKA activity and the maintenance of a strong inward gradient for Na$^+$.

In the core region of an ischemic stroke, intracellular ATP levels fall, and astrocytes suffer from a breakdown of ionic homeostasis, including an increase in their $[Na^+]_i$ [7–9]. Prominent, albeit more transient, changes in astrocytic $[Na^+]_i$ are observed in the ischemic penumbra during the passage of a wave of spreading depolarization [9–11]. Earlier work has shown that an increase in $[Na^+]_i$ results in a reduced capacity for glutamate uptake or even in the reversal of glutamate transporters and reduced glial uptake of K^+ [12–14]. Moreover, Na^+ loading promotes the reversal of the plasma membrane Na^+/Ca^{2+}-exchanger (NCX), thereby contributing to the deleterious Ca^{2+} loading of astrocytes [10,15–18].

Another Na^+-dependent transporter whose driving force is affected by changes in astrocytic $[Na^+]_i$ is the electrogenic sodium-bicarbonate cotransporter 1 (NBCe1; *SLC4A4*) [19]. Outward transport of HCO_3^- via the NBCe1 is a major mechanism for regulation of intracellular pH (pH_i) in astrocytes when pH_i is above 6.8 [20–22]. Depending on the cellular membrane potential and the respective ion concentrations, it can mediate either the influx of Na^+ and HCO_3^- (forward/inward mode) or their efflux (reverse/outward mode). While the forward operation of NBCe1 is activated, for example, in response to neuronal activity and an increase in the extracellular K^+ concentration ($[K^+]_o$) [23–26], its reverse operation can be induced by a decrease in extracellular HCO_3^- ($[HCO_3^-]_o$) or by inhibition of the NKA and the concomitant increase in astrocytic $[Na^+]_i$ [27,28]. In addition to ionic driving forces, NBCe1 is modulated via several intracellular signaling pathways that influence its activity [29,30].

The mode of operation of NBCe1 in ischemic conditions, however, is unclear. In the ischemic brain, cells not only undergo a membrane depolarization and an increase in $[Na^+]_i$, but also a long-lasting acidification [7–9,11,31,32], and all of these processes will strongly affect the driving forces for NBCe1. It is therefore unclear whether NBCe1 activity promotes (forward mode) or dampens (reverse mode) astrocytic Na^+ loading. Furthermore, it is unclear if (and how) NBCe1 activity contributes to astrocytic pH_i changes in ischemic conditions.

To address these questions, we performed fluorescence imaging with the chemical indicator dyes SBFI and BCECF to measure changes in astrocytic $[Na^+]_i$ and pH_i, respectively, in organotypic slice cultures of the murine neocortex. Astrocytic ATP levels were analyzed using the genetically encoded fluorescent sensor ATeam1.03^{YEMK} [33]. The involvement of NBCe1 activity was probed using pharmacological tools and by comparing slice preparations derived from wild-type and *Nbce1*-deficient animals. Experimental results allowed for a comprehensive mathematical simulation of ion fluxes and NBCe1 activity during metabolic failure and enabled the prediction of NBCe1-related ATP consumption in astrocytes. Our experimental results and computational modeling demonstrate that a 2 min transient energy failure stimulates inwardly directed NBCe1 activity in astrocytes. This increases Na^+ influx and astrocytic ATP consumption while at the same time dampening their acidification.

2. Materials and Methods

2.1. Preparation of Organotypic Slice Cultures

In the present study, we used wild-type mice as well as transgenic, *Nbce1*-deficient mice (NBCe1 KO [34]; originally received from Dr. Gary E. Shull, UC College of Medicine, University of Cincinnati, Cincinnati, OH, USA). For the preparation of organotypic hippocampal slice cultures, mice (*Mus musculus*, Balb/C; both sexes) ranging from postnatal days (P) 6–9 were used, on which most experiments were performed. In addition, neonatal NBCe1 KO mice were used from P5 to 6 [34]. C57BL/6N mice (both sexes) served as wild-type controls for these animals.

Acute slices of 250 µm thickness were prepared using the methods previously published [35]. Briefly, mice were quickly decapitated, and brains were immediately placed in ice-cold standard artificial cerebrospinal fluid (aCSF) containing (in mM): 130 NaCl, 2.5 KCl, 2 $CaCl_2$, 1 $MgCl_2$, 1.25 NaH_2PO_4, 26 $NaHCO_3$, and 10 glucose, which was bubbled with 5% CO_2/95% O_2, resulting in a pH of 7.35–7.4. Brains were separated into hemispheres and cut parasagittally using a vibratome (HM 650 V; Thermo Fisher Scientific, Waltham, MA, USA). Acute slices were cultured following a procedure by Stoppini [36] and as described in more detail recently [37]. Slice cultures were kept in an incubator at 36 °C at the interface between humidified carbogen (5% CO_2/95% O_2) and culture medium containing minimum essential medium (MEM; M7M278), 20% heat-inactivated horse serum (Origin, Brazil; Thermo Fisher Scientific, Waltham, MA, USA), 1 mM of L-glutamine, 0.01 mg mL^{-1} of insulin, 14.5 mM of NaCl, 2 mM of $MgSO_4$, 1.44 mM of $CaCl_2$, 0.00125% ascorbic acid, and 13 mM of D-glucose. The medium was replaced every 3 days.

If not stated otherwise, experiments were performed on organotypic slices cultured between 10 and 21 days. For astrocyte identification, slices were exposed to 0.5–1 μM of sulforhodamine 101 (SR101) in aCSF for 30–60 min in the incubator. Experiments were carried out at room temperature (21 ± 1 °C). Throughout experiments, slices were perfused with aCSF containing 0.5 μM of tetrodotoxin (TTX, Biotrend, Cologne, Germany) to block action potential discharges, typical for organotypic slice preparations [38]. Pharmacological substances were diluted in aCSF and bath-applied via the perfusion system for 15 min before the beginning of and throughout the recordings. Brief chemical ischemia was induced via a 2 min bath application of glucose-free aCSF containing sodium azide (NaN_3, 5 mM; inhibitor of cytochrome C oxidase and hence mitochondrial respiration) as well as 2-deoxyglucose (2-DG, 2 mM; inhibitor of hexokinase and thus glycolysis). If not stated otherwise, chemicals were purchased from Sigma-Aldrich (Munich, Germany).

2.2. Imaging of Intracellular pH and Na^+

For the determination of changes in pH_i or $[Na^+]_i$, we performed wide-field imaging using an epifluorescence microscope (Nikon Eclipse FN 1, Nikon Europe, Düsseldorf, Germany), equipped with a Fluor 40×/0.8 NA water immersion objective (Nikon) coupled to a Poly-V monochromator (Thermo Scientific/FEI, Planegg, Germany). After staining for astrocytes with SR101, organotypic slice cultures were bolus-loaded with fluorescent indicators for the respective ion, as reported before [35].

Changes in pH_i were determined by loading the slices with BCECF (BCECF-AM, 125 μM, A.G. Scientific, San Diego, CA, USA). Standard dual excitation ratiometric imaging of BCECF was performed by alternating excitation at 458 (isosbestic wavelength) and 488 nm (pH-sensitive wavelength) and recording fluorescence emission ranging between 518 and 563 nm using a 518 beam splitter and a 537/26 band pass filter.

Changes in $[Na^+]_i$ were determined using SBFI (SBFI-AM, 116.7 μM, ION Biosciences, San Marcos, TX, USA). SBFI was excited at 400 nm, and its emission was detected above ~430 nm using a 409 beam splitter and a band pass 510/84 nm emission filter [10]. Images were obtained at 0.5–1 Hz with an ORCA FLASH 4.0LT camera (Hamamatsu Photonics, Herrsching, Germany). Emission was collected from regions of interest (ROIs) representing the cell bodies of SR101-positive astrocytes and analyzed using OriginPro 2019 (OriginLab Corporation, Northampton, MA, USA). Fluorescence emission from individual ROIs was background-corrected as reported earlier [35]. Afterwards, the fluorescence ratio (F_{458}/F_{488}) was calculated for BCECF.

Changes in SBFI emission and BCECF ratio were converted into mM (Na^+) and pH units, respectively, using established in situ calibration procedures [10,39,40]. In brief, for the calibration of SBFI, SBFI -loaded slices were perfused with calibration salines containing $[Na^+]$ ranging from 0 to 150 mM, as well as the NKA inhibitor ouabain (100 μM), the Na^+/H^+ exchanger monensin (10 μM), and the Na^+ ionophore gramicidin (3 μM) to equilibrate intra- and extracellular Na^+. Calibration was started by perfusing the slices with nominally Na^+-free calibration saline. Subsequently, slices were exposed to calibration solutions containing 5–150 mM Na^+ before going back to nominally Na^+-free calibration saline. Detected changes in SBFI fluorescence were normalized to the initial baseline in nominally Na^+-free calibration saline and plotted against the respective Na^+ concentration. The data followed a Michaelis-Menten relationship (R^2 = 0.975), revealing an apparent K_D of about 32 mM (not illustrated). This calibration curve enabled a calculation of changes in astrocytic $[Na^+]_i$, assuming a baseline $[Na^+]_i$ of 12.1 mM as determined for neocortical astrocytes in a recent study from our laboratory (baseline $[Na^+]_i$: 12.1 ± 0.5 mM [15]).

For calibration of BCECF, we used calibration salines from pH 5.5 to 8.5 (0.5 increments) containing the K^+/H^+ exchanger nigericin (10 μM) to equilibrate intra- and extracellular pH. Calibration was started by perfusing BCECF-loaded slices with a calibration saline at pH 5.5 and then switching to other calibration salines. Mean changes in the astrocytic BCECF ratio (F_{458}/F_{488}) were recorded, normalized to pH 5.5, and then plotted against the respective pH. A linear fit between pH 6.5 and 8.0 (R^2 = 0.982; not illustrated) enabled the calculation of changes in astrocytic pH_i, assuming a baseline pH_i of 7.33 as determined for neocortical astrocytes in a recent study from our laboratory (baseline pH_i: 7.33 ± 0.04 [41]). The data were analyzed further offline using OriginPro 2019.

Of note, our in situ calibrations allow a reliable and reproducible calculation of absolute changes in ion concentrations ($[Na^+]_i$, pH_i) at a given rig at which experiments are performed. However, initial tissue swelling and movement upon perfusion with calibration salines prevent a direct readout of cellular baseline ion concentrations in physiological saline. Determination of the latter thus requires alternative strategies and experimental approaches as described before (e.g., [15,41]). Importantly, the range of the near-linear behavior of both dyes (SBFI and BCECF) fully covers and even exceeds the range ± S.D. of baseline $[Na^+]_i$ (12.1 ± 0.5 mM) and pH_i (7.33 ± 0.04) determined earlier (e.g., [15,41]). This means that even if baseline ion concentrations differ between different cells to some extent, this will not distort the calculation made of the absolute changes in ion concentrations presented throughout the present manuscript.

2.3. FRET-Based Imaging of Intracellular ATP

For the determination of changes in astrocytic ATP levels in organotypic slice cultures, we employed the genetically encoded, fluorescence resonance energy transfer (FRET)-based sensor ATeam1.03^{YEMK} (ATeam) [33]. To this end, 0.5 μL of a vector (AAV 2/5) carrying the coding sequence for ATeam under the transcriptional control of the human glial fibrillary acid protein (*hGFAP*) promoter fragment ABC1D was applied to the top of a cultured slice at 1–3 days in vitro as described before [42]. After transduction, slices were maintained in the incubator for at least 10 more days before being used for experiments.

Slices expressing ATeam were imaged using an upright microscope (Nikon Eclipse FN-I, Nikon GmbH Europe, Düsseldorf, Germany) equipped with an Achroplan 40×/0.8 NA water immersion objective (Nikon). ATeam was excited using a Poly-V monochromator (Thermo Fisher Scientific/FEI, Planegg, Germany) at 434 nm, and images were taken at 0.5 Hz with a CMOS camera (Orca 4 LT Plus, Hamamatsu Photonics, Herrsching, Germany). Fluorescence emission was split at ~500 nm using a 505 beam splitter (WVIEW GEMINI optic system; Hamamatsu Photonics, Herrsching, Germany) onto two band pass filters (483/32 nm, imaging of the enhanced cyan fluorescent protein (*eCFP*)/donor; and 542/27 nm, imaging of Venus/acceptor). After background correction, the fluorescence ratio (Venus/eCFP) was calculated for individual ROIs, representing the cell bodies of astrocytes. Subsequent analysis was performed using OriginPro 2021 software (OriginLab Corporation, Northampton, MA, USA). Normalized changes in the Venus/eCFP fluorescence ratio are given as a percentage change thereof (ΔATeam ratio (%)).

2.4. Measurement of Extracellular K^+, pH, and Na^+

Measurements of extracellular ion shifts were obtained by employing double-barreled ion-sensitive microelectrodes as described before (e.g., [41,43]). In brief, two thin-walled borosilicate glass capillaries with filament were glued and pulled out together. The tip of one capillary was silanized via exposure to vaporized hexamethyldisilazane (Fluka, Buchs, Switzerland) and filled with a liquid neutral ion carrier based on valinomycin for measurement of $[K^+]_o$ (Ionophore I, Cocktail B, Fluka, Buchs, Switzerland), Hydrogen Ionophore I for pH (Cocktail A, 95291, Merck, Darmstadt, Germany), or ETH 157 for $[Na^+]_o$ (Ionophore II, Cocktail A, Fluka, Buchs, Switzerland). Afterwards, the capillary was backfilled with 100 mM NaCl or 100 mM KCl. The reference electrode was filled with 150 mM NaCl/1 mM HEPES (titrated to pH 7.0 with NaOH).

Electrodes were calibrated directly before and after each individual experiment. Calibration of K^+-sensitive electrodes was performed in HEPES-based saline containing a total of 150 mM NaCl and KCl, in which KCl was 0–10 mM and NaCl adjusted accordingly to maintain osmolarity. For calibration of Na^+-sensitive electrodes, solutions with Na^+ concentrations ranging from 70 to 160 mM, to which N-methyl-D-glucamine chloride (NMDG-Cl) was added to adjust osmolarity (326 ± 5 mOsm l^{-1}), were used. pH-sensitive microelectrodes were calibrated with salines titrated to a pH of 7.0 or 7.6, containing (in mM): 144.25 NaCl (pH 7.0)/108.48 NaCl (pH 7.6), 2.5 KCl, 1.25 NaH_2PO_4, and 12 $NaHCO_3$ (pH 7.0)/47.77 $NaHCO_3$ (pH 7.6), bubbled with carbogen. The data were processed in OriginPro 2019.

2.5. Patch-Clamp Recordings

Changes in astrocytic membrane potential were monitored using cell-attached patch-clamp recordings. These were carried out at an upright microscope (E600FN, Nikon, Tokyo, Japan), which was equipped with infrared differential interference contrast optics. The latter included a 60× water immersion objective (Fluor 60×/1.00 W, DIC H/N2, ∞/0 WD 2.0, Nikon) and an infrared video camera (XC ST70CE, Hamamatsu Photonics, Herrsching, Germany). Patch pipettes with a resistance of 3.5–4.0 MΩ (when filled with external solution) were pulled from borosilicate glass capillaries (GB150(F) 8P, Science Products, Hofheim am Taunus, Germany) using a vertical puller (PC-10 Puller, Narishige International, London, UK). Electrophysiological measurements were performed using an EPC10 amplifier and "PatchMaster" software (Harvard Bioscience/HEKA Elektronik, Lambrecht, Germany). Recordings in cell-attached configuration were conducted as reported recently [41]. In brief, pipettes were filled with standard aCSF, and the offset potential was corrected. After achieving a seal higher than 1 GΩ, cells were recorded for 30–45 min to ensure reliable measurement of the membrane potential [44]. The data were analyzed offline using OriginPro 2021.

2.6. Modeling NBCe1 Activity

We build on our previous work, where we modeled dynamic changes in astrocytic Na^+, K^+, Cl^-, and Ca^{2+} concentrations [45,46] by incorporating the main pathways regulating intra- and extracellular pH (pH_i and pH_o, respectively). The equations for intra- and extracellular Na^+ concentrations ($[Na^+]_i$ and $[Na^+]_o$, respectively) are modified accordingly. The rate equations for various state variables and related fluxes are detailed in Appendix A. A schematic of the model is shown in the results section (see Figure 4). Here, we only discuss equations related to pH.

pH_o and astrocyte pH_i are controlled with fluxes through NBCe1 (J_{NBCe1}), sodium/proton exchanger (J_{NHE}), and diffusion between the ECS and bath solution, that is

$$\frac{dpH_o}{dt} = \frac{1}{VR_{sa}\beta}(-J_{NHE} - J_{NBCe1}) + diff(pH_{bath} - pH_o) \quad (1)$$

$$\frac{dpH_i}{dt} = \frac{J_{NHE} + J_{NBCe1}}{\beta}, \quad (2)$$

where diff is the rate at which pH_o equilibrates with the bath solution (pH_{bath}) and VR_{sa} is the ECS to astrocytic volume ratio.

$\beta = \beta_{i/o} + \frac{2.3[HCO_3^-]_{i/o}}{[pH_{rest}]_{i/o}}$ is the total pH buffering capacity, where $\beta_{i/o} = \Delta[HCO_3^-]_{i/o}/\Delta pH$ is the intrinsic buffering capacity. $[HCO_3^-]_{i/o}$ is the intra- and extracellular bicarbonate concentration, respectively, and $[pH_{rest}]_{i/o}$ is the resting pH of the intra- (7.33) and extracellular (7.35) regions in our experiments. ΔpH is the maximum observed change in pH with respect to the resting state inside or outside the cell. In our experiments, the highest change occurred in pH_i (that is, the maximum change in pH_o is lower), thus we used the maximum observed change in pH_i as ΔpH. We observed that the peak change in $[HCO_3^-]_i$, $[HCO_3^-]_o$, and pH_i with respect to their resting values was 12 mM, 5 mM, and 0.48 pH units, respectively. All these considerations led to intra- and extracellular intrinsic buffering powers of 25 mM/pH and 10.5 mM/pH, respectively. Notice that $\beta_{i/o}$ is the slope of the $[HCO_3^-]$ versus pH curve. Thus, our estimate of $\beta_{i/o}$ is an average approximation as we look at the maximum changes in pH and $[HCO_3^-]$ in our experiments. To measure the value $\beta_{i/o}$ as a continuous function of pH, one would have to measure $[HCO_3^-]$ at multiple pH values and find the slope of the curve at different pH values.

The second expression in β represents the buffering power of the intra- and extracellular regions in an open system at a steady-state pH due to CO_2. Our model assumes fixed CO_2, so carbonic acid (H_2CO_3) concentration does not change. We remark that replacing $[pH_{rest}]_{i/o}$ in the second term in β with pH unit (i.e., removing $[pH_{rest}]_{i/o}$ from the equation) does not change our results significantly. Thus, we leave the equation unperturbed. Furthermore, the effect of surrounding cells other than the astrocyte modeled on pH_o is incorporated in the diffusion term. However, some pathways regulating pH_i, including H^+ and HCO_3^- leaks, background acid loading, and CO_2 dissociation and respiration, are not included, as the goal was to develop a simple model capable of reproducing our observations. However, as more data on the pH regulation mechanism of astrocytes emerge, incorporating these and other pH-regulating pathways in the model will be crucial, which is the subject of our future research.

The equations for J_{NBCe1} and J_{NHE} are similar to those used in [47]. That is,

$$J_{NBCe1} = G_{NBCe1}(v_i - E_{NBCe1}), \tag{3}$$

where E_{NBCe1}, v_i, and G_{NBCe1} are the reversal potential for Na^+ and HCO_3^- flux through NBCe1, the membrane potential of astrocytes, and the whole-cell conductance of NBCe1, respectively. E_{NBCe1} is calculated using the Nernst equation.

$$E_{NBCe1} = \frac{V_T}{z_{NBCe1}} \ln\left(\frac{[Na^+]_o}{[HCO_3^-]_i^2} \frac{[HCO_3^-]_o^2}{[Na^+]_i} \right). \tag{4}$$

$V_T = \frac{RT}{F}$, where R, T, and F represent the gas constant, temperature, and Faraday's constant and Z_{NBCe1} represents the net charge transported. $[HCO_3^-]_i$ is the intracellular bicarbonate concentration. J_{NHE} is given as

$$J_{NHE} = G_{NHE}(v_i - E_{NHE}), \tag{5}$$

where G_{NHE} is the whole-cell conductance of NHE and E_{NHE} is its reversal potential.

$$E_{NHE} = \frac{V_T}{z_{NHE}} \ln\left(\frac{[Na^+]_i [H^+]_o}{[Na^+]_o [H^+]_i} \right). \tag{6}$$

$[H^+]_o$ and $[H^+]_i$ represent extra- and intracellular hydrogen concentrations, respectively. Z_{NBCe1} is the net charge transported using NHE.

$[HCO_3^-]_o$ is given using the Henderson-Hasselbalch equation [48],

$$[HCO_3^-]_o = 10^{(pH_o - pK_a)} \left[CO_{2(aq)}\right], \tag{7}$$

where pK_a is the negative logarithm (base = 10) of the acid dissociation constant of carbonic acid, and $[CO_{2(aq)}]$ is the product of solubility (s) in aqueous solution or water and partial pressure of carbon dioxide (P_{CO2}). Similarly, $[HCO_3^-]_i$ is given as

$$[HCO_3^-]_i = 10^{(pH_o - pH_i)} [HCO_3^-]_o. \tag{8}$$

$[H^+]_o$ and $[H^+]_i$ are calculated using the Kassirer–Bleich approximation [48],

$$[H^+]_{i/o} = \frac{sK_h P_{CO_2}}{[HCO_3^-]_{i/o}}, \quad (9)$$

where K_h is the dissociation constant of carbonic acid.

The equations for NKA are modified to simulate 2 min chemical ischemia as explained in Appendix A.

2.7. Numerical Methods

The rate equations are solved in Fortran 90 using the Euler method with a time step of 0.1 μs. The system of equations is allowed to reach a steady state before imposing chemical ischemia. The data are visualized using MATLAB (MATLAB Version: 9.14.0.2286388 (R2023a) Update 3, The MathWorks Inc., Natick, MA, USA). The codes reproducing the main results are available from the authors upon request.

2.8. Data Analysis and Statistics

Each series of experiments on tissue slices was prepared from at least three different animals; "n" represents the number of cells analyzed and "N" the number of individual experiments/slices. In the results section, numbers for individual experiments are stated as follows: number of cells investigated/number of different slices/number of different animals. Power analysis was conducted using G*Power 3.1.9.6 [49]. The effect size was calculated as the difference of means divided by the standard deviation of the control group. For a significance level of 0.001, it revealed a minimum power of 0.8. Results are given as the mean ± standard deviation (SD). Data are illustrated in box plots indicating the mean (black square), interquartile range (box), median (middle line), and SD (whiskers). Additionally, all individual data points are shown underneath box plots as dots. The data were statistically analyzed using OriginPro 2019. The normality of the data was assessed using a Shapiro-Wilk test. In the case of a normal distribution, the data were statistically analyzed using either a Student's *t*-test ("*t*-test") or a one-way ANOVA ("ANOVA"), followed by a post hoc Bonferroni test. Otherwise, statistical analysis was conducted using a U-test (Mann-Whitney, "MWU"). *p*-values below 0.05 were considered to indicate a significant difference. The following symbols are used to illustrate the results of statistical tests in the figures (*p* represents the error probability): * $p < 0.05$, ** $p < 0.01$, and *** $p < 0.001$.

3. Results

3.1. Probing NBCe1 Activity in Organotypic Slice Cultures

The goal of the present study was to reveal the mode of operation of NBCe1 in astrocytes of the mouse neocortex during short periods of energy failure and to analyze its effects on ischemia-induced changes in astrocytic pH_i and $[Na^+]_i$. To test whether NBCe1 is functionally active in organotypic slice cultures and can be addressed pharmacologically, we first probed for a Depolarization-Induced *A*lkalinization ("DIA") of astrocytes, which is based on stimulation of inward NBCe1 activity [50,51].

To depolarize astrocytes, we elevated the K^+ concentration of the aCSF from 2.5 to 10 mM for 2 min. Recordings with K^+-sensitive microelectrodes showed that this resulted in a transient increase in the $[K^+]_o$ within the slice preparation from 2.5 to 8.6 mM (3/3/3; not illustrated). Raising $[K^+]_o$ indeed resulted in an alkalinization of astrocytes, amounting to on average 0.26 ± 0.07 pH units (81/7/4) (Figure 1a). The amplitude of this alkalinization remained stable upon a second elevation of $[K^+]_o$, performed 30 min after the first one (amplitude at second application: 0.27 ± 0.09 pH units; 81/7/4; MWU, $p = 0.221$) (Figure 1a).

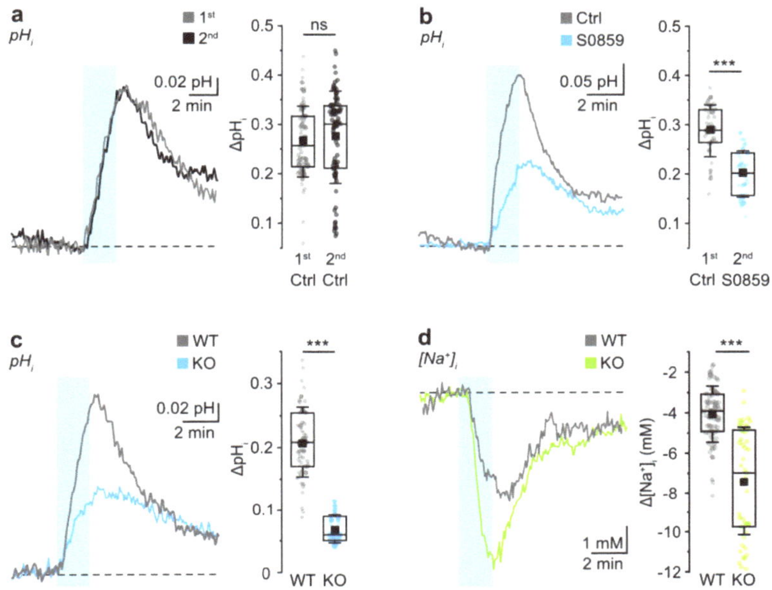

Figure 1. K^+-induced changes in astrocytic pH_i and $[Na^+]_i$. The left panels of (**a–d**) show representative traces from single experiments; the right panels show data for all the experiments. (**a**) Left: Change in pH_i evoked by two consecutive increases in $[K^+]_o$ for 2 min each. (**b**) Left: Change in pH_i evoked by two consecutive increases in $[K^+]_o$ for 2 min in control conditions (1st application, Ctrl) and in the presence of the NBC inhibitor S0859 (2nd application). (**c**) Left: Change in pH_i evoked by an increase in $[K^+]_o$ for 2 min in WT and in NBCe1 KO animals. (**d**) Left: Change in astrocytic $[Na^+]_i$ evoked by an increase in $[K^+]_o$ for 2 min in WT and in NBCe1 KO animals. Periods of increased $[K^+]_o$ are indicated with the light blue areas; upward traces represent alkalinizations. Dashed lines indicate baseline levels. Box plots on the right summarize the results from these experiments and show individual data points at maximum change (dots), means (squares), interquartile ranges (boxes), medians (lines), and SD (whiskers). ns: non-significant; *** $p < 0.001$.

To probe for the involvement of sodium-bicarbonate co-transport (NBC) in this alkalinization, we applied S0859 (30 μM), an inhibitor of the NBC transporter family [52]. At an assumed average baseline pH_i of 7.33 ([41], see methods), we found that pH_i was 0.01 ± 0.10 pH units lower after the addition of S0859 than in the control. Thus, S0859 did not alter pH_i significantly compared to the control prior to its addition (control: 7.33 ± 0.04, 8/5/4; S0589: 7.32 ± 0.09, 59/5/5; *t*-test, $p = 0.592$). However, in the presence of S0859, the K^+-induced astrocytic alkalinization was reduced from 0.29 ± 0.05 pH units (first application, control) to 0.20 ± 0.05 pH units (second application with blocker) (42/5/4; ANOVA, *** $p < 0.001$) (Figure 1b).

We next analyzed astrocytes in organotypic slice preparations derived from NBCe1 KO mice [34], for which slices from C57BL/6N mice served as wild-type controls (WT). In WT animals, the 2 min elevation in $[K^+]_o$ caused an alkalinization of 0.21 ± 0.05 pH units (57/6/3) (Figure 1c). Astrocytes in slices of NBCe1 KO displayed a significantly reduced alkalinization of 0.07 ± 0.02 pH units (75/5/3; MWU, *** $p > 0.001$) (Figure 1c).

Finally, we studied K^+-induced changes in astrocytic $[Na^+]_i$ in both WT and NBCe1 KO mice. The elevation of $[K^+]_o$ caused a decrease in $[Na^+]_i$ in WT mice by 4.1 ± 1.4 mM (67/5/3) (Figure 1d). In NBCe1 KO, this decrease was significantly larger (7.4 ± 2.7 mM; 50/5/4; MWU, *** $p < 0.001$) (Figure 1d), indicating reduced Na^+ influx under these conditions.

In summary, our data demonstrate functional NBCe1 activity in astrocytes from organotypic brain slices. They show that a moderate, transient elevation of $[K^+]_o$ induces a transient astrocytic alkalinization, which is significantly reduced by the NBC inhibitor S0859. Moreover, the amplitude of this alkalinization is significantly reduced in animals deficient in *Nbce1*. The latter also show a larger drop in $[Na^+]_i$ in response to elevation of $[K^+]_o$, suggesting less Na^+ influx than in WT animals. These effects are all in line with a stimulation of inward NBCe1 in response to elevation of $[K^+]_o$ and demonstrate that this transporter is functionally active in astrocytes of organotypic slices of mouse neocortex.

3.2. Role of NBCe1 Activity in Astrocytic Ion Changes during Brief Chemical Ischemia

After demonstrating functional activity of NBCe1 in astrocytes of organotypic slice cultures, we investigated its role in the generation of changes in astrocytic pH_i and $[Na^+]_i$ evoked by short periods of energy deprivation. The latter was induced by perfusing slices for 2 min with a glucose-free saline containing 5 mM of NaN_3 and 2 mM of 2-DG to block cellular ATP production by blocking mitochondrial respiration and glycolysis, respectively ("chemical ischemia") [10,41,53].

As expected from earlier work [7,41,54,55], brief chemical ischemia resulted in an acidification of astrocytes, decreasing their pH_i by 0.48 ± 0.07 pH units (79/5/3) (Figure 2a). With ATP production from glucose metabolism and oxidative respiration blocked, a likely mechanism contributing to astrocyte acidosis is e.g. the activation of glutamate transporters and the uptake of protons in conjunction with glutamate [56]. The peak acidification was reached within 183 ± 26 s, after which astrocytes slowly recovered back to their initial pH_i (full width at half maximum (FWHM): 399 ± 70 s). Inhibition of NBC via bath application of 30 µM of S0859 resulted in a significantly larger acidification by 0.60 ± 0.08 pH units upon chemical ischemia (70/4/3; MWU, *** $p < 0.001$) (Figure 2a). Notably, only half of the cells recovered to their initial baseline pH_i, indicating a compromised ability for the export of acid equivalents under these conditions.

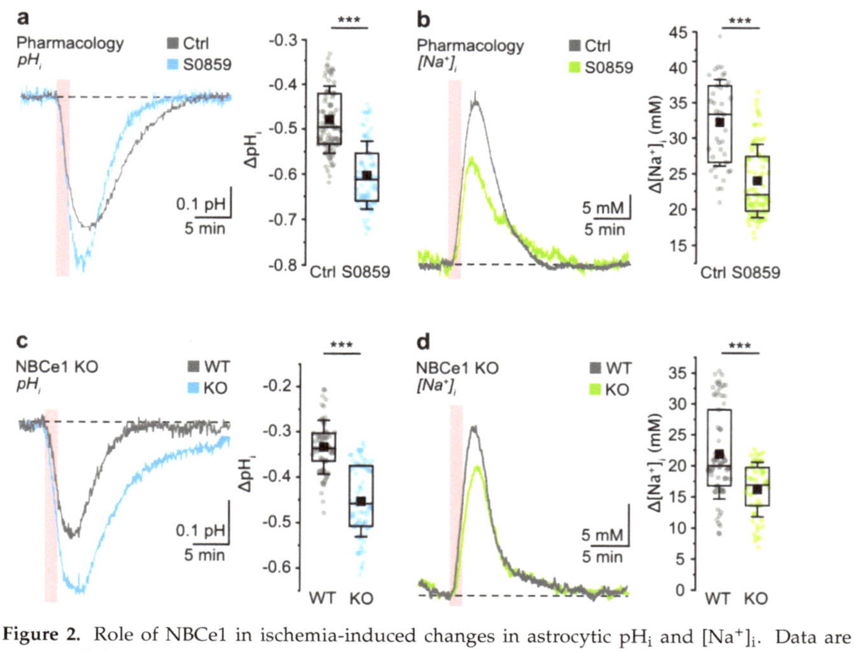

Figure 2. Role of NBCe1 in ischemia-induced changes in astrocytic pH_i and $[Na^+]_i$. Data are presented the same as in Figure 1. (**a**) Change in pH_i evoked by chemical ischemia for 2 min in control and with addition of S0859. (**b**) Change in $[Na^+]_i$ evoked by chemical ischemia in control (Ctrl) and with addition of S0859. (**c**) Change in pH_i evoked by chemical ischemia for 2 min in slices derived from wild-type mice (WT) and from NBCe1 KO. (**d**) Left: Change in $[Na^+]_i$ in wild-type and

NBCe1 KO. Periods of chemical ischemia are indicated with the light-red areas; downward traces represent acidification. Dashed lines indicate baseline levels. Box plots on the right summarize the results from these experiments and show individual data points (dots), means (squares), interquartile ranges (boxes), medians (lines), and SD (whiskers). *** $p < 0.001$.

In addition to the acidification, chemical ischemia evoked a transient increase in astrocytic $[Na^+]_i$ by 33.8 ± 8.7 mM, from which cells recovered within about 15 min (39/4/4) (Figure 2b). Assuming an average baseline $[Na^+]_i$ of 12.1 mM in neocortical astrocytes ([15], see methods), we found that addition of S0859 did not alter $[Na^+]_i$ significantly (S0589: 12.45 ± 0.97 mM (31/3/3); *t*-test, $p = 0.118$). Upon application of S0859, the peak amplitude of the chemical-ischemia-induced $[Na^+]_i$ increase was significantly smaller than in control, amounting to 22.1 ± 7.3 mM (76/6/3; MWU, *** $p < 0.001$) (Figure 2b). As in the control, and opposed to the pH changes induced, cells recovered fully from the $[Na^+]_i$ increase during pharmacological inhibition of NBCe1.

Next, we analyzed the effects of the 2 min chemical ischemia in organotypic slices from NBCe1 KO animals. Again, astrocytes in organotypic slices prepared from C57BL/6N wild-type mice were used as controls (see methods and above). We found that in the control, astrocytes acidified by 0.33 ± 0.06 pH units, from which cells fully recovered (FWHM: 410 ± 81 s) (57/6/3) (Figure 2c). Cells from *Nbce1*-deficient mice showed a significantly larger acidification of 0.45 ± 0.08 pH units upon metabolic inhibition (75/5/3; MWU, *** $p < 0.001$) (Figure 2c). Furthermore, the time to recover to baseline was significantly longer compared to WT (FWHM in NBCe1 KO: 513 ± 85 s; MWU, *** $p < 0.001$). In wild-type mice, chemical ischemia induced a $[Na^+]_i$ increase of 21.9 ± 7.2 mM (67/5/3). This $[Na^+]_i$ increase, from which cells fully recovered, was significantly smaller in NBCe1 KO astrocytes (16.2 ± 4.4 mM; 50/5/4; MWU, *** $p < 0.001$) (Figure 2d).

Taken together, these experiments show that pharmacological inhibition of NBC increases the amplitude of the ischemia-induced acidification in astrocytes while reducing the ischemia-induced $[Na^+]_i$ elevation. A similar result was obtained when comparing ion signals from astrocytes in slices from wild-type animals with those of NBCe1 KO animals. These data strongly suggest the activation of inwardly directed NBCe1 upon transient ischemia. While the resulting inward transport of HCO_3^- dampens the astrocytic reduction in pH_i, the accompanying Na^+ influx increases their Na^+ load.

3.3. Modeling Ion Dynamics and NBCe1 Activity during Transient Energy Deprivation

Based on the data on ischemia-induced changes in pH_i and $[Na^+]_i$ (see above), we aimed to develop a comprehensive model to simulate astrocytic NBCe1 activity during chemical ischemia. In addition to intracellular ion transients, chemical ischemia also depolarizes astrocytic membrane potential and changes extracellular ion concentrations, which will influence NBCe1 [10,41]. To make the simulation as realistic as possible, we additionally determined the most relevant of the latter parameters experimentally in our preparation.

Astrocytic membrane potentials were measured in cell-attached mode to avoid artifacts induced via a dialysis of cells. Chemical ischemia for 2 min resulted in a transient depolarization of astrocytes by 14.3 ± 3.3 mV from a baseline resting membrane potential of −83.0 ± 4.3 mV. After the washout of the metabolic inhibitors, cells hyperpolarized by 10.5 ± 6.0 mV below the initial baseline, after which membrane potentials recovered (5/5/3) (Figure 3a). Extracellular ion concentrations were analyzed using ion-selective microelectrodes. Baseline pH_o was 7.35 ± 0.06. Chemical ischemia caused a biphasic alkaline-acid shift: pH_o first increased briefly by 0.05 ± 0.02 pH units, followed by a long-lasting drop to pH_o 7.22 ± 0.05, after which pH_o fully recovered (5/5/3) (Figure 3b). The baseline $[K^+]_o$ was 2.5 ± 0.0 mM. The induction of chemical ischemia caused a $[K^+]_o$ increase by 1.0 ± 0.5 mM, followed by an undershoot of 0.2 ± 0.1 mM below the initial baseline, from which $[K^+]_o$ slowly recovered towards the baseline (5/5/3) (Figure 3c). Finally, the baseline $[Na^+]_o$ was 156.3 ± 1.2 mM and reversibly decreased by 1.7 ± 1.0 mM upon chemical ischemia (5/5/3) (Figure 3d).

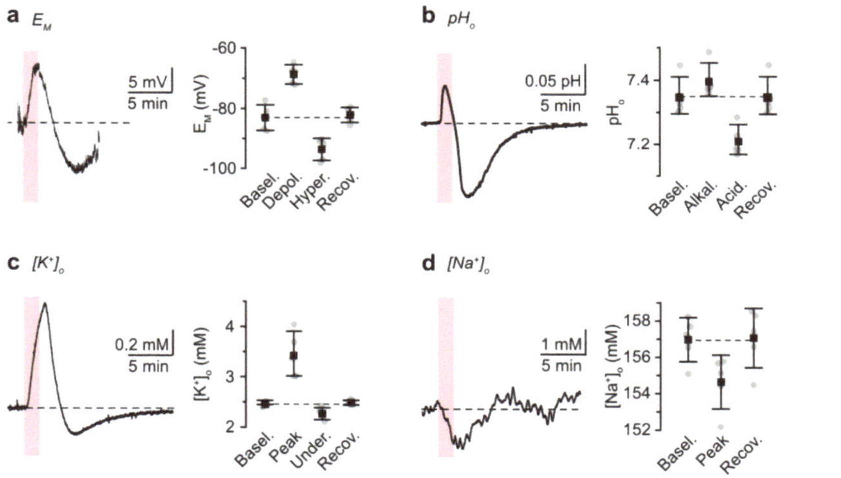

Figure 3. Ischemia-induced changes in astrocytic membrane potential and extracellular ions. (**a**) Change in astrocytic membrane potential (E_M), measured in cell-attached mode, evoked by chemical ischemia for 2 min. (**b**) Change in pH_o. Alkaline shifts are drawn upwards. (**c**,**d**) Change in $[K^+]_o$ and $[Na^+]_o$. Periods of chemical ischemia are indicated with the light-red areas. Dashed lines indicate baseline levels. Plots on the right summarize the results from these experiments and show individual data points (dots), means (squares), and SD (whiskers).

To incorporate the interaction of the modeled astrocyte with its environment (tissue and bath solution), the experimentally determined $[Na^+]_o$ and pH_o were fitted using polynomial functions of the form

$$[i]_{bath} = [i]\prime_{bath} \sum_j p_j (t - t_0)^{n-j}, \quad (10)$$

where i refers to Na^+, K^+, and pH; $i\prime_{bath}$ represents the baseline concentration in the bath; n represents the degree of a polynomial; p represents its coefficient; j ranges from 0 to n; and t is the time. A polynomial of degree 9 was used for fitting $[Na^+]_{bath}$ and pH_{bath}. Note that these functions representing concentrations or pH in the tissue are needed as the effects of other cells (neurons and other astrocytes) are not explicitly modeled and are different than the concentrations or pH immediately next to the cell. The concentrations or pH immediately next to the cell ($[Na^+]_o$, $[K^+]_o$, etc.) change dynamically in the model as the ions flow across the cell membrane (see Appendix A).

K^+ in the bath ($[K^+]_{bath}$) is modeled using exponential functions to incorporate baseline, rise, decay, and recovery phases as

$$\begin{aligned} [K^+]_{bath} = [K^+]'_{bath} \text{ for } t < t_o, \\ -[K^+]'_{bath}(\exp(k_1(t - t_o)) - k_2) \text{ for } t > t_0 \text{ and } t < t_0 + 240, \\ [K^+]'_{bath}\left(\frac{\exp(k_3(-t_0 - 240))}{k_4} + k_5\right) \text{ for } t > t_0 + 240 \text{ and } t < t_o + 535, \\ -[K^+]'_{bath}\left(\frac{\exp(k_6(t - t_0 - 535))}{k_7} + k_8\right) \text{ for } t > t_0 + 535 \end{aligned} \quad (11)$$

$[K^+]'_{bath}$ represents the baseline concentration of K^+ in the bath. The fitting parameters k_1, k_2, k_3, k_4, k_5, k_6, k_7, and k_8 are used to model the rise, decay, and recovery phases, and their values are provided in Appendix A. Fits to the observed extracellular Na^+, K^+, and pH are shown in Figure A1.

To model chemical ischemia, we decreased the activity of NKA by 2 min (see Equation (A4) in Appendix A). Furthermore, we assume that the local ischemia near the simulated astrocyte and recovery from ischemia set in slowly after the solution is switched from normal to chemical ischemia and vice versa. We mimicked this scenario in the model by decreasing and restoring the activity of NKA using sigmoidal functions (Appendix A). The activation of NBCe1 can lead to alterations in H^+ and HCO_3^- concentrations, resulting in changes in the pH_i of astrocytes in primary cell culture [57]. To simulate pH changes, we incorporated both NBCe1 and NHE into the model. In our simulations representing control mice, pH_i decreased from 7.33 to 6.92, while in simulations in which NBCe1 was inhibited, pH_i dropped to 6.77 (Figure 4b,c). These findings suggest that astrocytes in control conditions experience a smaller acidification in response to ischemia compared to cells lacking NBCe1. The smaller decrease in pH_i in cells expressing NBCe1 indicates that it acts as an acid extruder during chemical ischemia.

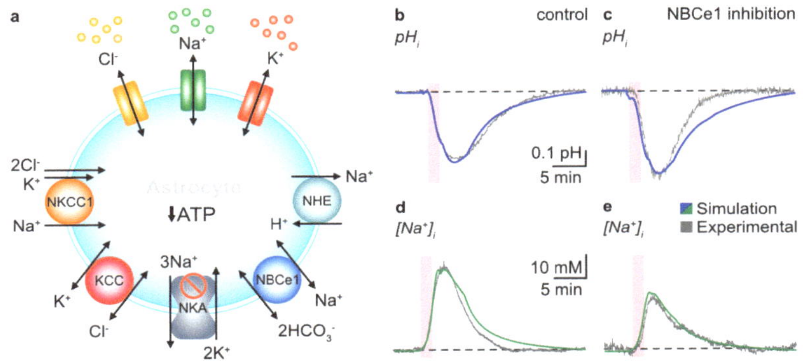

Figure 4. Computational modeling of ischemia-induced pH_i and $[Na^+]_i$ changes in astrocytes. (**a**) Schematic of the main pathways incorporated in the model. The arrow heads represent the direction of the flux. (**b–e**) Illustrating ischemia-induced pH_i and $[Na^+]_i$ changes in control (left panel) and upon inhibition of NBCe1 (right panel) in experimental and simulated conditions over a time course of 30 min. The blue and green traces represent simulated changes; the gray traces are experimental data taken from Figure 2a,b. Dashed lines indicate baseline levels. Periods of chemical ischemia are indicated with the light-red areas.

When chemical ischemia was active for 2 min, we also observed an increase in the $[Na^+]_i$. In control simulations, $[Na^+]_i$ increased by 34 mM from a baseline value of 12 mM (Figure 4d). However, in simulations mimicking an astrocyte with a 99% inhibition of NBCe1, $[Na^+]_i$ only rose by 26 mM (Figure 4e), indicating that NBCe1 activity is responsible for a 17% larger $[Na^+]_i$ increase. Notably, the results from these simulations are comparable to the experimental data, both concerning the kinetics and absolute amplitudes of the induced changes in pH_i and $[Na^+]_i$ (Figure 4).

3.4. NBCe1 Activity and ATP Consumption in Astrocytes

Given the prominent role of NBCe1 in Na^+ homeostasis, we next determined how inhibition of NBCe1 would affect ATP levels in astrocytes. We computed the consumption rate of ATP via NKA (knowing that NKA consumes one ATP molecule for transporting 3 Na^+ out of the cell in exchange for 2 K^+) by simulating the activity of NKA under 2 min of chemical ischemia in control and NBCe1-inhibited astrocytes. A comparison between the two conditions is shown in Figure 5a, where in both cases the NKA flux decreases (normalized to the resting state), reflecting the decreased activity of NKA and the resulting decline in its ATP consumption.

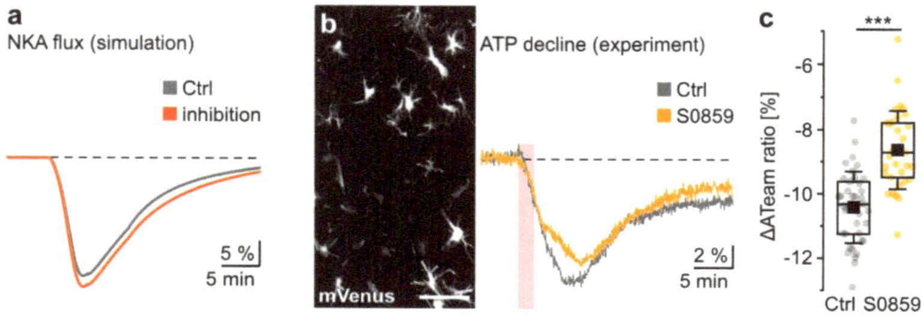

Figure 5. Role of NBCe1 in ischemia-induced astrocytic ATP depletion. (**a**) Computational modeling of reduction in the NKA flux in astrocytes. Gray represents the control flux, and light red represents the consumption rate upon inhibition of NBCe1. Note that NKA flux declines more with NBCe1 inhibition, indicating less NKA activity and, consequently, less consumption of ATP. (**b**) Left: Image of the ATeam fluorescence (mVenus) in cortical astrocytes. Scale: 50 μm. Right: Changes in the ATeam ratio during a 2 min induction of chemical ischemia in control and upon inhibition of NBC with S0859. Dashed lines indicate baseline levels. (**c**) Box plots summarizing the experimental results. Shown are individual data points (dots), means (squares), interquartile ranges (boxes), medians (lines), and SD (whiskers). *** $p < 0.001$. Periods of chemical ischemia are indicated with the light-red area.

Interestingly, the NKA flux (and NKA's ATP consumption rate) at the peak of chemical ischemia is lower in the astrocyte with NBCe1 inhibited compared to the control condition (Figure 5a). Thus, our model predicts that ATP levels should decline less during chemical ischemia in astrocytes with NBCe1 inhibited. Overall, this result suggests that NBCe1 activity promotes higher energy depletion during chemical ischemia.

To test the relation between NBCe1 and ATP depletion predicted using the model simulation, we conducted experiments in brain slices using FRET-based imaging with the ATP sensor ATeam1.03^{YEMK} ("ATeam") expressed in astrocytes. Consistent with our earlier work [41,42], we observed a transient decrease in the ATeam ratio by 10.4 ± 1.1% in astrocytes subjected to chemical ischemia for 2 min (39/4/3) (Figure 5b,c), indicating a decrease in intracellular ATP. Within the observation period of 15 min after the washout of the blockers, cells partially recovered from this decline to an ATeam ratio that was about 4% lower than before chemical ischemia. During inhibition of NBC with S0859, chemical ischemia induced a decline in the astrocytic ATeam ratio by 8.6 ± 1.2%, which was significantly lower than in the control (33/3/3; ANOVA, *** $p < 0.001$) (Figure 5b,c). As observed in the control, ATP levels did not fully recover to their initial baseline values within the observation period of 15 min.

Taken together, both our simulation and experimental results indicate that inhibition of NBCe1 reduces the astrocytic decline in ATP induced via metabolic inhibition. Our data thus strongly suggest that activation of inward NBCe1 and the accompanying Na^+ influx, respectively, promote energy depletion in ischemic conditions.

4. Discussion

In the present study, we demonstrate that a brief, 2 min period of chemical ischemia results in a decrease in the pH_i of astrocytes, which is accompanied by an increase in their $[Na^+]_i$. Pharmacological inhibition of NBC using S0859 caused a significant enlargement of the ischemia-induced astrocytic acidification and reduced the $[Na^+]_i$ elevation. Similar results were obtained when comparing ischemia-induced changes in astrocytic pH_i and $[Na^+]_i$ from wild-type animals with those of NBCe1 KO animals. Mathematical modeling based on our experimental data confirmed our experimental observations. Simulations furthermore predicted that activation of NBCe1 and the associated import of Na^+ result in a higher flux of NKA and, thus, a higher cellular ATP consumption. This prediction was tested and confirmed experimentally via imaging cellular ATP, showing that NBCe1 promotes the loss of astrocytic ATP in ischemic conditions.

4.1. NBCe1 Activity Influences Astrocytic pH_i and $[Na^+]_i$

NBCe1 (*SLC4A4*) is highly expressed in astrocytes and is the major plasma membrane transporter responsible for cellular pH regulation at a pH_i above ~6.8 [19–22,58,59]. At more severe decreases in pH_i; e.g., those accompanying severe ischemia, the Na^{+}/H^{+} exchanger NHE1 is central for the export of protons [18,60,61]. The stoichiometry of NBCe1 in the forward mode is $1Na^{+}$:$2HCO_3^{-}$ [57,62,63], while in the reverse mode, it has recently been demonstrated to transport $1Na^{+}$:$1HCO_3^{-}$:$1CO_3^{2-}$ [64,65]. In physiological conditions, the reversal potential of NBCe1 is close to the typical resting membrane potential of forebrain astrocytes (about −85 mV; see, e.g., [41]). NBCe1 thus readily switches operating directions, e.g., following changes in the astrocytic membrane potential in $pH_i/[HCO_3^{-}]_i$ and/or in $[Na^+]_i$ [24,27,30,56].

Inward NBCe1 is accompanied by an increase in the $[Na^+]_i$ and an alkalinization of astrocytes [27,28,66]. This mode of operation is stimulated via neuronal activity and is a central element in neuro-metabolic coupling in the forebrain [23,25,26]. Moreover, inward NBCe1 is an integral part of the chemosensory Ca^{2+} signaling of astrocytes in the brainstem [67]. Inward NBCe1 is promoted with a K^+-induced membrane depolarization of astrocytes, resulting in the so-called Depolarization-Induced Alkalinization (DIA; [50,51,62,68]). Reverse NBCe1, in contrast, serves the recovery from intracellular alkalosis [69] and may buffer extracellular acidifications [30]. It can be induced by switching from CO_2/HCO_3^{-}-buffered to HEPES-buffered saline [27,69]. Reverse operation has also been shown upon inhibition of the NKA and the concomitant increase in astrocytic $[Na^+]_i$ [28].

In the present study, we probed for NBCe1 activity, employing two major strategies. One strategy was the comparison of intracellular pH and Na^+ signals with and without the NBC blocker S0859. S0859 is an inhibitor of the NBC family and is not specific to NBCe1 [52]. Although it cannot be excluded that the effects obtained with S0859 might represent a mixture of inhibition of NBCe1 and other transporters such as NBCn1, it is well established that NBCe1 dominates in astrocytes [58]. In addition, S0859 has been reported to inhibit lactate transport via MCTs in astrocytes [70]. However, as the chemical ischemia protocol used in our study involves an inhibition of glycolysis, a significant production of lactate is unlikely under these conditions.

The other was a comparison of ion transients in preparations derived from wild-type mice with those from *Nbce1*-deficient mice. NBCe1 KO mice suffer from severe metabolic acidosis and other systemic effects and exhibit a sharp increase in mortality starting around postnatal day 8 [34,58]. To address the consequences of a deletion of NBCe1 on astrocytic ion changes, we therefore performed our study on organotypic tissue slice cultures prepared from neonatal animals. This is a well-established model system in which the cellular organization of neuronal networks is maintained and in which astrocytes continue to differentiate, similar to what is observed in vivo [71–73]. Notably, both strategies (application of S0589 to wild-type preparations as well as studying NBCe1 KO animals) produced similar results, indicating that S0859 indeed mainly targeted NBCe1 in astrocytes.

We first established that NBCe1 is functional in astrocytes in organotypic slices of the mouse neocortex by inducing a K^+-induced DIA, which has been shown to involve NBCe1 activity [57,69]. This conclusion is based on the following observations: (1) brief elevation of $[K^+]_o$ from 2.5 to about 8.6 mM transiently increased astrocytic pH_i; (2) the K^+-induced alkalinization was sensitive to the NBC inhibitor S0859; (3) the amplitude of the K^+-induced alkalinization was significantly smaller in slices prepared from animals deficient for NBCe1 compared to wild-type mice. (4) In addition to the described changes in pH_i, the $[K^+]_o$ elevation resulted in a decrease in astrocytic $[Na^+]_i$. This phenomenon was reported before and can most likely be ascribed to an activation of the astrocytic NKA [2,13,27]. The K^+-induced decrease in $[Na^+]_i$ was augmented in NBCe1 KO mice. The latter is indicative of a reduced Na^+ influx in *Nbce1*-deficient mice, pointing to an NBCe1-mediated influx of $[Na^+]_i$ in wild-type animals. The results from our experiments thus confirm that NBCe1 activity results in detectable changes in the astrocytic pH_i and $[Na^+]_i$, thereby dynamically coupling intracellular Na^+ homeostasis with cellular acid/base balance.

4.2. Operation of NBCe1 during Brief Chemical Ischemia

Brain ischemia results in a rapid decline in cellular ATP levels and in a reduction or failure of plasma membrane transporters, first and foremost of the NKA [7,74–76]. While cells of the ischemic core are doomed to die, the surrounding ischemic penumbra may recover if perfusion is restored in time. Recovery, however, is compromised with repeated waves of spreading depolarizations, which promote further damage [77]. Initially, spreading depolarizations are fully reversible and accompanied with a transient reduction in cellular ATP, an increase in $[K^+]_o$ and extracellular glutamate, as well as an increase in $[Na^+]_i$ and $[Ca^{2+}]_i$ and an intracellular acidification [9–11,32,78,79].

To mimic such conditions, we exposed tissue slices to blockers of cellular ATP production for 2 min ("chemical ischemia"). Earlier work has shown that chemical ischemia induces an immediate reduction in cellular ATP levels as well as changes in ion concentrations reminiscent of those observed during a spreading depolarization in the ischemic penumbra [10,41,42]. Here, we confirm these observations, demonstrating that brief periods of metabolic inhibition cause transient fluctuations in $[K^+]_o$, pH_o, and $[Na^+]_o$, accompanied with a transient membrane depolarization, a decrease in pH_i, and an increase in the $[Na^+]_i$ of astrocytes.

While results obtained in our study thus generally confirm earlier observations made using other in vitro models (oxygen–glucose deprivation or other strategies to mimic ischemic conditions; e.g., [7,10,55,80,81]) or using mouse models of ischemia in vivo [9–11,32,78,79], it is noteworthy that both the absolute amplitudes as well as the kinetics of ion changes can differ depending on the model employed. For example, it has been demonstrated earlier that the degree of extracellular acidification depends on the depth and duration of ischemia [80,82–84]. Compared to these studies, the pH_o changes determined in the present study were moderate. The same holds true for increases in $[K^+]_o$ determined here, which are lower in amplitude than those observed during spreading depolarizations in situ or in vivo [7,11]. A major reason for this discrepancy is the rather short duration of chemical ischemia employed. Moreover, brain slice preparations allow for more rapid extracellular diffusion and, thereby, may dampen extracellular ionic changes compared to the in vivo situation. In contrast, and as reported before [10], the amplitudes and time courses of $[Na^+]_i$ increases observed upon chemical ischemia were similar to those induced during a spreading depolarization in the ischemic penumbra of the mouse neocortex in vivo [10].

The cellular pathways of the ischemia-induced Na^+ influx into astrocytes are still not fully understood. Notably, while moderate elevations in $[K^+]_o$ stimulate NKA and therefore cause a decrease in $[Na^+]_i$ (present work and [2,13,27]), spreading depolarizations and metabolic inhibition are additionally characterized by an increase in extracellular glutamate, as mentioned above. Indeed, blocking glutamate transporters strongly reduced Na^+ increases in neocortical astrocytes exposed to transient energy depletion, suggesting a prominent Na^+ influx into astrocytes via this pathway, temporarily overriding the capacity of NKA to export Na^+ [10]. Inhibition of glutamate transporters reduced Na^+ influx by only about 50%, implying that additional mechanisms for Na^+ entry must exist. Our results obtained in the present study provide clear evidence for a stimulation of inward NBCe1 during chemical ischemia, showing that the latter represents such a pathway contributing to astrocytic Na^+ loading. This conclusion is supported by the observation that blocking NBC with S0859 enlarged the acidification while at the same time reducing the $[Na^+]_i$ increase evoked by the metabolic inhibition. A similar observation was made using tissue from NBCe1 KO animals, in which the ischemia-induced acidification was enlarged and the ischemia-induced Na^+ loading was reduced compared to wild-type animals. Stimulation of inward operation of NBCe1 in neocortical astrocytes upon brief chemical ischemia is in line with the proposed inward NBCe1 activity in cardiomyocytes upon myocardial ischemia [85] or in astrocytes exposed to long-lasting extracellular acidification to mimic cerebral ischemia [86].

Our experimental results are in line with the increased inward activity of NBCe1 transporters. As our experimental setting focused on the acute effects of brief metabolic inhibition (that is, on effects seen within a period of about 30 min), changes in expression levels are unlikely to play a role. Additionally, a substrate-mediated increase in transport activity, fast and efficient regulation of NBCe1 transport may occur via post-translational modifications such as phosphorylation or by changes in transporter trafficking to the plasma membrane, without changes in transcript or protein expression [29]. In the context of the present work, it is noteworthy that brain ischemia causes regulation of mTOR signaling (reviewed by [87]), which is known to be able to phosphorylate several residues of NBCe1, thereby altering its transport activity [88,89].

To further investigate the role of NBCe1 in astrocytic function, we integrated known physiological parameters and experimental data into a detailed biophysical model for ion concentrations and pH dynamics in the astrocyte and extracellular space. Our model predicted the inward activity of NBCe1 during chemical ischemia, which is consistent with our experimental findings. Notably, our modeling approach replicated increased inward NBCe1 activity in response to changes in ionic driving forces only, indicating that the latter will suffice to explain our experimental results. Furthermore, the model predicted that inhibition of NBCe1 activity results in a decrease in ATP consumption. This result suggests that the inward activity of NBCe1 exacerbates the depletion of ATP under metabolic stress. We validated these predictions in brain slices using FRET-based imaging with the ATP-sensor ATeam1.3^{YEMK}. Overall, our mathematical model offers a complementary approach to exploring the complex dynamics of NBCe1-mediated Na^+ and HCO_3^- fluxes and how they modulate astrocytic $[Na^+]_i$, pH_i, and ATP under different conditions.

4.3. Possible Consequences of NBCe1 Activity in the Ischemic Brain

The pathophysiological role of NBCe1 in the ischemic brain in vivo is still unclear. Several studies have reported an upregulation of sodium-bicarbonate cotransporters (NBC) in ischemic conditions or after induction of ischemic stroke in vivo. Jung et al. [90] showed significantly increased protein expression of NBC in the rat brain 3 and 6 h after a permanent middle cerebral artery occlusion (pMCAO) in rats. Moreover, Sohn and colleagues [91] demonstrated increased NBC protein levels in the hippocampal CA1 region of gerbils 12 h after exposure to transient cerebral ischemia, which, however, returned to control levels after 4 days. Astrocytes in culture, derived from human-induced pluripotent stem cells (hiPSCs), showed increased NBCe1 protein expression 24 h after exposure to a saline mimicking ischemic conditions in the brain [83].

Earlier work addressing the role of NBC transporters in brain injury concluded that their increased expression causes an increased vulnerability to extracellular acidosis by promoting the accumulation of intracellular Na^+, thereby contributing to cellular damage in cerebral ischemia in vivo [86,90]. The study by Yao et al. [83], in contrast, showed that NBC activity protected astrocytes from cell death when exposed to ischemic conditions. Moreover, a recently posted manuscript [92] reports that loss of astrocytic NBCe1 increased the infarct volume after ischemic stroke in a mouse model in vivo. Results also indicate that NBCe1 dampens the breakdown of pH homeostasis in the ischemic brain [92].

The protective effect of NBCe1 might be related to its inward activity, increasing cell survival by reducing a harmful intracellular acidosis [18,86]. The latter idea is in line with the present study, in which we demonstrate that inward NBCe1 ameliorates the intracellular acidification in acute ischemic conditions. Intracellular acidosis is a harmful event, exacerbating cytotoxic edema and cellular damage through mechanisms including free radical generation, impaired protein synthesis, and/or intracellular Ca^{2+} accumulation [55,80,81,93–95]. Moreover, suppressing acidosis in astrocytes using optogenetic tools has been demonstrated to suppress glial glutamate release and ischemic brain damage in mouse models in vivo, presumably by reducing glutamate-induced excitotoxicity [96,97].

In another study, loss of NBCe1 was shown to cause familial hemiplegic migraine, and the authors suggested that a resulting lack of glial acid secretion into the ECS increased neuronal excitability [98].

The above-mentioned studies thus indicate that stimulation of inward NBCe1 and the resulting reduction in astrocytic acidosis and secretion of acid into the ECS, respectively, are neuroprotective. Its upregulation following ischemic stroke reported from in vivo mouse models [76,77] will support such a neuroprotective role. Regulation of pH, however, comes at a high price. Inward NBCe1 can drive water influx, astrocyte swelling, and shrinkage of the ECS [99–102], thereby exacerbating neuronal damage. Importantly, and as shown here, it also results in increased Na^+ influx, aggravating the ischemia-induced Na^+ loading of astrocytes. Besides weakening the driving force of Na^+-dependent transporters, our present work demonstrates that NBCe1-related Na^+ influx results in a larger decline in cellular ATP levels. The latter aggravates the loss of cellular energy upon chemical ischemia, most likely representing a harmful action of NBCe1, which might promote cellular damage in the ischemic brain. Clearly, further work, including studies in vivo, is needed to clarify the relevance of NBCe1 and its involvement in generating brain damage seen after stroke in animal models as well as in human patients.

Author Contributions: Conceptualization, K.E., P.T., G.U. and C.R.R.; methodology, K.E., P.T., E.R., G.U. and C.R.R.; validation, investigation, and data curation, K.E., P.T., N.P., S.D., S.E., E.R., G.U. and C.R.R.; writing—original draft preparation, K.E., P.T. and C.R.R.; writing—review and editing, K.E., P.T., N.P., S.E., E.R., G.U. and C.R.R.; visualization, K.E. and P.T.; supervision, G.U. and C.R.R.; project administration, C.R.R.; funding acquisition, G.U. and C.R.R. All authors have read and agreed to the published version of the manuscript.

Funding: Research performed at the Institute of Neurobiology, HHU, and the Department of Physics, USF, was supported by the Deutsche Forschungsgemeinschaft (DFG), Research Unit 2795 "Synapses under Stress" (Ro2327/13-2 to C.R.R. and Mercator fellowship to G.U.). Research performed at the Department of Physics, USF, was supported by National Institutes of Health through grant number R01NS130916 awarded to G.U.

Institutional Review Board Statement: The study was carried out in accordance with the institutional guidelines of the Heinrich Heine University Düsseldorf as well as the European Community Council Directive (2010/63/EU). All experiments using brain slices were communicated to and approved by the Animal Welfare Office at the Animal Care and Use Facility of the Heinrich Heine University Düsseldorf (institutional act no. O50/05). Neonatal animals were killed via rapid decapitation before preparation of brain slices in accordance with the recommendations of the European Commission [103].

Informed Consent Statement: Not applicable.

Data Availability Statement: The experimental data are fully available from the authors (C.R.R./HHU) upon reasonable request. The codes reproducing the main results are available from the authors (G.U./USF) upon request.

Acknowledgments: The authors wish to thank Claudia Roderigo (HHU) and Simone Durry (HHU) for their expert technical assistance.

Conflicts of Interest: The authors declare no conflict of interest.

Appendix A. Model Equations and Parameters

The model used in this study builds on our previous work and uses the following equations [1]. K^+ concentration in the ECS ($[K^+]_o$) depends upon the fluxes through Na^+/K^+-ATPase (J_{NKA}), K^+ channels (J_K), $Na^+/K^+/Cl^-$ co-transporter (J_{NKCC1}), K^+/Cl^- co-transporter (J_{KCC1}), as well as K^+ exchange with the bath solution (J_{Kdiff}). The rate equation for $[K^+]_o$ is

$$\frac{d[K^+]_o}{dt} = \frac{1}{VR_{sa}}(J_K - 2J_{NKA} - J_{NKCC1} - J_{KCC1}) + J_{Kdiff} \quad (A1)$$

where VR_{sa} is the ratio of the ECS to the astrocytic volume. The flux through the K^+ channels ($\mu M\ s^{-1}$) is

$$J_K = G_K(v_i - E_K) \tag{A2}$$

where G_K is whole-cell conductance of K^+ channels and v_i is the membrane potential. Strictly speaking, $\mu M\ s^{-1}$ is not a standard flux unit; however, it can be converted to current density ($pA/\mu m^2$) using a conversion factor of $10^3 \frac{A}{F\times V}$, where *A*, *F*, and *V* is the area of the cell, Faraday's constant, and volume of the cell, respectively. E_K is the reversal potential of K^+ channels given using the Nernst equation (mV).

$$E_K = \frac{V_T}{z_K}\ln\left(\frac{[K^+]_o}{[K^+]_i}\right) \tag{A3}$$

z_K is the valency of K^+. J_{NKA} exports three Na^+ and imports two K^+. J_{NKA} ($\mu M\ s^{-1}$) is given as

$$\begin{aligned} J_{NKA} = J_{NKA_{max}}(&-I_1(a_1, b_1, t, t_0, c_1)I_2(a_2, b_2, t, t_0, c_2) \\ &+d)H_{1.5}([Na^+]_i, K_{Na_i})H([K^+]_o, K_{K_o}) \end{aligned} \tag{A4}$$

where J_{NKAmax} is the maximum flux through Na^+/K^+-ATPase and $H_n(x,K)$ is from $\frac{x^n}{x^n+K^n}$, where n is the Hill coefficient, x is the concentration of Na^+ or K^+, and K in the function $H_n(x,K)$ is the dissociation constant of the respective ion to the pump.

$$I(a, b, t, t_0, c) = \frac{a}{1 + a\exp(b(t - t_0) + c)} \tag{A5}$$

where a, b, c, and d are constants, t represents time during the simulation and t_0 represents the time at which ischemia is initiated. The sigmoidal forms are used to mimic the scenario where local ischemia near the cell settles in slowly or normal oxygen and glucose supply restore slowly after the solution is switched to ischemic condition and back to normal, respectively.

K^+ diffusion between the ECS and bath solution ($\mu M\ s^{-1}$) is given as

$$J_{Kdiff} = diff([K^+]_{bath} - [K^+]_o) \tag{A6}$$

where *diff* is the diffusion constant. One Na^+, one K^+, and two Cl^- ions move in inward direction through NKCC1. The flux through NKCC1 ($\mu M\ s^{-1}$) is

$$J_{NKCC1} = G_{NKCC1}\varnothing\ln\left(\frac{[Na^+]_o[K^+]_o[Cl^-]_o^2}{[Na^+]_o[K^+]_i[Cl^-]_i^2}\right) \tag{A7}$$

G_{NKCC} is the whole-cell conductance of NKCC1. Cl^- and K^+ flux through KCC1 channels ($\mu M\ s^{-1}$) is

$$J_{KCC1} = G_{KCC1}\varnothing\ln\left(\frac{[K^+]_o[Cl^-]_o}{[K^+]_i[Cl^-]_i}\right), \tag{A8}$$

where G_{KCC1} is the whole-cell conductance of KCC1.

The rate equation for $[Na^+]_o$ depends on the flux through Na^+ channels (J_{Na}), NKA, NBCe1, Na^+/H^+ exchanger, and Na^+ exchange with the bath solution (J_{Nadiff}).

$$\frac{d[Na^+]_o}{dt} = \frac{1}{VR_{sa}}(J_{Na} + 3J_{NKA} - J_{NKCC1} - J_{NBCe1} + J_{NHE}) + J_{Nadiff}, \tag{A9}$$

Na^+ flux through Na^+ channels ($\mu M\ s^{-1}$) is

$$J_{Na} = G_{Na}(v_i - E_{Na}) \tag{A10}$$

where G_{Na} is the whole-cell conductance of Na^+ channels and E_{Na} is the reversal potential for Na^+.

$$E_{Na} = \frac{V_T}{z_{Na}} \ln\left(\frac{[Na^+]_o}{[Na^+]_i}\right) \tag{A11}$$

Fluxes through NBCe1 and NHE ($\mu M\ s^{-1}$) are given in Section 2.6.

Na^+ exchange with the bath solution ($\mu M\ s^{-1}$) is given as

$$J_{Nadiff} = diff\left([Na^+]_{bath} - [Na^+]_o\right) \tag{A12}$$

where diff is the diffusion constant of Na^+.

K^+ concentration in the astrocyte ($[K^+]_i$) depends on the fluxes due to K^+ channels, NKA, NKCC1, and KCC1. That is,

$$\frac{d[K^+]_i}{dt} = -J_K + 2J_{NKA} + J_{NKCC1} + J_{KCC1}. \tag{A13}$$

$[Na^+]_i$ depends on the fluxes through Na^+ channels, NKA, NBCe1, and NHE. These fluxes are already described above.

$$\frac{d[Na^+]_i}{dt} = -J_{Na} - 3J_{NKA} + J_{NKCC1} + J_{NBCe1} - J_{NHE} \tag{A14}$$

Cl^- concentration in the astrocyte ($[Cl^-]_i$) and ECS ($[Cl^-]_o$) is given via electroneutrality (μM).

$$\frac{d[Cl^-]_i}{dt} = \frac{d[Na^+]_i}{dt} + \frac{d[K^+]_i}{dt} - J_{NBCe1} \tag{A15}$$

$$[Cl^-]_o = [Na^+]_o + [K^+]_o - [HCO_3^-]_o \tag{A16}$$

Membrane potential of the astrocyte (mV) is given as

$$\frac{dv_i}{dt} = \gamma_v(-J_K - J_{Na} - J_{Cl} - J_{NKA} + J_{NBCe1} - J_{NHE}) \tag{A17}$$

where γ_v converts flux from concentration unit to current unit. Cl^- flux through leak channels ($\mu M\ s^{-1}$) is

$$J_{Cl} = G_{Cl}(v_i - E_{Cl}) \tag{A18}$$

where G_{Cl} is the maximum conductance of Cl^- channels. E_{Cl} is the reversal potential of Cl^- and is given using

$$E_{Cl} = \frac{v_T}{z_{Cl}} \ln\left(\frac{[Cl^-]_o}{[Cl^-]_i}\right) \tag{A19}$$

where z_{Cl} is the valence of Cl^-.

The parameters used in the model are given in Tables A1–A4 (see below).

Table A1. Parameters used in the equations for ion dynamics in the astrocyte and ECS as well as membrane potential of the astrocyte.

Parameter	Description	Value
γ_v	Scaling factor for membrane potential	1970 mV μM^{-1}
VR_{sa}	Volume ratio between the ECS and astrocyte	3
V_T	Voltage constant in Nernst equation	26.7 mV
G_K	Peak conductance of K^+ channels	2072.3 $\mu M\ mV^{-1}\ s^{-1}$
G_{Na}	Peak conductance of Na^+ channels	68.08 $\mu M\ mV^{-1}\ s^{-1}$
G_{NBCe1}	Peak conductance of NBCe1	392.22 $\mu M\ mV^{-1}\ s^{-1}$

Table A1. *Cont.*

Parameter	Description	Value
G_{NHE}	Peak conductance of NHE	$\frac{1}{3}G_{NBCe1}$
J_{NaKmax}	Maximum flux through NKA	4.26×10^4 μM s^{-1}
K_{Na}	Association constant for Na^+ to NKA	10×10^3 μM
K_{Ko}	Association constant for K^+ to NKA	1.5×10^3 μM
diff	Diffusion constant of Na^+, K^+, and pH	0.5 s^{-1}
β_i	Intracellular intrinsic buffering capacity	25 mM/pH unit
β_o	Extracellular intrinsic buffering capacity	10.5 mM/pH unit
pK_a	Negative logarithm of the acid dissociation constant of carbonic acid	6.1
s	Solubility of CO_2	2.25×10^{-4} mM Pa^{-1}
P_{CO2}	Partial pressure of CO_2	5332.9 Pa
K_h	Dissociation constant of CO_2	800 nmol L^{-1}

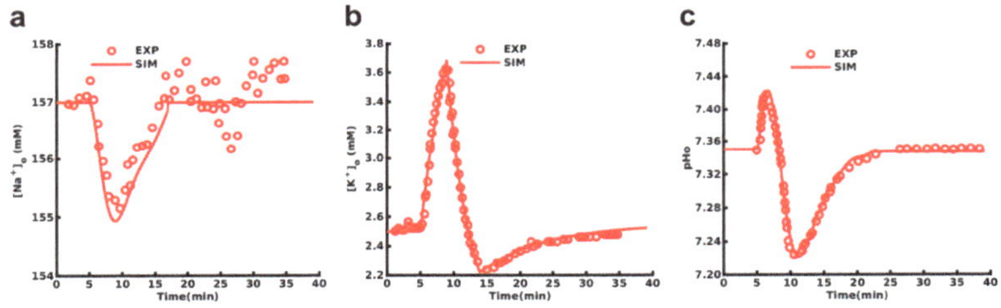

Figure A1. Fits to observed changes in $[Na^+]_o$ (**a**), $[K^+]_o$ (**b**), and pH_o (**c**). Solid lines and markers represent simulated and experimental values, respectively.

Table A2. The initial values of state variables.

Parameter	Description	Value
$[Na^+]_o$	Extracellular Na^+ concentration	157 mM
$[Na^+]_i$	Intracellular Na^+ concentration	12 mM
$[K^+]_o$	Extracellular K^+ concentration	2.5 mM
$[K^+]_i$	Intracellular K^+ concentration	146 mM
V_i	Intracellular membrane potential	−83 mV
pH_o	Extracellular pH value	7.35
pH_i	Intracellular pH value	7.32

Table A3. Parameters used in functional forms of ions in the bath.

Parameter	Description	Value
$[K^+]'_{bath}$	Baseline value of $[K^+]_{bath}$	2.5 mM
	k_1	−0.002675 s^{-1}
	k_2	−2
	k_3	−0.0052 s^{-1}
	k_4	1.350
	k_5	0.73
	k_6	−0.00169
	k_7	7.6
	k_8	−1.02
$[Na^+]'_{bath}$	Baseline value of $[Na^+]_{bath}$	157 mM
	p_1	-1.1906×10^{-26} s^{-1}
	p_2	5.045×10^{-23} s^{-1}
	p_3	-8.122×10^{-20} s^{-1}

Table A3. *Cont.*

Parameter	Description	Value
	p_4	5.6637×10^{-17} s^{-1}
	p_5	-7.006×10^{-15} s^{-1}
	p_6	-1.3156×10^{-11} s^{-1}
	p_7	7.521×10^{-9} s^{-1}
	p_8	-1.326×10^{-6} s^{-1}
	p_9	3.51×10^{-6} s^{-1}
	p_{10}	1.0005
pH'_{bath}	Baseline value of pH_{bath}	$7.35 - 5.4275 \times 10^{-19}$ s^{-1}
	p_1	-4.7273×10^{-26} s^{-1}
	p_2	2.47264×10^{-22} s^{-1}
	p_3	-5.4275×10^{-19} s^{-1}
	p_4	6.4515×10^{-16} s^{-1}
	p_5	-4.4486×10^{-13} s^{-1}
	p_6	1/7487 s^{-10}
	p_7	−3.4571 s^{-8}
	p_8	2.0687×10^{-6} s^{-1}
	p_9	1.0069×10^{-4} s^{-1}
	p_{10}	0.9999

Table A4. Parameters used in the NKA flux.

Parameter	Description	Value
a_1	Parameter in sigmoidal function	1000
b_1	Parameter in sigmoidal function	0.0022134 s^{-1}
c_1	Parameter in sigmoidal function	0
t_0	Beginning of pump inhibition	300 s
a_2	Parameter in sigmoidal function	1.52
b_2	Parameter in sigmoidal function	−0.02014 s^{-1}
c_2	Parameter in sigmoidal function	5.738
d	Intercept parameter in sigmoidal functions	1.13

References

1. Kofuji, P.; Newman, E.A. Potassium buffering in the central nervous system. *Neuroscience* **2004**, *129*, 1045–1056. [CrossRef]
2. Larsen, B.R.; Stoica, A.; MacAulay, N. Managing Brain Extracellular K^+ during Neuronal Activity: The Physiological Role of the Na^+/K^+-ATPase Subunit Isoforms. *Front. Physiol.* **2016**, *7*, 141. [CrossRef]
3. Silver, I.A.; Erecinska, M. Energetic demands of the Na^+/K^+ ATPase in mammalian astrocytes. *Glia* **1997**, *21*, 35–45. [CrossRef]
4. Rose, C.R.; Verkhratsky, A. Principles of sodium homeostasis and sodium signalling in astroglia. *Glia* **2016**, *64*, 1611–1627. [CrossRef]
5. Danbolt, N.C. Glutamate uptake. *Prog. Neurobiol.* **2001**, *65*, 1–105. [CrossRef]
6. Schousboe, A.; Sarup, A.; Bak, L.K.; Waagepetersen, H.S.; Larsson, O.M. Role of astrocytic transport processes in glutamatergic and GABAergic neurotransmission. *Neurochem. Int.* **2004**, *45*, 521–527. [CrossRef]
7. Somjen, G.G. *Ions in the Brain: Normal Function, Seizures, and Stroke*; Oxford University Press: New York, NY, USA, 2004.
8. Dirnagl, U.; Iadecola, C.; Moskowitz, M.A. Pathobiology of ischaemic stroke: An integrated view. *Trends Neurosci.* **1999**, *22*, 391–397. [CrossRef]
9. Van Putten, M.J.A.M.; Fahlke, C.; Kafitz, K.W.; Hofmeijer, J.; Rose, C.R. Dysregulation of astrocyte ion homeostasis and its relevance for stroke-induced brain damage. *Int. J. Mol. Sci.* **2021**, *22*, 5679. [CrossRef]
10. Gerkau, N.J.; Rakers, C.; Durry, S.; Petzold, G.C.; Rose, C.R. Reverse NCX Attenuates Cellular Sodium Loading in Metabolically Compromised Cortex. *Cereb. Cortex* **2018**, *28*, 4264–4280. [CrossRef]
11. Pietrobon, D.; Moskowitz, M.A. Chaos and commotion in the wake of cortical spreading depression and spreading depolarizations. *Nat. Rev. Neurosci.* **2014**, *15*, 379–393. [CrossRef]
12. Rossi, D.J.; Oshima, T.; Attwell, D. Glutamate release in severe brain ischaemia is mainly by reversed uptake. *Nature* **2000**, *403*, 316–321. [CrossRef]
13. Karus, C.; Mondragao, M.A.; Ziemens, D.; Rose, C.R. Astrocytes restrict discharge duration and neuronal sodium loads during recurrent network activity. *Glia* **2015**, *63*, 936–957. [CrossRef]
14. Kelly, T.; Kafitz, K.W.; Roderigo, C.; Rose, C.R. Ammonium-evoked alterations in intracellular sodium and pH reduce glial glutamate transport activity. *Glia* **2009**, *57*, 921–934. [CrossRef]

15. Ziemens, D.; Oschmann, F.; Gerkau, N.J.; Rose, C.R. Heterogeneity of activity-induced sodium transients between astrocytes of the mouse hippocampus and neocortex: Mechanisms and consequences. *J. Neurosci.* **2019**, *39*, 2620–2634. [CrossRef]
16. Rose, C.R.; Ziemens, D.; Verkhratsky, A. On the special role of NCX in astrocytes: Translating Na^+-transients into intracellular Ca^{2+} signals. *Cell Calcium* **2020**, *86*, 102154. [CrossRef]
17. Annunziato, L.; Boscia, F.; Pignataro, G. Ionic transporter activity in astrocytes, microglia, and oligodendrocytes during brain ischemia. *J. Cereb. Blood Flow Metab.* **2013**, *33*, 969–982. [CrossRef]
18. Song, S.; Luo, L.; Sun, B.; Sun, D. Roles of glial ion transporters in brain diseases. *Glia* **2020**, *68*, 472–494. [CrossRef]
19. Parker, M.D.; Boron, W.F. The divergence, actions, roles, and relatives of sodium-coupled bicarbonate transporters. *Physiol. Rev.* **2013**, *93*, 803–959. [CrossRef]
20. Bevensee, M.O.; Boron, W.F. Effects of acute hypoxia on intracellular-pH regulation in astrocytes cultured from rat hippocampus. *Brain Res.* **2008**, *1193*, 143–152. [CrossRef]
21. Deitmer, J.W.; Rose, C.R. Ion changes and signalling in perisynaptic glia. *Brain Res. Rev.* **2010**, *63*, 113–129. [CrossRef]
22. Chesler, M. Regulation and modulation of pH in the brain. *Physiol. Rev.* **2003**, *83*, 1183–1221. [CrossRef]
23. Chatton, J.Y.; Magistretti, P.J.; Barros, L.F. Sodium signaling and astrocyte energy metabolism. *Glia* **2016**, *64*, 1667–1676. [CrossRef]
24. Deitmer, J.W.; Theparambil, S.M.; Ruminot, I.; Noor, S.I.; Becker, H.M. Energy Dynamics in the Brain: Contributions of Astrocytes to Metabolism and pH Homeostasis. *Front. Neurosci.* **2019**, *13*, 1301. [CrossRef]
25. Barros, L.F.; Ruminot, I.; Sotelo-Hitschfeld, T.; Lerchundi, R.; Fernandez-Moncada, I. Metabolic Recruitment in Brain Tissue. *Annu. Rev. Physiol.* **2023**, *85*, 115–135. [CrossRef]
26. Allaman, I.; Belanger, M.; Magistretti, P.J. Astrocyte-neuron metabolic relationships: For better and for worse. *Trends Neurosci.* **2011**, *34*, 76–87. [CrossRef]
27. Rose, C.R.; Ransom, B.R. Intracellular sodium homeostasis in rat hippocampal astrocytes. *J. Physiol.* **1996**, *491*, 291–305. [CrossRef]
28. Noor, Z.N.; Deitmer, J.W.; Theparambil, S.M. Cytosolic sodium regulation in mouse cortical astrocytes and its dependence on potassium and bicarbonate. *J. Cell. Physiol.* **2018**, *234*, 89–99. [CrossRef]
29. Thornell, I.M.; Bevensee, M.O. Regulators of Slc4 bicarbonate transporter activity. *Front. Physiol.* **2015**, *6*, 166. [CrossRef]
30. Theparambil, S.M.; Hosford, P.S.; Ruminot, I.; Kopach, O.; Reynolds, J.R.; Sandoval, P.Y.; Rusakov, D.A.; Barros, L.F.; Gourine, A.V. Astrocytes regulate brain extracellular pH via a neuronal activity-dependent bicarbonate shuttle. *Nat. Commun.* **2020**, *11*, 5073. [CrossRef]
31. Kraig, R.P.; Pulsinelli, W.A.; Plum, F. Carbonic acid buffer changes during complete brain ischemia. *Am. J. Physiol.* **1986**, *250*, R348–R357. [CrossRef]
32. Menyhart, A.; Zolei-Szenasi, D.; Puskas, T.; Makra, P.; Orsolya, M.T.; Szepes, B.E.; Toth, R.; Ivankovits-Kiss, O.; Obrenovitch, T.P.; Bari, F.; et al. Spreading depolarization remarkably exacerbates ischemia-induced tissue acidosis in the young and aged rat brain. *Sci. Rep.* **2017**, *7*, 1154. [CrossRef]
33. Imamura, H.; Huynh Nhat, K.P.; Togawa, H.; Saito, K.; Iino, R.; Kato-Yamada, Y.; Nagai, T.; Noji, H. Visualization of ATP levels inside single living cells with fluorescence resonance energy transfer-based genetically encoded indicators. *Proc. Natl. Acad. Sci. USA* **2009**, *106*, 15651–15656. [CrossRef]
34. Gawenis, L.R.; Bradford, E.M.; Prasad, V.; Lorenz, J.N.; Simpson, J.E.; Clarke, L.L.; Woo, A.L.; Grisham, C.; Sanford, L.P.; Doetschman, T.; et al. Colonic anion secretory defects and metabolic acidosis in mice lacking the NBC1 Na^+/HCO_3^- cotransporter. *J. Biol. Chem.* **2007**, *282*, 9042–9052. [CrossRef]
35. Langer, J.; Rose, C.R. Synaptically induced sodium signals in hippocampal astrocytes in situ. *J. Physiol.* **2009**, *587*, 5859–5877. [CrossRef]
36. Stoppini, L.; Buchs, P.A.; Muller, D. A simple method for organotypic cultures of nervous tissue. *J. Neurosci. Methods* **1991**, *37*, 173–182. [CrossRef]
37. Lerchundi, R.; Kafitz, K.W.; Farfers, M.; Beyer, F.; Huang, N.; Rose, C.R. Imaging of intracellular ATP in organotypic tissue slices of the mouse brain using the FRET-based sensor ATeam1.03^{YEMK}. *J. Vis. Exp. JoVE* **2019**, *154*, e60294. [CrossRef]
38. Gutierrez, R.; Heinemann, U. Synaptic reorganization in explanted cultures of rat hippocampus. *Brain Res.* **1999**, *815*, 304–316. [CrossRef]
39. Meier, S.D.; Kovalchuk, Y.; Rose, C.R. Properties of the new fluorescent Na^+ indicator CoroNa Green: Comparison with SBFI and confocal Na^+ imaging. *J. Neurosci. Methods* **2006**, *155*, 251–259. [CrossRef]
40. Kelly, T.; Rose, C.R. Ammonium influx pathways into astrocytes and neurones of hippocampal slices. *J. Neurochem.* **2010**, *115*, 1123–1136. [CrossRef]
41. Eitelmann, S.; Stephan, J.; Everaerts, K.; Durry, S.; Pape, N.; Gerkau, N.J.; Rose, C.R. Changes in Astroglial K^+ upon Brief Periods of Energy Deprivation in the Mouse Neocortex. *Int. J. Mol. Sci.* **2022**, *23*, 4836. [CrossRef]
42. Lerchundi, R.; Kafitz, K.W.; Winkler, U.; Farfers, M.; Hirrlinger, J.; Rose, C.R. FRET-based imaging of intracellular ATP in organotypic brain slices. *J. Neurosci. Res.* **2019**, *97*, 933–945. [CrossRef]
43. Haack, N.; Rose, C.R. Preparation, calibration and application of potassium-selective microelectrodes. In *Microelectrodes*; Lei, K.F., Ed.; Nova Science Publishers: New York, NY, USA, 2014; pp. 87–105.
44. Perkins, K.L. Cell-attached voltage-clamp and current-clamp recording and stimulation techniques in brain slices. *J. Neurosci. Methods* **2006**, *154*, 1–18. [CrossRef]

45. Thapaliya, P.; Pape, N.; Rose, C.R.; Ullah, A. Modeling the heterogeneity of sodium and calcium homeostasis between cortical and hippocampal astrocytes and its impact on bioenergetics. *Front. Cell. Neurosci.* **2023**, *17*, 1035553. [CrossRef]
46. Kenny, A.; Plank, M.J.; David, T. Macro scale modelling of cortical spreading depression and the role of astrocytic gap junctions. *J. Theor. Biol.* **2018**, *458*, 78–91. [CrossRef]
47. Salameh, A.I.; Hubner, C.A.; Boron, W.F. Role of Cl^-–HCO_3^- exchanger AE3 in intracellular pH homeostasis in cultured murine hippocampal neurons, and in crosstalk to adjacent astrocytes. *J. Physiol.* **2017**, *595*, 93–124. [CrossRef]
48. Haddad, N.J.; Agarwal, A.K. Metabolic Acidosis and Metabolic Alkalosis. In *Textbook of Nephrology*, 3rd ed.; Mandal, A.K., Prakash, J., Eds.; Jaypee Brothers Medical Publishers: New Delhi, India, 2014.
49. Faul, F.; Erdfelder, E.; Lang, A.G.; Buchner, A. G*Power 3: A flexible statistical power analysis program for the social, behavioral, and biomedical sciences. *Behav. Res. Methods* **2007**, *39*, 175–191. [CrossRef]
50. Siebens, A.W.; Boron, W.F. Depolarization-induced alkalinization in proximal tubules. I. Characteristics and dependence on Na^+. *Am. J. Physiol.* **1989**, *256*, F342–F353. [CrossRef]
51. Pappas, C.A.; Ransom, B.R. Depolarization-induced alkalinization (DIA) in rat hippocampal astrocytes. *J. Neurophysiol.* **1994**, *72*, 2816–2826. [CrossRef]
52. Ch'en, F.F.; Villafuerte, F.C.; Swietach, P.; Cobden, P.M.; Vaughan-Jones, R.D. S0859, an N-cyanosulphonamide inhibitor of sodium-bicarbonate cotransport in the heart. *Br. J. Pharmacol.* **2008**, *153*, 972–982. [CrossRef]
53. Meyer, J.; Gerkau, N.J.; Kafitz, K.W.; Patting, M.; Jolmes, F.; Henneberger, C.; Rose, C.R. Rapid Fluorescence Lifetime Imaging Reveals That TRPV4 Channels Promote Dysregulation of Neuronal Na^+ in Ischemia. *J. Neurosci.* **2022**, *42*, 552–566. [CrossRef]
54. Leng, T.; Shi, Y.; Xiong, Z.G.; Sun, D. Proton-sensitive cation channels and ion exchangers in ischemic brain injury: New therapeutic targets for stroke? *Prog. Neurobiol.* **2014**, *115*, 189–209. [CrossRef]
55. Chesler, M. Failure and function of intracellular pH regulation in acute hypoxic-ischemic injury of astrocytes. *Glia* **2005**, *50*, 398–406. [CrossRef]
56. Rose, C.R.; Ransom, B.R. Mechanisms of H^+ and Na^+ changes induced by glutamate, kainate, and D-Aspartate in rat hippocampal astrocytes. *J. Neurosci.* **1996**, *16*, 5393–5404. [CrossRef]
57. Theparambil, S.M.; Ruminot, I.; Schneider, H.P.; Shull, G.E.; Deitmer, J.W. The electrogenic sodium bicarbonate cotransporter NBCe1 is a high-affinity bicarbonate carrier in cortical astrocytes. *J. Neurosci.* **2014**, *34*, 1148–1157. [CrossRef]
58. Majumdar, D.; Bevensee, M.O. Na-coupled bicarbonate transporters of the solute carrier 4 family in the nervous system: Function, localization, and relevance to neurologic function. *Neuroscience* **2010**, *171*, 951–972. [CrossRef]
59. Rickmann, M.; Orlowski, B.; Heupel, K.; Roussa, E. Distinct expression and subcellular localization patterns of Na^+/HCO_3^- cotransporter (SLC 4A4) variants NBCe1-A and NBCe1-B in mouse brain. *Neuroscience* **2007**, *146*, 1220–1231. [CrossRef]
60. Kintner, D.B.; Su, G.; Lenart, B.; Ballard, A.J.; Meyer, J.W.; Ng, L.L.; Shull, G.E.; Sun, D. Increased tolerance to oxygen and glucose deprivation in astrocytes from Na^+/H^+ exchanger isoform 1 null mice. *Am. J. Physiol. Cell Physiol.* **2004**, *287*, C12–C21. [CrossRef]
61. Begum, G.; Song, S.; Wang, S.; Zhao, H.; Bhuiyan, M.I.H.; Li, E.; Nepomuceno, R.; Ye, Q.; Sun, M.; Calderon, M.J.; et al. Selective knockout of astrocytic Na^+/H^+ exchanger isoform 1 reduces astrogliosis, BBB damage, infarction, and improves neurological function after ischemic stroke. *Glia* **2018**, *66*, 126–144. [CrossRef]
62. Bevensee, M.O.; Apkon, M.; Boron, W.F. Intracellular pH regulation in cultured astrocytes from rat hippocampus. II. Electrogenic Na/HCO_3 cotransport. *J. Gen. Physiol.* **1997**, *110*, 467–483. [CrossRef]
63. Brune, T.; Fetzer, S.; Backus, K.H.; Deitmer, J.W. Evidence for electrogenic sodium-bicarbonate cotransport in cultured rat cerebellar astrocytes. *Pflügers Arch.* **1994**, *429*, 64–71. [CrossRef]
64. Wu, H.; Liu, S.; Su, P.; Xie, Z.D.; Gui, T.X.; Zhao, L.; Liu, Y.; Chen, L.M. Molecular insight into coordination sites for substrates and their coupling kinetics in Na^+/HCO_3^- cotransporter NBCe1. *J. Physiol.* **2022**, *600*, 3083–3111. [CrossRef]
65. Romero, M.F. Who's on first ... Na^+, HCO_3^- or CO_3^{2-}? *J. Physiol.* **2022**, *600*, 3005–3006. [CrossRef]
66. Turovsky, E.; Theparambil, S.M.; Kasymov, V.; Deitmer, J.W.; Del Arroyo, A.G.; Ackland, G.L.; Corneveaux, J.J.; Allen, A.N.; Huentelman, M.J.; Kasparov, S.; et al. Mechanisms of CO_2/H^+ Sensitivity of Astrocytes. *J. Neurosci.* **2016**, *36*, 10750–10758. [CrossRef]
67. Gourine, A.V.; Dale, N. Brain H^+/CO_2 sensing and control by glial cells. *Glia* **2022**, *70*, 1520–1535. [CrossRef]
68. Grichtchenko, I.I.; Chesler, M. Depolarization-induced alkalinization of astrocytes in gliotic hippocampal slices. *Neuroscience* **1994**, *62*, 1071–1078. [CrossRef]
69. Theparambil, S.M.; Naoshin, Z.; Thyssen, A.; Deitmer, J.W. Reversed electrogenic sodium bicarbonate cotransporter 1 is the major acid loader during recovery from cytosolic alkalosis in mouse cortical astrocytes. *J. Physiol.* **2015**, *593*, 3533–3547. [CrossRef]
70. Heidtmann, H.; Ruminot, I.; Becker, H.M.; Deitmer, J.W. Inhibition of monocarboxylate transporter by N-cyanosulphonamide S0859. *Eur. J. Pharmacol.* **2015**, *762*, 344–349. [CrossRef]
71. Del Rio, J.A.; Heimrich, B.; Soriano, E.; Schwegler, H.; Frotscher, M. Proliferation and differentiation of glial fibrillary acidic protein-immunoreactive glial cells in organotypic slice cultures of rat hippocampus. *Neuroscience* **1991**, *43*, 335–347. [CrossRef]
72. Frotscher, M.; Zafirov, S.; Heimrich, B. Development of identified neuronal types and of specific synaptic connections in slice cultures of rat hippocampus. *Prog. Neurobiol.* **1995**, *45*, 143–164. [CrossRef]
73. Benediktsson, A.M.; Schachtele, S.J.; Green, S.H.; Dailey, M.E. Ballistic labeling and dynamic imaging of astrocytes in organotypic hippocampal slice cultures. *J. Neurosci. Methods* **2005**, *141*, 41–53. [CrossRef]

74. Moskowitz, M.A.; Lo, E.H.; Iadecola, C. The science of stroke: Mechanisms in search of treatments. *Neuron* **2010**, *67*, 181–198. [CrossRef]
75. Daniele, S.G.; Trummer, G.; Hossmann, K.A.; Vrselja, Z.; Benk, C.; Gobeske, K.T.; Damjanovic, D.; Andrijevic, D.; Pooth, J.S.; Dellal, D.; et al. Brain vulnerability and viability after ischaemia. *Nat. Rev. Neurosci.* **2021**, *22*, 553–572. [CrossRef]
76. Erecinska, M.; Silver, I.A. Ions and energy in mammalian brain. *Prog. Neurobiol.* **1994**, *43*, 37–71. [CrossRef]
77. Hartings, J.A.; Shuttleworth, C.W.; Kirov, S.A.; Ayata, C.; Hinzman, J.M.; Foreman, B.; Andrew, R.D.; Boutelle, M.G.; Brennan, K.C.; Carlson, A.P.; et al. The continuum of spreading depolarizations in acute cortical lesion development: Examining Leao's legacy. *J. Cereb. Blood Flow Metab.* **2017**, *37*, 1571–1594. [CrossRef]
78. Mies, G.; Paschen, W. Regional changes of blood flow, glucose, and ATP content determined on brain sections during a single passage of spreading depression in rat brain cortex. *Exp. Neurol.* **1984**, *84*, 249–258. [CrossRef]
79. Rossi, D.J.; Brady, J.D.; Mohr, C. Astrocyte metabolism and signaling during brain ischemia. *Nat. Neurosci.* **2007**, *10*, 1377–1386. [CrossRef]
80. Kraig, R.P.; Chesler, M. Astrocytic acidosis in hyperglycemic and complete ischemia. *J. Cereb. Blood Flow Metab.* **1990**, *10*, 104–114. [CrossRef]
81. Mutch, W.A.; Hansen, A.J. Extracellular pH changes during spreading depression and cerebral ischemia: Mechanisms of brain pH regulation. *J. Cereb. Blood Flow Metab.* **1984**, *4*, 17–27. [CrossRef]
82. Hansen, A.J. Effect of anoxia on ion distribution in the brain. *Physiol. Rev.* **1985**, *65*, 101–148. [CrossRef]
83. Yao, H.; Azad, P.; Zhao, H.W.; Wang, J.; Poulsen, O.; Freitas, B.C.; Muotri, A.R.; Haddad, G.G. The Na^+/HCO_3^- co-transporter is protective during ischemia in astrocytes. *Neuroscience* **2016**, *339*, 329–337. [CrossRef]
84. Bondarenko, A.; Chesler, M. Rapid astrocyte death induced by transient hypoxia, acidosis, and extracellular ion shifts. *Glia* **2001**, *34*, 134–142. [CrossRef]
85. Lu, M.; Jia, M.; Wang, Q.; Guo, Y.; Li, C.; Ren, B.; Qian, F.; Wu, J. The electrogenic sodium bicarbonate cotransporter and its roles in the myocardial ischemia-reperfusion induced cardiac diseases. *Life Sci.* **2021**, *270*, 119153. [CrossRef]
86. Giffard, R.G.; Papadopoulos, M.C.; van Hooft, J.A.; Xu, L.; Giuffrida, R.; Monyer, H. The electrogenic sodium bicarbonate cotransporter: Developmental expression in rat brain and possible role in acid vulnerability. *J. Neurosci.* **2000**, *20*, 1001–1008. [CrossRef]
87. Villa-Gonzalez, M.; Martin-Lopez, G.; Perez-Alvarez, M.J. Dysregulation of mTOR Signaling after Brain Ischemia. *Int. J. Mol. Sci.* **2022**, *23*, 2814. [CrossRef]
88. Khakipoor, S.; Giannaki, M.; Theparambil, S.M.; Zecha, J.; Kuster, B.; Heermann, S.; Deitmer, J.W.; Roussa, E. Functional expression of electrogenic sodium bicarbonate cotransporter 1 (NBCe1) in mouse cortical astrocytes is dependent on S255–257 and regulated by mTOR. *Glia* **2019**, *67*, 2264–2278. [CrossRef]
89. Giannaki, M.; Ludwig, C.; Heermann, S.; Roussa, E. Regulation of electrogenic Na^+/HCO_3^- cotransporter 1 (NBCe1) function and its dependence on m-TOR mediated phosphorylation of Ser(245). *J. Cell. Physiol.* **2022**, *237*, 1372–1388. [CrossRef]
90. Jung, Y.W.; Choi, I.J.; Kwon, T.H. Altered expression of sodium transporters in ischemic penumbra after focal cerebral ischemia in rats. *Neurosci. Res.* **2007**, *59*, 152–159. [CrossRef]
91. Sohn, Y.; Yoo, K.Y.; Park, O.K.; Kwon, S.H.; Lee, C.H.; Choi, J.H.; Hwang, I.K.; Seo, J.Y.; Cho, J.H.; Won, M.H. Na^+/HCO_3^- cotransporter immunoreactivity changes in neurons and expresses in astrocytes in the gerbil hippocampal CA1 region after ischemia/reperfusion. *Neurochem. Res.* **2011**, *36*, 2459–2469. [CrossRef]
92. Ye, Q.; Jo, J.; Wang, C.-Y.; Oh, H.; Choy, T.J.; Kim, K.; D'Alessandro, A.; Reshetnyak, Y.K.; Jung, S.Y.; Chen, Z.; et al. Astrocytic Slc4a4 regulates blood-brain barrier integrity in healthy and stroke brains via a NO-CCL2-CCR2 pathway. *bioRxiv* **2023**. *preprint*. [CrossRef]
93. Toth, M.O.; Menyhart, A.; Frank, R.; Hantosi, D.; Farkas, E.; Bari, F. Tissue Acidosis Associated with Ischemic Stroke to Guide Neuroprotective Drug Delivery. *Biology* **2020**, *9*, 460. [CrossRef]
94. Simon, R.P. Acidotoxicity trumps excitotoxicity in ischemic brain. *Arch. Neurol.* **2006**, *63*, 1368–1371. [CrossRef]
95. Shono, Y.; Kamouchi, M.; Kitazono, T.; Kuroda, J.; Nakamura, K.; Hagiwara, N.; Ooboshi, H.; Ibayashi, S.; Iida, M. Change in intracellular pH causes the toxic Ca^{2+} entry via NCX1 in neuron- and glia-derived cells. *Cell. Mol. Neurobiol.* **2010**, *30*, 453–460. [CrossRef]
96. Beppu, K.; Sasaki, T.; Tanaka, K.F.; Yamanaka, A.; Fukazawa, Y.; Shigemoto, R.; Matsui, K. Optogenetic countering of glial acidosis suppresses glial glutamate release and ischemic brain damage. *Neuron* **2014**, *81*, 314–320. [CrossRef]
97. Sloan, S.A.; Barres, B.A. The Detrimental Role of Glial Acidification during Ischemia. *Neuron* **2014**, *81*, 221–223. [CrossRef]
98. Suzuki, M.; Van Paesschen, W.; Stalmans, I.; Horita, S.; Yamada, H.; Bergmans, B.A.; Legius, E.; Riant, F.; De Jonghe, P.; Li, Y.; et al. Defective membrane expression of the $Na^+–HCO_3^-$ cotransporter NBCe1 is associated with familial migraine. *Proc. Natl. Acad. Sci. USA* **2010**, *107*, 15963–15968. [CrossRef]
99. Larsen, B.R.; MacAulay, N. Activity-dependent astrocyte swelling is mediated by pH-regulating mechanisms. *Glia* **2017**, *65*, 1668–1681. [CrossRef]
100. Ostby, I.; Oyehaug, L.; Einevoll, G.T.; Nagelhus, E.A.; Plahte, E.; Zeuthen, T.; Lloyd, C.M.; Ottersen, O.P.; Omholt, S.W. Astrocytic mechanisms explaining neural-activity-induced shrinkage of extraneuronal space. *PLoS Comput. Biol.* **2009**, *5*, e1000272. [CrossRef]

101. Florence, C.M.; Baillie, L.D.; Mulligan, S.J. Dynamic volume changes in astrocytes are an intrinsic phenomenon mediated by bicarbonate ion flux. *PLoS ONE* **2012**, *7*, e51124. [CrossRef]
102. MacAulay, N. Molecular mechanisms of K^+ clearance and extracellular space shrinkage-Glia cells as the stars. *Glia* **2020**, *68*, 2192–2211. [CrossRef]
103. Close, B.; Banister, K.; Baumans, V.; Bernoth, E.M.; Bromage, N.; Bunyan, J.; Erhardt, W.; Flecknell, P.; Gregory, N.; Hackbarth, H.; et al. Recommendations for euthanasia of experimental animals: Part 2. DGXT of the European Commission. *Lab. Anim.* **1997**, *31*, 137–161. [CrossRef]

cells MDPI

Article

The Matrix Receptor CD44 Is Present in Astrocytes throughout the Human Central Nervous System and Accumulates in Hypoxia and Seizures

Osama Al-Dalahmah [1], Alexander A. Sosunov [2], Yu Sun [1], Yang Liu [3], Nacoya Madden [1], E. Sander Connolly [2], Carol M. Troy [1,4,5], Guy M. McKhann II [2] and James E. Goldman [1,5,*]

[1] Department of Pathology and Cell Biology, Vagelos College of Physicians and Surgeons, Columbia University Irving Medical Center and the New York Presbyterian Hospital, New York, NY 10032, USA
[2] Department of Neurosurgery, Vagelos College of Physicians and Surgeons, Columbia University Irving Medical Center and the New York Presbyterian Hospital, New York, NY 10032, USA; esc5@cumc.columbia.edu (E.S.C.)
[3] Department of Pathology, Albany Medical Center, Albany, NY 12208, USA
[4] Department of Neurology, Vagelos College of Physicians and Surgeons, Columbia University Irving Medical Center and the New York Presbyterian Hospital, New York, NY 10032, USA
[5] The Taub Institute, Columbia University Irving Medical Center, New York, NY 10032, USA
* Correspondence: jeg5@cumc.columbia.edu

Abstract: In the mammalian isocortex, CD44, a cell surface receptor for extracellular matrix molecules, is present in pial-based and fibrous astrocytes of white matter but not in protoplasmic astrocytes. In the hominid isocortex, CD44+ astrocytes comprise the subpial "interlaminar" astrocytes, sending long processes into the cortex. The hippocampus also contains similar astrocytes. We have examined all levels of the human central nervous system and found CD44+ astrocytes in every region. Astrocytes in white matter and astrocytes that interact with large blood vessels but not with capillaries in gray matter are CD44+, the latter extending long processes into the parenchyma. Motor neurons in the brainstem and spinal cord, such as oculomotor, facial, hypoglossal, and in the anterior horn of the spinal cord, are surrounded by CD44+ processes, contrasting with neurons in the cortex, basal ganglia, and thalamus. We found CD44+ processes that intercalate between ependymal cells to reach the ventricle. We also found CD44+ astrocytes in the molecular layer of the cerebellar cortex. Protoplasmic astrocytes, which do not normally contain CD44, acquire it in pathologies like hypoxia and seizures. The pervasive and inducible expression of CD44 in astrocytes is a novel finding that lays the foundations for functional studies into the significance of CD44 in health and disease.

Keywords: CD44; astrocytes; hypoxia; seizures; interlaminar astrocytes

Citation: Al-Dalahmah, O.; Sosunov, A.A.; Sun, Y.; Liu, Y.; Madden, N.; Connolly, E.S.; Troy, C.M.; McKhann, G.M., II; Goldman, J.E. The Matrix Receptor CD44 Is Present in Astrocytes throughout the Human Central Nervous System and Accumulates in Hypoxia and Seizures. *Cells* **2024**, *13*, 129. https://doi.org/10.3390/cells13020129

Academic Editors: Vladimir Parpura, H. Rheinallt Parri and Michael Brenner

Received: 17 March 2023
Revised: 20 November 2023
Accepted: 22 December 2023
Published: 10 January 2024

1. Introduction

The interlaminar astrocytes of the mammalian isocortex were first described well over one hundred years ago [1] and further characterized in recent years [2–4]. These astrocytes display marked differences from the better-known, bushy, protoplasmic astrocytes of the cortex. They project multiple, long, unbranched cell processes, usually oriented orthogonally to the pial surface, through several cortical layers, forming cascades of parallel processes. A subset of astrocytes in the deep cortical layers near the subcortical white matter also projects similar, radially oriented processes into the deeper cortical layers. Immunocytochemical studies reveal that these long-process astrocytes contain CD44 [5,6], a cell surface receptor for extracellular matrix molecules, including osteopontin, hyaluronan, laminin, and integrins [7]. They are different from the protoplasmic astrocytes, which do not contain detectable levels of CD44 [6,8], not only in morphology but also by being low in glutamine synthetase and the two plasmalemmal glutamate transporters, EAAT1 and

EAAT2 [6]. The astrocytes in white matter are also CD44+ [6,8–11]. However, CD44 can be increased in all types of astrocytes in response to pathological conditions [6,8].

We have now examined other parts of the human central nervous system (CNS) and found that CD44+ astrocytes are common in every part of the CNS, including cortices, striatum, thalamus, brainstem, cerebellum, and spinal cord. In all areas, we found CD44 associated with long-process astrocytes. CD44+ astrocytes populate white matter, subependymal zones, and subpial areas. We found that the proximity of CD44+ processes to neuronal cell bodies varies considerably. In particular, unlike neurons in the isocortex and diencephalon, the motor neurons of the brainstem and spinal cord show a close apposition of CD44+ processes and are even encircled by these processes. We raise the issue of whether this apposition may suggest a novel, intimate relationship between neurons and CD44+ astrocytes for specific neuronal classes.

In this study, we have also examined biopsy and autopsy material from individuals who had experienced hypoxic events and seizures and found that protoplasmic astrocytes accumulate CD44 in these pathological conditions. Furthermore, we have investigated rat models of ischemia and seizures, finding that in those brains, protoplasmic astrocytes also acquire CD44.

2. Materials and Methods

2.1. Human Autopsy and Biopsy Material

Eight autopsy brains, fixed in 10% formalin for 10 days after removal, were sectioned coronally and samples from all areas of the CNS, including superior frontal, cingulate, and striate cortices with subcortical white matter, hippocampus with temporal isocortex, basal ganglia, thalamus, midbrain, pons, medulla, cerebellum with cortex and dentate nucleus, and spinal cord were removed and embedded in paraffin blocks, cut at 7 μm thickness, and mounted on glass slides. Patients' ages ranged from 27 to 74; 6 were female, and 2 were male. None of these individuals displayed neurological symptoms, and none of the brains showed evidence of neuropathology except for mild atherosclerosis and/or arteriolosclerosis with acute but mild hypoxic/ischemic changes. In addition, we sampled sections of anterior striatum from 3 more autopsies (ages in years and sex 1M (male), 76F (female), and 94F), sections of hippocampus from 8 more autopsies (ages in years and sex: 1M, 2M, 5M, 7M, 73M, 71M, 76M, and 66M), and sections of spinal cord from 4 more autopsies (ages in years and sex: 27F, 66M, 80M, and 80M). We also sampled sets of sections of cerebral hemispheres containing the subependymal zone at the lateral ventricle from 8 fetal and neonatal brains, ranging in age from 19 to 40 weeks of gestation and 1 day to 7 weeks of postnatal life. All autopsies were performed with the consent of the next of kin, and all protocols were approved by the Institutional Review Board of Columbia University Medical Center.

2.2. Epilepsy Surgical Specimens

We examined 9 samples of temporal isocortex of patients with mesial temporal sclerosis (ages ranged from 12 to 55; 6M and 3F). We also examined 5 samples of hippocampi from patients with mesial temporal sclerosis (ages ranged from 9 to 66; 3M and 2F) obtained from surgical resections from patients with medically intractable epilepsy. We did not include specimens from seizure patients with brain tumors, vascular malformations, cortical dysplasias, and inflammatory and infectious disorders, although we found that many of them also show CD44+ astrocytes.

2.3. Hypoxia Specimens

Eight neurosurgical specimens that represented hypoxic/ischemic insults were selected. We used both the time between the initial clinical presentation and the biopsy and the neuropathology to classify these as acute, subacute, or chronic hypoxic/ischemic changes, acute within a few days, and showing eosinophilic neuronal change but no reactivities of blood vessels, astrocytes, or microglia, subacute between 10 and 14 days and showing proliferative vasculature and reactive astrocytes and microglia with macrophages, and chronic more than 14 days and showing tissue necrosis with foamy macrophages. Some of the biopsies showed acute and subacute or subacute and chronic changes, indicating progressive changes in the evolution of the lesions. The age range was 19–82, 3M and 5F. CD44-positive astrocytes were counted in three 20X (objective numerical aperture 0.40) power fields from each case. Sample *t*-tests were performed between pairs of the three groups.

2.4. Rodent Hypoxia/Ischemia and Seizure Models

Adult male rats were housed in standard cages with free access to food and water on a 12 h light/dark cycle. All procedures performed on animals were approved by Columbia University's Institutional Animal Care and Use Committee and conducted according to Institutional and Federal guidelines.

2.5. Stroke/Transient Middle Cerebral Artery Occlusion (tMCAO)

Wistar rats (275–300 g) were subjected to unilateral tMCAO using intraluminal vascular occlusion. Animals were anesthetized with halothane in a mix of 70% nitrous oxide/30% oxygen, and core temperatures were maintained at 37 °C throughout the entire procedure and for 60 min after reperfusion. The right common carotid artery, the right external carotid artery, and the right internal carotid artery were exposed and isolated. MCAO was accomplished by advancing a 25 mm 4–0 nylon suture with a blunted silicone tip (outer diameter, 0.38 mm) through an incision in the external carotid artery until the suture was 18 mm past the carotid bifurcation. MCAO was confirmed by transcranial measurements of cerebral blood flow via laser Doppler flowmetry (Periflux System 5000; Perimed, Inc., Järfälla, Sweden). After 120 min, the suture was removed, and reperfusion was confirmed by laser Doppler flowmetry. After 7 days, animals were deeply anesthetized with an overdose of ketamine/xylazine and perfused with 4% (v/v). After perfusion, brains were removed and additionally fixed in 4% PFA in phosphate-buffered saline (PBS) for 14–18 h (4 °C). Forty-micron sections were prepared with a vibratome (Leica VT1000S, Wetzlar, Germany) and stored in cryoprotectant solution at −20 °C. After perfusion, other brains were dissected and incubated at 4 °C with 4% (w/v) PFA (24 h), PBS (24 h), and 30% sucrose (v/v) in PBS, then embedded in O.C.T. compound and 20 μm cryosections were prepared.

2.6. Pilocarpine-Induced Status Epilepticus

Seizures were produced in rats with pilocarpine, as described elsewhere [12]. After premedication with scopolamine (5 mg/kg, i.p.) to prevent the effects of peripheral cholinergic stimulation, pilocarpine (330 mg/kg, i.p.) was administered to Sprague–Dawley rats (100–150 g) to induce seizures. Seizures were graded on the modified Racine scale [13], and only animals with grade 4–5 seizures for 2 h were used in experiments. After 2 h of continuous seizures, ketamine (80 mg/kg, i.p.) was administered to stop seizures, and a second dose (40 mg/kg, i.p.) was administered if seizures did not stop in 10 min after the first. Animals were deeply anesthetized 3 days post-seizures and perfused as above. After perfusion, brains were removed and processed as above.

2.7. Immunostaining and Antibodies

Formalin-fixed paraffin-embedded (FFPE) sections were stained using hematoxylin and eosin (H&E). Immunohistochemistry with positive and negative controls was performed using an antibody to CD44 (Roche, mouse monoclonal antibody pre-diluted at 1:100 and run on the Ventana platform using the Ultraview DAB kit). For immunofluorescence staining on tissue sections from the rat MCAO occlusions and pilocarpine-induced seizures, we used a CD44 mouse monoclonal antibody (1:500, OX49/ab238464, Abcam, Cambridge, UK) and a glial fibrillary acidic protein (GFAP) rabbit polyclonal (1:1000, Z 0334, Dako, Carpinteria, CA, USA). Secondary antibodies conjugated to fluorophores: goat anti-mouse Alexa Fluor™ 488, 594, and 633, goat anti-rabbit Alexa Fluor™ 488 and

594, and donkey anti-goat Alexa Fluor^{TM} 488, 594, and 633 (1:300, ThermoFisher Scientific, Eugene, OR, USA). After blocking with 10% (v/v) normal goat (or donkey) serum (30 min, RT = 20–22 °C), free-floating sections were incubated in a mixture of primary antibodies overnight (4 °C). Alexa Fluor^{TM}-conjugated secondary antibodies were used for 1 h at RT = 20–22 °C). For visualization of nuclei, DAPI (5 μg/mL; D9542, Sigma-Aldrich, St. Louis, MI, USA) was applied with the secondary antibodies. Blocking serum, primary, and secondary antibodies were applied in 0.2% Triton X-100 in PBS. Sections for fluorescence microscopy were mounted on slides in Vectashield (Vector Laboratories, Burlingame, CA, USA). To control for the specificity of immunostaining, primary antibodies were omitted and substituted with appropriate normal serum. Slides were viewed using a confocal microscope (Nikon Ti Eclipse, Tokyo, Japan). For immunofluorescence staining of cerebellar sections, we used 5–7 μm thick FFPE sections, which were immunostained on a Leica Bond RXm™ with the following modifications: After blocking in 10% (v/v) donkey serum, sections were incubated with rabbit anti-CD44 monoclonal antibody (1:100, Abcam #ab101531) and chicken anti-GFAP (1:1000, #b4674, Abcam) using antigen retrieval with Leica ER2 antigen retrieval buffer for 20 min. The secondary antibodies were Alexa Fluor^{TM} 488 donkey anti-chicken and 568 donkey anti-rabbit (1:500). Images were acquired at 40X (objective numerical aperture 0.65) on a Leica DMi8 Thunder system.

2.8. Quantification of CD44 in Hypoxia Sections

All CD44+ astrocytes were counted in 20X (numerical aperture 0.40) objective fields on a Nikon BX43 microscope. Three fields were counted per slide. Statistical analysis was performed using ANOVA. Data are shown as mean ± SEM. For quantification of the proportion of astrocytes that are CD44+ in different brain regions, we used QuPath to first quantify GFAP+ astrocytes using positive cell detection, then trained a random tree classifier to classify astrocytes as CD44+ or CD44−. The training data for the classifier included many examples of positive and negative cells derived from all of the slides examined.

2.9. Quantification of CD44 in Pilocarpine Sections

GFAP+ astrocytes were counted as either CD44+ or CD44− in the cortical area that contained double-labeled astrocytes. All GFAP+ astrocytes were counted in an area 221 × 221 μm^2 with a 20X objective lens (objective numerical aperture 0.40) of the confocal microscope, and the percentage of all GFAP+ astrocytes that were also CD44+ was quantitated.

2.10. Statistics

We performed all statistics in GraphPad Prism 9 (GraphPad Software, Boston, MA, USA) or Miscrosoft Excel. We did not perform a power analysis because the experimental studies are well known. The post hoc statistical analysis examined a large number of astrocytes in several animals and showed a highly significant effect. Unpaired Student t-tests were used to compare test groups, hypoxia vs. control for the tMCAO study and pilocarpine vs. controls for the seizure study, as indicated in the text. For reporting the percentage of astrocytes that were CD44+, we did not perform statistical tests. All data are presented as mean ± standard error of the mean unless otherwise indicated.

3. Results

3.1. Cell Specificity of CD44 Immunostaining for Astrocytes

CD44 co-localized with GFAP in the isocortex and hippocampus [6]. Neurons, oligodendrocytes, myelin, and microglia were CD44-negative, although macrophages and classes of lymphocytes were positive. None of the autopsy brains we examined for the characterization of CD44+ astrocytes in non-pathological tissues had any neuropathological findings, including areas of macrophage accumulation, lymphocyte infiltration, or astrocyte scars.

3.2. Isocortex

The cellular architecture of CD44+ astrocytes has been described in the isocortex [5,6,14] and matches that of the well-known interlaminar astrocytes with cell bodies at or near the pial surface and also astrocytes around large blood vessels. Note that the CD44+ astrocyte cell bodies sit at or near the pial surface (Figure 1). Isocortical neurons are not apposed or surrounded by CD44+ processes, including the Betz cells of the primary motor cortex (see below, in contrast with brainstem and spinal cord motor neurons, which are surrounded by CD44+ processes).

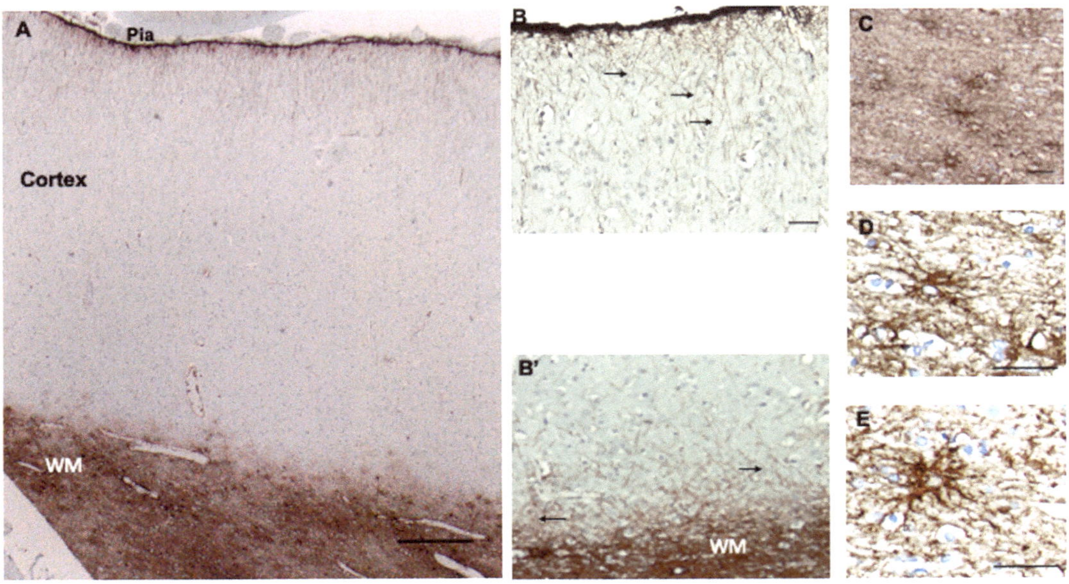

Figure 1. Superior frontal cortex shows CD44 positivity in subpial (Pia), interlaminar astrocytes, and white matter (WM) astrocytes, but not in the protoplasmic astrocytes of the cortex (**A**). Higher magnification of the pial surface and upper cortex shows the long interlaminar processes, radial to the pial surface (arrows) (**B**). Higher magnification of the white matter and lower cortex shows long astrocyte processes emerging into the cortex (arrows) (**B′**). (**B**,**B′**) are not from (**A**) but from other specimens, but all cortical specimens show the same pattern. The subcortical white matter contains dense astrocyte processes (**C**). Higher magnification of white matter astrocytes, which extend long processes in many directions, although the longest processes are roughly parallel and in the direction of axonal tracts (**D**,**E**). All specimens are counterstained with hematoxylin. Scale bars: (**A**) 250 μm, (**B**) 50 μm, (**C**) 25 μm, (**D**,**E**) 20 μm.

3.3. Subcortical White Matter

The fibrous astrocytes of white matter were CD44+ in every area of the CNS. These astrocytes were complex in shape, extending long processes along axonal tracts and shorter processes in other orientations as well (Figure 1C–E). As we have noted previously [6], some of the subcortical white matter astrocytes projected long, unbranched, radially oriented processes into the upper cortical levels (Figure 1B′). A number of other CNS areas also showed the projection of long CD44+ processes from white matter into gray matter (e.g., see dentate nucleus below).

3.4. Striatum

CD44+ astrocytes occupied the subependymal layer, forming a dense matrix (Figure 2A–F). From there, long processes of CD44+ astrocytes were projected into the caudate nucleus. In examining the ependymal lining, we found that CD44+ processes intercalated between the CD44-negative ependymal cells (Figure 2B–F). Confocal imaging confirmed this finding (Figure 2E,F). In the ependymal lining over the caudate nucleus, these processes were more frequent dorsally than ventrally (Figure 2C,D). We observed this dorsal–ventral gradient in 11 of 11 sections of anterior caudate in patients ranging in age from 1 year to 94 years. In addition, we examined the germinal matrix at the angle of the lateral ventricle in eight fetal brains, aged 19 to 40 weeks of gestation, and four postnatal brains, aged 1 day to 7 weeks. All showed CD44+ processes in between ependymal cells (Supplemental Figure S1). In some cases, it was possible to trace these processes into the underlying subependymal zone. CD44+ astrocytes also appeared in the striatal white matter in both the pencil fiber bundles and the internal capsule, as expected. Protoplasmic astrocytes of the striatal gray matter were CD44-negative. No caudate or putamen neurons were surrounded by CD44+ processes.

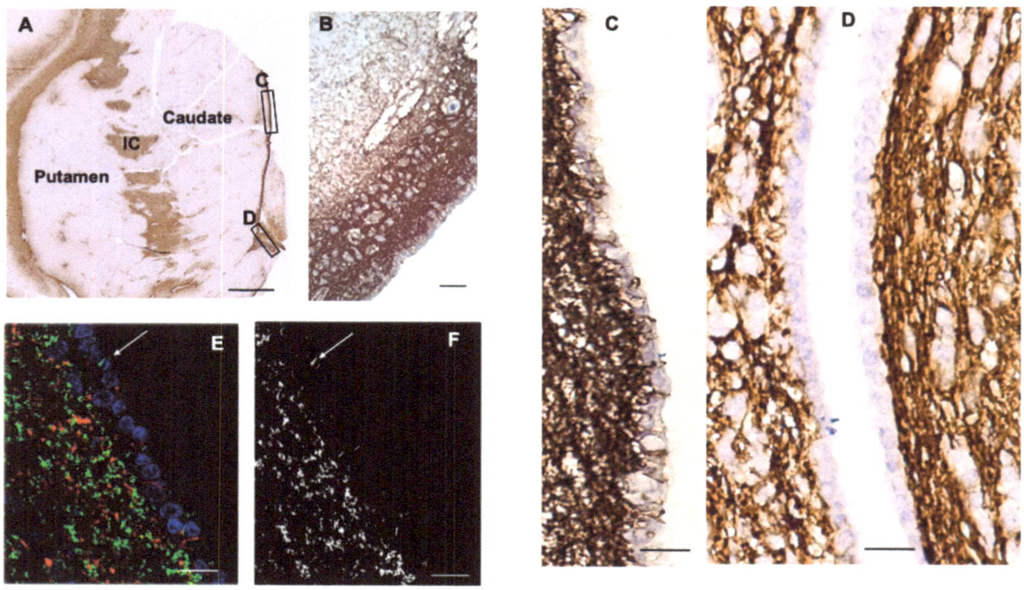

Figure 2. Caudate and putamen show CD44 positivity in the subependymal region, the internal capsule (IC), and the white matter fibers of the striatum (**A**). Higher magnification of the subependymal regions shows a dense network of CD44+ processes, some of which extend into the caudate (**B**). Higher magnification of the subependymal network (**C,D**). CD44+ processes extend between ependymal cells over the dorsal caudate to the ventricle in the dorsal area (**C**), but far fewer of these are present in more caudal areas (**D**). Confocal imaging also reveals CD44+ processes in between ependymal cells (arrows) (**E**, CD44—green, GFAP—red, DAPI—blue, (**F**, CD44+ only). Specimens (**A–D**) are counterstained with hematoxylin. Scale bars: (**A**) 5 mm, (**B**) 100 μm, (**C–F**) 20 μm.

3.5. Thalamus

In paraventricular thalamic nuclei, there was a dense network of CD44+ processes in the subependymal zone (Figure 3A,B). CD44+ processes also intercalated between ependymal cells (Figure 3B). Astrocytes surrounding large thalamic blood vessels appeared CD44+ and sent long processes into the thalamic neuropil (Figure 3C). There were more CD44+ astrocytes dispersed throughout thalamic nuclei than in the striatum. This may correlate with the large numbers of groups of myelinated axons that course through the thalamus, although some CD44+ astrocytes appear to be unassociated with white matter or blood vessels (Figure 3D). The anterior and lateral dorsal thalamic nuclei were relatively poor in CD44+ astrocytes, whereas the lateral nuclei and the subthalamic nucleus contained more of them (Figure 3A). Thalamic neurons were not surrounded by CD44+ processes, however. In the lateral geniculate nucleus (LGN), the white matter between layers contained CD44+ astrocytes, which extended long processes into the neuronal layers (Figure 3E). The large neurons of the magnocellular layers I and II were surrounded by CD44+ staining, while the small neurons in the other layers were not (Figure 3F,G).

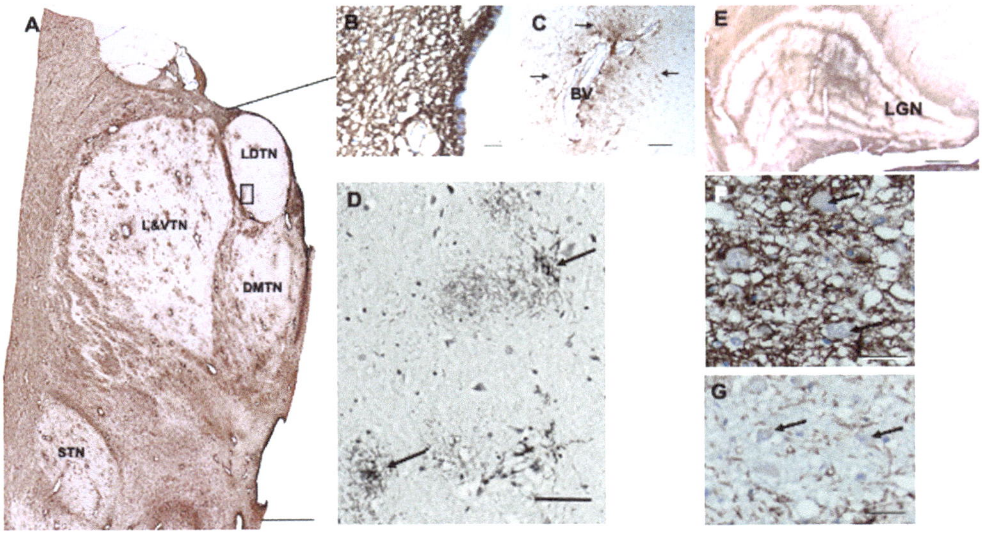

Figure 3. The anterior thalamus, including the latero-dorsal (LDTN), dorso-medial (DMTN), and latero-ventral (L&VTN) nuclei and the subthalamic nucleus (STN) show CD44 positivity around blood vessels and in white matter striae (**A**). CD44+ processes intercalate between ependymal cells (the line from A to B denotes the ependymal and subependymal area shown in (**B**)). Higher magnification of the boxed area in (**A**) shows a large blood vessel surrounded by CD44+ astrocytes, which send long unbranched processes into the parenchyma (arrows indicate some of the many processes) (**C**). Some of the CD44+ astrocytes in the L&VTN do not apparently contact blood vessels (arrows, (**D**)). Myelinated fiber bands of the lateral geniculate nucleus (LGN) are CD44+ (**E**). CD44+ surrounds magnocellular neurons (arrows, (**F**)) but not parvicellular neurons (arrows, (**G**)). All specimens are counterstained with hematoxylin. Scale bars: (**A**) 5 mm, (**B**) 10 μm, (**C**,**D**) 50 μm, (**F**,**G**) 20 μm.

3.6. *Hypothalamus, Third Ventricle*

As in other paraventricular areas, there was a dense subependymal network of CD44+ processes and an extension of long, unbranched processes from this into the parenchyma (Supplemental Figure S2A). As with the lateral ventricle, there were apparently more CD44+ processes intercalating between ependymal cells in the dorsal aspect compared with the ventral aspect of the third ventricle (Supplemental Figure S2B,C). Neurons of the paraventricular nuclei of the hypothalamus were completely or partially surrounded by CD44+ processes (Supplemental Figure S2D).

3.7. *Hippocampus*

As described [6], CD44+ astrocytes in the stratum oriens projected long processes through the depth of the pyramidal layer of CA1–CA3 and subiculum, ending at the border of the stratum pyramidale, which contains the cell bodies of the pyramidal neurons (Figure 4A,B). The pyramidal neurons were not surrounded by CD44+ processes. CD44+ astrocytes were localized to the molecular layer of the dentate gyrus, in which processes of astrocytes in the subgranular layer extended through the molecular layer (Figure 4B′).

CD44+ astrocytes occupied the stratum lacunosum, the white matter tract from the entorhinal cortex. Note that both white matter tracts, the alveus (AL) and the stratum lacunosum (SL), are CD44+ labeled, as are all other white matter tracts in the CNS.

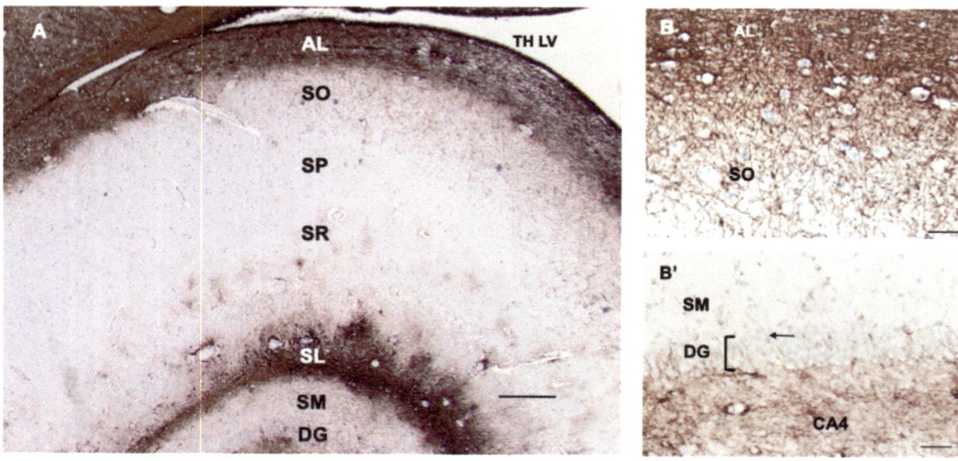

Figure 4. A section through the layers of the hippocampus (**A**) show CD44 positivity in the alveus (AL) and stratum lacunosum (SL). Thin, unbranched CD44+ processes run through the stratum oriens (SO) toward the pyramidal layer (stratum pyramidale, SP) (**A**,**B**). CD44+ processes do not surround pyramidal cells. Astrocytes in the stratum pyramidale, stratum radiatum (SR), and stratum moleculare (SM) show little CD44 reactivity. Dentate gyrus (DG). A higher magnification (**B**) shows the radially oriented processes in the SO. A higher magnification also shows the CD44+ processes (arrow) emanating from the subgranular zone of the CA4 sector through the dentate gyrus (DG, bracket) into the stratum moleculare (**B′**). (**B**,**B′**) are from a different specimen than (**A**), but all hippocampal specimens show the same patterns. Temporal horn of the lateral ventricle (TH LV). All specimens are counterstained with hematoxylin. Scale bars: (**A**) 250 μm, (**B**,**B′**) 200 μm.

3.8. Cerebellum

A. Cortex—We found cell bodies of CD44+ cells located either at the pial surface or just below it (Figure 5A,C–F). Many processes extended radially toward the Purkinje cell layer (Figure 5A and Supplemental Figure S3D), although we found that many cells also showed processes that were parallel to the pia (Figure 5C–F). The granule cell layer itself contained a network of CD44+ processes from the velate astrocytes, the astrocytes that populate this layer, and surrounding groups of granule cells, but not individual granule cells (Figure 5A and Supplemental Figure S3E,F). Finally, the spatial relationship of Purkinje cells to CD44+ processes seemed to vary. Some were not surrounded by CD44+ processes, although a few processes closely traveled by the lateral aspects of Purkinje cells, while others appeared to be surrounded (Supplemental Figure S3A–C). CD44+ processes extended radially from the granule cell layer into the molecular layer (Supplemental Figure S3D), indicating that some of the thin processes in the molecular layer had their origin in the granule cell layer. Moreover, in the Purkinje cell layer, a small subset of Bergmann glia, which sends long GFAP+ processes all the way to the pial surface, was CD44+ (Supplemental Figure S4D–F).

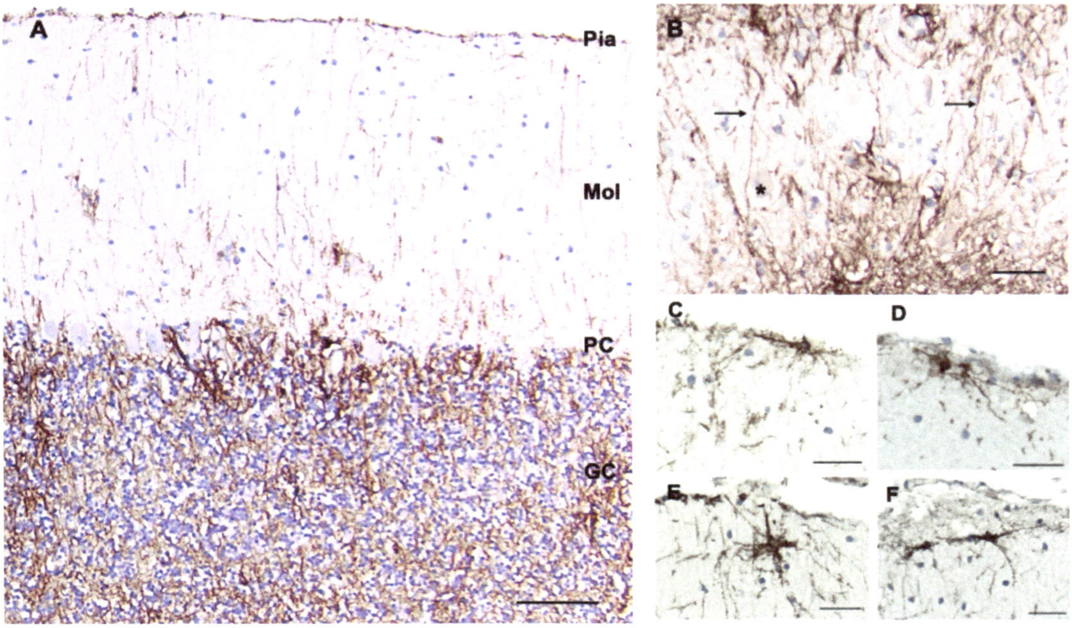

Figure 5. A section through the cerebellar cortex shows CD44+ long processes arising from the pial surface and entering the molecular layer (**A**). Some appear to begin in the granule cell layer and project into the molecular layer (see Supplemental Figure S3D). Fine CD44+ processes course through the granule cell layer, surrounding groups of granule cells, but do not surround individual granule cells. CD44+ processes (arrows) run through the dentate nucleus but do not surround dentate neurons (one labeled with *) (**B**). CD44+ astrocytes at or near the pial surface of the molecular layer extend processes parallel to the pia or into the molecular layer (**C–F**). Molecular layer (Mol), purkinje cell layer (PC), granule cell layer (GC). All specimens are counterstained with hematoxylin. Scale bars: (**A**) 100 µm, (**B–F**) 20 µm.

B. Dentate nucleus—The white matter surrounding the dentate nucleus, like other white matter in the CNS, was CD44+. Thin CD44+ processes entered and sometimes crossed the dentate nucleus and occasionally enwrapped some dentate neurons but not others (Figure 5B and Supplemental Figure S4A–C).

3.9. Brainstem

Many areas of the brainstem contained CD44+ astrocyte cell bodies and processes, in part due to the abundance of white matter pathways. In the midbrain, white matter pathways, such as the cerebral peduncles and the medial lemniscus, were all CD44+ (Figure 6). The red nucleus contained many CD44+ astrocytes, likely correlating with the myriad of myelinated axon bundles, but neurons were not surrounded by CD44+ processes. The cell bodies of motor neurons of the oculomotor nucleus were in close contact with CD44+ processes, in some cases enwrapped by them (Figure 6B). The substantia nigra contained few CD44+ astrocytes, and nigral neurons were not enwrapped by CD44+ processes (Figure 6C).

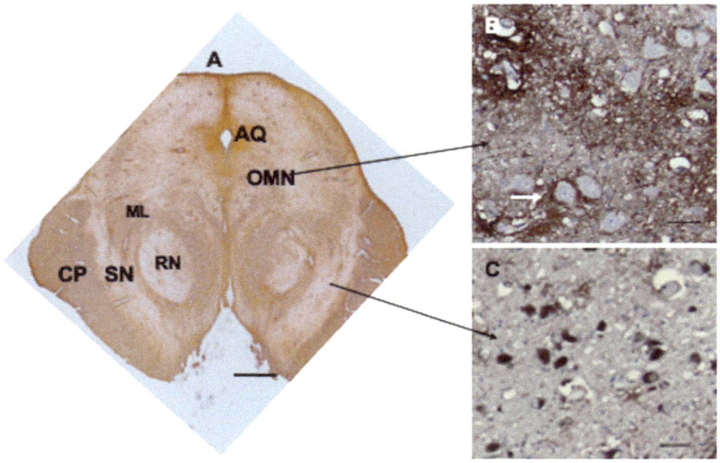

Figure 6. The midbrain appears complex, with CD44 staining of white matter tracts such as the cerebral peduncles (CP) and medial lemniscus (CP). The red nucleus (RN) contains CD44+ astrocytes. The substantia nigra (SN) appears relatively free of CD44+ staining (**A**). Neurons of the oculomotor nucleus (OMN) are surrounded by CD44+ staining (arrow shows one of these neurons) (**B**), but neurons of the substantia nigra are not (**C**). Aqueduct of Sylvius (AQ). Scale bars: (**A**) 5 mm, (**B**,**C**) 25 µm.

The medulla also contained CD44+ astrocytes in white matter tracts (Figure 7A). Neurons of the hypoglossal nucleus (Figure 7B), the dorsal motor nucleus of the vagus (Figure 7C), the nucleus ambiguous, but not the motor neurons of the primary notor cortex (Betz ells) (Figure 7D) or the solitary tract nucleus or trigeminal nuclei, were surrounded by CD44+ processes (see Table 1 for a list of neurons that are and are not surrounded).

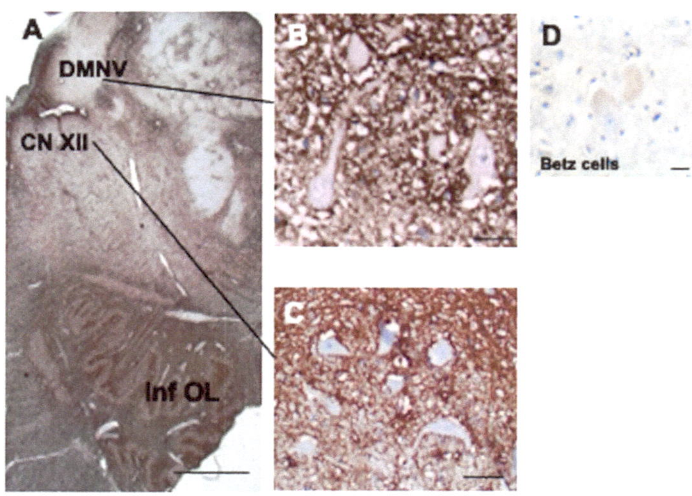

Figure 7. The medulla contains many CD44+ processes, including white matter tracts (**A**). Motor neurons of the dorsal motor nucleus of the vagus (**B**) and hypoglossal nucleus (**C**) are surrounded by CD44+ processes. In contrast, Betz cells of the motor cortex are not surrounded by CD44+ processes (**D**). Hypoglossal nucleus (CN XII), dorsal motor nucleus of the vagus (DMNV), inferior olivary nucleus (Inf OL). All specimens are counterstained with hematoxylin. Scale bars: (**A**), 2 mm, (**B**,**C**) 10 µm, (**D**) 50 µm.

Table 1. A list of neurons that are or are not surrounded by CD44+ processes.

Not Surrounded by CD44+	**Surrounded by CD44+**
Isocortex, including Betz cells	Oculomotor nucleus
Thalamus	Facial nucleus
Basal ganglia	Hypoglossal nucleus
Subthalamic nucleus	Dorsal motor nucleus of vagus
Hippocampal dentate gyrus	Lateral vestibular nucleus
Hippocampal and para-hippocampal pyramidal neurons	Inferior olivary nucleus
Locus ceruleus	Spinal anterior motor neurons
Red nucleus	Onufrowitz nucleus
Substantia nigra	Clarke's column
Colliculi	Intermediolateralis nucleus
Pontine base	
Cerebellar granule cells	
Dentate nucleus +/−	
Purkinje cells +/−	
Solitary nucleus	

3.10. Spinal Cord

The CD44 immunostaining of the spinal cord revealed CD44+ astrocytes in gray and white matter (Figure 8A–D). Motor neurons of the anterior horn were in close contact with CD44+ processes, in some cases enwrapped by them (Figure 8B). We found these contacts in all cord levels: cervical, thoracic, and lumbar. Neurons of the nucleus intermediolateralis (Figure 8C) and Clarke's column (Figure 8D) were also surrounded by CD44+ processes. Double immunofluorescence labeling for CD44 and GFAP shows that CD44+ overlaps extensively with GFAP+, indicating that these processes belong to astrocytes (Figure 8 E,F). As a quantitative measure of neurons surrounded by CD44+ processes, we counted the numbers of anterior horn motor neurons in the CD44-immunostained spinal cords in eight separate autopsies (two anterior horns per section), each slide counterstained with hematoxylin to reveal cell nuclei and cytoplasm and found that all were surrounded by CD44+ processes (461/461).

To quantitate the proportion of GFAP+ cells that are also CD44+, we counted the double-labeled cells in the caudate nucleus, including the parenchyma (gray matter only) subependymal region, the pencil fibers (white matter tracts), and around small arteries and then in the hypoglossal nucleus (Figure 9). Examples of double-labeled cells are provided in Figure 9A–E, and the quantitation of the proportion of GFAP+ cells that are also CD44+ in Figure 9F).

Finally, as a summary of all neurons that are or are not surrounded by CD44+ processes, we provide Table 1.

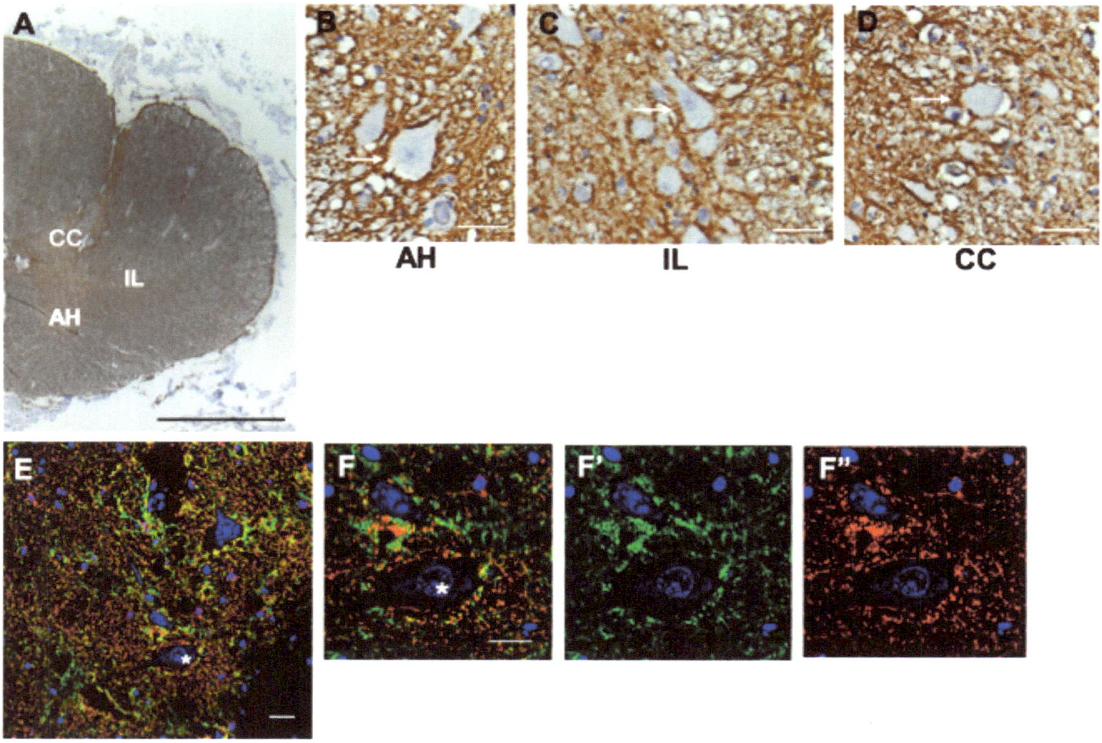

Figure 8. A hemi-section through the thoracic spinal cord, dorsal at top, ventral at bottom, shows many CD44+ processes, including white matter tracts (**A**). Motor neurons of the anterior horn (**B**), intermediolateralis (**C**), and Clarke's column (**D**) are surrounded by CD44+ processes. One neuron in each panel is shown by an arrow. Anterior horn (AH), nucleus intermediolateralis (IL). Clarke's column (CC). Specimens in (**A–D**) are counterstained with hematoxylin. Double immunolabeling for CD44—green and GFAP—red, showing low magnification of ventral horn (**E**, motor neuron *) and high magnification of the same anterior horn motor neuron (**F**, double stain, **F'**, CD44, **F''**, GFAP). Scale bars: (**A**), 5 mm, (**B–F**) 20 µm.

3.11. CD44+ Astrocytes in the Hypoxic Cortex

Protoplasmic astrocytes of the isocortex are not normally CD44+. To see if the protoplasmic astrocytes of the human cortex increase CD44 after hypoxia/ischemia, we examined by immunohistochemistry a series of neurosurgical and autopsy isocortical specimens from patients with well-defined clinical histories for the onset of hypoxic/ischemic events, allowing us to gauge the timing of CD44 accumulation (Figure 10). Very few CD44+ astrocytes were present in acute infarcts (1–2 days) (Figure 10A,B). However, within several days after the insult, some cortical astrocytes began to show weak staining for CD44 (Figure 10C,D), and by 12–14 days, many astrocytes in the affected cortex stained strongly (Figure 10E,F and Supplemental Figure S5). The CD44+ astrocytes resided in the penumbra of infarcts or within hypoxic lesions that did not contain necrosis.

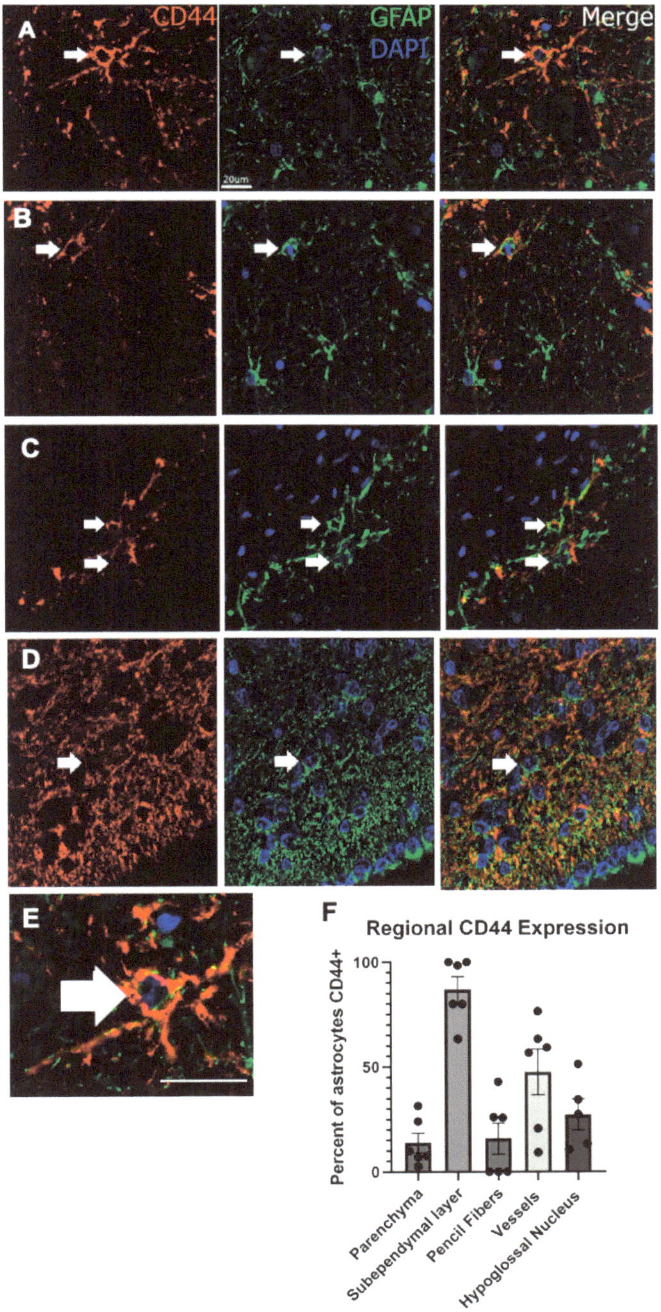

Figure 9. Examples of GFAP+/CD44+ double-labeled astrocytes in the hypoglossal nucleus (**A**), in the caudate parenchyma, excluding the pencil fibers (**B**), around small arteries (**C**), and the subependymal lining of the caudate nucleus (**D**). A higher magnification of A shows the double labeling (**E**). Note that

the GFAP label localizes inside the CD44 label, the former being cytoplasmic and the latter membraneous. Note also that the CD44 extends beyond the GFAP since GFAP normally does not extend into the smaller distal processes of astrocytes. Quantitative measurements of double-labeled cells displayed as the percentage of GFAP+ cells that were also CD44+ (**F**). Arrows in all images show double-labeled cells. A total of 8045 astrocytes were counted in the parenchyma, 4482 in the subependymal layer, 21 in the pencil fibers, 670 next to vessels, and 1522 in the hypoglossal nucleus in a total of 6 different brains. Scale bar for panels (**A–E**) 20 μm (shown in (**A**) for the GFAP stain), but applicable for all images.

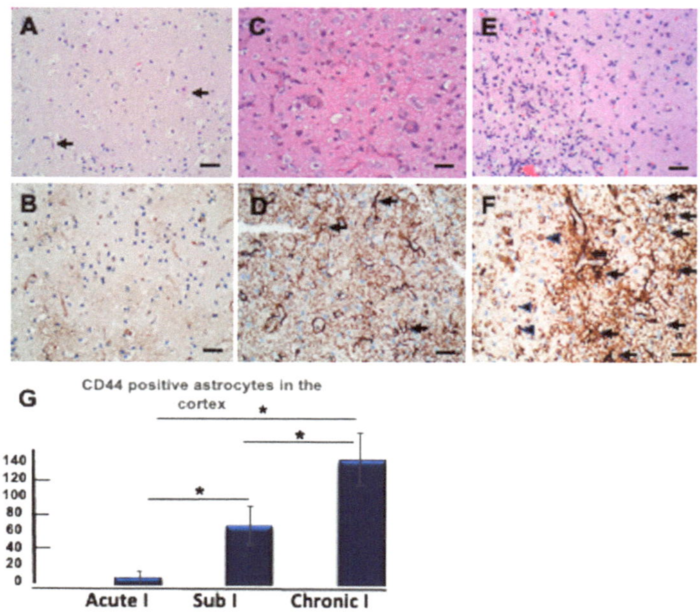

Figure 10. Protoplasmic astrocytes in the isocortex stain for CD44 following hypoxic/ischemic conditions. (**A–F**) represent acute, subacute, and chronic infarcts, respectively. Eosinophilic neurons (arrows) are seen in (**A**). Proliferative vasculature is present in (**C**). Foamy macrophages are more pronounced in (**E**) in an area of necrosis (left side of the panel). (**B**,**D**,**F**) are immunohistochemical stains for CD44 in the same tissue samples and hypoxic areas as shown in (**A**,**C**,**E**), respectively, but not the same fields. CD44+ astrocytes are highlighted (black arrows). Macrophages (arrowheads) are also positive for CD44. The vessels are outlined by CD44 stain. Specimens in (**A**,**C**,**E**) are stained with hematoxylin and eosin. Specimens in (**B**,**D**,**F**) are counterstained with hematoxylin. Scale bars: 20 μm. (**G**) The numbers of CD44+ astrocytes increase as infarcts evolve (average numbers in three 20X fields/specimen). Acute I: acute infarct; Sub I: subacute infarct; Chronic I: chronic infarct. (* $p < 0.01$, Sub I vs. Acute I, * $p < 0.01$, Chronic I vs. AI; * $p < 0.01$, Chronic I vs. Sub I). ANOVA with Tukey post hoc analysis. Values are shown as means ± SEM.

We also examined a rat model of brain ischemia generated by unilateral transient occlusion of the middle cerebral artery (tMCAO). After 7 days, many of the cortical and striatal astrocytes in the cortex and striatum ipsilateral to the MCAO, which had previously been GFAP and CD44-negative, became GFAP+, as expected after a hypoxic insult. Many of these also acquired CD44+. We show the hemispheres ipsilateral and contralateral to the tMCAO (Figure 11A, left vs. right, respectively) to illustrate the difference. The majority of GFAP+ cortical and striatal astrocytes on the contralateral side remained CD44-negative with the exception of the GFAP+ subpial astrocytes, which are normally CD44+, and some striatal astrocytes, as well as the astrocytes of the white matter. Quantitative counts of GFAP+/CD44+ cells show a significant difference (Figure 11B,C). Higher magnification of astrocytes illustrates details (Figure 11D,E).

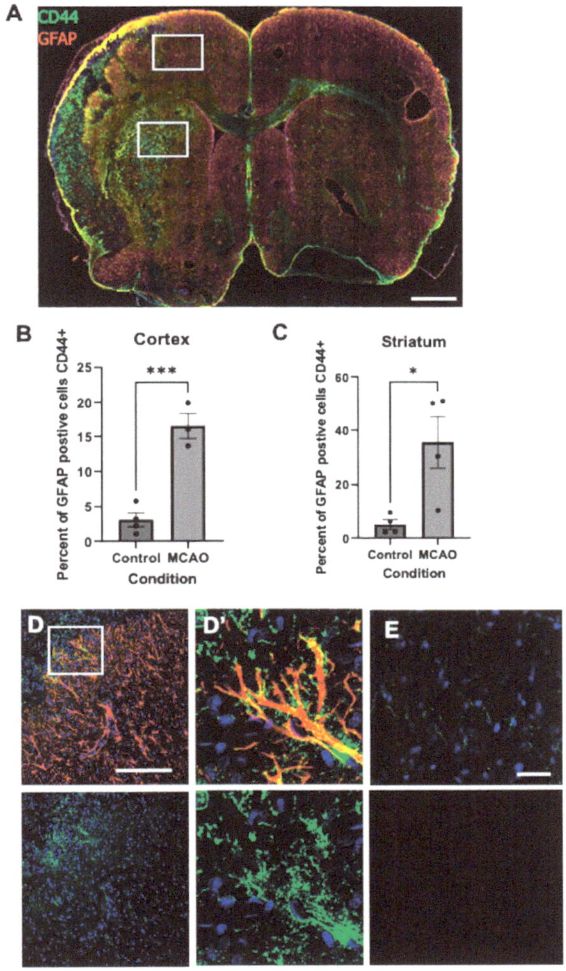

Figure 11. (**A**) Section of rat brain 7 days after left-sided tMCAO. Note the large number of CD44+-green signals on the left cortex and striatum and far fewer on the right side. This is a section from one of four separate experiments. The white rectangles delineate the areas in which GFAP+/CD44+ cells were counted for quantitation. Quantitative measurements of double-labeled cells are shown in (**B**,**C**), displayed as the percentages of GFAP+ cells that are also CD44+. * $p = 0.021$, *** $p = 0.001$, bars are SEM. Lower magnification (**D**) and higher magnification of boxed area (**D′**) show CD44+/GFAP+ astrocytes in the ipsilateral cortex. CD44—green, GFAP—red. The contralateral cortex only shows small numbers of lightly stained GFAP+ astrocytes; in this field, none is CD44+ (**E**). In (**E**), GFAP—green, CD44—red. In (**D**,**E**), the top row shows GFAP and CD44, and the bottom row shows only CD44 of the same field. A total of 6457 and 3004 astrocytes were counted in the hypoxic cortex and striatum, respectively, and 6212 and 1790 in the control cortex and striatum, in aggregate from 4 separate brains, each point is one brain. Scale bars: (**A**) 2 mm, (**D**) 240 μm, (**E**) 25 μm. The optically empty spaces in the right hemisphere in (**A**) are sectioning artifacts.

3.12. CD44+ Astrocytes in Epilepsy

Because of the reports of CD44 increase in epilepsy [15,16] and in tuberous sclerosis, which is often accompanied by seizures [14,17], we examined a group of neurosurgical specimens removed from patients with temporal lobe epilepsy, both the hippocampi and the adjacent temporal isocortex, for CD44+ astrocytes. The hippocampi showed an apparent increase in CD44 immunostaining in all sectors (Figure 12A,B) compared with controls (Figure 4). CD44 staining of all four resections for epileptic foci that contained the hippocampus revealed that neurons of the pyramidal cell layer were surrounded by CD44+ processes (Figure 12C–E), a morphology not present in the normal brain (Figure 4). The temporal isocortex showed a variety of intensities of CD44 staining in cortical astrocytes (Figure 13), although more appeared severe than moderate or low. We do not have the full clinical histories of all of these patients, however, so we cannot correlate the extent of CD44 staining with clinical behavior. Furthermore, sampling is an issue since we do not know how far away the cortical samples were taken from the hippocampus. Thus, we can only conclude that in some of these individuals, there has been a transformation of CD44-negative protoplasmic astrocytes to CD44+ astrocytes. Interestingly, some of the CD44+ cortical astrocytes had extended long processes (Figure 13D), an abnormal morphology for protoplasmic astrocytes, which normally do not extend such processes.

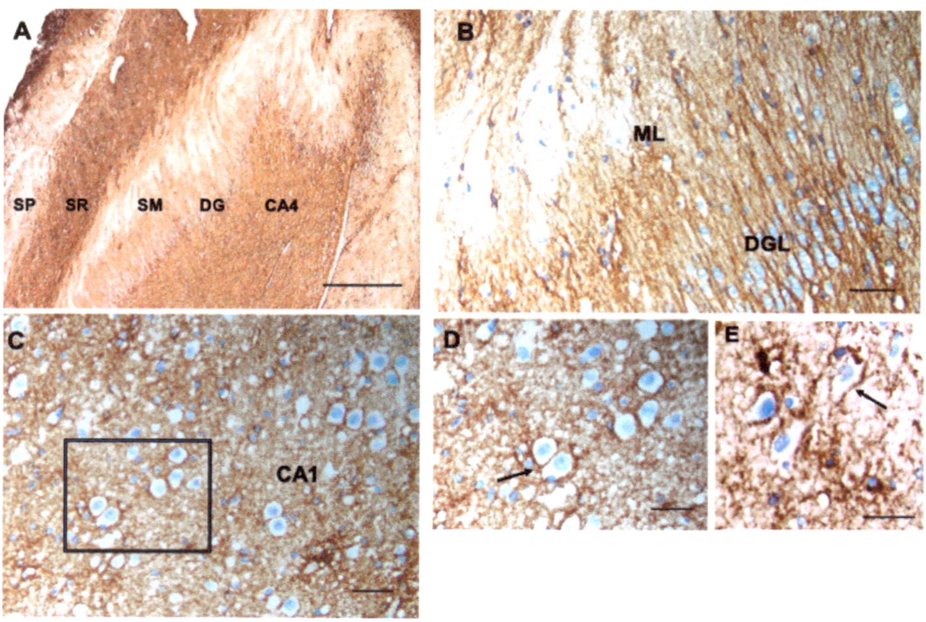

Figure 12. Changes in CD44 in the sclerotic hippocampus in human mesial temporal lobe epilepsy. Compare with Figure 4. There is an apparent increase in CD44 in all layers (abbreviations as in Figure 4) (**A**). The radially oriented astrocyte processes in the dentate granule layer (DGL), which extend into the molecular layer (SM), are CD44+ (**B**). Pyramidal neurons, which are normally not surrounded by CD44+ processes, have become so (**C**,**D**). (**D**) is the boxed area in (**C**). Pyramidal neurons are surrounded by CD44+ processes in another epilepsy specimen (**E**). Arrows represent pyramidal neurons surrounded by CD44+ processes. CA1, cornu ammonis 1, CA4, cornu ammonis 4. All specimens are counterstained with hematoxylin. Scale bars: (**A**) 200 μm, (**B**) 25 μm, (**C**) 25 μm, (**D**,**E**) 20 μm.

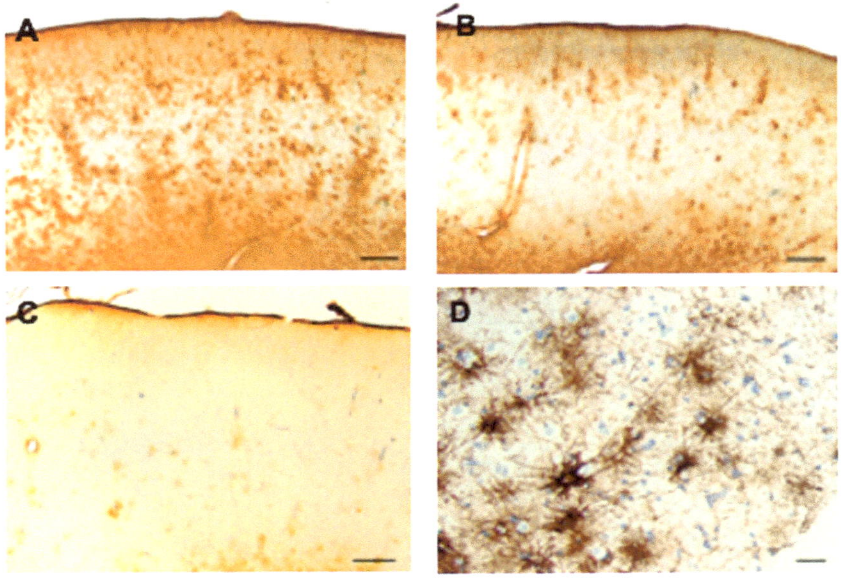

Figure 13. Temporal isocortex in individuals with temporal lobe epilepsy. In some resections, many astrocytes have become CD44+ (**A**). In others, fewer are CD44+ (**B**,**C**). The pial surface of the cortex is at the top. A higher magnification of CD44+ astrocytes in the temporal isocortex reveals that many of them have extended long processes, an abnormal morphology for protoplasmic astrocytes (**D**). The specimen in (**D**) is counterstained with hematoxylin. Scale bars: (**A–C**) 250 µm, (**D**) 20 µm.

We also examined the entorhinal cortex of rats that had been subjected to pilocarpine-induced seizures. The normal rat entorhinal cortex contains CD44+ astrocytes at the pial surface and in the subcortical white matter but not in the mid-cortex itself (Figure 14B). However, after the pilocarpine-induced seizures, a band of necrosis forms under the subpial area (delineated by "n" in Figure 14A and by dotted lines in Figure 14C). Note that there are no astrocytes in the necrotic zone. In the deeper cortical layers, there are many GFAP+/CD44+ astrocytes (Figure 14 C). The round CD44+ cells are inflammatory cells, and blood vessels in this area are also CD44+ (Figure 14C,C'). Higher magnification Figure 14D,D') shows GFAP+ astrocytes, both CD44+ and CD44−. Note, as above, that the CD44 stain reveals the fine processes of astrocytes, not shown by GFAP since the filament does not extend into fine processes. Quantitative counts of GFAP/CD44+ astrocytes in the cortex show a highly significant difference between control and seizures (Figure 14F).

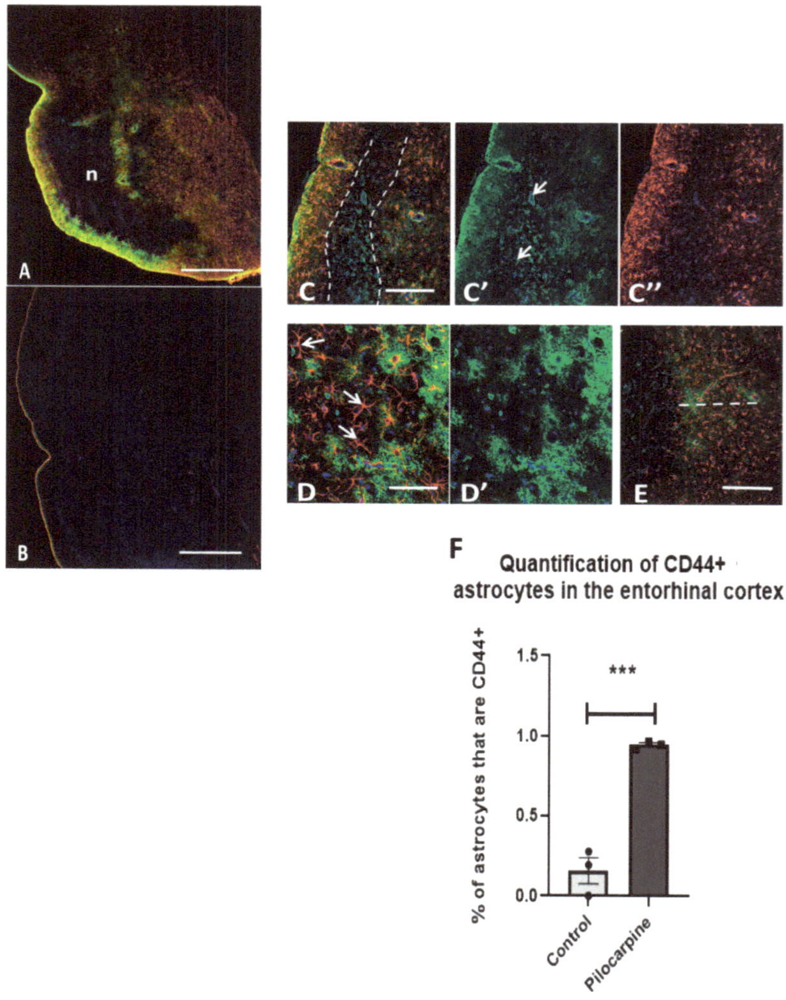

Figure 14. Cortical astrocytes become CD44+ in the rat entorhinal cortex 3 days after pilocarpine-induced seizures. A low magnification image shows strong GFAP+/CD44+ immunostaining in the subpial layer at the left of the image, no GFAP immunostaining in the necrotic layer (n) beneath the subpial layer, and strong GFAP+/CD44+ immunostaining in the layer beneath that (**A**). The normal entorhinal cortex in the control age-matched rat shows subpial astrocytes that are GFAP+/CD44+, but otherwise, there are very few double-labeled astrocytes (**B**). A higher magnification of the cortex shows the subpial gliosis, the necrotic layer outlined by the dotted line, and the gliotic layer beneath that (**C**). The necrotic zone contains CD44+ blood vessels (arrows) and round, CD44+ inflammatory cells (**C′**) but no astrocytes (**C″**). The gliotic zone under the necrotic layer (right side of the image) contains astrocytes, many of which are GFAP+/CD44+ (**D**,**D′**). Arrows show GFAP+/CD44− astrocytes. Many are also double labeled, the CD44 showing the usually fine processes of the astrocyte domains. The width of the reactive astrocyte part of the cortex is denoted by a 500 μm dotted line in (**E**). In all images, GFAP—red, CD44—green, DAPI—blue. (**F**) Quantitative measurements of double-labeled cells displayed as the proportions of GFAP+ cells that were also CD44+. A total of 637 GFAP+ astrocytes were counted in 3 pilocarpine and 3 control brains. *** $p = 0.001$, bars are SEM. Scale bars: (**A**,**B**) 440 μm, (**C**) 220 μm, (**D**) 65 μm, (**E**) 220 μm.

4. Discussion

4.1. CD44+ Astrocytes Populate the Entire Human CNS

In examining CD44+ astrocytes throughout the human CNS, we found not only that they populate every area but also that they are morphologically remarkably heterogeneous. Previous studies have reported CD44 immunostaining in the CNS, some of which localized it to astrocytes, but there has not been a systematic investigation of the entire human neuraxis in cellular detail [5,6,9–11,14,18]. Furthermore, we report novel observations of CD44+ processes intercalated between ependymal cells, subpial CD44+ astrocytes in the cerebellum, and CD44+ astrocyte processes that surround motor neurons and other large neurons.

4.2. Interlaminar Astrocytes

Long-process astrocytes in the isocortex have been known for over a century in human and other large mammalian brains. The smaller brains, such as those of rodents, do not show these, although astrocytes at the pial surface are also CD44+, as are white matter astrocytes. That the morphology is species dependent is implied by a study that produced long-process astrocytes in the mouse after transplanting human-induced pluripotent stem cells into the mouse cortex [19], suggesting that human long-process astrocytes are intrinsically programmed to assume this morphology.

Our findings in the isocortex and hippocampus confirm and extend those of our previous work [6]; we now show that CD44+ astrocytes are found throughout the human CNS. Long, unbranched processes lacking a bushy appearance distinguish the CD44+ astrocytes from the protoplasmic astrocytes of gray matter, which are CD44-negative. The presence of CD44, however, does not imply that all of these astrocytes represent a single population. For example, the "fibrous" astrocytes of the white matter may be different from the subpial, long-process astrocytes and the astrocytes that occupy subependymal locations or the astrocytes that wrap around brainstem and spinal cord neurons. Single-cell and single-nucleus RNA sequencing of the human CNS may indeed separate these populations, now defined by location and morphology, into multiple groups with different transcriptional profiles.

The pial-based CD44+ interlaminar astrocytes and the white matter-based astrocytes with processes entering the deeper cortical layers may represent remnants of the outer radial glia and ventricular radial glia that develop late in gestation in the human hemispheres after neuronal migration has taken place [20].

4.3. Neurons Surrounded by CD44+ Processes

One of the unexpected findings was that in the brainstem and spinal cord, many neurons are surrounded by CD44+ astrocyte processes. The presence of CD44+ processes surrounding motor neurons of the brainstem and spinal cord has not been previously reported. We did not see neurons surrounded by CD44+ processes in the isocortex, basal ganglia, thalamus, or hippocampus in the normal brain. Other large neurons that are surrounded by CD44+ astrocyte processes are the neurons of Clarke's column, which are large, Nissl-rich neurons that receive input from large synapses from tendon organs and muscle spindles, and the magnocellular neurons of the LGN. Also, some Purkinje cells appear to be partially surrounded by CD44+ processes. Whether these contacts between neurons and CD44+ astrocytes are functionally important awaits further investigation. Studies of carcinoma cells may provide one clue to this interaction [21]. The intracellular domain of CD44 (CD44ICD) acts as a transcription factor for genes that contain a CD44ICD response element and promotes the expression of pyruvate dehydrogenase kinase (*PDK1*), 6-phosphofructo-2-kinase (*PFKFB4*), and aldolase C (*ALDOC*). Miletti-Gonzalez et al. [21] suggest that upregulation of these genes implies that CD44 may play a role in glycolysis since PDK1 promotes the conversion of pyruvate to lactate, a molecule known to be transferred from astrocytes to neurons to promote energy. Since these studies were performed in cancer cell lines, it would be interesting to test what genes the CD44ICD promotes in CD44+ astrocytes.

We found that hippocampal pyramidal neurons in temporal lobe epilepsy were also surrounded by CD44+ processes. We interpret this as the acquisition of CD44 by normal protoplasmic astrocytes whose processes normally contact neurons. Whether this represents a new process formation by astrocytes will await further investigation. What functional significance this enwrapping has will be interesting to examine.

4.4. Subependymal Astrocytes

Another unexpected finding was the dense matrix of CD44+ processes in the subependymal region. Because of the density, it was difficult to follow individual astrocyte morphologies, even with confocal microscopy. Future experiments injecting dyes into individual astrocytes would provide a better view of their morphologies. We did notice that CD44+ processes appeared to intercalate between ependymal cells, which themselves were CD44-negative. This morphology and location are reminiscent of neural stem cells in the adult mouse brain and subependymal astrocytes in the human brain, which can function as neural stem cells [22–24]. However, further comparison between these human CD44+ subependymal cells and neural stem cells will have to await further analysis.

4.5. Thalamic Astrocytes

Many of the CD44+ astrocytes in the thalamus and subthalamic nucleus are attached to large blood vessels, as seen elsewhere in the CNS. Other CD44+ astrocytes may be associated with myelinated fiber bundles not previously reported with CD44 immunostains.

4.6. Cerebellar Astrocytes

The cerebellar cortex contains several types of CD44+ astrocytes. Velate astrocytes, which have unique morphologies, send out many processes to enwrap clusters of granule cells [25]. Astrocytes in the upper part of the molecular layer, near the pial surface, are not well described to our knowledge. There are thin, radial, CD44+ processes passing through the molecular layer not previously described with CD44 immunostains. The source of these is not clear, although some are contiguous with CD44+ processes in the granule cell layer. Such thin processes were described by Santiago Ramón y Cajal, who thought, from his Weigert stains, that they originated from astrocytes in the granule cell layer or the underlying white matter [26]. These processes are far less dense than those of Bergmann glia, but some may represent a subpopulation of Bergmann astrocytes.

4.7. Astrocytes with Vascular Contacts

We also observed that CD44+ astrocytes are in contact with large blood vessels, sending long processes radially into the neuropil, as illustrated in Akiyama et al. [5]. In contrast, astrocytes that contact capillaries are CD44-negative. Thus, astrocyte contacts with arterioles may promote the CD44+ phenotype, or contact with capillaries may repress the phenotype and promote the characteristics of protoplasmic astrocytes. If CD44 is involved in the interaction, it may bind to the matrix surrounding large vessels, which is different from that around capillaries.

4.8. Hypoxia

We found that hypoxic/ischemic insults caused the protoplasmic astrocytes of the isocortex to become CD44+. Studies of tumor cell lines suggest mechanisms underlying this change. Thus, hypoxia regulates *CD44* expression via HIF-1α in breast cancer [27] and gastric cancer [28]. Furthermore, the CD44ICD is released in hypoxia by gamma secretase cleavage [29] and binds to HIF-2α [30]. This activates HIF-regulated genes. This latter study, which was performed with human glioblastoma cell lines, suggests a positive feedback mechanism in which hypoxia regulates *CD44*, which, in turn, may regulate more hypoxia-related genes. We found that cortical astrocytes in the rat brain become CD44+

after a tMCAO. In a previous study, *CD44* was found to be increased in the cortex and striatum of rats after a tMCAO, and localized macrophages and microglia [31], but the authors did not describe CD44+ in astrocytes.

4.9. Seizures

Seizures provoke an increase in CD44 in astrocytes, at least in the isocortex and hippocampus, the two areas we examined in the human CNS and in the mouse entorhinal cortex following pilocarpine-induced seizures. Furthermore, transcriptomics of the reactive astrocytosis accompanying temporal lobe epilepsy show upregulated *CD44* [16]. The mechanisms underlying this are not known. Hypoxia may be involved, or inflammatory responses. A previous study found CD44 in the mouse hippocampus after pilocarpine-induced seizures, but astrocyte localization was not described [15].

4.10. Regulation of CD44 That Might Be Relevant to Studies in the Human CNS

Although the transcriptional regulation of *CD44* has not been studied in human astrocytes, studies with other cultured cell types show that the upstream promoter region of CD44 is complex, containing binding sites for both positive and negative transcriptional regulators, including sequences that bind AP-1, NF-kB, Sp1, Egr1, TCF4, ETS-1, p53, KLF4, and Foxp3, the latter three as repressors [32,33]. As noted above, hypoxia regulates *CD44* through HIF-1α. These studies were carried out in non-CNS cancer cell lines, and it is not known whether similar regulations occur in astrocytes. Mouse astrocytes in culture upregulate *CD44* when treated with phorbol-12-myristate 13-acetate or TNFα plus gamma interferon [34], suggesting that inflammatory conditions may upregulate *CD44*. Which of these transcription factors acts to regulate *CD44* in human astrocytes is unknown.

In addition, there is evidence for epigenetic regulation of *CD44*. Thus, a 5′-CpG island of *CD44* is methylated in prostate cancer [35]. CpGs are highly methylated in the CD44-negative T cell lymphoma line AKR1, although transfection with *c-jun* can activate transcription without major changes in methylation [36]. It will be of interest to examine the methylation status of *CD44* in CD44+ and CD44-negative astrocytes and investigate if these changes in hypoxia or seizures allow *CD44* transcription.

4.11. There Is a Negative Correlation between CD44 and Protoplasmic Astrocyte Genes/Proteins

We previously determined that the CD44+ interlaminar astrocytes in the normal human isocortex have lower levels of glutamate transporters and glutamine synthetase than CD44-negative protoplasmic astrocytes [6]. In a mouse model of Alexander disease, a CNS degenerative disorder caused by mutations in *GFAP*, we found that protoplasmic astrocytes of the hippocampus and isocortex became CD44+ as the disease progressed [37]. The acquisition of CD44 correlated strongly with the loss of the plasmalemmal glutamate transporter, Glt-1. In a study of the human cingulate gyrus in Huntington disease, RNA sequencing revealed an upregulation of *CD44* during the disease [38]. This study also revealed that astrocytes lost the expression of normal protoplasmic astrocyte genes. Thus, there appears to be a negative correlation between CD44 and several protoplasmic astrocyte genes and proteins. Whether this correlation is caused by the presence of CD44 is not known, although it is testable in future studies. The expression of *CD44* caused by hypoxia and seizures may well alter the physiology of protoplasmic astrocytes in ways that are deleterious to neurons.

4.12. Is There Any Value in Downregulating CD44 in Pathological Situations?

If CD44 plays a role in changing the phenotypes of protoplasmic astrocytes, would there be any value in downregulating *CD44* or preventing its upregulation? Would this preserve a normal protoplasmic astrocyte state? This could be tested in mouse models. In studies of human cancers, CD44 is associated with STAT3, which may lead to STAT3 activation [39]. STAT3, in turn, activates other astrocyte genes, such as *GFAP* [40,41].

So et al. [39] found that a vitamin D analog represses *CD44* expression in breast cancer cells. It would be interesting to determine if astrocytes show the same response.

4.13. Limitations

Although this is the most thorough study of CD44+ astrocytes in the human CNS, and we show the conversion of protoplasmic astrocytes to being CD44+ in hypoxia and epilepsy, our observations do not imply specific molecular mechanisms. Such data on the mechanisms of *CD44* expression appear largely in the cancer biology literature and have not been investigated in astrocytes. The present study provides an important background that will promote further investigations by our labs and others.

5. Conclusions

This study includes a number of novel observations, further characterizing the heterogeneity of astrocytes in the human CNS.

1. CD44+ astrocytes are present at all levels of the human CNS.
2. Some neurons are surrounded by CD44+ processes, including motor neurons in the brain stem and spinal cord, while isocortical, hippocampal, and most diencephalic neurons are not.
3. CD44+ processes intercalate between ependymal cells at ventricular surfaces.
4. CD44 immunostaining revealed astrocytes in the upper part of the cerebellar molecular layer. These send processes parallel to the pia and radially toward the Purkinje cell layer.
5. Velate astrocytes of the cerebellar granule cell layer are CD44+.
6. Hypoxia and seizures induce CD44-negative protoplasmic astrocytes to become CD44+.

Supplementary Materials: The following supporting information can be downloaded at: https://www.mdpi.com/article/10.3390/cells13020129/s1, Figure S1: Sections through the ependymal lining overlying the subependymal zone in fetal brains to show CD44+ processes between ependymal cells, Figure S2: A coronal section through the lateral and third ventricles shows CD44+ in the anterior commissure (AC), optic tracts (OT), and internal capsule (IC), Figure S3: Sections through the cerebellar cortex, Figure S4: CD44+-positive astrocytes in the dentate nucleus (left panels, A–C) and cerebellar cortex (right panels, D–F), Figure S5: In an individual who died with acute hypoxia, isocortical neurons are eosinophilic and shrunken (arrows, A), but astrocytes are not CD44+ (B). In individuals with chronic hypoxia, cortical astrocytes become CD44+ (C–E).

Author Contributions: J.E.G. and O.A.-D. conceived of the studies and wrote the paper. Y.S., Y.L., N.M., O.A.-D., A.A.S., G.M.M.II, E.S.C., C.M.T. and J.E.G. carried out the studies and edited the manuscript. All authors have read and agreed to the published version of the manuscript.

Funding: The ADRC REC grant (P30AG066462) (OAD), NINDS RO1 NS081333 (CMT), K08-NS048064 (GM), and the Tuberous Sclerosis Alliance (GM).

Institutional Review Board Statement: All investigations were carried out following the rules of the Declaration of Helsinki of 1975. Human brain tissues from patient brain biopsies and autopsies, which were diagnosed based on accepted neuropathological criteria, were obtained from the Department of Pathology and Cell Biology of Columbia University and the New York Brain Bank. All surgeries were performed with the consent of the patients, and studies from biopsy tissues are provided without any patient identifying information. All autopsy brains were donated after consent from the next of kin or an individual with legal authority to grant such permission. The use of postmortem brain tissues for research was approved by the Columbia University Institutional Review Board with informed consent from patients or their families. The Institutional Review Board has determined that clinicopathologic studies on de-identified postmortem tissue samples are exempt from Human Subject Research according to Exemption 45 CFR 46.104(d)(2).

Informed Consent Statement: Not applicable. The study was performed on autopsy tissue donated after consent from the next of kin. The Institutional Review Board of Columbia University has determined that clinicopathologic studies on de-identified postmortem tissue samples are exempt from Human Subject Research according to Exemption 45 CFR 46.104(d)(2).

Data Availability Statement: There are no research data or data files other than the text and images presented in this paper.

Acknowledgments: The authors greatly appreciate the efforts of members of the Immuno-staining Laboratory of the Department of Pathology and Cell Biology at Columbia Irving Medical Center, Michael Miller, and the staff of the DPCL of the Department of Pathology and Cell Biology for digitizing images, Juncheng Li for technical assistance, Nsikan Akpan for early studies on hypoxia, and the Neuropathology staff of the Division of Neuropathology for their interest and for many discussions. The Graphic Abstract was created with BioRender.com.

Conflicts of Interest: The authors declare no conflicts of interest.

References

1. Andriezen, W.L. The Neuroglia Elements in the Human Brain. *Br. Med. J.* **1893**, *2*, 227–230. [CrossRef] [PubMed]
2. Colombo, J.A.; Gayol, S.; Yanez, A.; Marco, P. Immunocytochemical and Electron Microscope Observations on Astroglial Interlaminar Processes in the Primate Neocortex. *J. Neurosci. Res.* **1997**, *48*, 352–357. [CrossRef]
3. Oberheim, N.A.; Takano, T.; Han, X.; He, W.; Lin, J.H.; Wang, F.; Xu, Q.; Wyatt, J.D.; Pilcher, W.; Ojemann, J.G.; et al. Uniquely Hominid Features of Adult Human Astrocytes. *J. Neurosci.* **2009**, *29*, 3276–3287. [CrossRef]
4. Falcone, C.; Wolf-Ochoa, M.; Amina, S.; Hong, T.; Vakilzadeh, G.; Hopkins, W.D.; Hof, P.R.; Sherwood, C.C.; Manger, P.R.; Noctor, S.C.; et al. Cortical Interlaminar Astrocytes across the Therian Mammal Radiation. *J. Comp. Neurol.* **2019**, *527*, 1654–1674. [CrossRef] [PubMed]
5. Akiyama, H.; Tooyama, I.; Kawamata, T.; Ikeda, K.; McGeer, P.L. Morphological diversities of CD44 positive astrocytes in the cerebral cortex of normal subjects and patients with Alzheimer's disease. *Brain Res.* **1993**, *632*, 249–259. [CrossRef]
6. Sosunov, A.A.; Wu, X.; Tsankova, N.M.; Guilfoyle, E.; McKhann, G.M., 2nd; Goldman, J.E. Phenotypic heterogeneity and plasticity of isocortical and hippocampal astrocytes in the human brain. *J. Neurosci.* **2014**, *34*, 2285–2298. [CrossRef] [PubMed]
7. Goodison, S.; Urquidi, V.; Tarin, D. CD44 cell adhesion molecules. *Mol. Pathol.* **1999**, *52*, 189–196. [CrossRef]
8. Sosunov, A.A.; McGovern, R.A.; Mikell, C.B.; Wu, X.; Coughlin, D.G.; Crino, P.B.; Weiner, H.L.; Ghatan, S.; Goldman, J.E.; McKhann, G.M., 2nd. Epileptogenic but MRI-normal perituberal tissue in Tuberous Sclerosis Complex contains tuber-specific abnormalities. *Acta Neuropathol. Commun.* **2015**, *3*, 17. [CrossRef]
9. Girgrah, N.; Letarte, M.; Becker, L.E.; Cruz, T.F.; Theriault, E.; Moscarello, M.A. Localization of the CD44 glycoprotein to fibrous astrocytes in normal white matter and to reactive astrocytes in active lesions in multiple sclerosis. *J. Neuropathol. Exp. Neurol.* **1991**, *50*, 779–792. [CrossRef]
10. Pinner, E.; Gruper, Y.; Ben Zimra, M.; Kristt, D.; Laudon, M.; Naor, D.; Zisapel, N. CD44 Splice Variants as Potential Players in Alzheimer's Disease Pathology. *J. Alzheimer's Dis.* **2017**, *58*, 1137–1149. [CrossRef]
11. Vogel, H.; Butcher, E.C.; Picker, L.J. H-CAM expression in the human nervous system: Evidence for a role in diverse glial interactions. *J. Neurocytol.* **1992**, *21*, 363–373. [CrossRef]
12. Sosunov, A.A.; Wu, X.; McGovern, R.A.; Coughlin, D.G.; Mikell, C.B.; Goodman, R.R.; McKhann, G.M., 2nd. The mTOR pathway is activated in glial cells in mesial temporal sclerosis. *Epilepsia* **2012**, *53* (Suppl. S1), 78–86. [CrossRef]
13. McKhann, G.M., 2nd; Wenzel, H.J.; Robbins, C.A.; Sosunov, A.A.; Schwartzkroin, P.A. Mouse strain differences in kainic acid sensitivity, seizure behavior, mortality, and hippocampal pathology. *Neuroscience* **2003**, *122*, 551–561. [CrossRef] [PubMed]
14. Arai, Y.; Takashima, S.; Becker, L.E. CD44 expression in tuberous sclerosis. *Pathobiology* **2000**, *68*, 87–92. [CrossRef] [PubMed]
15. Borges, K.; McDermott, D.L.; Dingledine, R. Reciprocal changes of CD44 and GAP-43 expression in the dentate gyrus inner molecular layer after status epilepticus in mice. *Exp. Neurol.* **2004**, *188*, 1–10. [CrossRef] [PubMed]
16. Pai, B.; Tome-Garcia, J.; Cheng, W.S.; Nudelman, G.; Beaumont, K.G.; Ghatan, S.; Panov, F.; Caballero, E.; Sarpong, K.; Marcuse, L.; et al. High-resolution transcriptomics informs glial pathology in human temporal lobe epilepsy. *Acta Neuropathol. Commun.* **2022**, *10*, 149. [CrossRef] [PubMed]
17. Boer, K.; Crino, P.B.; Gorter, J.A.; Nellist, M.; Jansen, F.E.; Spliet, W.G.; van Rijen, P.C.; Wittink, F.R.; Breit, T.M.; Troost, D.; et al. Gene expression analysis of tuberous sclerosis complex cortical tubers reveals increased expression of adhesion and inflammatory factors. *Brain Pathol.* **2010**, *20*, 704–719. [CrossRef]
18. Kaaijk, P.; Pals, S.T.; Morsink, F.; Bosch, D.A.; Troost, D. Differential expression of CD44 splice variants in the normal human central nervous system. *J. Neuroimmunol.* **1997**, *73*, 70–76. [CrossRef]
19. Padmashri, R.; Ren, B.; Oldham, B.; Jung, Y.; Gough, R.; Dunaevsky, A. Modeling human-specific interlaminar astrocytes in the mouse cerebral cortex. *J. Comp. Neurol.* **2021**, *529*, 802–810. [CrossRef]
20. Nowakowski, T.J.; Pollen, A.A.; Sandoval-Espinosa, C.; Kriegstein, A.R. Transformation of the Radial Glia Scaffold Demarcates Two Stages of Human Cerebral Cortex Development. *Neuron* **2016**, *91*, 1219–1227. [CrossRef]
21. Miletti-González, K.E.; Murphy, K.; Kumaran, M.N.; Ravindranath, A.K.; Wernyj, R.P.; Kaur, S.; Miles, G.D.; Lim, E.; Chan, R.; Chekmareva, M.; et al. Identification of function for CD44 intracytoplasmic domain (CD44-ICD): Modulation of matrix metalloproteinase 9 (MMP-9) transcription via novel promoter response element. *J. Biol. Chem.* **2012**, *287*, 18995–19007. [CrossRef]

22. Mirzadeh, Z.; Merkle, F.T.; Soriano-Navarro, M.; Garcia-Verdugo, J.M.; Alvarez-Buylla, A. Neural stem cells confer unique pinwheel architecture to the ventricular surface in neurogenic regions of the adult brain. *Cell Stem Cell* **2008**, *3*, 265–278. [CrossRef] [PubMed]
23. Sanai, N.; Tramontin, A.D.; Quiñones-Hinojosa, A.; Barbaro, N.M.; Gupta, N.; Kunwar, S.; Lawton, M.T.; McDermott, M.W.; Parsa, A.T.; Manuel-García Verdugo, J.; et al. Unique astrocyte ribbon in adult human brain contains neural stem cells but lacks chain migration. *Nature* **2004**, *427*, 740–744. [CrossRef]
24. Quiñones-Hinojosa, A.; Sanai, N.; Soriano-Navarro, M.; Gonzalez-Perez, O.; Mirzadeh, Z.; Gil-Perotin, S.; Romero-Rodriguez, R.; Berger, M.S.; Garcia-Verdugo, J.M.; Alvarez-Buylla, A. Cellular composition and cytoarchitecture of the adult human subventricular zone: A niche of neural stem cells. *J. Comp. Neurol.* **2006**, *494*, 415–434. [CrossRef]
25. Chan-Palay, V.; Palay, S.L. The form of velate astrocytes in the cerebellar cortex of monkey and rat: High voltage electron microscopy of rapid Golgi preparations. *Z. Anat. Entwicklungsgesch.* **1972**, *138*, 1–19. [CrossRef]
26. Ramon y Cajal, S. *Histology of the Nervous System*; Swanson, N.; Swanson, L.W., Translators; Oxford University Press: New York, NY, USA, 1995; Volume 2.
27. Krishnamachary, B.; Penet, M.F.; Nimmagadda, S.; Mironchik, Y.; Raman, V.; Solaiyappan, M.; Semenza, G.L.; Pomper, M.G.; Bhujwalla, Z.M. Hypoxia regulates CD44 and its variant isoforms through HIF-1α in triple negative breast cancer. *PLoS ONE* **2012**, *7*, e44078. [CrossRef]
28. Liang, G.; Li, S.; Du, W.; Ke, Q.; Cai, J.; Yang, J. Hypoxia regulates CD44 expression via hypoxia-inducible factor-1α in human gastric cancer cells. *Oncol. Lett.* **2017**, *13*, 967–972. [CrossRef] [PubMed]
29. Pietras, A.; Katz, A.M.; Ekström, E.J.; Wee, B.; Halliday, J.J.; Pitter, K.L.; Werbeck, J.L.; Amankulor, N.M.; Huse, J.T.; Holland, E.C. Osteopontin-CD44 signaling in the glioma perivascular niche enhances cancer stem cell phenotypes and promotes aggressive tumor growth. *Cell Stem Cell* **2014**, *14*, 357–369. [CrossRef]
30. Johansson, E.; Grassi, E.S.; Pantazopoulou, V.; Tong, B.; Lindgren, D.; Berg, T.J.; Pietras, E.J.; Axelson, H.; Pietras, A. CD44 Interacts with HIF-2α to Modulate the Hypoxic Phenotype of Perinecrotic and Perivascular Glioma Cells. *Cell Rep.* **2017**, *20*, 1641–1653. [CrossRef] [PubMed]
31. Wang, H.; Zhan, Y.; Xu, L.; Feuerstein, G.Z.; Wang, X. Use of suppression subtractive hybridization for differential gene expression in stroke: Discovery of CD44 gene expression and localization in permanent focal stroke in rats. *Stroke* **2001**, *32*, 1020–1027. [CrossRef]
32. Zhang, C.; Xu, Y.; Hao, Q.; Wang, S.; Li, H.; Li, J.; Gao, Y.; Li, M.; Li, W.; Xue, X.; et al. FOXP3 suppresses breast cancer metastasis through downregulation of CD44. *Int. J. Cancer* **2015**, *137*, 1279–1290. [CrossRef] [PubMed]
33. Chen, C.; Zhao, S.; Karnad, A.; Freeman, J.W. The biology and role of CD44 in cancer progression: Therapeutic implications. *J. Hematol. Oncol.* **2018**, *11*, 64. [CrossRef] [PubMed]
34. Haegel, H.; Tölg, C.; Hofmann, M.; Ceredig, R. Activated mouse astrocytes and T cells express similar CD44 variants. Role of CD44 in astrocyte/T cell binding. *J. Cell Biol.* **1993**, *122*, 1067–1077. [CrossRef]
35. Lou, W.; Krill, D.; Dhir, R.; Becich, M.J.; Dong, J.T.; Frierson, H.F., Jr.; Isaacs, W.B.; Isaacs, J.T.; Gao, A.C. Methylation of the CD44 metastasis suppressor gene in human prostate cancer. *Cancer Res.* **1999**, *59*, 2329–2331. [PubMed]
36. Hyman, R. Lack of a consistent relationship between demethylation of the CD44 promoter and CD44 expression. *Immunogenetics* **2002**, *53*, 914–924. [CrossRef]
37. Sosunov, A.A.; Guilfoyle, E.; Wu, X.; McKhann, G.M., 2nd; Goldman, J.E. Phenotypic conversions of "protoplasmic" to "reactive" astrocytes in Alexander disease. *J. Neurosci.* **2013**, *33*, 7439–7450. [CrossRef]
38. Al-Dalahmah, O.; Sosunov, A.A.; Shaik, A.; Ofori, K.; Liu, Y.; Vonsattel, J.P.; Adorjan, I.; Menon, V.; Goldman, J.E. Single-nucleus RNA-seq identifies Huntington disease astrocyte states. *Acta Neuropathol. Commun.* **2020**, *8*, 19. [CrossRef]
39. So, J.Y.; Smolarek, A.K.; Salerno, D.M.; Maehr, H.; Uskokovic, M.; Liu, F.; Suh, N. Targeting CD44-STAT3 signaling by Gemini vitamin D analog leads to inhibition of invasion in basal-like breast cancer. *PLoS ONE* **2013**, *8*, e54020. [CrossRef] [PubMed]
40. Xu, Z.; Xue, T.; Zhang, Z.; Wang, X.; Xu, P.; Zhang, J.; Lei, X.; Li, Y.; Xie, Y.; Wang, L.; et al. Role of signal transducer and activator of transcription-3 in up-regulation of GFAP after epilepsy. *Neurochem. Res.* **2011**, *36*, 2208–2215. [CrossRef] [PubMed]
41. O'Callaghan, J.P.; Kelly, K.A.; VanGilder, R.L.; Sofroniew, M.V.; Miller, D.B. Early activation of STAT3 regulates reactive astrogliosis induced by diverse forms of neurotoxicity. *PLoS ONE* **2014**, *9*, e102003. [CrossRef] [PubMed]

cells

MDPI

Article

Astrocytic CD44 Deficiency Reduces the Severity of Kainate-Induced Epilepsy

Patrycja K. Kruk [1], Karolina Nader [2], Anna Skupien-Jaroszek [1], Tomasz Wójtowicz [1], Anna Buszka [1], Gabriela Olech-Kochańczyk [3], Grzegorz M. Wilczynski [3,†], Remigiusz Worch [1], Katarzyna Kalita [2], Jakub Włodarczyk [1] and Joanna Dzwonek [1,*]

1 Laboratory of Cell Biophysics, Nencki Institute of Experimental Biology, Polish Academy of Sciences, 3 Pasteura St, 02-093 Warsaw, Poland
2 Laboratory of Neurobiology, Nencki-EMBL Partnership for Neural Plasticity and Brain Disorders-Braincity, 3 Pasteura St, 02-093 Warsaw, Poland
3 Laboratory of Molecular and Structural Neuromorphology, Nencki Institute of Experimental Biology, Polish Academy of Sciences, 3 Pasteura St, 02-093 Warsaw, Poland
* Correspondence: j.dzwonek@nencki.edu.pl
† Deceased on 13 July 2020.

Abstract: Background: Epilepsy affects millions of people worldwide, yet we still lack a successful treatment for all epileptic patients. Most of the available drugs modulate neuronal activity. Astrocytes, the most abundant cells in the brain, may constitute alternative drug targets. A robust expansion of astrocytic cell bodies and processes occurs after seizures. Highly expressed in astrocytes, CD44 adhesion protein is upregulated during injury and is suggested to be one of the most important proteins associated with epilepsy. It connects the astrocytic cytoskeleton to hyaluronan in the extracellular matrix, influencing both structural and functional aspects of brain plasticity. Methods: Herein, we used transgenic mice with an astrocyte CD44 knockout to evaluate the impact of the hippocampal CD44 absence on the development of epileptogenesis and ultrastructural changes at the tripartite synapse. Results: We demonstrated that local, virally-induced CD44 deficiency in hippocampal astrocytes reduces reactive astrogliosis and decreases the progression of kainic acid-induced epileptogenesis. We also observed that CD44 deficiency resulted in structural changes evident in a higher dendritic spine number along with a lower percentage of astrocyte-synapse contacts, and decreased post-synaptic density size in the hippocampal molecular layer of the dentate gyrus. Conclusions: Overall, our study indicates that CD44 signaling may be important for astrocytic coverage of synapses in the hippocampus and that alterations of astrocytes translate to functional changes in the pathology of epilepsy.

Keywords: epilepsy; CD44; astrocytes; synapse; seizures

Citation: Kruk, P.K.; Nader, K.; Skupien-Jaroszek, A.; Wójtowicz, T.; Buszka, A.; Olech-Kochańczyk, G.; Wilczynski, G.M.; Worch, R.; Kalita, K.; Włodarczyk, J.; et al. Astrocytic CD44 Deficiency Reduces the Severity of Kainate-Induced Epilepsy. *Cells* **2023**, *12*, 1483. https://doi.org/10.3390/cells12111483

Academic Editors: Michael Brenner and Vladimir Parpura

Received: 1 February 2023
Revised: 5 May 2023
Accepted: 24 May 2023
Published: 26 May 2023

1. Introduction

Epilepsy is one of the most common chronic neurological disorders, affecting around 50 million people worldwide. It is characterized by recurrent seizures which are a result of excessive electrical discharges in neurons [1]. Most of the currently available antiepileptic drugs work by suppressing neuronal hyperexcitability, but they also cause several side effects, negatively affecting patients' quality of life [2]. Additionally, one-third of patients suffer from drug-resistant types of epilepsy [3], of which temporal lobe epilepsy (TLE) is the most common one. A target alternative to neurons, such as astrocytes able to modulate neuronal activity, could be a promising next step in the treatment of epilepsy. Reactive astrocytes in the epileptic brain are well known to undergo extensive morphological and biochemical changes that modify their physiological functions, including glutamate reuptake, regulation of the ionic environment and interstitial volume, and blood-brain barrier permeability control [4,5]. Both the protective and detrimental functions of reactive astrocytes in seizure development have been documented [6,7]. A better understanding of the pro- and anticonvulsant actions of reactive astrocytes could lead to novel strategies for epilepsy treatment.

Seizure onset in TLE can begin in the interconnected structures within the temporal lobe and adjoining regions, such as the hippocampus [8]. Epileptic seizures induce morphological rearrangements of neuronal circuits that are accompanied by pruning, remodeling, and de novo formation of synapses in the hippocampus [9,10]. The loss of neurons in the hippocampus has been described in both human specimens [11] and animal models [12–15] of TLE. Experimental animal models of *status epilepticus* (SE) showed that hippocampal dendritic spines are dynamically altered during the establishment and maintenance of spontaneous seizures. In addition to the extensive loss of spines, the remaining spines are of increased size and often appear as multisynaptic giant spines, with numerous postsynaptic densities (PSDs), making contact with different presynaptic boutons [9,16–18].

Astrocytes possess thin terminal processes, also called perisynaptic astrocyte processes (PAPs) or leaflets, that surround synapses and regulate synaptic transmission by bilateral communication (reviewed in [19,20]). PAPs located close to a synaptic cleft undergo structural rearrangements upon neuronal activation, which results in altering synaptic transmission and plasticity [21–24]. Interestingly, astrocytes also undergo seizure-induced alterations characterized by overall hypertrophy [25,26], and their ensheathment around presynaptic and postsynaptic elements is enhanced in CA3-CA1 synapses after kainate-induced SE [27]. However, little is known about the functional relevance of morphological changes in astrocytes following SE.

CD44 protein, the main transmembrane receptor for hyaluronan (HA), is expressed in the central nervous system (CNS). The localization of CD44 protein in the brain varies depending on cell type and brain region and is developmentally regulated. A number of reports have pointed to the astrocytic expression of CD44 in the nervous system [28–31], while others have also described its expression in neurons [32–36]. The diversity of the CD44 cell type and developmental expression should be considered when interpreting data concerning its function. While the extracellular N-terminal domain of CD44 is responsible for the binding of HA, the C-terminal cytoplasmatic region is involved in interactions with different effectors, along with cytoskeletal proteins [37]. CD44-mediated signaling can additionally influence the function of small Rho GTPases (Rac1-Rac Family Small GTPase 1, Cdc42-Cell Division Cycle 42, RhoA-Ras Homolog Family Member A) [32,38,39] and Src kinases [33]. CD44 is upregulated during injury and other pathological conditions, including epilepsy [40–43]. Its strong reactivity is seen in both pilocarpine- and kainate-induced SE, as well as various epilepsy-related neurological disorders [42,44]. On the other hand, a huge increase in the number of CD44-positive astrocytes was observed in the post-mortem brains of Alzheimer's disease patients, which was not related to epilepsy [43]. In humans, transcriptome analysis in epileptic patients identified CD44 as one of the five most important proteins associated with the pathogenesis of epilepsy [45].

CD44 expression correlates with the morphological characteristics of astrocytes. Our previous in vitro studies demonstrated that astrocytes without CD44 adhesion protein have distinctive stellate-like morphology [39]. Similarly, CD44-negative glial cells in the human brain show numerous branches and short processes and are often classified as protoplasmic astrocytes. In comparison, a high CD44 expression level is observed in more elongated cells with fewer long and unbranched processes; these cells have been categorized as fibrous astrocytes [39,46]. Nevertheless, the role of CD44 in astrocytes in physiological and pathological conditions in the brain, especially during epileptogenesis, remains largely unknown.

In this study, we employed molecular biology techniques, simultaneous EEG and video recording (vEEG), fluorescence immunolabelling, serial block-face scanning electron microscopy (SBEM), and 3D reconstruction [47] to evaluate the impact of the CD44 absence in hippocampal astrocytes on the development of epileptogenesis and consecutive ultrastructural alterations in the tripartite synapse. The presented results point out significant changes in the ultrastructural and functional remodeling of astrocytes and PSD, together with altered seizure progression during epileptogenesis. We show that CD44 deficiency in astrocytes decreases the number of behavioral seizures, increases the number of grade 0 seizures, and reduces reactive astrogliosis during epileptogenesis triggered by kainic acid (KA) administration. Our findings demonstrate that CD44 astrocytic knockout mice show a decreased percentage of astrocyte–synapse contacts, and increased dendritic spine number, along with a smaller PSD area and volume in the molecular layer (ML) of the hippocampal dentate gyrus (DG) of kainic acid-treated mice. Overall, our work suggests that CD44 signaling is involved in astrocyte coverage of neuronal synapses in the ML of the hippocampus and that astrocyte morphological modifications may translate to functional changes in the pathology of epilepsy.

2. Materials and Methods

2.1. Animals

Adult mice (female and male, 12 to 14 weeks old at the start of the experiments) carrying the conditional deletion of exon 3 of the CD44 gene ($CD44^{fl/fl}$; "floxed" mice) were initially acquired from Professor Veronique Orian-Rousseau (Karlsruhe Institute of Technology, Germany) [48]. Animals were bred in the Animal House of the Nencki Institute of Experimental Biology and were treated according to the ethical standards of European and Polish regulations. All mice had free access to food and water, and were kept in individually ventilated cages (at temperatures 21–23 °C, 50–60% humidity) under a 12-h light/12-h dark cycle. All experimental procedures were performed following the European Communities Council Directive of 24 November 1986 (86/609/EEC), Animal Protection Act of Poland, and approved by the 1st Local Ethics Committee in Warsaw (permission no. 860/2019).

2.2. Generation of CD44 Knockout Mice

Before the surgery, mice were anesthetized with a mix of ketamine + medetomidine (80 mg/kg + 1 mg/kg, i.p.) and were given anti-inflammatory and analgesic drugs (butorphanol 3.3 mg/kg, tolfenamic acid 2.0 mg/kg). Then, they were placed in a stereotaxic frame on a heating pad to maintain a constant body temperature. To produce animals with the astrocytic knockout of CD44 (AsKO), one group was stereotaxically injected with 500 nL of AAV5 vector with Cre recombinase–AAV5.*gfaABC1D*::GFP-Cre (construct: Bryan Roth, AAV Serotype 5, UNC Vector Core, Chapel Hill, NC, USA, AAV vector with ITR2 where the human *gfaABC1D* promoter [49] drives the expression of eGFP attached to the N-terminus of Cre recombinase), whereas the control (CTRL) group was injected with 500 nL of AAV5.*gfaABC1D*::GFP vector (Construct: Bryan Roth, Addgene plasmid #50473; AAV Serotype 5, UNC Vector Core, AAV vector with ITR2 where the human *gfaABC1D* promoter drives the expression of eGFP). Injections were performed at coordinates (AP −2.0 and ML ±1.4 from bregma, DV −1.5 below the dura) corresponding to the molecular layer of the DG of both hippocampi. The proper injection side was confirmed post-mortem before histochemical analysis. After-surgery recovery time lasted 4 weeks.

2.3. Western Blot

The dentate gyrus part of the hippocampus was dissected and homogenized in RIPA lysis buffer with protease inhibitor and phosphatase inhibitor cocktails (Sigma-Aldrich, St. Louis, MO, USA) using TissueLyser II (Qiagen, Hilden, Germany). Protein concentration was measured with a BCA Protein Assay Kit (Pierce, Appleton, WI, USA). Proteins were separated and transferred using 10% TGX FastCast Acrylamide Gels and Trans-Blot Turbo Mini PVDF Transfer Packs (Bio-Rad, Hercules, CA, USA). Western blot was performed according to a standard procedure, with anti-CD44 antibody (R&D Systems, Minneapolis, MN, USA, AF6127; 1:2500) and anti-GAPDH antibody (Abcam, Cambridge, UK, ab9485; 1:5000), with the latter being used as a loading control. For band visualization, the chemiluminescence detection method was used. For densitometric calculations, a scan of the X-ray film was analyzed using the Band/Peak Quantification plugin [50] in Fiji 2.3.0 software [29] and normalized to GAPDH.

2.4. Induction of Status Epilepticus

To examine the changes in the CD44 expression level in kainate-induced SE, CD44$^{fl/fl}$ mice underwent intrahippocampal administration of KA. This model reproduces many of the morphological and functional features similar to human patients with temporal lobe epilepsy [51]. Before the surgery, mice were anesthetized with a mix of ketamine + medetomidine (80 mg/kg + 1 mg/kg, i.p.) and were given anti-inflammatory and analgesic drugs (butorphanol 3.3 mg/kg and tolfenamic acid 2.0 mg/kg). Then, animals were placed in a stereotaxic frame on a heating pad to maintain a constant body temperature. Animals were stereotaxically injected with 70 nL of 20 mM solution of KA (Tocris Bioscience, Bristol, UK) in 0.9% w/v NaCl saline or saline alone (SA) to the left CA1 region of the dorsal hippocampus (coordinates: AP −1.8, ML +1.7, DV −2.1, relative to bregma). Injections were administered at a 50 nL/min flow rate. Mice were sacrificed 4 days post-surgery.

To induce SE for electroencephalographic monitoring, 4 weeks after induction of the conditional mutation, mice underwent a second surgery. Animals from both AsKO and CTRL groups were stereotaxically injected with KA (AsKO+KA, CTRL+KA) or SA (AsKO, CTRL) as described above. After the surgery, they were given atipamezole hydrochloride (0.5 mg/kg, i.p.) to facilitate awakening from anesthesia.

2.5. Electrode Implantation and EEG-Video Monitoring

The implantation of five electrodes was carried out immediately after kainate administration during the same surgery session. Two stainless-steel screw electrodes (1.6 mm Ø; Bilaney Consultants GmbH, Dusseldorf, Germany) were bilaterally positioned over the frontal cortex, with two additional ones over the cerebellum (as the right cerebellar electrode served as the ground electrode, there was a total of three reference electrodes). A bipolar hippocampal electrode (Bilaney Consultants GmbH) was implanted into the left hippocampus (coordinates: AP −2.0 and ML +1.3 from bregma, DV −1.7 below the dura). The placement of the electrode was confirmed post-mortem and corresponded to the granular layer of the DG. All electrodes were attached to a pedestal (Bilaney Consultants GmbH) and secured with dental acrylate (Duracryl Plus, Spofa Dental, Czech Republic). Before vEEG monitoring, animals were separately placed in plexiglass cages and connected to a digital acquisition system with commutators (SL6C, Plastic One). vEEG was continuously carried out with TWin Clinical Software for EEG (Grass Technologies, Middleton, WI, USA) and I-PRO WV SC385 digital cameras (Panasonic, Kadoma, Japan) for 4 weeks (24/7).

2.6. Analysis of Video EEG Recordings

The occurrence of spontaneous seizures was analyzed by manually browsing through EEG recordings on a computer screen. An encephalographic episode was defined as a high amplitude (over 2× baseline) discharge lasting ≥10 s. The behavioral severity of each detected spontaneous seizure was analyzed from the matching video recordings. Seizure severity was estimated according to Racine [52], Schauwecker and Steward [53], and Ndode-Ekane and Pitkanen [54] with minor modifications: grade 0—encephalographic seizure with no detectable motor manifestation; grade 1—rigid posture or immobility; grade 2—stiffened, extended, and often arched tail; grade 3—partial body clonus, head bobbing, or whole-body clonic seizures; grade 4—rearing, severe whole-body continuous clonic seizures while retaining posture; grade 5—continuous rearing and falling, with severe clonic seizures; grade 6—generalized tonic-clonic seizures with jumping. Spontaneous seizures were defined as episodes appearing after 24 h post-KA administration. During the analysis, various parameters were characterized: seizure latency (time to the first spontaneous seizure), number of seizures per day, seizure score (behavioral severity), and average seizure duration.

2.7. Immunofluorescence

Mice were transcardially perfused with 4% *w/v* paraformaldehyde (PFA; Sigma-Aldrich) in phosphate-buffered saline (PBS), and brains were dissected and post-fixed by immersion in the same buffer overnight at 4 °C. Then, they were cryoprotected in 30% *w/v* sucrose in PBS for 24 h at 4 °C, snap-frozen in isopentane, and stored at −70 °C. After cutting 40-μm-thick coronal brain slices, they were subjected to antigen retrieval for 30 min at 80 °C in a preheated sodium citrate buffer (10 mM sodium citrate, 0.05% Tween 20, pH 6.0). Subsequently, slices were washed in PBS, blocked in 5% *v/v* NDST (normal donkey serum in PBS + 0.2% Triton X-100) for 1 h at room temperature (20–24 °C, RT), and incubated with primary antibody diluted in 5% NDST overnight at 4 °C. This step was followed by a PBS wash and incubation with species-specific secondary antibodies in 5% NDST for 2 h at RT. After incubation, slices were washed, co-stained with Hoechst (Invitrogen, Carlsbad, CA, USA, H3570; 1:500), a cell-permeant nuclear stain, for 10 min at RT, washed again, and coverslipped on glass slides with Vectashield (Vector Laboratories, Newark, CA, USA). Primary antibodies used: sheep anti-CD44 (R&D Systems, AF6577; 1:500), chicken anti-glial fibrillary acidic protein (GFAP) (Abcam, ab4674; 1:500), and rabbit anti-s100β (Abcam, ab52642; 1:100). Secondary antibodies used: donkey anti-sheep conjugated to Alexa Fluor^{TM}488 (Invitrogen; 1:500), donkey anti-sheep conjugated to Alexa Fluor^{TM}555 (Invitrogen; 1:500), donkey anti-rabbit conjugated to Alexa Fluor^{TM}647 (Jackson ImmunoResearch; 1:500).

2.8. Fluorescence Microscopy Analysis

Microscopic imaging of fluorescent specimens was carried out using a Leica AF 7000 fluorescence microscope (objective lenses: HCX PL APO 10.0×/0.40 DRY, HCX PL APO 20.0×/0.70 DRY) and a Zeiss LSM 800 with Airyscan confocal microscope (objective lenses: EC Plan-Neofluar 10×/0.30, Plan-Apochromat 63×/1.40 Oil DIC M27). All scans were analyzed with Fiji software. Fractions of virally transfected astrocytes were calculated from s100β-immunostained slices. The area of astrocytes was determined by outlining individual astrocytes observed as GFAP immunosignal. Because GFAP protein is mostly localized in the major primary and secondary branches [55,56], undetectable terminal astrocyte processes were excluded from this analysis. Comparative analyses were only performed in the case of images taken with the same acquisition parameters.

2.9. SBEM Sample Preparation

Mice were transcardially perfused with 4% PFA + 2.5% e/v glutaraldehyde (GA; EM grade, Sigma-Aldrich) in PBS. Brains were dissected and post-fixed by immersion in the same buffer overnight at 4 °C, then cut on a vibratome into 100 μm-thick coronal slices. The ML of the DG region in the hippocampus was cut out from each slice, along with a small piece of granular layer for orientation, with a razor blade.

SBEM staining was performed based on a previously published protocol [57] with slight modifications. Specimens were washed in cold 0.1 M phosphate buffer pH 7.4, post-fixed in 1.5% potassium ferrocyanide (Sigma-Aldrich) and 2% osmium tetroxide (Electron Microscopy Sciences, Hatfield, PA, USA) aqueous solution for 1 h on ice. All later steps were performed at room temperature. Samples were washed with double distilled (dd) H_2O, immersed in 1% aqueous thiocarbohydrazide (TCH; Sigma) solution for 20 min, and washed again. Subsequently, they were postfixed in a 2% osmium tetroxide aqueous solution for 30 min, rinsed with dd H_2O, and incubated in a 1% aqueous solution of uranyl acetate overnight. The next morning, samples were washed and immersed in a freshly prepared lead aspartate aqueous solution pH 5.5 (0.066 g lead nitrate in 10 mL dd H_2O + 0.04 g aspartic acid pH 3.8) for 20 min at 60 °C, then washed again. Next, they were dehydrated using reversely graded dilutions of ethanol (20%, 50%, 70%, 100%, 100% anhydrous) for 5 min each. After the last ethanol incubation, fragments were infiltrated with a 1:1 (*v/v*) solution of Durcupan resin (Sigma-Aldrich) with anhydrous ethanol for 30 min, and 100% Durcupan resin for another 1 h. The resin was then replaced with a fresh one and left to incubate overnight. The next day, Durcupan resin was replaced again for an additional 1-h infiltration, after which samples were flat embedded between Aclar sheets (Electron Microscopy Sciences), and put in an oven at 65 °C for at least 48 h for the resin to polymerize. After the resin hardened, Aclar sheets were separated and the resin-embedded ML pieces were taken out, glued to aluminum pins (Gatan 3View system scanning electron microscopy pin stubs, Micro to Nano), and trimmed on an ultramicrotome (Leica Ultracut R). Samples were covered with conductive silver paint (Ted Pella) and mounted into the 3View chamber (Gatan).

2.10. SBEM Imaging

Imaging of samples was conducted with a SigmaVP (Zeiss, Oberkochen, Germany) scanning electron microscope equipped with the 3View2 chamber (Gatan, Pleasanton, CA, USA) using a backscatter electron detector. Scans were taken in the inner molecular layer of the hippocampus. From every sample, 200 sections were collected (thickness 60 nm). Imaging settings: 2.5 kV EHT, 20 μm aperture, 7–9 Pa variable pressure, 15,000× magnification, 7 μs pixel dwell time, 5 nm × 5 nm pixel size (3072 × 3072 camera array).

2.11. SBEM Analysis

Scans were aligned using the TrakEM plugin in Fiji software and normalized with Microscopy Image Browser 2.82 software [58]. The reconstruction of synaptic connections and astrocytic leaflets was carried out using Reconstruct 1.1.0.0 software [59]. For 3D reconstruction, areas from each sample taken for analysis were randomly picked and processed using the unbiased brick method [60]. For every specimen, reconstructed structures were limited to a brick of size 6 μm × 6 μm × 6 μm. Dendritic spines were defined as dendritic protrusions having an electron-dense region, the approximate core of post-synaptic density (PSD), close to the membrane adjacent to an axonal bouton with synaptic vesicles. Astrocytes and their leaflets were identified by characteristic features: an irregular stellate shape, numerous glycogen granules, bundles of intermediate filaments, or a relatively clear cytoplasm. After identifying and reconstructing PSD areas, dendritic spines and axonal boutons, the parts of astrocytic leaflets located in their proximity, were drafted. Dendritic spines were classified as filopodia (protrusions without enlargement at the tip), stubby spines (short protrusions without a neck and in direct continuity with the dendritic surface), and mushroom spines (protrusions with a neck and an enlargement at the tip). The PSD surface area and volume were measured by outlining darkened electron-dense regions on every section with a particular PSD. Following the work of Ostroff et al. [61], an astrocytic leaflet localized in the proximity ≤20 nm from a PSD was considered as one having contact with the synapse. Reconstruct software was used to extract PSD parameters and PSD to leaflet distance.

2.12. Statistical Analysis and Figures

Data statistical analysis was performed using GraphPad Prism 9 software (GraphPad Software Inc., San Diego, CA, USA). Distributions were tested with the Shapiro-Wilk normality test. Differences in semiquantitative scores between groups were analyzed using Fisher's test. For other comparisons, the parametric Student's *t*-test (groups with equal variances), Student's *t*-test with Welch's correction (groups without equal variances), two-way ANOVA, or the nonparametric Mann–Whitney test (groups with equal variances), the Kolmogorov-Smirnov test (groups without equal variances) were used. Data on graphs are presented as mean ± standard errors of the mean (SEM) in parametric analyses or as median with interquartile range (IQR) in nonparametric analyses. Differences between groups were considered significant if $p < 0.05$ (* $p < 0.05$, ** $p < 0.01$, *** $p < 0.001$). The Pearson correlation coefficient was calculated to measure the correlation between the intensity of CD44 and the area of GFAP immunoreactivities.

Apart from the information on the statistical parameters included in the figure captions, additional parameters (Supplementary Table S1 and Figure S1) were calculated in Python (ver. 3.7.4) using the following libraries: numpy (1.19.2) and scipy (1.5.2). The linear regression was performed using the linregress module (1.10.1). The power analysis was performed using a corresponding function from the statsmodels module (0.12.1). It was calculated for a pair of datasets based on the number of data points of a larger set. The effect size was calculated as a difference of means divided by the minimum value of standard deviations. The significance level was 0.05.

3. Results

3.1. Increased Level of CD44 in Astrocytes following Kainate-Induced Status Epilepticus

To shed light on the role of CD44 in the development and progression of epilepsy, we characterized the expression of CD44 in the mouse hippocampus after kainic acid-induced seizures. *Status epilepticus* was induced in CD44$^{fl/fl}$ mice by intrahippocampal KA injections ($n = 4$ per group). This model reproduces many of the morphological and functional features observed in human patients with temporal lobe epilepsy [51]. The immunohistochemistry with anti-CD44 and anti-glial fibrillary acidic (GFAP) antibodies was performed to visualize CD44 protein expression in astrocytes (Figure 1a,b). We compared the CD44 level in the hippocampus of control animals and animals 4 days after KA-induced SE. Sham injections with saline (SA) were used to give sufficient control for KA administration in terms of the mechanical injury caused by needle insertion and fluid infusion. The astrocytic GFAP profile area increased 4 days after KA-induced SE (Figure 1c). Furthermore, in comparison to a low level of CD44 expression in SA-treated controls, we observed a visible upregulation of CD44 in hippocampal astrocytes in the ML of the DG (Figure 1d).

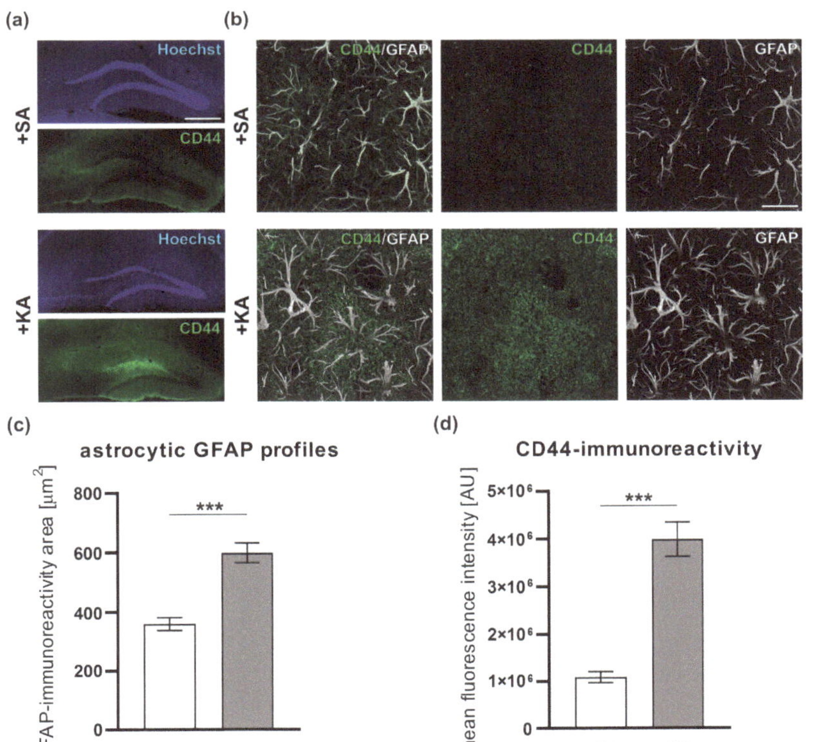

Figure 1. Upregulation of the CD44 expression in hippocampal astrocytes of CD44$^{fl/fl}$ mice following kainic acid (KA)-induced *status epilepticus*. (**a**) Representative immunofluorescence micrographs of

CD44 expression (green) and nuclei (Hoechst, purple) in the hippocampus of SA- or KA-injected mice. Scale bar: 500 μm. (**b**) High-power view of the DG neuropil (molecular layer) immunostained for GFAP (white) and CD44 (green) in SA- or KA-injected mice. Scale bar: 20 μm. (**c**) Quantitative analysis of astrocytic GFAP area in CD44$^{fl/fl}$ mice 96 h after SA or KA administration ($n = 4$ independent animals in each group, Student's t-test, *** $p < 0.001$, data presented as mean ± SEM) (**d**) CD44-immunoreactivity in astrocytic GFAP profiles 96 h after SA or KA administration ($n = 4$ independent animals in each group, Student's t-test, *** $p < 0.001$, data presented as mean ± SEM).

3.2. CD44 Depletion in Astrocytes by AAV5-Mediated Expression of gfaABC1D::GFP-Cre Recombinase in the Hippocampi of CD44$^{fl/fl}$ Mice

To further investigate the role of CD44 in astrocytes during epileptogenesis, we implemented a genetic approach to delete CD44 exclusively in the hippocampal astrocytes. We performed stereotaxic delivery of an adeno-associated viral vector (AAV5) encoding Cre-recombinase into the molecular layer of the dentate gyrus of the hippocampi of CD44$^{fl/fl}$ mice (Figure 2a). Cre recombinase was expressed in astrocytic cell nuclei and was N-terminally appended with green fluorescent protein (GFP), a fluorescent reporter, under the control of an astrocyte-specific GFAP promoter. These mice are indicated in the figures as astrocytic CD44 knockouts (AsKO). As a control (indicated as CTRL), we used CD44fl/fl mice injected with the AAV5 virus encoding GFP under the control of the GFAP promoter, resulting in GFP expression in the cell cytoplasm of astrocytes. To estimate the efficiency of viral transduction, we calculated the percentage of s100β-positive cells that were also GFP-positive. The analysis revealed that the efficiency of transduction was 84.1 ± 3.51% for AAV5.gfaABC1D::GFP and 74.22 ± 3.07% for AAV5.gfaABC1D::GFP-Cre (data presented as mean ± SEM). Next, we examined CD44 expression levels by Western blot analysis, 4–5 weeks after AAV delivery. We found a strong reduction of CD44 in the DG of mice expressing Cre recombinase compared to the controls (Figure 2b,c). Immunohistochemical analysis revealed that, unlike the CTRL group, AsKO animals had decreased CD44 expression in the infected astrocytes (detected with the anti-s100β antibody) in the molecular layer of the DG (Figure 2d,e).

3.3. CD44 Deficiency Reduces KA-Induced Epileptogenesis

To investigate the role of CD44 in the process of epileptogenesis, mice with CD44 deficiency in hippocampal astrocytes and CTRL animals ($n = 12$ per group) were intrahippocampally injected with KA, and continuously vEEG monitored for the next 4 weeks (Figure 3a). Both groups developed seizures during the recording period. When analyzing seizures with behavioral manifestations (grades 1–6; for scale description see Section 2.6 of Materials and Methods), no significant differences in the latency to the first spontaneous seizure (Figure 3b), the average duration of spontaneous seizures (Figure 3c), or seizure severity (Figure 3d) were observed. However, the average number of behavioral seizures per day during the fourth week was significantly lower in CD44 AsKO relative to the CTRL animals (Figure 3e). In contrast, the average number of grade 0 seizures (detected in EEG but without any behavioral symptoms) was significantly higher in knockout mice than in the controls (Figure 3f); 84% of all seizures in the CD44 knockout group were of grade 0, compared to only 33% in the control mice (Figure 3g). The difference in the total number of all seizures (including non-behavioral) between AsKO+KA and CTRL+KA animals (Figure 3g) mainly results from a significantly greater number of seizures of grade 0 observed in AsKO+KA mice (Figure 3f).

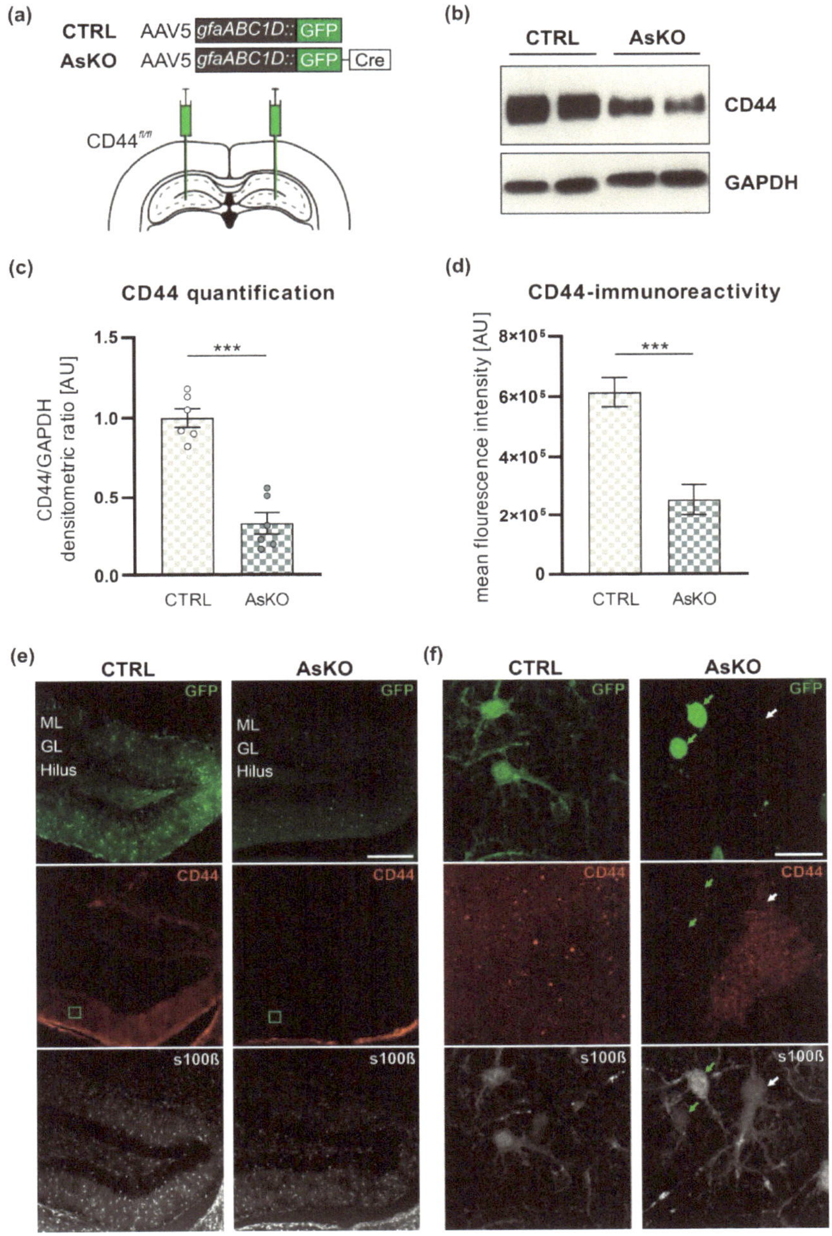

Figure 2. GFAP-Cre recombinase viral approach to target CD44 in adult dentate gyrus astrocytes. (**a**) Scheme of bilateral intrahippocampal administration of AAV vectors (CTRL: AAV5.*gfaABC1D*::GFP; GFP localizes in the cytoplasm of astrocytes, and AsKO: AAV5.*gfaABC1D*::GFP-Cre; GFP-Cre localizes in cell nuclei of astrocytes) into the brains of $CD44^{fl/fl}$ mice. (**b**) CD44 level in protein extracts from DG

of CTRL and AsKO animals (Western blot). Glyceraldehyde 3-phosphate dehydrogenase (GAPDH) was used as the loading control protein. (**c**) Quantitative analysis of Western blot results ($n = 6$ independent animals, Student's t-test, *** $p < 0.001$, data presented as mean ± SEM) (**d**) CD44-immunoreactivity in astrocytic s100β profiles from the ML of the DG in CTRL and AsKO mice (see panel f for representative immunostaining) (Student's t-test, *** $p < 0.001$, data presented as mean ± SEM). Representative immunofluorescence micrographs of GFP (green), astrocyte marker s100β (white), and CD44 (red) expression in the hippocampus of CTRL and AsKO mice for matching triple stain images shown at (**e**) a lower magnification or (**f**) the close-ups reflecting green squares on panel (**e**). Arrows point to astrocytes (s100β staining) that were or were not transfected (green arrow for GFP-expressing and white arrow for non-expressing, respectively). ML—molecular layer, GL—granular layer. Scale bar: (**e**) 200 μm and (**f**) 20 μm.

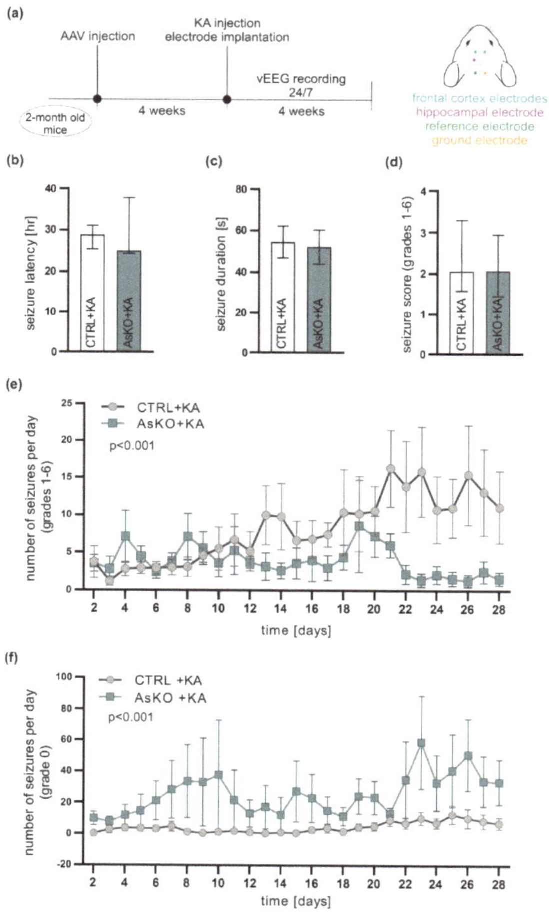

Figure 3. *Cont.*

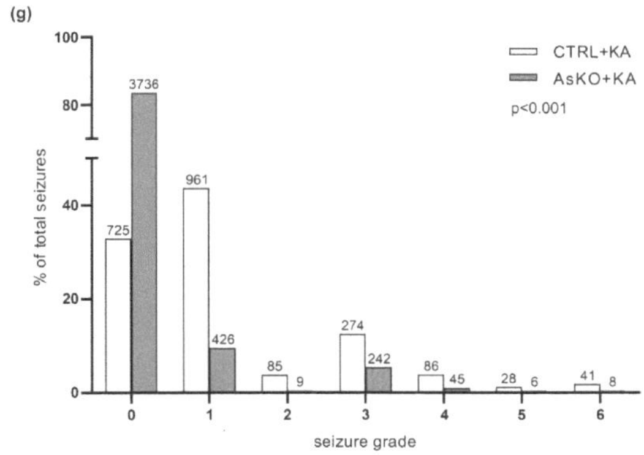

Figure 3. Astrocytic knockout of CD44 in the dentate gyrus results in fewer spontaneous behavioral seizures after intrahippocampal KA administration. (**a**) Schematic representation of the experimental design. (**b**) Seizure latency to the first spontaneous seizure was not affected by CD44 knockout (Mann-Whitney test, $p = 0.22$, data presented as median with IQR). (**c**) The mean duration of spontaneous seizures was not different in the AsKO compared to the CTRL group (Student's t-test, $p = 0.83$, data presented as mean ± SEM). (**d**) The average seizure score (for seizures of grades 1–6) was not different in the AsKO compared to the CTRL group (Mann-Whitney test, $p = 0.38$, data presented as median with IQR). (**e**) Knockout animals developed a decreased number of behavioral seizures (grades 1–6) per day when compared to CTRL animals (two-way ANOVA, genotype effect $p < 0.001$, data presented as mean ± SEM). (**f**) CD 44 knockout animals developed a greater average number of seizures of grade 0 per day than CTRL mice (two-way ANOVA, genotype effect $p < 0.001$, data presented as mean ± SEM). (**g**) The severity of behavioral seizures presented as a percentage of each seizure type (grades 0–6) out of all episodes per group. Most seizures in AsKO mice were less severe than the ones occurring in the CTRL group (Fisher's exact test, $p < 0.001$).

3.4. *CD44 Deficiency Reduces Reactive Astrogliosis after Kainate-Induced Status Epilepticus*

It is well known that reactive astrogliosis occurs in the mouse hippocampus in response to seizures (rev. in [62]). Having demonstrated that CD44 is upregulated in astrocytes following SE, we then evaluated how CD44-deficient astrocytes respond to kainate-induced SE. To assess reactive astrogliosis, we performed GFAP immunohistochemical staining in CTRL and AsKO mice 4 weeks after SA (Figure 4a) or KA injection (Figure 4b). For quantification, the mean area occupied by GFAP immunoreactivity (GFAP-ir) for a single astrocyte, which reflects the extent of reactive astrogliosis, was measured. The results showed that CD44 deletion had no significant effect on the GFAP-ir area in naive, saline-treated mice (Figure 4c), whereas seizure-induced astrogliosis was significantly reduced in CD44 AsKO+KA versus CTRL+KA mice (Figure 4e). Alongside the GFAP immunostaining, we additionally assessed the CD44 expression level in astrocytes. CD44 immunoreactivity (CD44-ir) was lowered in AsKO vs. CTRL mice (Figure 4d). The seizure-induced upregulation of CD44 at this time point was still well observed in the CTRL+KA group (Figure 4b,f) and was even more robust than the one seen 4 days post-kainate administration (presented in Figure 1). There was no apparent CD44-ir upregulation in AsKO+KA compared to untreated AsKO (Figure 4d,f). Moreover, the intensity of CD44-ir positively correlated with the area of GFAP-ir in both the CTRL ($R = 0.901$, Pearson correlation coefficient) and in CTRL+KA ($R = 0.919$, Pearson correlation coefficient) groups. The correlation was not observed in the AsKO group ($R = 4.8 \times 10^{-5}$, Pearson correlation coefficient) or was very low in the AsKO+KA group ($R = 0.52$, Pearson correlation coefficient) (see Supplementary Figure S1b,d for more details of the analysis), probably due to the low CD44 expression in the AsKO.

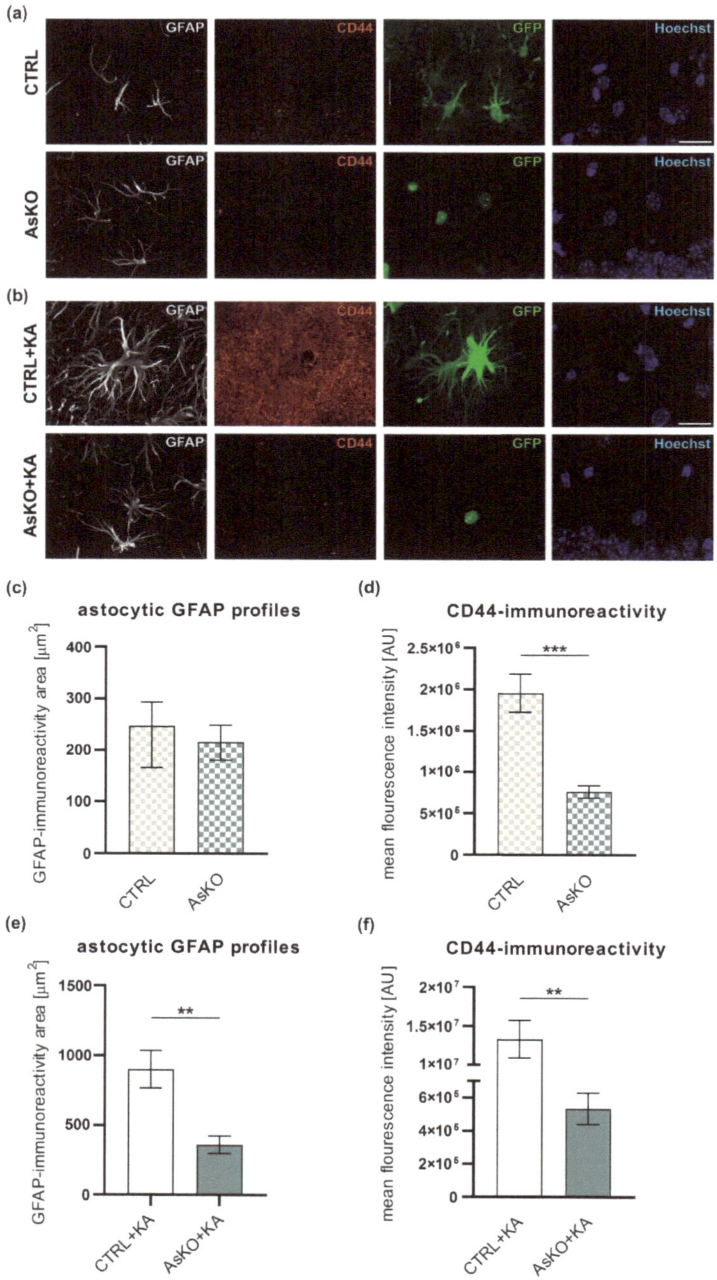

Figure 4. CD44 deletion in astrocytes reduces reactive astrogliosis following kainate-induced epilepsy. Representative images of astrocytes (GFAP, white), CD44 expression (red), GFP (green), and nuclei (Hoechst, purple) in the hippocampal ML of the DG of CTRL and CD44 AsKO mice before (**a**) and after

(**b**) 4 weeks post-KA administration. Scale bar: 20 µm. Quantitative analysis of GFAP-immunoreactive area (µm^2) in the ML of the DG in CTRL and CD44 AsKO mice 4 weeks after administration of (**c**) SA (Kolmogorov-Smirnov test, p = 0.41, data presented as median with IQR) or (**e**) KA (n = 6 independent animals in each group, Student's t-test with Welch's correction, ** p = 0.004, data presented as mean ± SEM). Quantitative analysis of CD44-immunoreactivity in astrocytic GFAP profiles in the ML of the DG in CTRL and CD44 AsKO mice 4 weeks after administration of (**d**) SA (Student's t-test with Welch's correction, *** p < 0.001, data presented as mean ± SEM) or (**f**) KA (Student's t-test with Welch's correction, ** p = 0.001, data presented as mean ± SEM).

3.5. Ultrastructural Changes in Astrocyte-Synapse Interactions and Dendritic Spines in CD44 AsKO Mice Hippocampi upon Seizures

To establish if alterations in seizure severity in CD44 astrocytic knockouts in the hippocampus are accompanied by structural changes in the astrocyte-neuron relationship, we employed electron microscopy and characterized synapses and adjacent astrocytes at the ultrastructural level. We used SBEM to three-dimensionally reconstruct synaptic contacts together with surrounding astrocytic terminal processes from the ML of the DG. All structures were analyzed from tissue bricks of 6 µm × 6 µm × 6 µm size per animal following the unbiased brick method [60]. We performed a volumetric analysis of PSDs representing the postsynaptic part of the synapse. We reconstructed 192 synapses with adjacent astrocyte leaflets from the control group (CTRL+KA) and 472 from the AsKO mice (AsKO+KA) upon KA-induced seizures. The distances between reconstructed PSDs and adjacent astrocytic leaflets in the ML of the hippocampi of CTRL+KA and AsKO+KA mice were measured (Figure 5a,b). Adopting the methodology of Ostroff et al. [61], we assumed that a leaflet-PSD distance ≤20 nm represents synapses with astrocyte contacts. As a result, we observed a lower percentage of such contacts in AsKO+KA mice in comparison to the CTRL+KA group (Figure 5c). Since extensive spine loss in the dendrites of the DG granule cells has been previously shown upon kainate-induced seizures [16], we also analyzed changes in dendritic spine density in mice after SA (CTRL and AsKO) or KA (CTRL+KA and AsKO+KA) injection. The density of dendritic spines in the ML was reduced upon seizures in CTRL, but not in AsKO mice (Figure 5d–f). At the same time, the PSD mean volume and area were significantly diminished in AsKO+KA mice in comparison to CTRL+KA animals (Figure 5h,i). To more precisely define the characteristics of the dendritic spines, we classified them into three main categories: filopodia, stubby spines, and mushroom spines. The ratio between these different types of dendritic spines did not vary in the CD44 AsKO+KA vs. CTRL+KA groups (Figure 5g). Together, these data show that the depletion of CD44 in astrocytes leads to a reduced interaction of astrocytic leaflets with the synaptic cleft, an increased density of dendritic spines, and a decreased size of postsynaptic densities in the epileptic ML within the dentate gyrus of the hippocampus.

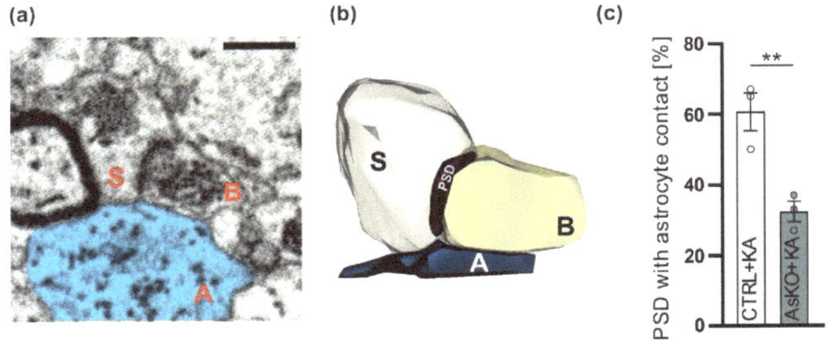

Figure 5. *Cont.*

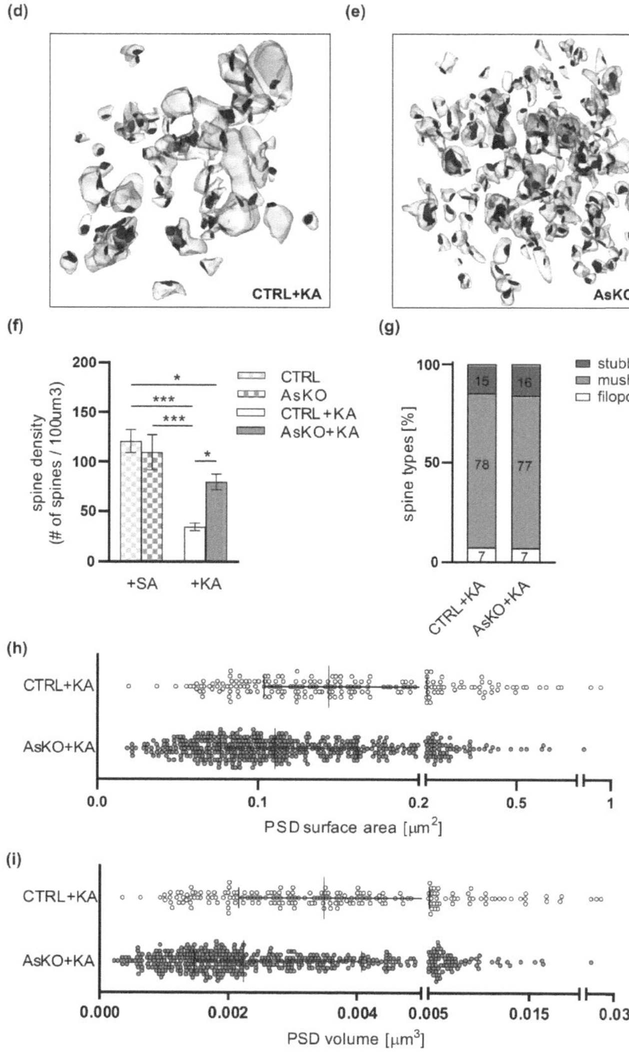

Figure 5. Astrocytic knockout of CD44 in the ML of the dentate gyrus prevents KA-induced spine loss and results in reduced PSD volume and surface area and diminished astrocyte–synapse interaction in the epileptic brain. (**a**) Representative image of a synaptic structure from SBEM. Astrocytic leaflet (A, blue) contacting presynaptic bouton (B) and postsynaptic dendritic spine (S). Scale bar: 500 nm. (**b**) Example of a 3D reconstruction of the tripartite synapse structure with PSD. (**c**) The proportion of synaptic clefts with astrocyte contacts. Shown as a percentage of PSDs with astrocytic leaflets in a ≤20 nm proximity (Fisher's exact test, ** $p < 0.001$, data presented as mean ± SEM). (**d**,**e**) Examples of 3D reconstructions of spines (white) with PSDs (red) from CTRL+KA and AsKO+KA mice (brick size: 6 μm × 6 μm × 6 μm). (**f**) Spine density was decreased in CTRL mice following KA treatment. In CD44 AsKO animals the KA-induced reduction in spine density is diminished (two-way ANOVA, * $p < 0.05$, *** $p < 0.001$ data presented as mean ± SEM). (**g**) The proportion of protrusion types (mushroom spines, stubby spines, and filopodia) detected on dendritic segments imaged with SBEM (two-way ANOVA). (**h**,**i**) Both PSD area and volume were decreased in AsKO+KA mice (Mann-Whitney test, *** $p < 0.001$, data presented as median with IQR).

4. Discussion

In the present study, we report that mice with selective deletion of CD44 in hippocampal astrocytes developed less severe epileptogenesis and gliosis upon intrahippocampal KA administration. Furthermore, we characterized ultrastructural seizure-induced changes in dendritic spines and thin astrocytic processes that contact synapses in the DG of the hippocampus of the CTRL and CD44 AsKO animals. Our work revealed that, compared to the control mice, astrocytic deficiency of CD44 in the epileptic hippocampus leads to a diminished astrocytic coverage of synapses, an increased dendritic spine number, and a decreased size of postsynaptic densities.

CD44 has been previously shown to be predominantly expressed in astrocytes in rodents and the human brain, and it has been demonstrated to be important in the regulation of astrocyte morphology [39,58]. CD44 expression increases in the brain after injury [29,63]. However, there are inconsistent data concerning its expression in animal models of epilepsy. CD44 was strongly upregulated in the dentate gyrus molecular layer 3 days post-pilocarpine-induced SE in mice, then it declined over the next 4 weeks [42]. This adhesion molecule was also identified in a proteomic approach as one of the most robustly induced proteins in the mouse brain in the kainate model of epileptogenesis [64]. Contrary to the above observation, studies in in vitro hippocampal organotypic cultures have shown that CD44 expression in the molecular layer was reduced following kainic acid treatment [40]. In our model of TLE, CD44 was overexpressed in hippocampal astrocytes and this enhanced expression persisted for the next four weeks. A potentially critical difference between studies may arise from the use of different models of epilepsy. The cited study was conducted in cell culture under strictly defined conditions [40], whereas ours was carried out in the in vivo mouse model where a variety of systemic players may influence the observed alterations. Although it is important to take into consideration both in vitro and in vivo results, the latter in many cases better represents the complicated changes happening in a living organism. Increased CD44 immunoreactivity in the astrocytic processes was also shown in human patients with tuberous sclerosis [65] and the mouse model of Alexander disease [58], which are conditions accompanied by seizures. Recently, CD44 was identified in genome-wide transcriptome analysis as one of the treatment targets for epilepsy in rats [44] and humans [45].

A growing body of evidence indicates that the elevated expression of CD44 is associated with epilepsy-related brain injuries [42,44,45,58,64,66]. However, the exact role of CD44 in activated astrocytes in epileptogenesis remains to be determined. Astrocytes are increasingly recognized as playing a fundamental role in bidirectional communication with neurons. They modulate neuronal excitability, synaptic transmission and plasticity [67], although the molecular mechanisms underlying these processes are largely unknown. Thus, we employed a viral infection approach to specifically target CD44 in hippocampal astrocytes to unravel its function in epileptogenesis. Based on the existing literature data [68,69] and our preliminary tests, we chose the AAV5 viral vector serotype and the astrocyte-specific GFAP promoter to obtain cell-specific CD44 deletion. Our results indicate that injection of the AAV5 vector expressing Cre recombinase appended at its N-terminus by GFP under the control of GFAP promoter into the molecular layer of the DG of CD44$^{fl/fl}$ mice effectively reduces the expression of CD44 in hippocampal astrocytes. However, we observed that upon kainate treatment, GFP was also expressed in granule neurons of the DG. These results suggest that AAV5 can also infect DG neurons and, what is more important and surprising, is that kainic acid influences neuronal gene expression so strongly that it can induce expression controlled by a GFAP promoter in granule neurons. Our observation is in line with previous data showing the expression of GFAP-driving transgenes in hippocampal neuronal cells [70]. An alternative explanation of KA-induced GFAP promoter-driven expression in neurons could be the enhanced neurogenesis in the DG and survival of the new granule neurons. It has been previously shown that seizure activity leads to transiently increased cell proliferation in the DG during the first 2 weeks after *status epilepticus* [71]. New neurons in the DG arise from a population of radial neural stem cells that, like astrocytes, express GFAP [72,73] and survive for at least 6 months [74]. Therefore, we cannot completely exclude a neuronal contribution to the results observed in our study. However, taking into account that the level of CD44 in neurons of the adult brain is low and that KA induces CD44 expression in astrocytes but not in neurons, we can still assume that the observed effects are predominantly caused by the CD44 knockout in astrocytes rather than neurons.

Recently, the non-specific actions of Cre recombinase in the brain have been reported but they regard mainly neurons, not astrocytes. The mechanism of the possible noxious effects of AAV-mediated Cre expression described in these reports points to defects in DNA repair machinery that lead to a massive decrease in the neuronal population [75]. Adding WT mice with a Cre injection as an additional control could shed more light on this phenomenon. Since we focused our project on astrocytic CD44 and we did not observe massive cell death, we decided not to include the WT control injected with AAV5.*gfaABC1D*::GFP-Cre in our experiments to minimize the number of sacrificed animals. It has been previously shown that AAV viruses expressing eGFP controlled by the *gfaABC1D* promoter induce a titer-dependent reactivity of astrocytes [76]. Following low viral-titer injection (10^{10} genome copies (GC) per injection), eGFP-positive cells had normal astrocytic morphology and low levels of GFAP staining, whereas eGFP-positive cells labeled by high-titer injection (3×10^{10} genome copies per injection) of the virus had a hypertrophic morphology typical of reactive astrocytes. In our experiments, we used a low viral titer injection (CTRL: 4×10^9 and Cre: 2.2×10^9 genome copies (GC) per injection) that does not induce astrocytic gliosis

In the present study, CD44 AsKO mice developed fewer spontaneous behavioral seizures following KA-induced *status epilepticus* than CTRL mice. The majority of the seizures observed in CD44 knockouts were of grade 0; i.e., were detected with EEG recording but expressed no behavioral symptoms. Thus, our findings demonstrate that CD44 depletion in astrocytes leads to defects in behavioral seizure development, but the molecular mechanisms of CD44 action are still unknown. Epilepsy is linked to a robust aberrant synaptic plasticity occurring in the ML of the DG [77]. Recently, it was determined that astrocytic deletion of one of the CD44 binding partners, the actin-associated protein Ezrin, resulted in smaller astrocyte territories and reduced astrocytic coverage of excitatory synapses. This led to a decrease in the synaptic glutamate level and increased extrasynaptic glutamate diffusion, and had a significant impact on contextual fear memory [78]. Since the C-terminal cytoplasmic domain of CD44 interacts with Ezrin (reviewed in [79,80]), the observed changes in seizure severity in CD44 knockout animals might be explained by alterations in glutamate uptake and diffusion. This hypothesis is also supported by the results showing that increased CD44 in astrocytes correlates with a loss of plasmalemmal glutamate transporter GLT-1 and diminished glutamate transporter current [58]. It is possible that astrocytes with increased expression of CD44 upon KA-induced *status epilepticus* do not buffer glutamate as efficiently as non-reactive astrocytes.

Reactive astrocytes play a crucial role in CNS pathophysiology. Astrogliosis, the process of astrocyte activation upon brain injury (including epilepsy), is connected with changes in astrocytic morphology. We have previously shown that CD44 regulates astrocyte shape in primary astrocytic cultures [39]. In the present study, we observed reduced GFAP-ir profiles of astrocytes upon seizures in CD44 AsKO mice, indicating that the enhanced CD44 expression supports the development of astrogliosis. These results are in line with the previously published observation showing that phenotypic conversion from regular protoplasmic to hypertrophic astrocytes in Alexander disease is associated with CD44 acquisition by normally CD44-negative protoplasmic astrocytes. This suggests that CD44 can play a role in astrogliosis; e.g., by mediating alterations in astrocyte shape [58]. There is also evidence that astrogliosis can be a cause of seizures. Genetically- or virally-induced astrogliosis was shown to be sufficient to cause epileptic seizures in mice without any other CNS pathologies [76,81]. Thus, CD44 might also regulate seizure severity by contributing to the development of astrogliosis.

Epilepsy results from an imbalance between neuronal excitation and inhibition. Among effective mechanisms for decreasing epilepsy occurrence are (i) the conversion of polyamines to GABA and (ii) the subsequent release of GABA from astrocytes [82–84]. There is no direct evidence that CD44 function is related to GABA release. Nevertheless, the observed beneficial effect of CD44 deficiency on seizure progression may be associated with changes in the interplay between either CD44 itself or CD44-dependent signaling pathways and astrocytic GABA release or mechanisms regulating polyamine conversion to GABA in astrocytes. In addition to excitatory synapses, the astrocytic deletion of CD44 may as well alter inhibitory synapses made by GABAergic neurons that are also present in the molecular layer of the hippocampus.

Astrocytic leaflets are linked to synaptic ensheathment, and interaction with dendritic spines and synaptic terminals, which regulate synaptic structure and functions [21,85,86]. The interplay between neurons and astrocytes is one of the mechanisms that contribute to the generation of epileptic discharges [87]. Our results revealed that astrocytic deletion of CD44 during epileptogenesis has an impact on the structural relationship between astrocyte processes and synapses, leading to a decreased number of synapses with astrocyte contacts. This may be explained by the possibility that AsKO results in decreased astrogliosis. In line with this, the loss of Ezrin reduces astrocyte-neuron proximity and the number of astrocyte contacts at synapses, and these structural changes are accompanied by increased glutamate spillover and NMDA receptor activation [78]. Thus, it is likely that CD44-dependent changes in astrocytic synaptic coverage in CD44 AsKO mice might be associated with a reduced glutamate concentration at synapses and in turn lead to reduced seizure progression. On the other hand, in CTRL+KA animals, the glutamate concentration at the synapse can be increased by (i) the reduced level of GLT-1 in reactive astrocytes (reviewed in [88]), leading to an inhibition of glutamate reuptake by astrocytes at the synaptic cleft, and (ii) an increased number of astrocyte-synapse contacts, resulting in a decrease in glutamate spillover outside the synaptic cleft.

We showed that spine density significantly decreases in the ML of the dentate gyrus of CTRL mice upon KA administration. These results are in line with previously shown extensive spine loss in the dendrites of granule cells upon kainate-induced seizures [16]. In CD44 AsKO, we did not observe KA-induced spine loss. This may suggest a protective effect of the astrocytic CD44 deletion on the KA-induced loss of existing synapses in the hippocampus of knockout mice. Moreover, electron microscopy ultrastructural observations of the epileptic brain uncovered the presence of degenerated presynaptic profiles partially or fully enveloped in astrocytic leaflets [18,27,89], suggesting that seizures can induce active synaptic phagocytosis by reactive astrocytes. Therefore, the observed elevated number of synapses in the hippocampus of CD44 knockouts might be associated with reduced phagocytosis by CD44-depleted astrocytes. Increased dendritic spine density and decreased PSD size in CD44 AsKO animals upon seizures indicate that CD44 deletion from astrocytes can have beneficial effects on brain structure after epilepsy-related injury. One possible target is the prevention of the formation of aberrant giant spines associated with epileptic trauma [16,60], which are believed to be a compensatory mechanism for a reduced number of axonal boutons as a way to recover some of the lost excitatory drive [90]. This possibility is unlikely in our animal model because we did not observe such giant spines with sufficient frequency to be statistically significant (Figure 5g). Another possible target is the transition of newly formed spines, which are usually thin and often called immature or learning spines, to become mature mushroom spines, also known as memory spines [91–94]. Such morphological changes in dendritic spines are prerequisites of synaptic plasticity [95,96]. However, our study revealed that the proportion of dendritic spines of different shapes is not changed in the ML of the DG in the CD44 AsKO brain upon KA-induced seizures, indicating that all investigated types of dendritic spines are equally susceptible to the beneficial effect of astrocytic CD44 deletion.

Supplementary Materials: The following supporting information can be downloaded at: https://www.mdpi.com/article/10.3390/cells12111483/s1, Figure S1. Correlation between the area of GFAP-ir and mean CD44 fluorescence for: (a): CTRL, (b) AsKO, (c) CTRL+KA, and (d) AsKO+KA groups. Statistical parameters in each panel correspond to the following parameters, respectively: slope and intercept of the fitted lines, R^2 (coefficient of determination), the p-value for a hypothesis test whose null hypothesis is that the slope is zero, standard error of the estimated slope. Lines represent linear regression given by the equation (GFAP-ir intensity (y) = slope∗ CD-44-ir intensity (x) + intercept). Table S1. Additional statistical parameters.

Author Contributions: P.K.K.: Investigation, Methodology, Formal analysis, Writing—original draft, Writing—review & editing. K.N.: Investigation, Formal analysis. A.S.-J.: Investigation, Formal analysis. T.W.: Investigation, Formal analysis. A.B.: Formal analysis. G.O.-K.: Methodology. G.M.W.: Conceptualization. R.W.: Formal analysis. K.K.: Conceptualization, Writing—review & editing. J.W.: Writing—review & editing. J.D.: Conceptualization, Writing—original draft, Writing—review & editing, Supervision, Funding acquisition. All authors have read and agreed to the published version of the manuscript.

Funding: This work was funded by a grant from the National Science Centre (2015/17/B/NZ4/02540, UMO-2015/17/B/NZ4/02540). PK and ASJ were supported by the National Science Centre grant (UMO-2018/ 29/B/NZ4/01473).

Institutional Review Board Statement: The animal study protocol was approved by the 1st Local Ethics Committee in Warsaw (permission no. 860/2019).

Informed Consent Statement: Not applicable.

Data Availability Statement: Raw Data that support the findings of this study are available from the corresponding author upon request.

Acknowledgments: The transgenic CD44$^{fl/fl}$ mouse line was kindly provided by Veronique Orian-Rousseau from Karlsruhe Institute of Technology in Germany. Optical microscopy and 3D electron microscopy experiments were performed at the Laboratory of Imaging Tissue Structure and Function which serves as an imaging core facility at the Nencki Institute of Experimental Biology.

Conflicts of Interest: The authors declare no conflict of interest. The funders had no role in the design of the study; in the collection, analyses, or interpretation of data; in the writing of the manuscript; or in the decision to publish the results.

References

1. WHO. Epilepsy. 2019. Available online: https://www.who.int/news-room/fact-sheets/detail/epilepsy (accessed on 15 January 2023).
2. Chen, B.; Choi, H.; Hirsch, L.J.; Katz, A.; Legge, A.; Buchsbaum, R.; Detyniecki, K. Psychiatric and behavioral side effects of antiepileptic drugs in adults with epilepsy. *Epilepsy Behav.* **2017**, *76*, 24–31. [CrossRef] [PubMed]
3. Tian, N.; Boring, M.; Kobau, R.; Zack, M.M.; Croft, J.B. Active Epilepsy and Seizure Control in Adults—United States, 2013 and 2015. *MMWR Morb. Mortal. Wkly. Rep.* **2018**, *67*, 437–442. [CrossRef] [PubMed]
4. Hayatdavoudi, P.; Hosseini, M.; Hajali, V.; Hosseini, A.; Rajabian, A. The role of astrocytes in epileptic disorders. *Physiol. Rep.* **2022**, *10*, e15239. [CrossRef] [PubMed]
5. Wetherington, J.; Serrano, G.; Dingledine, R. Astrocytes in the epileptic brain. *Neuron* **2008**, *58*, 168–178. [CrossRef]
6. Myer, D.J.; Gurkoff, G.G.; Lee, S.M.; Hovda, D.A.; Sofroniew, M.V. Essential protective roles of reactive astrocytes in traumatic brain injury. *Brain* **2006**, *129*, 2761–2772. [CrossRef]
7. Chan, F.; Lax, N.Z.; Voss, C.M.; Aldana, B.I.; Whyte, S.; Jenkins, A.; Nicholson, C.; Nichols, S.; Tilley, E.; Powell, Z.; et al. The role of astrocytes in seizure generation: Insights from a novel in vitro seizure model based on mitochondrial dysfunction. *Brain* **2019**, *142*, 391–411. [CrossRef]
8. Gonzalez Otarula, K.A.; Schuele, S. Networks in Temporal Lobe Epilepsy. *Neurosurg. Clin. N. Am.* **2020**, *31*, 309–317. [CrossRef]
9. Isokawa, M. Remodeling dendritic spines of dentate granule cells in temporal lobe epilepsy patients and the rat pilocarpine model. *Epilepsia* **2000**, *41* (Suppl. S6), S14–S17. [CrossRef]
10. Ben-Ari, Y. Cell death and synaptic reorganizations produced by seizures. *Epilepsia* **2001**, *42* (Suppl. S3), 5–7. [CrossRef]
11. Magloczky, Z.; Wittner, L.; Borhegyi, Z.; Halasz, P.; Vajda, J.; Czirjak, S.; Freund, T.F. Changes in the distribution and connectivity of interneurons in the epileptic human dentate gyrus. *Neuroscience* **2000**, *96*, 7–25. [CrossRef]
12. Buckmaster, P.S.; Dudek, F.E. Neuron loss, granule cell axon reorganization, and functional changes in the dentate gyrus of epileptic kainate-treated rats. *J. Comp. Neurol.* **1997**, *385*, 385–404. [CrossRef]

13. Rao, M.S.; Hattiangady, B.; Reddy, D.S.; Shetty, A.K. Hippocampal neurodegeneration, spontaneous seizures, and mossy fiber sprouting in the F344 rat model of temporal lobe epilepsy. *J. Neurosci. Res.* **2006**, *83*, 1088–1105. [CrossRef]
14. Bouilleret, V.; Loup, F.; Kiener, T.; Marescaux, C.; Fritschy, J.M. Early loss of interneurons and delayed subunit-specific changes in GABA(A)-receptor expression in a mouse model of mesial temporal lobe epilepsy. *Hippocampus* **2000**, *10*, 305–324. [CrossRef]
15. Borges, K.; Gearing, M.; McDermott, D.L.; Smith, A.B.; Almonte, A.G.; Wainer, B.H.; Dingledine, R. Neuronal and glial pathological changes during epileptogenesis in the mouse pilocarpine model. *Exp. Neurol.* **2003**, *182*, 21–34. [CrossRef] [PubMed]
16. Xie, L.; Li, T.; Song, X.; Sun, H.; Liu, J.; Yang, J.; Zhao, W.; Cheng, L.; Chen, H.; Liu, B.; et al. Dynamic alteration of dendrites and dendritic spines in the hippocampus and microglia in mouse brain tissues after kainate-induced status epilepticus. *Int. J. Neurosci.* **2021**, *131*, 1045–1057. [CrossRef] [PubMed]
17. Flores-Soto, M.; Romero-Guerrero, C.; Vazquez-Hernandez, N.; Tejeda-Martinez, A.; Martin-Amaya-Barajas, F.L.; Orozco-Suarez, S.; Gonzalez-Burgos, I. Pentylenetetrazol-induced seizures in adult rats are associated with plastic changes to the dendritic spines on hippocampal CA1 pyramidal neurons. *Behav. Brain Res.* **2021**, *406*, 113198. [CrossRef]
18. Witcher, M.R.; Park, Y.D.; Lee, M.R.; Sharma, S.; Harris, K.M.; Kirov, S.A. Three-dimensional relationships between perisynaptic astroglia and human hippocampal synapses. *Glia* **2010**, *58*, 572–587. [CrossRef]
19. Perea, G.; Araque, A. GLIA modulates synaptic transmission. *Brain Res. Rev.* **2010**, *63*, 93–102. [CrossRef]
20. Durkee, C.A.; Araque, A. Diversity and Specificity of Astrocyte–neuron Communication. *Neuroscience* **2019**, *396*, 73–78. [CrossRef]
21. Bernardinelli, Y.; Muller, D.; Nikonenko, I. Astrocyte-synapse structural plasticity. *Neural Plast.* **2014**, *2014*, 232105. [CrossRef]
22. Henneberger, C.; Bard, L.; Panatier, A.; Reynolds, J.P.; Kopach, O.; Medvedev, N.I.; Minge, D.; Herde, M.K.; Anders, S.; Kraev, I.; et al. LTP Induction Boosts Glutamate Spillover by Driving Withdrawal of Perisynaptic Astroglia. *Neuron* **2020**, *108*, 919–936.e11. [CrossRef] [PubMed]
23. Lushnikova, I.; Skibo, G.; Muller, D.; Nikonenko, I. Synaptic potentiation induces increased glial coverage of excitatory synapses in CA1 hippocampus. *Hippocampus* **2009**, *19*, 753–762. [CrossRef]
24. Perez-Alvarez, A.; Navarrete, M.; Covelo, A.; Martin, E.D.; Araque, A. Structural and functional plasticity of astrocyte processes and dendritic spine interactions. *J. Neurosci.* **2014**, *34*, 12738–12744. [CrossRef]
25. Roch, C.; Leroy, C.; Nehlig, A.; Namer, I.J. Magnetic resonance imaging in the study of the lithium-pilocarpine model of temporal lobe epilepsy in adult rats. *Epilepsia* **2002**, *43*, 325–335. [CrossRef] [PubMed]
26. Oberheim, N.A.; Tian, G.F.; Han, X.; Peng, W.; Takano, T.; Ransom, B.; Nedergaard, M. Loss of astrocytic domain organization in the epileptic brain. *J. Neurosci.* **2008**, *28*, 3264–3276. [CrossRef] [PubMed]
27. Clarkson, C.; Smeal, R.M.; Hasenoehrl, M.G.; White, J.A.; Rubio, M.E.; Wilcox, K.S. Ultrastructural and functional changes at the tripartite synapse during epileptogenesis in a model of temporal lobe epilepsy. *Exp. Neurol.* **2020**, *326*, 113196. [CrossRef] [PubMed]
28. McKenzie, J.L.; Dalchau, R.; Fabre, J.W. Biochemical characterisation and localization in brain of a human brain-leucocyte membrane glycoprotein recognised by a monoclonal antibody. *J. Neurochem.* **1982**, *39*, 1461–1466. [CrossRef]
29. Girgrah, N.; Letarte, M.; Becker, L.E.; Cruz, T.F.; Theriault, E.; Moscarello, M.A. Localization of the CD44 glycoprotein to fibrous astrocytes in normal white matter and to reactive astrocytes in active lesions in multiple sclerosis. *J. Neuropathol. Exp. Neurol.* **1991**, *50*, 779–792. [CrossRef]
30. Vogel, H.; Butcher, E.C.; Picker, L.J. H-CAM expression in the human nervous system: Evidence for a role in diverse glial interactions. *J. Neurocytol.* **1992**, *21*, 363–373. [CrossRef]
31. Jones, L.L.; Liu, Z.; Shen, J.; Werner, A.; Kreutzberg, G.W.; Raivich, G. Regulation of the cell adhesion molecule CD44 after nerve transection and direct trauma to the mouse brain. *J. Comp. Neurol.* **2000**, *426*, 468–492. [CrossRef]
32. Roszkowska, M.; Skupien, A.; Wojtowicz, T.; Konopka, A.; Gorlewicz, A.; Kisiel, M.; Bekisz, M.; Ruszczycki, B.; Dolezyczek, H.; Rejmak, E.; et al. CD44: A novel synaptic cell adhesion molecule regulating structural and functional plasticity of dendritic spines. *Mol. Biol. Cell.* **2016**, *27*, 4055–4066. [CrossRef] [PubMed]
33. Skupien, A.; Konopka, A.; Trzaskoma, P.; Labus, J.; Gorlewicz, A.; Swiech, L.; Babraj, M.; Dolezyczek, H.; Figiel, I.; Ponimaskin, E.; et al. CD44 regulates dendrite morphogenesis through Src tyrosine kinase-dependent positioning of the Golgi. *J. Cell. Sci.* **2014**, *127 Pt 23*, 5038–5051.
34. Glezer, I.; Bittencourt, J.C.; Rivest, S. Neuronal expression of Cd36, Cd44, and Cd83 antigen transcripts maps to distinct and specific murine brain circuits. *J. Comp. Neurol.* **2009**, *517*, 906–924. [CrossRef] [PubMed]
35. Kaaijk, P.; Pals, S.T.; Morsink, F.; Bosch, D.A.; Troost, D. Differential expression of CD44 splice variants in the normal human central nervous system. *J. Neuroimmunol.* **1997**, *73*, 70–76. [CrossRef]
36. Naruse, M.; Shibasaki, K.; Yokoyama, S.; Kurachi, M.; Ishizaki, Y. Dynamic changes of CD44 expression from progenitors to subpopulations of astrocytes and neurons in developing cerebellum. *PLoS ONE* **2013**, *8*, e53109. [CrossRef] [PubMed]
37. Ponta, H.; Sherman, L.; Herrlich, P.A. CD44: From adhesion molecules to signalling regulators. *Nat. Rev. Mol. Cell Biol.* **2003**, *4*, 33–45. [CrossRef]
38. Bourguignon, L.Y.; Gilad, E.; Rothman, K.; Peyrollier, K. Hyaluronan-CD44 interaction with IQGAP1 promotes Cdc42 and ERK signaling, leading to actin binding, Elk-1/estrogen receptor transcriptional activation, and ovarian cancer progression. *J. Biol. Chem.* **2005**, *280*, 11961–11972. [CrossRef]

39. Konopka, A.; Zeug, A.; Skupien, A.; Kaza, B.; Mueller, F.; Chwedorowicz, A.; Ponimaskin, E.; Wilczynski, G.M.; Dzwonek J. Cleavage of Hyaluronan and CD44 Adhesion Molecule Regulate Astrocyte Morphology via Rac1 Signalling. *PLoS ONE* **2016**, *11*, e0155053. [CrossRef]
40. Bausch, S.B. Potential roles for hyaluronan and CD44 in kainic acid-induced mossy fiber sprouting in organotypic hippocampal slice cultures. *Neuroscience* **2006**, *143*, 339–350. [CrossRef]
41. Dzwonek, J.; Wilczynski, G.M. CD44: Molecular interactions, signaling and functions in the nervous system. *Front. Cell. Neurosci* **2015**, *9*, 175. [CrossRef]
42. Borges, K.; McDermott, D.L.; Dingledine, R. Reciprocal changes of CD44 and GAP-43 expression in the dentate gyrus inner molecular layer after status epilepticus in mice. *Exp. Neurol.* **2004**, *188*, 1–10. [CrossRef] [PubMed]
43. Akiyama, H.; Tooyama, I.; Kawamata, T.; Ikeda, K.; McGeer, P.L. Morphological diversities of CD44 positive astrocytes in the cerebral cortex of normal subjects and patients with Alzheimer's disease. *Brain Res.* **1993**, *632*, 249–259. [CrossRef] [PubMed]
44. Lipponen, A.; Paananen, J.; Puhakka, N.; Pitkanen, A. Analysis of Post-Traumatic Brain Injury Gene Expression Signature Reveals Tubulins, Nfe2l2, Nfkb, Cd44, and S100a4 as Treatment Targets. *Sci. Rep.* **2016**, *6*, 31570. [CrossRef] [PubMed]
45. Li, S.; Shao, H.; Chang, L. The Important Role of Perituberal Tissue in Epileptic Patients with Tuberous Sclerosis Complex by the Transcriptome Analysis. *Biomed. Res. Int.* **2020**, *2020*, 4980609. [CrossRef] [PubMed]
46. Sosunov, A.A.; Wu, X.; Tsankova, N.M.; Guilfoyle, E.; McKhann, G.M., II; Goldman, J.E. Phenotypic heterogeneity and plasticity of isocortical and hippocampal astrocytes in the human brain. *J. Neurosci.* **2014**, *34*, 2285–2298. [CrossRef] [PubMed]
47. Denk, W.; Horstmann, H. Serial block-face scanning electron microscopy to reconstruct three-dimensional tissue nanostructure *PLoS Biol.* **2004**, *2*, e329. [CrossRef]
48. Shatirishvili, M.; Burk, A.S.; Franz, C.M.; Pace, G.; Kastilan, T.; Breuhahn, K.; Hinterseer, E.; Dierich, A.; Bakiri, L.; Wagner, E.F.; et al. Epidermal-specific deletion of CD44 reveals a function in keratinocytes in response to mechanical stress. *Cell Death Dis.* **2016**, *7*, e2461. [CrossRef]
49. Lee, Y.; Messing, A.; Su, M.; Brenner, M. GFAP promoter elements required for region-specific and astrocyte-specific expression *Glia* **2008**, *56*, 481–493. [CrossRef]
50. Ohgane, K.; Yoshioka, H. *Quantification of Gel Bands by an Image J Macro, Band/Peak Quantification Tool*; Protocol IO: San Francisco, CA, USA, 2019.
51. Bouilleret, V.; Ridoux, V.; Depaulis, A.; Marescaux, C.; Nehlig, A.; Le Gal La Salle, G. Recurrent seizures and hippocampal sclerosis following intrahippocampal kainate injection in adult mice: Electroencephalography, histopathology and synaptic reorganization similar to mesial temporal lobe epilepsy. *Neuroscience* **1999**, *89*, 717–729. [CrossRef]
52. Racine, R.J. Modification of seizure activity by electrical stimulation. II. Motor seizure. *Electroencephalogr. Clin. Neurophysiol.* **1972**, *32*, 281–294. [CrossRef]
53. Schauwecker, P.E.; Steward, O. Genetic determinants of susceptibility to excitotoxic cell death: Implications for gene targeting approaches. *Proc. Natl. Acad. Sci. USA* **1997**, *94*, 4103–4108. [CrossRef]
54. Ndode-Ekane, X.E.; Pitkanen, A. Urokinase-type plasminogen activator receptor modulates epileptogenesis in mouse model of temporal lobe epilepsy. *Mol. Neurobiol.* **2013**, *47*, 914–937. [CrossRef] [PubMed]
55. Kimelberg, H.K. The problem of astrocyte identity. *Neurochem. Int.* **2004**, *45*, 191–202. [CrossRef] [PubMed]
56. Torres-Ceja, B.; Olsen, M.L. A closer look at astrocyte morphology: Development, heterogeneity, and plasticity at astrocyte leaflets. *Curr. Opin. Neurobiol.* **2022**, *74*, 102550. [CrossRef]
57. Deerinck, T.J.; Bushong, E.A.; Ellisman, M.H.; Thor, A. NCMIR methods for 3D EM: A new protocol for preparation of biological specimens for serial block face scanning electron microscopy. *Microscopy* **2010**, *1*, 6–8.
58. Sosunov, A.A.; Guilfoyle, E.; Wu, X.; McKhann, G.M., II; Goldman, J.E. Phenotypic conversions of "protoplasmic" to "reactive" astrocytes in Alexander disease. *J. Neurosci.* **2013**, *33*, 7439–7450. [CrossRef] [PubMed]
59. Utkina-Sosunova, I.V.; Niatsetskaya, Z.V.; Sosunov, S.A.; Ratner, V.I.; Matsiukevich, D.; Ten, V.S. Nelfinavir inhibits intra-mitochondrial calcium influx and protects brain against hypoxic-ischemic injury in neonatal mice. *PLoS ONE* **2013**, *8*, e62448. [CrossRef]
60. Fiala, J.C.; Harris, K.M. Extending unbiased stereology of brain ultrastructure to three-dimensional volumes. *J. Am. Med. Inform. Assoc.* **2001**, *8*, 1–16. [CrossRef]
61. Ostroff, L.E.; Manzur, M.K.; Cain, C.K.; Ledoux, J.E. Synapses lacking astrocyte appear in the amygdala during consolidation of Pavlovian threat conditioning. *J. Comp. Neurol.* **2014**, *522*, 2152–2163. [CrossRef]
62. Devinsky, O.; Vezzani, A.; Najjar, S.; De Lanerolle, N.C.; Rogawski, M.A. Glia and epilepsy: Excitability and inflammation. *Trends Neurosci.* **2013**, *36*, 174–184. [CrossRef]
63. Stylli, S.S.; Kaye, A.H.; Novak, U. Induction of CD44 expression in stab wounds of the brain: Long term persistence of CD44 expression. *J. Clin. Neurosci.* **2000**, *7*, 137–140. [CrossRef]
64. Tse, K.; Hammond, D.; Simpson, D.; Beynon, R.J.; Beamer, E.; Tymianski, M.; Salter, M.W.; Sills, G.J.; Thippeswamy, T. The impact of postsynaptic density 95 blocking peptide (Tat-NR2B9c) and an iNOS inhibitor (1400W) on proteomic profile of the hippocampus in C57BL/6J mouse model of kainate-induced epileptogenesis. *J. Neurosci. Res.* **2019**, *97*, 1378–1392. [CrossRef] [PubMed]
65. Arai, Y.; Takashima, S.; Becker, L.E. CD44 expression in tuberous sclerosis. *Pathobiology* **2000**, *68*, 87–92. [CrossRef] [PubMed]

66. Pai, B.; Kuo, C.-C.; Chen, Y.-C.; Lin, C.-H.; Chen, C.-C.; Chang, Y.-J.; Hsu, Y.-C.; Huang, Y.-S.; Lin, Y.-C.; Li, H.-M.; et al. High-resolution transcriptomics informs glial pathology in human temporal lobe epilepsy. *Acta Neuropathol. Commun.* **2022**, *10*, 149. [CrossRef] [PubMed]
67. Pannasch, U.; Rouach, N. Emerging role for astroglial networks in information processing: From synapse to behavior. *Trends Neurosci.* **2013**, *36*, 405–417. [CrossRef]
68. Papouin, T.; Dunphy, J.M.; Tolman, M.; Dineley, K.T.; Haydon, P.G. Septal Cholinergic Neuromodulation Tunes the Astrocyte-Dependent Gating of Hippocampal NMDA Receptors to Wakefulness. *Neuron* **2017**, *94*, 840–854.e847. [CrossRef]
69. Garcia-Caceres, C.; Quarta, C.; Varela, L.; Gao, Y.; Gruber, T.; Legutko, B.; Jastroch, M.; Johansson, P.; Ninkovic, J.; Yi, C.X.; et al. Astrocytic Insulin Signaling Couples Brain Glucose Uptake with Nutrient Availability. *Cell* **2016**, *166*, 867–880. [CrossRef]
70. Su, M.; Hu, H.; Lee, Y.; d'Azzo, A.; Messing, A.; Brenner, M. Expression specificity of GFAP transgenes. *Neurochem. Res.* **2004**, *29*, 2075–2093. [CrossRef]
71. Jessberger, S.; Parent, J.M. Epilepsy and Adult Neurogenesis. *Cold Spring Harb. Perspect. Biol.* **2015**, *7*, a020677. [CrossRef]
72. Fukuda, S.; Kato, F.; Tozuka, Y.; Yamaguchi, M.; Miyamoto, Y.; Hisatsune, T. Two distinct subpopulations of nestin-positive cells in adult mouse dentate gyrus. *J. Neurosci.* **2003**, *23*, 9357–9366. [CrossRef]
73. Sierra, A.; Martin-Suarez, S.; Valcarcel-Martin, R.; Pascual-Brazo, J.; Aelvoet, S.A.; Abiega, O.; Deudero, J.J.; Brewster, A.L.; Bernales, I.; Anderson, A.E.; et al. Neuronal hyperactivity accelerates depletion of neural stem cells and impairs hippocampal neurogenesis. *Cell Stem Cell* **2015**, *16*, 488–503. [CrossRef] [PubMed]
74. Bonde, S.; Ekdahl, C.T.; Lindvall, O. Long-term neuronal replacement in adult rat hippocampus after status epilepticus despite chronic inflammation. *Eur. J. Neurosci.* **2006**, *23*, 965–974. [CrossRef] [PubMed]
75. Rezai Amin, S.; Gruszczynski, C.; Guiard, B.P.; Callebert, J.; Launay, J.M.; Louis, F.; Betancur, C.; Vialou, V.; Gautron, S. Viral vector-mediated Cre recombinase expression in substantia nigra induces lesions of the nigrostriatal pathway associated with perturbations of dopamine-related behaviors and hallmarks of programmed cell death. *J. Neurochem.* **2019**, *150*, 330–340. [CrossRef] [PubMed]
76. Ortinski, P.I.; Dong, J.; Mungenast, A.; Yue, C.; Takano, H.; Watson, D.J.; Haydon, P.G.; Coulter, D.A. Selective induction of astrocytic gliosis generates deficits in neuronal inhibition. *Nat. Neurosci.* **2010**, *13*, 584–591. [CrossRef]
77. Wenzel, H.J.; Woolley, C.S.; Robbins, C.A.; Schwartzkroin, P.A. Kainic acid-induced mossy fiber sprouting and synapse formation in the dentate gyrus of rats. *Hippocampus* **2000**, *10*, 244–260. [CrossRef]
78. Badia-Soteras, A.; Llorens-Martin, M.; Jurado-Arjona, J.; Avila, J.; Hernandez, F. Retraction of Astrocyte Leaflets From the Synapse Enhances Fear Memory. *Biol. Psychiatry* **2022**, *in press*. [CrossRef]
79. Pure, E.; Cuff, C.A. A crucial role for CD44 in inflammation. *Trends Mol. Med.* **2001**, *7*, 213–221. [CrossRef]
80. Tsukita, S.; Yonemura, S. ERM (ezrin/radixin/moesin) family: From cytoskeleton to signal transduction. *Curr. Opin. Cell Biol.* **1997**, *9*, 70–75. [CrossRef]
81. Robel, S.; Buckingham, S.C.; Boni, J.L.; Campbell, S.L.; Danbolt, N.C.; Riedemann, T.; Sutor, B.; Sontheimer, H. Reactive astrogliosis causes the development of spontaneous seizures. *J. Neurosci.* **2015**, *35*, 3330–3345. [CrossRef]
82. Heja, L.; Barabas, P.; Nyitrai, G.; Kekesi, K.A.; Lasztoczi, B.; Toke, O.; Tarkanyi, G.; Madsen, K.; Schousboe, A.; Dobolyi, A.; et al. Glutamate uptake triggers transporter-mediated GABA release from astrocytes. *PLoS ONE* **2009**, *4*, e7153. [CrossRef]
83. Kovacs, Z.; Skatchkov, S.N.; Veh, R.W.; Szabo, Z.; Nemeth, K.; Szabo, P.T.; Kardos, J.; Heja, L. Critical Role of Astrocytic Polyamine and GABA Metabolism in Epileptogenesis. *Front. Cell Neurosci.* **2021**, *15*, 787319. [CrossRef] [PubMed]
84. Kovacs, Z.; Skatchkov, S.N.; Szabo, Z.; Qahtan, S.; Mendez-Gonzalez, M.P.; Malpica-Nieves, C.J.; Eaton, M.J.; Kardos, J.; Heja, L. Putrescine Intensifies Glu/GABA Exchange Mechanism and Promotes Early Termination of Seizures. *Int. J. Mol. Sci.* **2022**, *23*, 8191. [CrossRef] [PubMed]
85. Araque, A.; Carmignoto, G.; Haydon, P.G.; Oliet, S.H.; Robitaille, R.; Volterra, A. Gliotransmitters travel in time and space. *Neuron* **2014**, *81*, 728–739. [CrossRef]
86. Halassa, M.M.; Haydon, P.G. Integrated brain circuits: Astrocytic networks modulate neuronal activity and behavior. *Annu. Rev. Physiol.* **2010**, *72*, 335–355. [CrossRef]
87. Gomez-Gonzalo, M.; Losi, G.; Chiavegato, A.; Zonta, M.; Cammarota, M.; Brondi, M.; Vetri, F.; Uva, L.; Pozzan, T.; de Curtis, M.; et al. An excitatory loop with astrocytes contributes to drive neurons to seizure threshold. *PLoS Biol.* **2010**, *8*, e1000352. [CrossRef]
88. Price, B.R.; Johnson, L.A.; Norris, C.M. Reactive Astrocytes: The Nexus of Pathological and Clinical Hallmarks of Alzheimer's Disease. *Ageing Res. Rev.* **2021**, *68*, 101335. [CrossRef]
89. Zhvania, M.G.; Ksovreli, M.; Japaridze, N.J.; Lordkipanidze, T.G. Ultrastructural changes to rat hippocampus in pentylenetetrazol- and kainic acid-induced status epilepticus: A study using electron microscopy. *Micron* **2015**, *74*, 22–29. [CrossRef] [PubMed]
90. Fiala, J.C.; Spacek, J.; Harris, K.M. Dendritic spine pathology: Cause or consequence of neurological disorders? *Brain Res. Brain Res. Rev.* **2002**, *39*, 29–54. [CrossRef]
91. Chidambaram, S.B.; Rathipriya, A.G.; Bolla, S.R.; Bhat, A.; Ray, B.; Mahalakshmi, A.M.; Manivasagam, T.; Thenmozhi, A.J.; Essa, M.M.; Guillemin, G.J.; et al. Dendritic spines: Revisiting the physiological role. *Prog. Neuropsychopharmacol. Biol. Psychiatry* **2019**, *92*, 161–193. [CrossRef]
92. Kayser, M.S.; Nolt, M.J.; Dalva, M.B. EphB receptors couple dendritic filopodia motility to synapse formation. *Neuron* **2008**, *59*, 56–69. [CrossRef]

93. Petrak, L.J.; Harris, K.M.; Kirov, S.A. Synaptogenesis on mature hippocampal dendrites occurs via filopodia and immature spines during blocked synaptic transmission. *J. Comp. Neurol.* **2005**, *484*, 183–190. [CrossRef] [PubMed]
94. Toni, N.; Teng, E.M.; Bushong, E.A.; Aimone, J.B.; Zhao, C.; Consiglio, A.; van Praag, H.; Martone, M.E.; Ellisman, M.H.; Gage, F.H. Synapse formation on neurons born in the adult hippocampus. *Nat. Neurosci.* **2007**, *10*, 727–734. [CrossRef] [PubMed]
95. Ziv, N.E.; Smith, S.J. Evidence for a role of dendritic filopodia in synaptogenesis and spine formation. *Neuron* **1996**, *17*, 91–102. [CrossRef]
96. Tonnesen, J.; Nagerl, U.V. Dendritic Spines as Tunable Regulators of Synaptic Signals. *Front. Psychiatry* **2016**, *7*, 101. [CrossRef] [PubMed]

cells MDPI

Review

Role of Impaired Astrocyte Gap Junction Coupling in Epileptogenesis

Peter Bedner * and Christian Steinhäuser *

Institute of Cellular Neurosciences, Medical Faculty, University of Bonn, Venusberg-Campus 1, 53127 Bonn, Germany
* Correspondence: peter.bedner@ukbonn.de (P.B.); cste@uni-bonn.de (C.S.)

Abstract: The gap-junction-coupled astroglial network plays a central role in the regulation of neuronal activity and synchronisation, but its involvement in the pathogenesis of neuronal diseases is not yet understood. Here, we present the current state of knowledge about the impact of impaired glial coupling in the development and progression of epilepsy and discuss whether astrocytes represent alternative therapeutic targets. We focus mainly on temporal lobe epilepsy (TLE), which is the most common form of epilepsy in adults and is characterised by high therapy resistance. Functional data from TLE patients and corresponding experimental models point to a complete loss of astrocytic coupling, but preservation of the gap junction forming proteins connexin43 and connexin30 in hippocampal sclerosis. Several studies further indicate that astrocyte uncoupling is a causal event in the initiation of TLE, as it occurs very early in epileptogenesis, clearly preceding dysfunctional changes in neurons. However, more research is needed to fully understand the role of gap junction channels in epilepsy and to develop safe and effective therapeutic strategies targeting astrocytes.

Keywords: astrocyte; gap junction coupling; connexin 43; connexin 30; hippocampus; epilepsy; epileptogenesis; temporal lobe epilepsy

Citation: Bedner, P.; Steinhäuser, C. Role of Impaired Astrocyte Gap Junction Coupling in Epileptogenesis. *Cells* **2023**, *12*, 1669. https://doi.org/10.3390/cells12121669

Academic Editors: Michael Brenner and Vladimir Parpura

Received: 27 January 2023
Revised: 25 May 2023
Accepted: 19 June 2023
Published: 20 June 2023

1. Introduction

Epilepsy is a disease of the brain that affects about 1% of the world's population and is characterised by the occurrence of recurrent unprovoked seizures [1,2]. Although more than twenty antiepileptic drugs (AEDs) are currently on the market, seizures cannot be fully controlled in about one-third of patients. Furthermore, available AEDs merely provide symptomatic treatment but are not able to prevent or reverse the development of the disorder [3]. These drugs act primarily on neurons through modulating voltage-activated ion channels, enhancing GABAergic inhibition or attenuating excitatory neurotransmission [3,4]. However, as the efficacy and tolerability of AEDs have not substantially improved in recent decades, and drugs with disease-modifying (antiepileptogenic) effects are still not available, this neurocentric strategy in the development of AEDs is increasingly being challenged. Indeed, it seems plausible that not only defective neurons but also a disruption of the control mechanisms that regulate neuronal excitability and synaptic transmission can cause epilepsy [5,6]. Astrocytes not only have crucial homeostatic and supportive functions, such as the supply of energy substrates to neurons, the control of extracellular ion and water homeostasis, the clearance of neurotransmitters from the synapse or the regulation of blood–brain barrier (BBB) permeability, but also actively modulate synaptic transmission and neuronal plasticity through the release of gliotransmitters [5,7]. Synapses interact with every physical component of an astrocyte, including endfeet and soma. An important prerequisite for the implementation and/or efficiency of most of these regulatory functions is the fact that astrocytes are extensively electrically and metabolically interconnected via low-resistance aqueous channels termed gap junctions (GJs). This leads to the formation of large syncytium-like functional networks that overlap with neuronal synaptic networks and enable their coordinated regulation and synchronisation [8,9]. Astrocytic gap junctions consist mainly of the channel proteins connexin43 (Cx43) and connexin30 (Cx30) whose relative expression levels vary considerably across developmental stages and brain regions [9–11]. In adult mouse and human hippocampi, astrocyte coupling appears to be predominately mediated by Cx43 [10,12,13], while mouse thalamic astrocytes couple mainly through Cx30 [11]. The two astrocytic Cx isoforms possess different biophysical properties (unitary conductance, selectivity, transjunctional voltage sensitivity and gating properties) and are differentially regulated [14,15]. The gating properties of Cx43-based channels as well as the trafficking, assembly and degradation of Cx43 proteins are tightly regulated by phosphorylation [16]. In contrast, there are no reports on Cx30 phosphorylation. Astrocytic Cxs can form in addition to homotypic GJ channels also functional heterotypic channels with each other and with oligodendrocytic Cxs [11,17,18]. In addition, under certain (primarily pathological) conditions, undocked Cx hemichannels (HCs) might also function as transmembrane channels, allowing the bidirectional exchange of ions and small molecules between the cytoplasm and the extracellular space [8,19]. Finally, there is growing evidence that Cx proteins play an important role in several other fundamental processes, including adhesion, migration, cellular life cycle, neurogenesis, cell death and signal transduction [8,20,21].

2. Role of Glial Cx Channels in Epilepsy

In view of the essential role of the astrocyte network for neuronal function and brain homeostasis, it is not surprising that various neurological disorders including stroke, migraine, gliomas, Alzheimer's disease and epilepsy have been linked to alterations in astrocyte coupling [8]. The role of astrocytic network disturbances in the development and progression of epilepsy is not yet conclusively understood, but there is growing evidence that they are fundamentally involved [5,22–25]. In principle, disruption of GJ coupling between astrocytes can affect neuronal excitability and synchronisation in many ways. An important epilepsy-promoting functional consequence of such a disruption would be the reduced astrocytic capacity to buffer elevated extracellular K^+ concentrations ($[K^+]_o$) during high neuronal activity. In the process known as spatial K^+ buffering, the excess neuronally released $[K^+]_o$ is passively taken up by astrocytes via Kir4.1 channels, distributed in the GJ-coupled network and released at sites of lower $[K^+]_o$ [26]. This process, which is driven by the difference between the local K^+ equilibrium potential and the more negative membrane potential of the glial syncytium, cannot be sustained in the absence of coupling, resulting in stronger local $[K^+]_o$ increases and consequently in a more pronounced neuronal depolarisation and enhanced neuronal excitability [26–30]. In addition to K^+, the redistribution of Na^+ ions through the GJ-coupled astrocytic network might also play an important role in preventing neuronal hyperexcitability. In fact, astrocytic plasmalemmal glutamate transporters are mainly responsible for the removal of excess glutamate during synaptic activity, a process driven by the Na^+ gradient and involving massive co-uptake of the ion [31–33]. Reduced GJ coupling could promote accumulation of cytosolic Na^+ in astrocytes, resulting in a reduced driving force for uptake of synaptically released glutamate and thus increased excitatory neurotransmission [34]. Furthermore, such increased cytosolic Na^+ can trigger the reversal of the Na^+/Ca^{2+} exchanger, resulting in effective astrocytic Ca^{2+} uptake and therefore defective Ca^{2+}-dependent processes such as aberrant gliotransmitter release, which in turn also affects network excitability [31,33,35,36]. Finally, cytosolic K^+ and Na^+ accumulation enhances water influx and thus the extent of activity-dependent astrocytic swelling. The resulting reduction in extracellular space volume further increases the concentration of extracellular ions and neurotransmitters, exacerbating the seizure-promoting effect of impaired spatial K^+ and glutamate buffering [22,37]. Experimental confirmation for these theoretical considerations is provided via experiments on brain slices from transgenic mice with coupling-deficient astrocytes or after pharmacological disruption of GJ communication. Uncoupling, in fact, resulted in impaired spatial K^+ and Na^+ buffering [29,30,32,38,39], reduced astrocytic glutamate clearance [37] and hypertrophic astrocytes [37,40]. Together, these findings point to an anti-epileptic function of the astroglial network. On the other hand, it was shown that the astrocytic network is crucial for the effective supply of energetic metabolites from blood vessels to sites of high neuronal activity [41,42]. It can be assumed that this function of the network has less effect on the initiation than on the maintenance of seizure activity. Thus, an acute reduction of astrocytic GJ coupling may include rapid seizure-promoting consequences due to reduced K^+ and glutamate buffering but delayed seizure-suppressing effects due to insufficient energy supply [22,23,31]. Reduced coupling may also restrict the propagation of Na^+ and Ca^{2+} through the astrocyte network and thus its ability to synchronise large populations of neurons. This effect may counteract the hypersynchronous neuronal firing that characterises epilepsy [5,22,31,43].

In addition to intercellular GJ channels, functional Cx HCs have also been reported to play an important role in the pathology of various diseases, including epilepsy, al-though it is unclear how cells can maintain their integrity upon opening of such large non-specific pores [8,44]. In the healthy brain, Cx HCs do not appear to have a significant open probability, but this increases in response to various stress conditions such as metabolic inhibition, hypoxia/ischemia, inflammation, strong depolarisation or altered intra- and extracellular Ca^{2+} concentrations [8,19]. In epilepsy, opened HCs were proposed to release neuroactive molecules such as glutamate, ATP and/or D-serine, which in turn enhance neuronal excitability and synchronisation. Accordingly, Cx HC activation has been proposed to have pro-epileptic effects [8,22,45–47].

Finally, non-channel functions might also be involved in the pathogenesis of epilepsy. For example, Cx43 was shown to alter the expression and dose–response curve of glial purinergic $P2Y_1$ receptors even in the absence of functional channels [8,48]. Due to its involvement in glial Ca^{2+} mobilisation and the propagation of astrocytic Ca^{2+} waves, the metabotropic P2Y1 receptor has been attributed an important role in epilepsy [49,50]. Another example of an epilepsy-relevant non-channel function of Cx proteins was provided by Pannasch and colleagues, who showed that Cx30 proteins control the intrusion of astroglial processes into the synaptic cleft, thereby indirectly modulating the efficiency of synaptic glutamate clearance and, thus, the strength of excitatory synaptic transmission [51] (Figure 1).

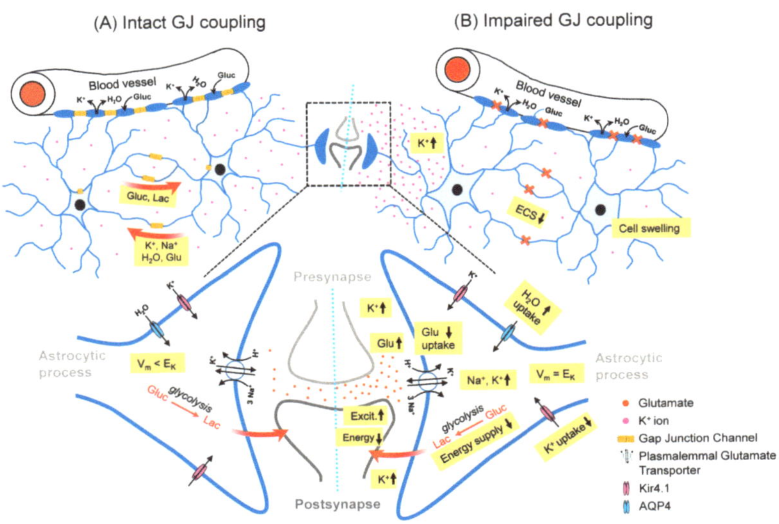

Figure 1. Schematic illustration of the role of the GJ-coupled astrocytic network in the healthy brain (left) and consequences of its disconnection (right) on astrocyte homeostatic functions and neuronal

activity. (**A**) In the intact astrocytic network, excessive extracellular K^+ released during high neuronal activity is taken up by astrocytes, redistributed through the astrocytic network and released at regions of lower extracellular K^+ concentrations. This process is energy-independent, as the uptake and release of K^+ through Kir4.1 channels are driven by the difference between the dynamic local K^+ equilibrium potential (E_K) and a relatively constant syncytial voltage (V_m). Astrocytic synaptic glutamate (Glu) buffering leads to massive Na^+ co-uptake, which is distributed in the astrocytic network Astrocytic GJ coupling mediates activity-dependent trafficking of glucose (Gluc) and its metabolic product lactate (Lac) from blood vessels to sites of high energy demand. (**B**) In the absence of GJ coupling, an increased extracellular K^+ concentration leads to depolarisation of nearby astrocytes ($V_m = E_K$), resulting in cessation of K^+ uptake and re-distribution through spatial buffering. This leads to increased/prolonged extracellular K^+ transients and increased/prolonged neuronal depolarisation, lowering the threshold for seizure generation. Loss of coupling also leads to intracellular accumulation of Na^+ taken up together with glutamate. Na^+ accumulation and depolarisation triggered by elevated extracellular K^+ reduce the driving force for glutamate uptake, resulting in a seizure-promoting prolongation of excitatory synapse activation. In addition, intracellular accumulation of K^+ and Na^+ trigger astrocytic water uptake through aquaporin-4 (AQP4) channels and a reduction of the extracellular space (ECS) volume due to astrocytic swelling, which in turn further increases the concentration of extracellular ions and neurotransmitters, exacerbating the seizure-promoting effect of impaired K^+ and glutamate buffering. On the other hand, uncoupled astrocytes are less effective at providing energy substrates, which counteracts neuronal hyperactivity.

3. Connexin Expression and GJ Coupling in Human Epilepsy

The most relevant information about astrocytic changes in human epilepsy has been gained from epileptogenic brain tissue specimens surgically resected from patients with drug-resistant temporal lobe epilepsy (TLE). Several studies compared epilepsy-related changes in Cx expression, for example in surgical specimens from TLE patients with (HS) vs. without hippocampal sclerosis (non-HS), or in samples from epilepsy patients vs. non-epileptic patients [12,52–58]. Except for one study [56], all found upregulation of Cx43 transcripts or protein in epileptic or sclerotic tissue. A detailed analysis was carried out by Deshpande and co-workers, who compared not only whole cell protein levels of the two astrocytic Cx isoforms but also the plasma membrane-associated protein fractions, subcellular distribution and phosphorylation status of Cx43 between sclerotic and non-sclerotic human tissue from TLE patients [12]. Neither total nor membrane-associated Cx30 protein levels nor subcellular distribution of Cx30 differed between both conditions. Interestingly, in HS, increased Cx43 protein levels were found only in the total protein fraction but not in the plasma membrane fraction. Importantly, the authors also reported a pronounced subcellular redistribution towards the perivascular endfeet as well as an altered phosphorylation pattern of Cx43 in HS, a finding suggestive of altered astrocytic coupling [12]. Information on functional astrocytic changes in human epileptic tissue is sparse due to the limited availability of live tissue, the technical difficulties of recording from them and the lack of healthy control tissue. The first evidence of a disturbance in the ability of astrocytes to spatially buffer K^+ in human TLE with HS was gained more than 2 decades ago, although it remained unclear whether changes in GJ coupling contributed to the phenomenon [59–62]. Characterisation of functional GJ coupling between human astrocytes in the sclerotic and non-sclerotic hippocampus was performed several years later via tracer diffusion analysis [24]. Intriguingly, in non-HS control-like tissue (obtained from patients with lesion-associated TLE, [62,63]) the extent of astrocyte coupling was in the same range as in the adult rodent hippocampus, indicating that GJ coupling has no influence on the pathogenesis of this type of epilepsy [13,24]. In contrast, complete loss of bona fide astrocytes and GJ coupling was found in hippocampal slices from TLE-HS [24]. Together, these data suggest that Cx proteins are abundantly expressed in the sclerotic hippocampus from TLE patients, but do not form functional GJ channels, most likely due to subcellular redistribution, morphological alterations and/or abnormal phosphorylation of Cx43.

4. Connexin Expression, Communication and Regulation in Experimental Epilepsy

The loss of astrocytic coupling observed in human tissue raises the question of whether this is causally involved in the genesis of epilepsy or whether it represents merely a consequence of the disease. To answer this question, information is required about the early pathophysiological changes that ultimately culminate in chronic epilepsy. As human tissue specimens are only available from the late chronic state of the disorder, animal models that closely mimic the human situation are needed. Most experimental TLE models involve the chemical or electrical induction of status epilepticus (SE) that triggers chronic epilepsy after a latent period of days to weeks [64–66]. Astrocyte changes occurring during SE or the latency period can be considered as possible causative factors in epileptogenesis. In accordance with human Cx expression studies, Cx43 and Cx30 mRNA and/or protein were found either unchanged or upregulated in rodent TLE models [12,67–75]. Intriguingly, in the unilateral intracortical kainate injection model of TLE-HS [24], the changes in astrocytic Cx expression, subcellular distribution and phosphorylation closely resembled those described in human epileptic tissues. Consistently, of the two astrocytic Cxs, only total protein levels of Cx43 were upregulated in the sclerotic hippocampus, while the membrane-associated levels were unaltered. Also in accordance with the situation in human tissue, Cx43 exhibited subcellular redistribution towards perivascular endfeet and an altered phosphorylation pattern in HS [12]. Interestingly, the model also recapitulated the complete loss of GJ coupling between astrocytes in chronic human TLE and revealed that uncoupling and the associated disruption of K^+ buffering started at a very early stage of epileptogenesis, even before the onset of neurodegeneration and chronic seizure activity [24]. This finding implies that the uncoupling of astrocytes plays a causal role in the initiation of TLE. A similar reduction in GJ coupling was also observed 5 days after hyperthermia-induced febrile seizures [76], a finding that might provide a mechanistic explanation for the increased risk of developing human TLE following prolonged juvenile febrile seizures [77]. In contrast, Takahashi and colleagues reported increased coupling between hippocampal astrocytes one week after systemic injection of kainate into rats [74]. Differences between models (e.g., in the time course of histopathological and functional changes associated with epileptogenesis) may account for these contradictory findings.

The signalling pathway underlying loss of astrocyte coupling in TLE remains unknown, but there are good reasons to assume that increased levels of pro-inflammatory cytokines are involved. Indeed, inflammation is a hallmark of both human and experimental epilepsy, and there is evidence that the process is not only a consequence but also a causal factor in the aetiology of the disease [78]. Initial evidence for the regulation of astrocytic GJ coupling by inflammatory mediators came from cell culture studies, which showed that the pro-inflammatory cytokines interleukin-1β (IL-1β) and tumour necrosis factor α (TNFα) released from microglia inhibit coupling between cultured astrocytes [79–81]. Interestingly, the cytokines exerted an opposite effect on the Cx HCs, which were activated in their presence [80]. In later work, inhibition of astrocytic coupling could also be induced in situ after incubation of acute hippocampal slices from mice with IL-1β and TNFα, as well as in vivo by means of systemic injection of the inflammation-inducing molecule lipopolysaccharide (LPS) [24]. Direct demonstration of the involvement of microglia-derived cytokines in astrocytic uncoupling in epileptic tissue was recently provided by Henning and co-workers, who showed that seizure activity triggered by intracortical kainate injections decreased GJ coupling between hippocampal astrocytes only in acute slices from wild-type mice, but not in slices from transgenic mice with microglia-specific TNFα deletion or from mice with depleted microglia [82]. These data suggest that at this early time point (4 h after kainate injection), astrocytic coupling is inhibited by microglial TNFα. The cytokine probably exerts its effects through binding to astrocytic TNF receptors, which in turn might mediate the observed changes in Cx expression and/or phosphorylation [12,76] or even trigger necroptotic astrocytic cell death as recently described in the same model [83]. Depletion of microglia attenuated the severity of SE [82], a finding that is in line with the proposed seizure-promoting role of astrocytic uncoupling.

Another pathological mechanism that may be involved in GJ regulation arises from the fact that epilepsy is associated with a breakdown of the BBB, allowing serum proteins such as albumin to enter the brain parenchyma [84]. Extravasated serum albumin was proposed to be endocytosed by astrocytes via binding to transforming growth factor beta receptors (TGFβR) to reduce Cx expression and GJ coupling [85–87]. However, in a recent study, no major astrocytic albumin uptake could be detected 4 h and 1 d after intracortical kainate injection, and inhibition of TGFβR1 kinase did not prevent seizure-induced GJ uncoupling, arguing against an involvement of this mechanism in the regulation of GJs during early epileptogenesis [88]. Nevertheless, as uncoupling and albumin extravasation was observed at the chronic stage in this TLE model [12,24], a later contribution of the albumin-TGFβR1 signalling pathway to astrocytic dysfunction cannot be ruled out.

5. Impact of Glial Cx Gene Knockout on Neuronal Excitability and Epileptogenesis

Important insights into the role of astrocytic GJ channels in the regulation of neuronal excitability and susceptibility to seizures have been gained from transgenic mice with coupling-deficient astrocytes, which were generated by Wallraff and colleagues via crossing astrocyte-specific Cx43-deficient mice (Cx43$^{fl/fl}$:*hGFAP*-Cre mice [89]) with mice globally lacking Cx30 (Cx30$^{-/-}$ [30,90]). The resulting double knockout (dKO) mice showed not only a complete disruption of astrocytic coupling and, as mentioned above, impaired K^+ and glutamate buffering, but also spontaneous epileptiform activity in acute hippocampal slices [30]. In a subsequent study, dKO mice exhibited a higher number of but less severe pentylenetetrazol-induced acute seizures in vivo during short-term (30 min) electroencephalogram (EEG)/video recording, a finding the authors attributed to increased neuronal excitability but decreased neuronal release probability and synchronisation [91]. Possibly, the lower energy supply due to lack of GJ coupling accounted for less severe seizures Recently, dKO mice were subjected to the intracortical kainate injection model of TLE-HS to assess not only the consequences of Cx deficiency on seizure susceptibility but also on the whole process of epileptogenesis [92]. Remarkably, dKO mice showed substantially increased seizure and interictal spike activity during the chronic phase but less pronounced HS, indicating that the interruption of astrocytic GJ communication promotes chronic seizures but attenuates seizure-induced histopathological changes [92]. However, neither the constitutive dKO mice nor the recently generated inducible, astrocyte-specific dKO mice (Cx30$^{fl/fl}$:Cx43$^{fl/fl}$:GLAST^{CreERT2}; [40]) showed spontaneous seizures or abnormal EEG activity in vivo [40,91]. Of course, one must bear in mind that in dKO mice, not only intercellular coupling but also the Cx proteins are missing, which does not correspond to the situation in human and experimental epilepsy where the proteins are even upregulated [12]. Hence, potential pro-epileptic non-channel and HC functions of Cxs are not considered in these mice. Moreover, astrocyte uncoupling in human and experimental TLE is spatially restricted to the sclerotic area and the epileptic foci, while Cx deletion in dKO mice involves the entire brain. Finally, compensatory developmental changes in the constitutive mouse line and residual coupling due to incomplete Cre recombination in the inducible mouse line limit the conclusions that can be drawn from these experiments.

Mice lacking only Cx30 developed less severe behavioural convulsions during the first 2 h after systemic kainate injection, a phenomenon that was dependent only on the presence of the protein but not on functional channels [72]. The authors attributed this to more effective astrocytic glutamate clearance in the absence of Cx30, which is consistent with the above-mentioned non-channel function of the protein in controlling the invasion of astrocytic processes into the synaptic cleft [51,72].

6. Therapeutic Potential of Targeting Glial Cx Channels in Epilepsy

The low efficacy and tolerability of currently available AEDs and the lack of antiepileptogenic, disease-modifying drugs underscore the urgent need for new cellular and molecular targets and therapeutic strategies in drug development. The evidence from the literature described above suggests that Cx channels, especially those composed of Cx43, play an important role in epilepsy and may therefore constitute promising targets for new AEDs. Interestingly, it has been shown that impaired GJ coupling in cytokine- or LPS-incubated astroglia/microglia co-cultures or in acute hippocampal slices from LPS-injected mice could be improved through treatment with the commonly used AED levetiracetam, leading to the assumption that the antiepileptogenic properties of the drug are at least partly attributable to its effect on coupling [24,81]. The suitability of substances that directly inhibit or increase GJ coupling as antiepileptic drugs is questionable, as Cxs are ubiquitously expressed in almost all tissues and cell types [93] and, therefore, systemic administration likely has side effects. For example, Cx43 is the most abundant GJ protein in the heart, where it plays a key role in the coordinated spread of electrical activity. Thus, its inhibition or activation in the heart would have deleterious effects on heart function. Another problem is the lack of specificity of substances that modulate GJs. For instance, carbenoxolone (CBX), the most commonly used GJ blocker, is not only unable to distinguish between different Cx isoforms or between GJ channels and HCs, but was also shown to inhibit voltage-gated Ca^{2+} channels, purinergic P2X7 receptors and tonic $GABA_A$ receptor currents [94–96]. Accordingly, several studies have shown that CBX inhibits synaptic transmission and neuronal network excitability in a GJ-independent manner, questioning its suitability for assessing the role of GJs in epilepsy [94,97–100]. This should be kept in mind when drawing conclusions from the extensive literature on antiepileptic effects of CBX treatment in epilepsy models (for review see [101]). Similarly, no substances are known to specifically increase astrocytic coupling. Regarding systemic side effects, it would probably make more sense not to search for more specific GJ blockers/activators but instead to search for substances that prevent/reverse the molecular mechanisms or signalling pathways that lead to impaired coupling under pathological conditions. For example, it was recently reported that epileptiform activity inhibits astrocytic coupling via $Na^+/HCO3^-$ cotransporter-mediated intracellular alkalization and that blockers of this cotransporter have antiepileptic effects [102]. In this context, polyamines such as spermine have been shown to accumulate in astrocytes and to prevent or reverse the proton- or calcium-mediated inhibition of Cx43 [103,104]. As data from human and experimental TLE indicate that GJ inhibition is mediated by phosphorylation-mediated closure of Cx43 channels [12,76], compounds that disrupt aberrant Cx phosphorylation may have therapeutic potential in the treatment of epilepsy. In this regard, antiarrhythmic peptides such as the Cx43-targeting peptide danegaptide (ZP1609), developed for the treatment of reperfusion injury after acute myocardial infarction [105], were investigated [8,46]. Indeed, danegaptide was found to normalize pathological phosphorylation-induced closure of Cx43 channels not only in the heart, but also between astrocytes in a mouse brain ischemia/reperfusion model [106,107]. However, the peptide exhibited no effect on astrocyte uncoupling induced by intracortical kainate injections (unpublished data from our group), indicating that the mechanism of Cx43 inhibition differs between epilepsy and ischemia.

In addition to GJ channels, specific Cx HC blockade has also been proposed as a promising strategy for treating epilepsy [45,47]. In accordance with the proposed seizure-promoting consequences of HC opening, inhibition of Cx43 HCs with the Cx mimetic peptide TAT-Gap19 or the small molecule D4, which were described to block HC activity without affecting GJ coupling, yielded anticonvulsant effects in mice and rats in different models of seizures and epilepsy [45,47]. The advantage of this approach is that HC activation occurs only under pathological conditions, so systemic administration of HC blockers should not have major side effects [45,47]. However, more research is needed to understand the biophysical properties and specific role of these channels in the healthy and diseased brain before the therapeutic potential of Cx HC blockers can be established.

7. Conclusions

Evidence is accumulating showing that astrocyte Cx channels are critically implicated in epilepsy, but the underlying mechanisms and therapeutic importance are still poorly understood. In the sclerotic hippocampus of TLE patients, astrocytic GJ coupling is completely lost, despite the fact that expression of the two astrocytic Cx isoforms does not decrease. This might simply be due to loss of physical contact between glial processes as a result of the dramatic morphological alterations astrocytes undergo in sclerosis (Hinterkeuser et al., 2000; Bedner et al., 2015) [24,62]. However, it can also not be excluded that besides altered intercellular GJ coupling, increased Cx HC activity or non-channel properties of Cxs affect neuronal excitability and synchronisation. Maybe, only a combination of several of these factors is sufficient to generate neuronal hyperexcitability, which would explain why mice with genetic deletion of astrocytic Cxs do not exhibit spontaneous behavioural convulsions. Identifying the exact mechanisms underlying generation and progression of this complex disease remains a challenge. Nevertheless, since astrocytic uncoupling is one of the earliest detectable cellular changes in experimental TLE that precedes the onset of neurodegeneration and chronic seizures, it can be considered as a causative factor. To validate the causality of the dysfunction, it would be necessary to prove that its suppression prevents the genesis of the disease. However, this requires detailed knowledge of the mechanism of uncoupling and the identification of substances that specifically inhibit it. The growing recognition of the importance of the astrocytic network in epilepsy, together with the increasingly sophisticated technical tools and methods available today, give rise to optimism that the mechanisms governing epilepsy will soon be better understood. This opens new avenues for future research and therapeutic interventions.

Author Contributions: Both authors wrote the manuscript and agreed to its final version. All authors have read and agreed to the published version of the manuscript.

Funding: This research was funded by BMBF, grant numbers 01DN20001 and 16GW0182, and EU, grant number H2020 MSCA-ITN project 722053.

Institutional Review Board Statement: Not applicable.

Informed Consent Statement: Not applicable.

Data Availability Statement: Data sharing not applicable.

Conflicts of Interest: The authors declare no conflict of interest.

Abbreviations

AED: antiepileptic drug, BBB: blood–brain barrier, CBX: carbenoxolone, Cx: connexin, Cx30: connexin30, Cx43: connexin43, dKO: double knockout, EEG: electroencephalogram, GJ: gap junction, HC: hemichannel, HS: hippocampal sclerosis, [K+]o: extracellular K+ concentration, IL-1β: interleukin-1β, LPS: lipopolysaccharide, SE: status epilepticus, TGFβR: transforming growth factor beta receptor, TLE: temporal lobe epilepsy, TNFα: tumor necrosis factor α.

References

1. Fisher, R.S.; Acevedo, C.; Arzimanoglou, A.; Bogacz, A.; Cross, J.H.; Elger, C.E.; Engel, J.; Forsgren, L.; French, J.A.; Glynn, M.; et al. ILAE Official Report: A Practical Clinical Definition of Epilepsy. *Epilepsia* **2014**, *55*, 475–482. [CrossRef] [PubMed]
2. Fiest, K.M.; Sauro, K.M.; Wiebe, S.; Patten, S.B.; Kwon, C.-S.; Dykeman, J.; Pringsheim, T.; Lorenzetti, D.L.; Jetté, N. Prevalence and Incidence of Epilepsy: A Systematic Review and Meta-Analysis of International Studies. *Neurology* **2017**, *88*, 296–303. [CrossRef] [PubMed]
3. Löscher, W.; Potschka, H.; Sisodiya, S.M.; Vezzani, A. Drug Resistance in Epilepsy: Clinical Impact, Potential Mechanisms, and New Innovative Treatment Options. *Pharmacol. Rev.* **2020**, *72*, 606–638. [CrossRef] [PubMed]
4. Stafstrom, C.E. Mechanisms of Action of Antiepileptic Drugs: The Search for Synergy. *Curr. Opin. Neurol.* **2010**, *23*, 157–163. [CrossRef]
5. Boison, D.; Steinhäuser, C. Epilepsy and Astrocyte Energy Metabolism. *Glia* **2018**, *66*, 1235–1243. [CrossRef]

6. Binder, D.K.; Steinhäuser, C. Astrocytes and Epilepsy. *Neurochem. Res.* **2021**, *46*, 2687–2695. [CrossRef]
7. Verkhratsky, A.; Nedergaard, M. Physiology of Astroglia. *Physiol. Rev.* **2018**, *98*, 239–389. [CrossRef]
8. Giaume, C.; Naus, C.C.; Sáez, J.C.; Leybaert, L. Glial Connexins and Pannexins in the Healthy and Diseased Brain. *Physiol. Rev.* **2020**, *101*, 93–145. [CrossRef]
9. Giaume, C.; Koulakoff, A.; Roux, L.; Holcman, D.; Rouach, N. Astroglial Networks: A Step Further in Neuroglial and Gliovascular Interactions. *Nat. Rev. Neurosci.* **2010**, *11*, 87–99. [CrossRef]
10. Gosejacob, D.; Dublin, P.; Bedner, P.; Hüttmann, K.; Zhang, J.; Tress, O.; Willecke, K.; Pfrieger, F.; Steinhäuser, C.; Theis, M. Role of Astroglial Connexin30 in Hippocampal Gap Junction Coupling. *Glia* **2011**, *59*, 511–519. [CrossRef]
11. Griemsmann, S.; Höft, S.P.; Bedner, P.; Zhang, J.; von Staden, E.; Beinhauer, A.; Degen, J.; Dublin, P.; Cope, D.W.; Richter, N.; et al. Characterization of Panglial Gap Junction Networks in the Thalamus, Neocortex, and Hippocampus Reveals a Unique Population of Glial Cells. *Cereb. Cortex* **2015**, *25*, 3420–3433. [CrossRef] [PubMed]
12. Deshpande, T.; Li, T.; Herde, M.K.; Becker, A.; Vatter, H.; Schwarz, M.K.; Henneberger, C.; Steinhäuser, C.; Bedner, P. Subcellular Reorganization and Altered Phosphorylation of the Astrocytic Gap Junction Protein Connexin43 in Human and Experimental Temporal Lobe Epilepsy. *Glia* **2017**, *65*, 1809–1820. [CrossRef]
13. Bedner, P.; Jabs, R.; Steinhäuser, C. Properties of Human Astrocytes and NG2 Glia. *Glia* **2020**, *68*, 756–767. [CrossRef] [PubMed]
14. Valiunas, V.; Manthey, D.; Vogel, R.; Willecke, K.; Weingart, R. Biophysical Properties of Mouse Connexin30 Gap Junction Channels Studied in Transfected Human HeLa Cells. *J. Physiol.* **1999**, *519*, 631–644. [CrossRef] [PubMed]
15. Valiunas, V.; Weingart, R.; Brink, P.R. Formation of Heterotypic Gap Junction Channels by Connexins 40 and 43. *Circ. Res.* **2000**, *86*, E42–E49. [CrossRef] [PubMed]
16. Solan, J.L.; Lampe, P.D. Connexin43 Phosphorylation: Structural Changes and Biological Effects. *Biochem. J.* **2009**, *419*, 261–272. [CrossRef]
17. Manthey, D.; Banach, K.; Desplantez, T.; Lee, C.G.; Kozak, C.A.; Traub, O.; Weingart, R.; Willecke, K. Intracellular Domains of Mouse Connexin26 and -30 Affect Diffusional and Electrical Properties of Gap Junction Channels. *J. Membr. Biol.* **2001**, *181*, 137–148. [CrossRef]
18. Magnotti, L.M.; Goodenough, D.A.; Paul, D.L. Functional Heterotypic Interactions between Astrocyte and Oligodendrocyte Connexins. *Glia* **2011**, *59*, 26–34. [CrossRef]
19. Giaume, C.; Leybaert, L.; Naus, C.; Sáez, J.-C. Connexin and Pannexin Hemichannels in Brain Glial Cells: Properties, Pharmacology, and Roles. *Front. Pharmacol.* **2013**, *4*, 88. [CrossRef]
20. Vinken, M.; Decrock, E.; Leybaert, L.; Bultynck, G.; Himpens, B.; Vanhaecke, T.; Rogiers, V. Non-Channel Functions of Connexins in Cell Growth and Cell Death. *Biochim. Biophys. Acta BBA-Biomembr.* **2012**, *1818*, 2002–2008. [CrossRef]
21. Kunze, A.; Congreso, M.R.; Hartmann, C.; Wallraff-Beck, A.; Hüttmann, K.; Bedner, P.; Requardt, R.; Seifert, G.; Redecker, C.; Willecke, K.; et al. Connexin Expression by Radial Glia-like Cells Is Required for Neurogenesis in the Adult Dentate Gyrus. *Proc. Natl. Acad. Sci. USA* **2009**, *106*, 11336–11341. [CrossRef]
22. Steinhäuser, C.; Seifert, G.; Bedner, P. Astrocyte Dysfunction in Temporal Lobe Epilepsy: K+ Channels and Gap Junction Coupling. *Glia* **2012**, *60*, 1192–1202. [CrossRef]
23. Bedner, P.; Steinhäuser, C. Altered Kir and Gap Junction Channels in Temporal Lobe Epilepsy. *Neurochem. Int.* **2013**, *63*, 682–687. [CrossRef] [PubMed]
24. Bedner, P.; Dupper, A.; Hüttmann, K.; Müller, J.; Herde, M.K.; Dublin, P.; Deshpande, T.; Schramm, J.; Häussler, U.; Haas, C.A.; et al. Astrocyte Uncoupling as a Cause of Human Temporal Lobe Epilepsy. *Brain* **2015**, *138*, 1208–1222. [CrossRef] [PubMed]
25. Henning, L.; Unichenko, P.; Bedner, P.; Steinhäuser, C.; Henneberger, C. Overview Article Astrocytes as Initiators of Epilepsy. *Neurochem. Res.* **2022**, *48*, 1091–1099. [CrossRef]
26. Orkand, R.K. Introductory Remarks: Glial-Interstitial Fluid Exchange. *Ann. N. Y. Acad. Sci.* **1986**, *481*, 269–272. [CrossRef] [PubMed]
27. Walz, W. Role of Astrocytes in the Clearance of Excess Extracellular Potassium. *Neurochem. Int.* **2000**, *36*, 291–300. [CrossRef]
28. Kofuji, P.; Newman, E.A. Potassium Buffering in the Central Nervous System. *Neuroscience* **2004**, *129*, 1043–1054. [CrossRef]
29. Breithausen, B.; Kautzmann, S.; Boehlen, A.; Steinhäuser, C.; Henneberger, C. Limited Contribution of Astroglial Gap Junction Coupling to Buffering of Extracellular K+ in CA1 Stratum Radiatum. *Glia* **2020**, *68*, 918–931. [CrossRef]
30. Wallraff, A.; Köhling, R.; Heinemann, U.; Theis, M.; Willecke, K.; Steinhäuser, C. The Impact of Astrocytic Gap Junctional Coupling on Potassium Buffering in the Hippocampus. *J. Neurosci.* **2006**, *26*, 5438–5447. [CrossRef]
31. Henneberger, C. Does Rapid and Physiological Astrocyte–Neuron Signalling Amplify Epileptic Activity? *J. Physiol.* **2017**, *595*, 1917–1927. [CrossRef] [PubMed]
32. Langer, J.; Stephan, J.; Theis, M.; Rose, C.R. Gap Junctions Mediate Intercellular Spread of Sodium between Hippocampal Astrocytes in Situ. *Glia* **2012**, *60*, 239–252. [CrossRef]
33. Kirischuk, S.; Parpura, V.; Verkhratsky, A. Sodium Dynamics: Another Key to Astroglial Excitability? *Trends Neurosci.* **2012**, *35*, 497–506. [CrossRef] [PubMed]
34. Rose, C.R.; Felix, L.; Zeug, A.; Dietrich, D.; Reiner, A.; Henneberger, C. Astroglial Glutamate Signaling and Uptake in the Hippocampus. *Front. Mol. Neurosci.* **2017**, *10*, 451. [CrossRef] [PubMed]
35. Riquelme, J.; Wellmann, M.; Sotomayor-Zárate, R.; Bonansco, C. Gliotransmission: A Novel Target for the Development of Antiseizure Drugs. *Neuroscientist* **2020**, *26*, 293–309. [CrossRef] [PubMed]

36. Müller, J.; Timmermann, A.; Henning, L.; Müller, H.; Steinhäuser, C.; Bedner, P. Astrocytic GABA Accumulation in Experimental Temporal Lobe Epilepsy. *Front. Neurol.* **2020**, *11*, 614923. [CrossRef]
37. Pannasch, U.; Vargová, L.; Reingruber, J.; Ezan, P.; Holcman, D.; Giaume, C.; Syková, E.; Rouach, N. Astroglial Networks Scale Synaptic Activity and Plasticity. *Proc. Natl. Acad. Sci. USA* **2011**, *108*, 8467–8472. [CrossRef]
38. Bazzigaluppi, P.; Weisspapir, I.; Stefanovic, B.; Leybaert, L.; Carlen, P.L. Astrocytic Gap Junction Blockade Markedly Increases Extracellular Potassium without Causing Seizures in the Mouse Neocortex. *Neurobiol. Dis.* **2017**, *101*, 1–7. [CrossRef]
39. EbrahimAmini, A.; Bazzigaluppi, P.; Aquilino, M.S.; Stefanovic, B.; Carlen, P.L. Neocortical in Vivo Focal and Spreading Potassium Responses and the Influence of Astrocytic Gap Junctional Coupling. *Neurobiol. Dis.* **2021**, *147*, 105160. [CrossRef] [PubMed]
40. Hösli, L.; Binini, N.; Ferrari, K.D.; Thieren, L.; Looser, Z.J.; Zuend, M.; Zanker, H.S.; Berry, S.; Holub, M.; Möbius, W.; et al. Decoupling Astrocytes in Adult Mice Impairs Synaptic Plasticity and Spatial Learning. *Cell Rep.* **2022**, *38*, 110484. [CrossRef]
41. Rouach, N.; Koulakoff, A.; Abudara, V.; Willecke, K.; Giaume, C. Astroglial Metabolic Networks Sustain Hippocampal Synaptic Transmission. *Science* **2008**, *322*, 1551–1555. [CrossRef]
42. Philippot, C.; Griemsmann, S.; Jabs, R.; Seifert, G.; Kettenmann, H.; Steinhäuser, C. Astrocytes and Oligodendrocytes in the Thalamus Jointly Maintain Synaptic Activity by Supplying Metabolites. *Cell Rep.* **2021**, *34*, 108642. [CrossRef] [PubMed]
43. Gómez-Gonzalo, M.; Losi, G.; Chiavegato, A.; Zonta, M.; Cammarota, M.; Brondi, M.; Vetri, F.; Uva, L.; Pozzan, T.; de Curtis, M.; et al. An Excitatory Loop with Astrocytes Contributes to Drive Neurons to Seizure Threshold. *PLoS Biol.* **2010**, *8*, e1000352. [CrossRef]
44. Nielsen, B.S.; Hansen, D.B.; Ransom, B.R.; Nielsen, M.S.; MacAulay, N. Connexin Hemichannels in Astrocytes: An Assessment of Controversies Regarding Their Functional Characteristics. *Neurochem. Res.* **2017**, *42*, 2537–2550. [CrossRef] [PubMed]
45. Walrave, L.; Pierre, A.; Albertini, G.; Aourz, N.; Bundel, D.D.; Eeckhaut, A.V.; Vinken, M.; Giaume, C.; Leybaert, L.; Smolders, I. Inhibition of Astroglial Connexin43 Hemichannels with TAT-Gap19 Exerts Anticonvulsant Effects in Rodents. *Glia* **2018**, *66*, 1788–1804. [CrossRef] [PubMed]
46. Walrave, L.; Vinken, M.; Leybaert, L.; Smolders, I. Astrocytic Connexin43 Channels as Candidate Targets in Epilepsy Treatment. *Biomolecules* **2020**, *10*, 1578. [CrossRef]
47. Guo, A.; Zhang, H.; Li, H.; Chiu, A.; García-Rodríguez, C.; Lagos, C.F.; Sáez, J.C.; Lau, C.G. Inhibition of Connexin Hemichannels Alleviates Neuroinflammation and Hyperexcitability in Temporal Lobe Epilepsy. *Proc. Natl. Acad. Sci. USA* **2022**, *119*, e2213162119. [CrossRef]
48. Scemes, E. Modulation of Astrocyte P2Y1 Receptors by the Carboxyl Terminal Domain of the Gap Junction Protein Cx43. *Glia* **2008**, *56*, 145–153. [CrossRef]
49. Nikolic, L.; Nobili, P.; Shen, W.; Audinat, E. Role of Astrocyte Purinergic Signaling in Epilepsy. *Glia* **2020**, *68*, 1677–1691. [CrossRef]
50. Vezzani, A.; Ravizza, T.; Bedner, P.; Aronica, E.; Steinhäuser, C.; Boison, D. Astrocytes in the Initiation and Progression of Epilepsy. *Nat. Rev. Neurol.* **2022**, *18*, 707–722. [CrossRef]
51. Pannasch, U.; Freche, D.; Dallérac, G.; Ghézali, G.; Escartin, C.; Ezan, P.; Cohen-Salmon, M.; Benchenane, K.; Abudara, V.; Dufour, A.; et al. Connexin 30 Sets Synaptic Strength by Controlling Astroglial Synapse Invasion. *Nat. Neurosci.* **2014**, *17*, 549–558. [CrossRef]
52. Aronica, E.; Gorter, J.A.; Jansen, G.H.; Leenstra, S.; Yankaya, B.; Troost, D. Expression of Connexin 43 and Connexin 32 Gap-Junction Proteins in Epilepsy-Associated Brain Tumors and in the Perilesional Epileptic Cortex. *Acta Neuropathol.* **2001**, *101*, 449–459. [CrossRef] [PubMed]
53. Collignon, F.; Wetjen, N.M.; Cohen-Gadol, A.A.; Cascino, G.D.; Parisi, J.; Meyer, F.B.; Marsh, W.R.; Roche, P.; Weigand, S.D. Altered Expression of Connexin Subtypes in Mesial Temporal Lobe Epilepsy in Humans. *J. Neurosurg.* **2006**, *105*, 77–87. [CrossRef]
54. Naus, C.C.G.; Bechberger, J.F.; Paul, D.L. Gap Junction Gene Expression in Human Seizure Disorder. *Exp. Neurol.* **1991**, *111*, 198–203. [CrossRef]
55. Das, A.; Wallace, G.C.; Holmes, C.; McDowell, M.L.; Smith, J.A.; Marshall, J.D.; Bonilha, L.; Edwards, J.C.; Glazier, S.S.; Ray, S.K.; et al. Hippocampal Tissue of Patients with Refractory Temporal Lobe Epilepsy Is Associated with Astrocyte Activation, Inflammation, and Altered Expression of Channels and Receptors. *Neuroscience* **2012**, *220*, 237–246. [CrossRef]
56. Elisevich, K.; Rempel, S.A.; Smith, B.J.; Edvardsen, K. Hippocampal Connexin 43 Expression in Human Complex Partial Seizure Disorder. *Exp. Neurol.* **1997**, *145*, 154–164. [CrossRef] [PubMed]
57. Fonseca, C.G.; Green, C.R.; Nicholson, L.F.B. Upregulation in Astrocytic Connexin 43 Gap Junction Levels May Exacerbate Generalized Seizures in Mesial Temporal Lobe Epilepsy. *Brain Res.* **2002**, *929*, 105–116. [CrossRef]
58. Garbelli, R.; Frassoni, C.; Condorelli, D.F.; Salinaro, A.T.; Musso, N.; Medici, V.; Tassi, L.; Bentivoglio, M.; Spreafico, R. Expression of Connexin 43 in the Human Epileptic and Drug-Resistant Cerebral Cortex. *Neurology* **2011**, *76*, 895–902. [CrossRef] [PubMed]
59. Heinemann, U.; Gabriel, S.; Jauch, R.; Schulze, K.; Kivi, A.; Eilers, A.; Kovacs, R.; Lehmann, T.-N. Alterations of Glial Cell Function in Temporal Lobe Epilepsy. *Epilepsia* **2000**, *41*, S185–S189. [CrossRef]
60. Jauch, R.; Windmüller, O.; Lehmann, T.-N.; Heinemann, U.; Gabriel, S. Effects of Barium, Furosemide, Ouabaine and 4,4′-Diisothiocyanatostilbene-2,2′-Disulfonic Acid (DIDS) on Ionophoretically-Induced Changes in Extracellular Potassium Concentration in Hippocampal Slices from Rats and from Patients with Epilepsy. *Brain Res.* **2002**, *925*, 18–27. [CrossRef]
61. Kivi, A.; Lehmann, T.-N.; Kovács, R.; Eilers, A.; Jauch, R.; Meencke, H.J.; Deimling, A.V.; Heinemann, U.; Gabriel, S. Effects of Barium on Stimulus-Induced Rises of [K+]o in Human Epileptic Non-Sclerotic and Sclerotic Hippocampal Area CA1. *Eur. J. Neurosci.* **2000**, *12*, 2039–2048. [CrossRef] [PubMed]

62. Hinterkeuser, S.; Schröder, W.; Hager, G.; Seifert, G.; Blümcke, I.; Elger, C.E.; Schramm, J.; Steinhäuser, C. Astrocytes in the Hippocampus of Patients with Temporal Lobe Epilepsy Display Changes in Potassium Conductances. *Eur. J. Neurosci.* **2000**, *12*, 2087–2096. [CrossRef] [PubMed]
63. Blümcke, I.; Beck, H.; Lie, A.A.; Wiestler, O.D. Molecular Neuropathology of Human Mesial Temporal Lobe Epilepsy. *Epilepsy Res.* **1999**, *36*, 205–223. [CrossRef] [PubMed]
64. Löscher, W.; Schmidt, D. Modern Antiepileptic Drug Development Has Failed to Deliver: Ways out of the Current Dilemma. *Epilepsia* **2011**, *52*, 657–678. [CrossRef]
65. Jefferys, J.; Steinhäuser, C.; Bedner, P. Chemically-Induced TLE Models: Topical Application. *J. Neurosci. Methods* **2016**, *260*, 53–61. [CrossRef]
66. Rusina, E.; Bernard, C.; Williamson, A. The Kainic Acid Models of Temporal Lobe Epilepsy. *eNeuro* **2021**, *8*, 1–24. [CrossRef]
67. Condorelli, D.F.; Mudò, G.; Trovato-Salinaro, A.; Mirone, M.B.; Amato, G.; Belluardo, N. Connexin-30 MRNA Is Up-Regulated in Astrocytes and Expressed in Apoptotic Neuronal Cells of Rat Brain Following Kainate-Induced Seizures. *Mol. Cell. Neurosci.* **2002**, *21*, 94–113. [CrossRef]
68. Condorelli, D.F.; Trovato-Salinaro, A.; Mudò, G.; Mirone, M.B.; Belluardo, N. Cellular Expression of Connexins in the Rat Brain: Neuronal Localization, Effects of Kainate-Induced Seizures and Expression in Apoptotic Neuronal Cells. *Eur. J. Neurosci.* **2003**, *18*, 1807–1827. [CrossRef]
69. Hussein, A.M.; Abbas, K.M.; Abulseoud, O.A.; El-Hussainy, E.-H.M.A. Effects of Ferulic Acid on Oxidative Stress, Heat Shock Protein 70, Connexin 43, and Monoamines in the Hippocampus of Pentylenetetrazole-Kindled Rats. *Can. J. Physiol. Pharmacol.* **2017**, *95*, 732–742. [CrossRef]
70. Kinjo, E.R.; Higa, G.S.V.; Morya, E.; Valle, A.C.; Kihara, A.H.; Britto, L.R.G. Reciprocal Regulation of Epileptiform Neuronal Oscillations and Electrical Synapses in the Rat Hippocampus. *PLoS ONE* **2014**, *9*, e109149. [CrossRef]
71. McCracken, C.B.; Roberts, D.C.S. A Single Evoked Afterdischarge Produces Rapid Time-Dependent Changes in Connexin36 Protein Expression in Adult Rat Dorsal Hippocampus. *Neurosci. Lett.* **2006**, *405*, 84–88. [CrossRef] [PubMed]
72. Pannasch, U.; Dossi, E.; Ezan, P.; Rouach, N. Astroglial Cx30 Sustains Neuronal Population Bursts Independently of Gap-Junction Mediated Biochemical Coupling. *Glia* **2019**, *67*, 1104–1112. [CrossRef]
73. Söhl, G.; Güldenagel, M.; Beck, H.; Teubner, B.; Traub, O.; Gutiérrez, R.; Heinemann, U.; Willecke, K. Expression of Connexin Genes in Hippocampus of Kainate-Treated and Kindled Rats under Conditions of Experimental Epilepsy. *Mol. Brain Res.* **2000**, *83*, 44–51. [CrossRef] [PubMed]
74. Takahashi, D.K.; Vargas, J.R.; Wilcox, K.S. Increased Coupling and Altered Glutamate Transport Currents in Astrocytes Following Kainic-Acid-Induced Status Epilepticus. *Neurobiol. Dis.* **2010**, *40*, 573–585. [CrossRef] [PubMed]
75. Wu, X.L.; Tang, Y.C.; Lu, Q.Y.; Xiao, X.L.; Song, T.B.; Tang, F.R. Astrocytic Cx 43 and Cx 40 in the Mouse Hippocampus during and after Pilocarpine-Induced Status Epilepticus. *Exp. Brain Res.* **2015**, *233*, 1529–1539. [CrossRef] [PubMed]
76. Khan, D.; Dupper, A.; Deshpande, T.; Graan, P.N.E.D.; Steinhäuser, C.; Bedner, P. Experimental Febrile Seizures Impair Interastrocytic Gap Junction Coupling in Juvenile Mice. *J. Neurosci. Res.* **2016**, *94*, 804–813. [CrossRef]
77. Choy, M.; Dubé, C.M.; Ehrengruber, M.; Baram, T.Z. Inflammatory Processes, Febrile Seizures, and Subsequent Epileptogenesis. *Epilepsy Curr.* **2014**, *14*, 15–22. [CrossRef]
78. Vezzani, A.; Balosso, S.; Ravizza, T. Neuroinflammatory Pathways as Treatment Targets and Biomarkers in Epilepsy. *Nat. Rev. Neurol.* **2019**, *15*, 459–472. [CrossRef]
79. Même, W.; Calvo, C.-F.; Froger, N.; Ezan, P.; Amigou, E.; Koulakoff, A.; Giaume, C. Proinflammatory Cytokines Released from Microglia Inhibit Gap Junctions in Astrocytes: Potentiation by β-Amyloid. *FASEB J.* **2006**, *20*, 494–496. [CrossRef]
80. Retamal, M.A.; Froger, N.; Palacios-Prado, N.; Ezan, P.; Sáez, P.J.; Sáez, J.C.; Giaume, C. Cx43 Hemichannels and Gap Junction Channels in Astrocytes Are Regulated Oppositely by Proinflammatory Cytokines Released from Activated Microglia. *J. Neurosci.* **2007**, *27*, 13781–13792. [CrossRef]
81. Haghikia, A.; Ladage, K.; Hinkerohe, D.; Vollmar, P.; Heupel, K.; Dermietzel, R.; Faustmann, P.M. Implications of Antiinflammatory Properties of the Anticonvulsant Drug Levetiracetam in Astrocytes. *J. Neurosci. Res.* **2008**, *86*, 1781–1788. [CrossRef] [PubMed]
82. Henning, L.; Antony, H.; Breuer, A.; Müller, J.; Seifert, G.; Audinat, E.; Singh, P.; Brosseron, F.; Heneka, M.T.; Steinhäuser, C.; et al. Reactive Microglia Are the Major Source of Tumor Necrosis Factor Alpha and Contribute to Astrocyte Dysfunction and Acute Seizures in Experimental Temporal Lobe Epilepsy. *Glia* **2023**, *71*, 168–186. [CrossRef] [PubMed]
83. Wu, Z.; Deshpande, T.; Henning, L.; Bedner, P.; Seifert, G.; Steinhäuser, C. Cell Death of Hippocampal CA1 Astrocytes during Early Epileptogenesis. *Epilepsia* **2021**, *62*, 1569–1583. [CrossRef] [PubMed]
84. Löscher, W.; Friedman, A. Structural, Molecular, and Functional Alterations of the Blood-Brain Barrier during Epileptogenesis and Epilepsy: A Cause, Consequence, or Both? *Int. J. Mol. Sci.* **2020**, *21*, 591. [CrossRef]
85. David, Y.; Cacheaux, L.P.; Ivens, S.; Lapilover, E.; Heinemann, U.; Kaufer, D.; Friedman, A. Astrocytic Dysfunction in Epileptogenesis: Consequence of Altered Potassium and Glutamate Homeostasis? *J. Neurosci.* **2009**, *29*, 10588–10599. [CrossRef]
86. Cacheaux, L.P.; Ivens, S.; David, Y.; Lakhter, A.J.; Bar-Klein, G.; Shapira, M.; Heinemann, U.; Friedman, A.; Kaufer, D. Transcriptome Profiling Reveals TGF-β Signaling Involvement in Epileptogenesis. *J. Neurosci.* **2009**, *29*, 8927–8935. [CrossRef]
87. Braganza, O.; Bedner, P.; Hüttmann, K.; von Staden, E.; Friedman, A.; Seifert, G.; Steinhäuser, C. Albumin Is Taken up by Hippocampal NG2 Cells and Astrocytes and Decreases Gap Junction Coupling. *Epilepsia* **2012**, *53*, 1898–1906. [CrossRef]

88. Henning, L.; Steinhäuser, C.; Bedner, P. Initiation of Experimental Temporal Lobe Epilepsy by Early Astrocyte Uncoupling Is Independent of TGFβR1/ALK5 Signaling. *Front. Neurol.* **2021**, *12*, 660591. [CrossRef]
89. Theis, M.; Jauch, R.; Zhuo, L.; Speidel, D.; Wallraff, A.; Döring, B.; Frisch, C.; Söhl, G.; Teubner, B.; Euwens, C.; et al. Accelerated Hippocampal Spreading Depression and Enhanced Locomotory Activity in Mice with Astrocyte-Directed Inactivation of Connexin43. *J. Neurosci. Off. J. Soc. Neurosci.* **2003**, *23*, 766–776. [CrossRef]
90. Teubner, B.; Michel, V.; Pesch, J.; Lautermann, J.; Cohen-Salmon, M.; Söhl, G.; Jahnke, K.; Winterhager, E.; Herberhold, C.; Hardelin, J.-P.; et al. Connexin30 (Gjb6)-Deficiency Causes Severe Hearing Impairment and Lack of Endocochlear Potential. *Hum. Mol. Genet.* **2003**, *12*, 13–21. [CrossRef]
91. Chever, O.; Dossi, E.; Pannasch, U.; Derangeon, M.; Rouach, N. Astroglial Networks Promote Neuronal Coordination. *Sci. Signal.* **2016**, *9*, ra6. [CrossRef] [PubMed]
92. Deshpande, T.; Li, T.; Henning, L.; Wu, Z.; Müller, J.; Seifert, G.; Steinhäuser, C.; Bedner, P. Constitutive Deletion of Astrocytic Connexins Aggravates Kainate-Induced Epilepsy. *Glia* **2020**, *68*, 2136–2147. [CrossRef] [PubMed]
93. Bedner, P.; Steinhäuser, C.; Theis, M. Functional Redundancy and Compensation among Members of Gap Junction Protein Families? *Biochim. Biophys. Acta BBA-Biomembr.* **2012**, *1818*, 1971–1984. [CrossRef] [PubMed]
94. Vessey, J.P.; Lalonde, M.R.; Mizan, H.A.; Welch, N.C.; Kelly, M.E.M.; Barnes, S. Carbenoxolone Inhibition of Voltage-Gated Ca Channels and Synaptic Transmission in the Retina. *J. Neurophysiol.* **2004**, *92*, 1252–1256. [CrossRef]
95. Suadicani, S.O.; Pina-Benabou, M.H.D.; Urban-Maldonado, M.; Spray, D.C.; Scemes, E. Acute Downregulation of Cx43 Alters P2Y Receptor Expression Levels in Mouse Spinal Cord Astrocytes. *Glia* **2003**, *42*, 160–171. [CrossRef]
96. Ransom, C.B.; Ye, Z.; Spain, W.J.; Richerson, G.B. Modulation of Tonic GABA Currents by Anion Channel and Connexin Hemichannel Antagonists. *Neurochem. Res.* **2017**, *42*, 2551–2559. [CrossRef]
97. Rekling, J.C.; Shao, X.M.; Feldman, J.L. Electrical Coupling and Excitatory Synaptic Transmission between Rhythmogenic Respiratory Neurons in the PreBötzinger Complex. *J. Neurosci.* **2000**, *20*, RC113. [CrossRef]
98. Rouach, N.; Segal, M.; Koulakoff, A.; Giaume, C.; Avignone, E. Carbenoxolone Blockade of Neuronal Network Activity in Culture Is Not Mediated by an Action on Gap Junctions. *J. Physiol.* **2003**, *553*, 729–745. [CrossRef]
99. Chepkova, A.N.; Sergeeva, O.A.; Haas, H.L. Carbenoxolone Impairs LTP and Blocks NMDA Receptors in Murine Hippocampus. *Neuropharmacology* **2008**, *55*, 139–147. [CrossRef]
100. Tovar, K.R.; Maher, B.J.; Westbrook, G.L. Direct Actions of Carbenoxolone on Synaptic Transmission and Neuronal Membrane Properties. *J. Neurophysiol.* **2009**, *102*, 974–978. [CrossRef]
101. Volnova, A.; Tsytsarev, V.; Ganina, O.; Vélez-Crespo, G.E.; Alves, J.M.; Ignashchenkova, A.; Inyushin, M. The Anti-Epileptic Effects of Carbenoxolone In Vitro and In Vivo. *Int. J. Mol. Sci.* **2022**, *23*, 663. [CrossRef] [PubMed]
102. Onodera, M.; Meyer, J.; Furukawa, K.; Hiraoka, Y.; Aida, T.; Tanaka, K.; Tanaka, K.F.; Rose, C.R.; Matsui, K. Exacerbation of Epilepsy by Astrocyte Alkalization and Gap Junction Uncoupling. *J. Neurosci.* **2021**, *41*, 2106–2118. [CrossRef] [PubMed]
103. Skatchkov, S.N.; Bukauskas, F.F.; Benedikt, J.; Inyushin, M.; Kucheryavykh, Y.V. Intracellular Spermine Prevents Acid-Induced Uncoupling of Cx43 Gap Junction Channels. *Neuroreport* **2015**, *26*, 528–532. [CrossRef] [PubMed]
104. Kucheryavykh, L.Y.; Benedikt, J.; Cubano, L.A.; Skatchkov, S.N.; Bukauskas, F.F.; Kucheryavykh, Y.V. Polyamines Preserve Connexin43-Mediated Gap Junctional Communication during Intracellular Hypercalcemia and Acidosis. *Neuroreport* **2017**, *28*, 208–213. [CrossRef]
105. Engstrøm, T.; Nepper-Christensen, L.; Helqvist, S.; Kløvgaard, L.; Holmvang, L.; Jørgensen, E.; Pedersen, F.; Saunamaki, K.; Tilsted, H.-H.; Steensberg, A.; et al. Danegaptide for Primary Percutaneous Coronary Intervention in Acute Myocardial Infarction Patients: A Phase 2 Randomised Clinical Trial. *Heart* **2018**, *104*, 1593–1599. [CrossRef] [PubMed]
106. Dhein, S.; Hagen, A.; Jozwiak, J.; Dietze, A.; Garbade, J.; Barten, M.; Kostelka, M.; Mohr, F.-W. Improving Cardiac Gap Junction Communication as a New Antiarrhythmic Mechanism: The Action of Antiarrhythmic Peptides. *Naunyn. Schmiedebergs Arch. Pharmacol.* **2010**, *381*, 221–234. [CrossRef]
107. Freitas-Andrade, M.; Bechberger, J.; Wang, J.; Yeung, K.K.C.; Whitehead, S.N.; Shultz Hansen, R.; Naus, C.C. Danegaptide Enhances Astrocyte Gap Junctional Coupling and Reduces Ischemic Reperfusion Brain Injury in Mice. *Biomolecules* **2020**, *10*, 353. [CrossRef]

cells

MDPI

Review

Astrocytes Are a Key Target for Neurotropic Viral Infection

Maja Potokar [1,2], Robert Zorec [1,2,*] and Jernej Jorgačevski [1,2,*]

1 Laboratory of Neuroendocrinology–Molecular Cell Physiology, Institute of Pathophysiology, Faculty of Medicine, University of Ljubljana, Zaloška 4, 1000 Ljubljana, Slovenia
2 Celica Biomedical, Tehnološki Park 24, 1000 Ljubljana, Slovenia
* Correspondence: robert.zorec@mf.uni-lj.si (R.Z.); jernej.jorgacevski@mf.uni-lj.si (J.J.)

Abstract: Astrocytes are increasingly recognized as important viral host cells in the central nervous system. These cells can produce relatively high quantities of new virions. In part, this can be attributed to the characteristics of astrocyte metabolism and its abundant and dynamic cytoskeleton network. Astrocytes are anatomically localized adjacent to interfaces between blood capillaries and brain parenchyma and between blood capillaries and brain ventricles. Moreover, astrocytes exhibit a larger membrane interface with the extracellular space than neurons. These properties, together with the expression of various and numerous viral entry receptors, a relatively high rate of endocytosis, and morphological plasticity of intracellular organelles, render astrocytes important target cells in neurotropic infections. In this review, we describe factors that mediate the high susceptibility of astrocytes to viral infection and replication, including the anatomic localization of astrocytes, morphology, expression of viral entry receptors, and various forms of autophagy.

Keywords: viruses; astrocytes; endosome; lysosome; autophagy; cell metabolism; oxidant species

Citation: Potokar, M.; Zorec, R.; Jorgačevski, J. Astrocytes Are a Key Target for Neurotropic Viral Infection. *Cells* **2023**, *12*, 2307. https://doi.org/10.3390/cells12182307

Academic Editors: Vladimir Parpura and Michael Brenner

Received: 23 February 2023
Revised: 28 August 2023
Accepted: 13 September 2023
Published: 19 September 2023

1. Introduction

Infectious diseases of the central nervous system (CNS) are most commonly caused by various types of viruses [1,2]; most of the 25 virus families implicated in human disease include representatives that have been associated with CNS disorders in humans [3,4]. Viruses involved in infection of the CNS can cause inflammation of distinct anatomic regions; inflammation of the meninges is known as meningitis, inflammation of the brain is called encephalitis, and inflammation of the spinal cord is called myelitis; simultaneous inflammation of multiple regions is referred to as either meningoencephalitis or encephalomyelitis [5]. Virus-induced diseases of the CNS may lead to various acute changes in mental and motor functions and cause chronic neurological dysfunctions with long-term consequences that have severe implications for the quality of life of affected individuals (despite an often-mild acute phase) and are associated with substantial mortality. However, viral infections and their long-term consequences are still frequently overlooked, under-reported and under-diagnosed and thus represent a significant burden to human health worldwide [6].

2. Virus Spread and Infection of the CNS

The CNS is typically not the site of initial virus entry, except for extrinsic viral contamination of the CNS during head trauma, neurosurgical procedures, via medical implants, and congenital malformations. Primary infections generally occur at more remote anatomic sites. Arboviruses, a group of viruses transmitted to humans by infected arthropod vectors (e.g., ticks and mosquitoes), enter the skin after an infectious insect inoculates the virus into the dermis of a human host (Figure 1A). After cell entry, virus replication may remain localized to the infected cell, or the virus may start spreading beyond the primary site of infection; the infection becomes systemic when it is not contained by the immune system [7]. Factors that restrict an infection from spreading beyond an epithelial surface have been described in detail previously [7]. When spreading occurs, virus progeny produced within the dermis have access to afferent lymphatic vessels through which viruses gain direct access to the permeable lymphatic capillaries of lymph nodes (Figure 1A) or target Langerhans cells [8], which then migrate to and accumulate in the draining lymph nodes [9]. Ingested and inhaled viruses, on the other hand, must first overcome the intrinsic barriers of the mucosal surfaces [10] and establish infection in oropharyngeal or small bowel lymphoid tissues [5]. Once in secondary lymphoid tissues, viruses can enter veins via the efferent lymphatic vessel (Figure 1A) [7]. Circulating viruses are then disseminated via a hematogenous route. During the initial entry of the virus into the bloodstream, which is known as primary viremia, the quantity of circulating virus is typically low, but the ensuing virus replication in target organs may result in the production of higher concentrations of virus in the blood during secondary viremia [7]. In addition to a hematogenous route, viruses can also spread from the primary site of infection to the CNS by retrograde transit within peripheral or olfactory nerve axons after entry into local nerve endings [11–13]. Some viruses rely on a specific pathway to enter the CNS (e.g., direct entry of the rabies virus into the CNS using neuronal spread [14]), but many viruses are capable of infecting the CNS in more than one way [15,16].

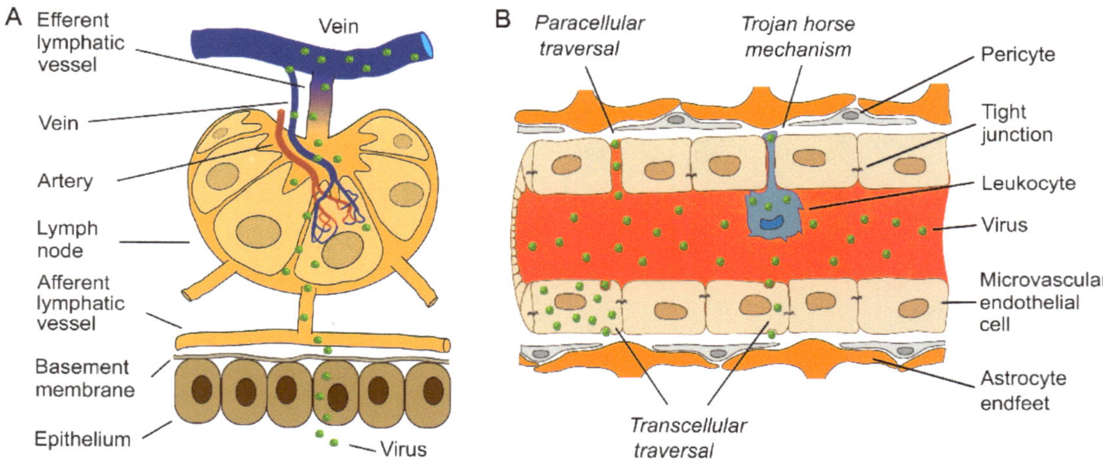

Figure 1. Entry of viruses through the epithelial barrier, the lymphatic spread of viruses and neuroinvasion through the blood–brain barrier. (**A**) After entry through the epithelium, viruses may continue with subepithelial invasion by exploiting the network of lymphatics to enter lymph nodes. Viruses enter porous lymphatic capillaries and/or efferent lymphatic vessels to access the venous system. Circulating blood is the fastest and most effective way of viral dissemination to different tissues, including the central nervous system (CNS). (**B**) Entry into the delicate CNS parenchyma is restricted by an elaborate barrier network, most notably by the blood–brain barrier (BBB). Certain viruses can pass the BBB by the transcellular route by transferring through the lining of microvascular endothelial cells, where they may or may not replicate. In addition, viruses can pass the BBB by paracellular traversal between adjacent endothelial cells, with or without disruption of tight junctions. Viruses can also penetrate the BBB by transmigration within infected haematopoietic cells, known as the "Trojan horse" mechanism.

Viruses circulating in the blood must successfully cross the protective barriers of the brain to establish infection of the CNS. A restricted environment is maintained in the CNS by strict regulation of the entry and exit of substances through the blood–brain barrier (BBB; Figure 1B) and the blood-cerebrospinal fluid barrier (BCSFB), which mediate the immune-privileged status of the CNS [17]. The BBB consists primarily of endothelial cells of microvascular blood vessels, compressed closely to each other by tight junctions

(Figure 1B). In contrast, the BCSFB is formed by the network of fenestrated microvessels in the ventricular system and the subarachnoid space [11,18,19].

The BBB was first demonstrated by Paul Ehrlich in the late 1800s when he noticed that acidic dyes injected intravenously stained the tissues of the body but not the brain [19]. Although gases and fat-soluble molecules can diffuse across these barriers, entry of water-soluble molecules, immune cells, and pathogens is limited; even water requires special transporter proteins [19]. Approximately 50 years after the groundbreaking work of Ehrlich, the term "Barrière hématoencéphalique" or BBB was introduced in 1921 by Lina Stern [20] (for an interesting article on the history of BBB permeability studies, see the review by Saunders et al. [17]). Even though the BBB is arguably the most selective and tightly controlled of the CNS barriers, most viruses invade the CNS by this pathway because the BBB is the largest surface area between the brain and the blood, with an estimated combined area of 12–18 m^2 [7,21,22]. The structural components that form the BBB are endothelial cells, which, unlike other vascular endothelial cells, express tight junction (TJ) proteins that seal the paracellular space between adjoining endothelial cells. These proteins include claudins, occludin, tricelllin, lipolysis-stimulated lipoprotein receptor, junctional adhesion molecules, and zonula occludens proteins; the latter form a structural link to the actin cytoskeleton and actin-binding proteins. In addition to endothelial cells, astrocytes and pericytes are critical for the development and maintenance of the BBB by regulating the TJs [23–25] (see also Figure 1B). To cross the BBB, viruses have developed diverse strategies that can be classified into three main categories: transcellular, paracellular and/or the "Trojan horse" mechanism(s) (Figure 1B) [26]. Transcellular traversal involves infection of endothelial cells, in which viruses may or may not replicate, followed by direct passage into the CNS [27,28]. In the paracellular crossing of the BBB, viruses migrate between the proteins comprising the TJs (occludin and claudin proteins that use scaffolding and regulatory proteins to link to the cytoskeleton) [11,26]. Systemic viral infection can lead to increased cytotoxic effects and induce inflammation, mediated by various cytokines, chemokines, free radicals and matrix metalloproteinases, which results in disruption of the TJs and the consequent breakdown of the BBB, allowing viruses unimpeded access to the CNS [11,29,30]. The Trojan horse mechanism of viral infection of the CNS refers to the haematopoietic cells (e.g., leukocytes; Figure 1) infected with a virus that transports the virus into the CNS [26,28,31,32]. After crossing the BBB, viruses may disperse through the CNS [33].

3. Astrocytes as the Key Cell Type of CNS Viral Infections

Once a virus enters the CNS, it is considered neuroinvasive. In contrast, neurotropic viruses can infect and successfully replicate in cells of the CNS, and neurovirulence indicates the ability of a viral infection to cause CNS pathology [34,35]. A high degree of cellular heterogeneity of the CNS enables neuroinvasive viruses to infect a variety of cell types, including but not limited to different subtypes of neurons, astrocytes, oligodendrocytes, microglia, pericytes, and ependymal cells. Due to their abundance, role in communicating information and their inherent poor ability to regenerate, most studies investigating neuroinvasive viruses have focused on mature neurons of the CNS. In the last decade, interest in other cell types, especially astrocytes, has grown considerably ([19,21,36–40] and references within), but the pathophysiology of viral infection in astrocytes remains poorly understood. However, several lines of evidence support the importance of astrocytes in viral infections of the CNS: astrocytes (i) are one of the first cell types to be infected after viruses cross the BBB; (ii) are almost as numerous as neurons [41] and have a surface area-to-volume ratio that is fourfold higher than that of neurons [42]; (iii) express several viral receptors (as discussed in detail later); (iv) are highly metabolically active and are closely coupled metabolically with neurons [35,43–45]; and (v) can produce and release high amounts of viral progeny [38,40,46]. The relatively high level of virus production in astrocytes likely relies on the properties of astroglial aerobic glycolysis [47]. Here, the end product of aerobic glycolysis is L-lactate, despite adequate oxygen levels, a phenomenon typically present in cancer cells known as "the Warburg effect" [48].

As discussed, astrocytes are one of the first cell types that a virus encounters when crossing the BBB. The basic morphological backbone of astrocytes is similar to that of neurons. Yet, detailed morphology reveals cell expansions that may branch to thousands of smaller processes [42], enabling astrocytes to have a much larger membrane interface with the extracellular space. Extracellular space, estimated to occupy ~20% of brain tissue [49], during infections contains complete viral particles and extracellular vesicles containing infective viral genomes and quasi-enveloped viruses [45]. In addition, astrocytes have been shown to express numerous viral entry receptors important for the preferential targeting certain viruses to particular cell types.

4. Viral Attachment to Entry Receptors in Astrocytes

Viral entry into astrocytes, like other cell types, is mediated by the attachment of viral surface proteins to receptors at the plasma membrane. These receptors are involved in various cellular physiological processes; however, viruses exploit them to enter their host cell. Potential viral entry receptors in astrocytes are composed of diverse proteins (Figure 2); a given virus may use several different receptors, as evident from Table 1. Concomitantly, individual viruses can exploit several receptors for entry. Most potential entry receptors summarized in Figure 2 have already been confirmed to mediate astrocyte entry (Table 1).

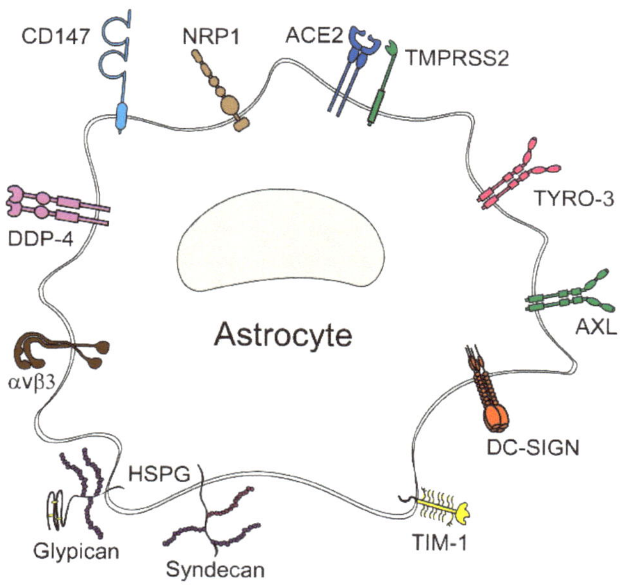

Figure 2. Viral entry receptors in astrocytes. Diverse plasma membrane receptors can mediate virus attachment to and entry into astrocytes. These include neuropilin-1 (NRP1), angiotensin-converting enzyme 2 (ACE2), type II transmembrane serine protease (TMPRSS2), TAM family receptors (TYRO-3 and AXL), dendritic cell (DC)-specific intercellular adhesion molecule-3-grabbing nonintegrin (DC-SIGN), T-cell Ig and mucin domain 1 (TIM-1), heparan sulphate proteoglycans (HSPGs) comprising syndecans and glypicans, αvβ3 integrin, dipeptidylpeptidase 4 (DPP4), and cluster of differentiation 147 (CD147).

Table 1. Confirmed entry receptors of neurotropic viruses in astrocytes.

Entry Receptor	Virus	References
HSPGs, $\alpha v\beta 3$	HSV-1	[50]
ACE2	SARS-CoV-2	[36]
NRP1	SARS-CoV-2	[51]
TMPRSS2	SARS-CoV-2	[52]
AXL	ZIKV	[53]
$\alpha 2\beta 1$	EV1	[54]
DC-SIGN	HIV-1	[55]

HSPGs, heparan sulphate proteoglycans; ACE2, angiotensin-converting enzyme 2; NRP1, neuropilin-1; TMPRSS2, type II transmembrane serine protease; AXL, a cell surface receptor tyrosine kinase; $\alpha v\beta 3$, $\alpha 2\beta 1$, class-I transmembrane proteins integrins; DC-SIGN, dendritic cell-specific intercellular adhesion molecule-3-grabbing nonintegrin; HSV-1, herpes simplex virus type-1; SARS-CoV-2, severe acute respiratory syndrome coronavirus 2; ZIKV, Zika virus; EV1, echovirus-1; HIV-1, human immunodeficiency virus type 1.

4.1. Confirmed Viral Entry Receptors in Astrocytes

4.1.1. Heparan Sulphate Proteoglycans

Heparan sulphate proteoglycans (HSPGs) are a prominent group of receptors and have been confirmed to mediate viral entry into astrocytes. These receptors comprise a core protein covalently linked to glycosaminoglycan chains formed by unbranched sulphated anionic polysaccharide heparan sulphates [56]. HSPGs comprise syndecans (the core protein is composed of an extracellular domain, a single transmembrane domain, and a short cytoplasmic domain that interacts with the cell cytoskeleton) and glypicans (the core protein is anchored in the cell membrane via glycosylphosphatidylinositol [GPI]) [56,57]. HSPGs are present at the plasma membrane of almost all eukaryotic cells and are involved in many cellular processes, e.g., they mediate interactions with extracellular factors regulating cellular adhesion, play a role in endocytosis, lysosomal degradation, transcellular transport, regulation of cell motility and growth, and affect intercellular signalling and development [56–60]. On the one hand, stress can lead to proteolytic cleavage and shedding of HSPGs.

Conversely, conditions mimicking injury response promote higher sulphation and upregulation of their expression [57,61]. The latter has been shown after viral infection [56]. In astrocytes, the involvement of HSPGs in the entry of viruses has been corroborated extensively. In the case of herpes simplex virus type-1 (HSV-1) infection of astrocytes, the interaction between HSV-1 and HSPGs leads to Ca^{2+}-dependent release of ATP. Released ATP activates purinergic P2 receptors (P2YR) at the plasma membrane of astrocytes and neighbouring neurons alike, evoking Ca^{2+}-dependent activation of glycogen synthase kinase (GSK)-3 [50]. GSK-3 is a versatile enzyme involved in many functions, including glycogen metabolism, insulin signalling, inflammatory response, and innate immunity [62]. GSK-3 was shown to be essential for phosphorylation of the severe acute respiratory syndrome coronavirus (SARS-CoV) SARS-CoV-2 nucleocapsid protein, and GSK-3 inhibition blocked SARS-CoV-2 infection in human lung epithelial cells [63]. A similar GSK-3-dependent molecular pathway underlying HSV-1 infection is likely shared by neurons and astrocytes, given that the blockade of GSK-3 activation inhibited infection of both cell types [50]. Of note, in human neuronal cells, HSV-1 upregulates GSK-3, which hyperphosphorylates tau protein [64]. In addition to being implicated in the accumulation of amyloid beta peptide (Aβ), dysregulation of calcium homeostasis, and impaired autophagy, this is one of the mechanisms by which HSV-1 could act as a causative agent in the pathogenesis of sporadic Alzheimer's disease.

The finding that HSPGs are expressed more abundantly in astrocytes in comparison with neurons [50] is consistent with a previous proposal that astrocytes are among the first cells to get infected by viruses that enter the CNS and that astrocytes are a major producer of certain viruses, as was confirmed for Zika virus (ZIKV) [38]. An increased presence of HSPGs at the plasma membrane of HSV-1-infected astrocytes [50] is in line with the prediction that HSPGs are also involved in attachment and entry of tick-borne encephalitis virus (TBEV) into astrocytes; increased time-dependent entry of pre-labelled TBEV and ZIKV was demonstrated, which suggests a positive feedback loop for viral entry [38,40].

Although HSPGs have an important role as a primary attachment molecule that concentrates viral particles on the cell surface and may facilitate subsequent binding to more specific receptor molecules, HSPGs may also mediate viral entry directly [60]. Nevertheless, HSPGs have been confirmed to serve as direct entry receptors only in rare cases, as described for HSV-1 [65].

4.1.2. Angiotensin Converting Enzyme 2

In recent years, angiotensin converting enzyme 2 (ACE2) has gained significant attention because it enables the binding and entry of severe acute respiratory syndrome coronaviruses SARS-CoV and SARS-CoV-2. ACE2 is a type I transmembrane protein composed of an extracellular heavily N-glycosylated N-terminal domain containing the carboxypeptidase site and a short intracellular C-terminal cytoplasmic tail (Figure 2) [66]. ACE2 is an ectoenzyme located on the surface of endothelial and other cells. Although its primary binding partner and substrate appears to be the hormone angiotensin II, ACE2 can hydrolyse other physiological substrates, including inactivation of bradykinin receptor as well as modulate various physiological processes, such as regulation of renal amino acid transport, pancreatic insulin secretion, control of the transport of intestinal neutral amino acid transporters of the B0AT1 family, mediation of intestinal inflammation and diarrhoea. In the CNS, it mediates cleavage of Aβ, and acts as the receptor for SARS viruses [67,68]. In the brain, ACE2 is expressed predominantly in the cerebral microvasculature, in the medulla oblongata, the hypothalamus, the subventricular zones, and the meninges around the medulla oblongata and hypothalamus [36]. At the cellular level, ACE2 is expressed primarily in astrocytes, rendering them highly susceptible to SARS-CoV-2 infection [36,69]. The S protein of SARS-CoV-2, essentially a crown-like (corona) trimer, facilitates the binding of the virus to ACE2 as well as membrane fusion [70]. Formation of a fully assembled infectious SARS-CoV-2 virion in its prior host or the extracellular space requires proteolytic cleavage of the S protein by furin or furin-like proprotein convertase into the S1 and S2 subunits. This cleavage occurs during biosynthesis and maturation in the Golgi apparatus of infected cells. Following cleavage, the S1 and S2 subunits of SARS-CoV-2 remain non-covalently associated. During the infection of a target cell, the receptor binding domain within the N-terminal end of the S1 subunit binds ACE2, and the S2 subunit anchors the S protein to the membrane and mediates membrane fusion [70–72]. Specifically, binding a SARS-CoV-2 virion to ACE2 exposes a site of the S2 subunit termed the S2′ site. This site of the S2 subunit is then further cleaved (acid-dependent proteolytic cleavage) by type II transmembrane serine protease (TMPRSS2) at the cell surface [73] or by cathepsin L in the endolysosomal compartments [74]. As a result, a fusion peptide is exposed and inserted into the membrane. In this process, an antiparallel six-helix bundle is formed, which allows the mixing of viral and cellular membranes, resulting in the fusion and consequent release of the viral genome into the cytoplasm [52,75,76]. Even though ACE2 expression has been confirmed in astrocytes on several occasions and is believed to be the prime receptor for SARS-CoV-2 [36,77,78], other proteins mediate SARS-CoV-2 entry.

4.1.3. Neuropilin-1

Neuropilins (NRPs) are cell surface transmembrane glycoprotein co-receptors for class 3 semaphorins, polypeptides with key roles in axonal guidance, and members of the vascular endothelial growth factor family of angiogenic cytokines [79]. The primary physiological role of NRP-1 is essential for neuronal and cardiovascular development; in the case of disease, it initiates signal transduction pathways involved in cell proliferation, migration, survival, and cancer invasiveness (of various cancer cells, including melanoma, breast cancer and glioblastoma [80,81]).

Moreover, NRP-1 was recently identified as a receptor for SARS-CoV-2 [82], with confirmed high expression levels in astrocytes and neurons [51]. The current understanding of NRP-1 mechanisms in viral infections (especially with SARS-CoV-2) includes (1) recognition of S protein, (2) facilitation of the membrane fusion, (3) possible modulation of endocytosis, and (4) modulation of immune responses. NRP-1 has been identified as a co-receptor for certain enveloped viruses, including the SARS-CoV-2. Namely, the S protein of the virus interacts with NRP-1 in addition to the primary receptor ACE2 [51]. NRP-1 not only promotes viral entry through S protein binding but was also suggested to aid in the fusion of the viral membrane with the host cell membrane, possibly even in the absence of ACE2 [83,84]. Following viral entry, it is likely that NRP-1 plays a role in directing the viral cargo to specific cellular compartments [85]. NRP-1 is also implicated in the immune response to viral infection, as NRP-1 is expressed on regulatory T-lymphocytes, and the therapeutic manipulation of these cells depends on the surface receptors, including NRP-1 [86,87].

Like ACE2, NRP1 is known to bind furin-cleaved substrates and has recently been demonstrated to function as one of the primary viral entry receptors in astrocytes in the case of SARS-CoV-2 [51]. While no studies addressed detailed mechanisms of NRP-1 mediated SARS-CoV-2 entry into astrocytes, binding of S protein to NRP-1 is enabled by aforementioned furin cleavage [51]. Namely, cleavage of S generates a polybasic Arg-Arg-Ala-Arg carboxyl-terminal sequence on S1, which conforms to a C-end rule motif that binds to cell surface NRP1 and enhances SARS-CoV-2 infection, as demonstrated in Caco-2 cells, a human colon adenocarcinoma cell line endogenously expressing ACE2 [88]. As Kong et al. [51] pointed out, measuring the absorption rate and dissociation constants for the interaction of different spike proteins with ACE2 versus NRP1 will better define the molecular basis of apparent differences in receptor utilization. Further research of NRP-1-induced modulations of specific signal transduction pathways in viral infection is warranted, as these pathways (e.g., the Akt and the MAPK/ERK signalling pathways) are associated with cell proliferation, survival, and migration [80,81].

4.1.4. TIM and TAM Phosphatidylserine Receptors

Various RNA and DNA-enveloped viruses exploit phosphatidylserine (PtdSer) receptors to enhance their entry into host cells. In the case of enveloped viruses, the nucleic acid and the capsid are enclosed within the envelope, derived from virus-encoded membrane-associated proteins and partly from modified host cell membranes acquired by the virus during its egress from the host cell [89]. The plasma membrane and a wide variety of intracellular membranes can serve as platforms for virus budding and egress, e.g., the nuclear envelope, rough and smooth endoplasmic reticulum (ER), endosomes, and Golgi compartment [89]. Hence, PtdSer in these membranes ends up in the viral envelope, enhancing virus binding and entry into host cells via PtdSer-mediated virus entry-enhancing receptors (PVEERs) [90,91]. PVEERs are broadly expressed across various tissues, including the CNS, and comprise TIM family proteins, TAM family receptors (AXL, TYRO3, Mertk), and MFG-E8/integrin $\alpha v\beta 3$ or $\alpha v\beta 5$ complexes [90].

All these receptors have been identified in astrocytes [30,92], but not all have been linked to viral entry. Among these, the receptor tyrosine kinase AXL and its ligands, growth-arrest-specific six and protein S, are implicated in multiple cellular responses, including cell survival, proliferation, migration, and adhesion [93,94]. In astrocytes, AXL has been identified as the primary entry receptor for ZIKV [92]. Compared with neural progenitor cells and neurons, where AXL expression is largely absent during developmental stages in humans, AXL becomes abundantly expressed in astrocytes and radial glia in the mid-gestation period, and it may play an important role in the replication and spread of ZIKV in the fetal CNS [95–97]. Astrocytes internalize significantly higher amounts of ZIKV, replicate more ZIKV and tolerate higher viral loads than neurons [38]. AXL has been proposed to promote ZIKV infection in astrocytes by antagonizing the type I interferon signalling [53]. In astrocytes, β-catenin, the central mediator of canonical Wnt signalling, negatively regulates the transcription of AXL but not TYRO3, thus limiting ZIKV internalization [92]. However, ZIKV bypasses this line of defence by downregulating β-catenin to facilitate AXL expression [98]. In comparison with AXL, other PtdSer receptors in astrocytes, such as TYRO3, TIM-1 and dendritic cell (DC)-specific intercellular adhesion molecule-3-grabbing nonintegrin (DC-SIGN) (Figure 2), have limited potential in mediating ZIKV entry due to their minuscule expression [92].

4.1.5. Integrins

Integrins, which are cell surface extracellular matrix receptors, can also be exploited by viruses and bacteria to invade cells [54]. Astrocytes express several integrins, e.g., αvβ3 [99], αvβ5 and αvβ8 [100], which may participate in mediating viral entry in these cells. Viruses from the Picornaviridae and Herpesviridae families, some of which were shown to infect astrocytes, interact with several members of the integrin family, including αvβ3, αvβ6, α2β1, α5β1 in the case of picornaviruses and β1, αvβ3, α3β1, α2β1 in the case of herpesviruses [54]. These interactions may affect downstream signalling events and activate an innate immune response; e.g., an interaction between αvβ3 with toll-like receptor 2 has been demonstrated in a 293T human kidney cell line infected with HSV-1 [101]. Integrin receptors are also exploited by viruses that infect the respiratory epithelium because they facilitate virus entry into cells and modify the induction of intracellular signalling, which has a profound effect on disease pathogenesis [54]. The participation of integrins as potential viral entry receptors in astrocytes has not yet been addressed.

4.1.6. Cluster of Differentiation 147 and Dipeptidylpeptidase 4

Transmembrane glycoproteins from the immunoglobulin superfamily cluster of differentiation 147 (CD147) and dipeptidylpeptidase 4 (DPP4, also known as CD26) are expressed in the human cortex and have been demonstrated as MERS-CoV and SARS-CoV-2 receptors [102–104]. The spike RBD of coronaviruses binds with high affinity to CD147 and DPP4 [104,105]. Expression of DPP4 and CD147 is known to increase in response to inflammation in reactive astrocytes. Both proteins appear to be upregulated in infected astrocytes [102,106,107], and as such, they might significantly contribute to viral attachment and entry into astrocytes. Therefore, the upregulation of DPP4 and CD147 may serve as a positive feedback mechanism during viral infection.

5. Viral Entry and Remodelling of Intracellular Organelle Traffic in Infected Astrocytes

Viral infection of astrocytes affects several cellular processes, altering their morphologic and physiologic properties, including rearrangement and reorganization of the cytoskeleton, modulation of vesicular traffic with endocytosis and cellular organelles, affecting the transcription of genes with an altered expression profile of cytokines, affecting autophagy and resulting in damage to cellular DNA [40]. Numerous neurotropic viruses may infect astrocytes and indirectly alter the cellular functions of neighbouring neurons and the BBB-forming endothelial cells ([108] and references within).

The viral replication cycle is comprised of four major steps: (1) attachment and entry into a target cell, (2) replication of the viral genome, (3) synthesis of viral proteins and genome packaging into infectious progeny, and (4) egress to the extracellular space and dissemination to the next target cell. All of these steps are affected by the dynamics of the transport and remodelling of intracellular membrane-bound compartments.

In general, viruses hijack the cytoskeletal macromolecular transport networks of the host cell to facilitate their movement across the cytoplasm [109]. Here, another positive feedback mechanism is present, in which, following virus entry, the remodelling of actin filaments and microtubules secondarily affects endocytosis and the ensuing viral invasion into cells. Several neurotropic viruses prominently exploit endocytosis in astrocytes as their entry mode [38,40]. In the case of certain flaviviruses (e.g., TBEV, ZIKV), an increase in the number of intracellular viral particles in a time-dependent manner is accompanied by the increase in speed and directionality of the endocytosed virus-laden vesicles [38,40].

In the case of endocytotic entry of ZIKV, binding to the AXL/Gas complex represents an important, although not exclusive, pathway of internalization [95]. Increased production of new infective virions in cells with increased internalization of endocytosed ZIKV and TBEV [38,40] implies that viruses mainly escape the degradation pathway, as the viral RNA is released from endocytic vesicles into the cytosol before the fusion with lysosomes. In contrast, after entry into astrocytes via the endocytotic pathway, HIV-1 gets predominantly degraded in lysosomes, and only a few virions are released to spread the infection [110].

In the case of infection of astrocytes with SARS-CoV-2, the two-pore cation channel 2 (TPC2) protein also plays a prominent role in the spread of the virus [51]. TPC2 is a member of the voltage-gated ion channel superfamily encoded by the *TPCN2* gene, which regulates the physiological functions of the endolysosomes [111]. TCP2 is localized at the endolysosomal membrane and controls the ionic homeostasis of these acidic organelles [112]. Physiologically, TPCs regulate several processes, including coupling endolysosomal function with the metabolic state of a cell and with autophagy [112]. During infection, TPC2 appears responsible for controlling the movement of viral particles through the endolysosomal system, thus relocating the virus from the endolysosomal network into the cell cytoplasm [51,113,114]. Namely, TPC2 mediates Ca^{2+} release from endosomes/lysosomes activated by nicotinic acid adenine dinucleotide phosphate. On the one hand, released Ca^{2+} regulates membrane fusion events through Ca^{2+} sensors (i.e., synaptotagmins) involving the soluble N-ethylmaleimide-sensitive fusion protein attachment protein receptors. On the other hand, it controls the direction and specificity of trafficking processes via Rab effectors and (Ca^{2+} sensors) synaptotagmins [115,116].

The pattern recognition receptor DC-SIGN also participates in viral endocytosis. DC-SIGN recognizes carbohydrate structures on viral glycoproteins and functions as an attachment factor for several enveloped viruses [117]. To date, only HIV-1 has been shown to be endocytosed in astrocytes via the DC-SIGN receptor, predominantly resulting in virus degradation [55]. In contrast, in dendritic cells, upon internalization and processing of HIV-1 via DC-SIGN, viral exogenous antigens are cross-presented on MHC class I, which induces anti-HIV cytotoxic T-cell responses. It is unclear how various viruses are engaged in DC-SIGN-mediated endocytosis in astrocytes.

6. Autophagy in Viral Infection of Astrocytes

Once in the cytoplasm, viruses commence their replication cycle by interacting with cellular membrane compartments to induce cytoplasmic membrane structures, in which genome replication and assembly occur [118]. Plus-stranded RNA viruses and large DNA viruses, in particular, cause large membrane structures originating from the ER or endosomes that support the replication of their genomes [118]. The ER is also associated with early autophagic structures [119,120] that have been proposed to promote viral replication in certain cases. The physiological significance of autophagy in astrocytes concerning the life cycle and transmission of viruses remains poorly investigated. Several flaviviruses (e.g., dengue virus (DENV), West Nile virus (WNV), chikungunya virus CHIKV and ZIKV) were assessed for their possible involvement in autophagy in various cell types, although rarely in astrocytes and neurons. Experimental evidence suggests that autophagy may act anti-viral or pro-viral, depending on the strain of the virus and the type of infected cell [121]. Judging by the limited number of studies conducted on astrocytes, autophagy may also have a dual role in this cell type.

Regarding the anti-viral role of autophagy, virus-induced autophagy helps host cells counteract arbovirus infection and replication by degrading the virus and enhancing anti-viral immunity by augmenting antigen presentation [121,122]. In virus-infected host cells, autophagy also initiates an innate immune response and production of antiviral cytokines. Viral pathogen-associated molecular patterns can be detected by host cell pattern recognition receptors, which initiate anti-viral responses by recruiting signalling molecules that promote the generation of type I interferon (IFN-I) and other pro-inflammatory cytokines [123]. IFN-I is a crucial anti-viral factor in astrocytes; it restricts viral growth via autocrine and paracrine signalling through the interferon-α/β receptors, as demonstrated after infection of astrocytes with TBEV, JEV, WNV, and ZIKV [124]. When autophagy was monitored in human astrocytes, induction of autophagy was detected after infection with TBEV and WNV (Figure 3) [46]. However, inhibition of the formation of autophagic structures does not affect viral replication or the production of infective viruses [46]. Thus in these cells, remodelling of the ER with the unfolded protein response pathway, not autophagic structures, appears to be a decisive factor for flavivirus replication [46]. Viruses have evolved to exploit evolutionary conserved stress responses, such as the ER unfolded protein response (UPR^{ER}), to manipulate cell stress and metabolic pathways, thereby enhancing infection and progeny formation or inducing cell death [125]. Several viruses have been shown to induce the UPR^{ER} by causing ER overload or by eliciting specific signals originating from the ER, as recently reviewed in [125].

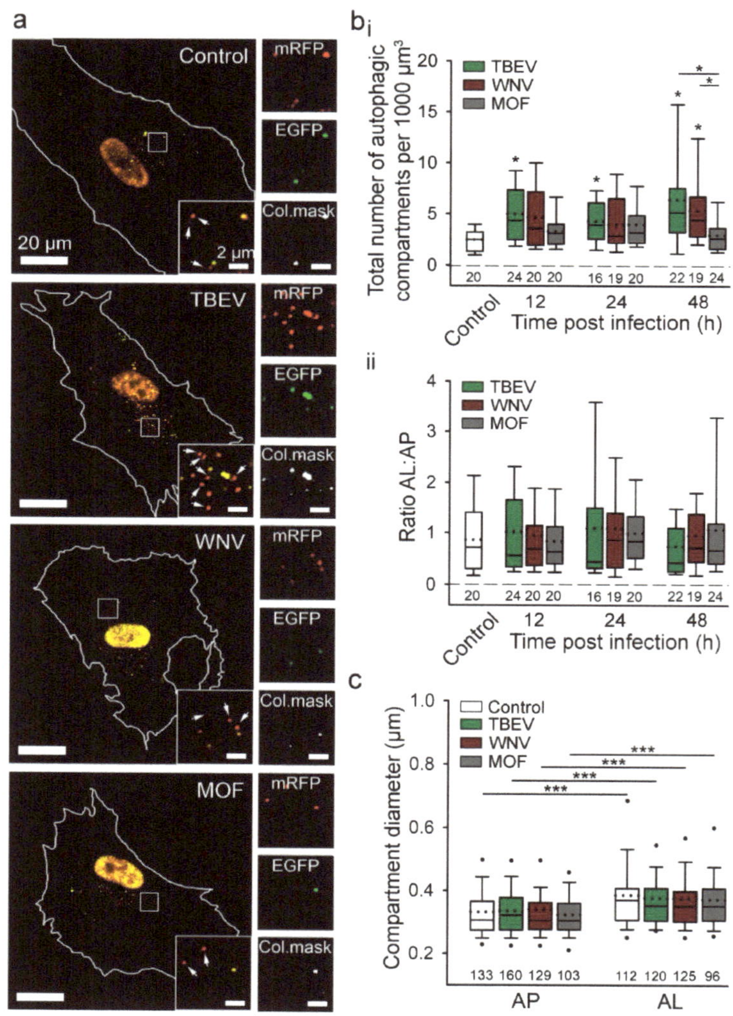

Figure 3. Tick-borne encephalitis virus (TBEV) and West Nile virus (WNV), but not mosquito-only flavivirus (MOF), increase autophagy in human astrocytes. (**a**) Representative fluorescence micrographs

documenting mock-infected human astrocytes (Control) and astrocytes after exposure to selected flaviviruses (TBEV, WNV, and MOF) for 48 h. Cells were expressing mRFP-EGFP-LC3, where LC3, a marker of autophagic compartments, is tandem fluorescent-tagged with mRFP (red fluorescence) and EGFP (green fluorescence). The pH of autophagosomes (AP) is close to neutral, which facilitates the fluorescence of both fluorophores, resulting in yellow objects. Fusion of AP with lysosomes yields autolysosomes (AL), i.e., organelles with acidic pH, where the EGFP fluorescence is quenched, resulting in red-only objects. Selected rectangular areas within the cells, enlarged at the corners (bottom, right), show superimposed images of mRFP and EGFP fluorescence. Arrows indicate AL. Adjacent, smaller panels display mRFP and EGFP fluorescence and co-localization masks (Col.mask) (co-localized objects correspond to AP) of the enlarged images. The white outlines in the large panels show the cell shape. (**b**) The total number of autophagic compartments (**i**) and the ratio of AL to AP (which is a measure of autophagic degradation activity) (**ii**) in mock-infected cells and after infection with TBEV, WNV, and MOF at 12, 24, and 48 h post-infection (hpi). Infection with TBEV increases the total number of autophagic compartments at all three time points tested, compared with mock-infected cells (i.e., controls at 48 hpi, which were confirmed to be comparable to controls at 12 and 24 hpi) (* $p < 0.05$, one-way ANOVA followed by Dunn's test), but does not affect the AL:AP ratio ($p > 0.05$, one-way ANOVA). WNV infection induces an increase in the total number of autophagic compartments at 48 hpi (* $p < 0.05$, one-way ANOVA followed by Dunn's test) and does not affect the ratio AL:AP ($p > 0.05$, one-way ANOVA). MOF infection does not affect the number of autophagic compartments or the AL: AP ratio compared with mock-infected cells at any time point tested ($p > 0.05$, one-way ANOVA). (**c**) Diameter of AP and AL in mock-infected cells and 48 hpi with selected flaviviruses. Infection with TBEV, WNV and MOF does not affect the size of the autophagic compartment compared with mock-infected cells ($p > 0.05$, one-way ANOVA). ALs are larger than APs in all experimental conditions (*** $p < 0.001$, Mann-Whitney U test). Full lines in the boxplots represent median values, and the dotted lines correspond to average values. The numbers below the boxplots are the number of cells (**b**) or compartments (**c**) analysed for each condition. Cells were infected with TBEV and MOF at an MOI of 0.1 and with WNV at an MOI of 1. Figure and figure legend reproduced from Tavcar Verdev et al. [46] with permission, licensed under a Creative Commons Attribution 4.0 International License (http://creativecommons.org/licenses/by/4.0 (accessed on 1 August 2023)).

In contrast to the induction of autophagy in astrocytes by TBEV and WNV, HIV-1 infection attenuates autophagy in astrocytes via the Nef protein, which blocks the last step of autophagy and thereby helps HIV-1 virus avoid autophagic degradation [126,127]. Modulation of autophagy in astrocytes by some viruses (e.g., TBEV and WNV), but not all (MOF), is evident from Figure 3.

6.1. Autophagy of Cellular Organelles in Viral Infection

In most cases, viral infection induces severe damage to various intracellular organelles, thus initiating selective autophagy to degrade dysfunctional organelles to maintain cell homeostasis. Depending on the organelle targeted for the elimination (e.g., mitochondria, peroxisomes, ER, lysosomes, lipid droplets), selective autophagy comprises mitophagy, pexophagy, reticulophagy, lysophagy, and lipophagy [128]. Selective autophagy can be utilized by a virus to promote its replication or help the host cell combat infection. Yet, the role of autophagy in the formation of virus-modified membranes that are involved in replication remains debated [118,128]. For example, mitophagy, which controls mitochondrial quality by inducing clearance of damaged mitochondria via the autophagic machinery, can be promoted by diverse viruses to favour their replication in host cells [129]. Various strategies have been revealed for how viruses exploit mitophagy to facilitate replication. For instance, degradation of autophagy-mediated mitochondrial antiviral-signalling protein (MAVS) has a pro-viral effect because MAVS-dependent activation of nuclear factor-kappa B (NF-κB) and IFN-regulatory factor (IRF3) is hampered and consequently type I IFN anti-viral action is impaired [108,129]. Also, mitophagy was proposed to, on the one hand, promote cell survival by inhibiting oxidative stress and apoptosis and, on the other hand, facilitate persistent viral infection [130,131]. So far, only one study has addressed the role of mitophagy in viral infection of astrocytes. During HIV-1 productive infection of astrocytes, mitophagy was shown to be crucial for cell death resistance and avoidance of inflammasome activation [132]. This study suggested that the inhibition of mitophagy in astrocytes may represent a potential therapeutic strategy aimed at reducing or eliminating HIV-1 reservoirs in the brain [132].

Although canonical macroautophagy is supposed to act pro-virally by viruses high jacking autophagic structures as viral replication sites [133], it also serves as a potent anti-viral pathway [120]. When a virus exploits the ER for its replication, newly synthesized viral proteins aggregate in the ER. Excessive protein processing in the ER inevitably culminates in reticulophagy, inhibiting viral protein synthesis and thus attenuating viral replication. As demonstrated in human brain microvascular endothelial cells, reticulophagy has been reported to assist resistance to infection and limit the replication of Ebola, dengue virus and ZIKV [134]. Yet, to promote their replication, viruses evolved to evade and subvert reticulophagy by inhibiting the ER-resided FAM134B receptor. This receptor binds to membrane-bound LC3 and promotes the docking of ER-derived vesicles to autophagosomes. Its inhibition thus impairs the reticulophagy-dependent clearance of viruses [120,135]. Relationships between reticulophagy and viral infections are poorly explored and thus far completely unexplored in astrocytes. Given that the ER is one of the major organelles that governs the replication and assembly of certain viruses (e.g., polyomaviruses) by penetrating the ER membrane to gain access to the cytosol and then the nucleus [136] and that astrocytes are major viral producers in the CNS, modifications and dynamics of ER inflicted by viral infection in astrocytes deserve attention in the future.

6.2. Biogenesis of Lipid Droplets and Lipophagy in Viral Infection

Lipid droplets (LDs) are dynamic intracellular lipid storage organelles that respond to the physiological state of cells. They participate in the regulation of cell metabolism and play a protective role in many cellular stressors, including oxidative stress [137,138]. In a pathological state of the CNS, the content of LDs increases, particularly in astrocytes and microglia, where LDs are speculated to support energy provision and to act neuroprotectively [137,139]. LDs could have both beneficial and detrimental effects in the context of neuroprotection, yet the exact mechanisms and overall impact are not yet fully understood. A potential neuroprotective action of glial LDs was proposed as a failure to form them resulted in more severe neurodegeneration [140,141]. For example, during hypoxia, the overproduction of saturated fatty acids generates ceramides and acylcarnitines that lead to lipid-mediated toxicity and contribute to cellular dysfunction and neuronal death [137]. The formation of LDs can buffer saturated fatty acid build-up in the cell, thereby reducing toxicity [137]. Understanding the roles of LDs biogenesis and lipophagy in viral infection is also elusive. While some studies report that LD build-up in virus-infected cells may reduce viral toxicity and that LD build-up can enhance viral replication, others report that LD build-up can reduce viral replication by activating antiviral innate pathways. For example, a study in baby hamster kidney fibroblasts (BHK cells) demonstrated that infection with DENV increases the number of LDs, some of which are enriched with dengue virus (DENV) viral capsid protein [142]. Thus, LDs are speculated to provide a platform for the formation of viral nucleocapsid and enhance viral replication, a speculation supported by the finding that pharmacological inhibition of LD formation reduced DENV replication [142].

In contrast to assisting viral replication, LDs were also suggested to play a pivotal role in protecting the host against viral infection [143]. A study by Monson et al. [144] in astrocytes showed that enhanced biogenesis and accumulation of LDs in HSV-1 and ZIKV-infected primary human foetal immortalized astrocytes corresponds with enhanced cellular type-I and -III IFN production, thereby decreasing viral replication. Contrary to LD biogenesis, lipophagy, the autophagic degradation of lipid droplets and an essential mechanism for maintaining homeostasis of LDs, can also be directly activated by viruses [145]. Lipophagy involves protein machinery common to macroautophagy, such as microtubule-associated proteins 1A/1B light chain 3B (LC3) [137,142]. As demonstrated in a human hepatoma cell line (HepG2 cells), DENV infection induces lipophagy via activation of 5′-AMP-activated protein kinase catalytic subunit α2 (AMPK), which in turn inhibits the mechanistic target of rapamycin 1 (mTORC1) [122]. These steps result in a depletion of stored triglycerides and an increase in β-oxidation and energy production [146,147]. An increase in ATP boosts the synthesis of the biomolecules needed for viral replication. ATP is required for the receptor-mediated attachment of viral particles and their entry into the cytoplasm [148]. Thus, the role of LDs in the viral cycle of astrocytes remains to be fully elucidated.

7. Conclusions

Newly emerging neurotropic viruses and the expanding geographic area of existing viruses represent a huge health burden and a major worldwide concern. It is becoming increasingly evident that to address the severe neurological conditions that can be elicited by neurotropic viruses, further research that significantly advances the boundaries of current knowledge is needed. With this in mind, attention is turning more and more to astrocytes. These glial cells are important as a viral reservoir, producers of new infective virions, and modulators of the immune response in the CNS. Despite the immense importance of astrocytes in neurotropic viral infections, only a handful of studies have systematically addressed the impact and exploitation of cellular processes by viruses in this cell type. Future research should focus on the possible targeting of entry receptors and specific cellular processes to limit virus-triggered malfunctions of astrocytes and the CNS. Of note, understanding the mechanisms of virus entry and replication in astrocytes may lead to therapeutic strategies that would reduce their capacity as a CNS viral reservoir and help prevent virus-associated astrocyte demise. This would be a major step forward, as currently, no effective antiviral drugs are available for many viral infections, nor are we in possession of astrocyte-specific therapies or drugs.

Author Contributions: Conceptualization: J.J., M.P. and R.Z.; Writing—original draft: J.J., M.P. and R.Z.; Writing—review & editing, J.J., M.P. and R.Z. All authors have read and agreed to the published version of the manuscript.

Funding: This work was supported by the Slovenian Research Agency grant no. P3 310 to R.Z. and the A4L_ACTIONS project funded by the European Union's Horizon 2020 research and innovation programme under grant agreement no. 964997 to J.J.

Institutional Review Board Statement: Not applicable.

Informed Consent Statement: Not applicable.

Data Availability Statement: No new data were created or analyzed in this study. Data sharing is not applicable to this article.

Conflicts of Interest: The funding sponsors had no role in the design of the study; in the collection, analyses, or interpretation of data; in the writing of the manuscript, and in the decision to publish the results.

References

1. Romero, J.R.; Newland, J.G. Viral meningitis and encephalitis: Traditional and emerging viral agents. *Semin. Pediatr. Infect. Dis.* **2003**, *14*, 72–82. [CrossRef]
2. Gora, H.; Smith, S.; Wilson, I.; Preston-Thomas, A.; Ramsamy, N.; Hanson, J. The epidemiology and outcomes of central nervous system infections in Far North Queensland, tropical Australia; 2000–2019. *PLoS ONE* **2022**, *17*, e0265410. [CrossRef]
3. Ludlow, M.; Kortekaas, J.; Herden, C.; Hoffmann, B.; Tappe, D.; Trebst, C.; Griffin, D.E.; Brindle, H.E.; Solomon, T.; Brown, A.S.; et al. Neurotropic virus infections as the cause of immediate and delayed neuropathology. *Acta Neuropathol.* **2016**, *131*, 159–184. [CrossRef]
4. Carroll, D.; Daszak, P.; Wolfe, N.D.; Gao, G.F.; Morel, C.M.; Morzaria, S.; Pablos-Mendez, A.; Tomori, O.; Mazet, J.A.K. The Global Virome Project. *Science* **2018**, *359*, 872–874. [CrossRef]
5. Swanson, P.A., II; McGavern, D.B. Viral diseases of the central nervous system. *Curr. Opin. Virol.* **2015**, *11*, 44–54. [CrossRef]
6. Valerio, F.; Whitehouse, D.P.; Menon, D.K.; Newcombe, V.F.J. The neurological sequelae of pandemics and epidemics. *J. Neurol.* **2021**, *268*, 2629–2655. [CrossRef]

7. Burrell, C.J.; Howard, C.R.; Murphy, F.A. Pathogenesis of Virus Infections. In *Fenner and White's Medical Virology*; Academic Press Cambridge, MA, USA, 2017; pp. 77–104. [CrossRef]
8. Wu, S.J.; Grouard-Vogel, G.; Sun, W.; Mascola, J.R.; Brachtel, E.; Putvatana, R.; Louder, M.K.; Filgueira, L.; Marovich, M.A.; Wong H.K.; et al. Human skin Langerhans cells are targets of dengue virus infection. *Nat. Med.* **2000**, *6*, 816–820. [CrossRef]
9. Byrne, S.N.; Halliday, G.M.; Johnston, L.J.; King, N.J. Interleukin-1beta but not tumor necrosis factor is involved in West Nile virus-induced Langerhans cell migration from the skin in C57BL/6 mice. *J. Investig. Dermatol.* **2001**, *117*, 702–709. [CrossRef]
10. Iwasaki, A.; Foxman, E.F.; Molony, R.D. Early local immune defences in the respiratory tract. *Nat. Rev. Immunol.* **2017**, *17*, 7–20 [CrossRef]
11. Cain, M.D.; Salimi, H.; Diamond, M.S.; Klein, R.S. Mechanisms of Pathogen Invasion into the Central Nervous System. *Neuron* **2019**, *103*, 771–783. [CrossRef]
12. Mori, I.; Nishiyama, Y.; Yokochi, T.; Kimura, Y. Olfactory transmission of neurotropic viruses. *J. Neurovirol.* **2005**, *11*, 129–137 [CrossRef]
13. Ren, R.; Racaniello, V.R. Poliovirus spreads from muscle to the central nervous system by neural pathways. *J. Infect. Dis.* **1992**, *166*, 747–752. [CrossRef]
14. Fooks, A.R.; Cliquet, F.; Finke, S.; Freuling, C.; Hemachudha, T.; Mani, R.S.; Muller, T.; Nadin-Davis, S.; Picard-Meyer, E.; Wilde, H.; et al. Rabies. *Nat. Rev. Dis. Primers* **2017**, *3*, 17091. [CrossRef]
15. Nathanson, N. The pathogenesis of poliomyelitis: What we don't know. *Adv. Virus Res.* **2008**, *71*, 1–50. [CrossRef]
16. Cho, H.; Diamond, M.S. Immune responses to West Nile virus infection in the central nervous system. *Viruses* **2012**, *4*, 3812–3830 [CrossRef]
17. Saunders, N.R.; Dreifuss, J.J.; Dziegielewska, K.M.; Johansson, P.A.; Habgood, M.D.; Mollgard, K.; Bauer, H.C. The rights and wrongs of blood-brain barrier permeability studies: A walk through 100 years of history. *Front. Neurosci.* **2014**, *8*, 404. [CrossRef]
18. Brown, P.D.; Davies, S.L.; Speake, T.; Millar, I.D. Molecular mechanisms of cerebrospinal fluid production. *Neuroscience* **2004**, *129*, 957–970. [CrossRef]
19. Kadry, H.; Noorani, B.; Cucullo, L. A blood-brain barrier overview on structure, function, impairment, and biomarkers of integrity *Fluids Barriers CNS* **2020**, *17*, 69. [CrossRef]
20. Sarikcioglu, L. Lina Stern (1878–1968): An outstanding scientist of her time. *Childs Nerv. Syst.* **2017**, *33*, 1027–1029. [CrossRef]
21. Dahm, T.; Rudolph, H.; Schwerk, C.; Schroten, H.; Tenenbaum, T. Neuroinvasion and Inflammation in Viral Central Nervous System Infections. *Mediat. Inflamm.* **2016**, *2016*, 8562805. [CrossRef]
22. Nag, S.; Begley David, J. Blood Brain Barrier, Exchange of metabolites and gases. In *Pathology and Genetics: Cerebrovascular Diseases*; ISN Neuropath Press: Basel, Switzerland, 2005.
23. Zhao, Y.; Xin, Y.; He, Z.; Hu, W. Function of Connexins in the Interaction between Glial and Vascular Cells in the Central Nervous System and Related Neurological Diseases. *Neural Plast.* **2018**, *2018*, 6323901. [CrossRef]
24. Alvarez, J.I.; Katayama, T.; Prat, A. Glial influence on the blood brain barrier. *Glia* **2013**, *61*, 1939–1958. [CrossRef]
25. Nikolakopoulou, A.M.; Zhao, Z.; Montagne, A.; Zlokovic, B.V. Regional early and progressive loss of brain pericytes but not vascular smooth muscle cells in adult mice with disrupted platelet-derived growth factor receptor-beta signaling. *PLoS ONE* **2017**, *12*, e0176225. [CrossRef]
26. Kim, K.S. Mechanisms of microbial traversal of the blood-brain barrier. *Nat. Rev. Microbiol.* **2008**, *6*, 625–634. [CrossRef]
27. Coyne, C.B.; Kim, K.S.; Bergelson, J.M. Poliovirus entry into human brain microvascular cells requires receptor-induced activation of SHP-2. *EMBO J.* **2007**, *26*, 4016–4028. [CrossRef]
28. Verma, S.; Lo, Y.; Chapagain, M.; Lum, S.; Kumar, M.; Gurjav, U.; Luo, H.; Nakatsuka, A.; Nerurkar, V.R. West Nile virus infection modulates human brain microvascular endothelial cells tight junction proteins and cell adhesion molecules: Transmigration across the in vitro blood-brain barrier. *Virology* **2009**, *385*, 425–433. [CrossRef]
29. Arsenio-Nunes, M.L.; Cerutti, I.; Farkas-Bargeton, E. Vascular and neuroglial changes in experimental herpes simplex encephalitis: Ultrastructural study. *Acta Neuropathol.* **1975**, *33*, 245–256. [CrossRef]
30. Daniels, B.P.; Holman, D.W.; Cruz-Orengo, L.; Jujjavarapu, H.; Durrant, D.M.; Klein, R.S. Viral pathogen-associated molecular patterns regulate blood-brain barrier integrity via competing innate cytokine signals. *mBio* **2014**, *5*, e01476-14. [CrossRef]
31. Tabor-Godwin, J.M.; Ruller, C.M.; Bagalso, N.; An, N.; Pagarigan, R.R.; Harkins, S.; Gilbert, P.E.; Kiosses, W.B.; Gude, N.A.; Cornell, C.T.; et al. A novel population of myeloid cells responding to coxsackievirus infection assists in the dissemination of virus within the neonatal CNS. *J. Neurosci.* **2010**, *30*, 8676–8691. [CrossRef]
32. Clay, C.C.; Rodrigues, D.S.; Ho, Y.S.; Fallert, B.A.; Janatpour, K.; Reinhart, T.A.; Esser, U. Neuroinvasion of fluorescein-positive monocytes in acute simian immunodeficiency virus infection. *J. Virol.* **2007**, *81*, 12040–12048. [CrossRef]
33. Suen, W.W.; Prow, N.A.; Hall, R.A.; Bielefeldt-Ohmann, H. Mechanism of West Nile virus neuroinvasion: A critical appraisal. *Viruses* **2014**, *6*, 2796–2825. [CrossRef] [PubMed]
34. Bauer, L.; Laksono, B.M.; de Vrij, F.M.S.; Kushner, S.A.; Harschnitz, O.; van Riel, D. The neuroinvasiveness, neurotropism, and neurovirulence of SARS-CoV-2. *Trends Neurosci.* **2022**, *45*, 358–368. [CrossRef]
35. Zorec, R.; Verkhratsky, A. Astrocytes in the pathophysiology of neuroinfection. *Essays Biochem.* **2022**, *67*, 131–145. [CrossRef]
36. Zhang, Y.; Archie, S.R.; Ghanwatkar, Y.; Sharma, S.; Nozohouri, S.; Burks, E.; Mdzinarishvili, A.; Liu, Z.; Abbruscato, T.J. Potential role of astrocyte angiotensin converting enzyme 2 in the neural transmission of COVID-19 and a neuroinflammatory state induced by smoking and vaping. *Fluids Barriers CNS* **2022**, *19*, 46. [CrossRef] [PubMed]

37. Tavcar, P.; Potokar, M.; Kolenc, M.; Korva, M.; Avsic-Zupanc, T.; Zorec, R.; Jorgacevski, J. Neurotropic Viruses, Astrocytes, and COVID-19. *Front. Cell Neurosci.* **2021**, *15*, 662578. [CrossRef]
38. Jorgačevski, J.; Korva, M.; Potokar, M.; Lisjak, M.; Avšič-Županc, T.; Zorec, R. ZIKV Strains Differentially Affect Survival of Human Fetal Astrocytes versus Neurons and Traffic of ZIKV-Laden Endocytotic Compartments. *Sci. Rep.* **2019**, *9*, 8069. [CrossRef]
39. Potokar, M.; Jorgacevski, J.; Zorec, R. Astrocytes in Flavivirus Infections. *Int. J. Mol. Sci.* **2019**, *20*, 691. [CrossRef]
40. Potokar, M.; Korva, M.; Jorgacevski, J.; Avsic-Zupanc, T.; Zorec, R. Tick-borne encephalitis virus infects rat astrocytes but does not affect their viability. *PLoS ONE* **2014**, *9*, e86219. [CrossRef]
41. von Bartheld, C.S.; Bahney, J.; Herculano-Houzel, S. The search for true numbers of neurons and glial cells in the human brain: A review of 150 years of cell counting. *J. Comp. Neurol.* **2016**, *524*, 3865–3895. [CrossRef]
42. Cali, C.; Agus, M.; Kare, K.; Boges, D.J.; Lehvaslaiho, H.; Hadwiger, M.; Magistretti, P.J. 3D cellular reconstruction of cortical glia and parenchymal morphometric analysis from Serial Block-Face Electron Microscopy of juvenile rat. *Prog. Neurobiol.* **2019**, *183*, 101696. [CrossRef]
43. Prebil, M.; Vardjan, N.; Jensen, J.; Zorec, R.; Kreft, M. Dynamic monitoring of cytosolic glucose in single astrocytes. *Glia* **2011**, *59*, 903–913. [CrossRef] [PubMed]
44. Prebil, M.; Jensen, J.; Zorec, R.; Kreft, M. Astrocytes and energy metabolism. *Arch. Physiol. Biochem.* **2011**, *117*, 64–69. [CrossRef] [PubMed]
45. Martins, S.T.; Alves, L.R. Extracellular Vesicles in Viral Infections: Two Sides of the Same Coin? *Front. Cell Infect. Microbiol.* **2020**, *10*, 593170. [CrossRef] [PubMed]
46. Tavcar Verdev, P.; Potokar, M.; Korva, M.; Resman Rus, K.; Kolenc, M.; Avsic Zupanc, T.; Zorec, R.; Jorgacevski, J. In human astrocytes neurotropic flaviviruses increase autophagy, yet their replication is autophagy-independent. *Cell Mol. Life Sci.* **2022**, *79*, 566. [CrossRef] [PubMed]
47. Magistretti, P.J.; Allaman, I. Lactate in the brain: From metabolic end-product to signalling molecule. *Nat. Rev. Neurosci.* **2018**, *19*, 235–249. [CrossRef]
48. Vander Heiden, M.G.; Cantley, L.C.; Thompson, C.B. Understanding the Warburg effect: The metabolic requirements of cell proliferation. *Science* **2009**, *324*, 1029–1033. [CrossRef]
49. Nicholson, C.; Hrabetova, S. Brain Extracellular Space: The Final Frontier of Neuroscience. *Biophys. J.* **2017**, *113*, 2133–2142. [CrossRef]
50. Li Puma, D.D.; Marcocci, M.E.; Lazzarino, G.; De Chiara, G.; Tavazzi, B.; Palamara, A.T.; Piacentini, R.; Grassi, C. Ca^{2+}-dependent release of ATP from astrocytes affects herpes simplex virus type 1 infection of neurons. *Glia* **2021**, *69*, 201–215. [CrossRef]
51. Kong, W.; Montano, M.; Corley, M.J.; Helmy, E.; Kobayashi, H.; Kinisu, M.; Suryawanshi, R.; Luo, X.; Royer, L.A.; Roan, N.R.; et al. Neuropilin-1 Mediates SARS-CoV-2 Infection of Astrocytes in Brain Organoids, Inducing Inflammation Leading to Dysfunction and Death of Neurons. *mBio* **2022**, *13*, e0230822. [CrossRef]
52. Xiao, L.; Sakagami, H.; Miwa, N. ACE2: The key Molecule for Understanding the Pathophysiology of Severe and Critical Conditions of COVID-19: Demon or Angel? *Viruses* **2020**, *12*, 491. [CrossRef]
53. Chen, J.; Yang, Y.F.; Yang, Y.; Zou, P.; Chen, J.; He, Y.; Shui, S.L.; Cui, Y.R.; Bai, R.; Liang, Y.J.; et al. AXL promotes Zika virus infection in astrocytes by antagonizing type I interferon signalling. *Nat. Microbiol.* **2018**, *3*, 302–309. [CrossRef] [PubMed]
54. Stewart, P.L.; Nemerow, G.R. Cell integrins: Commonly used receptors for diverse viral pathogens. *Trends Microbiol.* **2007**, *15*, 500–507. [CrossRef]
55. Chauhan, A.; Khandkar, M. Endocytosis of human immunodeficiency virus 1 (HIV-1) in astrocytes: A fiery path to its destination. *Microb. Pathog.* **2015**, *78*, 1–6. [CrossRef] [PubMed]
56. Cagno, V.; Tseligka, E.D.; Jones, S.T.; Tapparel, C. Heparan Sulfate Proteoglycans and Viral Attachment: True Receptors or Adaptation Bias? *Viruses* **2019**, *11*, 596. [CrossRef] [PubMed]
57. Hacker, U.; Nybakken, K.; Perrimon, N. Heparan sulphate proteoglycans: The sweet side of development. *Nat. Rev. Mol. Cell Biol.* **2005**, *6*, 530–541. [CrossRef]
58. Bishop, J.R.; Schuksz, M.; Esko, J.D. Heparan sulphate proteoglycans fine-tune mammalian physiology. *Nature* **2007**, *446*, 1030–1037. [CrossRef]
59. Bernfield, M.; Gotte, M.; Park, P.W.; Reizes, O.; Fitzgerald, M.L.; Lincecum, J.; Zako, M. Functions of cell surface heparan sulfate proteoglycans. *Annu. Rev. Biochem.* **1999**, *68*, 729–777. [CrossRef]
60. Kroschewski, H.; Allison, S.L.; Heinz, F.X.; Mandl, C.W. Role of heparan sulfate for attachment and entry of tick-borne encephalitis virus. *Virology* **2003**, *308*, 92–100. [CrossRef]
61. Properzi, F.; Lin, R.; Kwok, J.; Naidu, M.; van Kuppevelt, T.H.; Ten Dam, G.B.; Camargo, L.M.; Raha-Chowdhury, R.; Furukawa, Y.; Mikami, T.; et al. Heparan sulphate proteoglycans in glia and in the normal and injured CNS: Expression of sulphotransferases and changes in sulphation. *Eur. J. Neurosci.* **2008**, *27*, 593–604. [CrossRef]
62. Alfhili, M.A.; Alsughayyir, J.; McCubrey, J.A.; Akula, S.M. GSK-3-associated signaling is crucial to virus infection of cells. *Biochim. Biophys. Acta Mol. Cell Res.* **2020**, *1867*, 118767. [CrossRef]
63. Liu, X.; Verma, A.; Garcia, G., Jr.; Ramage, H.; Lucas, A.; Myers, R.L.; Michaelson, J.J.; Coryell, W.; Kumar, A.; Charney, A.W.; et al. Targeting the coronavirus nucleocapsid protein through GSK-3 inhibition. *Proc. Natl. Acad. Sci. USA* **2021**, *118*, e2113401118. [CrossRef] [PubMed]

64. Harris, S.A.; Harris, E.A. Molecular Mechanisms for Herpes Simplex Virus Type 1 Pathogenesis in Alzheimer's Disease. *Front. Aging Neurosci.* **2018**, *10*, 48. [CrossRef] [PubMed]
65. Shukla, D.; Liu, J.; Blaiklock, P.; Shworak, N.W.; Bai, X.; Esko, J.D.; Cohen, G.H.; Eisenberg, R.J.; Rosenberg, R.D.; Spear, P.G. A novel role for 3-O-sulfated heparan sulfate in herpes simplex virus 1 entry. *Cell* **1999**, *99*, 13–22. [CrossRef] [PubMed]
66. Lambert, D.W.; Yarski, M.; Warner, F.J.; Thornhill, P.; Parkin, E.T.; Smith, A.I.; Hooper, N.M.; Turner, A.J. Tumor necrosis factor-alpha convertase (ADAM17) mediates regulated ectodomain shedding of the severe-acute respiratory syndrome-coronavirus (SARS-CoV) receptor, angiotensin-converting enzyme-2 (ACE2). *J. Biol. Chem.* **2005**, *280*, 30113–30119. [CrossRef]
67. Evans, J.P.; Liu, S.L. Role of host factors in SARS-CoV-2 entry. *J. Biol. Chem.* **2021**, *297*, 100847. [CrossRef]
68. Unger, T.; Steckelings, U.M.; Santos, R.A.S.d. *The Protective Arm of the Renin Angiotensin System: Functional Aspects and Therapeutic Implications*; Academic Press: Amsterdam, The Netherlands, 2015; 312p.
69. Crunfli, F.; Carregari, V.C.; Veras, F.P.; Silva, L.S.; Nogueira, M.H.; Antunes, A.; Vendramini, P.H.; Valenca, A.G.F.; Brandao-Teles, C.; Zuccoli, G.D.S.; et al. Morphological, cellular, and molecular basis of brain infection in COVID-19 patients. *Proc. Natl. Acad. Sci. USA* **2022**, *119*, e2200960119. [CrossRef]
70. Jackson, C.B.; Farzan, M.; Chen, B.; Choe, H. Mechanisms of SARS-CoV-2 entry into cells. *Nat. Rev. Mol. Cell Biol.* **2022**, *23*, 3–20. [CrossRef]
71. Bestle, D.; Heindl, M.R.; Limburg, H.; Van Lam van, T.; Pilgram, O.; Moulton, H.; Stein, D.A.; Hardes, K.; Eickmann, M.; Dolnik, O.; et al. TMPRSS2 and furin are both essential for proteolytic activation of SARS-CoV-2 in human airway cells. *Life Sci. Alliance* **2020**, *3*, e202000786. [CrossRef]
72. Hoffmann, M.; Kleine-Weber, H.; Schroeder, S.; Kruger, N.; Herrler, T.; Erichsen, S.; Schiergens, T.S.; Herrler, G.; Wu, N.H.; Nitsche, A.; et al. SARS-CoV-2 Cell Entry Depends on ACE2 and TMPRSS2 and Is Blocked by a Clinically Proven Protease Inhibitor. *Cell* **2020**, *181*, 271–280.e278. [CrossRef]
73. Matsuyama, S.; Nagata, N.; Shirato, K.; Kawase, M.; Takeda, M.; Taguchi, F. Efficient activation of the severe acute respiratory syndrome coronavirus spike protein by the transmembrane protease TMPRSS2. *J. Virol.* **2010**, *84*, 12658–12664. [CrossRef]
74. Simmons, G.; Gosalia, D.N.; Rennekamp, A.J.; Reeves, J.D.; Diamond, S.L.; Bates, P. Inhibitors of cathepsin L prevent severe acute respiratory syndrome coronavirus entry. *Proc. Natl. Acad. Sci. USA* **2005**, *102*, 11876–11881. [CrossRef] [PubMed]
75. Li, W.; Zhang, C.; Sui, J.; Kuhn, J.H.; Moore, M.J.; Luo, S.; Wong, S.K.; Huang, I.C.; Xu, K.; Vasilieva, N.; et al. Receptor and viral determinants of SARS-coronavirus adaptation to human ACE2. *EMBO J.* **2005**, *24*, 1634–1643. [CrossRef] [PubMed]
76. Fehr, A.R.; Perlman, S. Coronaviruses: An overview of their replication and pathogenesis. *Methods Mol. Biol.* **2015**, *1282*, 1–23. [CrossRef] [PubMed]
77. Torices, S.; Cabrera, R.; Stangis, M.; Naranjo, O.; Fattakhov, N.; Teglas, T.; Adesse, D.; Toborek, M. Expression of SARS-CoV-2-related receptors in cells of the neurovascular unit: Implications for HIV-1 infection. *J. Neuroinflamm.* **2021**, *18*, 167. [CrossRef] [PubMed]
78. Hernandez, V.S.; Zetter, M.A.; Guerra, E.C.; Hernandez-Araiza, I.; Karuzin, N.; Hernandez-Perez, O.R.; Eiden, L.E.; Zhang, L. ACE2 expression in rat brain: Implications for COVID-19 associated neurological manifestations. *Exp. Neurol.* **2021**, *345*, 113837. [CrossRef]
79. Pellet-Many, C.; Frankel, P.; Jia, H.; Zachary, I. Neuropilins: Structure, function and role in disease. *Biochem. J.* **2008**, *411*, 211–226. [CrossRef]
80. Graziani, G.; Lacal, P.M. Neuropilin-1 as Therapeutic Target for Malignant Melanoma. *Front. Oncol.* **2015**, *5*, 125. [CrossRef]
81. Lee, J.; Chong, K.; Lee, J.; Kim, C.; Kim, J.H.; Choi, K.; Choi, C. Differential dependency of human glioblastoma cells on vascular endothelial growth factor-A signaling via neuropilin-1. *Int. J. Oncol.* **2022**, *61*, 122. [CrossRef]
82. Cantuti-Castelvetri, L.; Ojha, R.; Pedro, L.D.; Djannatian, M.; Franz, J.; Kuivanen, S.; van der Meer, F.; Kallio, K.; Kaya, T.; Anastasina, M.; et al. Neuropilin-1 facilitates SARS-CoV-2 cell entry and infectivity. *Science* **2020**, *370*, 856–860. [CrossRef]
83. Chapoval, S.P.; Keegan, A.D. Perspectives and potential approaches for targeting neuropilin 1 in SARS-CoV-2 infection. *Mol. Med.* **2021**, *27*, 162. [CrossRef]
84. Pang, H.B.; Braun, G.B.; Friman, T.; Aza-Blanc, P.; Ruidiaz, M.E.; Sugahara, K.N.; Teesalu, T.; Ruoslahti, E. An endocytosis pathway initiated through neuropilin-1 and regulated by nutrient availability. *Nat. Commun.* **2014**, *5*, 4904. [CrossRef] [PubMed]
85. Balistreri, G.; Yamauchi, Y.; Teesalu, T. A widespread viral entry mechanism: The C-end Rule motif-neuropilin receptor interaction. *Proc. Natl. Acad. Sci. USA* **2021**, *118*, e2112457118. [CrossRef] [PubMed]
86. Gudowska-Sawczuk, M.; Mroczko, B. The Role of Neuropilin-1 (NRP-1) in SARS-CoV-2 Infection: Review. *J. Clin. Med.* **2021**, *10*, 2772. [CrossRef] [PubMed]
87. De Vlaeminck, Y.; Bonelli, S.; Awad, R.M.; Dewilde, M.; Rizzolio, S.; Lecocq, Q.; Bolli, E.; Santos, A.R.; Laoui, D.; Schoonooghe, S.; et al. Targeting Neuropilin-1 with Nanobodies Reduces Colorectal Carcinoma Development. *Cancers* **2020**, *12*, 3582. [CrossRef]
88. Daly, J.L.; Simonetti, B.; Klein, K.; Chen, K.E.; Williamson, M.K.; Anton-Plagaro, C.; Shoemark, D.K.; Simon-Gracia, L.; Bauer, M.; Hollandi, R.; et al. Neuropilin-1 is a host factor for SARS-CoV-2 infection. *Science* **2020**, *370*, 861–865. [CrossRef]
89. Welsch, S.; Muller, B.; Krausslich, H.G. More than one door—Budding of enveloped viruses through cellular membranes. *FEBS Lett.* **2007**, *581*, 2089–2097. [CrossRef]
90. Moller-Tank, S.; Maury, W. Phosphatidylserine receptors: Enhancers of enveloped virus entry and infection. *Virology* **2014**, *468–470*, 565–580. [CrossRef]

91. Tsuji, T.; Cheng, J.; Tatematsu, T.; Ebata, A.; Kamikawa, H.; Fujita, A.; Gyobu, S.; Segawa, K.; Arai, H.; Taguchi, T.; et al. Predominant localization of phosphatidylserine at the cytoplasmic leaflet of the ER, and its TMEM16K-dependent redistribution. *Proc. Natl. Acad. Sci. USA* **2019**, *116*, 13368–13373. [CrossRef]
92. Nowakowski, T.J.; Pollen, A.A.; Di Lullo, E.; Sandoval-Espinosa, C.; Bershteyn, M.; Kriegstein, A.R. Expression Analysis Highlights AXL as a Candidate Zika Virus Entry Receptor in Neural Stem Cells. *Cell Stem Cell* **2016**, *18*, 591–596. [CrossRef]
93. Miner, J.J.; Daniels, B.P.; Shrestha, B.; Proenca-Modena, J.L.; Lew, E.D.; Lazear, H.M.; Gorman, M.J.; Lemke, G.; Klein, R.S.; Diamond, M.S. The TAM receptor Mertk protects against neuroinvasive viral infection by maintaining blood-brain barrier integrity. *Nat. Med.* **2015**, *21*, 1464–1472. [CrossRef]
94. Axelrod, H.; Pienta, K.J. Axl as a mediator of cellular growth and survival. *Oncotarget* **2014**, *5*, 8818–8852. [CrossRef] [PubMed]
95. Meertens, L.; Labeau, A.; Dejarnac, O.; Cipriani, S.; Sinigaglia, L.; Bonnet-Madin, L.; Le Charpentier, T.; Hafirassou, M.L.; Zamborlini, A.; Cao-Lormeau, V.M.; et al. Axl Mediates ZIKA Virus Entry in Human Glial Cells and Modulates Innate Immune Responses. *Cell Rep.* **2017**, *18*, 324–333. [CrossRef] [PubMed]
96. Hutterer, M.; Knyazev, P.; Abate, A.; Reschke, M.; Maier, H.; Stefanova, N.; Knyazeva, T.; Barbieri, V.; Reindl, M.; Muigg, A.; et al. Axl and growth arrest-specific gene 6 are frequently overexpressed in human gliomas and predict poor prognosis in patients with glioblastoma multiforme. *Clin. Cancer Res.* **2008**, *14*, 130–138. [CrossRef] [PubMed]
97. Retallack, H.; Di Lullo, E.; Arias, C.; Knopp, K.A.; Laurie, M.T.; Sandoval-Espinosa, C.; Mancia Leon, W.R.; Krencik, R.; Ullian, E.M.; Spatazza, J.; et al. Zika virus cell tropism in the developing human brain and inhibition by azithromycin. *Proc. Natl. Acad. Sci. USA* **2016**, *113*, 14408–14413. [CrossRef] [PubMed]
98. Jimenez, O.A.; Narasipura, S.D.; Barbian, H.J.; Albalawi, Y.A.; Seaton, M.S.; Robinson, K.F.; Al-Harthi, L. beta-Catenin Restricts Zika Virus Internalization by Downregulating Axl. *J. Virol.* **2021**, *95*, e0070521. [CrossRef]
99. Lagos-Cabre, R.; Alvarez, A.; Kong, M.; Burgos-Bravo, F.; Cardenas, A.; Rojas-Mancilla, E.; Perez-Nunez, R.; Herrera-Molina, R.; Rojas, F.; Schneider, P.; et al. $\alpha_V\beta_3$ Integrin regulates astrocyte reactivity. *J. Neuroinflamm.* **2017**, *14*, 194. [CrossRef]
100. Milner, R.; Huang, X.; Wu, J.; Nishimura, S.; Pytela, R.; Sheppard, D.; ffrench-Constant, C. Distinct roles for astrocyte alphavbeta5 and alphavbeta8 integrins in adhesion and migration. *J. Cell Sci.* **1999**, *112 Pt 23*, 4271–4279. [CrossRef]
101. Gianni, T.; Leoni, V.; Chesnokova, L.S.; Hutt-Fletcher, L.M.; Campadelli-Fiume, G. alphavbeta3-integrin is a major sensor and activator of innate immunity to herpes simplex virus-1. *Proc. Natl. Acad. Sci. USA* **2012**, *109*, 19792–19797. [CrossRef]
102. Andrews, M.G.; Mukhtar, T.; Eze, U.C.; Simoneau, C.R.; Ross, J.; Parikshak, N.; Wang, S.; Zhou, L.; Koontz, M.; Velmeshev, D.; et al. Tropism of SARS-CoV-2 for human cortical astrocytes. *Proc. Natl. Acad. Sci. USA* **2022**, *119*, e2122236119. [CrossRef]
103. Wang, K.; Chen, W.; Zhang, Z.; Deng, Y.; Lian, J.Q.; Du, P.; Wei, D.; Zhang, Y.; Sun, X.X.; Gong, L.; et al. CD147-spike protein is a novel route for SARS-CoV-2 infection to host cells. *Signal Transduct. Target. Ther.* **2020**, *5*, 283. [CrossRef]
104. Li, Y.; Zhang, Z.; Yang, L.; Lian, X.; Xie, Y.; Li, S.; Xin, S.; Cao, P.; Lu, J. The MERS-CoV Receptor DPP4 as a Candidate Binding Target of the SARS-CoV-2 Spike. *iScience* **2020**, *23*, 101400. [CrossRef] [PubMed]
105. Li, Y.; Xu, J.; Chen, L.; Zhong, W.D.; Zhang, Z.; Mi, L.; Zhang, Y.; Liao, C.G.; Bian, H.J.; Jiang, J.L.; et al. HAb18G (CD147), a cancer-associated biomarker and its role in cancer detection. *Histopathology* **2009**, *54*, 677–687. [CrossRef]
106. Kiraly, K.; Kozsurek, M.; Lukacsi, E.; Barta, B.; Alpar, A.; Balazsa, T.; Fekete, C.; Szabon, J.; Helyes, Z.; Bolcskei, K.; et al. Glial cell type-specific changes in spinal dipeptidyl peptidase 4 expression and effects of its inhibitors in inflammatory and neuropatic pain. *Sci. Rep.* **2018**, *8*, 3490. [CrossRef]
107. Jin, R.; Xiao, A.Y.; Chen, R.; Granger, D.N.; Li, G. Inhibition of CD147 (Cluster of Differentiation 147) Ameliorates Acute Ischemic Stroke in Mice by Reducing Thromboinflammation. *Stroke* **2017**, *48*, 3356–3365. [CrossRef] [PubMed]
108. Jorgacevski, J.; Potokar, M. Immune Functions of Astrocytes in Viral Neuroinfections. *Int. J. Mol. Sci.* **2023**, *24*, 3514. [CrossRef] [PubMed]
109. Walsh, D.; Naghavi, M.H. Exploitation of Cytoskeletal Networks during Early Viral Infection. *Trends Microbiol.* **2019**, *27*, 39–50. [CrossRef]
110. Chauhan, A.; Tikoo, A.; Patel, J.; Abdullah, A.M. HIV-1 endocytosis in astrocytes: A kiss of death or survival of the fittest? *Neurosci. Res.* **2014**, *88*, 16–22. [CrossRef]
111. She, J.; Zeng, W.; Guo, J.; Chen, Q.; Bai, X.C.; Jiang, Y. Structural mechanisms of phospholipid activation of the human TPC2 channel. *Elife* **2019**, *8*, e45222. [CrossRef]
112. Cang, C.; Zhou, Y.; Navarro, B.; Seo, Y.J.; Aranda, K.; Shi, L.; Battaglia-Hsu, S.; Nissim, I.; Clapham, D.E.; Ren, D. mTOR regulates lysosomal ATP-sensitive two-pore Na^+ channels to adapt to metabolic state. *Cell* **2013**, *152*, 778–790. [CrossRef]
113. Abboud-Jarrous, G.; Atzmon, R.; Peretz, T.; Palermo, C.; Gadea, B.B.; Joyce, J.A.; Vlodavsky, I. Cathepsin L is responsible for processing and activation of proheparanase through multiple cleavages of a linker segment. *J. Biol. Chem.* **2008**, *283*, 18167–18176. [CrossRef]
114. Sakurai, Y.; Kolokoltsov, A.A.; Chen, C.C.; Tidwell, M.W.; Bauta, W.E.; Klugbauer, N.; Grimm, C.; Wahl-Schott, C.; Biel, M.; Davey, R.A. Ebola virus. Two-pore channels control Ebola virus host cell entry and are drug targets for disease treatment. *Science* **2015**, *347*, 995–998. [CrossRef] [PubMed]
115. Zhu, M.X.; Ma, J.; Parrington, J.; Galione, A.; Evans, A.M. TPCs: Endolysosomal channels for Ca^{2+} mobilization from acidic organelles triggered by NAADP. *FEBS Lett.* **2010**, *584*, 1966–1974. [CrossRef] [PubMed]

116. Shen, D.; Wang, X.; Xu, H. Pairing phosphoinositides with calcium ions in endolysosomal dynamics: Phosphoinositides control the direction and specificity of membrane trafficking by regulating the activity of calcium channels in the endolysosomes. *Bioessays* **2011**, *33*, 448–457. [CrossRef] [PubMed]
117. Khoo, U.S.; Chan, K.Y.; Chan, V.S.; Lin, C.L. DC-SIGN and L-SIGN: The SIGNs for infection. *J Mol Med.* **2008**, *86*, 861–874. [CrossRef]
118. Miller, S.; Krijnse-Locker, J. Modification of intracellular membrane structures for virus replication. *Nat. Rev. Microbiol.* **2008**, *6*, 363–374. [CrossRef]
119. Hayashi-Nishino, M.; Fujita, N.; Noda, T.; Yamaguchi, A.; Yoshimori, T.; Yamamoto, A. A subdomain of the endoplasmic reticulum forms a cradle for autophagosome formation. *Nat. Cell Biol.* **2009**, *11*, 1433–1437. [CrossRef]
120. Lennemann, N.J.; Coyne, C.B. Dengue and Zika viruses subvert reticulophagy by NS2B3-mediated cleavage of FAM134B. *Autophagy* **2017**, *13*, 322–332. [CrossRef]
121. Echavarria-Consuegra, L.; Smit, J.M.; Reggiori, F. Role of autophagy during the replication and pathogenesis of common mosquito-borne flavi- and alphaviruses. *Open Biol.* **2019**, *9*, 190009. [CrossRef]
122. Choi, Y.; Bowman, J.W.; Jung, J.U. Autophagy during viral infection—A double-edged sword. *Nat. Rev. Microbiol.* **2018**, *16*, 341–354. [CrossRef]
123. Tian, Y.; Wang, M.L.; Zhao, J. Crosstalk between Autophagy and Type I Interferon Responses in Innate Antiviral Immunity. *Viruses* **2019**, *11*, 132. [CrossRef]
124. Lindqvist, R.; Mundt, F.; Gilthorpe, J.D.; Wolfel, S.; Gekara, N.O.; Kroger, A.; Overby, A.K. Fast type I interferon response protects astrocytes from flavivirus infection and virus-induced cytopathic effects. *J. Neuroinflamm.* **2016**, *13*, 277. [CrossRef] [PubMed]
125. Prasad, V.; Greber, U.F. The endoplasmic reticulum unfolded protein response—Homeostasis, cell death and evolution in virus infections. *FEMS Microbiol. Rev.* **2021**, *45*, fuab016. [CrossRef] [PubMed]
126. Mehla, R.; Chauhan, A. HIV-1 differentially modulates autophagy in neurons and astrocytes. *J. Neuroimmunol.* **2015**, *285*, 106–118. [CrossRef]
127. Saribas, A.S.; Khalili, K.; Sariyer, I.K. Dysregulation of autophagy by HIV-1 Nef in human astrocytes. *Cell Cycle* **2015**, *14*, 2899–2904. [CrossRef] [PubMed]
128. Ren, S.; Ding, C.; Sun, Y. Morphology Remodeling and Selective Autophagy of Intracellular Organelles during Viral Infections. *Int. J. Mol. Sci.* **2020**, *21*, 3689. [CrossRef] [PubMed]
129. Cho, D.H.; Kim, J.K.; Jo, E.K. Mitophagy and Innate Immunity in Infection. *Mol. Cells* **2020**, *43*, 10–22. [CrossRef] [PubMed]
130. Li, S.; Wang, J.; Zhou, A.; Khan, F.A.; Hu, L.; Zhang, S. Porcine reproductive and respiratory syndrome virus triggers mitochondrial fission and mitophagy to attenuate apoptosis. *Oncotarget* **2016**, *7*, 56002–56012. [CrossRef]
131. Zhu, L.; Mou, C.; Yang, X.; Lin, J.; Yang, Q. Mitophagy in TGEV infection counteracts oxidative stress and apoptosis. *Oncotarget* **2016**, *7*, 27122–27141. [CrossRef]
132. Ojeda, D.S.; Grasso, D.; Urquiza, J.; Till, A.; Vaccaro, M.I.; Quarleri, J. Cell Death Is Counteracted by Mitophagy in HIV-Productively Infected Astrocytes but Is Promoted by Inflammasome Activation Among Non-productively Infected Cells. *Front. Immunol.* **2018**, *9*, 2633. [CrossRef]
133. Heurich, A.; Hofmann-Winkler, H.; Gierer, S.; Liepold, T.; Jahn, O.; Pohlmann, S. TMPRSS2 and ADAM17 cleave ACE2 differentially and only proteolysis by TMPRSS2 augments entry driven by the severe acute respiratory syndrome coronavirus spike protein. *J. Virol.* **2014**, *88*, 1293–1307. [CrossRef]
134. Li, J.; Gao, E.; Xu, C.; Wang, H.; Wei, Y. ER-Phagy and Microbial Infection. *Front. Cell Dev. Biol.* **2021**, *9*, 771353. [CrossRef] [PubMed]
135. Mo, J.; Chen, J.; Zhang, B. Critical roles of FAM134B in ER-phagy and diseases. *Cell Death Dis.* **2020**, *11*, 983. [CrossRef] [PubMed]
136. Inoue, T.; Tsai, B. How viruses use the endoplasmic reticulum for entry, replication, and assembly. *Cold Spring Harb. Perspect. Biol.* **2013**, *5*, a013250. [CrossRef] [PubMed]
137. Ralhan, I.; Chang, C.L.; Lippincott-Schwartz, J.; Ioannou, M.S. Lipid droplets in the nervous system. *J. Cell Biol.* **2021**, *220*, e202102136. [CrossRef] [PubMed]
138. Zhang, S.; Peng, X.; Yang, S.; Li, X.; Huang, M.; Wei, S.; Liu, J.; He, G.; Zheng, H.; Yang, L.; et al. The regulation, function, and role of lipophagy, a form of selective autophagy, in metabolic disorders. *Cell Death Dis.* **2022**, *13*, 132. [CrossRef]
139. Smolic, T.; Tavcar, P.; Horvat, A.; Cerne, U.; Haluzan Vasle, A.; Tratnjek, L.; Kreft, M.E.; Scholz, N.; Matis, M.; Petan, T.; et al. Astrocytes in stress accumulate lipid droplets. *Glia* **2021**, *69*, 1540–1562. [CrossRef]
140. Liu, L.; Zhang, K.; Sandoval, H.; Yamamoto, S.; Jaiswal, M.; Sanz, E.; Li, Z.; Hui, J.; Graham, B.H.; Quintana, A.; et al. Glial lipid droplets and ROS induced by mitochondrial defects promote neurodegeneration. *Cell* **2015**, *160*, 177–190. [CrossRef]
141. Liu, L.; MacKenzie, K.R.; Putluri, N.; Maletic-Savatic, M.; Bellen, H.J. The Glia-Neuron Lactate Shuttle and Elevated ROS Promote Lipid Synthesis in Neurons and Lipid Droplet Accumulation in Glia via APOE/D. *Cell Metab.* **2017**, *26*, 719–737.e716. [CrossRef]
142. Samsa, M.M.; Mondotte, J.A.; Iglesias, N.G.; Assuncao-Miranda, I.; Barbosa-Lima, G.; Da Poian, A.T.; Bozza, P.T.; Gamarnik, A.V. Dengue virus capsid protein usurps lipid droplets for viral particle formation. *PLoS Pathog.* **2009**, *5*, e1000632. [CrossRef]
143. Monson, E.A.; Trenerry, A.M.; Laws, J.L.; Mackenzie, J.M.; Helbig, K.J. Lipid droplets and lipid mediators in viral infection and immunity. *FEMS Microbiol. Rev.* **2021**, *45*, fuaa066. [CrossRef]
144. Monson, E.A.; Whelan, D.R.; Helbig, K.J. Lipid Droplet Motility Increases Following Viral Immune Stimulation. *Int. J. Mol. Sci.* **2021**, *22*, 4418. [CrossRef] [PubMed]

145. Shin, D.W. Lipophagy: Molecular Mechanisms and Implications in Metabolic Disorders. *Mol. Cells* **2020**, *43*, 686–693. [CrossRef] [PubMed]
146. Heaton, N.S.; Randall, G. Multifaceted roles for lipids in viral infection. *Trends Microbiol.* **2011**, *19*, 368–375. [CrossRef] [PubMed]
147. Jordan, T.X.; Randall, G. Dengue Virus Activates the AMP Kinase-mTOR Axis To Stimulate a Proviral Lipophagy. *J. Virol.* **2017**, *91*, e02020-16. [CrossRef] [PubMed]
148. Sanchez-Garcia, F.J.; Perez-Hernandez, C.A.; Rodriguez-Murillo, M.; Moreno-Altamirano, M.M.B. The Role of Tricarboxylic Acid Cycle Metabolites in Viral Infections. *Front. Cell Infect. Microbiol.* **2021**, *11*, 725043. [CrossRef]

cells

MDPI

Article

Transcriptional and Chromatin Accessibility Profiling of Neural Stem Cells Differentiating into Astrocytes Reveal Dynamic Signatures Affected under Inflammatory Conditions

Maria Angeliki S. Pavlou [1,2,†], Kartikeya Singh [3,†], Srikanth Ravichandran [3], Rashi Halder [4], Nathalie Nicot [5,6], Cindy Birck [1], Luc Grandbarbe [1], Antonio del Sol [3,7,8,*,‡] and Alessandro Michelucci [2,4,*,‡]

1 Department of Life Sciences and Medicine, University of Luxembourg, L-4365 Esch-sur-Alzette, Luxembourg
2 Neuro-Immunology Group, Department of Cancer Research, Luxembourg Institute of Health, L-1210 Luxembourg, Luxembourg
3 Computational Biology Group, Luxembourg Centre for Systems Biomedicine, University of Luxembourg, L-4365 Esch-sur-Alzette, Luxembourg
4 Scientific Central Services, Luxembourg Centre for Systems Biomedicine, University of Luxembourg, L-4365 Esch-sur-Alzette, Luxembourg
5 Translational Medicine Operations Hub, Luxembourg Institute of Health, L-3555 Dudelange, Luxembourg
6 LuxGen Genome Center, Luxembourg Institute of Health & Laboratoire National de Santé, L-3555 Dudelange, Luxembourg
7 Computational Biology Group, CIC bioGUNE-BRTA (Basque Research and Technology Alliance), 48160 Derio, Spain
8 IKERBASQUE, Basque Foundation for Science, 48009 Bilbao, Spain
* Correspondence: antonio.delsol@uni.lu (A.d.S.); alessandro.michelucci@lih.lu (A.M.)
† These authors contributed equally to this work.
‡ These authors contributed equally to this work.

Citation: Pavlou, M.A.S.; Singh, K.; Ravichandran, S.; Halder, R.; Nicot, N.; Birck, C.; Grandbarbe, L.; del Sol, A.; Michelucci, A. Transcriptional and Chromatin Accessibility Profiling of Neural Stem Cells Differentiating into Astrocytes Reveal Dynamic Signatures Affected under Inflammatory Conditions. *Cells* **2023**, *12*, 948. https://doi.org/10.3390/cells12060948

Academic Editors: Michael Brenner and Vladimir Parpura

Received: 18 January 2023
Revised: 14 March 2023
Accepted: 17 March 2023
Published: 21 March 2023

Abstract: Astrocytes arise from multipotent neural stem cells (NSCs) and represent the most abundant cell type of the central nervous system (CNS), playing key roles in the developing and adult brain. Since the differentiation of NSCs towards a gliogenic fate is a precisely timed and regulated process, its perturbation gives rise to dysfunctional astrocytic phenotypes. Inflammation, which often underlies neurological disorders, including neurodevelopmental disorders and brain tumors, disrupts the accurate developmental process of NSCs. However, the specific consequences of an inflammatory environment on the epigenetic and transcriptional programs underlying NSCs' differentiation into astrocytes is unexplored. Here, we address this gap by profiling in mice glial precursors from neural tissue derived from early embryonic stages along their astrocytic differentiation trajectory in the presence or absence of tumor necrosis factor (TNF), a master pro-inflammatory cytokine. By using a combination of RNA- and ATAC-sequencing approaches, together with footprint and integrated gene regulatory network analyses, we here identify key differences during the differentiation of NSCs into astrocytes under physiological and inflammatory settings. In agreement with its role to turn cells resistant to inflammatory challenges, we detect *Nrf2* as a master transcription factor supporting the astrocytic differentiation under TNF exposure. Further, under these conditions, we unravel additional transcriptional regulatory hubs, including *Stat3*, *Smad3*, *Cebpb*, and *Nfkb2*, highlighting the interplay among pathways underlying physiological astrocytic developmental processes and those involved in inflammatory responses, resulting in discrete astrocytic phenotypes. Overall, our study reports key transcriptional and epigenetic changes leading to the identification of molecular regulators of astrocytic differentiation. Furthermore, our analyses provide a valuable resource for understanding inflammation-induced astrocytic phenotypes that might contribute to the development and progression of CNS disorders with an inflammatory component.

Keywords: neural stem cell; astrocyte; tumor necrosis factor; transcriptomics; chromatin accessibility; transcription factors

1. Introduction

In the developing brain, multipotent neural stem cells (NSCs) can give rise to neurons, oligodendrocytes, and astrocytes. The process of differentiation towards a neurogenic or gliogenic fate is finely timed and regulated through mammalian brain development. During the expansion phase, NSCs self-replicate via symmetric cell division to expand their own pool. Later on, during mid-gestation they undergo asymmetric cell division and receive cues to differentiate into neurons. Lastly, from late-gestation to perinatal periods they enter the gliogenic phase and differentiate into astrocytes and oligodendrocytes [1,2]. Importantly, undifferentiated NSCs still reside in two niches of the adult brain, namely the subventricular zone (SVZ) of the lateral ventricles and the subgranular zone (SGZ) of the dentate gyrus [3].

Astrocytes are active supporters of the neuronal network, providing neurons with nutrients and metabolites, modulating neuronal synaptic transmission, participating in the formation and functioning of the synapses as well as directly communicating with neurons [4]. Further, astrocytes support the immune system and contribute to the formation of the blood brain barrier. Additionally, they contribute to oligodendrocyte differentiation and maturation as well as support myelination [5]. Therefore, aberrant astrocytic phenotypes can give rise to neurological diseases, including neurodevelopmental disorders [6]. For example, dominant gain-of-function mutations in the gene coding for glial fibrillary acidic protein (GFAP), the hallmark intermediate filament in astrocytes, elicit the accumulation of cytoplasmic GFAP aggregates resulting in macrocephaly, white matter degeneration, and developmental delay in Alexander disease [7,8]. During inflammatory events in the CNS, several immune mediators, including cytokines and free radicals, affect intrinsic NSC properties and neurogenesis [9]. Injuries of the CNS largely elicit proliferation and migration of NSCs toward the injury site, suggesting the activation of a putative regenerative response. However, detrimental cues encountered within the inflammatory niche are responsible, at least partly, for an incomplete regeneration from NSCs [9]. Thus, in traumatic brain injuries, for example, attempts to promote a regenerative-favoring environment using endogenous or transplanted NSCs for reparative purposes represent critical challenges [10,11]. In a different context, perturbations of the fine-tuned process of NSCs' differentiation under inflammatory conditions can also give rise to brain tumors. Indeed, inflammation is linked to gliomagenesis [12] and evidence indicates that NSCs can give rise to gliomas [13,14]. Hence, defective differentiation abilities of NSCs and aberrant neoplastic astrocytic phenotypes in gliomas are attributed, at least partly, to an inflammatory microenvironment.

Tumor necrosis factor (TNF) is a crucial inflammatory cytokine mainly produced by microglia during neuroinflammatory processes. TNF induces various signaling responses within the cells that eventually result in apoptosis or necrosis [15], cell survival, differentiation, and inflammation in cells expressing the corresponding receptors [16]. Notably, the crosstalk between interferons and TNF affects the reprogramming of the macrophage epigenome and their inflammatory responses [17]. In this context, the role of epigenetic processes in the establishment, maintenance, and differentiation of NSCs is emerging [18]. For example, deficiency of the chromatin remodeler gene *Cdh5* in NSCs leads to their improper activation altering cell fate decisions and favoring an astroglial over a neuronal identity [19].

Here, we sought to investigate the dynamics of the transcriptional and chromatin accessibility changes underlying NSCs differentiation into astrocytes under normal and inflammatory conditions. For this, we adopted an integrated genome-wide approach combining RNA-sequencing (RNA-seq) and assay for transposase-accessible chromatin followed by deep sequencing (ATAC-seq) [20] to investigate the astrocytic differentiation process from murine neurospheres. We uncovered that normal differentiated astrocytes display mature cell-specific characteristics, whereas differentiating cells exposed to TNF do not reach the same level of maturity. In line with these observations and in agreement with a more restricted lineage-specific transcriptional program, we identified an overall reduction of chromatin accessibility in differentiated astrocytes compared to multipotent undifferentiated NSCs. Notably, this restriction was less pronounced under inflammatory conditions. Lastly, we reconstructed gene regulatory networks to identify transcription factors involved in the regulation of differentiation under normal and inflammatory settings.

Overall, our findings shed light on the molecular mechanisms underlying NSC differentiation into astrocytes and the effect of an inflammatory environment during this developmental process, which might pave the way to novel therapeutic targets restoring appropriate astrocytic phenotypes in CNS disorders with an inflammatory component including neurodevelopmental diseases and brain tumors.

2. Materials and Methods

2.1. Animals and Ethics

C57BL/6J mice, both wild type and transgenic mice expressing green fluorescent protein (GFP) under the control of glial fibrillary acidic protein (GFAP) promoter were housed in individually ventilated cages (IVC) in a conventional animal facility at the University of Luxembourg in accordance with the EU Directive 2010/63/EU and Commission recommendation 2007/526/EC. We kept mice in groups under a dark–light cycle with ad libitum access to water and food. All the animal work of the present study has been conducted and reported in accordance with the ARRIVE (Animal Research: Reporting of In Vivo Experiments) guidelines to improve the design, analysis, and reporting of research using animals, maximizing information published, and minimizing unnecessary studies. We conducted all animal procedures in accordance with the local Committee for Care and Use of Laboratory Animals.

2.2. Cell Culture

We isolated primary NSCs from the ventricular zone of single wild type mouse brains (C57BL/6J, Harlan, The Netherlands) at embryonic day 14 (E14) as described before [21,22]. Primary cultures of neurospheres (NSPs) were kept under proliferating conditions in neurobasal medium (DMEM F12; Lonza, Basel, Switzerland) supplemented with 1% B27 without vitamin A (Life Technologies, Carlsbad, CA, USA), penicillin (100 U/mL; Lonza), streptomycin (100 g/mL; Lonza), and 20 ng/mL EGF (Epidermal Growth Factor; Life Technologies). We differentiated NSPs into astrocytes on 6-well or 12-well plates coated with poly-L-ornithine and by exchanging the proliferation medium with DMEM containing 10% FBS (Fetal Bovine Serum; Gibco, Grand Island, NE, USA), penicillin (100 U/mL; Lonza), and streptomycin (100 g/mL; Lonza). We differentiated cells into astrocytes at different time points (24, 48, 72 h and 1 week) at 37 °C in 5% CO_2/95% air atmosphere, and we kept them under normal conditions or treated with TNF (50 ng/mL; R&D Systems, Minneapolis, MN, USA).

2.3. Selection of Astrocytes Based on Antibody-Coated Beads

We used NSPs and differentiating NSPs at different time points (24, 48 and 72 h) to isolate astrocytes by magnetic separation. We enriched astrocytic populations by magnetic separation using ACSA1+ (GLAST) beads (MACS Miltenyi Biotec) according to the manufacturer's recommendations.

2.4. Real-Time qPCR (RT-qPCR)

We kept murine NSPs under proliferating conditions or we differentiated them into astrocytes for 24, 48, and 72 h or 1 week with or without TNF treatment. We extracted total RNA by the innuPREP RNA Mini Kit (Westburg) according to manufacturer's recommendations. Complementary DNA (cDNA) was synthetized by using the ImProm-II Reverse Transcription System (Promega). We conducted cDNA synthesis for 5 min at 25 °C, followed by 1 h at 42 °C and then 15 min at 70 °C. We measured gene expression levels using SYBR Green Supermix (Promega) following the manufacturer's recommendations and by using CFX Connect Real-Time PCR Detection System (Bio-Rad). The expression levels of mRNA were expressed as $2^{-\Delta Ct}$ and normalized to beta actin levels [23]. Primer sequences were as follows: *Mki67* forward primer: 5′-TTCCTTCAGCAAGCCTGAG-3′; *Mki67* reverse primer: 5′-GTATTAGGAGGCAAGTT-3; *Ccnb1* forward primer: 5′-AGAGGTGGAACTTGCTGAGCCT-3′; *Ccnb1* reverse primer: 5′-GCACATCCAGATGTTTCCATCGG-3′; *Egr1* forward primer: 5′-AGCCGAGCGAACAAC CCTAT-3′; *Egr1* reverse primer: 5′-TGTCAGAAAAGGACTCTGTGGTCA-3′; *Gfap* forward primer: 5′-GGTTGAATCGCTGGAGGAG-3′; *Gfap* reverse primer: 5′-CTGTGAGGTCTGGC TTGG-3′; *Gpld1* forward primer: 5′-ACCCTAACCCAAGTCCTGCT-3′; *Gpld1* reverse primer: 5′-CAGGTCAGTCAGGTGCAGAA-3′; *Nfkbia* forward primer: 5′-GCCAGTGTAGC AGTCTTGAC-3′; *Nfkbia* reverse primer: 5′-GCCAGGTAGCCGTGAGTAG-3′; *Actb* forward primer: 5′-AGGGAAATCGTGCGTGACATCAAAGAG-3′; *Actb* reverse primer: 5′-GGAGGAAGAGGATGCGGCAGTGG-3′.

2.5. Immunocytochemistry

NSPs and differentiating cells were cultivated on poly-L-ornithine-coated coverslips and fixed with 4% paraformaldehyde in phosphate-buffered saline (PBS), followed by a permeabilization with 0.05% Triton X-100 in PBS. We performed a blocking step in PBS containing 3% bovine serum albumin (BSA) at 20–24 °C for 1 h. We conducted immunostainings by incubating cells with primary antibodies diluted in blocking solution overnight at 4 °C, followed by incubation with the corresponding secondary antibodies diluted in the blocking solution for 1 h at room temperature (1:1000; Jackson ImmunoResearch, West Grove, PA, USA). We used the following antibodies: rabbit anti-CD44 IgG (1:100; Abcam, Waltham, MA, USA), rabbit anti-Kir4.1 IgG (1:100; Alomone Labs, Jerusalem, Israel), rabbit anti-GLT1 (SLC1A2) IgG (1:100; Abcam), rabbit anti-GLAST (SLC1A3) IgG (1:100; Abcam), mouse anti-MAP2 IgG (1:200; Chemicon, Rolling Meadows, IL, USA) and mouse anti-O4 IgM (1:100; Bio-Techne). For GFAP staining, we used a cyanine 3-conjugated mouse anti-GFAP IgG antibody (1:400; Sigma, St. Louis, MO, USA). Cells were then washed and mounted with DAPI-Fluoromount G (SouthernBiotech, Birmingham, AL, USA). Images were collected by confocal microscopy using a LSM 510 confocal microscope (Carl Zeiss Micro Imaging, Jena, Germany) and analyzed on Adobe Photoshop (San Jose, CA, USA) software.

2.6. Preparation of Mouse Forebrain Cell Suspensions from hGFAP::eGFP Transgenic Mice and Flow Cytometry Analyses

hGFAP::eGFP transgenic mice (*hGFAP*::eGFP mice (FVB/N- Tg(GFAPGFP)14Mes/J) were purchased from The Jackson Laboratory (Jax stock #003257; Bar Harbor, ME, USA) [24]. The genetic identification of transgenic mice was determined by the analysis of DNA extracted from tails of 3-week old mice. DNA was extracted from tail samples (<5 mm) using "DirectPCR Lysis Reagent" (Viagen Biotech, Los Angeles, CA, USA) containing freshly prepared 20 mg/mL of proteinase K (Invitrogen, Merelbeke, Belgium). After a lysis incubation step at 55 °C for 5 h under gentle shaking (550 rpm), proteinase K was inactivated by incubating lysates at 85 °C for 45 min. DNA analysis was performed by Polymerase Chain Reaction (PCR). The PCR reaction mixture, with a total volume of 25 mL, contained 1 mL of DNA lysate, 12.5 mL of a PCR master mix (50 U/mL Taq DNA polymerase, 400 μM of each dNTP, 3 mM $MgCl_2$; Promega; Leiden, The Netherlands), and 11.5 mL of primers mix (Eurogentec, Seraing, Belgium). Primer sequences were as follows: GFP forward primer: 5′-AAGTTCATCTGCACCACCG-3′; GFP reverse primer: 5′-TCCTTGAAGAAGATGGTGCG-3′; internal positive control forward primer: 5′-CTAGGCCACAGAATTGAAAGATCT-3′; internal positive control reverse primer: 5′-GTAGGTGGAAATTCTAGCATCATCC-3′. Initial denaturation at 94 °C for 1 min was followed by 35 cycles of denaturation at 94 °C for 30 s, annealing at 60 °C for 1 min and extension at 72 °C for 1 min, with final extension at 72 °C for 2 min. PCR samples were analyzed on a 2% agarose gel (ThermoFisher Scientific, Merelbeke, Belgium), revealed on a ChemiDoc XRS+ System (Bio-Rad, Temse, Belgium) and visualized using ImageLab software (Bio-Rad).

We collected forebrains from *hGFAP*::eGFP transgenic mice in calcium/magnesium free HBSS at different postnatal developmental days (P4, P10, or P21). Tissue was diced and papain digested at 33 °C for 90 min (20 U/mL, Sigma) in dissociation buffer (EBSS (Sigma), D(+)-glucose 22.5 mM, $NaHCO_3$ 26 mM and DNaseI 125 U/mL with EDTA 0.5 mM and L-cysteine–HCl 1 mM (Sigma)) and washed 3 times in dissociation buffer with BSA (1 mg/mL, Sigma) and trypsin inhibitor (1 mg/mL, Sigma) before mechanical dissociation through fire-polished Pasteur pipettes to a single cell suspension. Cells were pelleted, re-suspended in cold PBS with DNaseI at 1×10^6 cell/mL, passed through a cell strainer (70-μm mesh; #352350, BD Falcon) and 7-aminoactinomycin D (7-AAD, Sigma) added. FITC-positive/PE-Cy5-negative cells were taken into account for quantification We performed flow cytometry analyses using a FACS Aria I SORP running FACS Diva6.3 software (BD Biosciences, San Jose, CA, USA).

2.7. RNA-Sequencing Analyses

For RNA-sequencing analyses, murine NSPs were kept under proliferation conditions or were differentiated into astrocytes for 24 and 72 h without or with TNF treatment and for 1 week under normal conditions. We extracted total RNA with TRIzol (Invitrogen) following the manufacturer's recommendations. We added a DNase treatment to the RNA extraction using DNA-free kit DNase Treatment and Removal Reagents (Ambion-Thermofisher cat #AM1906) following manufacturer's instructions. We checked the RNA quality using a Fragment Analyzer System (Agilent Technologies, Santa Clara, CA, USA), and RQN were between 3.4 and 10. We converted total RNA into Stranded Total RNA Library for RNA-sequencing by using the TruSeq Stranded Total RNA Library Prep workflow with Ribo-Zero Gold kit according to Illumina's instructions. Briefly, ribosomal RNA was removed from 300 ng of total RNA using biotinylated, target-specific oligos combined with Ribo-Zero Gold beads that deplete cytoplasmic and mitochondrial rRNA. Then, RNA was fragmented using divalent cations under elevated temperature of 94 °C. Cleaved RNA fragments are double-stranded cDNA converted, keeping strand-specificity. A single index adapter was ligated to each sample, and DNA fragments were enriched using a polymerase. All libraries have been quantified using a Qubit HS dsDNA kit (Thermofisher), and the library quality check has been performed using a High-Sensitivity NGS Fragment Analysis Kit on a Fragment Analyzer System (Agilent Technologies). We normalized indexed DNA libraries and pooled them at 10 nM. Libraries were paired-end sequenced at LuxGen Genome Center with 75 bp length on a NextSeq Illumina sequencer.

Fastq files contain raw sequenced data, and the FastQC tool was used for a quality check. The fastq data was trimmed to remove bad quality reads using Trim Galore and trimmed reads were then aligned to the transcriptome using STAR aligner. STAR produces unsorted BAM files as output, which were then sorted using Samtools. We used the sorted files as input for Feature Counts to obtain gene expression counts. We conducted differential gene expression analyses using DESeq2 package in R program (The R Foundation for Statistical Computing, version 4.0.4).

2.8. ATAC-Sequencing Analyses

Cells grown as either free-floating spheres (NSPs) or as astrocytes (for 24 h and 72 h non-treated or treated with TNF and for 1 week without TNF treatment) in poly-L-ornithine-coated plates were washed once with ice-cold PBS and were spun at 500× *g* for 5 min in a pre-chilled (4 °C) centrifuge. We then re-suspended the cells in 200 μL of lysis buffer (5 mM PIPES at pH 8, 85 mM KCl, 0.5% NP-40) and incubated them in a tube placed on ice for 20 min. Next, 800 μL of resuspension buffer (10 mM Tris pH 7.5, 10 mM NaCl, 3 mM $MgCl_2$, 0.1% Tween-20) were added to the cells. Following cell counting, we centrifuged 50,000 cells at 500× *g* for 10 min in a pre-chilled (4 °C) centrifuge. The cell pellet was re-suspended in 50 μL of transposition mix (25 μL 2× TD buffer, 2.5 μL transposase, 17 μL PBS, 0.5 μL 10% Tween-20 and 5 μL water). Transposition reactions were incubated at 37 °C for 40 min in a thermomixer with shaking at 1000 rpm. Reactions were cleaned up with

MinElute Reaction purification kit (Qiagen, Hilden, Germany) following manufacturer's recommendations. Following purification, library fragments were amplified using 1× NEBnext PCR master mix and 0.6 μM of custom PCR primers 1 and 2, using the following PCR conditions: 72 °C for 5 min; 98 °C for 30 s; and thermocycling at 98 °C for 10 s, 63 °C for 30 s and 72 °C for 1 min. Full libraries were amplified for five cycles, after which an aliquot of the PCR reaction was used together with 9 μL of the PCR cocktail with Sybr Green at a final dilution of 0.05×. We ran this reaction for 30 cycles to determine the additional number of cycles needed for the remaining reaction. Libraries were purified using AMPure XP magnetic beads (Beckman Coulter, Brea, CA, USA), quantified using qubit fluorimeter and diluted to make a 4-nM final concentration. We sequenced pooled libraries on NextSeq500 75 bp paired-end reads. We used the FastQC tool to perform a quality check on fastq files containing raw sequenced reads, and bad-quality reads were removed using Trim Galore. The trimmed fastq reads were aligned to the genome using Bowtie2. We used Samtools to convert the SAM output files from Bowtie2 into sorted BAM files and to remove reads aligned to mitochondrial chromosomes. We used Model-based Analysis of ChIP-seq data (MACS) to call chromatin accessibility peaks, and the peaks that were found in the blacklisted regions of the genome (ENCODE) were removed before further analyses. Peaks were annotated using annotate peaks tool from HOMER and differential accessibility analysis was performed using DiffBind package in R program.

2.9. Transcription Factor Footprint Analysis

We used the Regulatory Genomic Toolbox to study the changes in the transcription factor binding motifs in open chromatin regions of the genome. Using the HINT tool [25], we identified transcription factor motifs in the epigenomic landscape. We conducted differential motif analysis using the rgt-motif analysis tool to identify transcription factor binding motifs differentially present across two conditions, thus providing information of underlying differences in gene regulation.

2.10. Reconstruction of Gene Regulatory Networks

We re-constructed gene regulatory networks and identified hubs based on topology analyses. First, we inferred gene regulatory networks (GRNs) at each time point by using the GRN inference tool GENIE3 [26], which utilizes random forest-based ensemble methods to identify regulatory links between genes. We performed bootstrapping on these interactions and only selected interactions with significant value of more than zero in more than 50 percent of runs. To obtain high confidence networks, we further overlaid the re-constructed networks from GENIE3 with manually curated interactions from the Metacore database. We further pruned the GRN by contextualizing it using a tool developed in our group: https://gitlab.lcsb.uni.lu/andras.hartmann/GRNOptR (accessed on 15 November 2021). Briefly, the tool utilizes a parent network and differential gene expression data with respect to the preceding time-point to create contextualized gene expression networks. Contextualization removes nodes whose expression cannot be justified by the nodes regulating it as input. We then identified the regulatory hubs by performing topological analyses using the igraph package in R. We further used the clusterprofiler package in R to perform Gene Ontology pathways enrichment of transcription factors included in the re-constructed GRNs. We visualized the resultant networks using Cytoscape.

2.11. Raw Data Files

We deposited ATAC-seq and RNA-seq datasets in the Gene Expression Omnibus (GEO) database with accession number GSE225729.

2.12. Statistical Analyses

Data were analyzed using GraphPad Prism 8 software (GraphPad software, La Jolla, CA, USA) and R environment (R Core Team, Vienna, Austria). The number of replicates required for comparison was estimated using power analysis (set 80% and $\alpha = 0.05$) and guided by our previously published work [21]. We assessed normality using Kolmogorov-Smirnov and Shapiro–Wilk tests for normality. Depending on the number of averaged samples, the Central Limit Theorem was applicable, and even if data were not normal the mean uncertainty was normally distributed. Unless otherwise indicated, all data are presented as mean ± standard deviation (SD) or standard error of the mean (SEM) of at least three independent biological experiments. We performed statistical analyses using unpaired *t* test or two-way ANOVA. All differences were considered significantly different at *p* value < 0.05 and were annotated as follows: * $p < 0.05$, ** $p < 0.01$.

3. Results

3.1. Characterization of an In Vitro Model of Astrocytic Differentiation from Murine Neurospheres

To investigate the molecular mechanisms underlying NSCs differentiation into astrocytes under inflammatory conditions, we first characterized our in vitro cellular model of primary neurospheres (NSPs) derived from NSCs isolated from the ventricular zone at embryonic day-14 (E14) of the mouse embryos. Briefly, we differentiated NSPs into astrocytes in the presence of 10% fetal calf serum (FCS) for one or two weeks and analyzed them at specific intermediate stages, including 24, 48, and 72 h (Figure 1A). Under these conditions differentiating NSPs showed progressively increased expression levels of the astrocytic marker GFAP and a concomitant decreased expression of the stem cell marker CD44 (Figure 1B). After 2 weeks of differentiation, mature astrocytic markers, such as Kir4.1 (*Kcnj10*), GLT1 (*Slc1a2*), and GLAST (*Slc1a3*) were highly expressed (Figure S1A). Notably we did not detect neurons and oligodendrocytes in these culture conditions (Figure S1B,C) while multipotentiality of the NSCs was confirmed by the appearance of these cells upon addition of 1% B27 and 2 ng/mL EGF to 0.5% FCS medium (Figure S1D,E). At the transcriptional level, differentiating astrocytes at all stages showed significantly decreased expression levels of the proliferating (*Mki67* and *Ccnb1*) and stem cell (*Egr1*) markers compared to NSPs (Figure 1C). On the other hand, the expression levels of *Gfap* and the astrocyte-enriched gene *Gpld1* were mainly up-regulated in differentiating cells at the different time points. We detected high *Gfap* expression levels at 48 and 72 h, which dropped at 1 week, while *Gpld1* expression was not detectable in NSPs and showed progressively increased levels up to 1 week of differentiation (Figure 1D). In order to corroborate these results in vivo, we investigated the decrease of *Gfap* expression during astrocytic maturation by isolating GFP+ cells from *hGFAP*::eGFP transgenic mice. For this, we separated GFP+ cells by FACS at postnatal day 4 (P4), P10, and P21, with P4 representing an immature phase, P10 signifying an intermediate stage, while P21 corresponded to a mature astrocytic stage. In line with the in vitro model, the amount of *hGFAP*::eGFP+ cells gradually decreased along these time points (P4: 39.2 ± 3.6%; P10: 21.0 ± 0.9%; P21: 3.5 ± 1.3%) (Figure 1E). Notably, we confirmed the upregulation of *Gpld1* as a marker of astrocytic maturation in vivo from our previous transcriptomics analyses of GFP+ cells sorted from *Aldh1l1*::eGFP BAC transgenic mice when comparing cells isolated at P4, P10, and P21 (Figure S2A) [27]. We similarly verified their progressive exit from the cell cycle by detecting the decrease of *Ccnb1* expression levels (Figure S2B). Lastly, to verify the purity of our in vitro cultures, we sorted enriched astrocytic populations by MACS using an antibody against the pan-astrocytic marker GLAST (*Slc1a3*). GLAST+ cells exhibited similar expression levels of *Mki67*, *Ccnb1*, *Gfap*, and *Gpld1* to unsorted cells (Figure S2C,D). Hence, this common gene expression pattern observed in GLAST+ and GLAST- cells suggests that our culture conditions give rise to pure astrocytic populations expressing GLAST levels consistent with their maturation state.

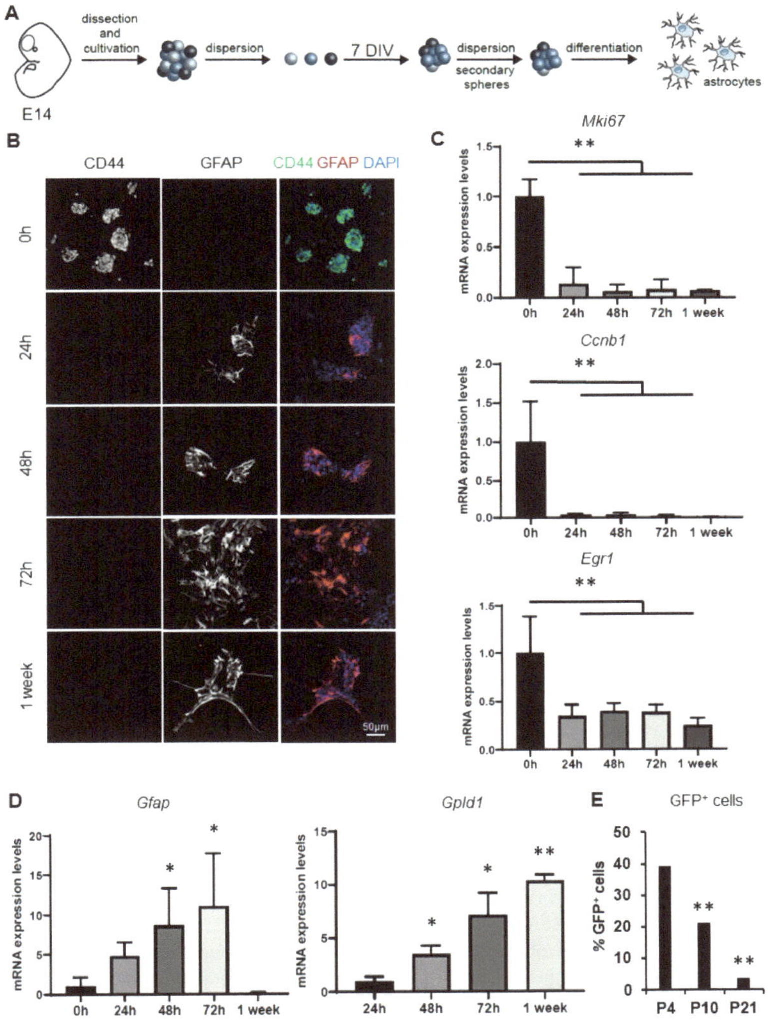

Figure 1. In vitro differentiation of neurospheres (NSPs) into astrocytes. (**A**) Schematic showing the in vitro cellular model of primary neurospheres (NSPs) derived from NSCs isolated from the ventricular zone at embryonic day-14 (E14) of the mouse embryos and the subsequent passages to differentiate them into astrocytes. (**B**) Immunostaining analyses of CD44 (green) and GFAP (red) of NSPs and differentiating astrocytes obtained from primary murine cultures of NSPs cultivated under proliferation conditions (DMEM F12, 1% B27 without vitamin A, antibiotics and 20 ng/mL EGF) or differentiated into astrocytes (DMEM, 10% FBS and antibiotics) for 24 h, 48 h, 72 h, and 1 week. Nuclei were counterstained with DAPI (in blue). Scale bar: 50 μm. (**C**,**D**) RT-qPCR analyses showing mRNA expression levels of (**C**) the cell cycle genes *Mki67* and *Ccnb1*, the NSC marker *Egr1*, (**D**) the astrocytic marker *Gfap*, and the astrocyte-enriched gene *Gpld1*. Results are expressed as mean ± standard deviation. Each time point is shown relative to a specific condition set to 1. * $p < 0.05$; ** $p < 0.01$, $n \geq 4$. (**E**) Quantification of GFP$^+$ cells from mouse forebrain cell suspensions of *hGFAP::*eGFP transgenic mice by FACS at postnatal day 4 (P4), P10, and P21. Bars represent the mean ± SEM. ** $p < 0.01$, $n = 3$.

Taken together, these results demonstrate that our in vitro model of NSPs differentiation into astrocytes represent a reproducible method enabling enrichment of astrocytic populations expressing markers of ex vivo-isolated astrocytes.

3.2. Exposure of NSPs to TNF under Differentiating Conditions Modulates the Expression of Specific NSC and Astrocytic Markers

Next, we took advantage of our model to investigate the effect of inflammation during the differentiation of NSPs into astrocytes. For this, we cultivated the NSPs in differentiation medium with or without TNF (50 ng/mL), as in our previous study [21]. As expected, TNF treatment induced NF-κB activation in differentiating NSPs as shown by enhanced *Nfkbia* mRNA levels compared to untreated cells at the different time points (Figure 2A). While we detected differences in the expression levels of the stem cell marker *Egr1* in the presence of TNF when compared to untreated samples at 24 h and 1 week (Figure 2B), TNF treatment did not cause significant changes in the expression levels of *Ccnb1* (Figure 2C) and *Mki67* (Figure 2D). Further, *Gfap* and *Gpld1* were down-regulated in TNF-treated cells at 48 and 72 h of differentiation for the former, and at 72 h and 1 week for the latter (Figure 2E,F).

These targeted gene expression studies, together with our recent analyses [21], show that exposure of differentiating NSPs to TNF induces the activation of the NF-κB pathway without affecting the exit of the differentiating astrocytes from the cell cycle. On the other hand, TNF treatment causes transcriptional changes in the expression levels of specific stem cell and astrocytic markers, thus affecting the physiologic astrocytic differentiation process.

3.3. Transcriptomics Analyses under Normal and Inflammatory Conditions Reveal Discrete Populations of Differentiating Astrocytes

As TNF affected the expression of specific NSC and astrocytic genes during differentiation of NSPs into astrocytes, we sought to expand our analyses by conducting a genome-wide RNA-sequencing analysis of the astrocytic transcriptome at different time points, both under normal and inflammatory conditions. Principal component analysis (PCA) using the top 1000 variable genes showed distinct clustering corresponding to the different conditions (Figure 3A). We identified differentially expressed genes ($p < 0.05$) between different conditions and compared them at 24 and 72 h under normal or inflammatory conditions. Specifically, we detected 4187 overexpressed genes and 4544 down-regulated genes at 24 h compared to undifferentiated NSPs, both with or without TNF. On the other hand, we detected 1005 and 756 genes, respectively, exclusively up- or down-regulated at 24 h under control conditions, while 1227 and 922 genes were correspondingly enhanced or decreased only under TNF treatments. We obtained similar results at 72 h (Figure 3B). Hierarchical gene clustering of the differentially expressed genes during normal differentiation (up- and down-regulated genes, adjusted p value < 0.05) revealed five major clusters differentially represented across the specific time points (Figure 3C). Corresponding Gene Ontology (GO) enrichment analyses indicated that proliferating NSPs were mainly associated with cluster 1, largely characterized by terms related to RNA processing. Clusters 2 and 4, largely represented in NSPs and 24 h-differentiated astrocytes, were characterized by processes associated with regulation of cell morphogenesis, motility, and migration. Clusters 3 and 5, mostly enriched in differentiating astrocytes at 72 h and 1 week, were characterized by terms linked to cell projection assembly and cilium organization (Figure 3D).

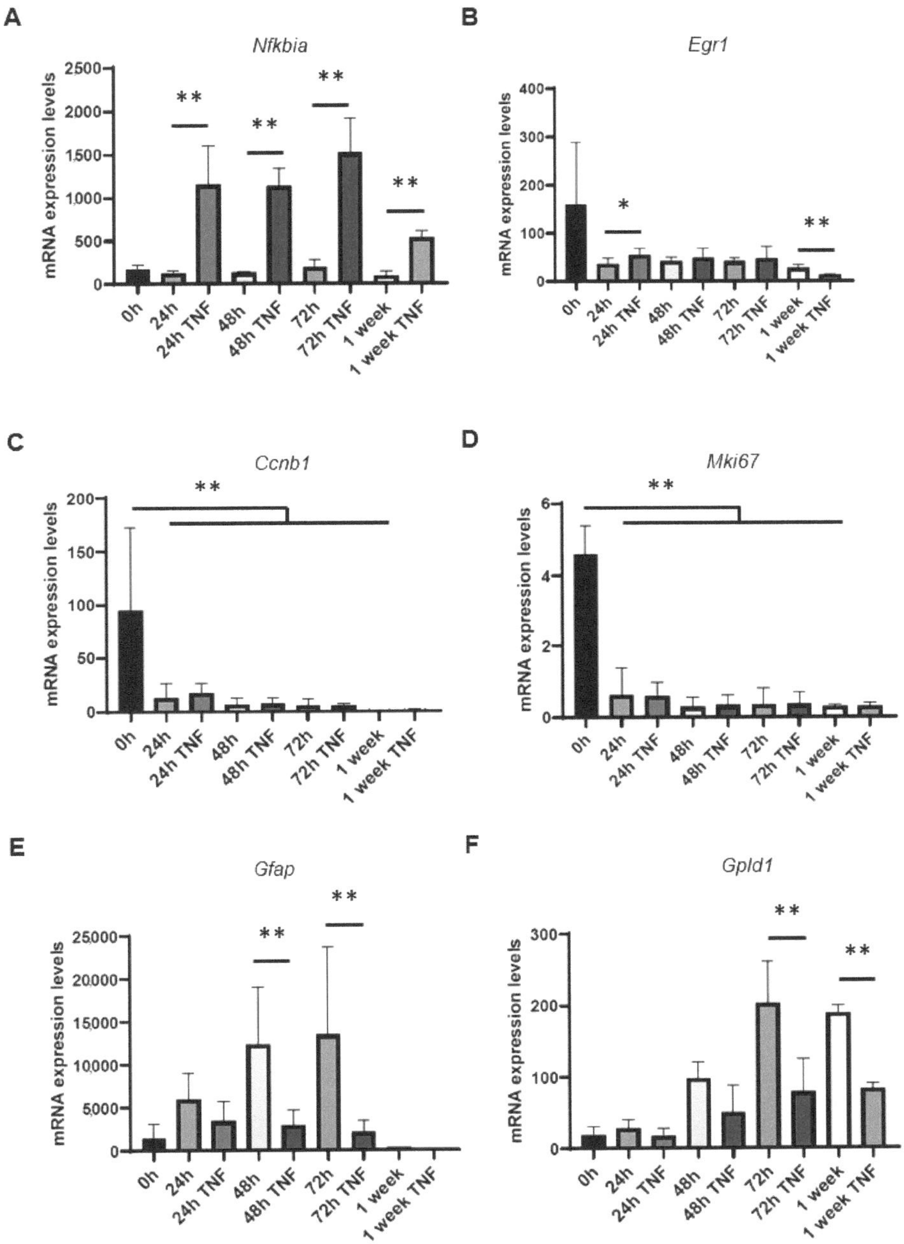

Figure 2. Gene expression levels of NSC and astrocyte markers following treatment of NSCs with TNF during astrocytic differentiation. NSPs were differentiated into astrocytes with or without exposure to TNF (50 ng/mL) for 24 h, 48 h, 72 h, and 1 week. RT-qPCR analyses show the mRNA expression levels of (**A**) *Nfkbia*, (**B**) *Egr1*, (**C**) *Ccnb1*, (**D**) *Mki67*, (**E**) *Gfap*, and (**F**) *Gpld1*. Results are expressed as mean ± standard deviation of $2^{-\Delta Ct}$ normalized to beta actin levels. * $p < 0.05$; ** $p < 0.01$, $n \geq 4$.

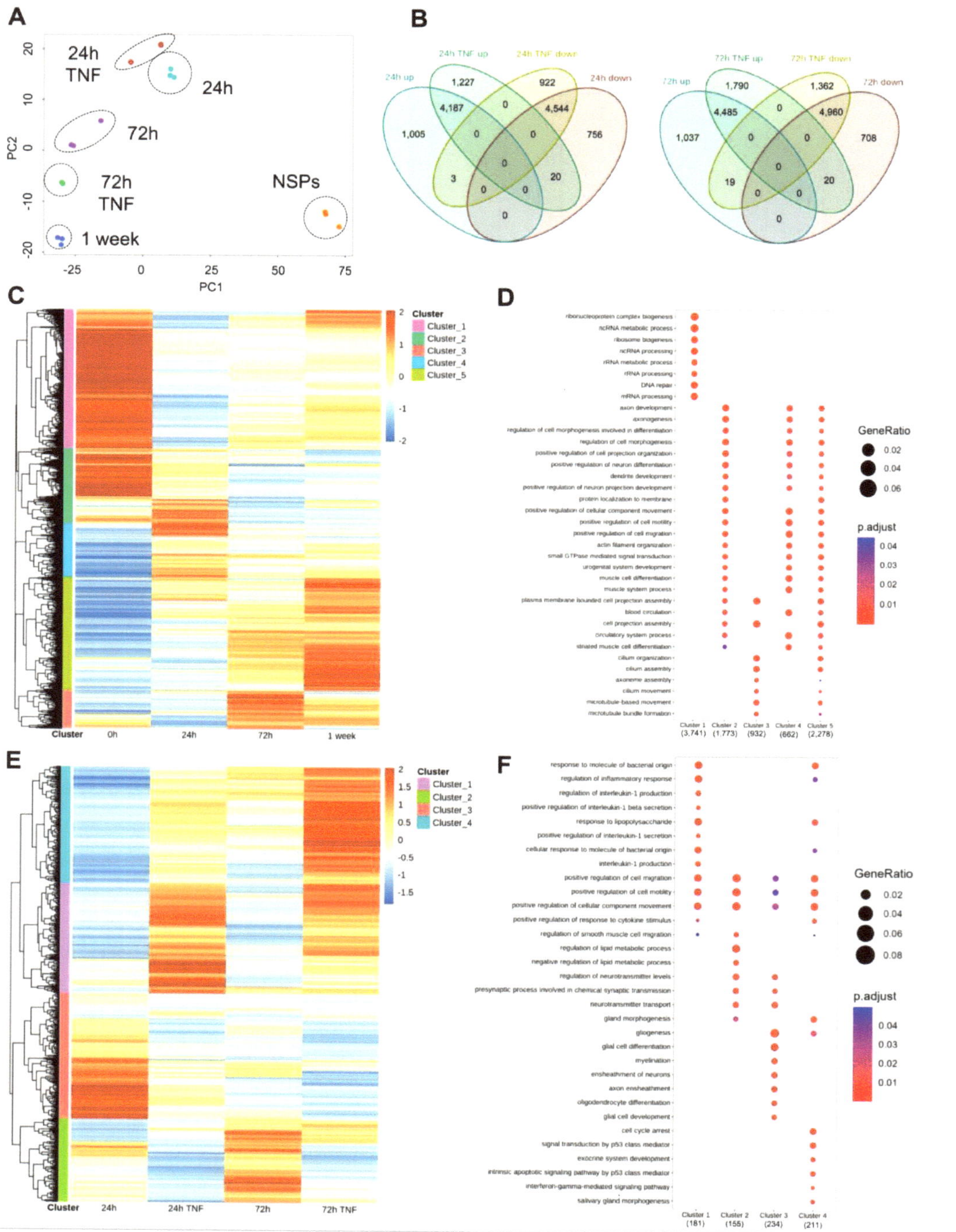

Figure 3. Transcriptomic signature of differentiating astrocytes under normal and inflammatory

conditions. (**A**) Representation of first two principal components (PC) analysis of the top 1000 most variable genes across all the conditions analyzed by RNA-seq. (**B**) Venn diagrams showing the number of significantly differentially expressed genes ($p < 0.05$) and their overlap across comparisons 24 h TNF vs. 24 h (left) and 72 h TNF vs. 72 h (right). (**C**) Heatmap showing normalized (z scores) raw gene expression counts of differentially expressed genes under normal conditions based on absolute log2 fold change from comparisons between the different time points (1 week vs. 72 h, 72 h vs. 24 h, and 24 h vs. 0 h). (**D**) Dot plot showing gene set enrichment analysis of genes in clusters identified in the heatmap (Figure 3C). (**E**) Heatmap showing normalized (z scores) raw gene expression counts under inflammatory conditions (24 h TNF vs. 24 h and 72 h TNF vs. 72 h) based on absolute log2 fold change. (**F**) Dot plot showing gene set enrichment analysis of genes in clusters identified in the heatmap (Figure 3E).

We conducted similar analyses to investigate the effect of TNF treatment at 24 and 72 h. Hierarchical gene clustering of the differentially expressed genes during differentiation under normal and inflammatory conditions (up- and down-regulated genes, adjusted p value < 0.05) identified four major clusters differentially represented across the specific settings (Figure 3E). Clusters 1 and 4, characterized by terms associated with response to molecules of bacterial origin and regulation of inflammatory response, were enriched under TNF treatment, at both 24 and 72 h. On the contrary, cluster 2 described by terms linked to regulation of cell motility and migration, was under-represented under inflammatory conditions. Notably, we observed that the presence of TNF, especially when comparing conditions at 24 h, decreased cluster 3, related to gliogenesis, glial cell differentiation and development (Figure 3F).

Next, we used heat maps to visualize gene expression levels across different time points between normal and inflammatory conditions of specific genes associated with different astrocytic categories. Specifically, we analyzed markers linked to "developing and differentiated astrocytes", "reactive astrocytes", "A1-neurotoxic-astrocytes", and "A2-neuroprotective-astrocytes" taking advantage of different available datasets [28–31] (Figure S3A–D). Notably, astrocytic gene markers, including *Gfap*, *Slc14a1*, and *Gja1* were up-regulated at 72 h in the absence of TNF (Figure S3A), while reactive astrocyte marker genes, including *Timp1*, *Lcn2*, *Icam1*, and *Ptx3*, were enhanced under inflammatory conditions, mainly at 24 h or *Sbno2*, *Cxcl10*, and *Cd109* [32–34] at 72 h (Figure S3B). Further, we showed that the obtained astrocytic populations express a mixture of A1 and A2 markers, both under normal and inflammatory conditions, although the specific markers differ between the two conditions (Figure S3C,D). Lastly, due to the importance of glycogen and glucose metabolism in astrocytes, we also looked at the expression levels of genes associated with "glycogen activity" and "glucose metabolism" (Figure S3E,F). Along the differentiation process, we detected variable expression levels of genes related to glycogen activity (*Ugp2*, *Pygb*, *Ppp1r3c*) and glucose metabolism (*Lmbrd1*, *Pid1*, *Esr1*, *Upk3b*, *Lep*), which were modulated under TNF exposure (Figure S3E,F).

Taken together, these results show that genes linked to gliogenesis and glial cell development are down-regulated in the presence of TNF, suggesting that an inflammatory environment alters the physiological astrocytic differentiation, thus affecting critical associated functional activities, including glycogen and glucose metabolic pathways.

3.4. Chromatin Accessibility Profiling Detects Extensive Chromatin Remodeling along the Astrocytic Differentiation Affected by TNF Exposure

To understand further the mechanisms underlying changes in gene expression under normal and inflammatory conditions, we sought to examine the corresponding chromatin accessibility profiles, which affect, at least partly, gene expression depending on their more open or closed states at specific loci. To do this, we performed ATAC-sequencing analyses under the same conditions as the transcriptome studies. A first PCA analysis using differential peaks discriminated the various time points and highlighted differences between treated and untreated samples. Of note, similarly to the PCA analyses of the RNA-seq samples, PC1 seemed to separate samples according to the developmental stage, while PC2 divided them according to stage and treatment (Figure 4A). Next, we conducted our analyses based on differentially accessible peaks in the promoter region nearest to the transcription start site. We identified differentially accessible peaks ($p < 0.05$) up- and down-regulated between different conditions and compared them at 24 and 72 h under normal or inflammatory conditions. Specifically, we detected 545 up-regulated and 9081 down-regulated accessible peaks at 24 h compared to undifferentiated NSPs, both with and without TNF. On the other hand, we detected 102 and 960 accessible peaks, respectively, exclusively up- or down-regulated at 24 h under control conditions, while 297 and 385 accessible peaks were correspondingly enhanced or decreased only under TNF treatments. We found similar proportions at 72 h (Figure 4B). Hierarchical gene clustering of the differentially accessible peaks across the analyzed normal differentiation time points (up- and down-accessible peaks, adjusted p value < 0.05) revealed five major clusters differentially represented across the specific time points (Figure 4C). Clusters 1, 3, and 5, mainly associated with proliferating NSPs, were the highest represented in the heatmap, thus indicating that multipotent NSPs have higher chromatin accessibility when compared to differentiated astrocytes (Figure 4C). NSPs were characterized by GO terms related to RNA processing, histone, and covalent chromatin modifications (clusters 1 and 3), synapse organization and axonogenesis (clusters 3 and 5) as well as lipid and monovalent inorganic cation transport (cluster 5). On the other hand, cluster 2, mainly enriched in NSPs and 1 week-differentiated astrocytes, was represented by GO terms related to response to chemical stress and external stimulus. Lastly, cluster 4, mostly enhanced in late-stage astrocytes, showed terms linked to lipid localization and transport (Figure 4D).

Under inflammatory conditions, we detected increased chromatin accessibility in TNF-treated cells when compared to untreated cells. More precisely, following TNF treatment, prevalent clusters 1, 2, and 3 were accessible at both 24 and 72 h (Figure 4E). Enrichment analysis of these clusters specifically showed involvement of the NF-κB signaling pathway (cluster 2), results that are in agreement with the RNA-seq data. Cluster 4, which was under-represented under inflammatory conditions, was associated with GO terms related to cell junction assembly, while cluster 5, mainly enriched in 24 h-differentiated astrocytes, was linked to regulation of developmental growth (Figure 4F). Lastly, we used volcano plots to analyze differentially accessible peaks across time points between normal and inflammatory conditions. Here, a comparison between TNF-treated and -untreated samples showed a skewed distribution towards peaks that were more open under inflammatory conditions, both at 24 and 72 h (Figure S3G).

Overall, we observed significant changes in chromatin accessibility upon astrocytic differentiation compared to undifferentiated cells, with genes associated, for example, with lipid localization and transport displaying increased accessible peaks in their promoter region. Upon TNF treatment, we detected a strong shift towards enhanced chromatin accessibility compared to normal conditions, with several gene clusters related to inflammation.

3.5. Transcription Factor Binding Analyses Uncover Dynamic Footprints Associated with Astrocyte Specification and Maturation under Normal and Inflammatory Conditions

Next, we investigated changes in the relative numbers of a given motif across accessible chromatin regions by footprint analyses using HINT-ATAC [25] to detect possible transcription factors (TFs) associated with astrocyte specification and maturation with or without TNF treatment.

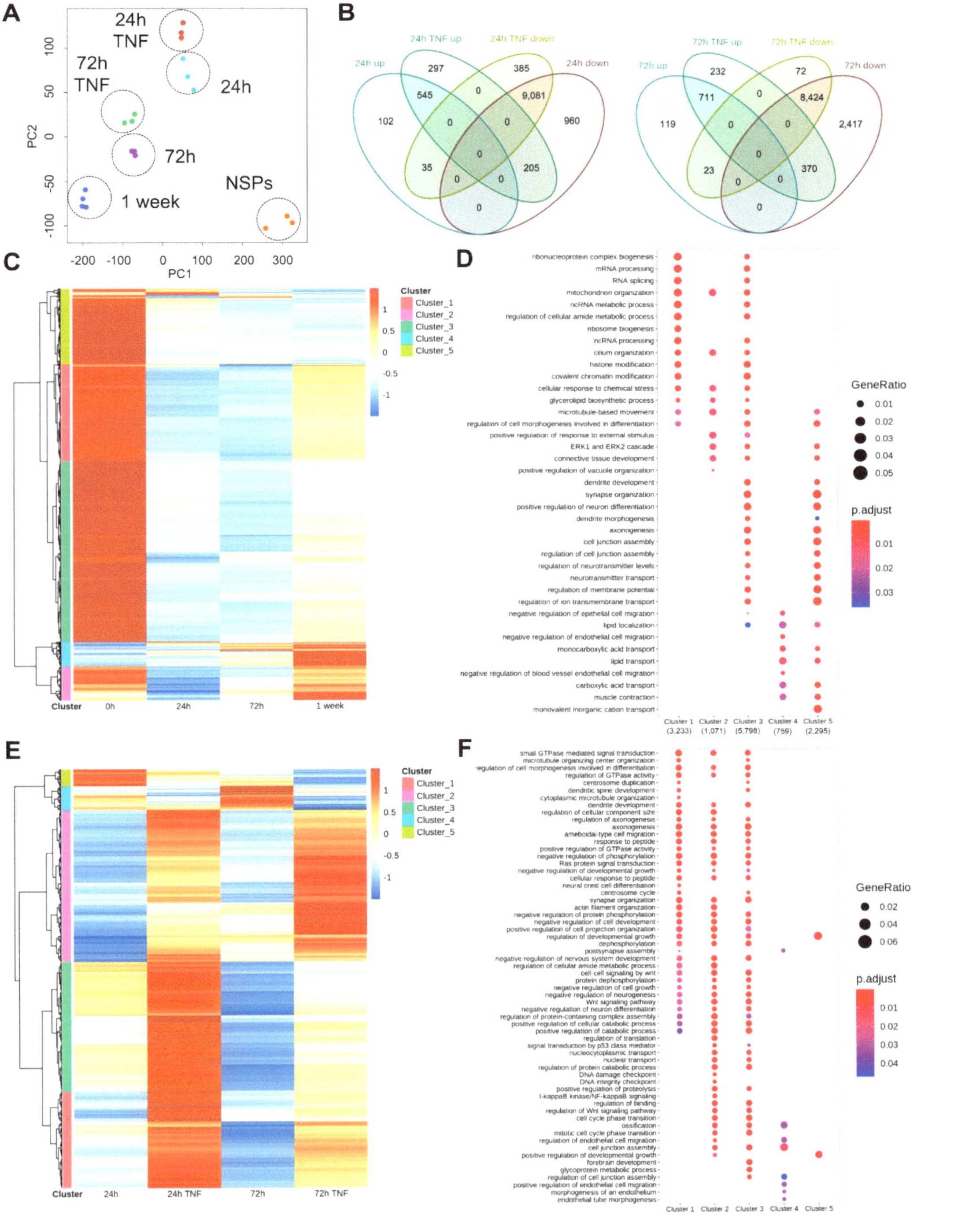

Figure 4. Chromatin accessibility profile of differentiating astrocytes under normal and inflammatory

conditions. (**A**) Representation of first two principal components (PC) analysis of the top 1000 most variable genes across all the conditions analyzed by ATAC-seq. (**B**) Venn diagrams showing the number of significantly differentially accessible peaks ($p < 0.05$) and their overlap across comparisons 24 h TNF vs. 24 h (left) and 72 h TNF vs. 72 h (right). (**C**) Heatmap showing normalized (z scores) raw expression of differential peaks under normal conditions based on absolute log2 fold change from comparisons between the different time points (1 week vs. 72 h, 72 h vs. 24 h, and 24 h vs. 0 h). (**D**) Dot plot showing gene set enrichment analysis of genes in clusters identified in the heatmap (Figure 4C). (**E**) Heatmap showing normalized (z scores) raw expression peak counts under inflammatory conditions (2 h TNF vs. 24 h and 72 h TNF vs. 72 h) based on absolute log2 fold change (**F**) Dot plot showing gene set enrichment analysis of genes in clusters identified in the heatmap (Figure 4E).

We first analyzed differences in the number of TF binding motifs between NSPs and differentiating astrocytes at 24 and 72 h under normal conditions. At 24 h, we detected an enriched activity score for *Atoh1*, *Dlx1*, *Rarg*, and *Arid3a* with a concomitant decreased score for *Dmbx1* and *Tcf21* TFs when compared to NSPs (Figure 5A). Similar analyses conducted at 72 h showed increased activity score for *Hoxd8*, *Lhx3*, *Rarb*, *Rarg*, and *Dlx1*, together with a decreased score for *Hes1*, *Tcf21*, and *Tcfl5* TFs compared to NSPs (Figure 5B) Hence, based on these analyses, ATOH1 and ARID3A may represent TFs associated with astrocytic specification (24 h), HOXD8, LHX3, and RARB may represent TFs linked to astrocyte maturation (72 h), and DLX1 and RARG are likely TFs linked to both processes (Figure 5A,B).

By conducting similar analyses under inflammatory conditions, we detected at 24 h increased activity score for *Nfe2l2 (Nrf2)* and a decreased score for *Dmbx1*, *Arid3a*, *Arid3b*, and *Atoh1* TFs when comparing to normal conditions (Figure 5C). Lastly, at 72 h, *Atoh1* and *Rarg* TFs showed an enhanced score by comparing TNF-treated versus untreated cells, while the score for *Tcf21* and *Dmbx1* was decreased (Figure 5D). Due to the importance of *Nfe2l2 (Nrf2)* in regulating inflammatory responses, mainly by inducing the expression of antioxidants and cytoprotective genes, we analyzed its binding motifs across accessible chromatin regions at the genome-wide level. We detected enriched footprints across the whole genome in TNF-treated cells at 24 h (Figure 5E), thus suggesting that NRF2 may represent a master TF driving the astrocytic differentiation in the presence of TNF.

3.6. *Transcriptional and Chromatin Accessibility States Positively Correlate at Specific NSC and Astrocytic Gene Loci*

Next, we sought to study the correlation between transcriptional and chromatin accessibility states at the genome-wide level to explore to which extent the chromatin states are linked to the transcriptional activities at specific gene loci. For this, we took advantage of our RNA-seq and ATAC-seq data to conduct correlation analyses between gene expression levels and their chromatin accessibility states at 24 h. We used scatterplots to show the number of genes differentially expressed and exhibiting differentially accessible peaks at their promoter regions under normal (Figure 6A) and inflammatory (Figure 6B) conditions. By using Pearson's correlation analysis for log-fold change in gene expression and accessibility, we identified similar correlation coefficients (normal: R = 0.38; inflammation: R = 0.51) to those detected in analogous studies comparing RNA-seq and ATAC-seq datasets (Ackermann et al., 2016, Starks et al., 2019). In a second step, we conducted a targeted correlation analysis at specific NSC, resting and reactive astrocytic gene loci [28–31]. Here, we calculated at 24 h a higher correlation score (R = 0.64) compared to genome-wide levels. Under normal conditions, we showed that specific NSC (e.g., *Ccnb1*, *Mki67*, *Egr1*) or astrocytic (e.g., *Gfap*, *Gpld1*, *Clcf1*) markers exhibited, respectively, decreased or increased gene expression levels and accessibility (Figure 6C). Under TNF exposure, we detected enhanced gene expression levels and accessibility for reactive astrocyte markers (e.g., *Lcn2*, *Cxcl10*, *Osmr*, *Sbno2*) (Figure 6D). Interestingly, *Nfkbia*, a pan-inflammatory marker, exhibited both enhanced gene expression levels and accessibility states (Figure 6D).

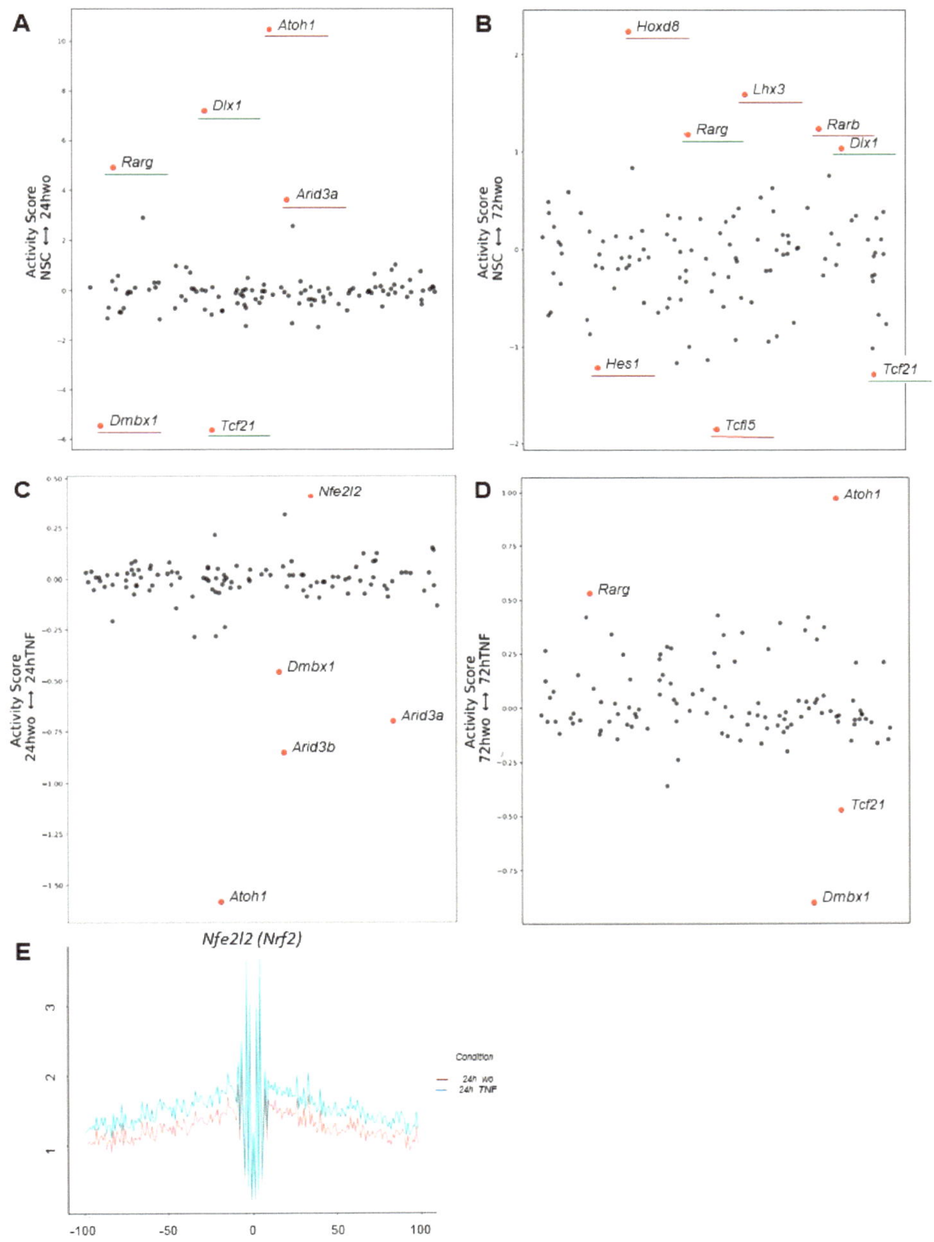

Figure 5. Chromatin footprinting activity under normal and inflammatory conditions. (**A–D**) Activity score plots at 24 and 72 h time points under (**A**,**B**) normal and (**C**,**D**) inflammatory conditions. In (**A**,**B**), shared TF binding motifs at 24 and 72 h are underlined in green, while unique TF binding motifs are in red. (**E**) Comparison of the activity of *Nfe2l2 (Nrf2)* gene based on motif enrichment analysis at 24 h between normal (red line) and inflammatory (blue line) conditions.

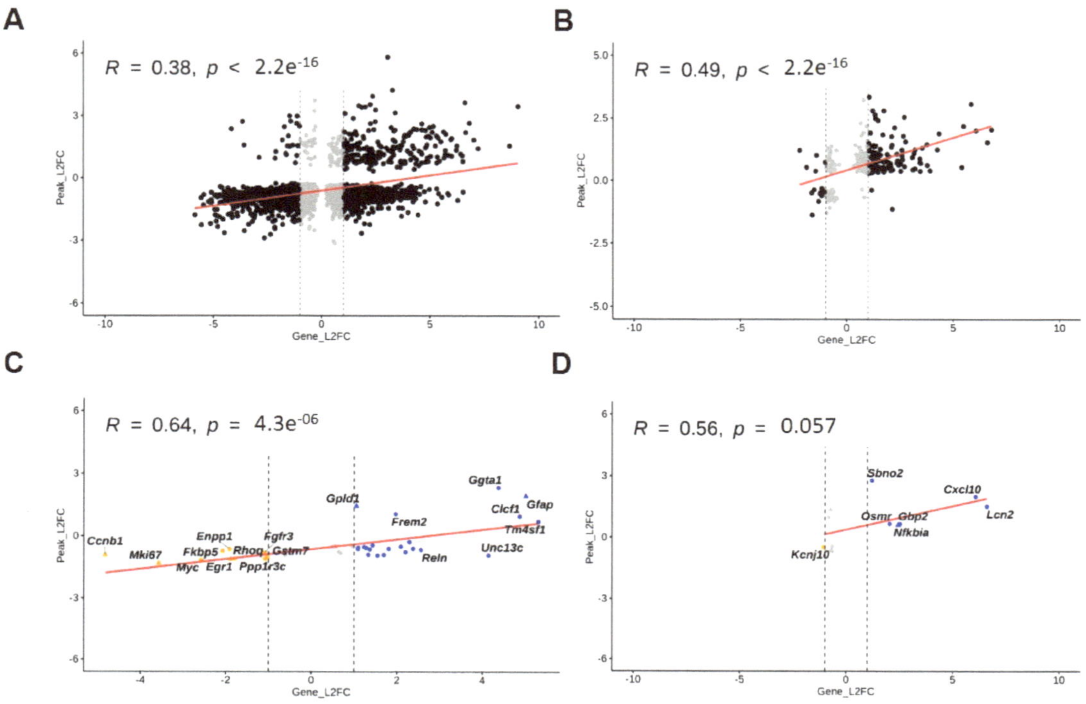

Figure 6. Correlation analyses of transcriptional and chromatin accessibility changes upon the differentiation process under normal and inflammatory conditions. Scatter plots for combined RNA-seq and ATAC-seq data (**A**,**B**) at the genome-wide level and (**C**,**D**) at specific NSC, resting and reactive astrocytic gene loci at 24 h under (**A**,**C**) normal and (**B**,**D**) inflammatory conditions. Plots display Pearson's correlation coefficient indicating correlation levels between RNA-seq and ATAC-seq data, along with corresponding *p*-values (considering differentially expressed genes with adjusted *p* value < 0.05 and differentially accessible peaks with false discovery rate (FDR) <0.05). Blue and orange dots represent marker genes, which are significantly differentially expressed (adjusted *p* value < 0.05), have more than 2-fold positive change (upregulation) or more than 2-fold negative change (downregulation), respectively, in gene expression and have a significantly differentially accessible peak (FDR < 0.05) in their promoter region. Red lines represent the regression line of correlation between fold change in gene expression and accessibility.

Taken together, these results indicate that at specific NSC and astrocytic gene loci, gene expression levels are associated with chromatin accessibility states, thus indicating a close link between chromatin remodeling and gene expression at lineage-specific gene loci.

3.7. Inference of Gene Regulatory Networks Enables the Identification of Key Transcription Factor Bindig Motifs Associated with Astrocytic Differentiation under Physiological and Inflammatory Conditions

To systematically identify regulators of astrocytic differentiation under physiological and inflammatory conditions, we reconstructed the corresponding gene regulatory network (GRN) for each analyzed time-point. The GRN after 24 h of physiological differentiation was represented by 262 interactions between 96 TFs, which, based on the topological analysis, include *Esr1*, *Nr3c1*, *Foxa1*, *Gata2*, *Stat5a*, *Fosl2*, *Stat1*, *Nfia*, and *Arid5b* (Figure 7A). Corresponding GO analyses highlighted terms associated to developmental processes, including "regulation of developmental process", "tissue development", and "regulation of cell differentiation" (Figure S4A). At 72 h, when compared to 24 h, the GRN consisted of 91 interactions between 46 TFs, thus reflecting a lineage-committed network made up of less than half the TFs when compared to earlier stages (Figure 7B). The top regulatory hubs include *Rxra*, *Rarg*, *Cebpb*, *Myc*, *Smad3*, *Rbpj*, and *Esr1*. Subsequent GO analyses showed enriched terms related to cell responses and communication, such as "cellular response to stimulus", "signaling", and "signal transduction" (Figure S4B), thus indicating that at earlier stages, differentiated cells went through the developmental process (astrocyte specification) and later acquired their functional astrocytic properties (astrocyte maturation). Interestingly, 10 TFs were shared between the two GRNs (Figure S4C). Out of these, we identified *Esr1* and *Foxp1* as regulatory hubs at both time points.

Similarly, we reconstructed GRNs under inflammatory conditions. Following a 24-h differentiation, the GRN consists of 107 TFs and 393 interactions (Figure 8A). We identified *Stat3*, *Smad3*, *Nfkb1*, *Runx1*, *Ar*, *Cebpb*, *Relb*, *Nfkb2*, and *Atf3* as regulatory hubs. Out of these TFs, 75 of them, such as *Esr1*, *Nr3c1*, *Gata2*, *Stat5a*, *Fosl2*, *Sat1*, *Junb*, *Mef2c*, *Arid5b*, *Mef2a*, and *Foxo3*, were shared with the corresponding GRN under normal conditions (Figure S5A), suggesting that the developmental process and the inflammatory response overlap during the acquisition of the astrocytic phenotype. GO analyses of the TFs underlying the GRN at 24 h under inflammatory conditions strengthen this observation, with terms associated to both developmental processes, such as "regulation of developmental process", and inflammatory responses, including "immune system process", being highly represented (Figure S5B). Lastly, following a 72-h differentiation under TNF treatment, the GRN is composed of 18 interactions between 15 TFs, including *Rxra*, *Myc*, *Stat5b*, and *Rorc* (Figure 8B). Corresponding GO analyses identified "response to chemical", "response to hormone" and "response to lipid" as main enriched terms (Figure S5C). Among the resultant TFs, 12 of them, including *Rxra*, *Rarg*, *Stat5b*, and *Myc*, were in common with the corresponding network under physiological conditions, while 3 TFs (*Thrb*, *Zfp217*, *Zbtb37*) were unique to the inflammatory condition (Figure S5D).

Taken together, reconstructed GRNs enabled the identification of regulatory hub TFs playing critical roles along the differentiation of astrocytes in the presence or absence of TNF. These analyses allowed establishing the effect of an inflammatory environment along the astrocytic development, indicating that it affects the underlying transcriptional signatures skewing astrocytes towards a discrete astrocytic phenotype when compared to physiological settings.

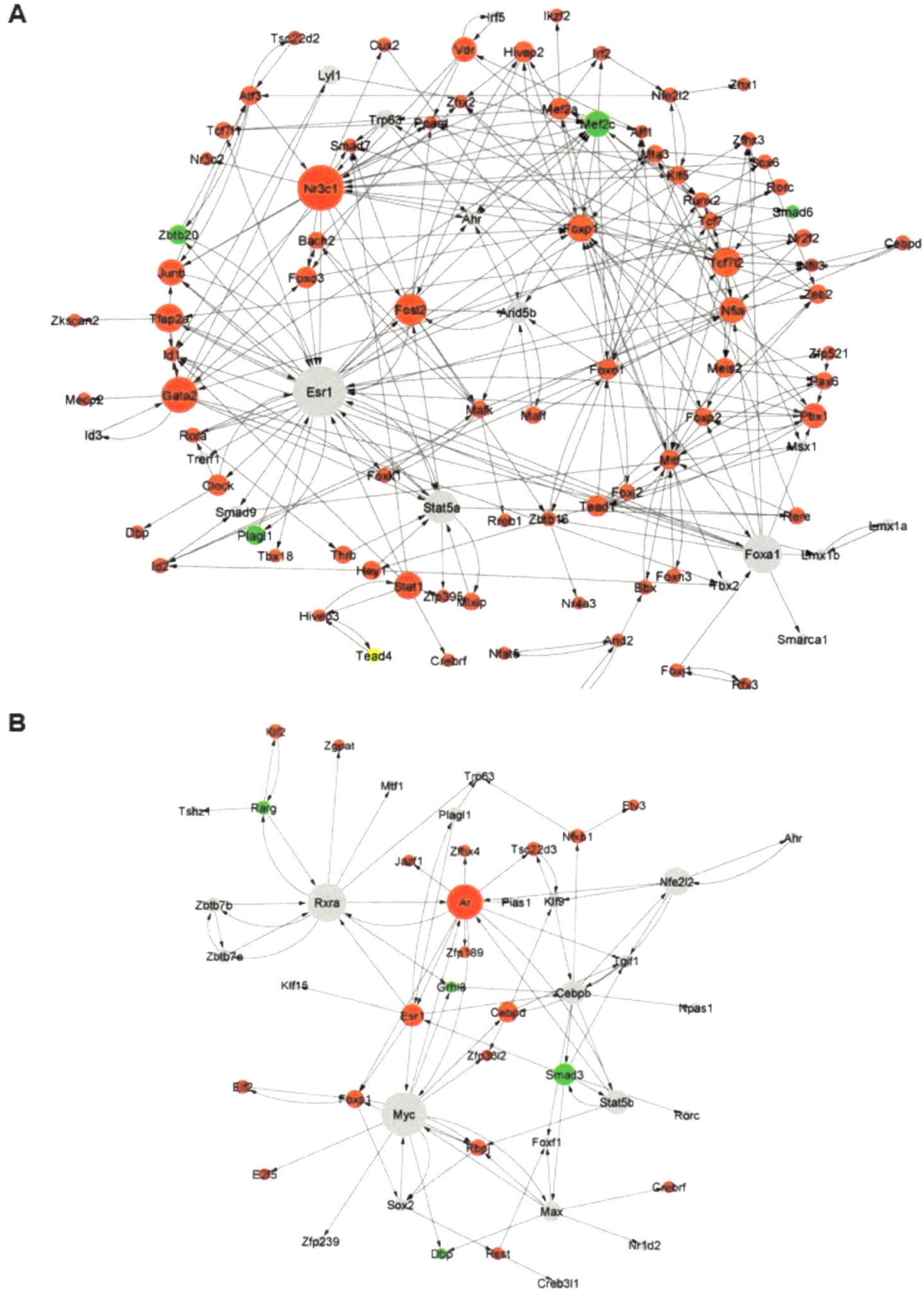

Figure 7. Reconstructed gene regulatory networks under normal conditions. TF-TF regulatory network at (**A**) 24-h and (**B**) 72-h time points. Color legend represents differential accessibility of the TFs (green: upregulated; red: downregulated; grey: not differentially accessible).

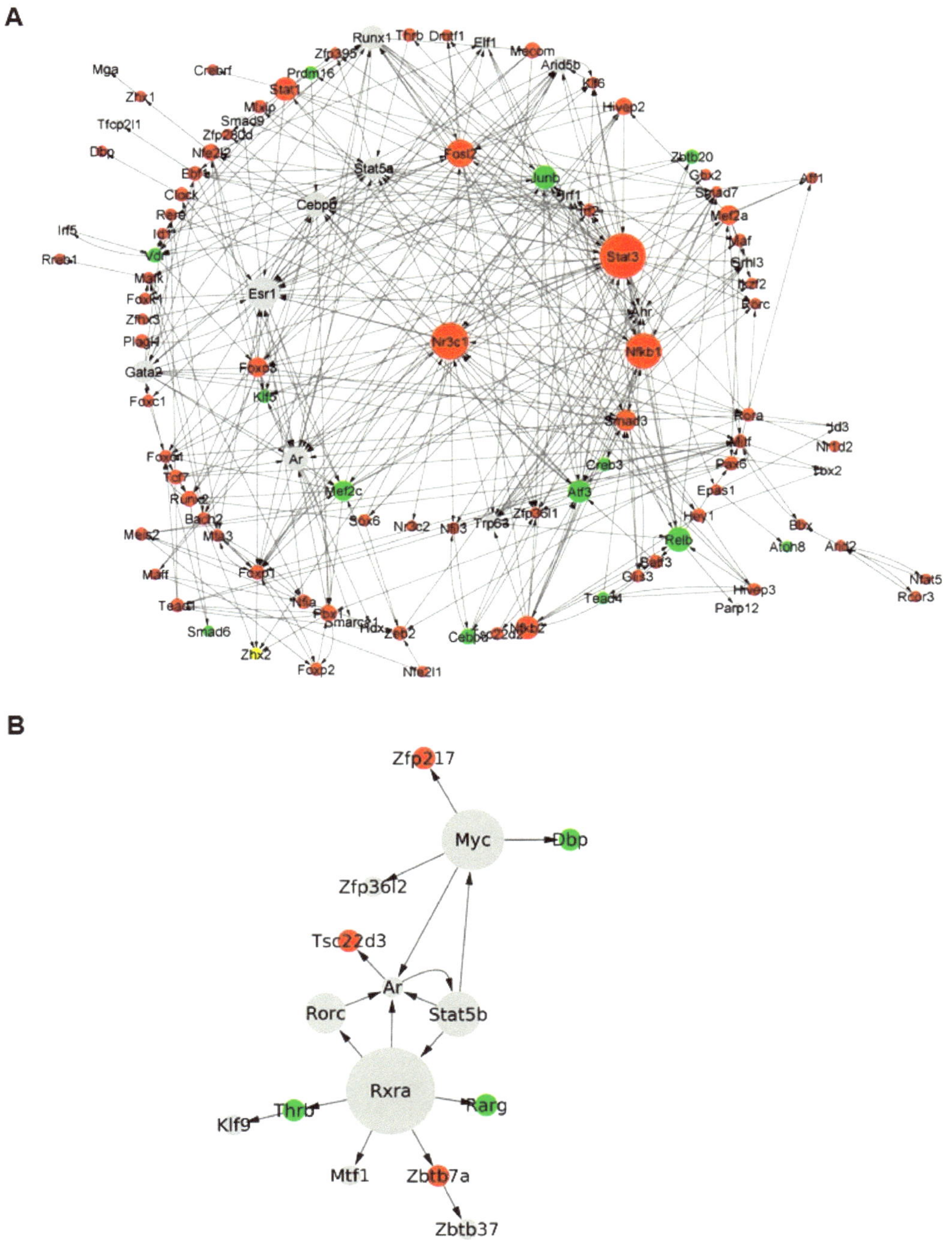

Figure 8. Reconstructed gene regulatory networks under inflammatory conditions. TF-TF regulatory network at (**A**) 24-h and (**B**) 72-h time points. Color legend represents differential accessibility of the TFs (green: upregulated; red: downregulated; grey: not differentially accessible).

4. Discussion

Astrogliogenesis depends on both extracellular cues and intrinsic dynamic changes of the epigenome. Perturbations during astrocytic development may result in neurological diseases, including neurodevelopmental disorders, such as Rett and fragile X syndromes [35] or Down syndrome and autism spectrum disorders [36]. In recent years, computational approaches to study astrocytes and their development using next-generation sequencing technologies have enabled a better understanding of their phenotypic and functional role in the CNS [37,38]. Further, the generation of novel in vitro models, such as astrocytes derived from induced pluripotent stem cells, has opened up a new area for studying neurological diseases in vitro. However, as the brain hosts a range of astrocyte populations, it is critical to develop standardized protocols for the in vitro generation of astrocytes with defined maturity states and phenotypic properties [39].

4.1. In Vitro Model Characterization

In this context, our in vitro NSP differentiation mouse model enabled us to conduct kinetic analyses along the process of astrocytic differentiation. In this model, under specific culture conditions, NSCs retain their multipotential properties, thus being able to give rise to neurons, astrocytes, and oligodendrocytes. For our experiments, we differentiated NSPs into astrocytic populations by cultivating them in the presence of serum, as previously described [21]. In these conditions, we detected the highest *Gfap* expression levels at 48 and 72 h, whereas its expression decreased after 1 week. This *Gfap* expression pattern is supported by previous in vivo studies, documenting that although GFAP is a reliable marker of reactive astrocytes, it is only variably detectable in mature astrocytes of the healthy mouse central nervous system [40]. Intriguingly, although GFAP is a reliable marker for reactive astrocytes, within our model we investigated the effect of inflammation in a different context than in mature astrocytes, as we treated NSCs with TNF and analyzed the resulting astrocytic phenotype. From our results, it appears that TNF dampens the expression of GFAP in differentiating astrocytes. This observation might be explained by various molecular insights as described below.

4.2. Targeted Gene Expression and Genome-Wide Transcriptional Analyses

We have previously shown that TNF is a key regulator of astrocytic reprogramming into neural progenitor-like cells [41]. In addition, TNF treatment of in vitro differentiated murine astrocytes resulted in modifications of their histone chromatin profile at the promoters of genes related to cell cycle, stemness or neuronal fate, thus indicating that inflammatory events intervene in such regulatory pathways [27]. In the present study, RNA-seq analyses showed major gene expression changes along the physiological process of astrocytic differentiation, which is modulated by TNF, suggesting the establishment of discrete astrocytic populations between normal and inflammatory conditions. More specifically, TNF-treated NSC-derived astrocytes exhibited significantly increased expression levels of classical reactive astrocytic marker genes, including *Timp1*, *Lcn2*, *Icam1*, and *Ptx3*, but not *Gfap*, which is the classic indicator of astrogliogenesis. In line with these observations, reactive astrocytes in inflammatory environments display increased expression levels of various proteins that are absent or weakly expressed in resting states. Among those proteins, LCN2 and TIMP1 are highly up-regulated in reactive astrocytes, with LCN2 considered as a marker of astrogliosis [28]. ICAM-1 is aberrantly expressed by astrocytes in CNS pathologies, such as multiple sclerosis, experimental allergic encephalomyelitis and Alzheimer's disease, suggesting a possible role for ICAM-1 in these disorders [42]. Lastly, *Ptx3* is considered as a marker of A2 reactive astrocytes [29,43]. Hence, TNF-treated NSC-derived astrocytes express markers of both A1 and A2 reactive types, either representing a distinct type of astrocytes that do not exclusively relate to the described dichotomous distribution or signifying a hybrid population.

Notably, the loss of normal astrocytic functions and not only reactivity contributes to the early pathologic events of neurological diseases, as described for example in multiple sclerosis [44,45]. In this context, we sought to examine genes associated with important astrocytic functions, including glucose import from blood, the primary energy source for generating ATP in the brain, and the metabolism of glycogen, which represents an important energy reserve synthesized from glucose in astrocytes [46]. Indeed, astrocytes represent an exclusive reserve of glycogen in the brain to produce lactate as an energy source transported to active neurons [47–50]. Thus, the detection of decreased expression levels of genes associated with glucose and glycogen metabolism under TNF exposure suggests an impairment of key functional activities when NSCs differentiate into astrocytes under inflammatory conditions. Of note, we have previously demonstrated that TNF treatment of primary mature astrocytes induces the decrease of genes related to glycogen metabolism. Specifically, the expression levels of glycogen phosphorylase and protein targeting to glycogen were strongly down-regulated following TNF exposure [41].

4.3. Genome-Wide Chromatin Accessibility Profiling

The ATAC-seq method enabled us to analyze the accessible chromatin landscape of differentiating astrocytes and the extent of its modifications upon inflammation, while providing insights into gene transcription complexity. We observed a general decrease of chromatin accessibility along NSC differentiation into astrocytes, thus suggesting a more lineage-restricted profile. However, upon TNF treatment, chromatin accessibility at the promoter region of specific NSC, astrocytic, and inflammatory markers was up-regulated, indicating that inflammation-induced astrocytes may retain some stemness vestige. The next step would be to identify which epigenetic mechanisms, such as DNA methylation or histone modifications, may be linked to the transcriptional and chromatin accessibility alterations that we identified either in the physiologic differentiating astrocytes or under inflammatory conditions. Intriguingly, a recent study showed that aberrant transcription in gliomas does not result from DNA methylation mechanisms, but rather from DNA methylation-independent modifications [51].

4.4. Footprint Analyses

Footprint analyses identified, among several modulated TF binding motifs across accessible chromatin regions, decreased levels of the downstream effector of the Notch pathway, HES1, when comparing differentiated astrocytes at 72 h with NSCs. Since HES1 is involved in the repression of pro-neural gene expression and maintenance of NSCs [52–54], decreased levels of its binding motifs at later time points are in line with a lineage-restricted fate of differentiated astrocytes when compared to NSCs. Further, we detected increased levels of NRF2 binding motifs across accessible chromatin regions at 24 h under inflammatory conditions. NRF2 is a master regulator of stress-induced antioxidant response inducing the expression of antioxidants and cytoprotective genes [55]. In the brain, NRF2 is involved in astrocytic neuroprotective pathways via their interactions with neurons [55,56]. Interestingly, in our dataset, the NRF2 target gene *Cebpb* [57] was both upregulated and accessible in the presence of TNF. Taken together, our results point towards NRF2 as a key TF counterbalancing inflammation and triggering self-protection in differentiating astrocytes exposed to TNF.

4.5. RNA-seq & ATAC-seq Correlation Analyses

Our correlation analyses are consistent with previous correlation studies addressing gene expression and chromatin accessibility, thus providing further support to our data [58]. Notably, correlation levels were enhanced when looking at NSC, astrocytic, and inflammatory markers. Further, our chromatin accessibility and gene expression analyses resulted in similar biological processes and pathways, thus indicating concordant modules of regulation at these molecular levels.

4.6. Reconstructed Gene Regulatory Networks

GRNs inferred from our transcriptional analyses enabled the identification of TFs underlying the astrocytic differentiation under physiological and inflammatory conditions Different pathways are known to be sequentially involved in the astrocytic differentiation process, with the Notch, STAT3, and BMP signaling pathways being the most important ones implicated in this process [59]. At 24 h of differentiation under normal conditions, the RNA of several TFs related to these pathways were highly represented in the GRN. These results validate once again the relevance of our in vitro model, which is often described in the literature as very different from the differentiation processes obtained in vivo [60]. Indeed, under our established differentiation conditions in the absence of growth factors and mitogen cocktails found in other studies [60,61], we obtained an efficient and reproducible process of astrocytic differentiation. Among the key detected regulators, *Esr1*, *Nr3c1*, *Foxp1*, *Runx2*, and *Nfia* appear particularly interesting. For example, *Esr1*-dependent signaling in astrocytes is involved in the regulation of several important pathways, such as PI3K-AKT, ERK, and JAK-STAT, and plays a crucial role in their metabolic control of providing nutrients to the neurons [62,63]. *Nr3c1*, also known as glucocorticoid receptor, is an important component of the metabolic functions of astrocytes and is required for memory formation [64]. NSCs express *Foxp1* promoting their maintenance and renewal [65], and is required for differentiation towards astrocytes [66]. *Runx2* promotes maturation-associated transcriptional changes in astrocytic cell culture models [61]. Lastly, *Nfia* contributes to the gliogenic switch enabling rapid derivation of functional human astrocytes from pluripotent stem cells [67]. Its expression is required for the maintenance of the astrocytic morphological complexity and its deficiency compromises the astrocyte–neuron communication [68]. Notably, most of the detected TFs show reduced chromatin accessibility after 24 h of differentiation, thus suggesting that they are implicated in the first step of differentiation, but not in the astrocytic maturation. The TFs identified at 72 h relate to the maturation and development of astrocytic functions. Compared to earlier time points, the network is less complex, characterized by the reduction of the number of TFs composing the GRN with a robust representation of TFs related to retinoic acid (RA). In astrocytes, GFAP expression induced by RA is mediated by the activation of RA receptor alpha (Rara), which downstream activates the PI3K pathway and binds to the STAT3-p300/CBP-SMAD complex [69]. The PI3K-AKT pathway also plays crucial roles in plasmalemmal glutamate transport [70] and synthesis of neuroprotective RANTES in astrocytes [71]. The presence of *Sox2* in the network reflects a certain degree of immaturity still at 72 h of differentiation. In fact, SOX2 acts upstream the Notch pathway to maintain cell proliferation potential, and its expression in the brain is restricted to neural stem and progenitor cells, glial precursors and proliferating astrocytes [72]. This observation is corroborated by the presence in the GRN of the effector of the Notch pathway, *Rbpj*, whose reduced accessibility suggests that the Notch pathway has lost part of its activity. Interestingly, reduction of Notch signaling in astrocytes promotes neurogenesis and absence of the Notch coactivator RBPJ activates neurogenesis in astrocytes [73]. Lastly, the TGFB1-SMAD signaling, with *Smad3* being among the TFs whose accessibility is increased, is involved in the fate commitment process of astrocytes [74].

Under inflammation, the TFs that are involved in the GRN at 24 h represent both astrocytic development and inflammatory responses. This is reflected by the overlap of regulatory hubs from the normal condition and TFs representing the inflammatory part of the process (e.g., *Stat3*, *Nfkb2*, *Runx1*, *Atf3*, *Relb*). STAT3 and NF-κB regulate the function and development of cells under inflammation [75–77]. The roles of NF-κB and STAT3 in colon, gastric, and liver cancers have been extensively investigated, and the activation and interaction between STAT3 and NF-κB plays a fundamental role in the control of the communication between cancer cells and inflammatory cells [78,79]. Specifically, STAT3 plays an important role in the development of astrocytes, as it binds to the promoter region of the *GFAP* gene inducing its expression, and its absence results in defects during astrogliogenesis [80]. The decrease of chromatin accessibility at the promoter region of *Stat3* may reflect the partial inhibition of the astrocytic differentiation induced by NF-κB activation. NFKB1 and NFKB2 are critical effectors of the canonical NF-κB pathway regulating expression of genes involved in inflammatory and cell-survival responses [81]. Further, *Runx1* codes for an important TF involved in the astrocytic development, and its overexpression is linked to differentiation along astrocytic and neuronal lineages [82]. Lastly, *Cebpb* is involved in the regulation of multiple pathways associated with inflammatory responses in astrocytes [83]. *Atf3* expression in astrocytes is induced by NRF2, a key regulator of stress-induced antioxidant response [84]. It is interesting to note that several TFs, such as *Esr1*, are present in the GRN both under normal and inflammatory conditions at 24 h of differentiation, thus suggesting that although inflammation affects the astrocytic differentiation, patterns of the normal process still drive the process of differentiation.

At 72 h under inflammation, the GRN is composed of TFs involved in multiple metabolic functions. For example, *Rarg*, *Rorc*, and *Rxra* are involved in the RA pathway, while *Thrb* is the thyroid hormone receptor. Astrocytes are the mediators of thyroid metabolism, which is linked to the proliferation and differentiation of astrocytes and also promotes neuronal development via astrocytes [85]. The corresponding GO terms support these processes related to metabolism (e.g., "response to steroid hormone", "response to corticosteroid", and "response to glucocorticoid"). Overall, the identified TFs support the tight interplay between development and inflammation pathways influencing astrocytic functional activities.

5. Conclusions

Taken together, our results show that an inflammatory environment affects NSC differentiation into astrocytes. Indeed, we here demonstrate that exposure of NSCs to TNF, a master pro-inflammatory cytokine, skews serum-differentiated astrocytes towards a specific phenotype characterized by impaired key functional activities ascribed to normal astrocytes, including glucose and glycogen metabolism. These divergent astrocytes may inadequately support the underlying neuronal network, would it be in its nascent phases, e.g., by affecting synaptogenesis; or in the adult brain, e.g., by inadequately providing nutrients to neurons, thus modifying the fine-tuned neuronal activities. Overall, these changes might contribute to the development and progression of specific neurological disorders, including neurodevelopmental diseases and brain tumors.

Supplementary Materials: The following supporting information can be downloaded at: https://www.mdpi.com/article/10.3390/cells12060948/s1, Figure S1. Generation of pure and mature astrocytic populations. (A) Primary murine neurospheres (NSPs) were differentiated into astrocytes (DMEM, 10% FBS) for 2 weeks. Pictures show immunostaining of differentiated astrocytes labeled with Kir4.1, SLC1A2, or SLC1A3 (in green) together with GFAP (in red). Nuclei were counterstained with DAPI (in blue). Scale bar: 50 µm. (B,C) Primary murine neurospheres (NSPs) were cultivated for 72 h in an altered differentiation medium (DMEM, 0.5% FBS, 1% B27 without vitamin A, 2 ng/mL EGF). Pictures show immunostaining of undifferentiated and differentiated NSPs at different time points (24 h, 48 h, and 72 h) labeled with antibodies against (B) GFAP (green) and MAP2 (red) or (C) O4 (green) and GFAP (red). Nuclei were counterstained with DAPI (in blue). Scale bar: 50 µm. (D,E) Magnification of (D) MAP2+ neurons (red) and (E) O4+ oligodendrocytes (green). Scale bar: 50 µm. Figure S2. Assessment of the validity of the in vitro model. (A,B) Gene expression levels of (A) *Gpld1* and (B) *Ccnb1* obtained from the transcriptomics analyses of GFP+ cells sorted from *Aldh1l1*::eGFP BAC transgenic mice comparing cells isolated at P4, P10 and P21 [25]. Results are shown as log FC relative to P4. (C,D) Primary murine neurospheres (NSPs) were cultivated under proliferation conditions or differentiated into astrocytes for 24 h, 48 h, and 72 h. We used GLAST magnetic beads to separate GLAST/SLC1A3+ and GLAST/SLC1A3- cells by MACS. Graphs show RT-qPCR results of *Mki67*, *Ccnb1*, *Gfap*, and *Gpld1* in (C) GLAST/SLC1A3+ and (D) GLAST/SLC1A3- cells. Bars represent the mean ± SD of 2 independent experiments. Figure S3. Gene expression analyses for specific astrocytic markers and chromatin peak distribution comparing normal and inflammatory conditions during NSP differentiation. (A–F) Heat maps showing gene expression of specific astrocytic markers classified in various categories: (A) "developing and differentiated astrocytes", (B) "reactive astrocytes", (C) "A1 astrocytes", (D) "A2 astrocytes", (E) "glycogen activity", and (F) "glucose metabolism". Heat maps represent normalized (z scores) raw gene expression counts of differentially expressed genes (adjusted p value < 0.05) in normal (24 h vs. 0 h, 72 h vs. 24 h) and inflammatory (24 h TNF vs. 24 h and 72 h TNF vs. 72 h) conditions. (G) Volcano plots showing the distribution of open and closed differentially accessible peaks at different time points upon astrocytic development between treated and untreated NSPs (24 h TNF vs. 24 h and 72 h TNF vs. 72 h); FDR, false discovery rate. Figure S4. Enrichment analysis of transcription factors in gene regulatory networks under normal conditions. (A,B) Gene Ontology terms for (A) 24 h and (B) 72 h time points. (C) Venn diagram showing the comparison between 24 h and 72 h. Figure S5. Enrichment analysis of transcription factors in gene regulatory networks under inflammatory conditions. (A) Venn diagram showing the comparison of astrocytic differentiation between physiological and inflammatory conditions at 24 h. (B,C) Gene Ontology terms for (B) 24 h and (C) 72 h time points. (D) Venn diagram showing the comparison of astrocytic differentiation between physiological and inflammatory conditions at 72 h.

Author Contributions: L.G., A.d.S. and A.M. designed project and obtained financial support; M.A.S.P., R.H., N.N., C.B. and A.M. performed experiments; M.A.S.P., K.S., S.R., L.G. and A.M. analyzed experiments; K.S. and S.R. conducted bioinformatics analyses; L.G., A.d.S. and A.M. supervised research; M.A.S.P., K.S. and A.M. wrote the manuscript. All authors have read and agreed to the published version of the manuscript.

Funding: M.A.S.P., L.G., A.d.S. and A.M. were supported by the University of Luxembourg (IRP ASTROSYS). A.M. was supported by the Luxembourg Institute of Health and the Luxembourg Centre for Systems Biomedicine (MIGLISYS). K.S. was supported by the Luxembourg National Research Fund (FNR) through the FNR-PRIDE program NEXTIMMUNE for doctoral education (PRIDE/11012546/NEXTIMMUNE).

Institutional Review Board Statement: Animal experiments were performed according to all applicable laws and regulations, after receiving approval by the institution's Animal Experimentation Ethics Committee at the University of Luxembourg (AEEC) and the Ministry of Agriculture, Viniculture and Rural Development.

Informed Consent Statement: Not applicable.

Data Availability Statement: Gene Expression Omnibus (GEO) database—Accession number GSE225729.

Acknowledgments: We thank Noel J. Buckley for the critical revision of the manuscript.

Conflicts of Interest: The authors declare no competing interest.

References

1. Hirabayashi, Y.; Gotoh, Y. Stage-dependent fate determination of neural precursor cells in mouse forebrain. *Neurosci. Res.* **2005**, *51*, 331–336. [CrossRef] [PubMed]
2. Takouda, J.; Katada, S.; Nakashima, K. Emerging mechanisms underlying astrogenesis in the developing mammalian brain. *Proc. Jpn. Acad. Ser. B Phys. Biol. Sci.* **2017**, *93*, 386–398. [CrossRef] [PubMed]
3. Gotz, M.; Nakafuku, M.; Petrik, D. Neurogenesis in the Developing and Adult Brain-Similarities and Key Differences. *Cold Spring Harb. Perspect. Biol.* **2016**, *8*, a018853. [CrossRef] [PubMed]
4. Pavlou, M.A.S.; Grandbarbe, L.; Buckley, N.J.; Niclou, S.P.; Michelucci, A. Transcriptional and epigenetic mechanisms underlying astrocyte identity. *Prog. Neurobiol.* **2019**, *174*, 36–52. [CrossRef]
5. Domingues, H.S.; Portugal, C.C.; Socodato, R.; Relvas, J.B. Corrigendum: Oligodendrocyte, Astrocyte and Microglia Crosstalk in Myelin Development, Damage, and Repair. *Front. Cell. Dev. Biol.* **2016**, *4*, 79. [CrossRef]
6. Pekny, M.; Pekna, M.; Messing, A.; Steinhauser, C.; Lee, J.M.; Parpura, V.; Hol, E.M.; Sofroniew, M.V.; Verkhratsky, A. Astrocytes: A central element in neurological diseases. *Acta Neuropathol.* **2016**, *131*, 323–345. [CrossRef]
7. Messing, A.; Brenner, M.; Feany, M.B.; Nedergaard, M.; Goldman, J.E. Alexander disease. *J. Neurosci.* **2012**, *32*, 5017–5023. [CrossRef]
8. Quinlan, R.A.; Brenner, M.; Goldman, J.E.; Messing, A. GFAP and its role in Alexander disease. *Exp. Cell Res.* **2007**, *313*, 2077–2087. [CrossRef]
9. Arvidsson, L.; Covacu, R.; Estrada, C.P.; Sankavaram, S.R.; Svensson, M.; Brundin, L. Long-distance effects of inflammation on differentiation of adult spinal cord neural stem/progenitor cells. *J. Neuroimmunol.* **2015**, *288*, 47–55. [CrossRef]

10. Lozano, D.; Gonzales-Portillo, G.S.; Acosta, S.; de la Pena, I.; Tajiri, N.; Kaneko, Y.; Borlongan, C.V. Neuroinflammatory responses to traumatic brain injury: Etiology, clinical consequences, and therapeutic opportunities. *Neuropsychiatr. Dis. Treat.* **2015**, *11*, 97–106. [CrossRef]
11. Hess, D.C.; Borlongan, C.V. Stem cells and neurological diseases. *Cell Prolif.* **2008**, *41* (Suppl. S1), 94–114. [CrossRef]
12. Galvao, R.P.; Zong, H. Inflammation and Gliomagenesis: Bi-Directional Communication at Early and Late Stages of Tumor Progression. *Curr. Pathobiol. Rep.* **2013**, *1*, 19–28. [CrossRef] [PubMed]
13. Sanai, N.; Alvarez-Buylla, A.; Berger, M.S. Neural stem cells and the origin of gliomas. *N. Engl. J. Med.* **2005**, *353*, 811–822. [CrossRef] [PubMed]
14. Alcantara Llaguno, S.R.; Parada, L.F. Cell of origin of glioma: Biological and clinical implications. *Br. J. Cancer* **2016**, *115*, 1445–1450. [CrossRef]
15. Idriss, H.T.; Naismith, J.H. TNFα and the TNF receptor superfamily: Structure-function relationship(s). *Microsc. Res. Tech.* **2000**, *50*, 184–195. [CrossRef]
16. Sun, M.; Fink, P.J. A new class of reverse signaling costimulators belongs to the TNF family. *J. Immunol.* **2007**, *179*, 4307–4312. [CrossRef] [PubMed]
17. Park, S.H.; Kang, K.; Giannopoulou, E.; Qiao, Y.; Kang, K.; Kim, G.; Park-Min, K.H.; Ivashkiv, L.B. Type I interferons and the cytokine TNF cooperatively reprogram the macrophage epigenome to promote inflammatory activation. *Nat. Immunol.* **2017**, *18*, 1104–1116. [CrossRef] [PubMed]
18. Adefuin, A.M.; Kimura, A.; Noguchi, H.; Nakashima, K.; Namihira, M. Epigenetic mechanisms regulating differentiation of neural stem/precursor cells. *Epigenomics* **2014**, *6*, 637–649. [CrossRef]
19. Hwang, D.W.; Jaganathan, A.; Shrestha, P.; Jin, Y.; El-Amine, N.; Wang, S.H.; Hammell, M.; Mills, A.A. Chromatin-mediated translational control is essential for neural cell fate specification. *Life Sci. Alliance* **2018**, *1*, e201700016. [CrossRef] [PubMed]
20. Buenrostro, J.D.; Giresi, P.G.; Zaba, L.C.; Chang, H.Y.; Greenleaf, W.J. Transposition of native chromatin for fast and sensitive epigenomic profiling of open chromatin, DNA-binding proteins and nucleosome position. *Nat. Methods* **2013**, *10*, 1213–1218. [CrossRef]
21. Birck, C.; Ginolhac, A.; Pavlou, M.A.S.; Michelucci, A.; Heuschling, P.; Grandbarbe, L. NF-κB and TNF Affect the Astrocytic Differentiation from Neural Stem Cells. *Cells* **2021**, *10*, 840. [CrossRef] [PubMed]
22. Grandbarbe, L.; Bouissac, J.; Rand, M.; Hrabe de Angelis, M.; Artavanis-Tsakonas, S.; Mohier, E. Delta-Notch signaling controls the generation of neurons/glia from neural stem cells in a stepwise process. *Development* **2003**, *130*, 1391–1402. [CrossRef] [PubMed]
23. Schmittgen, T.D.; Livak, K.J. Analyzing real-time PCR data by the comparative C_T method. *Nat. Protoc.* **2008**, *3*, 1101–1108. [CrossRef] [PubMed]
24. Zhuo, L.; Sun, B.; Zhang, C.-L.; Fine, A.; Chiu, S.-Y.; Messing, A. Live Astrocytes Visualized by Green Fluorescent Protein in Transgenic Mice. *Dev. Biol.* **1997**, *187*, 36–42. [CrossRef] [PubMed]
25. Li, Z.; Schulz, M.H.; Look, T.; Begemann, M.; Zenke, M.; Costa, I.G. Identification of transcription factor binding sites using ATAC-seq. *Genome Biol.* **2019**, *20*, 45. [CrossRef]
26. Huynh-Thu, V.A.; Irrthum, A.; Wehenkel, L.; Geurts, P. Inferring regulatory networks from expression data using tree-based methods. *PLoS ONE* **2010**, *5*, e12776. [CrossRef]
27. Michelucci, A.; Bithell, A.; Burney, M.J.; Johnston, C.E.; Wong, K.Y.; Teng, S.W.; Desai, J.; Gumbleton, N.; Anderson, G.; Stanton, L.W.; et al. The Neurogenic Potential of Astrocytes Is Regulated by Inflammatory Signals. *Mol. Neurobiol.* **2016**, *53*, 3724–3739. [CrossRef]
28. Escartin, C.; Galea, E.; Lakatos, A.; O'Callaghan, J.P.; Petzold, G.C.; Serrano-Pozo, A.; Steinhauser, C.; Volterra, A.; Carmignoto, G.; Agarwal, A.; et al. Reactive astrocyte nomenclature, definitions, and future directions. *Nat. Neurosci.* **2021**, *24*, 312–325. [CrossRef]
29. Liddelow, S.A.; Guttenplan, K.A.; Clarke, L.E.; Bennett, F.C.; Bohlen, C.J.; Schirmer, L.; Bennett, M.L.; Munch, A.E.; Chung, W.S.; Peterson, T.C.; et al. Neurotoxic reactive astrocytes are induced by activated microglia. *Nature* **2017**, *541*, 481–487. [CrossRef]
30. Cahoy, J.D.; Emery, B.; Kaushal, A.; Foo, L.C.; Zamanian, J.L.; Christopherson, K.S.; Xing, Y.; Lubischer, J.L.; Krieg, P.A.; Krupenko, S.A.; et al. A transcriptome database for astrocytes, neurons, and oligodendrocytes: A new resource for understanding brain development and function. *J. Neurosci.* **2008**, *28*, 264–278. [CrossRef]
31. Lovatt, D.; Sonnewald, U.; Waagepetersen, H.S.; Schousboe, A.; He, W.; Lin, J.H.; Han, X.; Takano, T.; Wang, S.; Sim, F.J.; et al. The transcriptome and metabolic gene signature of protoplasmic astrocytes in the adult murine cortex. *J. Neurosci.* **2007**, *27*, 12255–12266. [CrossRef] [PubMed]
32. Zamanian, J.L.; Xu, L.; Foo, L.C.; Nouri, N.; Zhou, L.; Giffard, R.G.; Barres, B.A. Genomic analysis of reactive astrogliosis. *J. Neurosci.* **2012**, *32*, 6391–6410. [CrossRef]
33. Rothhammer, V.; Quintana, F.J. Control of autoimmune CNS inflammation by astrocytes. *Semin. Immunopathol.* **2015**, *37*, 625–638. [CrossRef] [PubMed]
34. Grill, M.; Syme, T.E.; Nocon, A.L.; Lu, A.Z.; Hancock, D.; Rose-John, S.; Campbell, I.L. Strawberry notch homolog 2 is a novel inflammatory response factor predominantly but not exclusively expressed by astrocytes in the central nervous system. *Glia* **2015**, *63*, 1738–1752. [CrossRef] [PubMed]
35. Molofsky, A.V.; Krencik, R.; Ullian, E.M.; Tsai, H.H.; Deneen, B.; Richardson, W.D.; Barres, B.A.; Rowitch, D.H. Astrocytes and disease: A neurodevelopmental perspective. *Genes Dev.* **2012**, *26*, 891–907. [CrossRef] [PubMed]

36. Sloan, S.A.; Barres, B.A. Mechanisms of astrocyte development and their contributions to neurodevelopmental disorders. *Curr Opin. Neurobiol.* **2014**, *27*, 75–81. [CrossRef] [PubMed]
37. Magistri, M.; Khoury, N.; Mazza, E.M.; Velmeshev, D.; Lee, J.K.; Bicciato, S.; Tsoulfas, P.; Faghihi, M.A. A comparative transcriptomic analysis of astrocytes differentiation from human neural progenitor cells. *Eur. J. Neurosci.* **2016**, *44*, 2858–2870 [CrossRef]
38. Goodnight, A.V.; Kremsky, I.; Khampang, S.; Jung, Y.H.; Billingsley, J.M.; Bosinger, S.E.; Corces, V.G.; Chan, A.W.S. Chromatin accessibility and transcription dynamics during in vitro astrocyte differentiation of Huntington's Disease Monkey pluripotent stem cells. *Epigenetics Chromatin* **2019**, *12*, 67. [CrossRef] [PubMed]
39. Chandrasekaran, A.; Avci, H.X.; Leist, M.; Kobolak, J.; Dinnyes, A. Astrocyte Differentiation of Human Pluripotent Stem Cells: New Tools for Neurological Disorder Research. *Front. Cell. Neurosci.* **2016**, *10*, 215. [CrossRef]
40. Sofroniew, M.V.; Vinters, H.V. Astrocytes: Biology and pathology. *Acta Neuropathol.* **2010**, *119*, 7–35. [CrossRef]
41. Gabel, S.; Koncina, E.; Dorban, G.; Heurtaux, T.; Birck, C.; Glaab, E.; Michelucci, A.; Heuschling, P.; Grandbarbe, L. Inflammation Promotes a Conversion of Astrocytes into Neural Progenitor Cells via NF-κB Activation. *Mol. Neurobiol.* **2016**, *53*, 5041–5055 [CrossRef] [PubMed]
42. Lee, S.J.; Drabik, K.; Van Wagoner, N.J.; Lee, S.; Choi, C.; Dong, Y.; Benveniste, E.N. ICAM-1-induced expression of proinflammatory cytokines in astrocytes: Involvement of extracellular signal-regulated kinase and p38 mitogen-activated protein kinase pathways. *J. Immunol.* **2000**, *165*, 4658–4666. [CrossRef] [PubMed]
43. Neal, M.; Luo, J.; Harischandra, D.S.; Gordon, R.; Sarkar, S.; Jin, H.; Anantharam, V.; Desaubry, L.; Kanthasamy, A.; Kanthasamy, A. Prokineticin-2 promotes chemotaxis and alternative A2 reactivity of astrocytes. *Glia* **2018**, *66*, 2137–2157. [CrossRef]
44. Brosnan, C.F.; Raine, C.S. The astrocyte in multiple sclerosis revisited. *Glia* **2013**, *61*, 453–465. [CrossRef] [PubMed]
45. Parratt, J.D.; Prineas, J.W. Neuromyelitis optica: A demyelinating disease characterized by acute destruction and regeneration of perivascular astrocytes. *Mult. Scler.* **2010**, *16*, 1156–1172. [CrossRef]
46. Belanger, M.; Allaman, I.; Magistretti, P.J. Brain energy metabolism: Focus on astrocyte-neuron metabolic cooperation. *Cell Metab.* **2011**, *14*, 724–738. [CrossRef]
47. Brown, A.M.; Ransom, B.R. Astrocyte glycogen and brain energy metabolism. *Glia* **2007**, *55*, 1263–1271. [CrossRef]
48. Brunet, J.F.; Allaman, I.; Magistretti, P.J.; Pellerin, L. Glycogen metabolism as a marker of astrocyte differentiation. *J. Cereb. Blood Flow Metab.* **2010**, *30*, 51–55. [CrossRef]
49. Bak, L.K.; Walls, A.B.; Schousboe, A.; Waagepetersen, H.S. Astrocytic glycogen metabolism in the healthy and diseased brain. *J. Biol. Chem.* **2018**, *293*, 7108–7116. [CrossRef]
50. Iglesias, J.; Morales, L.; Barreto, G.E. Metabolic and Inflammatory Adaptation of Reactive Astrocytes: Role of PPARs. *Mol. Neurobiol.* **2017**, *54*, 2518–2538. [CrossRef]
51. Court, F.; Le Boiteux, E.; Fogli, A.; Muller-Barthelemy, M.; Vaurs-Barriere, C.; Chautard, E.; Pereira, B.; Biau, J.; Kemeny, J.L.; Khalil, T.; et al. Transcriptional alterations in glioma result primarily from DNA methylation-independent mechanisms. *Genome Res.* **2019**, *29*, 1605–1621. [CrossRef] [PubMed]
52. Imayoshi, I.; Sakamoto, M.; Yamaguchi, M.; Mori, K.; Kageyama, R. Essential roles of Notch signaling in maintenance of neural stem cells in developing and adult brains. *J. Neurosci.* **2010**, *30*, 3489–3498. [CrossRef]
53. Yun, S.J.; Byun, K.; Bhin, J.; Oh, J.H.; Nhung, L.T.H.; Hwang, D.; Lee, B. Transcriptional regulatory networks associated with self-renewal and differentiation of neural stem cells. *J. Cell. Physiol.* **2010**, *225*, 337–347. [CrossRef] [PubMed]
54. Ahmed, S.; Gan, H.T.; Lam, C.S.; Poonepalli, A.; Ramasamy, S.; Tay, Y.; Tham, M.; Yu, Y.H. Transcription factors and neural stem cell self-renewal, growth and differentiation. *Cell Adhes. Migr.* **2009**, *3*, 412–424. [CrossRef] [PubMed]
55. Habas, A.; Hahn, J.; Wang, X.; Margeta, M. Neuronal activity regulates astrocytic Nrf2 signaling. *Proc. Natl. Acad. Sci. USA* **2013**, *110*, 18291–18296. [CrossRef] [PubMed]
56. Liddell, J.R. Are Astrocytes the Predominant Cell Type for Activation of Nrf2 in Aging and Neurodegeneration? *Antioxidants* **2017**, *6*, 65. [CrossRef]
57. Hayes, J.D.; Dinkova-Kostova, A.T. The Nrf2 regulatory network provides an interface between redox and intermediary metabolism. *Trends Biochem. Sci.* **2014**, *39*, 199–218. [CrossRef]
58. Starks, R.R.; Biswas, A.; Jain, A.; Tuteja, G. Combined analysis of dissimilar promoter accessibility and gene expression profiles identifies tissue-specific genes and actively repressed networks. *Epigenetics Chromatin* **2019**, *12*, 16. [CrossRef]
59. Wen, S.; Li, H.; Liu, J. Dynamic signaling for neural stem cell fate determination. *Cell Adhes. Migr.* **2009**, *3*, 107–117. [CrossRef]
60. Lattke, M.; Goldstone, R.; Ellis, J.K.; Boeing, S.; Jurado-Arjona, J.; Marichal, N.; MacRae, J.I.; Berninger, B.; Guillemot, F. Extensive transcriptional and chromatin changes underlie astrocyte maturation in vivo and in culture. *Nat. Commun.* **2021**, *12*, 4335. [CrossRef]
61. Tiwari, N.; Pataskar, A.; Peron, S.; Thakurela, S.; Sahu, S.K.; Figueres-Onate, M.; Marichal, N.; Lopez-Mascaraque, L.; Tiwari, V.K.; Berninger, B. Stage-Specific Transcription Factors Drive Astrogliogenesis by Remodeling Gene Regulatory Landscapes. *Cell Stem Cell* **2018**, *23*, 557–571.e8. [CrossRef] [PubMed]
62. Fuente-Martin, E.; Garcia-Caceres, C.; Morselli, E.; Clegg, D.J.; Chowen, J.A.; Finan, B.; Brinton, R.D.; Tschop, M.H. Estrogen, astrocytes and the neuroendocrine control of metabolism. *Rev. Endocr. Metab. Disord.* **2013**, *14*, 331–338. [CrossRef] [PubMed]
63. Spence, R.D.; Hamby, M.E.; Umeda, E.; Itoh, N.; Du, S.; Wisdom, A.J.; Cao, Y.; Bondar, G.; Lam, J.; Ao, Y.; et al. Neuroprotection mediated through estrogen receptor-α in astrocytes. *Proc. Natl. Acad. Sci. USA* **2011**, *108*, 8867–8872. [CrossRef] [PubMed]

64. Tertil, M.; Skupio, U.; Barut, J.; Dubovyk, V.; Wawrzczak-Bargiela, A.; Soltys, Z.; Golda, S.; Kudla, L.; Wiktorowska, L.; Szklarczyk, K.; et al. Glucocorticoid receptor signaling in astrocytes is required for aversive memory formation. *Transl. Psychiatry* **2018**, *8*, 255. [CrossRef]
65. Pearson, C.A.; Moore, D.M.; Tucker, H.O.; Dekker, J.D.; Hu, H.; Miquelajauregui, A.; Novitch, B.G. Foxp1 Regulates Neural Stem Cell Self-Renewal and Bias Toward Deep Layer Cortical Fates. *Cell Rep.* **2020**, *30*, 1964–1981.e3. [CrossRef]
66. Braccioli, L.; Vervoort, S.J.; Adolfs, Y.; Heijnen, C.J.; Basak, O.; Pasterkamp, R.J.; Nijboer, C.H.; Coffer, P.J. FOXP1 Promotes Embryonic Neural Stem Cell Differentiation by Repressing Jagged1 Expression. *Stem Cell Rep.* **2017**, *9*, 1530–1545. [CrossRef]
67. Tchieu, J.; Calder, E.L.; Guttikonda, S.R.; Gutzwiller, E.M.; Aromolaran, K.A.; Steinbeck, J.A.; Goldstein, P.A.; Studer, L. NFIA is a gliogenic switch enabling rapid derivation of functional human astrocytes from pluripotent stem cells. *Nat. Biotechnol.* **2019**, *37*, 267–275. [CrossRef]
68. Zhang, J.; Xu, X.; Liu, H.; Jin, L.; Shen, X.; Xie, C.; Xiang, W.; Yang, D.; Feng, W.; Wang, J.; et al. Astrocytic YAP prevents the demyelination through promoting expression of cholesterol synthesis genes in experimental autoimmune encephalomyelitis. *Cell Death Dis.* **2021**, *12*, 907. [CrossRef]
69. Herrera, F.; Chen, Q.; Schubert, D. Synergistic effect of retinoic acid and cytokines on the regulation of glial fibrillary acidic protein expression. *J. Biol. Chem.* **2010**, *285*, 38915–38922. [CrossRef]
70. Wu, X.; Kihara, T.; Akaike, A.; Niidome, T.; Sugimoto, H. PI3K/Akt/mTOR signaling regulates glutamate transporter 1 in astrocytes. *Biochem. Biophys. Res. Commun.* **2010**, *393*, 514–518. [CrossRef]
71. Lin, M.S.; Sun, Y.Y.; Chiu, W.T.; Hung, C.C.; Chang, C.Y.; Shie, F.S.; Tsai, S.H.; Lin, J.W.; Hung, K.S.; Lee, Y.H. Curcumin attenuates the expression and secretion of RANTES after spinal cord injury in vivo and lipopolysaccharide-induced astrocyte reactivation in vitro. *J. Neurotrauma* **2011**, *28*, 1259–1269. [CrossRef] [PubMed]
72. Bani-Yaghoub, M.; Tremblay, R.G.; Lei, J.X.; Zhang, D.; Zurakowski, B.; Sandhu, J.K.; Smith, B.; Ribecco-Lutkiewicz, M.; Kennedy, J.; Walker, P.R.; et al. Role of Sox2 in the development of the mouse neocortex. *Dev. Biol.* **2006**, *295*, 52–66. [CrossRef]
73. Santopolo, G.; Magnusson, J.P.; Lindvall, O.; Kokaia, Z.; Frisen, J. Blocking Notch-Signaling Increases Neurogenesis in the Striatum after Stroke. *Cells* **2020**, *9*, 1732. [CrossRef] [PubMed]
74. Stipursky, J.; Gomes, F.C. TGF-β1/SMAD signaling induces astrocyte fate commitment in vitro: Implications for radial glia development. *Glia* **2007**, *55*, 1023–1033. [CrossRef] [PubMed]
75. Platanitis, E.; Decker, T. Regulatory Networks Involving STATs, IRFs, and NFκB in Inflammation. *Front. Immunol.* **2018**, *9*, 2542. [CrossRef]
76. Herrmann, J.E.; Imura, T.; Song, B.; Qi, J.; Ao, Y.; Nguyen, T.K.; Korsak, R.A.; Takeda, K.; Akira, S.; Sofroniew, M.V. STAT3 is a critical regulator of astrogliosis and scar formation after spinal cord injury. *J. Neurosci.* **2008**, *28*, 7231–7243. [CrossRef] [PubMed]
77. Lattke, M.; Reichel, S.N.; Baumann, B. NF-κB-mediated astrocyte dysfunction initiates neurodegeneration. *Oncotarget* **2017**, *8*, 50329–50330. [CrossRef] [PubMed]
78. Grivennikov, S.I.; Karin, M. Dangerous liaisons: STAT3 and NF-κB collaboration and crosstalk in cancer. *Cytokine Growth Factor Rev.* **2010**, *21*, 11–19. [CrossRef] [PubMed]
79. Fan, Y.; Mao, R.; Yang, J. NF-κB and STAT3 signaling pathways collaboratively link inflammation to cancer. *Protein Cell* **2013**, *4*, 176–185. [CrossRef]
80. Hong, S.; Song, M.R. STAT3 but not STAT1 is required for astrocyte differentiation. *PLoS ONE* **2014**, *9*, e86851. [CrossRef]
81. Lawrence, T. The nuclear factor NF-κB pathway in inflammation. *Cold Spring Harb. Perspect. Biol.* **2009**, *1*, a001651. [CrossRef] [PubMed]
82. Logan, T.T.; Rusnak, M.; Symes, A.J. Runx1 promotes proliferation and neuronal differentiation in adult mouse neurosphere cultures. *Stem Cell Res.* **2015**, *15*, 554–564. [CrossRef]
83. Fields, J.; Ghorpade, A. C/EBPβ regulates multiple IL-1beta-induced human astrocyte inflammatory genes. *J. Neuroinflamm.* **2012**, *9*, 177. [CrossRef]
84. Hunt, D.; Raivich, G.; Anderson, P.N. Activating transcription factor 3 and the nervous system. *Front. Mol. Neurosci.* **2012**, *5*, 7. [CrossRef] [PubMed]
85. Trentin, A.G. Thyroid hormone and astrocyte morphogenesis. *J. Endocrinol.* **2006**, *189*, 189–197. [CrossRef] [PubMed]

cells

MDPI

Review

The Glymphatic System May Play a Vital Role in the Pathogenesis of Hepatic Encephalopathy: A Narrative Review

Ali Sepehrinezhad [1,2], Fin Stolze Larsen [3], Rezan Ashayeri Ahmadabad [4], Ali Shahbazi [1,5] and Sajad Sahab Negah [2,4,*]

1 Department of Neuroscience, Faculty of Advanced Technologies in Medicine, Iran University of Medical Sciences, Tehran 1449614535, Iran; sepehrinezhad.a@iums.ac.ir (A.S.); shahbazi.a@iums.ac.ir (A.S.)
2 Neuroscience Research Center, Mashhad University of Medical Sciences, Mashhad 9919191778, Iran
3 Department of Gastroenterology and Hepatology, Rigshospitalet, Copenhagen University Hospital, 999017 Copenhagen, Denmark; fin.stolze.larsen@regionh.dk
4 Shefa Neuroscience Research Center, Khatam Alanbia Hospital, Tehran 1449614535, Iran; re.ashayeri@gmail.com
5 Cellular and Molecular Research Center, Iran University of Medical Sciences, Tehran 1449614535, Iran
* Correspondence: sahabnegahs@mums.ac.ir; Tel.: +98-51-38002473

Abstract: Hepatic encephalopathy (HE) is a neurological complication of liver disease resulting in cognitive, psychiatric, and motor symptoms. Although hyperammonemia is a key factor in the pathogenesis of HE, several other factors have recently been discovered. Among these, the impairment of a highly organized perivascular network known as the glymphatic pathway seems to be involved in the progression of some neurological complications due to the accumulation of misfolded proteins and waste substances in the brain interstitial fluids (ISF). The glymphatic system plays an important role in the clearance of brain metabolic derivatives and prevents aggregation of neurotoxic agents in the brain ISF. Impairment of it will result in aggravated accumulation of neurotoxic agents in the brain ISF. This could also be the case in patients with liver failure complicated by HE. Indeed, accumulation of some metabolic by-products and agents such as ammonia, glutamine, glutamate, and aromatic amino acids has been reported in the human brain ISF using microdialysis technique is attributed to worsening of HE and correlates with brain edema. Furthermore, it has been reported that the glymphatic system is impaired in the olfactory bulb, prefrontal cortex, and hippocampus in an experimental model of HE. In this review, we discuss different factors that may affect the function of the glymphatic pathways and how these changes may be involved in HE.

Keywords: glymphatic system; hepatic encephalopathy; astrocytes; brain fluids; waste substances

Citation: Sepehrinezhad, A.; Stolze Larsen, F.; Ashayeri Ahmadabad, R.; Shahbazi, A.; Sahab Negah, S. The Glymphatic System May Play a Vital Role in the Pathogenesis of Hepatic Encephalopathy: A Narrative Review. *Cells* **2023**, *12*, 979. https://doi.org/10.3390/cells12070979

Academic Editors: Michael Brenner and Vladimir Parpura

Received: 19 October 2022
Revised: 20 February 2023
Accepted: 21 March 2023
Published: 23 March 2023

1. Introduction

Hepatic encephalopathy (HE) is a neurological and neuropsychiatric complication following acute liver failure (ALF) or chronic liver disease (CLD) [1–3]. The exact pathophysiological mechanisms of HE are not fully understood. However, numerous studies have demonstrated that HE has complex pathology at the cellular molecular levels, e.g., astrocyte dysfunction plays a critical role in the pathogenesis of both forms of HE (i.e., ALF and CLD) [4–9]. Furthermore, substances including ammonia, glutamine, glutamate, aromatic amino acids, lactate, alanine, bile acids, and neurosteroids accumulate in the brain parenchyma during HE [10–12] and are directly implicated in the pathogenesis. Intracellular molecules are normally removed by two major protein degradation systems including the ubiquitin-proteasome system and autophagy in the neural cells [13]. Thereafter, cytosolic proteins and wastes are released into the interstitial space. These waste proteins and metabolic products are released into the cerebrospinal fluid (CSF) and finally are drained away by the glymphatic system [14]. This pathway provides a unidirectional fluid current to transport glucose and other nutrients, but the system is particularly known for the clearance of waste substances from the brain interstitial space [15]. Impairment of the glymphatic system, and the subsequent accumulation of waste soluble proteins has been proposed for the development of neurological disorders, such as Alzheimer's disease [16–18], and is implicated in stroke [19], and traumatic brain injury [20,21]. Recently, researchers have mentioned the disruption of the glymphatic system in an experimental model of HE [22,23]. In this review, we focus on the components of the glymphatic pathway that may be compromised in HE.

2. An Overview

The glymphatic system is a system for waste proteins and metabolic products clearance from interstitial fluids (ISF) in the brain parenchyma. The name "glymphatic" derives from the major role of astrocytes, the central nervous system's most abundant glial cells [24] in this system that operates like a lymphatic system. Its unidirectional flow initiates from Virchow-Robin spaces or periarterial spaces, where CSF flows from the subarachnoid space to the brain tissue along the penetrating arterioles (Figure 1). The driving force for the influx of CSF to the ISF and movement of CSF-ISF into the interstitial space is arterial pulsation [25,26]. Astrocytic end-feet express aquaporin 4 (AQP4) water channels, where astrocytes and endothelial cells constitute barriers between CSF and ISF [27,28]. AQP4 channels are bidirectional and facilitate water flux between blood circulation, periarterial space, interstitial space, and perivenous space. These channels decrease the resistance of CSF flow from the periarterial space into the ISF and facilitate the movement of ISF toward the perivenous space due to the expression on astrocyte processes [29]. CSF-ISF fluid subsequently enters the perivenous space and releases interstitial metabolic agents, soluble proteins, and neurotoxic factors, including lactate, glutamine, glutamate, tau, and amyloid peptides. Finally, waste products are released into the meningeal lymphatic vessels in the dura mater and are transported to the deep cervical lymph nodes (Figure 1) [20]. Enhancing the drainage of CSF from brain tissue ensures the proper functioning of the glymphatic pathway as a powerful driving force for CSF-ISF flow.

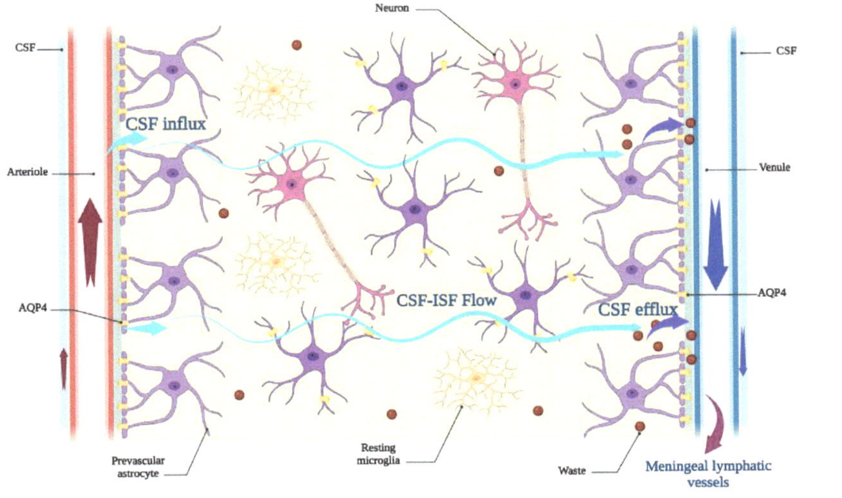

Figure 1. The components of the glymphatic pathway. In the region of Virchow-Robin spaces, CSF along with penetrating arterioles flow into the brain parenchyma. Arterial pulsation provides influx of CSF into the periarterial interstitial space. The system eliminates most neurotoxic agents and waste substances from ISF. AQP4 water channels that are localized in perivascular astrocyte end-feet mediate fluid movements throughout the system. AQP4: aquaporin 4; CSF: cerebrospinal fluid; ISF: interstitial fluid. Created with BioRender.com.

3. Factors Affecting the Function of the Glymphatic System

The normal function of the glymphatic pathway is to eliminate misfolded proteins completely, maintain water and ions equilibrium, and regulate the CSF, ISF, and intracranial pressure. If any element of the glymphatic pathway is impaired, waste accumulates in the brain ISF. Such impairments occur in the development of neurological disorders. Several factors can affect the glymphatic function in the central nervous system (CNS), such as circadian cycle, body posture, aging, and autonomic innervation. The function of the glymphatic system decreases during wakefulness and becomes better during sleep and under anesthesia conditions [25]. Sleep and anesthesia lead to interstitial space expansion and promote the movement of CSF flow in the perivascular spaces in mice [25]. Another study of rodents showed that the glymphatic system is controlled under circadian rhythms, and AQP4 localization on perivascular astrocyte end-feet is highest in sleepy mice [26]. In a human study using intrathecal injection of gadobutrol as a CSF tracer, glymphatic function was also highest over-night compared to daytime, which indicates the role of sleep for maintenance of an efficient glymphatic pathway [30]. Anesthesia following injection of pentobarbital also improved glymphatic activity in mice [31]. In terms of other factors, the effect of body posture was also evaluated in that latter study. Humans and animals display different body postures at every moment and during sleep and wakefulness. Previous studies have shown that body position affects cerebral blood circulation and intracranial pressure [32,33]. In one study, dynamic-contrast-enhanced magnetic resonance imaging (MRI) and a radioactive CSF tracer revealed that CSF-ISF flow and waste clearance of ISF improved in lateral position in rats [34]. Aging is another factor that has negative effects on the function of the glymphatic system. Aging decreases vascular elasticity and, therefore, reduces the power of arterial pulsation [35,36]. Studies in mice have shown that aging suppresses CSF flow, CSF-ISF flow, and clearance of waste substances [37,38]. Decreasing the levels of collagen type IV and other basement membrane proteins is associated with vascular stiffness and suppression of the amplitude of arterial pulsation, which leads to impairment of CSF influx into the interstitial space [39]. Aging also affects the morphogenesis and cellular behavior of astrocytes, which are fundamental to the structure of glymphatic corridors [40,41]. Furthermore, the distribution of AQP4 throughout the brain was changed with aging [42,43]. The autonomic nervous system (ANS) is another factor that affects the glymphatic system. Activation of the sympathetic nervous system (SNS) can exert negative effects on the glymphatic system through the main neurotransmitter norepinephrine. Reciprocally, activation of the parasympathetic nervous system (PNS) through the vagus nerve positively regulates and improves glymphatic function [44]. Administration of dexmedetomidine, a suppressor of norepinephrine release from locus coeruleus neurons, enhanced glymphatic function in the hippocampus of lightly anesthetized rats [45]. Likewise, intracisternal injection of adrenergic receptor antagonists increased CSF influx into the ISF in conscious mice, as evaluated by a CSF tracer [25]. These studies suggest that norepinephrine and basal tone of the SNS are responsible for suppression of glymphatic transports during wakefulness. Stimulation of the vagus nerve through an implanted device enhanced CSF influx into the ISF and improved CSF-ISF flow in the brain of mice [44].

4. Direct Evidence of Glymphatic System Impairment in HE

The first evidence of a dysfunctional glymphatic system in HE was reported by Hadjihambi and her colleagues at University College London [22]. In this study, they measured the glymphatic function in brain parenchyma in bile duct ligation (BDL) rats as chronic liver disease-induced HE model, using dynamic contrast-enhanced MRI and mass-spectroscopy. Decreased penetration of the contrast agent gadolinium indicated reduced CSF-ISF flow in the olfactory bulb and prefrontal cortex of BDL rats. These findings correlated with reduced expression of AQP4 in these brain regions [22]. Thus, they concluded that reduced expression of AQP4 mediated glymphatic impairment and behavioral deficits in HE rats [22]. Furthermore, Jung Hsu et al. [46] demonstrated that the distribution of fluorescent ovalbumin dye as an indicator of glymphatic function was impaired in the brain parenchyma and later in deep cervical lymph nodes in BDL rats compared to the control animals. In addition, the BDL rats displayed microglia activation and neuroinflammation. This study indicated that dysfunctional glymphatic clearance was caused by impaired meningeal lymphatic drainage in HE [46]. A review paper by Claeys et al. [47] explained that impaired glymphatic clearance of waste substances may contribute to the pathophysiology of HE. An editorial letter by Gallina et al. [23] also hypothesized that several abnormalities (i.e., cardiomyopathy-heart failure, pulmonary hypertension, alternation of intracranial pressure, hyperdynamic circulation, and decreased venous outflow) might contribute to the impairment of glymphatic function following liver disease and HE. In this review we describe in detail the links between HE pathogenesis and malfunction of the glymphatic system. The presence of such links suggests areas for future research.

5. Factors That May Indirectly Affect the Efficacy of Glymphatic System in Hepatic Encephalopathy

5.1. Sleep Disturbances in HE

As mentioned earlier, the sleep–wake cycle can affect the glymphatic system. Sleep improves the function of glymphatic system, while wakefulness reduces it. Sleep disturbance is quite common in patients with liver insufficiency well before HE becomes clinically overt [48,49]. A neuroimaging study suggested that sleep disturbance is a valuable diagnostic sign for subclinical HE in cirrhotic patients [50]. Moreover, decreased sleep quality and daily drowsiness measured by polysomnography and sleep indexes have been reported in HE patients [48,51,52]. Long-lasting sleep abnormalities were also revealed in patients with portosystemic shunt encephalopathy [53]. In an experimental model of this type of HE, disruption of sleep patterns and significant reduction of both rapid eye movement sleep and non-rapid eye movement compared to control rats were shown [54]. Similar findings, such as increased total duration of wakefulness and decreased slow-wave sleep and paradoxical sleep times, have been reported in rats with hyperammonemia [55]. Therefore, manipulating the sleep–wake cycle and correcting sleep disturbances can be an effective effort for the recovery of glymphatic system function and consequently result in ameliorating HE.

5.2. Cardiomyopathy, Arterial Hypotension, and Impaired Cerebral Blood Flow in Cirrhosis and HE

Starling forces regulate the fluid flux across the cerebral microvessels, but are severely imbalanced in liver failure, especially if the normal function of the BBB is compromised [56,57]. The main driving forces for CSF influx into the interstitial space and CSF-ISF flow in the brain parenchyma are arterial pulsation induced by the amplitude of the cerebral perfusion pressure (mean arterial pressure (MAP) minus intracranial pressure) [58]. Decreased inotropic and chronotropic properties of heart muscles, reduced peripheral vascular resistance, and reduced systemic blood pressure are commonly seen in patients with insufficient liver function [59–61]. Ventricular diastolic dysfunction, left atrial enlargement [62], and both systolic and diastolic cardiac failures [63,64] have been reported in cirrhotic patients. Furthermore, an autopsy study has shown ventricular hypertrophy, cardiac valve calcification, interstitial edema, fibrosis in myocytes, and coronary artery atherosclerosis in cirrhotic patients [65]. Cardiomyopathy as systolic and diastolic abnormalities, inotropic and chronotropic dysfunction, increased plasma levels of myocardial enzymes, myocyte swelling and hypertrophy, and gross abnormalities in cardiomyocytes in histological examinations have also been reported in animal models of cirrhosis and HE [66–69]. Reduced response to vasoconstrictors and accumulation of circulatory vasodilators, such as nitric oxide, carbon monoxide, and endocannabinoids explain the suppression of systemic vascular resistance in liver failure [70–73]. In patients with acute liver failure and HE, arterial hypotension is very common [74]. Similar findings have also been shown in the experimental model of HE induced by bile duct ligation in rats, where the systolic blood pressure, measured by tail-cuff, significantly decreased compared to the sham group. Furthermore, injection of thioacetamide as a hepatotoxic agent for induction of HE resulted in a decrease in systolic blood pressure [75]. Dysregulation of cerebral blood flow (CBF) is another important contributing factor that may affect glymphatic transports. Normal regulation of CBF plays a critical role in the supply of vital nutrients to the neuronal cells and elimination of waste agents from brain parenchyma. The cerebral microcirculation is controlled by the partial carbon dioxide tension in the blood, the ANS and cerebral perfusion pressure; i.e., by the CBF autoregulation [76,77]. Numerous human and animal studies have reported that the normal regulation of CBF is impaired in liver failure and HE [12,78–82]. The intracranial pressure increases in severe HE due to astrocyte swelling [83–89]. Intracranial hypertension also leads to a reduction in cerebral perfusion pressure in patients with acute liver failure [86,90–94]. The combination of arterial hypotension, intracranial hypertension, and impaired regulation of CBF will tend to compromise all the transition pathways of the glymphatic system, aggravating the accumulation of waste metabolic products in the ISF as previously observed in patients with acute liver failure [95–97]. This further underscores the importance of the current clinical practice to promptly correct arterial hypotension and hyperammonemia to avoid brain edema and high intracranial pressure, thereby improving CSF distribution in para-arterial spaces and CSF-ISF flow in brain interstitial space.

5.3. Alteration in the Function of Autonomic Nervous System in HE

As mentioned in Section 3, ANS affects function of the glymphatic system. Hyperactivity of sympathetic tone and suppression of vagal tone is present in cirrhosis [98–101]. A recent study showed ANS imbalance in patients with chronic liver disease [102]. Activation of the SNS and elevated levels of norepinephrine in the systemic circulation of cirrhotic patients have been reported [98,103]. Furthermore, PNS activity, as the vagal tone, is strongly suppressed in patients with end-stage liver disease and candidates for liver transplantation [101]. Another study showed an increase in the tonicity of the SNS in patients with high grades of HE compared to healthy control individuals [104]. There was also a considerable change in the function of ANS in patients with HE in another research [105]. Hyperactivity of the SNS is also seen in experimental models of HE induced by thioacetamide in rats [106]. Plasma levels of norepinephrine were found to be increased in rat models of cirrhosis [107,108]. Regarding these findings, modulation of ANS function to restore parasympathetic balance could be of importance to preserve the normal functioning of the glymphatic system, but further study is needed.

5.4. Astrocyte Dysfunction and AQP4 Mislocalization in HE

The normal functioning of the glymphatic system in the clearance of ISF requires the presence of healthy astrocytes as a main component. Astrocytes are the most abundant brain glial cells, especially in the cerebral cortex, which regulate ion equilibrium in the extracellular space, provide nutrient agents for their surrounding cells, particularly neuronal cells, release neurotransmitters at the location of astrocyte-neuron and astrocyte-microglia interactions, repair injured regions of CNS, support cerebral microvessels endothelial cells, contribute to the BBB, and finally detoxify many neurotoxic substances (e.g., ammonia and osmotic agents) from the brain interstitial space [10,109–111]. The function and morphology of astrocytes are strongly impaired following HE [10]. Glial fibrillary acidic protein (GFAP) is an intermediate filament protein that is intensely expressed in astrocyte cells. GFAP plays an important role in maintaining cell structure, cell proliferation, astrocyte-neuron communication, and in the formation of glial scars following brain injury [112,113]. Postmortem analysis of brain samples from HE patients has revealed decreased protein levels of GFAP in the cerebral cortex and basal ganglia [114]. In the same way, experimental studies on animal models of HE have reported decreased expression of GFAP in several brain areas, such as substantia nigra, ventral tegmental area, hippocampus [115], sensorimotor cortex, thalamus [116], and subcortical white matter [117,118]. In hyperammonemic conditions, astrocytes are primarily the cells that detoxify ammonia from brain parenchyma through their glutamine synthetase (GS) enzyme [119]. The enzyme catalyzes the conversion of ammonia and glutamate to glutamine [119]. Glutamine, which is an active osmolyte, triggers the influx of water into the cell. In hyperammonemic conditions, accumulation of glutamine following ammonia detoxification leads to a hypertonic state, and the influx of water into the astrocytes results in astrocyte swelling [119–122]. A set of pathological agents and processes, such as inflammation, oxidative stress, ATP depletion, cell senescence, osmolyte accumulation, mitochondrial dysfunction, impaired GFAP, and AQP4 expression triggers and even exacerbates astrocyte swelling following hyperammonemic conditions [10]. Microglia activation-induced neuroinflammation is another factor by which production of proinflammatory cytokines and production of oxidative stress worsens astrocyte swelling following hyperammonemia [123–125]. Normal homeostasis of the internal environment of the brain tissue requires an intact BBB in the region of cerebral microvasculature [126]. The BBB is made up of several components: endothelial cells that are joined together with tight junctions, pericytes that are located on the abluminal side and incompletely cover the endothelial cells, and astrocyte end-feet that act as linkers between neuronal networks and blood vessels [127]. This barrier is semipermeable and allows some ions and crucial nutrients (i.e., glucose and amino acids) to transport between blood and brain parenchyma, which ensures the survival of neuronal cells. However, it should be mentioned that the BBB is strongly impermeable to substances for which it does not have a transporter due to tight junctions composed of occludin, claudins, and junctional adhesion molecules; thus, it prevents the entry of circulatory substances into the brain parenchyma [128]. Impairment of this barrier can make it permeable to many circulatory neurotoxic agents, such as ammonia, bile acids, proinflammatory cytokines, and bacterial toxins like lipopolysaccharides [129,130]. Disruption of the BBB may also cause penetration of inflammatory cells such as phagocytic cells, lymphocytes, and antigen-presenting cells into the brain tissue and result in the induction of neuroinflammation and production of reactive oxygen species [131–135]. Therefore, increasing the permeability of the BBB can exacerbate astrocyte damage/swelling, as the main component of glymphatic system, through induction of neuroinflammation and oxidative stress and activation of microglia cells. BBB disruption has been reported in the pathogenesis of HE from experimental to clinical studies. For example, treatment of the human cerebral microvessel endothelial cell (hCMEC/D3) line as an in vitro model of the BBB with plasma samples from cirrhotic patients with HE increases trans-endothelial migration of leukocytes [136]. Increased extravasation of Evans blue dye, as an indicator of BBB disruption, in mice models of HE has been reported [137,138]. Ultrastructural analysis of cerebral cortex, using transmission electron microscopy and [^{3}H] inulin, in a rat model of HE has revealed increased leakage in cerebral capillaries [139]. Moreover, the protein level of S100 calcium-binding protein B, as a marker of BBB disruption, increased in serum samples of HE patients [140,141]. As a result, BBB dysfunction in HE is linked with astrocyte swelling and cerebral edema [139,142–154]. Microglia activation and neuroinflammation are triggered by astrogliosis and induction of the reactive inflammatory A1 phenotype subtype of astrocytes [155]. These reactive phenotypes release pro-inflammatory cytokines such as TNFα and IL-1β which induce BBB disruption, neurodegeneration, and gliopathy [156–159].

AQP4 water channels are integral membrane proteins that are localized on astrocytic end-feet surrounding the capillaries and facilitate the maintenance of cerebral water flow through their bidirectional properties [160]. However, AQP4 is expressed on plasma membrane instead of perivascular end-feet in swollen and reactive astrocytes [161–166]. This AQP4 mislocalization may be one of the main reasons for glymphatic impairment in several neurological diseases (Figure 2) [21,163,167,168]. For instance, the loss of perivascular AQP4 localization as a main part of the glymphatic system was assessed on clearance of amyloid β (Aβ) in human postmortem cases and the α-syntrophin knockout mouse model, which lacks perivascular AQP4 localization. Data indicated that the perivascular localization of AQP4 is significantly decreased in frontal cortical gray matter of AD cases compared to cognitively intact subjects. In the α-syntrophin knockout mouse, CSF influx,

CSF-ISF flow, and perivenous interstitial fluids efflux were significantly decreased, while Aβ was increased [167]. Mislocalization of perivascular astrocyte AQP4 and impairment of the glymphatic system have been well identified in AD [16,169–171]. Postmortem cerebral cortex analyses from AD patients found that the loss of immunofluorescence for perivascular AQP4 was associated with misfolded amyloid-β [167,172]. The AQP4 channels were also decreased in perivascular reactive astrocytes of the APP/PS1 mice model of AD and aged mice, and the mislocalization was accompanied by a decrease in the transport of a fluorescent tracer as an indicator of lymphatic clearance in brain interstitial space [37,169] Mislocalization of AQP4 channels on reactive perivascular astrocytes was also reported in mice models of traumatic brain injury (TBI) [173,174]. Moreover, localization of AQP4 on perivascular reactive astrocytes was impaired in hippocampus of TBI mice together with accumulation of hyperphosphorylated tau proteins in this region [163]. Immunofluorescence and tracer tracking techniques demonstrated that perivascular AQP4 mislocalized on reactive astrocytes was associated with a 60% decrease in glymphatic function in TBI mice [21]. Impairment of glymphatic clearance was also reported in the limbic system of TBI rats using dye tracing and MRI [175]. Astrocyte swelling and astrogliosis along with mislocalization of perivascular astrocytic AQP4 as well as glymphatic impairment were also revealed in stroke [162,176–179]. It is concluded that mislocalization of AQP4 from perivascular end-feet to the cell body in reactive astrocytes is associated with the reduction of efficacy of the glymphatic system and the onset of pathological cerebral findings following AD, TBI, and stroke (Figure 2). This raises the question, does the same mechanism occur in HE? In the context of liver diseases, it seems reasonable to consider similar mechanisms for HE onset. Astrocyte swelling, Alzheimer type II astrocytes, and reactive astrocytes are also predominant cerebral changes in HE that are associated with neurotoxicity [9,180–183]. These predominant phenotypes of astrocytes may negatively impact glymphatic function in HE as seen in AD, TBI, and stroke. Mislocalization of AQP4 channels in reactive perivascular astrocytes have been previously reported in liver diseases and HE [22,184–187]. A postmortem analysis of different brain regions (i.e., striatum, cerebellum, thalamus, hippocampus, and cortex) from cirrhotic patients indicated that the expression of AQP4 channels localized to the cell body's plasma membrane was significantly increased [184]. Furthermore, decreased expression of AQP4 in the olfactory bulb and prefrontal cortex of BDL rats along with the impairment of the glymphatic system, using dynamic contrast-enhanced MRI and mass-spectroscopy, was also observed [22]. In these rats, penetration of gadolinium as an indicator of subarachnoid CSF-ISF flow as visualized by dynamic contrast-enhanced MRI was significantly decreased in prefrontal cortex and olfactory bulbs as compared to the control [22]. Moreover, microarray expression analyses of astrocytes isolated from hyperammonemic mice showed that the expression of the gap junction proteins connexin-43 and AQP4 was decreased compared to the control cells [187]. In addition to that, astrocytes exposed to ammonia in culture medium decreased the expression of AQP4 and changed the arrangement of AQP4 channels on the plasma membrane [186]. Based on the existing data, it is suggested that mislocalization of AQP4 water channels on astrocytes in HE may impair glymphatic clearance and waste aggregation (Figure 2; Table 1.). Further research is needed to confirm the role of the glymphatic system dysfunction in the context of HE.

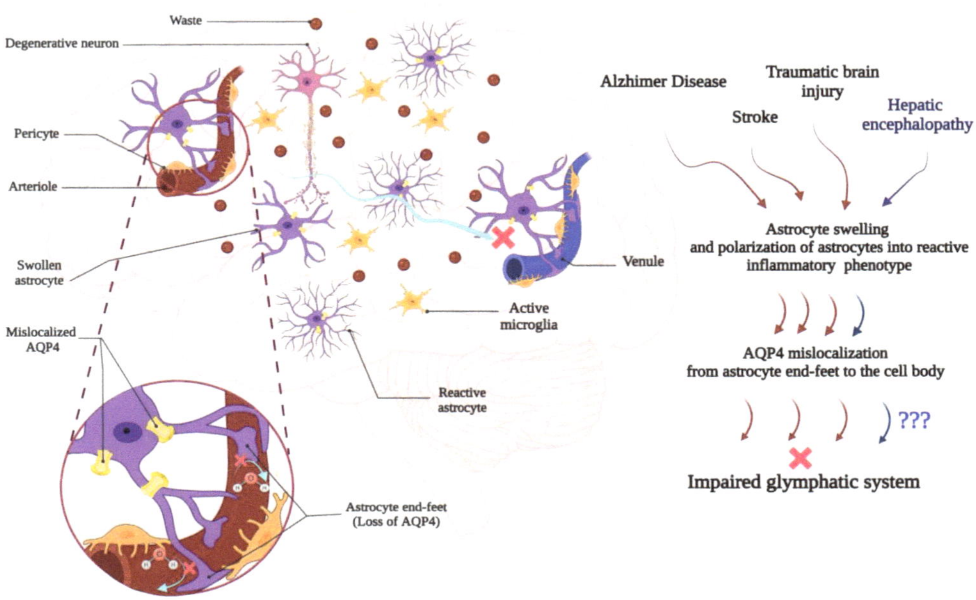

Figure 2. Injured astrocytes differentially express AQP4, impairing the glymphatic system. In several neurological disorders, AQP4 is primarily expressed on the cell body of swollen and reactive astrocytes (A1 phenotype). The AQP4 mislocalization disrupts the flow through the ISF and is responsible for impairment of the glymphatic system, resulting in accumulation of waste metabolites, microglial activation, neuroinflammation, and neuronal toxicity in the brain parenchyma. Similar astrogliosis and mislocalization of AQP4 have been observed in HE. These pathological changes in astrocytes, the main cellular component of the glymphatic pathway, may compromise clearance of ISF in HE. Created with BioRender.com.

Table 1. Main effective factors on the normal function of glymphatic system following liver diseases and HE.

Studies	Affected Factor	Key Results	References
Clinical human cases	Sleep disorders	Sleep disturbances are defined as common features of liver diseases and indicators of subclinical HE in cirrhosis, evaluated by Pittsburgh Sleep Quality Index, Epworth Sleepiness Scale, and polysomnography.	[48,50,53,188–191]
	Arterial hypotension	Portosystemic shunt and accumulation of vasodilator mediators in systemic circulation, shock, decrease systemic vascular resistance, and blood pressure in HE and cirrhosis.	[70,71,73,74,192–194]
	Impaired cerebral blood flow	Decrease in the cerebral perfusion pressure, intracranial hypertension, and impairment of CBF following liver failure and HE.	[83–94]
	Altered autonomic nervous system	General imbalance in ANS tone, sympathetic tone hyperactivity, increase in plasma levels of norepinephrine and inhibition of vagal tone in cirrhosis, chronic liver disease and HE.	[98–105]

Table 1. *Cont.*

Studies	Affected Factor	Key Results	References
Clinical human cases	Astrocyte dysfunction	Decrease in GFAP expression in the frontal cortex and basal ganglia in postmortem analysis of HE patients, solutes accumulation in astrocytes, brain edema, and astrocyte swelling in patients with fulminant hepatic failure.	[114,144,150,153]
	Dysregulation of AQP4	Postmortem brain analysis of cirrhosis patients indicated an increase in the expression of AQP4 in cell body.	[184]
In Vivo	Sleep disorders	Paradoxical sleep, disruption of sleep patterns, suppression in the duration of rapid eye movement sleep, non-rapid eye movement sleep, and alternation of circadian rhythms in rat models of HE.	[54,55,190,195,196]
	Cardiomyopathy	Cardiac hypertrophy, myocyte swelling, gross abnormalities in cardiomyocytes in histological examinations, increase plasma levels of myocardial enzymes, systolic and diastolic abnormalities, inotropic and chronotropic dysfunction, and QT interval prolongation in rat models of HE and cirrhosis.	[66–69]
	Arterial hypotension	Decrease in the systemic arterial blood pressure in animal models of HE and liver failure.	[75,197–202]
	Impaired cerebral blood flow	Raising intracranial pressure, cerebral hyperemia, vasogenic edema, and impairment of CBF in hyperammonemic HE rats.	[82,201,203–206]
	Altered autonomic nervous system	Sympathetic tone hyperactivity and increase in circulatory norepinephrine in rats model of cirrhosis and HE.	[106–108,202,207]
	Astrocyte dysfunction	Decrease in GFAP expression in substantia nigra, ventral tegmental area, hippocampus, sensorimotor cortex, thalamus, cerebral edema, astrogliosis, and astrocyte swelling in animal models of HE.	[115–118,142,208,209]
	Dysregulation of AQP4	Decreased expression of AQP4 in olfactory bulb and prefrontal cortex accompanied with glymphatic impairment in BDL rats.	[22]
	Meningeal lymphatic drainage	Improvement in the severity of HE in an experimental rat model following enhanced meningeal lymphatic drainage (requires further experiments)	[46]
In Vitro	Astrocyte dysfunction	Cell swelling and AQP4 rearrangement in ammonia-exposed astrocytes.	[186,210–216]
	Dysregulation of AQP4	Decreased AQP4 expression in astrocytes isolated from hyperammonemic mice and AQP4 rearrangement in ammonia-exposed astrocytes.	[187]

6. Summary

The glymphatic system is an astrocyte-dependent unidirectional transport pathway that empties macromolecules and waste substances from the brain interstitial space. HE mainly evolves and progresses due to hyperammonemia and inflammation [217]. Accumulation of some neurotoxic and osmotic agents such as ammonia, glutamine, reactive oxygen species, glutamate, alanine, aromatic amino acids, lactate, and proinflammatory cytokines, into the brain parenchyma and interstitial space is considered a central issue in the development and progression of HE [10]. Therefore, an efficient glymphatic pathway that can help remove these waste agents from the brain parenchyma may be of potential clinical interest. For instance, it has been shown that an increase in meningeal lymphangiogenesis through injection of adeno-associated virus 8-vascular endothelial growth factor C (AAV8-VEGF-C) into the cisterna magna of bile duct ligation-induced rats with HE reduced microglial activation, improved the glymphatic function, decreased the expression of proinflammatory cytokines in the cerebral cortex, and improved the motor function [46]. However, there remains a paucity of evidence on the impairment of the glymphatic system HE, or on the effect of treatments such as liver transplantation on its function. Further systematic studies are thus warranted. For the first time, Hadjihambi et al. showed evidence of glymphatic system impairment in the BDL rats model of HE using neuroimaging and immunofluorescence techniques [22]. Here we discussed different levels of pathological changes in HE that can be related to the glymphatic pathway, including the sleep–wake cycle, arterial pulsation, astrocyte morphology and function, AQP4 expression patterns, and meningeal lymphangiogenesis, to draw attention to this topic for future studies (Figure 3).

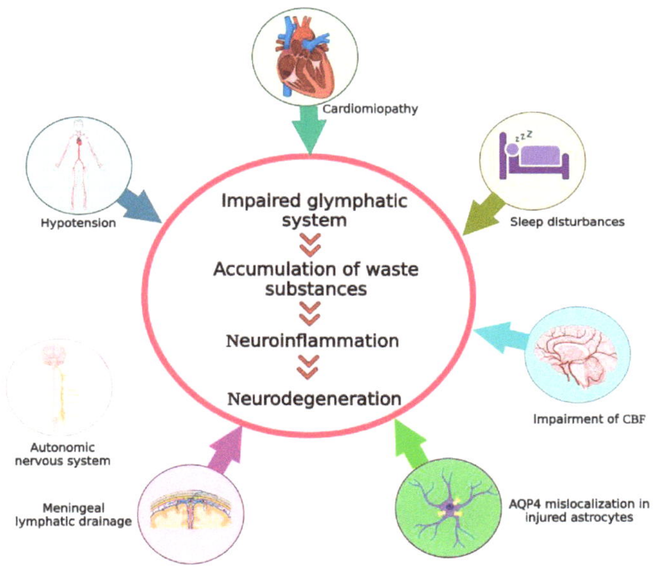

Figure 3. Schematic representations of some pathological processes which may affect normal function of the glymphatic system following HE. All these factors can impair the function of the glymphatic system throughout the cerebral tissue. This induces accumulation of waste substances in the brain interstitial space and triggers neuroinflammation, resulting in neuronal injury and progression of hepatic encephalopathy. CBF: cerebral blood flow; HE: hepatic encephalopathy. Created with BioRender.com.

Author Contributions: A.S. (Ali Sepehrinezhad) designed the study, performed the literature review, drafted the manuscript, and drew and prepared the figures. S.S.N. performed the literature review, supervised the work, corrected grammatical errors, and drafted the manuscript. F.S.L. drafted the manuscript, critically and scientifically edited the manuscript. R.A.A. clinically edited the whole manuscript, corrected grammatical errors, and added extra information where needed. A.S. (Ali Shahbazi) critically edited the final version. All authors have read and agreed to the published version of the manuscript.

Funding: This research received no external funding.

Institutional Review Board Statement: Not applicable.

Informed Consent Statement: Not applicable.

Data Availability Statement: Not applicable.

Conflicts of Interest: The authors declare no conflict of interest.

Abbreviations

AAV8-VEGF-C: adeno-associated virus 8-vascular endothelial growth factor C; AQP4: aquaporin 4; BBB: blood–brain barrier; CBF: cerebral blood flow; CNS: central nervous system; CSF cerebrospinal fluid; GFAP: glial fibrillary acidic protein; hCMEC/D3: human cerebral microvessel endothelial cells; HE: hepatic encephalopathy; ICP: intracranial pressure; ISF: interstitial fluid; MAP mean arterial pressure; mLVs: meningeal lymphatic vessels; MRI: magnetic resonance imaging.

References

1. Wijdicks, E.F.M. Hepatic Encephalopathy. *N. Engl. J. Med.* **2016**, *375*, 1660–1670. [CrossRef] [PubMed]
2. Stewart, C.A.; Malinchoc, M.; Kim, W.R.; Kamath, P.S. Hepatic encephalopathy as a predictor of survival in patients with end-stage liver disease. *Liver Transplant.* **2007**, *13*, 1366–1371. [CrossRef] [PubMed]
3. Bloom, P.P.; Tapper, E.B.; Young, V.B.; Lok, A.S. Microbiome therapeutics for hepatic encephalopathy. *J. Hepatol.* **2021**, *75*, 1452–1464. [CrossRef] [PubMed]
4. Vigo, M.B.; Pérez, M.J.; De Fino, F.; Gómez, G.; Martínez, S.A.; Bisagno, V.; Di Carlo, M.B.; Scazziota, A.; Manautou, J.E. Ghanem, C.I. Acute acetaminophen intoxication induces direct neurotoxicity in rats manifested as astrogliosis and decreased dopaminergic markers in brain areas associated with locomotor regulation. *Biochem. Pharmacol.* **2019**, *170*, 113662. [CrossRef]
5. Tevethia, H.V.; Choudhury, A.; Maiwall, R.; Mitra, L.; Saluja, V.; Singh, P.; Trigatia, A.; Kumar, G.; Thapar, S.; Sarin, S.K Noninvasive diagnosis of cerebral edema in patients of acute liver failure. *Hepatol. Int.* **2018**, *12*, S614. [CrossRef]
6. Tamnanloo, F.; Ochoa-Sanchez, R.; Tremblay, M.; Rose, C.F. Repeated ammonia-induced episodes of hepatic encephalopathy leads to neuronal cell loss in rats with chronic liver disease. *Hepatology* **2022**, *76*, S1168–S1169. [CrossRef]
7. Claeys, W.; Van Hoecke, L.; Geerts, A.; Van Vlierberghe, H.; Lefere, S.; Van Imschoot, G.; Van Wonterghem, E.; Ghesquière, B. Vandenbroucke, R.E.; Van Steenkiste, C. A mouse model of hepatic encephalopathy: Bile duct ligation induces brain ammonia overload, glial cell activation and neuroinflammation. *Sci. Rep.* **2022**, *12*, 17558. [CrossRef] [PubMed]
8. Balzano, T.; Forteza, J.; Molina, P.; Giner, J.; Monzó, A.; Sancho-Jiménez, J.; Urios, A.; Montoliu, C.; Felipo, V. Lymphocyte infiltration, glial activation and neuronal loss in cerebellum of patients with different stages of chronic liver disease. *Glia* **2019**, *67*, E518–E519. [CrossRef]
9. Agarwal, A.N.; Mais, D.D. Sensitivity and Specificity of Alzheimer Type II Astrocytes in Hepatic Encephalopathy. *Arch. Pathol. Lab. Med.* **2019**, *143*, 1256–1258. [CrossRef]
10. Sepehrinezhad, A.; Zarifkar, A.; Namvar, G.; Shahbazi, A.; Williams, R. Astrocyte swelling in hepatic encephalopathy: Molecular perspective of cytotoxic edema. *Metab. Brain Dis.* **2020**, *35*, 559–578. [CrossRef]
11. Weiss, N.; Barbier Saint Hilaire, P.; Colsch, B.; Isnard, F.; Attala, S.; Schaefer, A.; Amador, M.D.; Rudler, M.; Lamari, F.; Sedel, F.; et al. Cerebrospinal fluid metabolomics highlights dysregulation of energy metabolism in overt hepatic encephalopathy *J. Hepatol* **2016**, *65*, 1120–1130. [CrossRef] [PubMed]
12. Bjerring, P.N.; Gluud, L.L.; Larsen, F.S. Cerebral blood flow and metabolism in hepatic encephalopathy—A meta-analysis. *J. Clin. Exp. Hepatol.* **2018**, *8*, 286–293. [CrossRef] [PubMed]
13. Nedelsky, N.B.; Todd, P.K.; Taylor, J.P. Autophagy and the ubiquitin-proteasome system: Collaborators in neuroprotection. *Biochim. Et Biophys. Acta* **2008**, *1782*, 691–699. [CrossRef]
14. Zhang, C.; Lin, J.; Wei, F.; Song, J.; Chen, W.; Shan, L.; Xue, R.; Wang, G.; Tao, J.; Zhang, G.; et al. Characterizing the glymphatic influx by utilizing intracisternal infusion of fluorescently conjugated cadaverine. *Life Sci.* **2018**, *201*, 150–160. [CrossRef] [PubMed]
15. Abbott, N.J.; Pizzo, M.E.; Preston, J.E.; Janigro, D.; Thorne, R.G. The role of brain barriers in fluid movement in the CNS: Is there a 'glymphatic' system? *Acta Neuropathol.* **2018**, *135*, 387–407. [CrossRef]
16. Reeves, B.C.; Karimy, J.K.; Kundishora, A.J.; Mestre, H.; Cerci, H.M.; Matouk, C.; Alper, S.L.; Lundgaard, I.; Nedergaard, M.; Kahle, K.T. Glymphatic System Impairment in Alzheimer's Disease and Idiopathic Normal Pressure Hydrocephalus. *Trends Mol. Med.* **2020**, *26*, 285–295. [CrossRef]
17. Silva, I.; Silva, J.; Ferreira, R.; Trigo, D. Glymphatic system, AQP4, and their implications in Alzheimer's disease. *Neurol. Res. Pract.* **2021**, *3*, 5. [CrossRef]
18. Harrison, I.F.; Ismail, O.; Machhada, A.; Colgan, N.; Ohene, Y.; Nahavandi, P.; Ahmed, Z.; Fisher, A.; Meftah, S.; Murray, T.K.; et al. Impaired glymphatic function and clearance of tau in an Alzheimer's disease model. *Brain* **2020**, *143*, 2576–2593. [CrossRef]
19. Goulay, R.; Flament, J.; Gauberti, M.; Naveau, M.; Pasquet, N.; Gakuba, C.; Emery, E.; Hantraye, P.; Vivien, D.; Aron-Badin, R.; et al. Subarachnoid Hemorrhage Severely Impairs Brain Parenchymal Cerebrospinal Fluid Circulation in Nonhuman Primate. *Stroke* **2017**, *48*, 2301–2305. [CrossRef]
20. Plog, B.A.; Dashnaw, M.L.; Hitomi, E.; Peng, W.; Liao, Y.; Lou, N.; Deane, R.; Nedergaard, M. Biomarkers of traumatic injury are transported from brain to blood via the glymphatic system. *J. Neurosci.* **2015**, *35*, 518–526. [CrossRef]
21. Iliff, J.J.; Chen, M.J.; Plog, B.A.; Zeppenfeld, D.M.; Soltero, M.; Yang, L.; Singh, I.; Deane, R.; Nedergaard, M. Impairment of glymphatic pathway function promotes tau pathology after traumatic brain injury. *J. Neurosci.* **2014**, *34*, 16180–16193. [CrossRef] [PubMed]

22. Hadjihambi, A.; Harrison, I.F.; Costas-Rodríguez, M.; Vanhaecke, F.; Arias, N.; Gallego-Durán, R.; Mastitskaya, S.; Hosford, P.S.; Olde Damink, S.W.M.; Davies, N.; et al. Impaired brain glymphatic flow in experimental hepatic encephalopathy. *J. Hepatol.* **2019**, *70*, 40–49. [CrossRef] [PubMed]
23. Gallina, P.; Gallo, O.; Nicoletti, C.; Romanelli, R.G. A hydrodynamic hypothesis for the pathogenesis of glymphatic system impairment in hepatic encephalopathy. *J. Hepatol.* **2019**, *71*, 228–229. [CrossRef] [PubMed]
24. Valenza, M.; Facchinetti, R.; Steardo, L.; Scuderi, C. Altered waste disposal system in aging and Alzheimer's disease: Focus on astrocytic aquaporin-4. *Front. Pharmacol.* **2020**, *10*, 1656. [CrossRef] [PubMed]
25. Xie, L.; Kang, H.; Xu, Q.; Chen, M.J.; Liao, Y.; Thiyagarajan, M.; O'Donnell, J.; Christensen, D.J.; Nicholson, C.; Iliff, J.J. Sleep drives metabolite clearance from the adult brain. *Science* **2013**, *342*, 373–377. [CrossRef]
26. Hablitz, L.M.; Plá, V.; Giannetto, M.; Vinitsky, H.S.; Stæger, F.F.; Metcalfe, T.; Nguyen, R.; Benrais, A.; Nedergaard, M. Circadian control of brain glymphatic and lymphatic fluid flow. *Nat. Commun.* **2020**, *11*, 4411. [CrossRef]
27. Camassa, L.M.A.; Lunde, L.K.; Hoddevik, E.H.; Stensland, M.; Boldt, H.B.; De Souza, G.A.; Ottersen, O.P.; Amiry-Moghaddam, M. Mechanisms underlying AQP4 accumulation in astrocyte endfeet. *Glia* **2015**, *63*, 2073–2091. [CrossRef]
28. De Bellis, M.; Pisani, F.; Mola, M.G.; Rosito, S.; Simone, L.; Buccoliero, C.; Trojano, M.; Nicchia, G.P.; Svelto, M.; Frigeri, A. Translational readthrough generates new astrocyte AQP4 isoforms that modulate supramolecular clustering, glial endfeet localization, and water transport. *Glia* **2017**, *65*, 790–803. [CrossRef]
29. Mestre, H.; Hablitz, L.M.; Xavier, A.L.; Feng, W.; Zou, W.; Pu, T.; Monai, H.; Murlidharan, G.; Rivera, R.M.C.; Simon, M.J. Aquaporin-4-dependent glymphatic solute transport in the rodent brain. *Elife* **2018**, *7*, e40070. [CrossRef]
30. Eide, P.K.; Ringstad, G. Delayed clearance of cerebrospinal fluid tracer from entorhinal cortex in idiopathic normal pressure hydrocephalus: A glymphatic magnetic resonance imaging study. *J. Cereb. Blood Flow Metab.* **2019**, *39*, 1355–1368. [CrossRef]
31. Hablitz, L.M.; Vinitsky, H.S.; Sun, Q.; Stæger, F.F.; Sigurdsson, B.; Mortensen, K.N.; Lilius, T.O.; Nedergaard, M. Increased glymphatic influx is correlated with high EEG delta power and low heart rate in mice under anesthesia. *Sci. Adv.* **2019**, *5*, eaav5447. [CrossRef] [PubMed]
32. Brosnan, R.J.; Steffey, E.P.; LeCouteur, R.A.; Imai, A.; Farver, T.B.; Kortz, G.D. Effects of body position on intracranial and cerebral perfusion pressures in isoflurane-anesthetized horses. *J. Appl. Physiol.* **2002**, *92*, 2542–2546. [CrossRef]
33. Kose, G.; Hatipoglu, S. Effect of head and body positioning on cerebral blood flow velocity in patients who underwent cranial surgery. *J. Clin. Nurs.* **2012**, *21*, 1859–1867. [CrossRef] [PubMed]
34. Lee, H.; Xie, L.; Yu, M.; Kang, H.; Feng, T.; Deane, R.; Logan, J.; Nedergaard, M.; Benveniste, H. The Effect of Body Posture on Brain Glymphatic Transport. *J. Neurosci.* **2015**, *35*, 11034–11044. [CrossRef] [PubMed]
35. Reddy, A.K.; Li, Y.-H.; Pham, T.T.; Ochoa, L.N.; Treviño, M.T.; Hartley, C.J.; Michael, L.H.; Entman, M.L.; Taffet, G.E. Measurement of aortic input impedance in mice: Effects of age on aortic stiffness. *Am. J. Physiol.-Heart Circ. Physiol.* **2003**, *285*, H1464–H1470. [CrossRef]
36. Avolio, A.P.; Kuznetsova, T.; Heyndrickx, G.R.; Kerkhof, P.L.M.; Li, J.K. Arterial Flow, Pulse Pressure and Pulse Wave Velocity in Men and Women at Various Ages. *Adv. Exp. Med. Biol.* **2018**, *1065*, 153–168. [CrossRef]
37. Kress, B.T.; Iliff, J.J.; Xia, M.; Wang, M.; Wei, H.S.; Zeppenfeld, D.; Xie, L.; Kang, H.; Xu, Q.; Liew, J.A.; et al. Impairment of paravascular clearance pathways in the aging brain. *Ann. Neurol.* **2014**, *76*, 845–861. [CrossRef]
38. Ma, Q.; Ineichen, B.V.; Detmar, M.; Proulx, S.T. Outflow of cerebrospinal fluid is predominantly through lymphatic vessels and is reduced in aged mice. *Nat. Commun.* **2017**, *8*, 1434. [CrossRef]
39. Chandra, P.K.; Cikic, S.; Rutkai, I.; Guidry, J.J.; Katakam, P.V.G.; Mostany, R.; Busija, D.W. Effects of aging on protein expression in mice brain microvessels: ROS scavengers, mRNA/protein stability, glycolytic enzymes, mitochondrial complexes, and basement membrane components. *GeroScience* **2022**, *44*, 371–388. [CrossRef]
40. Bronzuoli, M.R.; Facchinetti, R.; Valenza, M.; Cassano, T.; Steardo, L.; Scuderi, C. Astrocyte Function Is Affected by Aging and Not Alzheimer's Disease: A Preliminary Investigation in Hippocampi of 3xTg-AD Mice. *Front. Pharmacol.* **2019**, *10*, 644. [CrossRef]
41. Palmer, A.L.; Ousman, S.S. Astrocytes and Aging. *Front. Aging Neurosci.* **2018**, *10*, 337. [CrossRef] [PubMed]
42. Shen, X.Q.; Miyajima, M.; Ogino, I.; Arai, H. Expression of the water-channel protein aquaporin 4 in the H-Tx rat: Possible compensatory role in spontaneously arrested hydrocephalus. *J. Neurosurg.* **2006**, *105*, 459–464. [CrossRef]
43. Owasil, R.; O'Neill, R.; Keable, A.; Nimmo, J.; MacGregor Sharp, M.; Kelly, L.; Saito, S.; Simpson, J.E.; Weller, R.O.; Smith, C.; et al. The Pattern of AQP4 Expression in the Ageing Human Brain and in Cerebral Amyloid Angiopathy. *Int. J. Mol. Sci.* **2020**, *21*, 1225. [CrossRef]
44. Cheng, K.P.; Brodnick, S.K.; Blanz, S.L.; Zeng, W.; Kegel, J.; Pisaniello, J.A.; Ness, J.P.; Ross, E.; Nicolai, E.N.; Settell, M.L.; et al. Clinically-derived vagus nerve stimulation enhances cerebrospinal fluid penetrance. *Brain Stimul.* **2020**, *13*, 1024–1030. [CrossRef] [PubMed]
45. Benveniste, H.; Lee, H.; Ding, F.; Sun, Q.; Al-Bizri, E.; Makaryus, R.; Probst, S.; Nedergaard, M.; Stein, E.A.; Lu, H. Anesthesia with Dexmedetomidine and Low-dose Isoflurane Increases Solute Transport via the Glymphatic Pathway in Rat Brain When Compared with High-dose Isoflurane. *Anesthesiology* **2017**, *127*, 976–988. [CrossRef]
46. Hsu, S.J.; Zhang, C.; Jeong, J.; Lee, S.I.; McConnell, M.; Utsumi, T.; Iwakiri, Y. Enhanced Meningeal Lymphatic Drainage Ameliorates Neuroinflammation and Hepatic Encephalopathy in Cirrhotic Rats. *Gastroenterology* **2021**, *160*, 1315–1329.e13. [CrossRef] [PubMed]

47. Claeys, W.; Van Hoecke, L.; Lefere, S.; Geerts, A.; Verhelst, X.; Van Vlierberghe, H.; Degroote, H.; Devisscher, L.; Vandenbroucke, R.E.; Van Steenkiste, C. The neurogliovascular unit in hepatic encephalopathy. *JHEP Rep.* **2021**, *3*, 100352. [CrossRef]
48. Singh, J.; Sharma, B.C.; Puri, V.; Sachdeva, S.; Srivastava, S. Sleep disturbances in patients of liver cirrhosis with minimal hepatic encephalopathy before and after lactulose therapy. *Metab. Brain Dis.* **2017**, *32*, 595–605. [CrossRef]
49. Agrawal, S.; Umapathy, S.; Dhiman, R.K. Minimal Hepatic Encephalopathy Impairs Quality of Life. *J. Clin. Exp. Hepatol.* **2015**, *5*, S42–S48. [CrossRef]
50. Velissaris, D.; Solomou, E.; Kalogeropoulos, A.; Georgiopoulou, V.; Thomopoulos, C.; Karatza, C. Sleep disorders and brain MRI as early indicators of subclinical hepatic encephalopathy. *Hepatogastroenterology* **2006**, *53*, 51–54.
51. Labenz, C.; Baron, J.S.; Toenges, G.; Schattenberg, J.M.; Nagel, M.; Sprinzl, M.F.; Nguyen-Tat, M.; Zimmermann, T.; Huber, Y.; Marquardt, J.U.; et al. Prospective evaluation of the impact of covert hepatic encephalopathy on quality of life and sleep in cirrhotic patients. *Aliment. Pharmacol. Ther.* **2018**, *48*, 313–321. [CrossRef] [PubMed]
52. Bruyneel, M.; Sersté, T.; Libert, W.; van den Broecke, S.; Ameye, L.; Dachy, B.; Mulkay, J.P.; Moreno, C.; Gustot, T. Improvement of sleep architecture parameters in cirrhotic patients with recurrent hepatic encephalopathy with the use of rifaximin. *Eur. J. Gastroenterol. Hepatol.* **2017**, *29*, 302–308. [CrossRef] [PubMed]
53. Spahr, L.; Coeytaux, A.; Giostra, E.; Hadengue, A.; Annoni, J.M. Histamine H1 blocker hydroxyzine improves sleep in patients with cirrhosis and minimal hepatic encephalopathy: A randomized controlled pilot trial. *Am. J. Gastroenterol.* **2007**, *102*, 744–753. [CrossRef]
54. Llansola, M.; Cantero, J.L.; Hita-Yañez, E.; Mirones-Maldonado, M.J.; Piedrafita, B.; Ahabrach, H.; Errami, M.; Agusti, A.; Felipo, V. Progressive reduction of sleep time and quality in rats with hepatic encephalopathy caused by portacaval shunts. *Neuroscience* **2012**, *201*, 199–208. [CrossRef] [PubMed]
55. Beaubernard, C.; Salomon, F.; Bismuth, H. Experimental hepatic encephalopathy. Study of the organization of a diurnal sleep pattern in rats with portocaval anastomosis. *Biomedicine* **1980**, *32*, 76–80. [PubMed]
56. Larsen, F.S. Cerebral blood flow in hyperammonemia: Heterogeneity and starling forces in capillaries. *Metab. Brain Dis.* **2002**, *17*, 229–235. [CrossRef] [PubMed]
57. Larsen, F.S.; Wendon, J. Brain edema in liver failure: Basic physiologic principles and management. *Liver Transplant.* **2002**, *8*, 983–989. [CrossRef]
58. Iliff, J.J.; Wang, M.; Zeppenfeld, D.M.; Venkataraman, A.; Plog, B.A.; Liao, Y.; Deane, R.; Nedergaard, M. Cerebral arterial pulsation drives paravascular CSF-interstitial fluid exchange in the murine brain. *J. Neurosci.* **2013**, *33*, 18190–18199. [CrossRef]
59. Liu, H.; Gaskari, S.A.; Lee, S.S. Cardiac and vascular changes in cirrhosis: Pathogenic mechanisms. *World J. Gastroenterol.* **2006**, *12*, 837–842. [CrossRef]
60. Pudil, R.; Pelouch, R.; Praus, R.; Vašatová, M.; Hůlek, P. Heart failure in patients with liver cirrhosis. *Cor Et Vasa* **2013**, *55*, e391–e396. [CrossRef]
61. Berzigotti, A.; Bosch, J. Editorial: Increased cardiac output in cirrhosis—*Non*-invasive assessment of regional blood flow by magnetic resonance angiography. *Aliment. Pharmacol. Ther.* **2016**, *43*, 1340–1342. [CrossRef]
62. Li, X.; Yu, S.; Li, L.; Han, D.; Dai, S.; Gao, Y. Cirrhosis-related changes in left ventricular function and correlation with the model for end-stage liver disease score. *Int. J. Clin. Exp. Med.* **2014**, *7*, 5751–5757.
63. Sampaio, F.; Pimenta, J.; Bettencourt, N.; Fontes-Carvalho, R.; Silva, A.P.; Valente, J.; Bettencourt, P.; Fraga, J.; Gama, V. Systolic and diastolic dysfunction in cirrhosis: A tissue-Doppler and speckle tracking echocardiography study. *Liver Int.* **2013**, *33*, 1158–1165. [CrossRef] [PubMed]
64. Cazzaniga, M.; Salerno, F.; Pagnozzi, G.; Dionigi, E.; Visentin, S.; Cirello, I.; Meregaglia, D.; Nicolini, A. Diastolic dysfunction is associated with poor survival in patients with cirrhosis with transjugular intrahepatic portosystemic shunt. *Gut* **2007**, *56*, 869–875. [CrossRef]
65. Srinivasamurthy, B.C.; Saravanan, S.P.; Marak, F.K.; Manivel, P.; Bhat, R.V.; Mathiyazhagan, D. Morphological Cardiac Alterations in Liver Cirrhosis: An Autopsy Study. *Heart Views Off. J. Gulf Heart Assoc.* **2021**, *22*, 96–101. [CrossRef]
66. Gregolin, C.S.; do Nascimento, M.; Borges de Souza, S.L.; Ferreira Mota, G.A.; Bomfim, G.F.; de Azevedo Melo Luvizotto, R.; Sugizaki, M.M.; Zanati Bazan, S.G.; Salomé de Campos, D.H.; Dias, M.C.; et al. Myocardial Dysfunction in Cirrhotic Cardiomyopathy is Associated with Alterations of Phospholamban Phosphorylation and IL-6 Levels. *Arch. Med. Res.* **2021**, *52*, 284–293. [CrossRef]
67. Yu, S.; Sun, L.; Wang, H.; Jiang, J.; Zhou, Q. Autonomic regulation of imbalance-induced myocardial fibrosis and its mechanism in rats with cirrhosis. *Exp. Ther. Med.* **2021**, *22*, 1040. [CrossRef] [PubMed]
68. Yang, C.H.; Ting, W.J.; Day, C.H.; Ju, D.T.; Yeh, Y.L.; Chung, L.C.; Tsai, F.J.; Tsai, C.H.; Tsai, Y.; Huang, C.Y. SHSST cyclodextrin complex prevents the fibrosis effect on CCl_4-induced cirrhotic cardiomyopathy in rats through TGF-β pathway inhibition effects. *Int. J. Mol. Sci.* **2014**, *15*, 8037–8048. [CrossRef]
69. Farahnaz, J.; Razieh, A.; Alireza, A.; Giti, G.; Ahmad Reza, D. Evaluation of Chronic Losartan Treatment Effect on Cardiac Chronotropic Dysfunction in Biliary Cirrhotic Rats. *Acta Med. Iran.* **2018**, *56*, 4–13.
70. Wiest, R.; Groszmann, R.J. The paradox of nitric oxide in cirrhosis and portal hypertension: Too much, not enough. *Hepatology* **2002**, *35*, 478–491. [CrossRef]

71. Zardi, E.M.; Abbate, A.; Zardi, D.M.; Dobrina, A.; Margiotta, D.; Tassel, B.W.V.; Afeltra, A.; Sanyal, A.J. Cirrhotic Cardiomyopathy. *J. Am. Coll. Cardiol.* **2010**, *56*, 539–549. [CrossRef] [PubMed]
72. Liu, H.; Ma, Z.; Lee, S.S. Contribution of nitric oxide to the pathogenesis of cirrhotic cardiomyopathy in bile duct-ligated rats. *Gastroenterology* **2000**, *118*, 937–944. [CrossRef] [PubMed]
73. Bátkai, S.; Járai, Z.; Wagner, J.A.; Goparaju, S.K.; Varga, K.; Liu, J.; Wang, L.; Mirshahi, F.; Khanolkar, A.D.; Makriyannis, A.; et al. Endocannabinoids acting at vascular CB1 receptors mediate the vasodilated state in advanced liver cirrhosis. *Nat. Med.* **2001**, *7*, 827–832. [CrossRef]
74. Trewby, P.N.; Williams, R. Pathophysiology of hypotension in patients with fulminant hepatic failure. *Gut* **1977**, *18*, 1021–1026. [CrossRef]
75. Tsai, C.-Y.; Su, C.-H.; Chan, J.Y.H.; Chan, S.H.H. Nitrosative Stress-Induced Disruption of Baroreflex Neural Circuits in a Rat Model of Hepatic Encephalopathy: A DTI Study. *Sci. Rep.* **2017**, *7*, 40111. [CrossRef] [PubMed]
76. Tomiyama, Y.; Johnny, E.; Brian, J.; Todd, M.M. Plasma viscosity and cerebral blood flow. *Am. J. Physiol.-Heart Circ. Physiol.* **2000**, *279*, H1949–H1954. [CrossRef]
77. Odunayo, A. Chapter 80–Traumatic Brain Injury. In *August's Consultations in Feline Internal Medicine*; Little, S.E., Ed.; W.B. Saunders: St. Louis, PA, USA, 2016; Volume 7, pp. 799–802. [CrossRef]
78. Bjerring, P.N.; Bjerrum, E.J.; Larsen, F.S. Impaired cerebral microcirculation induced by ammonium chloride in rats is due to cortical adenosine release. *J. Hepatol.* **2018**, *68*, 1137–1143. [CrossRef]
79. Dam, G.; Keiding, S.; Munk, O.L.; Ott, P.; Vilstrup, H.; Bak, L.K.; Waagepetersen, H.S.; Schousboe, A.; Sørensen, M. Hepatic encephalopathy is associated with decreased cerebral oxygen metabolism and blood flow, not increased ammonia uptake. *Hepatology* **2013**, *57*, 258–265. [CrossRef]
80. Zheng, G.; Zhang, L.J.; Cao, Y.; Lu, G.M. Venous blood ammonia can be associated with cerebral blood flow in hepatic encephalopathy. *Hepatology* **2013**, *58*, 832–833. [CrossRef]
81. Strauss, G.; Hansen, B.A.; Kirkegaard, P.; Rasmussen, A.; Hjortrup, A.; Larsen, F.S. Liver function, cerebral blood flow autoregulation, and hepatic encephalopathy in fulminant hepatic failure. *Hepatology* **1997**, *25*, 837–839. [CrossRef]
82. Larsen, F.S.; Knudsen, G.M.; Paulson, O.B.; Vilstrup, H. Cerebral blood flow autoregulation is absent in rats with thioacetamide-induced hepatic failure. *J. Hepatol.* **1994**, *21*, 491–495. [CrossRef]
83. Donovan, J.P.; Schafer, D.F.; Shaw, B.W., Jr.; Sorrell, M.F. Cerebral oedema and increased intracranial pressure in chronic liver disease. *Lancet* **1998**, *351*, 719–721. [CrossRef] [PubMed]
84. Crippin, J.S.; Gross, J.B., Jr.; Lindor, K.D. Increased intracranial pressure and hepatic encephalopathy in chronic liver disease. *Am. J. Gastroenterol.* **1992**, *87*, 879–882. [PubMed]
85. Detry, O.; De Roover, A.; Honoré, P.; Meurisse, M. Brain edema and intracranial hypertension in fulminant hepatic failure: Pathophysiology and management. *WJG* **2006**, *12*, 7405. [CrossRef]
86. Bernal, W.; Murphy, N.; Brown, S.; Whitehouse, T.; Bjerring, P.N.; Hauerberg, J.; Frederiksen, H.J.; Auzinger, G.; Wendon, J.; Larsen, F.S. A multicentre randomized controlled trial of moderate hypothermia to prevent intracranial hypertension in acute liver failure. *J. Hepatol.* **2016**, *65*, 273–279. [CrossRef]
87. Bjerring, P.N.; Eefsen, M.; Hansen, B.A.; Larsen, F.S. The brain in acute liver failure. A tortuous path from hyperammonemia to cerebral edema. *Metab. Brain Dis.* **2009**, *24*, 5–14. [CrossRef]
88. Blei, A.T.; Larsen, F.S. Pathophysiology of cerebral edema in fulminant hepatic failure. *J. Hepatol.* **1999**, *31*, 771–776. [CrossRef] [PubMed]
89. Jalan, R.; Damink, S.W.O.; Deutz, N.E.; Hayes, P.C.; Lee, A. Moderate hypothermia in patients with acute liver failure and uncontrolled intracranial hypertension. *Gastroenterology* **2004**, *127*, 1338–1346. [CrossRef]
90. Tofteng, F.; Larsen, F.S. The Effect of Indomethacin on Intracranial Pressure, Cerebral Perfusion and Extracellular Lactate and Glutamate Concentrations in Patients with Fulminant Hepatic Failure. *J. Cereb. Blood Flow Metab.* **2004**, *24*, 798–804. [CrossRef]
91. Tofteng, F.; Hauerberg, J.; Hansen, B.A.; Pedersen, C.B.; Jørgensen, L.; Larsen, F.S. Persistent Arterial Hyperammonemia Increases the Concentration of Glutamine and Alanine in the Brain and Correlates with Intracranial Pressure in Patients with Fulminant Hepatic Failure. *J. Cereb. Blood Flow Metab.* **2006**, *26*, 21–27. [CrossRef]
92. Ranjan, P.; Mishra, A.M.; Kale, R.; Saraswat, V.A.; Gupta, R.K. Cytotoxic Edema Is Responsible for Raised Intracranial Pressure in Fulminant Hepatic Failure: In Vivo Demonstration Using Diffusion-Weighted MRI in Human Subjects. *Metab. Brain Dis.* **2005**, *20*, 181–192. [CrossRef]
93. Bjerring, P.N.; Hauerberg, J.; Jørgensen, L.; Frederiksen, H.-J.; Tofteng, F.; Hansen, B.A.; Larsen, F.S. Brain hypoxanthine concentration correlates to lactate/pyruvate ratio but not intracranial pressure in patients with acute liver failure. *J. Hepatol.* **2010**, *53*, 1054–1058. [CrossRef]
94. Davenport, A.; Will, E.J.; Davison, A.M. Early Changes in Intracranial Pressure During Haemofiltration Treatment in Patients with Grade 4 Hepatic Encephalopathy and Acute Oliguric Renal Failure. *Nephrol. Dial. Transplant.* **1990**, *5*, 192–198. [CrossRef] [PubMed]
95. Schmidt, L.E.; Tofteng, F.; Strauss, G.I.; Larsen, F.S. Effect of treatment with the Molecular Adsorbents Recirculating System on arterial amino acid levels and cerebral amino acid metabolism in patients with hepatic encephalopathy. *Scand. J. Gastroenterol.* **2004**, *39*, 974–980. [CrossRef] [PubMed]

96. Bjerring, P.N.; Dale, N.; Larsen, F.S. Acute hyperammonemia and systemic inflammation is associated with increased extracellular brain adenosine in rats: A biosensor study. *Neurochem. Res.* **2015**, *40*, 258–264. [CrossRef] [PubMed]
97. Bjerring, P.N.; Hauerberg, J.; Frederiksen, H.J.; Nielsen, H.B.; Clemmesen, J.O.; Larsen, F.S. The effect of fractionated plasma separation and adsorption on cerebral amino acid metabolism and oxidative metabolism during acute liver failure. *J. Hepatol.* **2012**, *57*, 774–779. [CrossRef]
98. Lenaerts, A.; Codden, T.; Meunier, J.-C.; Henry, J.-P.; Ligny, G. Effects of clonidine on diuretic response in ascitic patients with cirrhosis and activation of sympathetic nervous system. *Hepatology* **2006**, *44*, 844–849. [CrossRef]
99. Miyajima, H.; Nomura, M.; Muguruma, N.; Okahisa, T.; Shibata, H.; Okamura, S.; Honda, H.; Shimizu, I.; Harada, M.; Saito, K.; et al. Relationship among gastric motility, autonomic activity, and portal hemodynamics in patients with liver cirrhosis. *J. Gastroenterol. Hepatol.* **2001**, *16*, 647–659. [CrossRef]
100. Henriksen, J.H.; Ring-Larsen, H.; Christensen, N.J. Sympathetic nervous activity in cirrhosis: A survey of plasma catecholamine studies. *J. Hepatol.* **1985**, *1*, 55–65. [CrossRef]
101. Barron, H.V.; Alam, I.; Lesh, M.D.; Strunk, A.; Bass, N.M. Autonomic nervous system tone measured by baroreflex sensitivity is depressed in patients with end-stage liver disease. *Am. J. Gastroenterol.* **1999**, *94*, 986–989. [CrossRef]
102. Lara, C.; Gajardo, A.; Giubergia, F.; Bustos, N.; Roblero, J.P.; Urzúa, A.; Poniachik, J.; Cattaneo, M. P-38 Utility of pupillary reactivity in the functional assessment of the autonomous nervous system in patients with chronic liver disease: Preliminary results. *Ann. Hepatol.* **2021**, *24*, 100402. [CrossRef]
103. Esler, M.; Dudley, F.; Jennings, G.; Debinski, H.; Lambert, G.; Jones, P.; Crotty, B.; Colman, J.; Willett, I. Increased Sympathetic Nervous Activity and the Effects of Its Inhibition with Clonidine in Alcoholic Cirrhosis. *Ann. Intern. Med.* **1992**, *116*, 446–456. [CrossRef] [PubMed]
104. Lenz, K.; Hörtnagl, H.; Magometschnigg, D.; Kleinberger, G.; Druml, W.; Laggner, A. Function of the autonomic nervous system in patients with hepatic encephalopathy. *Hepatology* **1985**, *5*, 831–836. [CrossRef]
105. Rega, D.; Aiko, M.; Peñaranda, N.; Urios, A.; Gallego, J.J.; Giménez-Garzó, C.; Casanova, F.; Fiorillo, A.; Cabrera-Pastor, A.; San-Miguel, T.; et al. Patients with Minimal Hepatic Encephalopathy Show Altered Thermal Sensitivity and Autonomic Function. *J. Clin. Med.* **2021**, *10*, 239. [CrossRef] [PubMed]
106. Zhang, Y.; Tan, S.-L.; Du, J.; Chen, Y.; Jia, J.; Feng, J.-G.; Liu, K.-X.; Zhou, J. Dexmedetomidine alleviates neuroinflammation, restores sleep disorders and neurobehavioral abnormalities in rats with minimal hepatic encephalopathy. *Int. Immunopharmacol.* **2021**, *96*, 107795. [CrossRef]
107. Kirstetter, P.; Moreau, R.; Soupison, T.; Cailmail, S.; Hartleb, M.; Lebrec, D. Role of sympathetic cardiovascular tone in control of arterial pressure in rats with cirrhosis. *Liver* **1996**, *16*, 263–266. [CrossRef]
108. Dietrich, P.; Moleda, L.; Kees, F.; Müller, M.; Straub, R.H.; Hellerbrand, C.; Wiest, R. Dysbalance in sympathetic neurotransmitter release and action in cirrhotic rats: Impact of exogenous neuropeptide Y. *J. Hepatol.* **2013**, *58*, 254–261. [CrossRef]
109. Jaeger, V.; DeMorrow, S.; McMillin, M. The direct contribution of astrocytes and microglia to the pathogenesis of hepatic encephalopathy. *J. Clin. Transl. Hepatol.* **2019**, *7*, 352. [CrossRef]
110. Ott, P.; Larsen, F.S. Blood-brain barrier permeability to ammonia in liver failure: A critical reappraisal. *Neurochem. Int.* **2004**, *44*, 185–198. [CrossRef]
111. Siracusa, R.; Fusco, R.; Cuzzocrea, S. Astrocytes: Role and Functions in Brain Pathologies. *Front. Pharmacol.* **2019**, *10*, 1114. [CrossRef]
112. Broadhead, M.J.; Miles, G.B. Bi-Directional Communication Between Neurons and Astrocytes Modulates Spinal Motor Circuits. *Front. Cell. Neurosci.* **2020**, *14*, 30. [CrossRef]
113. Hol, E.M.; Pekny, M. Glial fibrillary acidic protein (GFAP) and the astrocyte intermediate filament system in diseases of the central nervous system. *Curr. Opin. Cell Biol.* **2015**, *32*, 121–130. [CrossRef]
114. Kretzschmar, H.A.; DeArmond, S.J.; Forno, L.S. Measurement of GFAP in Hepatic Encephalopathy by ELISA and Transblots. *J. Neuropathol. Exp. Neurol.* **1985**, *44*, 459–471. [CrossRef]
115. Hiba, O.E.; Elgot, A.; Ahboucha, S.; Gamrani, H. Differential regional responsiveness of astroglia in mild hepatic encephalopathy: An Immunohistochemical approach in bile duct ligated rat. *Acta Histochem.* **2016**, *118*, 338–346. [CrossRef] [PubMed]
116. Shuliatnikova, T. Region-dependent heterogeneity in GFAP expression in acute hepatic encephalopathy. *Interdiscip.Res. Sci. Horiz. Perspect.* **2021**, *2*, 56–57.
117. Chileski, G.S.; García, E.N.; Lértora, J.W.; Mussart, N.; Hernández, D.R.; Cholich, L.A. Hepatic encephalopathy in swine experimentally poisoned with Senna occidentalis seeds: Effects on astrocytes. *Toxicon* **2021**, *201*, 86–91. [CrossRef]
118. Bélanger, M.; Desjardins, P.; Chatauret, N.; Butterworth, R.F. Loss of expression of glial fibrillary acidic protein in acute hyperammonemia. *Neurochem. Int.* **2002**, *41*, 155–160. [CrossRef]
119. Jayakumar, A.; Rao, K.R.; Murthy, C.R.; Norenberg, M. Glutamine in the mechanism of ammonia-induced astrocyte swelling. *Neurochem. Int.* **2006**, *48*, 623–628. [CrossRef]
120. Strauss, G.I.; Knudsen, G.M.; Kondrup, J.; Møller, K.; Larsen, F.S. Cerebral metabolism of ammonia and amino acids in patients with fulminant hepatic failure. *Gastroenterology* **2001**, *121*, 1109–1119. [CrossRef]
121. Albrecht, J.; Norenberg, M.D. Glutamine: A Trojan horse in ammonia neurotoxicity. *Hepatology* **2006**, *44*, 788–794. [CrossRef] [PubMed]

122. Tanigami, H.; Rebel, A.; Martin, L.J.; Chen, T.Y.; Brusilow, S.W.; Traystman, R.J.; Koehler, R.C. Effect of glutamine synthetase inhibition on astrocyte swelling and altered astroglial protein expression during hyperammonemia in rats. *Neuroscience* **2005**, *131*, 437–449. [CrossRef]
123. Zemtsova, I.; Görg, B.; Keitel, V.; Bidmon, H.J.; Schrör, K.; Häussinger, D. Microglia activation in hepatic encephalopathy in rats and humans. *Hepatology* **2011**, *54*, 204–215. [CrossRef] [PubMed]
124. McMillin, M.; Frampton, G.; Thompson, M.; Galindo, C.; Standeford, H.; Whittington, E.; Alpini, G.; DeMorrow, S. Neuronal CCL2 is upregulated during hepatic encephalopathy and contributes to microglia activation and neurological decline. *J. Neuroinflammation* **2014**, *11*, 121. [CrossRef] [PubMed]
125. Rao, K.V.R.; Brahmbhatt, M.; Norenberg, M.D. Microglia contribute to ammonia-induced astrocyte swelling in culture. *Metab. Brain Dis.* **2013**, *28*, 139–143. [CrossRef] [PubMed]
126. Weiss, N.; Miller, F.; Cazaubon, S.; Couraud, P.O. The blood-brain barrier in brain homeostasis and neurological diseases. *Biochim. Et Biophys. Acta* **2009**, *1788*, 842–857. [CrossRef]
127. Abbott, N.J.; Patabendige, A.A.; Dolman, D.E.; Yusof, S.R.; Begley, D.J. Structure and function of the blood–brain barrier. *Neurobiol. Dis.* **2010**, *37*, 13–25. [CrossRef]
128. Hajal, C.; Le Roi, B.; Kamm, R.D.; Maoz, B.M. Biology and Models of the Blood–brain Barrier. *Annu. Rev. Biomed. Eng.* **2021**, *23*, 359–384. [CrossRef] [PubMed]
129. Huang, X.; Hussain, B.; Chang, J. Peripheral inflammation and blood–brain barrier disruption: Effects and mechanisms. *CNS Neurosci. Ther.* **2021**, *27*, 36–47. [CrossRef]
130. Galea, I. The blood–brain barrier in systemic infection and inflammation. *Cell. Mol. Immunol.* **2021**, *18*, 2489–2501. [CrossRef]
131. Profaci, C.P.; Munji, R.N.; Pulido, R.S.; Daneman, R. The blood–brain barrier in health and disease: Important unanswered questions. *J. Exp. Med.* **2020**, *217*, e20190062. [CrossRef]
132. Song, K.; Li, Y.; Zhang, H.; An, N.; Wei, Y.; Wang, L.; Tian, C.; Yuan, M.; Sun, Y.; Xing, Y. Oxidative stress-mediated blood-brain barrier (BBB) disruption in neurological diseases. *Oxidative Med. Cell. Longev.* **2020**, *2020*, 4356386. [CrossRef]
133. Marchetti, L.; Engelhardt, B. Immune cell trafficking across the blood-brain barrier in the absence and presence of neuroinflammation. *Vasc. Biol.* **2020**, *2*, H1–H18. [CrossRef] [PubMed]
134. Jayakumar, A.R.; Rama Rao, K.V.; Norenberg, M.D. Neuroinflammation in hepatic encephalopathy: Mechanistic aspects. *J. Clin. Exp. Hepatol.* **2015**, *5*, S21–S28. [CrossRef] [PubMed]
135. Chen, Z.; Ruan, J.; Li, D.; Wang, M.; Han, Z.; Qiu, W.; Wu, G. The Role of Intestinal Bacteria and Gut–Brain Axis in Hepatic Encephalopathy. *Front. Cell. Infect. Microbiol.* **2021**, *10*, 595759. [CrossRef] [PubMed]
136. Schaefer, A.; Journaux, M.; Mourabit, H.E.; Mouri, S.; Wendum, D.; Lasnier, E.; Couraud, P.-O.; Housset, C.; Thabut, D.; Rudler, M.; et al. A systemic mechanism of increased transendothelial migration of leukocytes through the blood-brain barrier in hepatic encephalopathy. *Clin. Res. Hepatol. Gastroenterol.* **2022**, *46*, 101801. [CrossRef]
137. Masago, K.; Kihara, Y.; Yanagida, K.; Hamano, F.; Nakagawa, S.; Niwa, M.; Shimizu, T. Lysophosphatidic acid receptor, LPA6, regulates endothelial blood-brain barrier function: Implication for hepatic encephalopathy. *Biochem. Biophys. Res. Commun.* **2018**, *501*, 1048–1054. [CrossRef]
138. McMillin, M.A.; Frampton, G.A.; Seiwell, A.P.; Patel, N.S.; Jacobs, A.N.; DeMorrow, S. TGFβ1 exacerbates blood–brain barrier permeability in a mouse model of hepatic encephalopathy via upregulation of MMP9 and downregulation of claudin-5. *Lab. Investig.* **2015**, *95*, 903–913. [CrossRef]
139. Kato, M.; Sugihara, J.; Nakamura, T.; Muto, Y. Electron microscopic study of the blood-brain barrier in rats with brain edema and encephalopathy due to acute hepatic failure. *Gastroenterol. Jpn.* **1989**, *24*, 135–142. [CrossRef]
140. Wiltfang, J.; Nolte, W.; Otto, M.; Wildberg, J.; Bahn, E.; Figulla, H.R.; Pralle, L.; Hartmann, H.; Rüther, E.; Ramadori, G. Elevated serum levels of astroglial S100beta in patients with liver cirrhosis indicate early and subclinical portal-systemic encephalopathy. *Metab. Brain Dis.* **1999**, *14*, 239–251. [CrossRef] [PubMed]
141. Saleh, A.; Kamel, L.; Ghali, A.; Ismail, A.; El Khayat, H. Serum levels of astroglial S100-beta and neuron-specific enolase in hepatic encephalopathy patients. *EMHJ-East. Mediterr. Health J.* **2007**, *13*, 1114–1123. [CrossRef]
142. Wright, G.; Davies, N.A.; Shawcross, D.L.; Hodges, S.J.; Zwingmann, C.; Brooks, H.F.; Mani, A.R.; Harry, D.; Stadlbauer, V.; Zou, Z.; et al. Endotoxemia produces coma and brain swelling in bile duct ligated rats. *Hepatology* **2007**, *45*, 1517–1526. [CrossRef] [PubMed]
143. Cauli, O.; Llansola, M.; Agustí, A.; Rodrigo, R.; Hernández-Rabaza, V.; Rodrigues, T.B.; López-Larrubia, P.; Cerdán, S.; Felipo, V. Cerebral oedema is not responsible for motor or cognitive deficits in rats with hepatic encephalopathy. *Liver Int.* **2014**, *34*, 379–387. [CrossRef]
144. Kale, R.A.; Gupta, R.K.; Saraswat, V.A.; Hasan, K.M.; Trivedi, R.; Mishra, A.M.; Ranjan, P.; Pandey, C.M.; Narayana, P.A. Demonstration of interstitial cerebral edema with diffusion tensor MR imaging in type C hepatic encephalopathy. *Hepatology* **2006**, *43*, 698–706. [CrossRef] [PubMed]
145. Córdoba, J.; Sanpedro, F.; Alonso, J.; Rovira, A. 1H Magnetic Resonance in the Study of Hepatic Encephalopathy in Humans. *Metab. Brain Dis.* **2002**, *17*, 415–429. [CrossRef]
146. Rai, R.; Ahuja, C.K.; Agrawal, S.; Kalra, N.; Duseja, A.; Khandelwal, N.; Chawla, Y.; Dhiman, R.K. Reversal of low-grade cerebral edema after lactulose/rifaximin therapy in patients with cirrhosis and minimal hepatic encephalopathy. *Clin. Transl. Gastroenterol.* **2015**, *6*, e111. [CrossRef] [PubMed]

147. Grant, S.; McMillin, M.; Frampton, G.; Petrescu, A.D.; Williams, E.; Jaeger, V.; Kain, J.; DeMorrow, S. Direct Comparison of the Thioacetamide and Azoxymethane Models of Type A Hepatic Encephalopathy in Mice. *Gene Expr.* **2018**, *18*, 171–185. [CrossRef]
148. Mardini, H.; Smith, F.E.; Record, C.O.; Blamire, A.M. Magnetic resonance quantification of water and metabolites in the brain of cirrhotics following induced hyperammonaemia. *J. Hepatol.* **2011**, *54*, 1154–1160. [CrossRef]
149. Häussinger, D.; Kircheis, G.; Fischer, R.; Schliess, F.; vom Dahl, S. Hepatic encephalopathy in chronic liver disease: A clinical manifestation of astrocyte swelling and low-grade cerebral edema? *J. Hepatol.* **2000**, *32*, 1035–1038. [CrossRef]
150. Martinez, A. Electron microscopy in human hepatic encephalopathy. *Acta Neuropathol.* **1968**, *11*, 82–86. [CrossRef]
151. Jiang, W.; Desjardins, P.; Butterworth, R.F. Hypothermia attenuates oxidative/nitrosative stress, encephalopathy and brain edema in acute (ischemic) liver failure. *Neurochem. Int.* **2009**, *55*, 124–128. [CrossRef]
152. Jayakumar, A.R.; Tong, X.Y.; Curtis, K.M.; Ruiz-Cordero, R.; Abreu, M.T.; Norenberg, M.D. Increased toll-like receptor 4 in cerebral endothelial cells contributes to the astrocyte swelling and brain edema in acute hepatic encephalopathy. *J Neurochem.* **2014**, *128*, 890–903. [CrossRef] [PubMed]
153. Kato, M.; Hughes, R.D.; Keays, R.T.; Williams, R. Electron microscopic study of brain capillaries in cerebral edema from fulminant hepatic failure. *Hepatology* **1992**, *15*, 1060–1066. [CrossRef] [PubMed]
154. Sepehrinezhad, A.; Shahbazi, A.; Sahab Negah, S.; Joghataei, M.T.; Larsen, F.S. Drug-induced-acute liver failure: A critical appraisal of the thioacetamide model for the study of hepatic encephalopathy. *Toxicol. Rep.* **2021**, *8*, 962–970. [CrossRef] [PubMed]
155. Liddelow, S.A.; Guttenplan, K.A.; Clarke, L.E.; Bennett, F.C.; Bohlen, C.J.; Schirmer, L.; Bennett, M.L.; Münch, A.E.; Chung, W.-S.; Peterson, T.C.; et al. Neurotoxic reactive astrocytes are induced by activated microglia. *Nature* **2017**, *541*, 481–487. [CrossRef] [PubMed]
156. Clark, D.P.Q.; Perreau, V.M.; Shultz, S.R.; Brady, R.D.; Lei, E.; Dixit, S.; Taylor, J.M.; Beart, P.M.; Boon, W.C. Inflammation in Traumatic Brain Injury: Roles for Toxic A1 Astrocytes and Microglial–Astrocytic Crosstalk. *Neurochem. Res.* **2019**, *44*, 1410–1424. [CrossRef]
157. Zha, Z.; Liu, Y.-J.; Liu, S.-S.; Zhang, N.; Li, J.-L.; Qi, F.; Jin, L.-Y.; Xue, B.; Yang, T.; Fan, Y.-P.; et al. Bu Shen Yi Sui capsule promotes myelin repair by modulating the transformation of A1/A2 reactive astrocytes In Vivo and In Vitro. *Oxidative Med. Cell. Longev.* **2022**, *2022*, 3800004. [CrossRef]
158. Zhang, Q.; Liu, C.; Shi, R.; Zhou, S.; Shan, H.; Deng, L.; Chen, T.; Guo, Y.; Zhang, Z.; Yang, G.Y.; et al. Blocking C3d(+)/GFAP(+) A1 Astrocyte Conversion with Semaglutide Attenuates Blood-Brain Barrier Disruption in Mice after Ischemic Stroke. *Aging Dis.* **2022**, *13*, 943–959. [CrossRef]
159. Song, N.; Zhu, H.; Xu, R.; Liu, J.; Fang, Y.; Zhang, J.; Ding, J.; Hu, G.; Lu, M. Induced Expression of kir6.2 in A1 Astrocytes Propagates Inflammatory Neurodegeneration via Drp1-dependent Mitochondrial Fission. *Front. Pharmacol.* **2021**, *11*, 618992. [CrossRef]
160. Manley, G.T.; Binder, D.K.; Papadopoulos, M.C.; Verkman, A.S. New insights into water transport and edema in the central nervous system from phenotype analysis of aquaporin-4 null mice. *Neuroscience* **2004**, *129*, 983–991. [CrossRef]
161. Yi, M.-H.; Lee, Y.S.; Kang, J.W.; Kim, S.J.; Oh, S.-H.; Kim, Y.M.; Lee, Y.H.; Lee, S.D.; Kim, D.W. NFAT5-Dependent Expression of AQP4 in Astrocytes. *Cell. Mol. Neurobiol.* **2013**, *33*, 223–232. [CrossRef]
162. Stokum, J.A.; Mehta, R.I.; Ivanova, S.; Yu, E.; Gerzanich, V.; Simard, J.M. Heterogeneity of aquaporin-4 localization and expression after focal cerebral ischemia underlies differences in white versus grey matter swelling. *Acta Neuropathol. Commun.* **2015**, *3*, 61. [CrossRef] [PubMed]
163. Zhao, Z.-A.; Li, P.; Ye, S.-Y.; Ning, Y.-L.; Wang, H.; Peng, Y.; Yang, N.; Zhao, Y.; Zhang, Z.-H.; Chen, J.-F.; et al. Perivascular AQP4 dysregulation in the hippocampal CA1 area after traumatic brain injury is alleviated by adenosine A2A receptor inactivation. *Sci. Rep.* **2017**, *7*, 2254. [CrossRef] [PubMed]
164. Binder, D.K. Astrocytes: Stars of the Sacred Disease. *Epilepsy Curr.* **2018**, *18*, 172–179. [CrossRef] [PubMed]
165. Potokar, M.; Stenovec, M.; Jorgačevski, J.; Holen, T.; Kreft, M.; Ottersen, O.P.; Zorec, R. Regulation of AQP4 surface expression via vesicle mobility in astrocytes. *Glia* **2013**, *61*, 917–928. [CrossRef]
166. Xu, Z.; Xiao, N.; Chen, Y.; Huang, H.; Marshall, C.; Gao, J.; Cai, Z.; Wu, T.; Hu, G.; Xiao, M. Deletion of aquaporin-4 in APP/PS1 mice exacerbates brain Aβ accumulation and memory deficits. *Mol. Neurodegener.* **2015**, *10*, 58. [CrossRef] [PubMed]
167. Simon, M.; Wang, M.X.; Ismail, O.; Braun, M.; Schindler, A.G.; Reemmer, J.; Wang, Z.; Haveliwala, M.A.; O'Boyle, R.P.; Han, W.Y.; et al. Loss of perivascular aquaporin-4 localization impairs glymphatic exchange and promotes amyloid β plaque formation in mice. *Alzheimer's Res. Ther.* **2022**, *14*, 59. [CrossRef] [PubMed]
168. Liu, E.; Sun, L.; Zhang, Y.; Wang, A.; Yan, J. Aquaporin4 Knockout Aggravates Early Brain Injury Following Subarachnoid Hemorrhage Through Impairment of the Glymphatic System in Rat Brain. *Acta Neurochir. Suppl.* **2020**, *127*, 59–64. [CrossRef] [PubMed]
169. Wang, T.; Chen, Y.; Zou, Y.; Pang, Y.; He, X.; Chen, Y.; Liu, Y.; Feng, W.; Zhang, Y.; Li, Q.; et al. Locomotor Hyperactivity in the Early-Stage Alzheimer's Disease-like Pathology of APP/PS1 Mice: Associated with Impaired Polarization of Astrocyte Aquaporin 4. *Aging Dis.* **2022**, *13*, 1504–1522. [CrossRef]
170. Kamagata, K.; Andica, C.; Takabayashi, K.; Saito, Y.; Taoka, T.; Nozaki, H.; Kikuta, J.; Fujita, S.; Hagiwara, A.; Kamiya, K. Association of MRI indices of glymphatic system with amyloid deposition and cognition in mild cognitive impairment and Alzheimer disease. *Neurology* **2022**, *99*, e2648–e2660. [CrossRef]

171. Taoka, T.; Masutani, Y.; Kawai, H.; Nakane, T.; Matsuoka, K.; Yasuno, F.; Kishimoto, T.; Naganawa, S. Evaluation of glymphatic system activity with the diffusion MR technique: Diffusion tensor image analysis along the perivascular space (DTI-ALPS) in Alzheimer's disease cases. *Jpn. J. Radiol.* **2017**, *35*, 172–178. [CrossRef]
172. Zeppenfeld, D.M.; Simon, M.; Haswell, J.D.; D'Abreo, D.; Murchison, C.; Quinn, J.F.; Grafe, M.R.; Woltjer, R.L.; Kaye, J.; Iliff, J.J. Association of Perivascular Localization of Aquaporin-4 With Cognition and Alzheimer Disease in Aging Brains. *JAMA Neurol.* **2017**, *74*, 91–99. [CrossRef] [PubMed]
173. Ren, Z.; Iliff, J.J.; Yang, L.; Yang, J.; Chen, X.; Chen, M.J.; Giese, R.N.; Wang, B.; Shi, X.; Nedergaard, M. 'Hit & Run' model of closed-skull traumatic brain injury (TBI) reveals complex patterns of post-traumatic AQP4 dysregulation. *J. Cereb. Blood Flow Metab.* **2013**, *33*, 834–845. [CrossRef] [PubMed]
174. Lu, H.; Zhan, Y.; Ai, L.; Chen, H.; Chen, J. AQP4-siRNA alleviates traumatic brain edema by altering post-traumatic AQP4 polarity reversal in TBI rats. *J. Clin. Neurosci.* **2020**, *81*, 113–119. [CrossRef] [PubMed]
175. Christensen, J.; Wright, D.K.; Yamakawa, G.R.; Shultz, S.R.; Mychasiuk, R. Repetitive Mild Traumatic Brain Injury Alters Glymphatic Clearance Rates in Limbic Structures of Adolescent Female Rats. *Sci. Rep.* **2020**, *10*, 6254. [CrossRef] [PubMed]
176. Banitalebi, S.; Skauli, N.; Geiseler, S.; Ottersen, O.P.; Amiry-Moghaddam, M. Disassembly and Mislocalization of AQP4 in Incipient Scar Formation after Experimental Stroke. *Int. J. Mol. Sci.* **2022**, *23*, 1117. [CrossRef] [PubMed]
177. Migliati, E.R.; Amiry-Moghaddam, M.; Froehner, S.C.; Adams, M.E.; Ottersen, O.P.; Bhardwaj, A. Na^+-K^+-$2Cl^-$ cotransport inhibitor attenuates cerebral edema following experimental stroke via the perivascular pool of aquaporin-4. *Neurocritical Care* **2010**, *13*, 123–131. [CrossRef]
178. Steiner, E.; Enzmann, G.U.; Lin, S.; Ghavampour, S.; Hannocks, M.J.; Zuber, B.; Rüegg, M.A.; Sorokin, L.; Engelhardt, B. Loss of astrocyte polarization upon transient focal brain ischemia as a possible mechanism to counteract early edema formation. *Glia* **2012**, *60*, 1646–1659. [CrossRef]
179. Gaberel, T.; Gakuba, C.; Goulay, R.; Martinez De Lizarrondo, S.; Hanouz, J.L.; Emery, E.; Touze, E.; Vivien, D.; Gauberti, M. Impaired glymphatic perfusion after strokes revealed by contrast-enhanced MRI: A new target for fibrinolysis? *Stroke* **2014**, *45*, 3092–3096. [CrossRef]
180. García-Lezana, T.; Oria, M.; Romero-Giménez, J.; Bové, J.; Vila, M.; Genescà, J.; Chavarria, L.; Cordoba, J. Cerebellar neurodegeneration in a new rat model of episodic hepatic encephalopathy. *J. Cereb. Blood Flow Metab.* **2017**, *37*, 927–937. [CrossRef]
181. Gelpi, E.; Rahimi, J.; Klotz, S.; Schmid, S.; Ricken, G.; Forcen-Vega, S.; Budka, H.; Kovacs, G.G. The autophagic marker p62 highlights Alzheimer type II astrocytes in metabolic/hepatic encephalopathy. *Neuropathology* **2020**, *40*, 358–366. [CrossRef]
182. Shulyatnikova, T.; Shavrin, V. Mobilisation and redistribution of multivesicular bodies to the endfeet of reactive astrocytes in acute endogenous toxic encephalopathies. *Brain Res.* **2021**, *1751*, 147174. [CrossRef] [PubMed]
183. Raviz, E.K.; Noormand, F.; Kermani, A.S.; Galedari, A.; Esmaeilpour, K.; Maneshian, M.; Kalantaripour, T.P.; Dabiri, S.; Asadi-Shekaari, M. Promising Effects of Naringenin and Melatonin against Hepatic Encephalopathy Impairments Induced by Bile Duct Ligation in Male Rats. *Cent. Nerv. Syst. Agents Med. Chem.* **2022**, *22*, 31–38. [CrossRef] [PubMed]
184. Shuliatnikova, T.; Tumanskyi, V. Expression of the water channel protein aquaporin-4 in the brain during human liver cirrhosis. *Grail Sci.* **2022**, *22*, 287–288. [CrossRef]
185. Eslimi Esfahani, D.; Oryan, S.; Nabiuni, M.; Hosseinynia, T.S. The effects of bile duct ligation on motor cortex region morphology and aquaporin 4 protein concentration in male Wistar rats. *Nova Biol. Reper.* **2019**, *6*, 1–9. [CrossRef]
186. Bodega, G.; Suárez, I.; López-Fernández, L.A.; García, M.I.; Köber, M.; Penedo, M.; Luna, M.; Juárez, S.; Ciordia, S.; Oria, M.; et al. Ammonia induces aquaporin-4 rearrangement in the plasma membrane of cultured astrocytes. *Neurochem. Int.* **2012**, *61*, 1314–1324. [CrossRef]
187. Lichter-Konecki, U.; Mangin, J.M.; Gordish-Dressman, H.; Hoffman, E.P.; Gallo, V. Gene expression profiling of astrocytes from hyperammonemic mice reveals altered pathways for water and potassium homeostasis in vivo. *Glia* **2008**, *56*, 365–377. [CrossRef]
188. De Rui, M.; Schiff, S.; Aprile, D.; Angeli, P.; Bombonato, G.; Bolognesi, M.; Sacerdoti, D.; Gatta, A.; Merkel, C.; Amodio, P.; et al. Excessive daytime sleepiness and hepatic encephalopathy: It is worth asking. *Metab. Brain Dis.* **2013**, *28*, 245–248. [CrossRef]
189. Velissaris, D.; Karamouzos, V.; Polychronopoulos, P.; Karanikolas, M. Chronotypology and melatonin alterations in minimal hepatic encephalopathy. *J. Circadian Rhythm.* **2009**, *7*, 6. [CrossRef]
190. Lozeva, V.; Tuomisto, L.; Sola, D.; Plumed, C.; Hippeläinen, M.; Butterworth, R. Increased density of brain histamine H1 receptors in rats with portacaval anastomosis and in cirrhotic patients with chronic hepatic encephalopathy. *Hepatology* **2001**, *33*, 1370–1376. [CrossRef]
191. Bajaj, J.S.; Saeian, K.; Schubert, C.M.; Franco, R.; Franco, J.; Heuman, D.M. Disruption of sleep architecture in minimal hepatic encephalopathy and ghrelin secretion. *Aliment Pharm.* **2011**, *34*, 103–105. [CrossRef]
192. Vora, R.S.; Subramanian, R.M. Hypotension in Cirrhosis. *Clin. Liver Dis.* **2019**, *13*, 149–153. [CrossRef] [PubMed]
193. Kim, E.; Choi, D.; Lim, H.K.; Lim, J.H. Multiple infarcted regenerative nodules in liver cirrhosis after systemic hypotension due to septic shock: Radiologic findings. *Abdom. Imaging* **2004**, *29*, 208–210. [CrossRef] [PubMed]
194. Sourianarayanane, A.; Barnes, D.S.; McCullough, A.J. Beneficial effect of midodrine in hypotensive cirrhotic patients with refractory ascites. *Gastroenterol Hepatol* **2011**, *7*, 132–134.
195. Lozeva, V.; Valjakka, A.; Lecklin, A.; Olkkonen, H.; Hippeläinen, M.; Itkonen, M.; Plumed, C.; Tuomisto, L. Effects of the histamine H(1) receptor blocker, pyrilamine, on spontaneous locomotor activity of rats with long-term portacaval anastomosis. *Hepatology* **2000**, *31*, 336–344. [CrossRef] [PubMed]

196. Ahabrach, H.; Piedrafita, B.; Ayad, A.; El Mlili, N.; Errami, M.; Felipo, V.; Llansola, M. Chronic hyperammonemia alters the circadian rhythms of corticosteroid hormone levels and of motor activity in rats. *J. Neurosci. Res.* **2010**, *88*, 1605–1614. [CrossRef] [PubMed]
197. Bomzon, A.; Binh, O.; Blendis, L.M. Hypotension in experimental cirrhosis: Is loss of vascular responsiveness to norepinephrine the cause of hypotension in chronic bile-duct-ligated dogs? *J. Hepatol.* **1993**, *17*, 116–123. [CrossRef]
198. Bomzon, A.; Rosenberg, M.; Gali, D.; Binah, O.; Mordechovitz, D.; Better, O.S.; Greig, P.D.; Blendis, L.M. Systemic hypotension and decreased pressor response in dogs with chronic bile duct ligation. *Hepatology* **1986**, *6*, 595–600. [CrossRef]
199. Clària, J.; Jiménez, W.; Ros, J.; Asbert, M.; Castro, A.; Arroyo, V.; Rivera, F.; Rodés, J. Pathogenesis of arterial hypotension in cirrhotic rats with ascites: Role of endogenous nitric oxide. *Hepatology* **1992**, *15*, 343–349. [CrossRef]
200. Caracuel, L.; Sastre, E.; Callejo, M.; Rodrigues-Díez, R.; García-Redondo, A.B.; Prieto, I.; Nieto, C.; Salaices, M.; Aller, M.Á.; Arias, J.; et al. Hepatic Encephalopathy-Associated Cerebral Vasculopathy in Acute-on-Chronic Liver Failure: Alterations on Endothelial Factor Release and Influence on Cerebrovascular Function. *Front. Physiol.* **2020**, *11*, 593371. [CrossRef] [PubMed]
201. Trewby, P.N.; Hanid, M.A.; Mackenzie, R.L.; Mellon, P.J.; Williams, R. Effects of cerebral oedema and arterial hypotension on cerebral blood flow in an animal model of hepatic failure. *Gut* **1978**, *19*, 999–1005. [CrossRef]
202. Estrela, H.F.G.; Damásio, E.S.; Fonseca, E.K.U.N.; Bergamaschi, C.T.; Campos, R.R. Differential Sympathetic Vasomotor Activation Induced by Liver Cirrhosis in Rats. *PLoS ONE* **2016**, *11*, e0152512. [CrossRef] [PubMed]
203. Rohde, J.; Pedersen, H.R.; Bjerring, P.N.; Larsen, F.S. Effects of Dexamethasone and Cox Inhibitors on Intracranial Pressure and Cerebral Perfusion in the Lipopolysaccharide Treated Rats with Hyperammonemia. *PLoS ONE* **2015**, *10*, e0117416. [CrossRef]
204. Larsen, R.H.; Kjær, M.S.; Eefsen, M.; Larsen, F.S.; Bjerring, P.N. Ciclosporin does not attenuate intracranial hypertension in rats with acute hyperammonaemia. *World J. Hepatol.* **2013**, *5*, 513–520. [CrossRef] [PubMed]
205. Master, S.; Gottstein, J.; Blei, A.T. Cerebral blood flow and the development of ammonia-induced brain edema in rats after portacaval anastomosis. *Hepatology* **1999**, *30*, 876–880. [CrossRef] [PubMed]
206. Cauli, O.; López–Larrubia, P.; Rodrigo, R.; Agusti, A.; Boix, J.; Nieto–Charques, L.; Cerdán, S.; Felipo, V. Brain Region-Selective Mechanisms Contribute to the Progression of Cerebral Alterations in Acute Liver Failure in Rats. *Gastroenterology* **2011**, *140*, 638–645. [CrossRef] [PubMed]
207. Lin, S.-Y.; Wang, Y.-Y.; Chuang, Y.-H.; Chen, C.-J. Skeletal muscle proteolysis is associated with sympathetic activation and TNF-α-ubiquitin-proteasome pathway in liver cirrhotic rats. *J. Gastroenterol. Hepatol.* **2016**, *31*, 890–896. [CrossRef]
208. Jia, W.; Liu, J.; Hu, R.; Hu, A.; Tang, W.; Li, L.; Li, J. Xiaochaihutang improves the cortical astrocyte edema in thioacetamide-induced rat acute hepatic encephalopathy by activating NRF2 pathway. *Front. Pharmacol.* **2020**, *11*, 382. [CrossRef]
209. Chen, J.-R.; Wang, B.-N.; Tseng, G.-F.; Wang, Y.-J.; Huang, Y.-S.; Wang, T.-J. Morphological changes of cortical pyramidal neurons in hepatic encephalopathy. *BMC Neurosci.* **2014**, *15*, 1–12. [CrossRef]
210. Ismail, F.S.; Faustmann, T.J.; Corvace, F.; Tsvetanova, A.; Moinfar, Z.; Faustmann, P.M. Ammonia induced microglia activation was associated with limited effects on connexin 43 and aquaporin 4 expression in an astrocyte-microglia co-culture model. *BMC Neurosci.* **2021**, *22*, 21. [CrossRef]
211. Rama Rao, K.V.; Chen, M.; Simard, J.M.; Norenberg, M.D. Increased aquaporin-4 expression in ammonia-treated cultured astrocytes. *Neuroreport* **2003**, *14*, 2379–2382. [CrossRef]
212. Pan, C.-F. Ammonia induces upregulation of aquaporin-4 in neocortical astrocytes of rats through the p38 mitogen-activated protein kinase pathway. *Chin. Med. J.* **2010**, *123*, 1888–1892. [CrossRef] [PubMed]
213. Stokum, J.A.; Kwon, M.S.; Woo, S.K.; Tsymbalyuk, O.; Vennekens, R.; Gerzanich, V.; Simard, J.M. SUR1-TRPM4 and AQP4 form a heteromultimeric complex that amplifies ion/water osmotic coupling and drives astrocyte swelling. *Glia* **2018**, *66*, 108–125 [CrossRef]
214. Sinke, A.P.; Jayakumar, A.R.; Panickar, K.S.; Moriyama, M.; Reddy, P.V.B.; Norenberg, M.D. NFκB in the mechanism of ammonia-induced astrocyte swelling in culture. *J. Neurochem.* **2008**, *106*, 2302–2311. [CrossRef]
215. Jayakumar, A.R.; Rama Rao, K.V.; Tong, X.Y.; Norenberg, M.D. Calcium in the mechanism of ammonia-induced astrocyte swelling. *J. Neurochem.* **2009**, *109*, 252–257. [CrossRef] [PubMed]
216. Reinehr, R.; Görg, B.; Becker, S.; Qvartskhava, N.; Bidmon, H.J.; Selbach, O.; Haas, H.L.; Schliess, F.; Häussinger, D. Hypoosmotic swelling and ammonia increase oxidative stress by NADPH oxidase in cultured astrocytes and vital brain slices. *Glia* **2007**, *55*, 758–771. [CrossRef] [PubMed]
217. Felipo, V.; Urios, A.; Montesinos, E.; Molina, I.; Garcia-Torres, M.L.; Civera, M.; Olmo, J.A.; Ortega, J.; Martinez-Valls, J.; Serra, M.A.; et al. Contribution of hyperammonemia and inflammatory factors to cognitive impairment in minimal hepatic encephalopathy. *Metab. Brain Dis.* **2012**, *27*, 51–58. [CrossRef] [PubMed]

cells

MDPI

Article

Heroin Self-Administration and Extinction Increase Prelimbic Cortical Astrocyte–Synapse Proximity and Alter Dendritic Spine Morphometrics That Are Reversed by N-Acetylcysteine

Benjamin M. Siemsen [1], Adam R. Denton [1,2], Jeffrey Parrila-Carrero [2], Kaylee N. Hooker [1], Eilish A. Carpenter [1], Meagan E. Prescot [1], Ashley G. Brock [1], Annaka M. Westphal [2], Mary-Nan Leath [1], John A. McFaddin [1], Thomas C. Jhou [2], Jacqueline F. McGinty [2,†] and Michael D. Scofield [1,2,*,†]

1 Department of Anesthesia and Perioperative Medicine, Medical University of South Carolina, Charleston, SC 29425, USA
2 Department of Neuroscience, Medical University of South Carolina, Charleston, SC 29425, USA
* Correspondence: scofield@musc.edu; Tel.: +1-843-792-1743
† These authors contributed equally to this work.

Abstract: Clinical and preclinical studies indicate that adaptations in corticostriatal neurotransmission significantly contribute to heroin relapse vulnerability. In animal models, heroin self-administration and extinction produce cellular adaptations in both neurons and astrocytes within the nucleus accumbens (NA) core that are required for cue-induced heroin seeking. Specifically, decreased glutamate clearance and reduced association of perisynaptic astrocytic processes with NAcore synapses allow glutamate release from prelimbic (PrL) cortical terminals to engage synaptic and structural plasticity in NAcore medium spiny neurons. Normalizing astrocyte glutamate homeostasis with drugs like the antioxidant N-acetylcysteine (NAC) prevents cue-induced heroin seeking. Surprisingly, little is known about heroin-induced alterations in astrocytes or pyramidal neurons projecting to the NAcore in the PrL cortex (PrL-NAcore). Here, we observe functional adaptations in the PrL cortical astrocyte following heroin self-administration (SA) and extinction as measured by the electrophysiologically evoked plasmalemmal glutamate transporter 1 (GLT-1)-dependent current. We likewise observed the increased complexity of the glial fibrillary acidic protein (GFAP) cytoskeletal arbor and increased association of the astrocytic plasma membrane with synaptic markers following heroin SA and extinction training in the PrL cortex. Repeated treatment with NAC during extinction reversed both the enhanced astrocytic complexity and synaptic association. In PrL-NAcore neurons, heroin SA and extinction decreased the apical tuft dendritic spine density and enlarged dendritic spine head diameter in male Sprague–Dawley rats. Repeated NAC treatment during extinction prevented decreases in spine density but not dendritic spine head expansion. Moreover, heroin SA and extinction increased the co-registry of the GluA1 subunit of AMPA receptors in both the dendrite shaft and spine heads of PrL-NAcore neurons. Interestingly, the accumulation of GluA1 immunoreactivity in spine heads was further potentiated by NAC treatment during extinction. Finally, we show that the NAC treatment and elimination of thrombospondin 2 (TSP-2) block cue-induced heroin relapse. Taken together, our data reveal circuit-level adaptations in cortical dendritic spine morphology potentially linked to heroin-induced alterations in astrocyte complexity and association at the synapses. Additionally, these data demonstrate that NAC reverses PrL cortical heroin SA-and-extinction-induced adaptations in both astrocytes and corticostriatal neurons.

Keywords: heroin; astrocytes; GFAP; N-acetylcysteine; prelimbic; dendritic spines

Citation: Siemsen, B.M.; Denton, A.R.; Parrila-Carrero, J.; Hooker, K.N.; Carpenter, E.A.; Prescot, M.E.; Brock, A.G.; Westphal, A.M.; Leath, M.-N.; McFaddin, J.A.; et al. Heroin Self-Administration and Extinction Increase Prelimbic Cortical Astrocyte–Synapse Proximity and Alter Dendritic Spine Morphometrics That Are Reversed by N-Acetylcysteine. *Cells* **2023**, *12*, 1812. https://doi.org/10.3390/cells12141812

Academic Editors: Michael Brenner and Vladimir Parpura

Received: 16 February 2023
Revised: 9 May 2023
Accepted: 6 July 2023
Published: 8 July 2023

1. Introduction

Relapse remains a major clinical obstacle in treating opioid use disorders and is typically brought on by a craving evoked by exposure to opioid-associated contexts or cues [1]. In humans, elevated activity in the corticostriatal circuit is a neural correlate of cue-evoked craving for heroin [2], and heroin-mediated dysfunction in corticostriatal circuity is faithfully recapitulated in rodent models of heroin self-administration (SA) and relapse [3]. A general consensus from rodent models is that the dysfunctional prefrontal cortical regulation of ventral striatal plasticity occurs as a result of chronic opioid exposure, allowing opioid-associated cues and contexts to drive drug seeking even after prolonged periods of abstinence [4].

The nucleus accumbens core (NAcore) is a ventral striatal region required for motivationally relevant external stimuli to evoke motor programs linked to seeking reinforcers or rewards [5]. It has been repeatedly demonstrated that heroin SA in rodents alters glutamatergic plasticity in the NAcore in a manner that is mechanistically linked to relapse [6,7]. For example, following the extinction of heroin [8] or cocaine [9] SA, an inability to induce long-term potentiation (LTP) and long-term depression (LTD) in the NAcore is observed following stimulation of the prelimbic (PrL) cortex in vivo. These data demonstrate drug-mediated dysfunction at PrL-NAcore synapses following withdrawal from addictive drugs that fundamentally alters synaptic plasticity. Akin to cocaine [10], heroin SA and extinction produce reduced glutamate clearance in the NAcore due to the downregulation of the high-affinity astrocytic glutamate transporter, GLT-1 [11], and concomitant reductions in the proximity of GLT-1-containing perisynaptic astrocytic processes (PAPs) to NAcore synapses [12]. Interestingly, decreased astrocytic association with NAcore synapses is also observed following the extinction of methamphetamine or cocaine SA [13,14]. Together, these data suggest that structural and cellular astrocyte dysfunction in the NAcore is a central aspect of the pathophysiology underlying relapse vulnerability [15]. As such, the reduced GLT-1 expression coupled with decreased proximity of astrocytic PAPs to synapses likely synergistically facilitate glutamate spillover to extrasynaptic sites following synaptic release. Consistent with dysfunctional plasticity at PrL-NAcore synapses, cue- and heroin-primed seeking promotes the sustained elevation of glutamate levels in the NAcore specifically due to the release from PrL cortical afferents [16]. This elevated glutamate release and spillover facilitates LTP-like increases in synaptic strength, including an upregulation of NR2B-containing NMDA receptors and enlargement of dendritic spine heads on NAcore medium spiny neurons (MSNs). Accordingly, it has been shown that normalizing drug-induced alterations in PrL-NAcore synaptic transmission, mediated in large part by astrocytic dysfunction, is a potent means of decreasing relapse vulnerability [17].

One drug that is particularly effective at targeting addictive drug-induced adaptations in astrocytes is the antioxidant, N-acetylcysteine (NAC). NAC is an L-cysteine pro-drug that has shown some promise clinically [18] but has been more extensively investigated preclinically as a research tool to probe glutamate homeostasis [15]. NAC has been shown to significantly limit heroin [19,20], cocaine [21,22], nicotine [23], or methamphetamine [14] seeking in rats with a drug self-administration history. Mechanistically, NAC is thought to prevent relapse by reversing the drug-induced downregulation of GLT-1 in the NAcore. This hypothesis has been confirmed in studies in which cocaine was used as a means to downregulate the expression of GLT-1 in the NAcore. In these studies, preventing the NAC-mediated restoration of GLT-1 expression in the NAcore following cocaine SA and extinction blocked NAC's ability to prevent relapse [10,24]. While NAC does indeed facilitate extinction and prevent cued heroin seeking [19], surprisingly, the ability of NAC to normalize heroin-induced decreases in GLT-1 in the NAcore remains to be directly evaluated. However, additional drugs that upregulate GLT-1 expression (i.e., ceftriaxone) have been shown to normalize heroin-induced decreases in glutamate clearance in the NAcore, inhibiting relapse [11]. Importantly, ceftriaxone has also been shown to normalize cocaine-induced decreases in PAPs' association with synapses in the NAcore [13], suggesting a significant structure–function relationship underlying drug-induced decreases in PAP–synapse associations, alterations in synaptic glutamate clearance, and relapse.

While there is a substantial body of work regarding the consequences of heroin SA and extinction on neuronal and astrocyte plasticity in the NAcore, little is known about the impact of heroin on neurons or astrocytes in the PrL cortex. This gap in our knowledge is particularly significant given that cue-induced glutamate release in the NAcore during heroin seeking requires THE activation of the PrL cortex [16]. While less is known about the relationship between cortical astrocytes and opiates, there is some literature describing opiate-induced adaptations in cortical astrocytes. For example, chronic morphine treatment increases the expression of glial fibrillary acidic protein (GFAP), a commonly used marker for astrocytes [25], in the frontal cortex [26]. These data suggest that morphine induces astrogliosis, yet the addition of morphine to astrocytes in vitro decreases the proliferation and increases the complexity of astrocyte processes, an effect prevented by naloxone pretreatment [27]. These data suggest that heroin may impact astrocytic function in the PrL cortex; however, such adaptations have yet to be shown in vivo or linked to the degree to which astrocytes are associated with synapses.

Here, we first employ electrophysiological recordings to interrogate functional adaptations in PrL astrocytes following heroin SA and extinction. Next, we employ immunohistochemical preparations in conjunction with confocal microscopy and 3D rendering to investigate the effects of heroin SA and extinction upon the GFAP cytoskeletal complexity of PrL astrocytes in addition to the PAP association with synaptic markers of NAcore projecting PrL neurons. We also investigate the therapeutic efficacy of NAC in reversing heroin SA-and-extinction-induced alterations in neuronal and astrocyte morphology. Finally, we investigate the effects of the inhibition of PrL thrombospondin 2 (TSP-2) in the context of NAC treatment following heroin SA, extinction, and cued-relapse to heroin seeking as a means to determine if NAC exerts its anti-relapse effects by engaging astrocyte-mediated synaptogenesis.

2. Materials and Methods

2.1. Animal Subjects and Surgery

Male Sprague–Dawley rats (Experiment 1: $N = 22$, Experiments 2–3: $N = 63$, Experiment 4: $N = 20$) were purchased from Charles River Laboratories (Wilmington, MA, USA) and were single-housed upon arrival in a temperature- and humidity-controlled vivarium with standard rat chow (Harlan; Indianapolis, IN, USA) and water available ad libitum. Rats were maintained on a 12 hr reverse light/dark cycle (lights off at 6 a.m.). All animal use protocols were approved by the Institutional Animal Care and Use Committee of the Medical University of South Carolina and were performed according to the National Institutes of Health Guide for the Care and Use of Laboratory Animals (8th ed., 2011). At the time of surgery, all rats weighed 275–300 g. On the day of surgery, rats were anesthetized with an intraperitoneal (i.p.) ketamine (66 mg/kg) and xylazine (1.33 mg/kg) injection, and received ketorolac (2.0 mg/kg, i.p.) for analgesia. A silastic catheter was implanted in the right jugular vein, attached to a cannula, which exited the animal's back. Rats in Experiment 1 received a bilateral intra-PrL cortical infusion of AAV5.*GfaABC1D*::tdTomato (0.75 μL/hemisphere, titer: 7×10^{12} viral genomes per mL (vg/mL); co-ordinates- AP: +2.8, ML: +/−0.6, DV: −3.8 mm from bregma). Rats in Experiment 2 received a bilateral intra-PrL cortical microinjection of AAV5.*GfaABC1D*::Lck-EGFP (0.75 μL/hemisphere, titer: 1×10^{13} vg/mL) to label the fine membrane processes of astrocytes [13,28]; here, the truncated Lck was used, i.e., only the first 26 amino acids of Lck lacking kinase activity but allowing plasma membrane targeting of appended EGFP. Rats in Experiment 3 received a bilateral microinjection of AAV1.*CAG*.Flex::Ruby2sm-FLAG (Titer: 5×10^{12} vg/mL) as well as intra-NAcore (co-ordinates-AP: +1.7, ML: +/−1.6, DV: −7 mm relative to bregma) injection of rgAAV.*hSyn*::Cre-WPRE.hGH (0.75 μL/hemisphere, titer: 7×10^{12}) to specifically target PrL cortical-NAcore neurons for dendritic spine analyses [29]. Rats in Experiment 4 received bilateral PrL injections of either AAV5 *GfaABC1D*::Lck-EGFP (titer: 1×10^{13} vg/mL) or AAV5. *GfaABC1D*::shTSP2-GFP. Injections (0.75 μL/hemisphere) were performed over a period of 5 min (0.15 μL/min) using a Nanoject II (Drummond Scientific, Broomall, PA, USA) and injectors were left in place for 5 min to facilitate viral diffusion from the injection site. Apart from the TSP2 RNAi construct, a gift from the Dong Lab [30], all viral constructs were purchased from Addgene (Cambridge, MA, USA).

Additionally, the pAAV5. *GfaABC1D*::Lck-EGFP construct was purchased as a plasmid from Addgene, then packaged to produce viral particles at the University of North Carolina Gene Therapy Center viral vector core (Chapel Hill, NC, USA). Rats were allowed at least 5 days of post-operative recovery during which catheters were flushed daily with taurolidine citrate solution (TCS, Access Technologies, Skokie, IL, USA) and food and water were available ad libitum. For each experiment, ≥2 independent cohorts were used. Figure 1 illustrates the experimental design and timeline for all experiments. Supplementary Table S1 contains names, descriptions, and sourcing information for all viral constructs used in the present experiments.

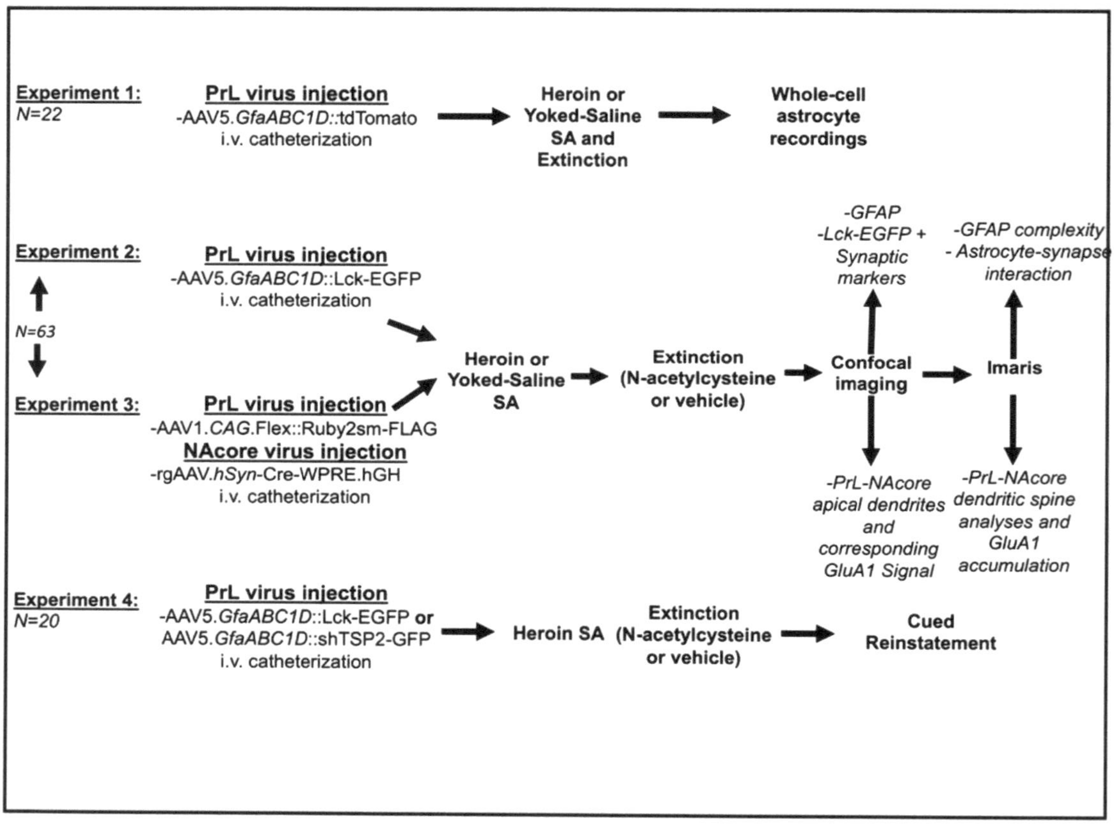

Figure 1. Diagram illustrating the experimental timeline for each experiment.

2.2. Drugs

Heroin hydrochloride was provided by the National Institute on Drug Abuse. N-acetylcysteine was purchased from Sigma-Aldrich (St. Louis, MO, USA, #A7250) and dissolved in sterile saline at 100 mg/mL.

2.3. Heroin Self-Administration, Extinction, and Cued-Reinstatement Procedures

All experiments were performed between the hours of 8 a.m. and 6 p.m. All SA experiments were performed in standard MedPC operant chambers equipped with two retractable levers, a tone generator, and a cue light (Med Associates, Fairfax, VT, USA). The beginning of each session was signaled by illumination of the house light and lever availability. Rats were trained for 14 days (3 h/day) to lever-press for heroin on a fixed ratio

1 (FR1) schedule of reinforcement, whereby each individual lever press on the active lever (ALP) resulted in the presentation of cue delivery and infusion of heroin followed by a 20 s timeout whereby ALP elicited no drug delivery. On days 1–2, active ALP resulted in a light and tone cue delivery followed by a 100 µg/50 µL i.v. infusion of heroin hydrochloride dissolved in sterile saline. On days 3–4, rats received a 50 µg/50 µL infusion followed by a 20 s. timeout period, and on days 5–14, rats received a 25 µg/50 µL infusion. Yoked-saline (YS) control animals were paired to a randomly selected heroin SA animal, receiving non-contingent saline infusions in a temporally paired manner. SA criteria were set at $\geq$1 infusion per session on days 1–4 whereas animals were required to receive $\geq$10 infusions per session on days 5–14. Following SA, all rats entered extinction whereby ALP resulted in no cue or drug delivery. All rats received 12–13 days of extinction training (3 h/day). Beginning on day 4, a subset of animals received N-acetylcysteine (100 mg/kg, i.p.) or vehicle (0.9% w/v NaCl) 30 min before each subsequent session, and all animals injected with N-acetylcysteine received 13 days of extinction training (Experiments 2–4). All YS animals received vehicle injections during extinction. This dose of NAC has previously been shown to reduce extinction responding on the active lever and provide an enduring reduction in cued and primed heroin seeking using the same heroin SA protocol [19]. Cued-reinstatement testing (Experiment 4) occurred following the final day of extinction training (3 h) in which ALP resulted in a light and tone cue delivery without the delivery of heroin.

2.4. Whole-Cell Recordings from Astrocytes

Twenty-four hours after the last extinction session, rats were anesthetized with isoflurane, decapitated, and their brains quickly removed and placed in an ice-cold and oxygenated artificial CSF (aCSF), slicing solution containing (in mM): 126 NaCl, 2.5 KCl, 1.2 $MgCl_2$, 1.4 NaH_2PO_4, 25 $NaHCO_3$, 11 D-glucose, and 0.4 ascorbate, supplemented with kynurenic acid (2.6 mM). Thick sections (~220 µM) containing the PrL region were prepared using a vibrating microtome (Leica Biosystems, Deer Park, IL, USA) and maintained in oxygenated aCSF in a dish in a 32 °C water bath until recording. During recording, sections were perfused at a flow-rate of ~2 mL/min with oxygenated aCSF at 33 °C. The prelimbic layer V/VI was identified initially using a 4× objective (0.1 numerical aperture (NA)), and later was magnified for patching with a 40 × (0.8 NA) water-immersion lens fitted to an upright microscope (Nikon Eclipse E600FN; Nikon Inc, Melville NY, USA) fitted with infrared differential interference contrast optics.

Whole-cell recordings were taken from astrocytes in PFC layers II/III and V by an experimenter blinded to treatment condition. Cell bodies of astrocytes were visualized using astrocyte-specific AAVs (AAV5.*GfaABC1D*::tdTomato) that were excited by wide-field illumination (555 nm LED, Thorlabs; coupled through a 40× objective). Light power and duration were controlled via the LED driver (LEDD1B, Thorlabs, Newton, NJ, USA) by external voltage modulation. Successful patching onto astrocytes was confirmed by electrophysiological features including their hyperpolarized resting membrane potential (RMP; Saline −84.11 ± 0.294, Heroin 205—83.43 ± 0.269 mV), their linear current–voltage relationship, their inability to generate APs, and their low membrane resistance (Rm; Saline 2.7 ± 0.2955, Heroin 2.473 ± 207 0.3709 MΩ). Recordings were taken with borosilicate glass pipettes (4 to 5 MΩ) containing the following internal solution: 115 mM K-gluconate, 6 mM KCl, 5 mM glucose, 7.8 mM Na-phosphocreatine, 4 mM Mg-ATP (adenosine triphosphate), 0.4 mM Na-GTP (guanosine triphosphate) (pH 7.25 with KOH; osmolarity, 295 mml/kg. Electrophysiological signals were amplified using the Multiclamp 700B amplifier (Molecular Devices, San Jose, CA, USA), low-pass filtered at 2 kHz, and digitized at 10 kHz). Data were collected using AxoGraph X and stored for offline analysis. For the recording of the threo-beta-benzyloxyaspartate (TBOA) (non-selective blocker of excitatory amino acid transporters)- and $BaCl_2$ (K^+ channel antagonist)-sensitive currents, the extracellular solution contained antagonists of n-methyl-D-aspartate, (NMDA) receptors (D-AP5; 50 µM), α-amino-3-hydroxyl-5-methyl-4-isoxazole-propionate (AMPA) receptors (NBQX; 10 μM), and γ-aminobutyric acid type A (GABAA) receptors (picrotoxin; 100 μM).

Astrocytes were held at −80 mV, and TBOA (100 μM)- and $BaCl_2$ (200 μM)-sensitive currents were evoked by single-pulse stimulation every 20 s using focal electrical stimulation with a bipolar tungsten electrode (FHC) positioned immediately medial (within ~100 μM) to the recorded astrocyte. The protocol was repeated 2 times per stimulation intensity and then averaged and analyzed. Access resistance was monitored (<20 MΩ), and recordings with an access resistance changing more than 20% between the beginning and the end of the recording were discarded. Resting membrane potential (RMP) and input resistances (Rm) were monitored for analysis of the electrophysiological properties of astrocytes.

2.5. Transcardial Perfusions and Immunohistochemistry

Twenty-four hours after the final extinction session, rats were transcardially perfused with 150 mL of 0.1 M phosphate buffer (PB), followed by 200 mL of 4% *w/v* paraformaldehyde (PFA) in 0.1 M PB. Brains were extracted and post-fixed in PFA for 24 h. Sections (80 μM) containing the PrL cortex and NAcore were collected in 0.1 M PBS containing 0.02% sodium azide with a vibrating microtome (Leica), and stored at 4 °C until processing.

Immunohistochemistry (IHC) was performed as previously described [13,14,31]. Briefly, 3 sections per animal were blocked and permeabilized by soaking with 0.1 M PBS containing 2% (*v/v*) TritonX-100 (PBST) containing 2% (*v/v*) normal goat serum (NGS). Sections were then incubated overnight with gentle agitation at 4 °C with the following primary antibodies: Experiment 2—chicken anti-GFP (Abcam, 1:1000, ab13970 RRID:AB_300798) + mouse anti-GluA2 (Millipore, 2.13 μ/mL, MAB397 RRID:AB_11212990) or rabbit anti-Synapsin I (Abcam, 1:500, ab64581 RRID:AB_1281135), and in a separate run, rabbit anti-GFAP (Abcam, 1:1000, ab7260 RRID:AB_305808); Experiment 3—mouse anti-FLAG (Sigma, 1:2000, F1804 RRID:AB_262044) and rabbit anti-GluA1 (Abcam, 1:500, ab31232 RRID:AB_2113447); Experiment 4—rabbit anti-NeuN (Millipore1:1000, MABN140 RRID:AB_2571567). Sections were then washed 3 × 5 min with PBST, and were incubated in species-specific secondary antibodies (Invitrogen, all 1:1000) conjugated to Alexa Fluor® 488 (RRID:AB_2534096, GFP), 594 (RRID:AB_2534091, FLAG), or 647 (RRID:AB_2535804, GFAP, Synapsin I, GluA1, and GluA2) diluted in PBST with 2% NGS for 5 h at room temperature (~20–25 °C) with gentle agitation. Sections were washed 3 × 5 min in PBST, then mounted on Superfrost plus slides with Prolong Gold antifade (ThermoFisher Scientific, Waltham, MA, USA). Slides were stored at 4 °C until imaging (<2 months).

2.6. Confocal Microscopy

For all high-resolution confocal microscopy experiments, a Leica SP8 confocal microscope (Leica Microsystems Inc., Deerfield, IL, USA) equipped with HyD detectors for enhanced sensitivity was used. Laser power, gain, and pinhole size (≤1 Airy Unit) were empirically determined to obtain optimal saturation values and subsequent deconvolution results. Laser power and gain values were then held relatively constant for the remainder of the experiment. For all experiments, Alexa® 488-mediated Lck-EGFP detection was carried out using an Argon 488 nm laser line, Alexa® 594-mediated FLAG detection was carried out using an OPSL 552 nm laser, and Alexa® 647-mediated GluA2, GFAP, and Synapsin I detection was carried out using a Diode 638 nm laser. In Experiment 2a, fields of GFAP + cells were imaged in layers II/III and V of the PrL cortex at a 4096 × 4096 frame size, 0.1 mm Z-step size, and 1× digital zoom, with a line average of 4 using a 63× oil-immersion objective (1.4 NA). In Experiment 2b, Lck-EGFP-expressing astrocytes that were visually isolated from neighboring astrocytes in layers II/III and V as well as GluA2/Synapsin I puncta were sampled using a 63× oil-immersion objective (1.4 N.A.) at a 2048 × 2048 frame size, 0.3 mm Z-step size, and 1.5× digital zoom, with a line average of 4. Care was taken to ensure that the majority of each astrocyte in the Z-plane was imaged. In Experiment 3, apical tuft dendrites of layer V PrL-NAcore neurons were imaged as previously described [31]. Briefly, dendritic spine segments were imaged only if they met the following criteria: 1: they could be traced back to the cell body of origin, 2: they were unobscured by neighboring dendritic segments, and 3: they were past the bifurcation of the proximal dendrite (terminating in layers I and II). Dendritic spine segments (~50 mm) were collected at 1024 × 512 frame size, 0.1 mm Z-step size, and 4.1× digital zoom, with a line average of 2. Each spine segment imaged contained an immunohistochemically labeled GluA1 channel in addition to FLAG for subsequent co-registry analyses.

2.7. Image Analysis

All image analyses were performed by an investigator blind to experimental groups. Following acquisition, confocal Z-stacks were deconvolved using Huygens software v.2 (Scientific Volume Imaging, Hilversum, NL, USA) and then exported to Bitplane Imaris v. 9.1 (Oxford Instruments, Concord, MA, USA) for digital rendering and analysis. Complexity of GFAP arbors were analyzed as previously described [14]. Briefly, Z-stacks were cropped to 12 mm in the Z plane to identify fields of cells with homogenous signal intensity, avoiding large areas of vasculature. We used a previously published semi-automated approach to skeletonize GFAP arbors belonging to individual, non-overlapping astrocytes. In doing so, we set the minimum diameter of a GFAP filament to 1 μM. Iterative processing of GFAP skeletons were performed to accurately skeletonize each astrocyte within the field. We exported and analyzed astrocyte structure using the following outputs from the Imaris filament function: the number of astrocytes in each image, the total sum of the GFAP filament length for each cell per image, the average GFAP diameter of each cell per image, the average number of branch points for each cell per image, and the average number of 3D Sholl intersections (1 μM radius) per cell, per image. The average number of Sholl intersections was then binned by 5 μM radii from 0 to 50 μM. Data are expressed as an average for each cell per image, or an overall image average as specified in the results.

The whole-cell astrocyte synaptic contact assay was performed as described previously [13,14], with minor modifications. Briefly, confocal Z-stacks were cropped in 3D to the border of each cell to isolate each astrocyte. A 3D rendering was generated from the GFP signal from each virally labeled astrocyte. We then isolated all GFP signal to each 3D rendering to remove background. The colocalization module of Imaris was used to perform co-registry analyses as previously described. A signal intensity threshold was automatically set according to the Costes method [32] for each channel (Channel 1: GFP, Channel 2: Synapsin I or GluA2). Images that could not be assigned automatic thresholds, due to uneven signal intensity distribution through the Z-plane of the data set, were not analyzed. Voxels that contained signal above threshold for each channel were considered colocalized. The greater number of co-registered voxels indicates a greater degree of astrocyte association at synapses on a whole-cell level. In a subset of images, we also digitally modeled each Synapsin I or GluA2 puncta to determine the effect of saline, heroin + vehicle, or heroin + NAC treatments on the number of puncta within each image, normalized to the data-set volume. Data exported included the percent of the data set colocalized normalized to the saline control group (% colocalized), and the number of Synapsin I/GluA2 puncta normalized to the data-set volume.

Dendritic spine morphometric feature and GluA1 co-registry were analyzed as previously described [31]. Briefly, the filament extension of Imaris was used to manually trace each dendritic shaft and automatically label corresponding dendritic spines under input from the experimenter. Care was taken to ensure that each dendritic spine head was modeled accurately. Variables exported included the average dendritic spine head diameter (d_H) of each segment and the dendritic spine density (spines/μm). We also binned the spine head diameters (0.2 μM bins) to generate a frequency distribution. For GluA1 co-registry analyses, a 3D rendering was then used to isolate the dendrite segment (shaft + spines), and GluA1 signal above background contained within the dendrite that was isolated for analysis. The colocalization module of Imaris was then used to generate a channel containing GluA1 and FLAG colocalization, and the 'filament analysis' MATLAB Xtension was used to split the channel into FLAG-GluA1 colocalization in dendritic subcompartments including the dendritic spine head and dendritic shaft for differential investigation of colocalization in respective subcompartments. Output variables included the percent of GluA1 volume normalized to the dendrite volume, the total volume of spine-head-specific GluA1 normalized to the dendrite volume, and the total volume of dendritic shaft GluA1 normalized to the dendrite volume, as well as signal intensity controls. Data are expressed as animal averages, collapsing each of the 3–5 dendrites sampled per animal into an average value of spine d_H and density per animal.

2.8. Statistical Analyses

All statistical analyses were performed with SPSS statistical software (v.28 Somer, NY USA) or Graphpad Prism software (v. 8.0, La Jolla, CA, USA). Evoked GLT-1-dependent current from whole-cell astrocyte recordings was modeled with non-linear regression techniques, and a linear discriminant analysis with bootstrapping procedure was performed to determine if evoked GLT-dependent current was predictive of initial drug treatment. A principal components analysis was performed to produce a lower-dimensional data set before a clustering analysis was performed for all cells recorded. Self-administration and extinction data across subsequent days were compared using a mixed-model ANOVA. When analyzing three groups across a single dependent variable, a one-way ANOVA was performed followed by Tukey's multiple comparison test if a significant effect of treatment was revealed. Normality of data within each group were analyzed using Shapiro–Wilk test. If one or more groups for each variable analyzed did not pass the Shapiro–Wilk test, a non-parametric Kruskal–Wallis test was used. In this instance, post hoc tests controlled for the false discovery rate (FDR) using the two-stage linear step-up procedure of Benjamini Krieger, and Yekutieli if a significant effect was revealed, in which case FDR-corrected q values are reported. If data were sampled from a normal distribution, but between-group standard deviations were not equal, a Brown–Forsythe-corrected one-way ANOVA was performed followed by Dunnett's T3 test when a significant effect of treatment was revealed. When three groups were compared across a repeated measure, a repeated-measure one-way ANOVA was applied followed by Tukey's multiple comparison test comparing the mean of the repeated measure between groups. Dendritic spine frequency distribution data and dendritic spine GluA1 colocalization data were analyzed with a two-way repeated-measure ANOVA followed by Tukey's multiple comparison test when a significant interaction was revealed. GluA1 immunoreactivity across dendritic spine head size was evaluated using linear regression, wherein all figures contain linear fits derived from the analysis. Unless otherwise stated, each data point represents an animal average derived from multiple images or SA/extinction sessions. Statistical outliers were detected according to the ROUT method (Q = 1%) and these data were removed from analyses as specified in the results. All data are expressed as the mean ± SEM, and significance was set at $p < 0.05$. A priori sample size and power estimates were derived from previously published literature [12,31].

3. Results

3.1. Experiment 1—Heroin SA and Extinction

In Experiment 1, we sought to evaluate the effects of heroin SA and extinction upon the physiological properties of PrL astrocytes using a strategy where an electrical-stimulation-induced inward current attributed to either the K^+ or glutamate current was evaluated. The current attributed to glutamate uptake was blocked with the excitatory amino transporter inhibitor TBOA, whereas K^+ conductance was blocked with $BaCl_2$, a K^+ channel antagonist. Of the 22 initial animals in Experiment 1, 1 animal did not finish SA due to catheter failure. Figure 2A shows a schematic of the AAV infusion in the PrL cortex and Figure 2B shows the acquisition and extinction data for yoked-saline (YS) (N = 11)- and heroin (N = 10)-treated animals. Comparing heroin and YS active lever presses across sessions, a mixed-model ANOVA reveals a significant drug-by-time interaction ($F(13,221) = 3.002$ $p < 0.001$). From sessions 9–14, heroin SA animals pressed the active lever significantly more than YS controls ($F(1,18) = 11.30$, $p = 0.003$).

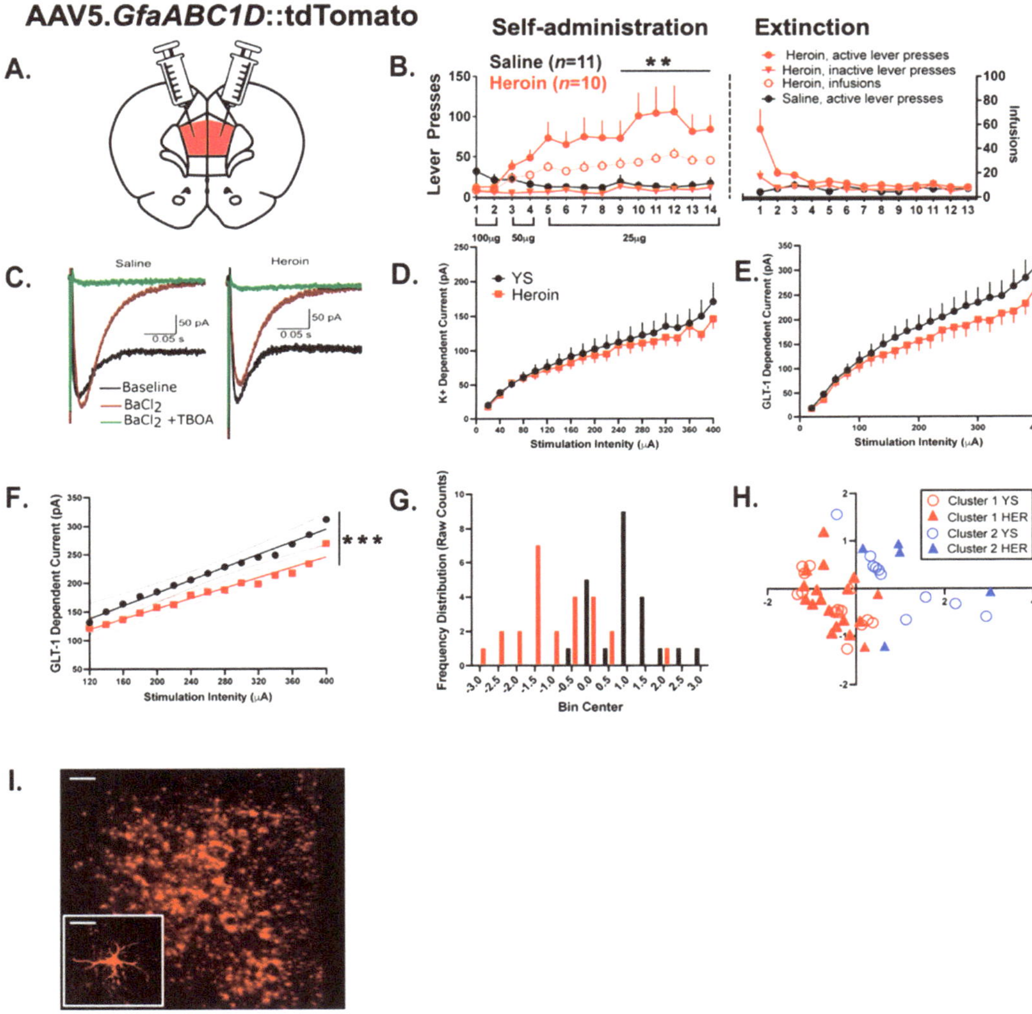

Figure 2. Evoked GLT−1-dependent current following heroin SA and extinction reveals functional clustering of PrL astrocyte populations. (**A**) Schematic indicating viral vector design. (**B**) Self-administration and extinction behavior for heroin and YS rats. Days of SA and extinction are displayed on the x-axis. Lever presses are shown on the left y-axis and infusions are shown on the right y-axis. Note that the infusion scale in panel B refers only to the self-administration data; no cue presentations of infusions occur during the extinction process. The statistical difference between active lever presses for rats receiving heroin and those receiving saline is indicated. (**C**) Representative traces of GLT−1 and K^+ current in PrL astrocytes following a single stimulation in saline- or heroin-treated rats. (**D**) Evoked K^+ current. (**E**) Evoked GLT−1 dependent current. (**F**) Isolation of area of interest and nonlinear modeling of YS (black)- vs. heroin (red)-receiving animals. (**G**) Discriminant

scores for YS (black)- versus heroin (red)-receiving animals. (**H**) Functional clustering of PrL astrocyte populations. (**I**) Representative low magnification image of AAV5.*GfaABC1D*::tdTomato expression in the prelimbic cortex with high magnification inset. Low magnification scale bar = 100 μM, high magnification scale bar = 5 μM. ** $p < 0.01$, *** $p < 0.001$.

3.2. Experiment 1—Heroin SA and Extinction Produced Alterations in Evoked Astrocytic GLT—1 Current

To examine functional adaptations in cortical astrocytes following heroin SA and extinction, heroin-treated animals ($N = 10$) and YS-treated animals ($N = 11$) were euthanized 24 h after the final extinction session and evoked K^+ and GLT−1-dependent currents were recorded from layer II/III and V astrocytes (N = 3–5 per animal). Figure 2C illustrates representative traces of the GLT-1 and K^+ current in PrL astrocytes following a single stimulation in saline- and heroin-treated rats. No difference was found for the evoked K^+ current (Figure 2D). Marginal evoked differences in the elicited GLT−1-dependent current were observed between treatment conditions, beginning at a stimulation intensity of 100 μA (Figure 2E). To further interrogate these differences, linear regression techniques were applied to model the responses following a 100 μA stimulation intensity using curve-fitting functions internal to GraphPad Prism, where it was determined that separate linear models were required to best characterize the data ($F(2,26) = 84.42$, $p < 0.001$, Figure 2F). In both treatment groups, model fit was evaluated using coefficients of determination (for YS $R^2 = 0.98$, for heroin, $R^2 = 0.96$). Next, to determine the degree to which drug treatment is predictive of the evoked GLT-1 response, the data were first bootstrapped (sample n = 1000) before a linear discriminant analysis was performed, where we show that the treatment condition can be predicted with 85.4% accuracy ($R_c = 0.687$, $\lambda = 0.893$, Wilk $\lambda = 0.528$). Obtained discriminant scores are displayed in Figure 2G, showing a clear separation between treatment conditions. To interrogate sources of variation in the evoked GLT-1-dependent current, a principal components analysis was first applied to the original data set to produce lower dimensional data while maintaining >90% of observed variance before a two-step cluster analysis was applied to the data. This analysis revealed the presence of at least two functional clusters of the PrL astrocyte response, largely independent of drug treatment (Figure 2H, silhouette coefficient > 0.5). Collectively, these results show that, while marginal, heroin SA and extinction lead to functional alterations in cortical astrocytes, though this effect may be somewhat obfuscated by drug-independent PrL astrocyte heterogeneity. Figure 2I illustrates representative viral expression in the PrL. Next, we interrogated heroin-dependent structural adaptations.

3.3. Experiments 2–3— Heroin Self-Administration and Extinction

Figure 3A shows self-administration and extinction data for heroin SA ($N = 31$) and YS ($N = 20$) rats used for Experiments 2–3. Of the 63 initial rats in Experiments 2–3, 12 animals did not finish SA due to catheter failure. When comparing heroin and YS active lever presses across SA sessions, a two-way repeated-measures ANOVA revealed a significant group-by-time interaction ($F(13,637) = 5.112$, $p < 0.001$). Heroin SA animals pressed the active lever significantly more than YS controls in sessions 9–14 ($F(1,49) = 10.168$, $p = 0.002$). There was no difference between groups (that were subsequently divided by treatment during extinction) in the average number of heroin infusions earned over the last 3 days of SA (Heroin + Vehicle: 29.38 ± 2.80, Heroin + NAC: 25.75 ± 4.87), indicating that the observed effects of NAC upon extinction were not dependent upon differences in initial heroin intake during SA. Figure 3B shows extinction training data for YS animals injected with vehicle ($N = 20$), and heroin SA rats injected with vehicle ($N = 15$) or NAC ($N = 16$) during the last 10 days of extinction. Figure 3C shows active lever presses averaged across the last 3 days of extinction for each group. A one-way ANOVA revealed a significant effect of treatment ($F(2,48) = 12.15$, $p < 0.0001$). Tukey's multiple comparison tests revealed that heroin SA animals showed significantly more lever pressing during the last 3 days of extinction than YS rats ($p < 0.0001$) and that NAC significantly decreased lever pressing to YS values ($p = 0.019$).

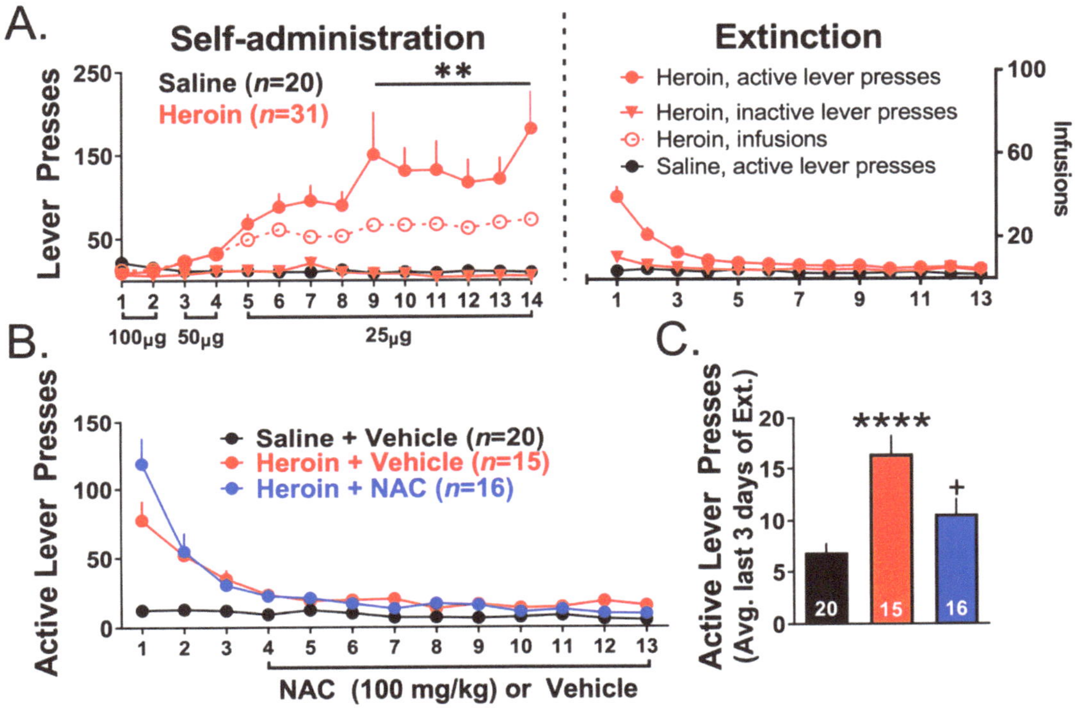

Figure 3. NAC reduces extinction responding. (**A**) Self-administration and extinction behavior for heroin and YS rats. Lever presses are shown on the left y-axis and infusions are shown on the right y-axis. ** $p < 0.01$ compared to saline active lever presses. (**B**) Active lever presses during extinction for YS, heroin animals injected with vehicle, and heroin animals injected with NAC. (**C**) Average active lever presses over the last 3 days of extinction for the three experimental groups. **** $p < 0.0001$ compared to YS, + $p < 0.05$ compared to Heroin + Vehicle. The number of animals within each group are shown within the bars.

3.4. Experiment 2A—Heroin SA and Extinction Increased Astrocyte GFAP Arbor Complexity, an Effect That Was Prevented by Repeated NAC Treatment during Extinction

Here, we used an astrocyte-specific membrane-targeted EGFP reporter virus to label the astrocyte plasma membrane (AAV5-*GfaABC1D*::Lck-EGFP) to evaluate GFAP arbor complexity following heroin SA, extinction, and NAC treatment. Figure 4 shows a representative immunohistochemical preparation indicating that EGFP expression is indeed specific to GFAP+ astrocytes. Of the initial 10 YS rats used in experiment 2A, 2 were not included due to technical issues related to IHC processing, 2 Heroin + Vehicle rats were not included due to technical issues related to IHC, and 1 Heroin + NAC animal was excluded due to lack of analyzable images. Figure 5A shows a representative field of GFAP+ cells in the PrL cortex (left); the same image is shown with an overlaid Imaris-based skeletonization (white) of each GFAP+ cell (middle), and an individual GFAP+ cell (yellow) isolated from the other cells (right). Using this high-throughput approach, we analyzed GFAP complexity in three to seven Z-series data sets from eight YS rats, seven Heroin + Vehicle rats, and nine Heroin + NAC rats using several different metrics. We first analyzed the average number of GFAP+ cells per field. A one-way ANOVA revealed no effect of treatment

(F(2,92) = 1.23, p = 0.30, Figure 5B). For complexity analyses, we first analyzed the total sum of GFAP filament length within each Z-series data set (Figure 5C). A Shapiro–Wilk test indicated that data in the YS (W = 0.9, p = 0.0005) and Heroin + NAC (W = 0.94, p = 0.04 groups were non-normally distributed. Accordingly, a Kruskal–Wallis non-parametric ANOVA was used, which indicated a significant difference between groups (H = 12.99 p = 0.002). The two-stage linear step-up procedure revealed that heroin increased total GFAP filament length per image when compared to YS (q = 0.0007), which was normalized by NAC treatment during extinction (q = 0.004). We next analyzed the average GFAP filament diameter for each cell per Z-series data set (Figure 5D). One Z-series data set from YS and two Z-series data sets from the Heroin + NAC group were identified as statistical outliers. A Brown–Forsythe test revealed that the standard deviations were significantly different between groups (F(2,89) = 3.26, p = 0.04). As such, these data were analyzed using a Brown–Forsythe-corrected ANOVA which revealed there was no significant effect of treatment (F(2,74.38) = 0.626, p = 0.54). We next analyzed the average number of branch points of each cell, within each Z-series data set (Figure 5E). One Z-series data set from the YS group was identified as a statistical outlier. A Shapiro–Wilk test indicated that data from the YS group were non-normally distributed. Accordingly, a Kruskal–Wallis test was used which revealed a significant difference between groups (H = 14.61, p = 0.0007). The two-stage linear step-up procedure revealed that heroin increased the average number of branch points within GFAP filament arbors compared to YS (q = 0.0002) which was normalized by NAC treatment during extinction (q = 0.003). We also performed a 3D Sholl analysis for each cell within each image using 1 μM 3D concentric sphere radii (which were then compiled into 5 μM bins) and analyzed the average number of intersections as a function of distance from the center of each cell (Figure 5F). A repeated-measures one-way ANOVA revealed a significant difference between group means (F(2,20) = 19.23, p < 0.0001) Tukey's multiple comparison test revealed that heroin increased the average number of Sholl intersections compared to YS (p = 0.0002), which was normalized by NAC treatment during extinction (p < 0.0001). In a subsequent analysis, we focused on the first 15 μM 3D sphere radius from the centroid of each cell to investigate the primary GFAP+ branches (Figure 5G). In these data, a one-way ANOVA revealed a significant difference between groups (F(2,92) = 8.41, p = 0.0004). Tukey's multiple comparison test revealed that heroin increased the number of Sholl intersections in this region when compared to YS (p = 0.003) which was normalized by NAC treatment during extinction (p = 0.0007). Figure 5H shows an individual cell with a representative skeletonized GFAP arbor from each treatment group.

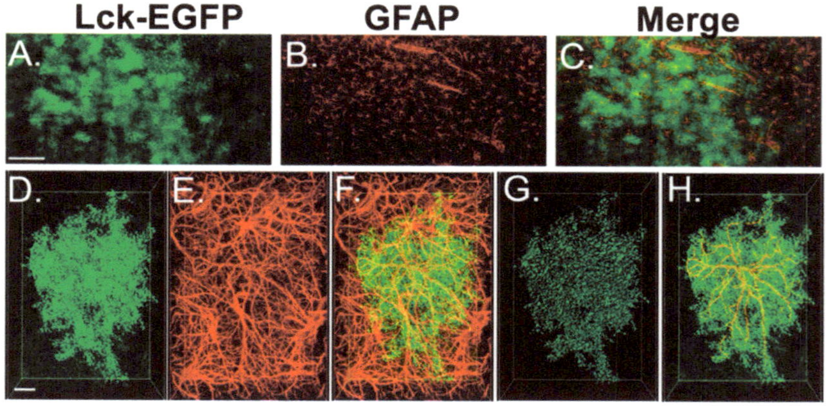

Figure 4. AAV5.*GfaABC1D*::Lck-EGFP expression is specific to GFAP+ astrocytes. (**A–C**) Low magnification view of AAV5.*GfaABC1D*::Lck-EGFP (green), GFAP (red), and the merge of the two signals.

(**D–F**) Raw deconvolved GFP, GFAP immunohistochemical detection, and the merge of the two signals. (**G**) 3D space-filling model (green) to isolate the EGFP+ astrocyte from background. (**H**) GFAP signal (red) confined to the isolated GFP+ astrocyte (green). Scale bars indicate either 150 μM (**A–C**) or 10 μM (**D–H**).

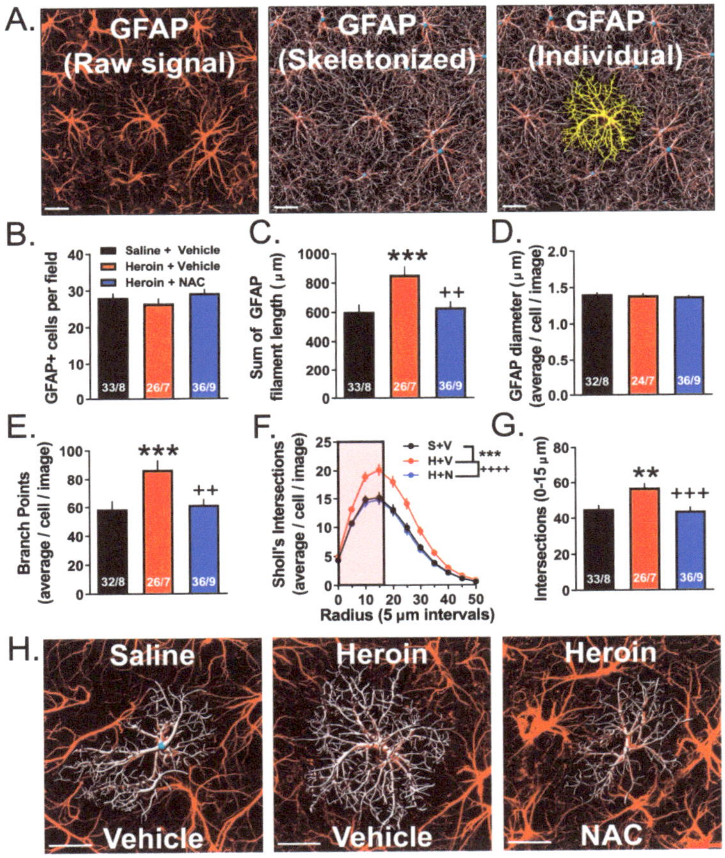

Figure 5. NAC normalizes heroin-induced alterations in PrL cortical astrocyte complexity. (**A**) Representative field of GFAP+ astrocytes skeletonized for analysis. Left—Raw GFAP signal from the confocal microscope. Middle—Skeletonized GFAP. Right—Isolated GFAP+ astrocyte from the field of cells. (**B**) There was no difference between groups in the number of GFAP+ cells per field. (**C**) Heroin increased the total length of the GFAP cytoskeletal filament which was normalized by repeated NAC treatment. (**D**) There was no difference between groups in the average diameter of the GFAP filament. (**E**) Heroin increased the average number of branch points which was normalized by repeated NAC treatment. (**F**) Frequency distribution of the number of Sholl intersections as a function of distance from the center of the cell. S + V = YS + Vehicle, H + V = Heroin + Vehicle, H + N = Heroin + NAC (**G**) Average number of Sholl intersections across the first 4 bins (0, 5, 10, and 15 μM). (**H**) Representative skeletonized GFAP+ cells for each treatment group. ** $p < 0.01$, *** $p < 0.001$ compared to YS, ++ $p < 0.01$, +++ $p < 0.001$, ++++ $p < 0.0001$ compared to Heroin + Vehicle. Numbers within each bar (**B–D**) denote the number of images taken/the number of animals used. Numbers within each bar (**E,G**) indicate the number of individual astrocytes processed/total number of animals in each condition. Scale bars indicate 20 μM.

3.5. Experiment 2B—Heroin SA and Extinction Increased Astrocyte Association with Multiple Synaptic Markers, an Effect That Was Prevented by Repeated NAC Treatment during Extinction

We performed two separate astrocyte–synaptic contact analyses using PrL cortical sections collected from the three experimental groups, YS ($N = 10$), Heroin + Vehicle ($N = 9$), and Heroin + NAC ($N = 10$), to examine the association of astrocytes with synaptic markers following heroin SA, extinction, and NAC treatment. In the first set, the astrocyte structure was visualized with viral-vector-mediated membrane-targeted EGFP expression achieved with the AAV5.*GfaABC1D*::Lck-EGFP viral vector and further amplified with GFP IHC whereas excitatory synapses were labeled with GluA2 IHC as described previously [31]. The second set of astrocyte–synaptic contact analyses was performed identically, except that synapsin I IHC was used as a general synaptic marker [13]. We imaged an average of 3–5 cells/animal for both GluA2 and synapsin I analyses. For the GluA2 analyses, the following were excluded: one YS animal for the missing localization of the Lck-EGFP virus expression, one YS animal for the lack of cells satisfying the Costes threshold assignment, one YS animal for issues related to IHC, one Heroin + Vehicle animal which had insufficient Lck-EGFP virus expression, one Heroin + NAC animal for the lack of cells satisfying the Costes threshold assignment, and one Heroin + NAC which had no more available tissue.

Figure 6A shows a representative PrL cortical Lck-EGFP-labeled astrocyte (green) and punctate signal corresponding to GluA2 (red). To aid in the visualization of the punctate GluA2 synaptic marker signal, an individual optical section from the many images that comprise each Z-series data set is shown. The left panel shows the raw signal, the middle panel shows the GluA2 signal contained within the 3D surface of the astrocyte membrane, and the right panel shows colocalized voxels. Within each panel, a smaller depiction of the full Z-series data set is visible in the top right corner, with an inset depicting the field seen within the single optical section view. Figure 6B shows the percent of the co-registry of the GFP and synaptic marker signals across the three treatment groups, a metric of astrocyte–synaptic interaction [13,14,33,34]. Since our experimental measurements utilized both a general synaptic marker and an excitatory synaptic marker, all data were normalized to YS animals and expressed as % of control. This was done for ease of comparison between markers and cohorts of animals. However, it is important to note that, although baseline values differed slightly from cohort to cohort, the effect of heroin was reproducible across both cohorts (Figure 7A–D). For GluA2, a Brown–Forsythe-corrected ANOVA revealed a significant effect of treatment ($F(2,9.743) = 4.74$, $p < 0.01$). The two-stage linear step-up procedure indicated that the Heroin + Vehicle group showed an increased co-registry of GluA2 and GFP compared to the YS group ($p = 0.01$), which was normalized by repeated NAC treatment during extinction ($p < 0.02$, Figure 6B). We next determined whether this effect was due to an increased number of GluA2 puncta within each data set. A one-way ANOVA revealed no significant effect of treatment ($F(2,20) = 2.24$, $p = 0.13$, Figure 6C). Figure 6D shows representative 3D-reconstructed astrocyte for each treatment group with GluA2 puncta (top) as well as the extent of synaptic contact as visualized by the white GFP and GluA2 co-registered voxels (bottom).

As described above, we also performed the same imaging and analysis for Lck-EGFP and Synapsin I, using tissue from the same animals to determine whether the observed effect was specific to synapses containing the GluA2 subunit of AMPA receptors. One YS, two Heroin + Vehicle animals, and two Heroin + NAC animals were removed from the analysis for the following reasons: YS + Vehicle: off-target Lck-EGFP virus expression ($N = 1$); Heroin + Vehicle: lack of available tissue ($N = 1$) or insufficient Lck-EGFP virus expression ($N = 1$); Heroin + NAC: lack of a sufficient number of analyzable cells after imaging ($N = 1$); and technical issues related to IHC ($N = 1$). A one-way ANOVA revealed a significant main effect of treatment ($F(2,21) = 6.51$; $p = 0.006$; Figure 6E). Tukey's multiple comparison test revealed that heroin SA and extinction increased astrocyte–synaptic interaction as assayed by Lck-EGFP-Synapsin I co-registration when compared to YS controls ($p = 0.012$), which was normalized by NAC treatment during extinction ($p = 0.012$). Akin to analyses performed for GluA2, we also analyzed the number of Synapsin I voxels in a subset of images from each animal (Figure 6F). A Shapiro–Wilk test indicated that data from YS controls (W = 0.83; $p = 0.04$) were non-normally distributed. A Kruskal–Wallis test indicated no difference between groups ($H = 2.04$; $p = 0.36$). There was no difference between groups in astrocyte complexity as measured by the ratio of the surface area to volume for cells used in either the GluA2 or Synapsin I data sets (Figure 7E,F).

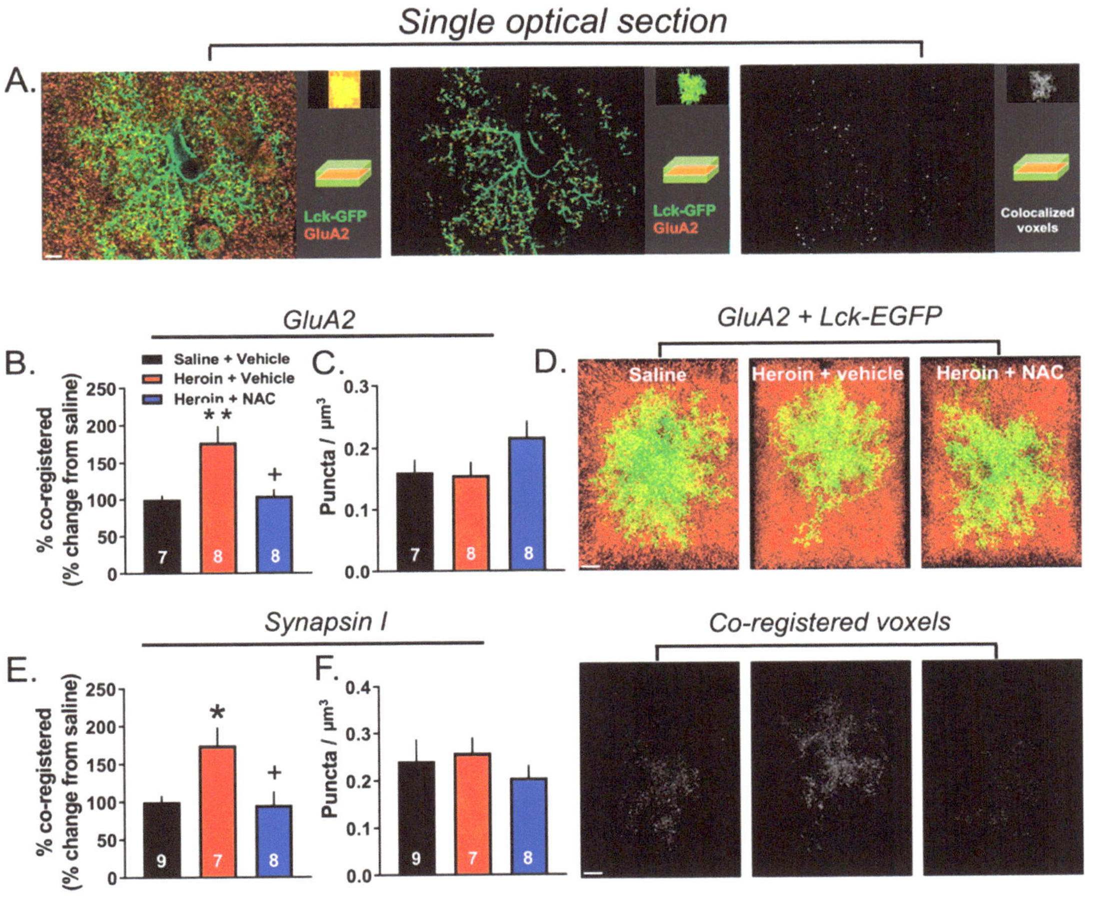

Figure 6. NAC normalized heroin-induced increases in PrL cortical astrocyte PAP–synapse association. (**A**) A single optical section from a representative Lck-EGFP-labeled astrocyte and GluA2 puncta. Left—Raw signal from confocal microscope. Middle—GluA2 confined within the proximity of the astrocyte membrane. Right—Co-registered voxels within the proximity of the astrocyte membrane. (**B**) Percent of co-registration of GFP and GluA2. (**C**) Number of GluA2 puncta normalized to the data-set volume. (**D**) Representative astrocytes from the three groups. Top—Raw signal from the confocal microscope. Bottom—Co-registered voxels within the region occupied by each astrocyte. (**E**) Heroin SA and extinction increased the percent co-registration of GFP and Synapsin I which was normalized by repeated NAC treatment (**F**) There was no difference between groups in the number of Synapsin I puncta normalized to the data-set volume. Numbers within each bar denote the number of animals used, black bars indicate Saline + Vehicle treatment, red bars indicate Heroin + Vehicle Treatment, blue bars indicate Heroin + NAC treatment. * $p < 0.05$ ** $p < 0.01$ compared to YS, + $p < 0.05$ compared to Heroin + Vehicle. Scale bars indicate 5 μM (**A**) and 10 μM (**D**).

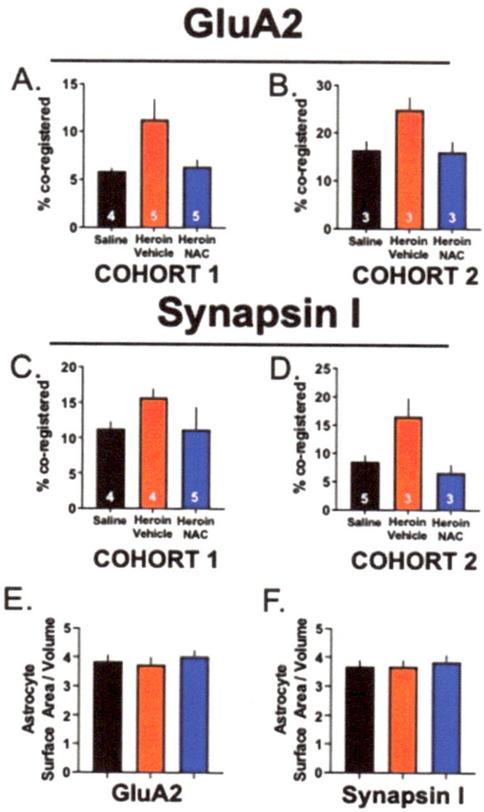

Figure 7. Between-cohort effects on GluA2 and Synapsin I co-registry with AAV5.*GfaABC1D*::LcK-EGFP. (**A**,**B**) Cohort 1 (**A**) and cohort 2 (**B**) showed the same effect of percent co-registry for GluA2 (**C**,**D**) Cohort 1 (**C**) and cohort 2 (**D**) showed the same effect of percent co-registry for Synapsin I. (**E**,**F**) 3D-rendered AAV5.*GfaABC1D*::LcK-EGFP-expressing astrocytes showed no difference between groups in the surface-area-to-volume ratio for GluA2 (**E**) or Synapsin I (**F**). Black bars indicate Saline + Vehicle treatment, red bars indicate Heroin + Vehicle Treatment, blue bars indicate Heroin + NAC treatment.

3.6. Experiment 3A—Heroin SA and Extinction Impacts Spine Density and Spine Head Diameter on PrL-NAcore Neurons, an Effect That Was Partially Reversed by Repeated NAC Treatment during Extinction

Others have shown that heroin SA decreases the number of dendritic spines on PrL cortical neurons after one month of abstinence [35]. Here, we sought to determine whether heroin SA and extinction altered dendritic spine morphometrics specifically in PrL cortical neurons projecting to the NAcore. We imaged 5–8 dendritic spine segments from YS (N = 10), Heroin + Vehicle (N = 6), and Heroin + NAC (N = 6) animals after extinction. Figure 8A shows a representative layer V PrL-NAcore neuron fully labeled with the highlighted area indicating the region sampled at the apical tuft. Figure 8B shows the dendritic spine segment sampled from the highlighted region in Figure 8A. Figure 8C shows the 3D rendering of the processed dendrite for spine detection and Figure 8D shows a color-coded map of the dendritic spine head diameter (d_H) with warmer colors indicating larger values. We first analyzed the average dendritic spine d_H across the three treatment groups. A one-way ANOVA revealed a significant effect of heroin treatment

($F(2,19) = 20.70$, $p < 0.0001$, Figure 8E). Tukey's multiple comparison test revealed that heroin SA and extinction increased the average dendritic spine d_H ($p < 0.0001$), which was not altered by NAC treatment during extinction. Heroin + NAC treatment likewise increased d_H relative to saline-treated animals ($p = 0.0007$). Next, we investigated whether this effect was specific to smaller or larger spines. To do so, we binned the spine d_H (0.2 µM bins) and generated a frequency distribution. A two-way repeated-measures ANOVA revealed a significant treatment-by-bin interaction ($F(8,76) = 8.144$, $p < 0.0001$, Figure 8F). Tukey's multiple comparison test revealed that heroin SA and extinction, regardless of NAC treatment, decreased the proportion of smaller spines (<0.2—$p < 0.0001$; 0.2–0.4—$p < 0.05$), but increased the proportion of spines with a larger d_H (0.4–0.6—$p < 0.01$) compared to YS rats. Interestingly, the proportion of spines with a d_H between 0.6–0.8 was increased in the Heroin SA + Vehicle group ($p = 0.008$), but there was no effect of NAC on this specific bin compared to the YS treatment group ($p = 0.17$). We next analyzed the average dendritic spine density in all three treatment groups from the same dendrite segments. A one-way ANOVA revealed a significant effect of treatment ($F(2,19) = 13.46$, $p = 0.0002$, Figure 8G). Tukey's multiple comparison test indicated that heroin SA and extinction decreased the average spine density compared to saline controls ($p = 0.0003$), which was prevented by NAC treatment during extinction ($p = 0.002$). To determine which spine head diameter bins accounted for the altered spine density, we performed a similar analysis as in Figure 8F, but instead normalized the number of spines within each bin to the length of the dendrite. A two-way repeated-measures ANOVA revealed a significant treatment-by-bin interaction ($F(8.76) = 8.722$, $p < 0.0001$, Figure 8H). Tukey's multiple comparison test indicated that heroin SA and extinction ($p < 0.0001$), regardless of NAC treatment ($p = 0.0003$), decreased the density of spines with a head diameter less than 0.2 µM. However, the Heroin + Vehicle group showed decreased spines in the 0.2–0.4 µM bin compared to YS ($p < 0.0001$), which was normalized in the Heroin + NAC group ($p < 0.0001$ compared to Heroin + Vehicle). Interestingly, the density of larger spines with head diameters between 0.4–0.6 µM were unaffected in the Heroin + Vehicle group compared to YS ($p = 0.194$), yet Heroin + NAC increased the density of these spines compared to the YS ($p = 0.0125$) and Heroin + Vehicle ($p = 0.0002$) groups. These effects on dendritic spine morphometric properties were likely not due to inherent differences in the FLAG signal intensity within the spine heads as there was no difference between groups in the average FLAG intensity within the spine head region when normalized to saline controls (Figure 9A). Representative spine segments (15 µM cropped segments) are shown in Figure 8I.

3.7. Experiment 3B—Heroin SA and Extinction Increased GluA1 Immunoreactivity in Dendritic Shafts and Spine Heads of PrL-NAcore Neurons, and NAC Potentiated Spine-Head-Specific GluA1

We previously showed that cocaine SA followed by one week of abstinence increases dendritic spine head diameter in the apical tuft dendrites of PrL-NAcore neurons. We also demonstrated that cocaine abstinence-duration-dependent structural plasticity occurs concomitantly with alterations in GluA1 immunoreactivity within the dendritic shaft and at dendritic spine heads. Here, we investigated whether the adaptations in dendritic spine morphometrics following the extinction of heroin SA (with or without repeated NAC treatment) were associated with altered GluA1 immunoreactivity in the discrete dendritic subcompartments of the PrL-NAcore apical tuft dendrites described above. When analyzing the total volume of GluA1 immunoreactivity in the dendritic spine heads and shaft combined, relative to the total dendrite volume, a one-way ANOVA revealed an omnibus significant effect of NAC treatment upon GluA1 immunoreactivity ($F(2,19) = 10.56$, $p = 0.0008$). Tukey's multiple comparison test indicated that the Heroin + Vehicle ($p = 0.02$) and Heroin + NAC ($p = 0.0009$) groups displayed increased GluA1 immunoreactivity compared to saline controls (Figure 10A). In the case of the Heroin + NAC, but not Heroin + Vehicle, group, increases in GluA1 volume could occur in part due to increased overall GluA1 signal intensity, as the average GluA1 intensity (spine heads + dendritic shaft) was significantly higher in this group (main effect of treatment: ($F(2,19) = 4.27$, $p = 0.029$, Tukey's post hoc comparing Heroin + NAC to Heroin + Vehicle—$p = 0.025$, Figure 9B).

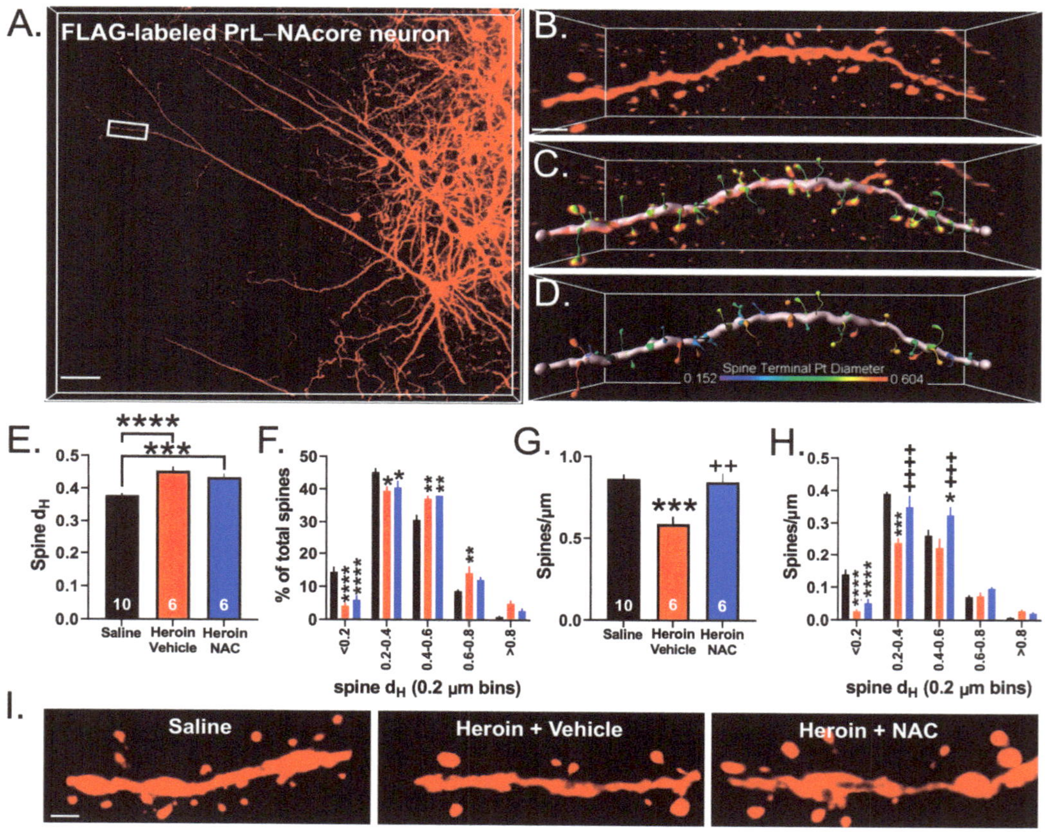

Figure 8. NAC normalized heroin-induced decreases in PrL–NAcore apical spine density, but not head diameter. (**A**) Left—Representative FLAG-labeled PrL–NAcore neuron. (**B**) Dendritic spine segment taken from the apical tuft (boxed region in **A**), (**C**) Filament-processed spine segment, (**D**) Heat map of dendritic spine head diameters along the segment—warmer colors indicate a greater head diameter. Note that most spines fall within the middle range of head diameters. (**E**) Average d_H across treatment groups. (**F**) Frequency distribution of the percent of total spines as a function of spine head diameter bin (0.2 µM). (**G**) Average dendritic spine density across treatment groups. (**H**) Frequency distribution of spine density normalized to spine d_H. (**I**) Representative spines from a 15 µM stretch of dendrite are shown. d_H = head diameter. Numbers within each bar denote the number of animals used. * $p < 0.05$, ** $p < 0.01$, *** $p < 0.001$, **** $p < 0.0001$ compared to YS. ++ $p < 0.01$, +++ $p < 0.001$, ++++ $p < 0.0001$ compared to Heroin + Vehicle. Scale bars indicate 70 µM (**A**), 3 µM (**B–D**), and 2 µM (**I**).

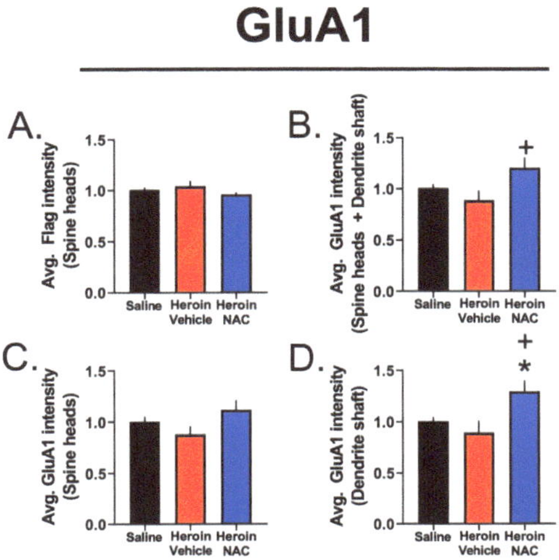

Figure 9. Signal intensity data pertaining to Experiment 3. (**A**) Average FLAG intensity in dendritic spine heads across treatment groups. (**B**) Average GluA1 intensity (dendritic shaft and spine head) across treatment groups. (**C**) Average GluA1 intensity isolated to dendritic spine heads. (**D**) Average GluA1 intensity isolated to dendritic shafts. * $p < 0.05$ compared to YS, + $p < 0.05$ compared to Heroin + Vehicle. Data are expressed as a fold change from control.

Next, we focused on dendritic-spine-head-specific GluA1 immunoreactivity. Here the volume of the signal in spine heads was expressed relative to the total dendrite volume, as a function of treatment. A Brown–Forsythe ANOVA revealed a significant effect of treatment ($F(2,9.028) = 20.22$, $p = 0.0005$). Dunnett's T3 multiple comparison test revealed that the Heroin + Vehicle ($p = 0.03$) and Heroin + NAC ($p = 0.004$) groups showed increased spine-head-specific GluA1 signal accumulation when compared to the Saline + Vehicle group (Figure 10B). However, in this instance, there was no difference between groups in the average GluA1 intensity in the co-registered volume (Figure 9C).

Finally, we analyzed dendritic-shaft-specific GluA1 immunoreactivity, relative to the total dendrite volume, as a function of treatment. A one-way ANOVA revealed a significant effect of treatment ($F(2,19) = 7.73$, $p = 0.004$). Tukey's multiple comparison test indicated that the Heroin + Vehicle ($p = 0.033$) and Heroin + NAC ($p = 0.005$) groups showed increased dendritic-shaft-specific GluA1 immunoreactivity compared to the Saline + Vehicle group (Figure 10C). The Heroin + NAC group displayed increased GluA1 intensity in the dendritic shaft subcompartment compared to both the Saline + Vehicle and Heroin + Vehicle groups (Figure 9D). This is to be expected given the results discussed above, as the shaft region constitutes a large portion of the total dendritic segment volume. Interestingly, correlation analyses revealed that % GluA1 volume in the total segment normalized to the total dendrite volume ($y = 34.10X - 8.093$, $R = 0.71$, $R^2 = 0.51$, $p = 0.0002$, Figure 10D), % GluA1 volume in the spine heads normalized to the total dendrite volume ($y = 4.457X - 1.231$, $R = 0.55$, $R^2 = 0.300$, $p = 0.008$, Figure 10E), and % GluA1 volume in the shaft normalized to the total dendrite volume ($y = 29.65X - 6.853$, $R = 0.71$, $R^2 = 0.504$, $p = 0.0002$, Figure 10F) showed a positive correlation with changes in the spine head diameter as a function of treatment. In contrast, the % GluA1 volume in the total dendrite, spine heads only, and shaft only did not correlate with changes in dendritic spine density as a function of treatment (Figure 11A–C). Representative dendritic spine segments with GluA1 immunoreactivity in the two subcompartments are shown in Figure 10G.

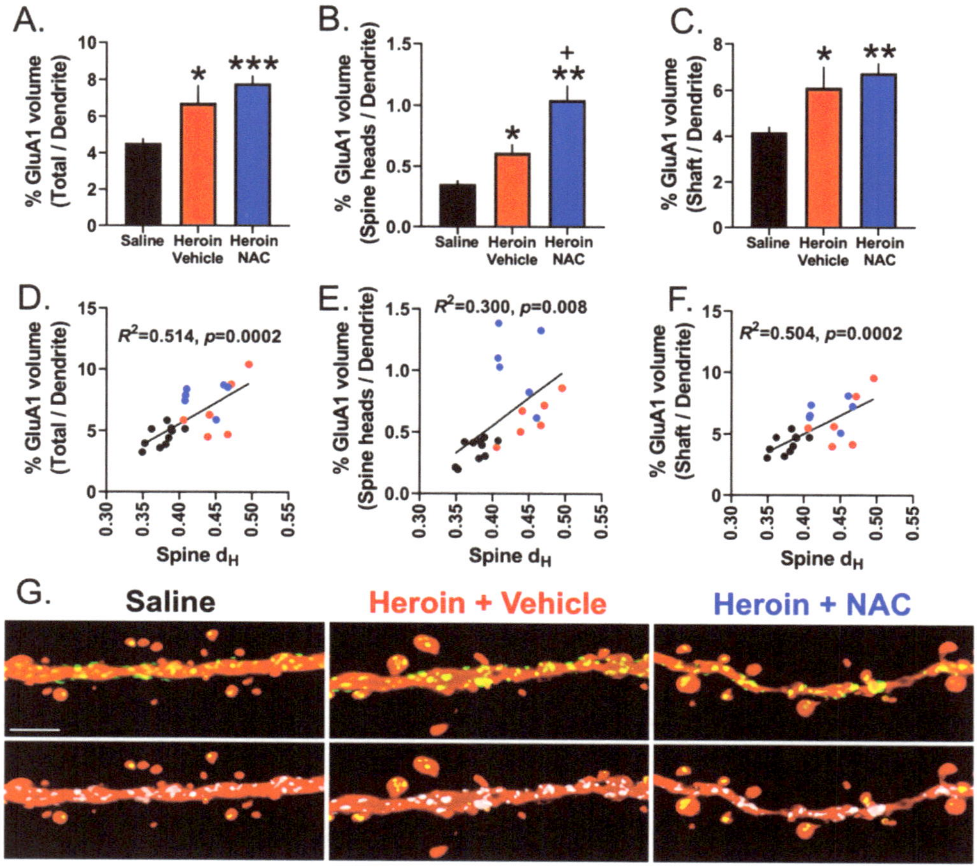

Figure 10. Heroin SA and extinction, regardless of NAC treatment, increased GluA1 immunoreactivity in dendritic shafts and spine heads of PrL-NAcore neurons. (**A**) Average GluA1 immunoreactivity in the total dendrite (shaft + spine head) normalized to dendrite volume. (**B**) Average GluA1 immunoreactivity in dendritic spine heads normalized to dendrite volume. (**C**) Average GluA1 immunoreactivity in the dendritic shaft normalized to dendrite volume. (**D**) Scatterplot depicting correlation between spine d_H and GluA1 immunoreactivity in the total dendrite (shaft + spine head) normalized to dendrite volume. (**E**) Scatterplot depicting correlation between spine d_H and GluA1 immunoreactivity in dendritic spine heads normalized to dendrite volume. (**F**) Scatterplot depicting correlation between spine d_H and GluA1 immunoreactivity in the dendritic shaft normalized to dendrite volume. Black lines indicate linear fit obtained from linear regression analyses. (**G**) Top panel—Representative dendritic spine segments showing Flag (red) and GluA1 (green). For clarity GluA1 is only shown in the volume occupied by the dendrite segment. Bottom panel—Co-registry between GluA1 and Flag in the dendritic shaft (white) and spine heads (yellow). * $p < 0.05$, ** $p < 0.01$, *** $p < 0.001$ compared to Saline; + $p < 0.05$ compared to Heroin + Vehicle. Scale bar indicates 4 µM (**D–F**) black points indicate Saline + Vehicle treatment, red points indicate Heroin + Vehicle Treatment, blue points indicate Heroin + NAC treatment.

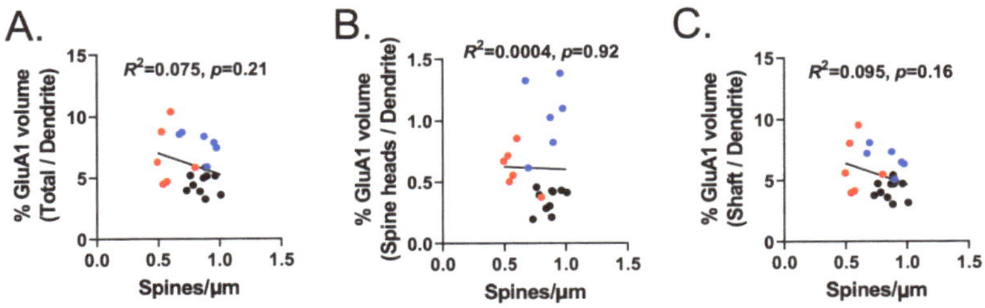

Figure 11. There was no correlation between the percent GluA1 volume in subcompartments normalized to the total dendrite volume and spine density. (**A**) % GluA1 volume in the co-registered volume within the total dendrite: y = −3.488X + 8.757, r = −0.27 r^2 = 0.075. (**B**) % GluA1 volume in the co-registered volume within the spine heads only: y = −0.043X − 0.647, r = −0.02 r^2 = 0.0004. (**C**) % GluA1 volume in the co-registered volume within the dendritic shafts only: y = −3.445X + 8.110, r = −0.31 r^2 = 0.095. Black lines indicate linear fit. Black points indicate Saline + Vehicle treatment, red points indicate Heroin + Vehicle Treatment, blue points indicate Heroin + NAC treatment.

3.8. Experiment 4—Both NAC Treatment and Delivery of shTSP-2 to PrL Astrocytes Blunt Cue-Induced Relapse to Heroin Seeking

Apart from the well-documented ability of NAC to normalize glutamate homeostasis [10,15,24], our data demonstrate that NAC treatment reversed heroin-mediated reductions in spine density at PrL-NAcore neurons. Accordingly, we hypothesized that NAC may exert its therapeutic effects through the restoration of dendritic spine populations at key subsets of cortical neurons following heroin SA. Astrocyte release of both TSP 1 and 2 mediate neuronal synaptogenesis, with astrocytic TSP signaling also required for the formation of silent synapses following cocaine exposure [30]. We thus hypothesized that the genetic elimination of TSP-2, and loss of an important signal for synaptogenesis, would disrupt the therapeutic efficacy of NAC treatment. To test the hypothesis that the delivery of shTSP-2 might blunt the effects of NAC treatment during extinction, rats received an infusion of either AAV5.*GfaABC1D*::Lck-EGFP (control, *N* = 11) or AAV5.*GfaABC1D*::shTSP2-GFP (shTSP, *N* = 14), an shRNA targeted to TSP-2 [30]. Figure 12A depicts a schematic of the AAV infusion and Figure 12B displays a schematic of viral placements in addition to representative low- and higher-magnification images illustrating isolation of expression of the viral constructs to astrocytes. Rats then underwent heroin SA and extinction (Figure 12C,D). During the last 10 days of extinction, rats received injections of either vehicle or NAC. Following extinction, rats underwent cued-relapse testing for heroin. PrL delivery of shTSP-2 was not found to alter heroin SA or extinction active lever pressing (Figure 12C,D). A 2 × 2 ANOVA revealed that both NAC treatment ($F(1,23) = 7.480$, $p = 0.013$) and the delivery of shTSP-2 ($F(1,23) = 5.005$, $p = 0.037$) significantly reduced active lever pressing during the last three days of extinction. Though NAC+ shTSP-2 treatment did result in a marginal decrease in active lever pressing during the last three days of extinction, the effect was not statistically significant. Similarly, both NAC ($F(1,23) = 6.155$, $p = 0.02$) and the delivery of shTSP-2 ($F(1,23) = 4.423$, $p = 0.04$) significantly blunted cued-relapse to heroin seeking. Similar to findings from extinction data, NAC + shTSP-2 treatment resulted in a marginal decrease in cued-relapse active lever pressing, though the effect was not statistically significant compared with either the LcK$^+$ NAC or shTSP2 + vehicle treatment groups (Figure 12E).

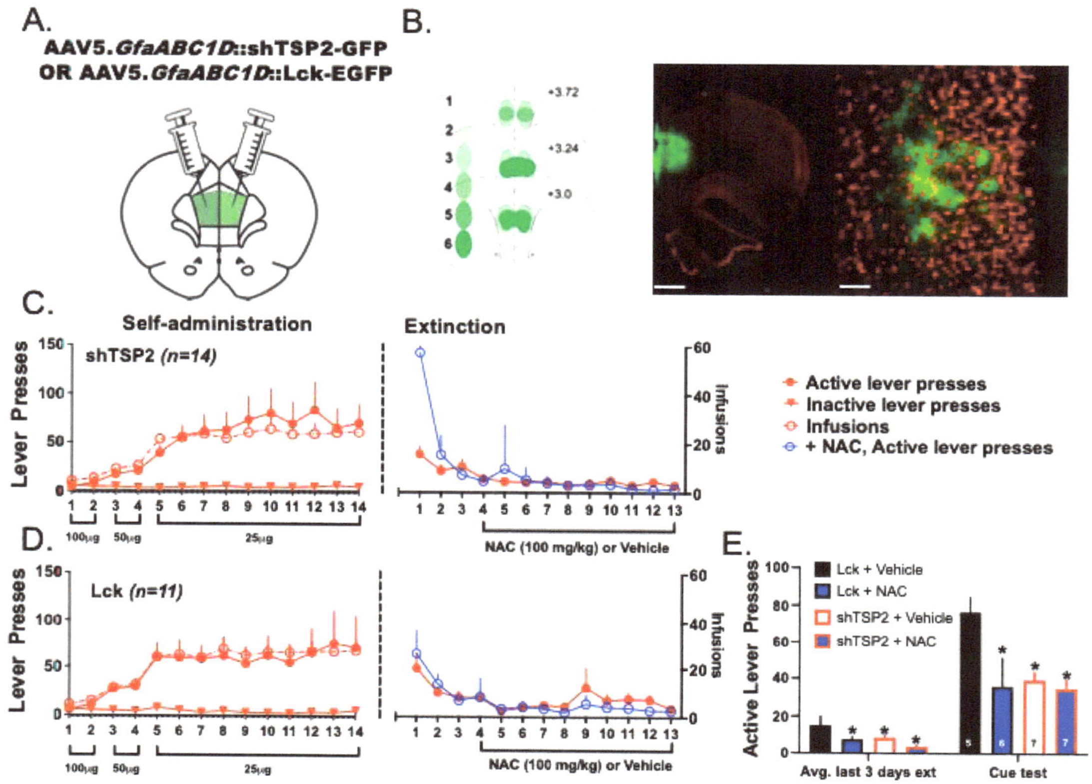

Figure 12. NAC treatment or loss of TSP2 on cortical astrocytes blunts cue-induced relapse to heroin. (**A**) Schematic indicating viral design. (**B**) Schematic indicating viral placements with approximate anterior/posterior viral spread from the injection site and representative low- and high-magnification images showing GFP expression in the PrL and viral labeling of astrocytes (green) versus neuN counterstain (red) (scale bars: low mag = 1500 μM, high mag = 100 μM). (**C**) Self-administration and extinction data for animals injected with shTSP-2. (**D**) Self-administration and extinction data for animals injected with Lck-EGFP. (**E**) Effects for the last three days of extinction and for cued-relapse to heroin seeking following extinction. * $p < 0.05$ compared to Lck + Vehicle.

4. Discussion

Here, we show that heroin SA and extinction induce functional adaptations in cortical astrocytes. Our data demonstrate that heroin SA followed by extinction enhances the fundamental cytoskeletal cellular complexity of astrocytes in the PrL cortex, a finding that was accompanied by the increased association of membranous astrocytic processes, virally labeled with membrane-targeted EGFP expression, with two distinct synaptic markers. Both of these effects, as well as heroin seeking during extinction, were normalized by repeated NAC treatment during the last 10 days of extinction training, confirming that NAC reduces responding during extinction as others have reported [19]. Apart from impacts on cortical astrocytes, heroin SA and extinction also increased the average spine d_H on apical tuft dendrites of PrL cortical neurons that project to the NAcore, an effect that was accompanied by decreased spine density. Interestingly, NAC treatment during extinction normalized heroin-induced decreases in dendritic spine density but did not impact dendritic spine d_H. Moreover, NAC treatment was shown to block cue-induced relapse to heroin. These results suggest that the ability of NAC to prevent heroin relapse in preclinical models may not only be due to its action in the NAcore but also include actions that normalize structural plasticity in astrocytes, as well as neurons within the PrL cortex that project to the NAcore. Further, our results indicate that shTSP-2 delivery to cortical astrocytes likewise decreases cue-induced reinstatement to heroin, thereby cementing the role of PrL astrocytes in the cycle of relapse to heroin addiction.

4.1. Heroin SA and Extinction Evoke Functional Alterations in PrL Astrocytes

Our data are among the first to show that heroin SA and extinction evoke functional alterations in evoked GLT-1-dependent current in cortical astrocytes, a finding that is consistent with our presently reported structural adaptations following heroin exposure. While the magnitude of these functional alterations is somewhat small, here, we show that these alterations can be used to predict prior heroin exposure with ~85% accuracy. The relevance of these alterations to cue-induced heroin reinstatement is an important future direction.

4.2. Evoked GLT-1-Dependent Current Reveals a Functional Clustering of Cortical Astrocytes, Independent of Heroin Treatment

Interestingly, our data provide evidence of a clustering of astrocyte subpopulations in the PrL that display characteristic functional properties that appear to occur independent of drug treatment. This finding mirrors existing data to support the existence of diverse heterogeneity in astrocyte populations. Despite being the most abundant cell type in the adult brain, early classification schemes of astrocyte subpopulations are inconsistent with emerging transcriptomic and morphological data that collectively suggest an incredible degree of diversity in astrocyte populations. Indeed, recent transcriptomic investigations have revealed considerable heterogeneity among astrocyte populations both across brain regions and within a single region, including the cortex [36]. Similarly, others have reported astrocyte-specific receptor heterogeneity across regions of the brain such as the thalamus [37]. Importantly, this diverse heterogeneity is maintained not only in healthy brains, but also across disease states [38]. The exact contributions of astrocyte heterogeneity to motivational and cognitive processes remain elusive, however. Given the distribution of astrocytes throughout the brain and their central importance to global neural functioning, it is clear that future studies of astrocyte contributions to heroin SA and extinction should include investigations into cell heterogeneity in an attempt to elucidate the role of these unique subpopulations of astrocytes. Moreover, the role that this functional heterogeneity plays in cue-induced relapse remains to be explored.

4.3. Repeated NAC Treatment Facilitated Extinction Training

Our data provide the first preclinical evidence for NAC acting to augment structural plasticity in a cortical region; we suspect that these adaptations play a role in NAC's ability to prevent cue-induced reinstatement. Although NAC's ability to prevent relapse has yet to be tested in the clinic for opiate use disorder, preclinical data indicate that daily NAC treatment during extinction reduces extinction responding as well as cue- and heroin-induced drug seeking for up to 40 days after the last injection [19]. Ultimately, NAC has shown promise preclinically in decreasing relapse for several drugs of abuse [14,20,22–24]. However, it is critical to note that NAC's ability to suppress relapse clinically, at least in the case of cocaine, is minimal to date [39]. The general consensus is that NAC is effective in increasing the abstinence duration in patients who are already abstinent but fails to decrease relapse rates in active users [39–41]. Despite its modest clinical efficacy, its well-described mechanism of action and ability to target drug-induced adaptations in non-neuronal cells substantiates its role as a powerful research tool for understanding astrocytic dysfunction in addiction.

4.4. Repeated NAC Treatment during Extinction Reversed Increases in Astrocyte Complexity and Association of PAPs with Synapses in PrL Cortex Induced by Heroin SA and Extinction

Here, we provide evidence that extinction from heroin SA increased GFAP cytoskeletal complexity without altering the number of GFAP+ cells in the PrL cortex. We conclude that this is not due to a transition of astrocytes to a reactive state as there was no enhancement of average GFAP filament diameter across groups, a phenomenon that serves as a canonical index of reactive astrocytes [42]. Given the profound impacts observed in PrL cortical astrocytes, the signal transduction required for this event is an immediately apparent question. It is possible that this effect may be due to heroin's direct impact on opioid receptors expressed on astrocytes. mRNA for all three opioid receptor subtypes is present in astrocytes, and PrL cortical astrocytes in particular express an abundance of mu opioid receptor mRNA [43], the primary receptor engaged by morphine [44]. Further the incubation of cortical astrocyte cultures with morphine enhances intracellular Ca^{2+} signaling [45], resulting in an increased area occupied by astrocytes [46]. In addition, astrocytic mu opioid receptor expression has been directly linked to the rewarding properties of drugs of abuse; e.g., activation of mu opioid receptors on hippocampal astrocytes induces conditioned place preference [47]. Thus, it is likely that one component of the observed increase in GFAP complexity in the PrL cortex of heroin-exposed rats is due to the chronic tone on astrocytic mu opioid receptors. In support, a study performed in the early 1990s showed that cultured astrocytes treated with morphine increase their branching complexity yet morphine decreased astrocyte proliferation [27]. Alternatively, chronic injections of morphine in rats have been shown to increase GFAP protein expression in the frontal cortex [26], yet this study did not have a prolonged withdrawal period. While we did not causally link increased GFAP complexity to relapse behavior, repeated NAC treatment prevented the action of heroin at cortical astrocyte GFAP arbors and is known to reverse drug-induced deficits in astrocyte function linked to relapse propensity [48,49]. Whether normalized cortical GFAP complexity is required for NAC's ability to suppress relapse remains to be determined. However, as an important caveat, future examinations should likewise consider astrocyte maturation in the context of these manipulations, which has been linked to enhanced GFAP arbor complexity at key developmental timepoints also associated with synaptogenesis [50,51].

Extinction from cocaine [13,34], heroin [12], or methamphetamine [14] SA leads to the decreased association of PAPs with synaptic markers in the NAcore. This finding can be interpreted as a withdrawal of PAPs, which are highly enriched in GLT-1 [52], from synapses [53,54]. Indeed, ezrin, a protein that links the astrocyte actin cytoskeleton with the astrocyte membrane and is highly enriched on peripheral astrocytic processes, has been implicated in this process. Specifically, knockdown of NAcore ezrin has been previously demonstrated to produce a similar withdrawal of PAPs from synapses comparable to the magnitude of withdrawal following extinction training [12]. In this same study, ezrin knockdown prevented the reassociation of PAPs with synaptic markers that occurs following cued-relapse and perpetuated cue-induced drug seeking [12]. This connection between astrocyte structure and function is readily made as drug-induced retraction of glial processes in the NAcore is often paired with decreased GLT-1 expression and reduced glutamate clearance [55]. While the specific mechanisms behind the downregulation of GLT-1 in the NAcore following drug SA remains elusive, possible candidates include epigenetic mechanisms [56] and potentially a disruption in GLT-1 recycling between the astrocyte plasma membrane and intracellular compartments, though more directed experimentation is required to confirm the latter hypothesis. We show here that extinction from heroin SA leads to the increased association of PAPs with synaptic markers in the PrL cortex, an effect that was specific to heroin and was not observed following the extinction of cocaine SA [34]. While further investigation would be required to determine if heroin SA and extinction impacts cortical glutamate clearance, we would not predict reduced glutamate clearance, as astrocyte retraction from synapses was not observed. The GFAP cytoskeleton represents a long-standing canonical maker for astrocytes and has been often utilized as a means to interrogate the astrocyte structure [57,58] and activation states [59]. However, it of course does not accurately recapitulate the membranous extremities of astrocytes [60] that extend beyond the GFAP skeleton. Accordingly, alterations in the cytoskeletal properties of astrocytes at the level of GFAP arbors do not necessarily mandate that adaptations in the astrocyte–synaptic interaction will be observed. As an example, methamphetamine SA and extinction decreased the PAP–synapse association, but not GFAP cytoskeletal complexity, in the NAcore [14]. Here, we do observe that heroin-induced increases in GFAP complexity were accompanied by a ~75% increase in the association of PAPs with synaptic markers in the PrL cortex following extinction from heroin SA. Interestingly, both adaptations in GFAP structure and astrocyte–synaptic interaction were normalized by repeated NAC treatment. One potential alternative explanation for our observations is that there was increased synaptogenesis in the PrL cortex after the extinction of heroin SA, with no alteration in the motility of astrocytes. However, this is unlikely given that we did not observe any difference between groups in the number of GluA2 or Synapsin I puncta.

It has been estimated that a single astrocyte makes contact with ~100 dendrites and ~100,000 synapses in the rodent brain [61]. Our astrocyte–synaptic co-registration data could be interpreted either as each astrocyte contacting a greater number of synapses in the PrL cortex after extinction or each astrocyte may be embracing individual synapses to a greater extent, though more sophisticated microscopy techniques such as electron microscopy would be required to address this issue. Additionally, some combination of these scenarios may also be occurring. Each of these potential outcomes would produce an overall increase in the co-registry of astrocyte membrane and synaptic marker signals. Astrocyte ensheathment of synapses shapes excitatory transmission by establishing the three-dimensional space in which neurotransmission occurs. More specifically, astrocyte ensheathment of synapses is a dynamic process that regulates the profiles of synaptic and non-synaptic receptor activation, allowing for astrocytic tuning of synaptic communication on a synapse-by-synapse basis. This modality of regulation has been directly explored by studies that examine how the degree to which astrocytes associate and physically interact with synapses is shaped by neuronal activity [62,63]. As an example, whisker stimulation has been shown to increase GLT-1 expression in the barrel cortex, which is associated with increased ensheathment of PAPs at enlarged dendritic spine heads [64]. Additionally, cue-induced reinstatement of heroin seeking has been associated with dynamic changes in the PAP–synapse association in the NAcore during the course of the reinstatement session that was directly linked to relapse behavior [12]. Together. these data demonstrate that astrocytes physically respond to changes in neurotransmission with physiologically and behaviorally relevant consequences. Moreover, the fidelity of synaptic transmission is tightly controlled by the PAP insulation of synapses; too little association is linked to glutamate overflow [12–14] whereas too much insulation can actually impede AMPA receptor-mediated excitatory transmission through enhanced glutamate uptake [65], a means of astrocyte regulation of synaptic function. Our data thus contribute to the evolving understanding of the astrocytic regulation of neuronal function, which was initially postulated to occur during sleep by Cajal almost 150 years ago [66]. Further experimentation will be needed to determine where astrocytic synapse ensheathment in the PrL cortex from heroin-experienced animals lies on this functional continuum. Whether PAP association with synapses has reached a "ceiling" and is impeding AMPA receptor transmission or whether the increased association leads to a suppression of volume transmission and a solidification of discrete excitatory inputs in a supraphysiological manner is yet to be determined.

4.5. Extinction from Heroin SA Altered Dendritic Spine Morphological Profiles and GluA1 Immunoreactivity in Discrete Dendritic Subcompartments of Prelimbic Cortical Neurons Projecting to the Nucleus Accumbens Core

Using similar viral-vector-based pathway-specific labeling of dendritic spines, we previously found that cocaine SA followed by one week of abstinence decreased the density of apical tuft dendritic spines of PrL-NAcore neurons, yet the remaining spines were enlarged and concomitantly displayed increased GluA1 and GluA2 immunoreactivity in the spine heads [31]. A previous study found that heroin SA and one month of abstinence decreased the density of dendritic spines of the PrL cortical layer V pyramidal neurons, with no reported alterations in the morphometric properties of the spines themselves [35]. To accompany the astrocyte–synapse interaction data discussed above, we investigated potential alterations in dendritic spine morphological profiles in PrL cortical neurons projecting to the NAcore. Akin to one week of abstinence from cocaine SA, we discovered that extinction from heroin SA decreased the density of dendritic spines along dendrites at the apical tuft of PrL-NAcore neurons, yet these neurons showed an increase in the average spine head size, as well as elevated GluA1 co-registry in the dendritic spine heads. These data are largely consistent with studies demonstrating that repeated treatment of cultured cortical neurons with morphine leads to destabilization of pre-existing spines [67]. In our data set, heroin-mediated decreases in dendritic spine density were prevented by repeated NAC treatment; yet the heroin-evoked increases in spine head diameter were not. In fact, to our surprise, we found that NAC-treated animals showed a potentiation of GluA1 immunoreactivity in dendritic spine heads relative to Heroin + Vehicle animals. Generally, an increase in the number of spines, the average spine head size, as well as AMPA receptor occupancy in spine heads, are interpreted as evidence for synaptic potentiation [68–71]. Thus, this finding is initially counter-intuitive given that NAC normalized the decrease in density but NAC-treated animals retained enlarged dendritic spines following heroin SA and extinction, suggesting that NAC is driving synaptic potentiation in the PrL-NAcore circuit. However, it should be noted that the apical tuft is a site of convergence of glutamatergic inputs from midline thalamic nuclei as well as dopamine afferents from the ventral tegmental area [72–74]. Thus, it remains possible that heroin-induced adaptations in spine morphology and AMPA receptor accumulation at this particular part of the dendritic arbor might not translate into whole-cell alterations in physiological indices of potentiation. Rather, this may reflect the potentiation of discrete inputs onto PrL-NAcore neurons which may or not may be counteracted by alterations in the synaptic strength of alternative inputs at other parts of the dendritic arbor. Future studies should examine whether heroin SA and extinction alter the electrophysiological properties of PrL-NAcore neurons and whether repeated NAC treatment during extinction has an impact on such alterations. Regardless, given that NAC is known to reverse drug-induced deficits in astrocyte function [18], we suspect that NAC's ability to restore dendritic spine density following heroin SA and extinction is mediated by augmented astrocyte function, perhaps through altering the release patterns of astrocyte-derived neuromodulators that regulate synaptogenesis.

4.6. NAC Treatment and Downregulation of Astrocyte TSP-2 Decreased Cued-Relapse to Heroin Seeking

NAC treatment and delivery of shTSP-2 on PrL astrocytes independently of NAC treatment decreased cued-relapse to heroin seeking. These novel findings extend our finding that NAC treatment blunted extinction responding following SA, and further suggest a therapeutic role for NAC in opioid use disorder. Although, as described previously, findings suggesting NAC's ability to suppress relapse in a clinical setting are minimal [18,39], and few, if any, existing studies directly investigated NAC's ability to block cued-relapse to heroin seeking. As discussed above, we predicted that NAC's ability to restore dendritic spine density following heroin SA and extinction may be mediated by augmented astrocytic function and the differential release of the astrocytic regulators of synaptogenesis. Thus, we initially hypothesized that the downregulation of astrocytic TSP-2 on PrL astrocytes would blunt the ability of NAC to both facilitate extinction training and to reduce cued-relapse following heroin SA. However, this was not the case. Despite this result, we did observe that the delivery of shTSP-2 to cortical astrocytes was sufficient to blunt cue-induced heroin seeking.

It is well-documented that astrocytes release a variety of neurotransmitters, neuromodulators, and growth factors that can have a profound impact on neuronal plasticity, (for review, see [75,76]). As an example, the astrocyte release of thrombospondin-1 (TSP—1) activates α2δ-1 receptors expressed on neurons to induce spinogenesis and excitatory synapse formation [77]. Dysfunction of this system has been linked to reduced dendritic spine density in Down Syndrome models [78], as well as elevated dendritic spine density in the NAc of animals receiving chronic cocaine injections [30]. Furthermore, the loss of astrocyte TSP-2 has been demonstrated to block the generation of silent synapses following cocaine training [30]. Presently, we report that the delivery of shTSP-2 to PrL astrocytes, independent of NAC treatment, is likewise sufficient to decrease cue-induced heroin relapse, which likewise suggests that cue-induced heroin relapse requires PrL astrocyte-mediated synaptogenesis, though more experimentation is required to directly test this hypothesis. Here, given that shTSP2 treatment did not block the ability of NAC to blunt cue-induced heroin relapse, an important alternative interpretation is that NAC may be protective with respect to heroin-induced reductions in PrL-NAcore dendritic spine density as opposed to our original hypothesis that NAC functions via synaptogenesis. However, additional experimentation is required to determine this distinction, and this research represents an important future direction of the present experiments.

It is possible that the heroin-induced decrease in spine density could be linked to dysfunctional gliotransmission following extinction, and that NAC's ability to drive spinogenesis could be linked to the normalization of such dysfunction. For example, repeated morphine administration in mice downregulates TSP expression in the cortex 24 h after the final injection, as well as in cultured astrocytes following repeated morphine treatment [79]. One mechanism by which heroin may impact TSP expression is through increased oxidative stress, as evidenced by the induction of oxidative stress in humans following chronic heroin exposure [80] and also in cultured cortical neurons [81,82]. Moreover, cultured primary astrocytes reduce TSP-1 gene expression under increased oxidative stress conditions, which is prevented by NAC [83]. With these considerations in mind, we were presently surprised that the delivery of shTSP-2 to PrL astrocytes did not block the ability of NAC to reduce both the extinction and cued-relapse to heroin, suggesting that their effects are mediated by independent mechanisms. It could be the case, however, that heroin exposure uniquely engages TSP-1 versus TSP-2. Echoing this hypothesis, it has previously been reported that cocaine administration elevates the presence of silent synapses in both wild-type and TSP-1 knockout mice, though cocaine administration was not shown to increase silent synapse formation in TSP-2 knockout mice [30].

Given that one well-documented mechanism by which NAC counteracts oxidative stress is increased glutathione synthesis [84], it is possible that heroin-induced adaptations in TSP-1 expression due to increased oxidative stress, and normalization of this by NAC, may be linked to the altered dendritic spine morphometric features observed herein. Overall, it is likely that NAC's ability to normalize heroin-induced adaptations in astrocyte–synapse proximity and GFAP cytoskeletal architecture has a direct influence on structural plasticity in PrL-NAcore neurons, perhaps through indirect or direct mechanisms at neurons. Thus, the degree to which PAPs associate with dendritic spines of PrL-NAcore neurons, whether this is altered following the extinction of heroin SA, the involvement of TSP-1 versus TSP-2, and whether NAC has an impact are areas of future investigation.

NAC has previously been demonstrated to regulate glutamate homeostasis, which has been consistently shown to be disrupted following drug self-administration [15–17,24]. Indeed, the normalization of glutamate transmission and related adaptations in astrocyte structural plasticity have been shown to be centrally important in preventing relapse to drugs of abuse. As an example, treatment with other compounds that restore glutamate homeostasis and astrocyte interaction with synapses such as ceftriaxone have shown the capacity to block cued-relapse to cocaine [10]. Overall, it is likely that NAC's ability to normalize heroin-induced adaptations in astrocyte–synapse proximity and GFAP cytoskeletal architecture have a direct influence on plasticity in PrL-NAcore neurons, perhaps through indirect or direct mechanisms at neurons mediated by extracellular glutamate concentrations and/or the levels of physical interaction of astrocytes with neurons. Accordingly, the degree to which heroin impacts PAPs' association with dendritic spines of PrL-NAcore neurons, whether this is altered following the extinction of heroin SA, the involvement of astrocyte signaling mechanisms, and whether NAC treatment reverses drug-induced adaptations in this pathway remain future directions for investigation.

4.7. Limitations and Methodological Considerations

The most obvious limitation to the present set of experiments is the omission of female rodents. These series of studies were initially planned as a follow-up to existing literature surrounding astrocyte–neuron interactions in the NAcore [12]. Because these initial experiments only used male animals, we omitted females from the present set of experiments as these earlier findings guided a priori power investigations and study planning. We fully recognize the importance and implications of sex differences in addiction, and plan to address this shortcoming by evaluating female rodents in identical experimental parameters.

Another limitation of the present findings is that they are largely correlative in nature and our present experimental manipulations do not allow us to infer causality between heroin adaptations to neurons and astrocytes, cued-relapse to heroin, and the modulatory role that NAC plays in this process. Future experiments will involve targeted manipulations of these specific features in an attempt to establish causal relationships between the relationships discussed.

While we show that the evoked astrocytic GLT-1-dependent current can be a predictive measurement of prior heroin exposure, it is unclear as to the functional ramifications this effect has owing to the relatively marginal differences observed between the evoked GLT-1 dependent current between YS- and heroin-treated animals. Moreover, as discussed above it is unclear if any such alterations play a role in cued-relapse to heroin.

Another limitation of the present experiments was the lack of YS control animals treated with NAC. In the current study, we chose to omit YS animals injected with NAC for several reasons. Most importantly, NAC has been shown to have no effect on NAcore glutamate concentrations or cystine uptake through Xc- in control animals [48]. Moreover NAC treatment normalized increased glutamate concentrations in the anterior cingulate of cocaine-addicted patients but had no effect in healthy controls [85].

Canonically, GFAP is thought to only be present in the major branches and endfeet of astrocytes and not to be present in all non-reactive astrocytes in adult healthy CNS tissue [86]. However, our data are potentially at odds with this view as we found consistent fields of astrocytes in all three experimental groups and did not observe gaps that could potentially represent GFAP-negative astrocytes. As described above, GFAP does not extend into the finest membranous processes, which are the primary domains of astrocyte-neuron interactions, the motility of which are regulated by actin cycling [87]. The ability to use membrane-targeted expression of EGFP driven by astrocyte-specific promoters in conjunction with the immunohistochemical detection of synaptic proteins has gained traction as an assay to determine the proximity of PAPs to synapses (see [53,54] for reviews) Moreover, GFAP comprises ~15% of total astrocyte volume, which has historically led to incorrect conclusions regarding the domains of individual astrocytes, which further suggests the necessity of this approach in the accurate estimation of the astrocyte–neuron contact [61]. In this study, we used two different synaptic markers, Synapsin I and GluA2 which label generic pre-synaptic terminals (Synapsin I) and excitatory post-synaptic AMPA receptors (GluA2), respectively. However, it is important to note that, with an approximate minimum resolution of 150 nm, we are not attempting to distinguish between pre- and post-synaptic elements, separated by an approximate distance of 10 nm [88]. We are merely using the IHC detection of these proteins as a means to label synapses. However, the fact that both antibodies yielded the increased co-registration of synapses with astrocytic membranes in heroin animals can be viewed as a means to exclude epiphenomenological or potentially off-target effects from each individual antibody.

Finally, in our shTSP2 experiments, we utilized the construct provided by the Dong Lab. Validation of TSP2 knockdown was performed by the Dong Lab in HEK293 cells and reduced cocaine-mediated synaptogenesis in vivo. [30]. Quantitative analysis of the magnitude of TSP2 knockdown would require alternative techniques such as RNAScope for TSP2 and an astrocyte marker, which is a limitation of our study. In addition, our experiment utilized the AAV5. *GfaABC1D*::Lck-EGFP virus as a negative control as opposed to an off-target shRNA construct which is also a limitation of the present study.

5. Conclusions

In conclusion, our data indicate that heroin SA and extinction elicit hypertrophic complexity and synapse association of astrocytes in the PrL cortex that are associated with altered dendritic spine morphological features in PrL-NAcore neurons. Our data fill a much-needed gap in the field by focusing on the upstream portion of the canonical relapse circuit to accompany what has been observed in the NAcore. Moreover, these findings point to an additional node of this circuitry whereby NAC can modify heroin-induced adaptations in the PrL cortex in addition to its well-defined modification of NAcore synaptic pathology. Ultimately, astrocytic and neuronal adaptations in the PrL cortex may be a target for informing future drug discovery for relapse prevention pharmacotherapies.

Supplementary Materials: The following supporting information can be downloaded at: https://www.mdpi.com/article/10.3390/cells12141812/s1, Table S1: Description of viral constructs.

Author Contributions: Conceptualization, M.D.S. and J.F.M.; methodology, B.M.S., A.R.D., J.P.-C. and T.C.J.; validation, B.M.S., A.R.D., J.F.M. and M.D.S.; formal analysis, B.M.S. and A.R.D.; investigation, B.M.S., A.R.D., J.P.-C., K.N.H., E.A.C., M.E.P., A.G.B., A.M.W., M.-N.L. and J.A.M.; resources, J.F.M. and M.D.S.; data curation, B.M.S., A.R.D. and J.P.-C.; writing—original draft preparation, B.M.S., A.R.D., J.P.-C., J.FM. and M.D.S.; writing—review and editing, B.M.S., A.R.D., J.F.M. and M.D.S.; visualization, B.M.S., A.R.D., J.P.-C. and M.D.S.; supervision, J.F.M. and M.D.S.; project administration, M.D.S.; funding acquisition, J.F.M. and M.D.S. All authors have read and agreed to the published version of the manuscript.

Funding: This was work was supported by RO1 DA054154, and MUSC TTS Award (M.D.S.), F32 DA050427 (B.M.S.), T32 DA007288 (J.F.M., B.M.S., A.R.D.), K01 DA053434 (J.P.-C.), R01 DA037327 (T.C.J.), U01 DA 044468 (T.C.J.), P50 DA046373 (J.F.M., M.D.S.), and R25 DA033680-11 (E.A.C., M.E.P.).

Institutional Review Board Statement: The animal study protocol was approved by the Institutional Review Board of the Medical University of South Carolina (protocol code 01081).

Informed Consent Statement: Not applicable.

Data Availability Statement: Not applicable.

Acknowledgments: The authors thank Giuseppe Giannotti for the excellent technical assistance and Yan Dong and Oliver M Schlüter for providing the astrocyte TSP2 RNAi construct and virus. The authors thank Antonieta Lavin and the MUSC Summer Undergraduate Research Program for the support of student coauthors.

Conflicts of Interest: The authors declare no conflict of interest.

References

1. Chen, L.; Li, N.; Ge, S.; Lozano, A.M.; Lee, D.J.; Yang, C.; Li, L.; Bai, Q.; Lu, H.; Wang, J.; et al. Long-term results after deep brain stimulation of nucleus accumbens and the anterior limb of the internal capsule for preventing heroin relapse: An open-label pilot study. *Brain Stimul.* **2019**, *12*, 175–183. [CrossRef] [PubMed]
2. Li, Q.; Wang, Y.; Zhang, Y.; Li, W.; Yang, W.; Zhu, J.; Wu, N.; Chang, H.; Zheng, Y.; Qin, W.; et al. Craving correlates with mesolimbic responses to heroin-related cues in short-term abstinence from heroin: An event-related fMRI study. *Brain Res.* **2012**, *1469*, 63–72. [CrossRef]
3. Lamontagne, S.J.; Olmstead, M.C. Animal models in addiction research: A dimensional approach. *Neurosci. Biobehav. Rev.* **2019**, *106*, 91–101. [CrossRef] [PubMed]
4. Hearing, M. Prefrontal-accumbens opioid plasticity: Implications for relapse and dependence. *Pharmacol. Res.* **2019**, *139*, 158–165. [CrossRef]

5. Hamel, L.; Thangarasa, T.; Samadi, O.; Ito, R. Caudal Nucleus Accumbens Core Is Critical in the Regulation of Cue-Elicited Approach-Avoidance Decisions. *eNeuro* **2017**, *4*, ENEURO.0330-16.2017. [CrossRef] [PubMed]
6. Bossert, J.M.; Ghitza, U.E.; Lu, L.; Epstein, D.H.; Shaham, Y. Neurobiology of relapse to heroin and cocaine seeking: An update and clinical implications. *Eur. J. Pharmacol.* **2005**, *526*, 36–50. [CrossRef]
7. Russo, S.J.; Dietz, D.M.; Dumitriu, D.; Morrison, J.H.; Malenka, R.C.; Nestler, E.J. The addicted synapse: Mechanisms of synaptic and structural plasticity in nucleus accumbens. *Trends Neurosci.* **2010**, *33*, 267–276. [CrossRef]
8. Shen, H.; Kalivas, P.W. Reduced LTP and LTD in prefrontal cortex synapses in the nucleus accumbens after heroin self-administration. *Int. J. Neuropsychopharmacol.* **2013**, *16*, 1165–1167. [CrossRef]
9. Moussawi, K.; Pacchioni, A.; Moran, M.; Olive, M.F.; Gass, J.T.; Lavin, A.; Kalivas, P.W. N-Acetylcysteine reverses cocaine-induced metaplasticity. *Nat. Neurosci.* **2009**, *12*, 182–189. [CrossRef]
10. Knackstedt, L.A.; Melendez, R.I.; Kalivas, P.W. Ceftriaxone restores glutamate homeostasis and prevents relapse to cocaine seeking. *Biol. Psychiatry* **2010**, *67*, 81–84. [CrossRef]
11. Shen, H.W.; Scofield, M.D.; Boger, H.; Hensley, M.; Kalivas, P.W. Synaptic glutamate spillover due to impaired glutamate uptake mediates heroin relapse. *J. Neurosci.* **2014**, *34*, 5649–5657. [CrossRef]
12. Kruyer, A.; Scofield, M.D.; Wood, D.; Reissner, K.J.; Kalivas, P.W. Heroin Cue-Evoked Astrocytic Structural Plasticity at Nucleus Accumbens Synapses Inhibits Heroin Seeking. *Biol. Psychiatry* **2019**, *86*, 811–819. [CrossRef] [PubMed]
13. Scofield, M.D.; Li, H.; Siemsen, B.M.; Healey, K.L.; Tran, P.K.; Woronoff, N.; Boger, H.A.; Kalivas, P.W.; Reissner, K.J. Cocaine Self-Administration and Extinction Leads to Reduced Glial Fibrillary Acidic Protein Expression and Morphometric Features of Astrocytes in the Nucleus Accumbens Core. *Biol. Psychiatry* **2016**, *80*, 207–215. [CrossRef]
14. Siemsen, B.M.; Reichel, C.M.; Leong, K.C.; Garcia-Keller, C.; Gipson, C.D.; Spencer, S.; McFaddin, J.A.; Hooker, K.N.; Kalivas, P.W.; Scofield, M.D. Effects of Methamphetamine Self-Administration and Extinction on Astrocyte Structure and Function in the Nucleus Accumbens Core. *Neuroscience* **2019**, *406*, 528–541. [CrossRef]
15. Scofield, M.D.; Kalivas, P.W. Astrocytic dysfunction and addiction: Consequences of impaired glutamate homeostasis. *Neuroscientist* **2014**, *20*, 610–622. [CrossRef] [PubMed]
16. LaLumiere, R.T.; Kalivas, P.W. Glutamate release in the nucleus accumbens core is necessary for heroin seeking. *J. Neurosci.* **2008**, *28*, 3170–3177. [CrossRef]
17. Kalivas, P.W.; Volkow, N.D. New medications for drug addiction hiding in glutamatergic neuroplasticity. *Mol. Psychiatry* **2011**, *16*, 974–986. [CrossRef]
18. Roberts-Wolfe, D.J.; Kalivas, P.W. Glutamate Transporter GLT-1 as a Therapeutic Target for Substance Use Disorders. *CNS Neurol. Disord. Drug Targets* **2015**, *14*, 745–756. [CrossRef] [PubMed]
19. Zhou, W.; Kalivas, P.W. N-acetylcysteine reduces extinction responding and induces enduring reductions in cue- and heroin-induced drug-seeking. *Biol. Psychiatry* **2008**, *63*, 338–340. [CrossRef]
20. Hodebourg, R.; Murray, J.E.; Fouyssac, M.; Puaud, M.; Everitt, B.J.; Belin, D. Heroin seeking becomes dependent on dorsal striatal dopaminergic mechanisms and can be decreased by N-acetylcysteine. *Eur. J. Neurosci.* **2019**, *50*, 2036–2044. [CrossRef]
21. LaRowe, S.D.; Kalivas, P.W. The Role of N-Acetylcysteine in Inhibiting Responding During Extinction in Rats Trained to Self-Administer Cocaine. *Open. Addict. J.* **2010**, *3*, 88–91. [CrossRef] [PubMed]
22. Kupchik, Y.M.; Moussawi, K.; Tang, X.C.; Wang, X.; Kalivas, B.C.; Kolokithas, R.; Ogburn, K.B.; Kalivas, P.W. The effect of N-acetylcysteine in the nucleus accumbens on neurotransmission and relapse to cocaine. *Biol. Psychiatry* **2012**, *71*, 978–986. [CrossRef] [PubMed]
23. Powell, G.L.; Leyrer-Jackson, J.M.; Goenaga, J.; Namba, M.D.; Pina, J.; Spencer, S.; Stankeviciute, N.; Schwartz, D.; Allen, N.P.; Del Franco, A.P.; et al. Chronic treatment with N-acetylcysteine decreases extinction responding and reduces cue-induced nicotine-seeking. *Physiol. Rep.* **2019**, *7*, e13958. [CrossRef] [PubMed]
24. Reissner, K.J.; Gipson, C.D.; Tran, P.K.; Knackstedt, L.A.; Scofield, M.D.; Kalivas, P.W. Glutamate transporter GLT-1 mediates N-acetylcysteine inhibition of cocaine reinstatement. *Addict. Biol.* **2015**, *20*, 316–323. [CrossRef]
25. Gomes, F.C.; Paulin, D.; Moura Neto, V. Glial fibrillary acidic protein (GFAP): Modulation by growth factors and its implication in astrocyte differentiation. *Braz. J. Med. Biol. Res.* **1999**, *32*, 619–631. [CrossRef]
26. Garrido, E.; Perez-Garcia, C.; Alguacil, L.F.; Diez-Fernandez, C. The alpha2-adrenoceptor antagonist yohimbine reduces glial fibrillary acidic protein upregulation induced by chronic morphine administration. *Neurosci. Lett.* **2005**, *383*, 141–144. [CrossRef]
27. Stiene-Martin, A.; Gurwell, J.A.; Hauser, K.F. Morphine alters astrocyte growth in primary cultures of mouse glial cells: Evidence for a direct effect of opiates on neural maturation. *Brain Res. Dev. Brain Res.* **1991**, *60*, 1–7. [CrossRef]
28. Shigetomi, E.; Kracun, S.; Sofroniew, M.V.; Khakh, B.S. A genetically targeted optical sensor to monitor calcium signals in astrocyte processes. *Nat. Neurosci.* **2010**, *13*, 759–766. [CrossRef]
29. Viswanathan, S.; Williams, M.E.; Bloss, E.B.; Stasevich, T.J.; Speer, C.M.; Nern, A.; Pfeiffer, B.D.; Hooks, B.M.; Li, W.P.; English, B.P.; et al. High-performance probes for light and electron microscopy. *Nat. Methods* **2015**, *12*, 568–576. [CrossRef] [PubMed]
30. Wang, J.; Li, K.L.; Shukla, A.; Beroun, A.; Ishikawa, M.; Huang, X.; Wang, Y.; Wang, Y.Q.; Yang, Y.; Bastola, N.D.; et al. Cocaine Triggers Astrocyte-Mediated Synaptogenesis. *Biol. Psychiatry* **2021**, *89*, 386–397. [CrossRef]
31. Siemsen, B.M.; Giannotti, G.; McFaddin, J.A.; Scofield, M.D.; McGinty, J.F. Biphasic effect of abstinence duration following cocaine self-administration on spine morphology and plasticity-related proteins in prelimbic cortical neurons projecting to the nucleus accumbens core. *Brain Struct. Funct.* **2019**, *224*, 741–758. [CrossRef] [PubMed]

32. Costes, S.V.; Daelemans, D.; Cho, E.H.; Dobbin, Z.; Pavlakis, G.; Lockett, S. Automatic and quantitative measurement of protein-protein colocalization in live cells. *Biophys. J.* **2004**, *86*, 3993–4003. [CrossRef] [PubMed]
33. Testen, A.; Ali, M.; Sexton, H.G.; Hodges, S.; Dubester, K.; Reissner, K.J.; Swartzwelder, H.S.; Risher, M.L. Region-Specific Differences in Morphometric Features and Synaptic Colocalization of Astrocytes During Development. *Neuroscience* **2019**, *400*, 98–109. [CrossRef]
34. Testen, A.; Sepulveda-Orengo, M.T.; Gaines, C.H.; Reissner, K.J. Region-Specific Reductions in Morphometric Properties and Synaptic Colocalization of Astrocytes Following Cocaine Self-Administration and Extinction. *Front. Cell. Neurosci.* **2018**, *12*, 246. [CrossRef]
35. Robinson, T.E.; Gorny, G.; Savage, V.R.; Kolb, B. Widespread but regionally specific effects of experimenter- versus self-administered morphine on dendritic spines in the nucleus accumbens, hippocampus, and neocortex of adult rats. *Synapse* **2002**, *46*, 271–279. [CrossRef]
36. John Lin, C.C.; Yu, K.; Hatcher, A.; Huang, T.W.; Lee, H.K.; Carlson, J.; Weston, M.C.; Chen, F.; Zhang, Y.; Zhu, W.; et al. Identification of diverse astrocyte populations and their malignant analogs. *Nat. Neurosci.* **2017**, *20*, 396–405. [CrossRef]
37. Hoft, S.; Griemsmann, S.; Seifert, G.; Steinhauser, C. Heterogeneity in expression of functional ionotropic glutamate and GABA receptors in astrocytes across brain regions: Insights from the thalamus. *Philos. Trans. R. Soc. Lond. B Biol. Sci.* **2014**, *369*, 20130602. [CrossRef] [PubMed]
38. Liddelow, S.A.; Guttenplan, K.A.; Clarke, L.E.; Bennett, F.C.; Bohlen, C.J.; Schirmer, L.; Bennett, M.L.; Munch, A.E.; Chung, W.S.; Peterson, T.C.; et al. Neurotoxic reactive astrocytes are induced by activated microglia. *Nature* **2017**, *541*, 481–487. [CrossRef]
39. LaRowe, S.D.; Kalivas, P.W.; Nicholas, J.S.; Randall, P.K.; Mardikian, P.N.; Malcolm, R.J. A double-blind placebo-controlled trial of N-acetylcysteine in the treatment of cocaine dependence. *Am. J. Addict.* **2013**, *22*, 443–452. [CrossRef]
40. Nocito Echevarria, M.A.; Andrade Reis, T.; Ruffo Capatti, G.; Siciliano Soares, V.; da Silveira, D.X.; Fidalgo, T.M. N-acetylcysteine for treating cocaine addiction—A systematic review. *Psychiatry Res.* **2017**, *251*, 197–203. [CrossRef]
41. Mardikian, P.N.; LaRowe, S.D.; Hedden, S.; Kalivas, P.W.; Malcolm, R.J. An open-label trial of N-acetylcysteine for the treatment of cocaine dependence: A pilot study. *Prog. Neuropsychopharmacol. Biol. Psychiatry* **2007**, *31*, 389–394. [CrossRef]
42. Schmidt-Kastner, R.; Wietasch, K.; Weigel, H.; Eysel, U.T. Immunohistochemical staining for glial fibrillary acidic protein (GFAP) after deafferentation or ischemic infarction in rat visual system: Features of reactive and damaged astrocytes. *Int. J. Dev. Neurosci.* **1993**, *11*, 157–174. [CrossRef] [PubMed]
43. Ruzicka, B.B.; Fox, C.A.; Thompson, R.C.; Meng, F.; Watson, S.J.; Akil, H. Primary astroglial cultures derived from several rat brain regions differentially express mu, delta and kappa opioid receptor mRNA. *Brain Res. Mol. Brain Res.* **1995**, *34*, 209–220. [CrossRef]
44. Kelly, E. Efficacy and ligand bias at the mu-opioid receptor. *Br. J. Pharmacol.* **2013**, *169*, 1430–1446. [CrossRef]
45. Hauser, K.F.; Stiene-Martin, A.; Mattson, M.P.; Elde, R.P.; Ryan, S.E.; Godleske, C.C. mu-Opioid receptor-induced Ca^{2+} mobilization and astroglial development: Morphine inhibits DNA synthesis and stimulates cellular hypertrophy through a Ca(2+)-dependent mechanism. *Brain Res.* **1996**, *720*, 191–203. [CrossRef] [PubMed]
46. Stiene-Martin, A.; Mattson, M.P.; Hauser, K.F. Opiates selectively increase intracellular calcium in developing type-1 astrocytes: Role of calcium in morphine-induced morphologic differentiation. *Brain Res. Dev. Brain Res.* **1993**, *76*, 189–196. [CrossRef]
47. Nam, M.H.; Han, K.S.; Lee, J.; Won, W.; Koh, W.; Bae, J.Y.; Woo, J.; Kim, J.; Kwong, E.; Choi, T.Y.; et al. Activation of Astrocytic mu-Opioid Receptor Causes Conditioned Place Preference. *Cell Rep.* **2019**, *28*, 1154–1166. [CrossRef] [PubMed]
48. Madayag, A.; Lobner, D.; Kau, K.S.; Mantsch, J.R.; Abdulhameed, O.; Hearing, M.; Grier, M.D.; Baker, D.A. Repeated N-acetylcysteine administration alters plasticity-dependent effects of cocaine. *J. Neurosci.* **2007**, *27*, 13968–13976. [CrossRef]
49. Kau, K.S.; Madayag, A.; Mantsch, J.R.; Grier, M.D.; Abdulhameed, O.; Baker, D.A. Blunted cystine-glutamate antiporter function in the nucleus accumbens promotes cocaine-induced drug seeking. *Neuroscience* **2008**, *155*, 530–537. [CrossRef]
50. Clavreul, S.; Abdeladim, L.; Hernandez-Garzon, E.; Niculescu, D.; Durand, J.; Ieng, S.H.; Barry, R.; Bonvento, G.; Beaurepaire, E.; Livet, J.; et al. Cortical astrocytes develop in a plastic manner at both clonal and cellular levels. *Nat. Commun.* **2019**, *10*, 4884. [CrossRef]
51. Stogsdill, J.A.; Ramirez, J.; Liu, D.; Kim, Y.H.; Baldwin, K.T.; Enustun, E.; Ejikeme, T.; Ji, R.R.; Eroglu, C. Astrocytic neuroligins control astrocyte morphogenesis and synaptogenesis. *Nature* **2017**, *551*, 192–197. [CrossRef] [PubMed]
52. Sullivan, R.; Rauen, T.; Fischer, F.; Wiessner, M.; Grewer, C.; Bicho, A.; Pow, D.V. Cloning, transport properties, and differential localization of two splice variants of GLT-1 in the rat CNS: Implications for CNS glutamate homeostasis. *Glia* **2004**, *45*, 155–169. [CrossRef]
53. Testen, A.; Kim, R.; Reissner, K.J. High-Resolution Three-Dimensional Imaging of Individual Astrocytes Using Confocal Microscopy. *Curr. Protoc. Neurosci.* **2020**, *91*, e92. [CrossRef] [PubMed]
54. Scofield, M.D. Exploring the Role of Astroglial Glutamate Release and Association With Synapses in Neuronal Function and Behavior. *Biol. Psychiatry* **2018**, *84*, 778–786. [CrossRef]
55. Scofield, M.D.; Heinsbroek, J.A.; Gipson, C.D.; Kupchik, Y.M.; Spencer, S.; Smith, A.C.; Roberts-Wolfe, D.; Kalivas, P.W. The Nucleus Accumbens: Mechanisms of Addiction across Drug Classes Reflect the Importance of Glutamate Homeostasis. *Pharmacol. Rev.* **2016**, *68*, 816–871. [CrossRef]
56. Kim, R.; Sepulveda-Orengo, M.T.; Healey, K.L.; Williams, E.A.; Reissner, K.J. Regulation of glutamate transporter 1 (GLT-1) gene expression by cocaine self-administration and withdrawal. *Neuropharmacology* **2018**, *128*, 1–10. [CrossRef] [PubMed]

57. Fattore, L.; Puddu, M.C.; Picciau, S.; Cappai, A.; Fratta, W.; Serra, G.P.; Spiga, S. Astroglial in vivo response to cocaine in mouse dentate gyrus: A quantitative and qualitative analysis by confocal microscopy. *Neuroscience* **2002**, *110*, 1–6. [CrossRef]
58. SheikhBahaei, S.; Morris, B.; Collina, J.; Anjum, S.; Znati, S.; Gamarra, J.; Zhang, R.; Gourine, A.V.; Smith, J.C. Morphometric analysis of astrocytes in brainstem respiratory regions. *J. Comp. Neurol.* **2018**, *526*, 2032–2047. [CrossRef]
59. Zhang, S.; Wu, M.; Peng, C.; Zhao, G.; Gu, R. GFAP expression in injured astrocytes in rats. *Exp. Ther. Med.* **2017**, *14*, 1905–1908. [CrossRef]
60. Haseleu, J.; Anlauf, E.; Blaess, S.; Endl, E.; Derouiche, A. Studying subcellular detail in fixed astrocytes: Dissociation of morphologically intact glial cells (DIMIGs). *Front. Cell. Neurosci.* **2013**, *7*, 54. [CrossRef]
61. Bushong, E.A.; Martone, M.E.; Jones, Y.Z.; Ellisman, M.H. Protoplasmic astrocytes in CA1 stratum radiatum occupy separate anatomical domains. *J. Neurosci.* **2002**, *22*, 183–192. [CrossRef] [PubMed]
62. Allen, N.J.; Eroglu, C. Cell Biology of Astrocyte-Synapse Interactions. *Neuron* **2017**, *96*, 697–708. [CrossRef]
63. Bernardinelli, Y.; Randall, J.; Janett, E.; Nikonenko, I.; Konig, S.; Jones, E.V.; Flores, C.E.; Murai, K.K.; Bochet, C.G.; Holtmaat, A.; et al. Activity-dependent structural plasticity of perisynaptic astrocytic domains promotes excitatory synapse stability. *Curr. Biol.* **2014**, *24*, 1679–1688. [CrossRef]
64. Genoud, C.; Quairiaux, C.; Steiner, P.; Hirling, H.; Welker, E.; Knott, G.W. Plasticity of astrocytic coverage and glutamate transporter expression in adult mouse cortex. *PLoS Biol.* **2006**, *4*, e343. [CrossRef] [PubMed]
65. Pannasch, U.; Freche, D.; Dallerac, G.; Ghezali, G.; Escartin, C.; Ezan, P.; Cohen-Salmon, M.; Benchenane, K.; Abudara, V.; Dufour, A.; et al. Connexin 30 sets synaptic strength by controlling astroglial synapse invasion. *Nat. Neurosci.* **2014**, *17*, 549–558. [CrossRef] [PubMed]
66. Tso, M.C.; Herzog, E.D. Was Cajal right about sleep? *BMC Biol.* **2015**, *13*, 67. [CrossRef] [PubMed]
67. Liao, D.; Lin, H.; Law, P.Y.; Loh, H.H. Mu-opioid receptors modulate the stability of dendritic spines. *Proc. Natl. Acad. Sci. USA* **2005**, *102*, 1725–1730. [CrossRef]
68. Gipson, C.D.; Kupchik, Y.M.; Shen, H.; Reissner, K.J.; Thomas, C.A.; Kalivas, P.W. Relapse induced by cues predicting cocaine depends on rapid, transient synaptic potentiation. *Neuron* **2013**, *77*, 867–872. [CrossRef] [PubMed]
69. Shi, S.H.; Hayashi, Y.; Petralia, R.S.; Zaman, S.H.; Wenthold, R.J.; Svoboda, K.; Malinow, R. Rapid spine delivery and redistribution of AMPA receptors after synaptic NMDA receptor activation. *Science* **1999**, *284*, 1811–1816. [CrossRef] [PubMed]
70. Matsuzaki, M.; Ellis-Davies, G.C.; Nemoto, T.; Miyashita, Y.; Iino, M.; Kasai, H. Dendritic spine geometry is critical for AMPA receptor expression in hippocampal CA1 pyramidal neurons. *Nat. Neurosci.* **2001**, *4*, 1086–1092. [CrossRef] [PubMed]
71. Matsuzaki, M.; Honkura, N.; Ellis-Davies, G.C.; Kasai, H. Structural basis of long-term potentiation in single dendritic spines. *Nature* **2004**, *429*, 761–766. [CrossRef] [PubMed]
72. Lambe, E.K.; Aghajanian, G.K. Hypocretin (orexin) induces calcium transients in single spines postsynaptic to identified thalamocortical boutons in prefrontal slice. *Neuron* **2003**, *40*, 139–150. [CrossRef] [PubMed]
73. Paspalas, C.D.; Wang, M.; Arnsten, A.F. Constellation of HCN channels and cAMP regulating proteins in dendritic spines of the primate prefrontal cortex: Potential substrate for working memory deficits in schizophrenia. *Cereb. Cortex* **2013**, *23*, 1643–1654. [CrossRef]
74. Kuroda, M.; Murakami, K.; Igarashi, H.; Okada, A. The convergence of axon terminals from the mediodorsal thalamic nucleus and ventral tegmental area on pyramidal cells in layer V of the rat prelimbic cortex. *Eur. J. Neurosci.* **1996**, *8*, 1340–1349. [CrossRef] [PubMed]
75. Parpura, V.; Zorec, R. Gliotransmission: Exocytotic release from astrocytes. *Brain Res. Rev.* **2010**, *63*, 83–92. [CrossRef]
76. Santello, M.; Cali, C.; Bezzi, P. Gliotransmission and the tripartite synapse. *Adv. Exp. Med. Biol.* **2012**, *970*, 307–331. [CrossRef]
77. Risher, W.C.; Kim, N.; Koh, S.; Choi, J.E.; Mitev, P.; Spence, E.F.; Pilaz, L.J.; Wang, D.; Feng, G.; Silver, D.L.; et al. Thrombospondin receptor alpha2delta-1 promotes synaptogenesis and spinogenesis via postsynaptic Rac1. *J. Cell Biol.* **2018**, *217*, 3747–3765. [CrossRef]
78. Garcia, O.; Torres, M.; Helguera, P.; Coskun, P.; Busciglio, J. A role for thrombospondin-1 deficits in astrocyte-mediated spine and synaptic pathology in Down's syndrome. *PLoS ONE* **2010**, *5*, e14200. [CrossRef]
79. Ikeda, H.; Miyatake, M.; Koshikawa, N.; Ochiai, K.; Yamada, K.; Kiss, A.; Donlin, M.J.; Panneton, W.M.; Churchill, J.D.; Green, M.; et al. Morphine modulation of thrombospondin levels in astrocytes and its implications for neurite outgrowth and synapse formation. *J. Biol. Chem.* **2010**, *285*, 38415–38427. [CrossRef]
80. Pereska, Z.; Dejanova, B.; Bozinovska, C.; Petkovska, L. Prooxidative/antioxidative homeostasis in heroin addiction and detoxification. *Bratisl. Lek. Listy* **2007**, *108*, 393–398.
81. Cunha-Oliveira, T.; Rego, A.C.; Garrido, J.; Borges, F.; Macedo, T.; Oliveira, C.R. Street heroin induces mitochondrial dysfunction and apoptosis in rat cortical neurons. *J. Neurochem.* **2007**, *101*, 543–554. [CrossRef]
82. Cunha-Oliveira, T.; Rego, A.C.; Garrido, J.; Borges, F.; Macedo, T.; Oliveira, C.R. Neurotoxicity of heroin-cocaine combinations in rat cortical neurons. *Toxicology* **2010**, *276*, 11–17. [CrossRef]
83. Chen, J.K.; Zhan, Y.J.; Yang, C.S.; Tzeng, S.F. Oxidative stress-induced attenuation of thrombospondin-1 expression in primary rat astrocytes. *J. Cell Biochem.* **2011**, *112*, 59–70. [CrossRef] [PubMed]
84. Flanagan, R.J.; Meredith, T.J. Use of N-acetylcysteine in clinical toxicology. *Am. J. Med.* **1991**, *30*, 5131–5139. [CrossRef] [PubMed]

85. Schmaal, L.; Veltman, D.J.; Nederveen, A.; van den Brink, W.; Goudriaan, A.E. N-acetylcysteine normalizes glutamate levels in cocaine-dependent patients: A randomized crossover magnetic resonance spectroscopy study. *Neuropsychopharmacology* **2012**, *37*, 2143–2152. [CrossRef]
86. Sofroniew, M.V.; Vinters, H.V. Astrocytes: Biology and pathology. *Acta Neuropathol.* **2010**, *119*, 7–35. [CrossRef]
87. Haber, M.; Zhou, L.; Murai, K.K. Cooperative astrocyte and dendritic spine dynamics at hippocampal excitatory synapses. *J. Neurosci.* **2006**, *26*, 8881–8891. [CrossRef] [PubMed]
88. Zuber, B.; Nikonenko, I.; Klauser, P.; Muller, D.; Dubochet, J. The mammalian central nervous synaptic cleft contains a high density of periodically organized complexes. *Proc. Natl. Acad. Sci. USA* **2005**, *102*, 19192–19197. [CrossRef]

cells

MDPI

Article

Loss of Astrocytic μ Opioid Receptors Exacerbates Aversion Associated with Morphine Withdrawal in Mice: Role of Mitochondrial Respiration

Kateryna Murlanova [1], Yan Jouroukhin [1], Ksenia Novototskaya-Vlasova [1], Shovgi Huseynov [1], Olga Pletnikova [2,3], Michael J. Morales [1], Yun Guan [4,5], Atsushi Kamiya [6], Dwight E. Bergles [7], David M. Dietz [8] and Mikhail V. Pletnikov [1,6,7,*]

1 Department of Physiology and Biophysics, Jacobs School of Medicine and Biomedical Sciences, University at Buffalo, Buffalo, NY 14203, USA
2 Department of Pathology and Anatomical Sciences, Jacobs School of Medicine and Biomedical Sciences, University at Buffalo, Buffalo, NY 14203, USA
3 Department of Pathology, Johns Hopkins University School of Medicine, Baltimore, MD 21205, USA
4 Department of Anesthesiology and Critical Care Medicine, Johns Hopkins University School of Medicine, Baltimore, MD 21205, USA
5 Department of Neurological Surgery, Johns Hopkins University School of Medicine, Baltimore, MD 21205, USA
6 Department of Psychiatry and Behavioral Sciences, Johns Hopkins University School of Medicine, Baltimore, MD 21287, USA
7 Solomon H. Snyder Department of Neuroscience, Johns Hopkins University School of Medicine, Baltimore, MD 21205, USA
8 Department of Pharmacology and Toxicology, Jacobs School of Medicine and Biomedical Sciences, University at Buffalo, Buffalo, NY 14203, USA
* Correspondence: mvpletni@buffalo.edu; Tel.: +1-716-829-2189

Abstract: Astrocytes express mu/μ opioid receptors, but the function of these receptors remains poorly understood. We evaluated the effects of astrocyte-restricted knockout of μ opioid receptors on reward- and aversion-associated behaviors in mice chronically exposed to morphine. Specifically, one of the floxed alleles of the *Oprm1* gene encoding μ opioid receptor 1 was selectively deleted from brain astrocytes in *Oprm1* inducible conditional knockout (icKO) mice. These mice did not exhibit changes in locomotor activity, anxiety, or novel object recognition, or in their responses to the acute analgesic effects of morphine. *Oprm1* icKO mice displayed increased locomotor activity in response to acute morphine administration but unaltered locomotor sensitization. *Oprm1* icKO mice showed normal morphine-induced conditioned place preference but exhibited stronger conditioned place aversion associated with naloxone-precipitated morphine withdrawal. Notably, elevated conditioned place aversion lasted up to 6 weeks in *Oprm1* icKO mice. Astrocytes isolated from the brains of *Oprm1* icKO mice had unchanged levels of glycolysis but had elevated oxidative phosphorylation. The basal augmentation of oxidative phosphorylation in *Oprm1* icKO mice was further exacerbated by naloxone-precipitated withdrawal from morphine and, similar to that for conditioned place aversion, was still present 6 weeks later. Our findings suggest that μ opioid receptors in astrocytes are linked to oxidative phosphorylation and they contribute to long-term changes associated with opioid withdrawal.

Keywords: astrocytes; μ opioid receptor; morphine withdrawal; oxidative phosphorylation; reward

Citation: Murlanova, K.; Jouroukhin, Y.; Novototskaya-Vlasova, K.; Huseynov, S.; Pletnikova, O.; Morales, M.J.; Guan, Y.; Kamiya, A.; Bergles, D.E.; Dietz, D.M.; et al. Loss of Astrocytic μ Opioid Receptors Exacerbates Aversion Associated with Morphine Withdrawal in Mice: Role of Mitochondrial Respiration. *Cells* **2023**, *12*, 1412. https://doi.org/10.3390/cells12101412

Academic Editors: Michael Brenner and Vladimir Parpura

Received: 15 January 2023
Revised: 19 April 2023
Accepted: 15 May 2023
Published: 17 May 2023

1. Introduction

Astrocytes, a major glial cell type in the brain, express opioid receptors. Exposure to an opioid dynamically alters the activity of astrocytes, resulting in adaptive structural and molecular changes that can be observed in vitro and in vivo [1–9]. However, it is not clear if astrocytes are active mediators of opioid-induced neuroadaptations.

Astrocytic glycolysis is a source of metabolic support for neuronal activity, but the role of astrocytic mitochondria in neuron–astrocyte interplay is understudied [10]. However, data from recent studies suggest that oxidative phosphorylation (OXPHOS) in astrocytes regulates brain bioenergetics and redox balance and could contribute to behavior [10–13]. Additionally, there is evidence of a link between μ opioid receptors (MORs) and astrocytic mitochondria in vitro [14,15], but there are no in vivo data implicating astrocytic OXPHOS in the molecular and cellular mechanisms of long-term opiate-induced neuroadaptations.

To assess the potential contribution of astrocytic MOR to the behavioral effects of opioids, we generated mice with astrocyte-specific knockout of the gene that encodes MOR-1, *Oprm1* [16], and then measured the responses of the mice to acute and chronic morphine treatments and morphine withdrawal. We also examined the effects of *Oprm1* deficiency on mitochondrial respiration and glycolysis in brain astrocytes in naïve and morphine-dependent mice. We found that astrocyte-specific knockout of *Oprm1* enhanced naloxone-precipitated conditioned place aversion that persisted for 6 weeks. Furthermore, *Oprm1* knockout produced an elevation in OXPHOS that was further enhanced by naloxone-precipitated morphine withdrawal. Our findings suggest that MORs may be linked to OXPHOS in astrocytes which contribute to long-term neuroadaptation produced by exposure to and/or withdrawal from opiates.

2. Materials and Methods

2.1. Mice

Mice with inducible astrocyte-specific knockout of *Oprm1* (inducible conditional knockout [icKO]) were generated by crossing *Aldh1l1*::Cre-ER^{T2} mice [17] with *Oprm1*$^{fl/fl}$ mice (a generous gift from Dr. Raja [Johns Hopkins University] and Dr. Kieffer [McGill University]) [18]. *Aldh1l1* is highly specific to astrocytes and is expressed in a broader population of astrocytes than other commonly used astrocyte markers in mice [17]. Male mice heterozygous for *Oprm1* icKO (*Aldh1l1*::Cre-ER^{T2}x*Oprm1*$^{fl/+}$) were the experimental group; male wild-type, *Aldh1l1*::Cre-ER^{T2}, and *Oprm1*$^{fl/fl}$, *Oprm1*$^{fl/+}$ mice were included as controls. All mice were treated with tamoxifen and were on the C57BL/6/J background. Mice were reared by their dams until postnatal day (PND) 21; afterwards, four same-sex mice were housed per cage.

Mice were given standard laboratory chow and water ad libitum and were housed in a temperature- (21 ± 2 °C), humidity- (55 ± 5%), and light-controlled room (reversed 12 h light/12 h dark cycle, light on at 6 p.m., light off at 6 a.m.).

2.2. Tamoxifen Injections

At PND 30, all mice were injected intraperitoneally (i.p.) for three consecutive days with 100 mg/kg tamoxifen (T5648, Sigma-Aldrich, St. Louis, MO, USA) dissolved in sunflower seed oil (S5007, Sigma-Aldrich), which induced conditional recombination in astrocytes of *Oprm1* icKO mice. The dose volume was 10 mL/kg body weight.

2.3. Morphine and/or Naloxone Injections

Morphine sulfate (1448005, USP, Rockville, MD, USA) dissolved in normal saline was injected i.p. for conditional place preference (10 mg/kg) and conditional place aversion (escalating doses of 20, 40, 60, 80, and 100 mg/kg) tests and subcutaneously (s.c.) for analgesic (5 mg/kg) and sensitization (10 mg/kg) tests. Naloxone (N7758, Sigma-Aldrich) dissolved in normal saline was injected s.c. for CPA test (0.25 mg/kg). For all treatments, the dose volume was 10 mL/kg body weight.

2.4. Behavioral Testing

At PND 60 (4 weeks after tamoxifen injections), open field test, elevated plus maze, and novel place recognition test were used to assess locomotor activity, anxiety, and recognition memory, respectively. Morphine-induced sensitization, conditioned place preference, conditioned place aversion, hot plate and Hargreaves test were tested in separate cohorts of *Oprm1* icKO and control mice (Table 1).

Table 1. Battery of behavioral tests used in the study.

PND 0 – – → PND 21 → PND 30-32 – – – → PND 60+
Weaning | Tamoxifen treatment | Behavioral tests

Cohort	Test	PND
	Open field test	60
Cohort 1	Elevated plus maze	80
	Novel place recognition test	90–92
Cohort 2	Morphine-induced sensitization	90–102
Cohort 3	Conditioned place preference	60–71
Cohort 4	Conditioned place aversion	105–110; 130; 150
Cohort 5	Hargreaves test	70
Cohort 6	Hot plate test	70

Mice were gently handled in the housing room for 1 min over the course of 3–5 days prior to behavioral testing. Mice were transported from the vivarium into the test rooms and were allowed to habituate for at least 30 min before testing. Body weight was monitored throughout the whole experimental period, and no genotype-dependent differences were observed. Mice were randomly and evenly allocated to each experimental group. To perform the group allocation in a blinded manner during data collection, animal preparation and experiments were performed by different investigators. All the experiments were carried out during the dark phase, and the red-light intensity was adjusted to ~20 lux during the behavioral testing.

2.4.1. Open Field Test

Locomotor activity, average speed and thigmotaxis were evaluated in an open field arena made of clear acrylic (40 × 40 × 30 cm) placed inside a frame containing evenly spaced photocells and receptors making a grid of infrared photobeams from the front to the back and from the left to the right (16 photobeams per axis) (Omnitech Electronics, Inc. Columbus, OH, USA). Any movement made by a mouse within the VersaMax monitor broke through the light beams, thus revealing the mouse's position in the (X-Y) plane. The mouse was placed in the center of the arena, and activity was recorded for a 30 min session. Data was analyzed using the VersaMax animal activity software (v.4.12-1AFE, AccuScan Instruments, Inc., Columbus, OH, USA). Distance moved and time spent in the outer zones (within 1 cm proximity to the walls of the cage) of the arena was interpreted as thigmotaxis, i.e., wall-hugging behavior.

2.4.2. Elevated Plus Maze

Anxiety-related behavior was assessed via an elevated plus maze. Mice were placed in the center of the maze consisting of two open arms and two closed arms 33 cm in length (San Diego Instruments Inc., San Diego, CA, USA) and allowed to explore the maze for 5 min. The elevated plus maze was raised 0.5 m above the floor. Any-maze Tracking Software (v.6.34, Stoelting, Co., Wood Dale, IL, USA) was used to automatically record and analyze preference to the open arms of the maze (% of time spent in the open arms of the maze). Criterion for arm entries was front two paws and snout inside of the arm.

2.4.3. Novel Place Recognition

The novel place recognition test was performed using a square arena (40 cm × 40 cm × 40 cm) made of white plastic. The test was divided into 3 sessions. During the habituation session (1 day before the training session), mice spent 10 min in the arena to decrease the exploration of the novel environment and anxiety during the training and test sessions. During the training session, mice became habituated to two identical objects placed at fixed locations in the arena. Each mouse began the training session at the arena's center, and the time spent in each object's immediate vicinity (within 2 cm) was recorded during a 10 min period, after which the mouse was placed back into the home cage. Long-term recognition memory was tested after 24 h, when mice revisited the arena, in which one of the object's location was changed. Time spent near the objects was again recorded for 5 min. Preference (%) for object in a novel location was calculated from the time mouse spent in each object's vicinity by dividing the exploration time of the object in a novel location (TN) by the total time spent by the mouse near objects in novel (TN) and familiar (TF) locations: Preference (%) (novel place) = (TN/(TN + TF)) × 100. Data were recorded and analyzed using the TopScan tracking software (v.2.0, Clever Sys., Inc., Reston, VA, USA).

2.4.4. Morphine-Induced Sensitization

Locomotor activity in response to acute effects of morphine and morphine sensitization were assessed in a separate cohort of *Oprm1* icKO and control mice as previously described [19] (Figure 1). The distance moved by each mouse in the activity box was detected by an infrared motion-sensor system (Omnitech Electronics, Inc., Columbus, OH, USA) fitted outside a transparent plastic cage and analyzed with Versa Max animal activity software (v.4.12-1AFE, AccuScan Instruments, Inc., Columbus, OH, USA).

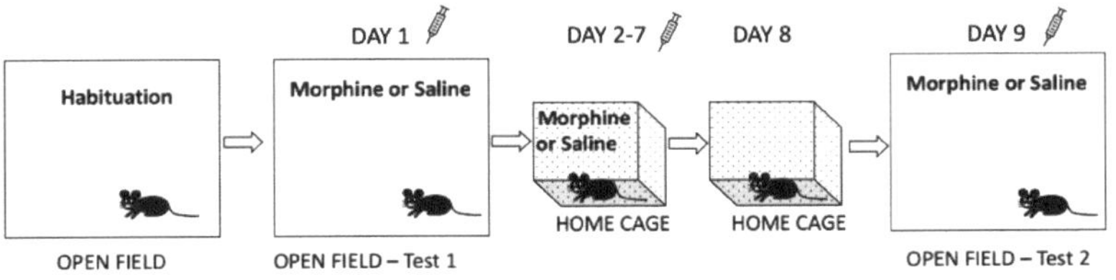

Figure 1. Schematic of the morphine sensitization procedure. Mice were individually habituated to the activity boxes for 30 min, 4 days before the test. On day 1 (test 1), mice were exposed to the activity box for 30 min, followed by a subcutaneous injection of morphine (10 mg/kg) or saline, and locomotor activity was recorded for 2 h. The same mice were administered morphine (10 mg/kg) or saline, respectively, once per day in their home cages for the next 6 days (total of 7 days of chronic daily morphine or saline treatment). On day 8, mice were not handled or treated. On day 9 (test 2), mice were treated the same as on day 1 (test 1), i.e., after a 30 min habituation period, morphine (10 mg/kg) or saline was administered, and activity was monitored for 2 h.

Each mouse was allowed to habituate to the activity box for 30 min 4 days before the test. On day 1 (test 1), mice were placed in the activity box 30 min before receiving an injection of morphine (10 mg/kg, s.c.) or saline, after which locomotor activity was recorded for 2 h. The mice then received similar injections once per day in home cages for the next 6 days (i.e., chronic daily treatment). No treatment or handling occurred on day 8. Locomotor activity in response to morphine or saline was again recorded for 2 h on day 9 (test 2) 30 min after they were placed in the activity boxes. Sensitization scores were calculated as [(distance traveled on day 9) − (distance on day 1)]/(distance on day 9) × 100.

2.4.5. Conditioned Place Preference

Morphine-induced conditioned place preference (CPP) was tested in a separate cohort of *Oprm1* icKO and control mice according to a previously published protocol [11]. CPP scores were calculated as the percent time spent in the morphine-paired compartment minus the percent time spent in the same compartment on the preconditioning day (Figure 2).

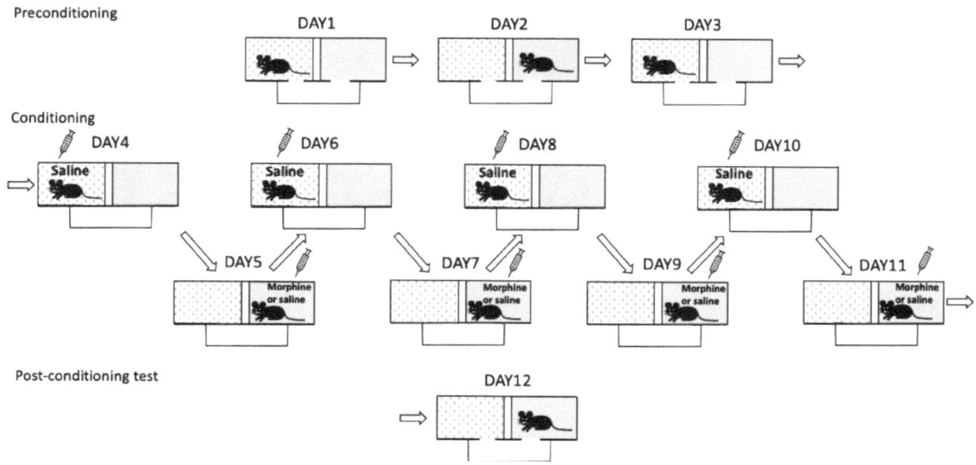

Figure 2. Schematic of the conditioned place preference (CPP) procedure. Preconditioning (habituation): The initial preference of a mouse for one of the compartments was tested. The mouse was placed in the bridge connecting two compartments with different floor textures and different wall patterns. The mouse was allowed to freely explore both compartments for 15 min daily for 3 days The compartment visited for less time on day 3 was designated the less preferred side and was paired with morphine during the conditioning session. Conditioning: Over the course of 8 days, the mouse was given a morphine or saline injection on alternate days (i.e., 4 days for each treatment) and kept in one of the compartments for 30 min. The mouse was confined to the less preferred side immediately after morphine injection (10 mg/kg; intraperitoneally) and was confined to the more preferred side immediately after saline injection. Half of the control and *Oprm1* icKO mice were given morphine/saline combinations in the opposite compartments, the other half received saline in both compartments. **Testing:** One day (24 h) after the last conditioning session, mice were tested for CPP by placing them in the bridge with open access to both compartments and recording the amount of time spent in each compartment over a 15 min period.

2.4.6. Conditioned Place Aversion

Conditioned place aversion (CPA) was tested in a separate cohort of *Oprm1* icKO and control mice as previously described [11] to evaluate the negative affective states induced by morphine withdrawal [20,21] (Figure 3). Mice were injected with escalating doses of morphine (20, 40, 60, and 80 mg/kg, i.p.) or saline twice a day for 4 days (at 9 a.m. and 5 p.m.). Three hours after the morning injection on day 3, mice were allowed to explore the CPA apparatus for 15 min (preconditioning). Place conditioning occurred 3 h after the morning injection on days 4 (saline, s.c.) and 5 (naloxone 0.25 mg/kg, s.c.). On day 5, mice received a single injection of morphine (100 mg/kg) or saline in their home cages. Place conditioning was tested on day 6, 24 h after the naloxone injection. The CPA score was calculated as the percent time spent in the naloxone-paired compartment minus the percent time spent in the same compartment on the preconditioning day.

2.4.7. Analgesic Effects

Analgesic effects of morphine were tested with a hot plate analgesia meter (Columbus Instruments, Columbus, OH, USA). Additionally, Hargreaves test was conducted using an IITC plantar analgesia meter (IITC Life Science Inc., Woodland Hills, CA, USA) as previously described [11]. For both tests, the maximal stimulus time was 30 s (cutoff time). The procedure was repeated 30 min after injection of morphine (5 mg/kg, s.c.), and the antinociceptive morphine response was calculated as a percentage of maximal possible effect (%MPE) using the formula 100 × (morphine response time − basal response time)/(cutoff time − basal response time).

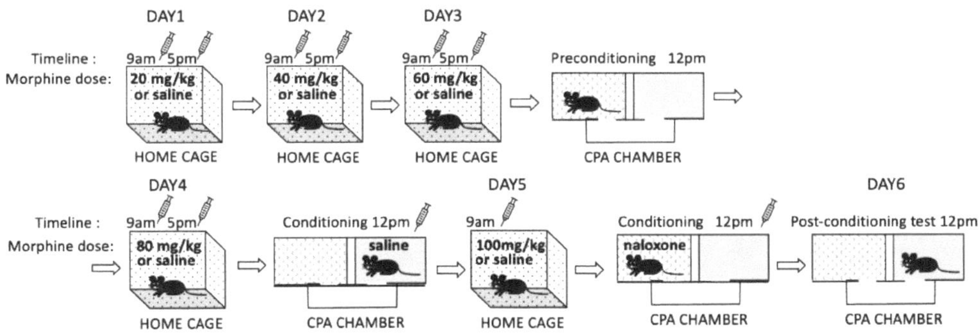

Figure 3. Schematic of the conditioned place aversion (CPA) procedure. A mouse received escalating doses of morphine (20, 40, 60, and 80 mg/kg; intraperitoneally) or saline twice a day for 4 days (at 9 a.m. and 5 p.m.). On day 5, the mouse received a single injection of morphine (100 mg/kg) or saline in the home cage. Three hours after the morning injection on day 3, the mouse was allowed to explore the CPA apparatus for 15 min (preconditioning). Place conditioning was performed on days 4 (saline) and 5 (naloxone 0.25 mg/kg; subcutaneously) 3 h after the morning injection. Place conditioning was tested on day 6, 24 h after the naloxone injection.

2.5. RNA In Situ Hybridization

In situ hybridization was performed using the "BaseScope" method [22] with a kit supplied by ACD (Newark, CA, USA) and 15 μm fresh-frozen sagittal mouse brain sections obtained from naïve *Oprm1* icKO and control mice (PND 90). Custom "3ZZ" probes were designed by ACD and were targeted to *Oprm1* mRNA (GenBank accession NM_001039652.2), nucleotides 845–976, and *Aldh1l1* mRNA (aldehyde dehydrogenase 1 family, member L1; NM_027406), nucleotides 1256–2112. *Aldh1l1* mRNA hybridization was used as a marker of astrocytes and was detected by precipitation of Fast Green. *Oprm1* mRNA hybridization was also detected by precipitation of Fast Red. All sections were additionally stained with 50% Gill's hematoxylin 1 (s209, Poly Scientific R&D Corp., Bay Shore, NY, USA) to disclose tissue morphology. Images were acquired using a Leica DM 6B upright microscope with a Leica DFC 7000T color camera and a 20× lens objective (numerical aperture 0.8) and a LED white light source (4500 K) using Leica LAS X acquisition software, v.3.7.3.2324.

Expression was evaluated using ImageJ 1.53c [23]. *Oprm1* expression was assessed by selecting random regions of interest (ROI; defined by a 15-μm circle) in a field that contained at least one green-turquoise punctum indicative of an astrocyte. Pixels containing red *Oprm1* staining were then counted within the ROI. Fifteen ROIs were evaluated within each 312 μm by 234 μm image from various brain regions. Images were uniformly adjusted for brightness, contrast, and color balance.

2.6. Biochemical Assays

Brain astrocytes were isolated at PND 150 from naïve *Oprm1* icKO and control mice (for Western blotting and the metabolic flux assay) and from mice that underwent the CPA test (for metabolic flux assay). Whole brains were dissected and placed in Dulbecco's phosphate-buffered saline containing Ca^{2+}, Mg^{2+}, D-glucose, and pyruvate (14287080, Thermo Fisher Scientific, Waltham, MA, USA). Astrocytes were isolated according to the "Isolation and cultivation of astrocytes from adult mouse brain" Miltenyi Biotec protocol for the Adult Brain Dissociation kit (130-107-677, Miltenyi Biotec, Bergisch Gladbach, Germany) and anti-ACSA-2 MicroBead kit (130-097-678, Miltenyi Biotec). After magnetic sorting, cells were centrifuged for 5 min at 300× *g* at 4 °C. Pellets were dissolved in cell lysis buffer (9803,

Cell Signaling Technology, Danvers, MA, USA) for Western blotting or in AstroMACS medium (130-117-031, Miltenyi Biotec) for subsequent cultivation.

2.6.1. Western Blotting

The expression of MORs by isolated astrocytes was evaluated using a standard Western blotting protocol [11]. The nitrocellulose membrane was incubated overnight at 4 °C with anti-MOR1 mouse monoclonal antibody (1:500, sc-515933, Santa Cruz Biotechnology, Dallas, TX, USA) and then with horseradish peroxidase-conjugated anti-mouse secondary antibody (1:1000, NA931, Cytiva, Marlborough, MA, USA). The optical density of protein bands on each digitized image was normalized to the optical density of the band for β-actin (A5441, Sigma-Aldrich) using ImageJ 1.53c.

2.6.2. Metabolic Flux Assays

Energy metabolism was evaluated in isolated astrocytes using an XF96 Extracellular Flux Analyzer (Agilent, Santa Clara, CA, USA). Specifically, mitochondrial OXPHOS was assessed on the basis of the oxygen consumption rate (OCR), and glycolysis was measured by analyzing the extracellular acidification rate (ECAR).

Isolated astrocytes were plated at a density of 50×10^3 per well in XF96 cell-culture microplates precoated with poly-D-lysine (P6407, Sigma-Aldrich). The cultures were incubated for 4 days in AstroMACS medium prior to the assay. On day 5, the medium was replaced with Seahorse XF medium supplemented with 10 mM D-glucose (G8769, Sigma-Aldrich), 1 mM L-glutamine (25030081, Life Technologies of Thermo Fisher Scientific), and 1 mM sodium pyruvate (11360070, Life Technologies,), and OCR and ECAR were measured using Seahorse XF Cell Mito Stress (103015-100, Agilent) and Glycolysis Stress (103020-100, Agilent) test kits.

OCR was determined at baseline and at each point after adding oligomycin (inhibits ATP synthase [complex V]), carbonyl cyanide-4 (trifluoromethoxy) phenylhydrazone (FCCP; uncoupling agent that collapses the proton gradient and disrupts the mitochondrial membrane potential), and rotenone (inhibits complex I). ECAR was determined at baseline and at each point after adding D-glucose, oligomycin, and 2-deoxy-glucose (2-DG; inhibits glycolysis by competitively binding to glucose hexokinase). The OCR and ECAR measurements were recorded after an 11 min equilibration, with a 1 min mix period and a 1 min wait period. All measurements were in pmol per min and were normalized to total protein concentration in each well.

2.7. Statistical Analysis

GraphPad Prism 9.4.1 was used for data analysis and graph generation. All data are presented as means ± SEMs and were analyzed with unpaired Student's t tests or two-way ANOVAs with or without corrections for repeated measures. Differences between the groups were further investigated using Bonferroni post hoc tests. The Bonferroni correction was used for multiple behavioral tests on the same mice. Normal distribution was checked using the Shapiro–Wilk test. Power analysis was conducted using G*Power 3.1.9.6 [24] based on previously published work [11] to determine the sample size. Significance was set at a p value of <0.05.

3. Results

3.1. Decreased Levels of Oprm1 mRNA and Protein in Astrocytes from icKO Mice

MORs are highly expressed in brain astrocytes [25,26] and Aldh1l1-Cre/ERT2 selectively targets brain astrocytes following tamoxifen induction [17,27].

We first validated astrocyte-specific knockout of *Oprm1*. BaseScope in situ hybridization was used to assess regional expression of *Oprm1* mRNA. In the *Oprm1* icKO mouse model, there was an approximately 50% decrease in expression in the medial prefrontal cortex, dentate gyrus and CA1-CA2 areas of the hippocampus, caudate putamen, nucleus accumbens, and the ventral tegmental area (Figure 4A). These regions were selected because they are where activity of the Aldh1l1 promoter is maximal [17] and also in these regions there is high expression of MORs [28]. Accordingly, protein levels of MOR were significantly lower in astrocytes from *Oprm1* icKO mice than in those from controls (Figure 4E,F). Astrocytes were acutely isolated from the whole brain the using magnetic beads technology, optimized to yield pure astrocytes as validated by the expression of astrocyte specific factors [29,30]. An approximately 50% reduction in astrocytic *Oprm1* mRNA and protein MOR expression were consistent with the mouse genotype being heterozygous.

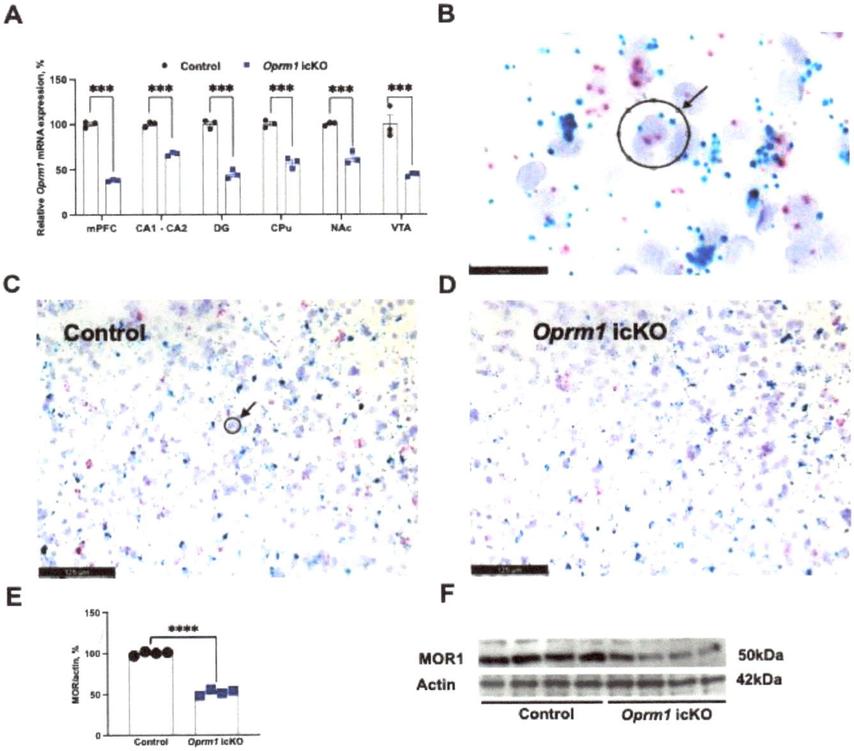

Figure 4. Decreased expression of *Oprm1* mRNA and protein with icKO. (**A**) Quantitative analysis of *Oprm1* mRNA expression in the medial prefrontal cortex (mPFC), CA1-CA2 areas and dentate gyrus (DG) of the hippocampus, caudate putamen (CPu), nucleus accumbens (NAc), and the ventral tegmental area (VTA). Two-way ANOVA revealed a significant effect of genotype [F (1, 24) = 519.3, $p < 0.001$]. A Bonferroni post hoc test showed that *Oprm1* icKO mice expressed significantly less *Oprm1* mRNA in all brain regions of interest. $n = 3$ mice per group; 2 sections per mouse; 15 regions of interest (ROI) per data point. ROI was defined by a 15 μm circle over a field that contained at least one green-turquoise (Fast Green; labeling *Aldh1* mRNA) punctum within an astrocyte, as seen in B. (**B**) Co-localization of *Oprm1* mRNA expression (red dots) surrounding randomly selected *Aldh1l1* mRNA expression loci (green-turquoise dots). The radius of the black circle (ROI) is 15 μm (707 μm^2). Pixels for red (Fast Red) *Oprm1* mRNA staining were counted within the ROI. (**C**,**D**) Representative images of sections counterstained with 50% Gill's hematoxylin 1 to disclose tissue morphology, i.e., the cell nuclei (purple). The black circle denotes the ROI shown in B. (**E**,**F**) Quantitative analysis of Western blots revealed significantly decreased expression of astrocytic MOR1 in *Oprm1* icKO mice (Student's *t* test; $t = 23.22$, $p < 0.0001$). $n = 4$ mice per group; Error bars indicate SEM. *** $p < 0.001$, **** $p < 0.0001$.

3.2. Normal Baseline Behavioral Phenotypes of Oprm1 icKO Mice

There were no genotype-dependent differences in general activity in the open field test, in anxiety-like behaviors in the elevated plus maze test, or in spatial memory in the novel place location recognition test (Figure 5).

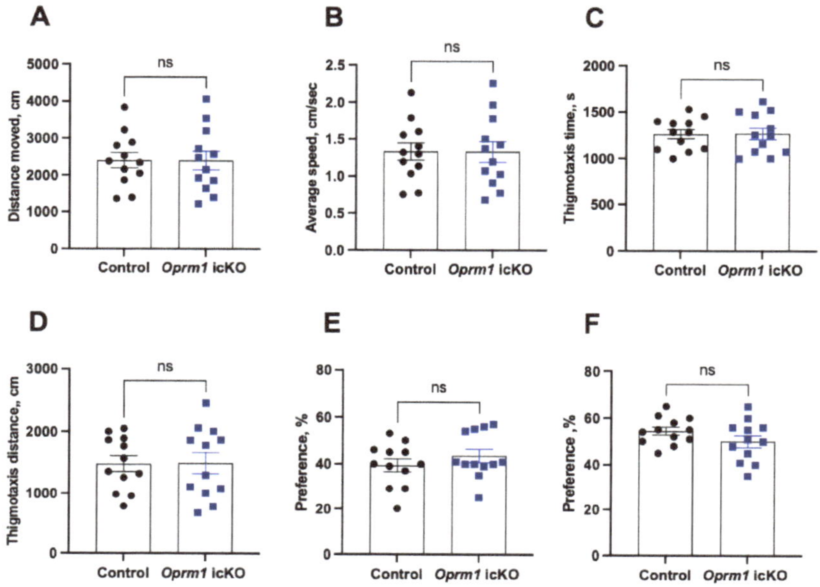

Figure 5. Locomotor activity, anxiety levels, and spatial memory in *Oprm1* icKO and control mice. There were no genotype differences in the distance traveled (Student's *t* test; $t = 0.009$, $p = 0.9928$) (**A**), average speed (Student's *t* test; $t = 0.009$, $p = 0.9928$) (**B**), thigmotaxis time (Student's *t* test; $t = 0.087$, $p = 0.9313$) (**C**), thigmotaxis distance (Student's *t* test; $t = 0.068$, $p = 0.9463$) (**D**) in the open field test, the preference for open arms in the elevated plus maze (Student's *t* test; $t = 1.05$, $p = 0.303$) (**E**), or the preference for a novel place in the novel place recognition test (Student's *t* test; $t = 1.42$, $p = 0.16$) (**F**) $n = 12$ mice per group. Error bars indicate SEM.

3.3. Effects of Oprm1 icKO on Drug-Induced Behaviors

3.3.1. Increased Morphine-Induced Locomotor Activity but Not Locomotor Sensitization with *Oprm1* icKO

Morphine administration (10 mg/kg, s.c.) increases locomotor activity in mice [19,31]. However, acute injections of morphine produced significantly greater locomotion in *Oprm1* icKO mice than in the controls (test 1) (Figure 6A,B). Because locomotor activation can be further enhanced after chronic administration of a drug in a process known as locomotor sensitization, or "reverse tolerance" [32,33], we tested for this in our mice. Six days of morphine treatment (10 mg/kg, s.c.) induced locomotor sensitization that was comparable in *Oprm1* icKO and control mice (test 2) (Figure 6C–E).

3.3.2. *Opmr1* icKO Mice Exhibit Normal CPP to Morphine

Morphine similarly produced CPP in *Oprm1* icKO and control mice (tested in a separate cohort of male 10-week-old mice) (Figure 7).

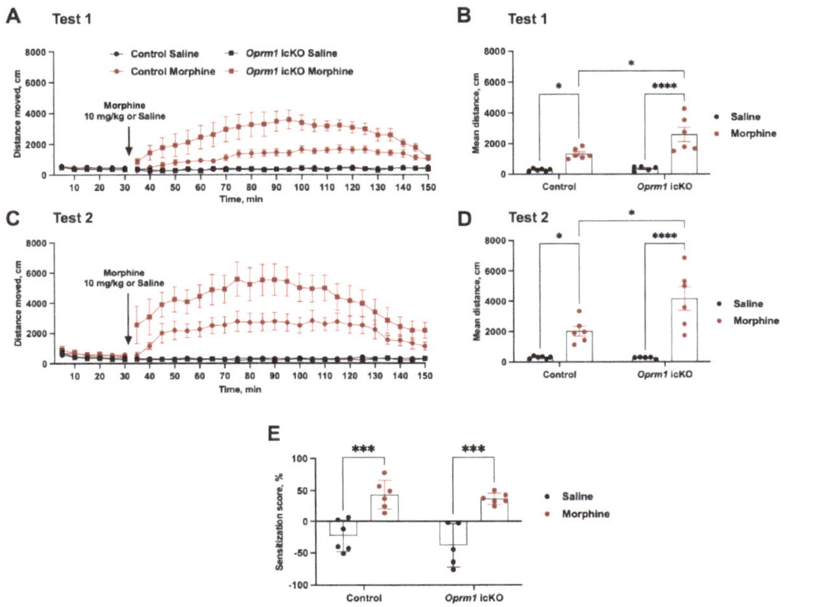

Figure 6. Astrocytic *Oprm1* deficiency leads to increased sensitivity to morphine but not locomotor sensitization to chronic morphine administration. (**A**,**C**) Mice were habituated to the open field (30 min) before locomotor activity in response to morphine (10 mg/kg, s.c.) or saline was assessed. Each data point represents the mean (±SEM) of the distance traveled during a 5 min interval. Both *Oprm1* icKO and control mice exhibited increased locomotion upon morphine injection, but the locomotor-stimulating effect of morphine was more pronounced in *Oprm1* icKO mice. (**B**,**D**) Mean distance traveled after morphine injection. Two-way AVOVA revealed significant effects of genotype [test 1: F (1, 19) = 5.34, $p = 0.03$; test 2: F (1, 19) = 5.59, $p = 0.02$] and morphine [test 1: F (1, 19) = 41.12, $p < 0.0001$; test 2: F (1, 19) = 40.22, $p < 0.0001$] and a genotype by morphine interaction [test 1: F (1, 19) = 5.34, $p = 0.032$; test 2: F (1, 19) = 6.04, $p = 0.02$] on distance traveled. (**E**) Sensitization scores did not differ between *Oprm1* icKO and control mice. Statistical significance between groups was assessed using two-way ANOVA and a Bonferroni post hoc test. $n = 5–6$ per group. Error bars indicate SEM. * $p < 0.05$, *** $p < 0.001$, **** $p < 0.0001$.

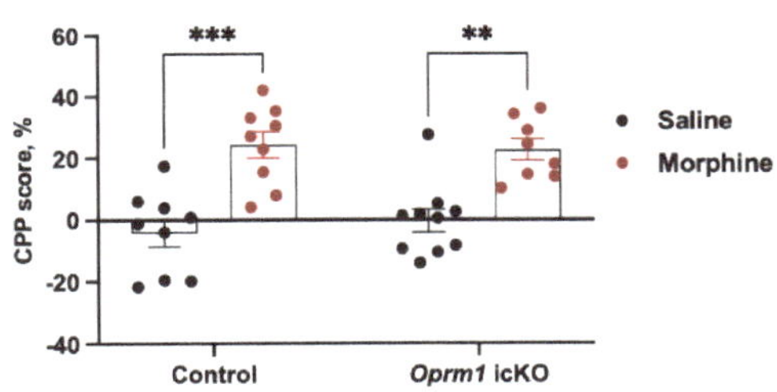

Figure 7. Conditioned place preference (CPP) in *Oprm1* icKO and control mice. Two-way ANOVA revealed a significant effect of morphine [F (1, 32) = 40.24, $p < 0.0001$]. No genotype effect was observed [F (1, 32) = 0.06, $p = 0.7934$]. A Bonferroni post hoc test showed that both morphine-treated control and *Oprm1* icKO mice exhibited significantly increased CPP scores compared to the respective saline-treated groups. $n = 8–10$ per group. Error bars indicate SEM. ** $p < 0.01$, *** $p < 0.001$.

3.3.3. *Oprm1* icKO Mice Exhibit Stronger CPA to Morphine Withdrawal

Another cohort of mice was tested for CPA to naloxone-precipitated morphine withdrawal (Figure 3). Relative to that of control mice, *Oprm1* icKO mice showed greater CPA (i.e., spent less time) in the naloxone-paired compartment 24 h after the naloxone injection (Figure 8A). When extinction of CPA was tested 3 or 6 weeks after naloxone treatment, morphine-treated *Oprm1* icKO mice but not control mice continued to show CPA (Figure 8B,C).

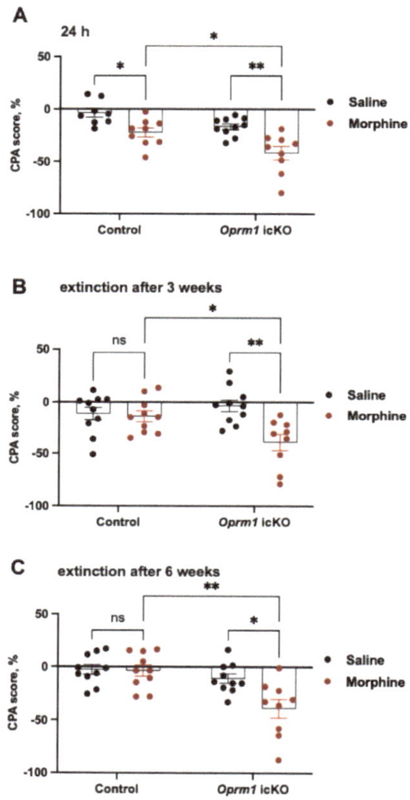

Figure 8. *Oprm1* icKO mice showed an exacerbated CPA and reduced CPA extinction in response to morphine withdrawal. (**A**) Morphine-dependent *Oprm1* icKO mice had stronger aversion (i.e. more negative CPA score) than control mice. Two-way ANOVA revealed significant effects of morphine [$F(1, 32) = 23.46$, $p < 0.0001$] and genotype [$F(1, 32) = 12.81$, $p = 0.001$]. A Bonferroni post hoc test also showed that both morphine-treated control and *Oprm1* icKO mice had significantly more negative CPA scores than saline-treated groups. Extinction of CPA was attenuated in morphine-dependent *Oprm1* icKO mice 3 (**B**) and 6 (**C**) weeks after naloxone-precipitated withdrawal. Two-way ANOVA of the CPA data revealed significant effects of morphine [extinction after 3 weeks: $F(1, 35) = 9.29$, $p = 0.004$; extinction after 6 weeks: $F(1, 35) = 6.10$, $p = 0.018$], genotype [extinction after 3 weeks: $F(1, 35) = 4.70$, $p = 0.03$; extinction after 6 weeks: $F(1, 35) = 13.55$, $p = 0.0008$] and a genotype by morphine interaction [extinction after 3 weeks: $F(1, 35) = 6.96$, $p = 0.012$; extinction after 6 weeks: $F(1, 35) = 5.36$, $p = 0.026$]. A Bonferroni post hoc test also showed that morphine-treated *Oprm1* icKO mice had significantly more negative CPA scores than saline-treated *Oprm1* icKO mice. $n = 8$–10 per group. Error bars indicate SEM. * $p < 0.05$, ** $p < 0.01$.

3.4. Oprm1 icKO Does Not Affect the Acute Analgesic Effects of Morphine

In separate cohorts of male 10-week-old mice, we evaluated the acute analgesic effects of morphine (5 mg/kg, s.c.) using the hot plate and Hargreaves tests. The effects were similar in *Oprm1* icKO and control mice (Figure 9).

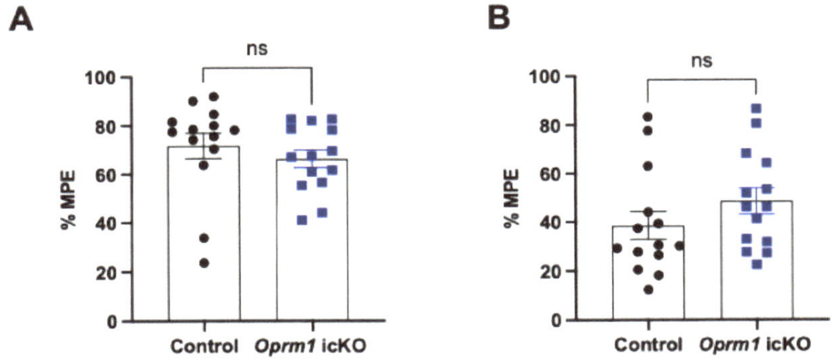

Figure 9. Morphine-induced analgesia in *Oprm1* icKO and control mice. %MPE (percentage of maximal possible effect) = 100 × (drug response time − basal response time)/(cutoff time − basal response time). There were no differences between the genotypes in the hot plate test (Student's *t* test; $t = 0.84$, $p = 0.407$) (**A**) or the Hargreaves test (Student's *t* test; $t = 1.27$, $p = 0.214$) (**B**). $n = 14$ mice per group. Error bars indicate SEM.

3.5. Increased OXPHOS in Oprm1-Deficient Astrocytes

Because in vitro studies suggest a link between opioid receptors and mitochondrial functions in astrocytes [14,15], we examined the effects of *Oprm1* knockout on mitochondrial function using the Seahorse XF Cell Mito Stress test. Astrocytes were isolated from the whole brain of naïve (no opioid treatment) *Oprm1* icKO and control mice. Figure 10A shows OCR of isolated astrocytes in response to compounds used in the mitochondrial stress test over time. Three basal rate measurements were taken prior to injection of pharmacological manipulators of mitochondrial respiratory chain proteins. Following the establishment of a basal OCR reading, oligomycin was applied to inhibit (H^+) flow through ATP synthase, essentially blocking all ATP-linked oxygen consumption. Maximal respiration was initiated by exposing cells to FCCP, which is a an ionophore that transports H^+ across the mitochondrial membrane leading to collapse of membrane potential and rapid consumption of oxygen. Rotenone prevented mitochondrial respiration by blocking complex I (NADH:ubiquinone oxidoreductase). Three measurements of OCR were obtained following the injection of each drug. Compared to controls, *Oprm1* icKO astrocytes exhibited significantly higher OCRs, indicating increased mitochondrial OXPHOS (Figure 10A). This was observed for basal respiration and the FCCP-induced maximal respiratory capacity.

No genotype-dependent changes were observed in the ECAR, suggesting comparable glycolytic function between the two groups (Figure 10B). *Oprm1* icKO and control astrocytes exhibited similar basal level of glycolysis at the beginning of the experiment. No difference was observed even after the addition of glucose, indicating the ability of *Oprm1* icKO and control astrocytes to perform increased glycolysis. Upon addition of oligomycin, a chemical that inhibits aerobic respiration causing the cells to rely on glycolysis for energy, the *Oprm1* icKO and control astrocytes showed similar acidification rates, also indicating their ability for increased glycolysis. ECAR, following treatment with 2-DG represented acidification associated with non-glycolytic activity was similar in both groups.

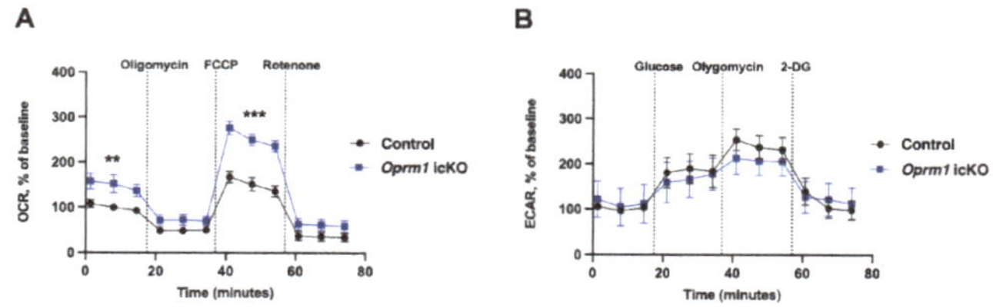

Figure 10. *Oprm1* deficiency increases mitochondrial respiration in astrocytes. (**A**) OCR for control and *Oprm1* icKO astrocytes. Two-way repeated measure ANOVA revealed a significant effect of *Oprm1* icKO for basal respiration [F (1, 9) = 6.04, $p = 0.036$] and maximal respiratory capacity following FCCP injection [F (1, 9) = 32.35, $p = 0.0003$]. Bonferroni post hoc tests showed that OCR at basal respiration and at maximal respiration (measured after FCCP addition) were significantly higher in *Oprm1* icKO astrocytes than in control cells. (**B**) ECAR for control and *Oprm1* icKO astrocytes Two-way repeated measure ANOVA revealed no significant effects. OCR and ECAR were normalized to the level of basal respiration in control astrocytes. $n = 5$ independent cultures for control astrocytes, $n = 6$ independent cultures for *Oprm1* icKO astrocytes. Each culture was assayed in duplicates. Error bars indicate SEM. ** $p < 0.01$, *** $p < 0.001$.

These data show that knockout of *Oprm1* significantly increases mitochondrial OXPHOS in astrocytes. Increased OXPHOS in *Oprm1* icKO astrocytes did not lead to compensatory decrease in glycolysis, as opposed to increased glycolysis in the absence of OXPHOS [34], suggesting a complex interplay between glycolysis and OXPHOS to adapt the mechanisms of energy production.

Because of the strong and long-lasting naloxone-precipitated aversion (up to 6 weeks), we isolated the whole brain astrocytes at PND 150 from *Oprm1* icKO and control mice exposed to CPA. Consistent with the OCR data obtained from astrocytes from naïve *Oprm1* icKO and control mice, a strong genotype effect was observed in basal respiration, ATP production (basal respiration—respiration after oligomycin addition), and maximal respiratory capacity (measured after FCCP addition) (Figure 11). OXPHOS was increased in *Oprm1* icKO and control mice that had been exposed to the escalating doses of morphine followed by naloxone-precipitated withdrawal compared to OXPHOS levels in saline-treated counterparts (Figures 8C and 11). Of note, was the increase in mitochondrial respiration in astrocytes measured from mice 6 weeks after their last exposure to morphine, suggesting that chronic exposure to morphine and pronounced withdrawal increases mitochondrial energy metabolism that persists for weeks.

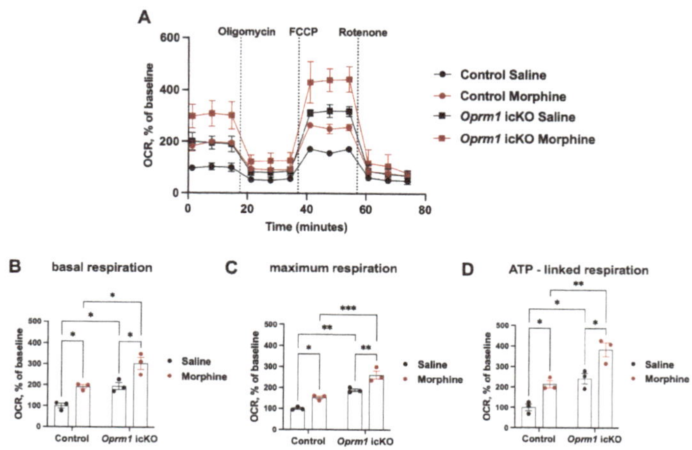

Figure 11. Astrocytic *Oprm1* deficiency leads to long-term increase in OXPHOS in mice exposed to

naloxone-precipitated morphine withdrawal. (**A**) OCR in astrocytes isolated from control and *Oprm1* icKO mice 6 weeks after last morphine (or saline) exposure. All mice were exposed to naloxone. The increase in astrocytic OXPHOS resulting from exposure to and withdrawal from morphine was more pronounced in the *Oprm1* icKO group. Basal (**B**), maximum (after FCCP injection; **C**), and ATP-linked (basal respiration—respiration after oligomycin addition; **D**) respiration in astrocytes isolated from morphine-treated *Oprm1* icKO and control mice exposed to naloxone-induced withdrawal and their saline-treated counterparts. Two-way ANOVA revealed significant effects of morphine exposure/withdrawal [basal: $F(1, 8) = 30.82$, $p = 0.0005$; maximum: $F(1, 8) = 32.80$, $p = 0.0004$; ATP-linked: $F(1, 8) = 26.81$, $p = 0.0008$] and genotype [basal: $F(1, 8) = 32.72$, $p = 0.0004$; maximum: $F(1, 8) = 79.24$, $p < 0.0001$; ATP-linked: $F(1, 8) = 39.76$, $p = 0.0002$] on mitochondrial respiration. Statistical significance between groups was assessed using two-way ANOVA and Bonferroni post hoc tests. OCR was normalized to basal respiration of astrocytes isolated from saline-treated control mice. $n = 3$ independent cultures for control and *Oprm1* icKO astrocytes. Each culture was assayed in duplicate. Error bars indicate SEM. * $p < 0.05$, ** $p < 0.01$, *** $p < 0.001$.

4. Discussion

Our study sheds light on the behavioral and metabolic functions of astrocytic MORs in a model of morphine dependency. Specifically, we showed that astrocyte conditional knockout of the *Oprm1* gene encoding MOR-1 increased locomotion in mice injected with morphine and enhanced and prolonged conditioned associations with morphine withdrawal. Furthermore, *Oprm1* deficiency increased astrocytic OXPHOS and further enhanced mitochondrial respiration long after morphine withdrawal in mice that were chronically treated with the drug. Taken together, our data support a functional in vivo link between astrocytic opioid signaling and OXPHOS.

It is tempting to speculate that MOR-mediated activation of astrocytic OXPHOS contributes to the negative affect behaviors (e.g., aversion as evidenced by CPA) in animals during protracted abstinence from chronic exposure to opiates [35–37]. Our previous work on mice with astrocyte-specific deficiency of *Cox10* (the gene encoding for the complex IV accessory protein COX10) similarly demonstrated a prolonged increase in naloxone-precipitated CPA in morphine-dependent *Cox10* icKO mice [11]. Although the exact reason(s) for the similarity in the CPA data and the similar striking difference in OXPHOS data between *Oprm1* and *Cox10* icKO mice remain unknown, it is not inconceivable that astrocyte mitochondrial respiration could play a significant role in both models. Future studies are warranted to address this issue by performing a more unbiased analysis of other potential changes in both mouse models.

The presence of mitochondria in the fine processes of astrocytes [12] raises the possibility that astrocytic OXPHOS supports neuronal activity [13]. Notably, impairment of OXPHOS in astrocytic mitochondria does not appreciably affect astrocytes themselves [34]. Our work supports data from in vitro studies linking MOR activation and mitochondria [14,15]. Nevertheless, the mechanism(s) by which astrocytic OXPHOS influences the behavioral effects of opioids remains unknown. Chronic exposure to opioids impairs glutamate uptake via downregulation of the plasma membrane glutamate transporter GLT-1 [38,39], and plasmalemmal glutamate transporters are in proximity to mitochondria in astrocytes [40–43] and interact with the Na^+/K^+ ATPase [44]. Thus, astrocytic OXPHOS might affect the activity of the Na^+/K^+ ATPase, thereby impacting the trafficking of glutamate transporters and glutamate uptake. Indeed, inhibition of mitochondrial function in astrocytes via fluorocitrate increases glutamate excitotoxicity in neuron–astrocyte co-cultures [45]. Additionally, activation of astrocytic MORs increases the release of glutamate by astrocytes in the hippocampus and nucleus accumbens [46]. Interestingly, activation of GLT-1 in the hippocampus via ceftriaxone reduces the signs of morphine withdrawal [47], indicating that glutamate may be a driver of the effects of morphine withdrawal.

Additionally, inflammatory signaling in astrocytes might influence negative emotional states associated with opioid withdrawal because genes involved in neuroinflammatory signaling (i.e., *Tnf*) are upregulated in glial cells in the amygdalae of rats experiencing morphine withdrawal [48].

There are several limitations of the study. First, we only examined male mice to avoid the influence of altered estrogenic responses in female mice as a result of our use of the selective estrogen receptor modulator tamoxifen to induce Cre recombinase [49]. Another reason was to avoid the variability introduced by the cyclical variation of female sex hormones during the estrous cycle which might affect basic and drug-related behaviors [50,51]. However, we plan to evaluate potential sex-dependent outcomes in future experiments. This is a critical issue because there is a growing appreciation of the influence of estrogens on glial function [52–55] and there are sex-dependent effects in non-contingent and contingent behavioral responses to drugs of abuse, including opioids [56,57]. Notably, pain sensitivity varies between men and women, with women reporting more chronic conditions that cause pain and are more likely to use prescription opioids than men [58].

Another limitation of the present work is that our MOR knockout affected astrocytes throughout the brain; thus, effects mediated from specific brain regions could not be resolved. For example, a recent work by Nam et al. [7] showed that activation of hippocampal astrocytic but not neuronal MORs was necessary for CPP. The selective MOR agonist, [D-Ala2, N-MePhe4, Gly-ol]-enkephalin (DAMGO), was injected either locally into the hippocampal CA1 region of mice or systemically to induce CPP. Manipulation of MOR expression using short hairpin RNA demonstrated that MOR expression in CA1 hippocampal astrocytes was both necessary and sufficient for DAMGO to induce CPP. In contrast to Nam et al. [7], we did not find any difference between control and *Oprm1* astrocyte icKO mice in CPP in our model with brain-wide knockout of the gene in astrocytes; more brain region-specific manipulations will be required to directly address brain region-dependent contributions of astrocyte MOR to the positive and negative valence behaviors. Another possibility for why our results differ from those of Nam et al. is the extent of MOR knockout, i.e., our mice had a 50% reduction in astrocytic MOR expression consistent with the mouse genotype being heterozygous. It is possible that the remaining 50% of MOR is sufficient for CPP. Complete elimination of astrocyte *Oprm1* expression, preferably in a circuit-related manner, would improve our understanding of the biology and behavioral roles of this gene. Future studies will utilize Cre-responsive viral reagents delivered postnatally within mesolimbic system to avoid other limitations associated with *Aldh1l1*::Cre-ER^{T2} -mediated recombination, e.g., promotor activity in peripheral organs and *Aldh1l1* expression in neural stem cells [59,60] that were not assessed in this work.

Future research should also assess the roles of astrocytic MORs in opioid self-administration and/or in seeking abstinence or extinction to determine if astrocytic MORs and bioenergetics are potential targets for treatments. Finally, there are three other types of opioid receptors [61] and thus additional work is needed to determine their roles to provide a more comprehensive picture of the molecular changes produced by the deletion of *Oprm1* in astrocytes.

In conclusion, our findings for the first time show that astrocytic MORs influence behavioral responses to morphine and to morphine withdrawal. These data also suggest a potential link between astrocytic MORs and mitochondrial OXPHOS that could be engaged by drugs of abuse to contribute to long-term neuroadaptations. Further research in this direction may yield new strategies for manipulating endogenous opioid signaling to combat opioid dependence.

Author Contributions: Conceptualization, K.M., Y.J., Y.G., A.K., D.E.B., D.M.D. and M.V.P.; Investigation, K.M., Y.J. and K.N.-V.; Investigation—BaseScope experiment, M.J.M., O.P. and S.H.; Writing—original draft, M.V.P. and K.M.; Writing—review and editing, Y.G., A.K. and D.M.D.; Supervision, M.V.P. All authors have read and agreed to the published version of the manuscript.

Funding: This research was funded by the Johns Hopkins University Drug Abuse Center (pilot grant to Y.J.), the National Institute of Neurological Disorders and Stroke (grant numbers R01NS110598 and R01NS117761 to Y.G.), the National Institute of Drug Abuse (R01DA041208 to A.K and M.V.P.) and the National Institute of Mental Health (grant numbers MH083728 to D.E.B. and M.V.P. and MH094268 to M.V.P.).

Institutional Review Board Statement: The animal study protocols were approved by the Institutional Animal Care and Use Committees (IACUCs) of the University at Buffalo (number 202000034, approved 20 July 2020) and Johns Hopkins University (number MO18M01, approved January 24, 2018). Animal experimentation was conducted in accordance with the US National Research Council's Guide for the Care and Use of Laboratory Animals and the US Public Health Service's Policy on Humane Care and Use of Laboratory Animals.

Informed Consent Statement: Not applicable.

Data Availability Statement: Data available on request from the authors.

Acknowledgments: We thank Srinivasa Raja (Johns Hopkins University) and Brigitte Kieffer (McGill University) for providing *Oprm1*$^{fl/fl}$ mice and Yavin Shaham (NIDA) for expert advice on testing morphine-produced behaviors.

Conflicts of Interest: The authors declare no conflict of interest.

References

1. Kruyer, A.; Scofield, M.D.; Wood, D.; Reissner, K.J.; Kalivas, P.W. Heroin Cue-Evoked Astrocytic Structural Plasticity at Nucleus Accumbens Synapses Inhibits Heroin Seeking. *Biol. Psychiatry* **2019**, *86*, 811–819. [CrossRef]
2. Avey, D.; Sankararaman, S.; Yim, A.K.Y.; Barve, R.; Milbrandt, J.; Mitra, R.D. Single-Cell RNA-Seq Uncovers a Robust Transcriptional Response to Morphine by Glia. *Cell Rep.* **2018**, *24*, 3619–3629.e3614. [CrossRef]
3. Dozio, V.; Daali, Y.; Desmeules, J.; Sanchez, J.C. Deep proteomics and phosphoproteomics reveal novel biological pathways perturbed by morphine, morphine-3-glucuronide and morphine-6-glucuronide in human astrocytes. *J. Neurosci. Res.* **2022**, *100*, 220–236. [CrossRef] [PubMed]
4. Liao, K.; Niu, F.; Hu, G.; Yang, L.; Dallon, B.; Villarreal, D.; Buch, S. Morphine-mediated release of miR-138 in astrocyte-derived extracellular vesicles promotes microglial activation. *J. Extracell. Vesicles* **2020**, *10*, e12027. [CrossRef]
5. Bachtell, R.K.; Jones, J.D.; Heinzerling, K.G.; Beardsley, P.M.; Comer, S.D. Glial and neuroinflammatory targets for treating substance use disorders. *Drug Alcohol Depend.* **2017**, *180*, 156–170. [CrossRef] [PubMed]
6. Rönnbäck, L.; Hansson, E. Are astroglial cells involved in morphine tolerance? *Neurochem. Res.* **1988**, *13*, 87–103. [CrossRef] [PubMed]
7. Nam, M.H.; Han, K.S.; Lee, J.; Won, W.; Koh, W.; Bae, J.Y.; Woo, J.; Kim, J.; Kwong, E.; Choi, T.Y.; et al. Activation of Astrocytic μ-Opioid Receptor Causes Conditioned Place Preference. *Cell Rep.* **2019**, *28*, 1154–1166.e1155. [CrossRef]
8. Kruyer, A.; Dixon, D.; Angelis, A.; Amato, D.; Kalivas, P.W. Astrocytes in the ventral pallidum extinguish heroin seeking through GAT-3 upregulation and morphological plasticity at D1-MSN terminals. *Mol. Psychiatry* **2022**, *27*, 855–864. [CrossRef]
9. Lv, Y.; Jing, M.Y.; Li, P.Y.; Zhao, T.Y.; Pang, C.; Lu, G.Y.; Wang, Z.Y.; Wu, N.; Hu, G.; Song, R.; et al. Aquaporin-4 deletion attenuates opioid-induced addictive behaviours associated with dopamine levels in nucleus accumbens. *Neuropharmacology* **2022**, *208*, 108986. [CrossRef]
10. Semyanov, A.; Verkhratsky, A. Astrocytic processes: From tripartite synapses to the active milieu. *Trends Neurosci.* **2021**, *44*, 781–792. [CrossRef]
11. Murlanova, K.; Jouroukhin, Y.; Huseynov, S.; Pletnikova, O.; Morales, M.J.; Guan, Y.; Baraban, J.M.; Bergles, D.E.; Pletnikov, M.V. Deficient mitochondrial respiration in astrocytes impairs trace fear conditioning and increases naloxone-precipitated aversion in morphine-dependent mice. *Glia* **2022**, *70*, 1289–1300. [CrossRef]
12. Agarwal, A.; Wu, P.H.; Hughes, E.G.; Fukaya, M.; Tischfield, M.A.; Langseth, A.J.; Wirtz, D.; Bergles, D.E. Transient Opening of the Mitochondrial Permeability Transition Pore Induces Microdomain Calcium Transients in Astrocyte Processes. *Neuron* **2017**, *93*, 587–605.e587. [CrossRef]
13. Rose, J.; Brian, C.; Pappa, A.; Panayiotidis, M.I.; Franco, R. Mitochondrial Metabolism in Astrocytes Regulates Brain Bioenergetics, Neurotransmission and Redox Balance. *Front. Neurosci.* **2020**, *14*, 536682. [CrossRef] [PubMed]
14. Zhang, C.; Wang, C.; Ren, J.; Guo, X.; Yun, K. Morphine Protects Spinal Cord Astrocytes from Glutamate-Induced Apoptosis via Reducing Endoplasmic Reticulum Stress. *Int. J. Mol. Sci.* **2016**, *17*, 1523. [CrossRef] [PubMed]
15. Lee, J.; Kim, M.S.; Park, C.; Jung, E.B.; Choi, D.H.; Kim, T.Y.; Moon, S.K.; Park, R. Morphine prevents glutamate-induced death of primary rat neonatal astrocytes through modulation of intracellular redox. *Immunopharmacol. Immunotoxicol.* **2004**, *26*, 17–28. [CrossRef] [PubMed]
16. Ravindranathan, A.; Joslyn, G.; Robertson, M.; Schuckit, M.A.; Whistler, J.L.; White, R.L. Functional characterization of human variants of the mu-opioid receptor gene. *Proc. Natl. Acad. Sci. USA* **2009**, *106*, 10811–10816. [CrossRef]
17. Srinivasan, R.; Lu, T.Y.; Chai, H.; Xu, J.; Huang, B.S.; Golshani, P.; Coppola, G.; Khakh, B.S. New Transgenic Mouse Lines for Selectively Targeting Astrocytes and Studying Calcium Signals in Astrocyte Processes In Situ and In Vivo. *Neuron* **2016**, *92*, 1181–1195. [CrossRef]
18. Weibel, R.; Reiss, D.; Karchewski, L.; Gardon, O.; Matifas, A.; Filliol, D.; Becker, J.A.; Wood, J.N.; Kieffer, B.L.; Gaveriaux-Ruff, C. Mu opioid receptors on primary afferent nav1.8 neurons contribute to opiate-induced analgesia: Insight from conditional knockout mice. *PLoS ONE* **2013**, *8*, e74706. [CrossRef]

19. Bohn, L.M.; Gainetdinov, R.R.; Sotnikova, T.D.; Medvedev, I.O.; Lefkowitz, R.J.; Dykstra, L.A.; Caron, M.G. Enhanced rewarding properties of morphine, but not cocaine, in beta(arrestin)-2 knock-out mice. *J. Neurosci.* **2003**, *23*, 10265–10273. [CrossRef]
20. Gómez-Milanés, I.; Almela, P.; García-Carmona, J.A.; García-Gutiérrez, M.S.; Aracil-Fernández, A.; Manzanares, J.; Milanés Maquilón, M.V.; Laorden, M.L. Accumbal dopamine, noradrenaline and serotonin activity after naloxone-conditioned place aversion in morphine-dependent mice. *Neurochem. Int.* **2012**, *61*, 433–440. [CrossRef]
21. Song, X.; Li, W.; Shi, Y.; Zhang, J.; Li, Y. Expression of protein kinase A and the kappa opioid receptor in selected brain regions and conditioned place aversion in morphine-dependent rats. *Oncotarget* **2017**, *8*, 82632–82642. [CrossRef]
22. Selvaraj, S.; Mezzano, V.; Loomis, C.; Hanniford, D. RNAscope® and BaseScope™: In-situ RNA analysis for formalin-fixed paraffin-embedded tissues and beyond. *J. Biomol. Tech.* **2020**, *31*, S31.
23. Schneider, C.A.; Rasband, W.S.; Eliceiri, K.W. NIH Image to ImageJ: 25 years of image analysis. *Nat. Methods* **2012**, *9*, 671–675. [CrossRef] [PubMed]
24. Faul, F.; Erdfelder, E.; Lang, A.G.; Buchner, A. G*Power 3: A flexible statistical power analysis program for the social, behavioral, and biomedical sciences. *Behav. Res. Methods* **2007**, *39*, 175–191. [CrossRef] [PubMed]
25. Nam, M.H.; Han, K.S.; Lee, J.; Bae, J.Y.; An, H.; Park, S.; Oh, S.J.; Kim, E.; Hwang, E.; Bae, Y.C.; et al. Expression of µ-Opioid Receptor in CA1 Hippocampal Astrocytes. *Exp. Neurobiol.* **2018**, *27*, 120–128. [CrossRef]
26. Corkrum, M.; Rothwell, P.E.; Thomas, M.J.; Kofuji, P.; Araque, A. Opioid-Mediated Astrocyte-Neuron Signaling in the Nucleus Accumbens. *Cells* **2019**, *8*, 586. [CrossRef] [PubMed]
27. Winchenbach, J.; Düking, T.; Berghoff, S.A.; Stumpf, S.K.; Hülsmann, S.; Nave, K.A.; Saher, G. Inducible targeting of CNS astrocytes in Aldh1l1-CreERT2 BAC transgenic mice. *F1000Research* **2016**, *5*, 2934. [CrossRef]
28. Le Merrer, J.; Becker, J.A.; Befort, K.; Kieffer, B.L. Reward processing by the opioid system in the brain. *Physiol. Rev.* **2009**, *89*, 1379–1412. [CrossRef]
29. Buenaventura, R.G.; Harvey, A.C.; Burns, M.P.; Main, B.S. Sequential Isolation of Microglia and Astrocytes from Young and Aged Adult Mouse Brains for Downstream Transcriptomic Analysis. *Methods Protoc.* **2022**, *5*, 77. [CrossRef] [PubMed]
30. Holt, L.M.; Olsen, M.L. Novel Applications of Magnetic Cell Sorting to Analyze Cell-Type Specific Gene and Protein Expression in the Central Nervous System. *PLoS ONE* **2016**, *11*, e0150290. [CrossRef]
31. Brase, D.A.; Loh, H.H.; Way, E.L. Comparison of the effects of morphine on locomotor activity, analgesia and primary and protracted physical dependence in six mouse strains. *J. Pharmacol. Exp. Ther.* **1977**, *201*, 368–374. [PubMed]
32. Robinson, T.E.; Berridge, K.C. The psychology and neurobiology of addiction: An incentive-sensitization view. *Addiction* **2000**, *95* (Suppl. S2), S91–S117. [CrossRef] [PubMed]
33. De Vries, T.J.; Shippenberg, T.S. Neural systems underlying opiate addiction. *J. Neurosci.* **2002**, *22*, 3321–3325. [CrossRef]
34. Supplie, L.M.; Düking, T.; Campbell, G.; Diaz, F.; Moraes, C.T.; Götz, M.; Hamprecht, B.; Boretius, S.; Mahad, D.; Nave, K.A. Respiration-Deficient Astrocytes Survive as Glycolytic Cells In Vivo. *J. Neurosci.* **2017**, *37*, 4231–4242. [CrossRef]
35. Kruyer, A.; Scofield, M.D. Astrocytes in Addictive Disorders. *Adv. Neurobiol.* **2021**, *26*, 231–254. [CrossRef]
36. Valentino, R.J.; Volkow, N.D. Untangling the complexity of opioid receptor function. *Neuropsychopharmacology* **2018**, *43*, 2514–2520. [CrossRef]
37. Darcq, E.; Kieffer, B.L. Opioid receptors: Drivers to addiction? *Nat. Rev. Neurosci.* **2018**, *19*, 499–514. [CrossRef] [PubMed]
38. Wang, W.; Zeng, F.; Hu, Y.; Li, X. A Mini-Review of the Role of Glutamate Transporter in Drug Addiction. *Front. Neurol.* **2019**, *10*, 1123. [CrossRef] [PubMed]
39. Roberts-Wolfe, D.J.; Kalivas, P.W. Glutamate Transporter GLT-1 as a Therapeutic Target for Substance Use Disorders. *CNS Neurol. Disord. Drug. Targets* **2015**, *14*, 745–756. [CrossRef] [PubMed]
40. Shevelkin, A.V.; Terrillion, C.E.; Hasegawa, Y.; Mychko, O.A.; Jouroukhin, Y.; Sawa, A.; Kamiya, A.; Pletnikov, M.V. Astrocyte DISC1 contributes to cognitive function in a brain region-dependent manner. *Hum. Mol. Genet.* **2020**, *29*, 2936–2950. [CrossRef]
41. Robinson, M.B.; Lee, M.L.; DaSilva, S. Glutamate Transporters and Mitochondria: Signaling, Co-compartmentalization, Functional Coupling, and Future Directions. *Neurochem. Res.* **2020**, *45*, 526–540. [CrossRef] [PubMed]
42. Genda, E.N.; Jackson, J.G.; Sheldon, A.L.; Locke, S.F.; Greco, T.M.; O'Donnell, J.C.; Spruce, L.A.; Xiao, R.; Guo, W.; Putt, M.; et al. Co-compartmentalization of the astroglial glutamate transporter, GLT-1, with glycolytic enzymes and mitochondria. *J. Neurosci.* **2011**, *31*, 18275–18288. [CrossRef] [PubMed]
43. Bauer, D.E.; Jackson, J.G.; Genda, E.N.; Montoya, M.M.; Yudkoff, M.; Robinson, M.B. The glutamate transporter, GLAST, participates in a macromolecular complex that supports glutamate metabolism. *Neurochem. Int.* **2012**, *61*, 566–574. [CrossRef]
44. Rose, E.M.; Koo, J.C.; Antflick, J.E.; Ahmed, S.M.; Angers, S.; Hampson, D.R. Glutamate transporter coupling to Na, K-ATPase. *J. Neurosci.* **2009**, *29*, 8143–8155. [CrossRef]
45. Voloboueva, L.A.; Suh, S.W.; Swanson, R.A.; Giffard, R.G. Inhibition of mitochondrial function in astrocytes: Implications for neuroprotection. *J. Neurochem.* **2007**, *102*, 1383–1394. [CrossRef] [PubMed]
46. Woo, D.H.; Bae, J.Y.; Nam, M.H.; An, H.; Ju, Y.H.; Won, J.; Choi, J.H.; Hwang, E.M.; Han, K.S.; Bae, Y.C.; et al. Activation of Astrocytic µ-opioid Receptor Elicits Fast Glutamate Release Through TREK-1-Containing K2P Channel in Hippocampal Astrocytes. *Front. Cell. Neurosci.* **2018**, *12*, 319. [CrossRef]
47. Saeedi, N.; Darvishmolla, M.; Tavassoli, Z.; Davoudi, S.; Heysieattalab, S.; Hosseinmardi, N.; Janahmadi, M.; Behzadi, G. The role of hippocampal glial glutamate transporter (GLT-1) in morphine-induced behavioral responses. *Brain Behav.* **2021**, *11*, e2323. [CrossRef]

48. O'Sullivan, S.J.; Malahias, E.; Park, J.; Srivastava, A.; Reyes, B.A.S.; Gorky, J.; Vadigepalli, R.; Van Bockstaele, E.J.; Schwaber, J.S. Single-Cell Glia and Neuron Gene Expression in the Central Amygdala in Opioid Withdrawal Suggests Inflammation With Correlated Gut Dysbiosis. *Front. Neurosci.* **2019**, *13*, 665. [CrossRef]
49. Zhang, Z.; Park, J.W.; Ahn, I.S.; Diamante, G.; Sivakumar, N.; Arneson, D.; Yang, X.; van Veen, J.E.; Correa, S.M. Estrogen receptor alpha in the brain mediates tamoxifen-induced changes in physiology in mice. *eLife* **2021**, *10*, e63333. [CrossRef]
50. Huhn, A.S.; Berry, M.S.; Dunn, K.E. Systematic review of sex-based differences in opioid-based effects. *Int. Rev. Psychiatry* **2018**, *30*, 107–116. [CrossRef]
51. Rocks, D.; Cham, H.; Kundakovic, M. Why the estrous cycle matters for neuroscience. *Biol. Sex. Differ.* **2022**, *13*, 62. [CrossRef]
52. Crespo-Castrillo, A.; Arevalo, M.A. Microglial and Astrocytic Function in Physiological and Pathological Conditions: Estrogenic Modulation. *Int. J. Mol. Sci.* **2020**, *21*, 3219. [CrossRef]
53. Martin-Jiménez, C.; Gaitán-Vaca, D.M.; Areiza, N.; Echeverria, V.; Ashraf, G.M.; González, J.; Sahebkar, A.; Garcia-Segura, L.M.; Barreto, G.E. Astrocytes Mediate Protective Actions of Estrogenic Compounds after Traumatic Brain Injury. *Neuroendocrinology* **2019**, *108*, 142–160. [CrossRef] [PubMed]
54. Chisholm, N.C.; Sohrabji, F. Astrocytic response to cerebral ischemia is influenced by sex differences and impaired by aging. *Neurobiol. Dis.* **2016**, *85*, 245–253. [CrossRef]
55. Wang, C.; Jie, C.; Dai, X. Possible roles of astrocytes in estrogen neuroprotection during cerebral ischemia. *Rev. Neurosci.* **2014**, *25*, 255–268. [CrossRef]
56. Kim, R.; Testen, A.; Harder, E.V.; Brown, N.E.; Witt, E.A.; Bellinger, T.J.; Franklin, J.P.; Reissner, K.J. Abstinence-Dependent Effects of Long-Access Cocaine Self-Administration on Nucleus Accumbens Astrocytes Are Observed in Male, But Not Female, Rats. *eNeuro* **2022**, *9*, 1–14. [CrossRef]
57. Hadschieff, V.; Drdla-Schutting, R.; Springer, D.N.; Siegert, A.S.M.; Schroeder, H.; Sandkühler, J. Fundamental sex differences in morphine withdrawal-induced neuronal plasticity. *Pain* **2020**, *161*, 2022–2034. [CrossRef]
58. Serdarevic, M.; Striley, C.W.; Cottler, L.B. Sex differences in prescription opioid use. *Curr. Opin. Psychiatry* **2017**, *30*, 238–246. [CrossRef] [PubMed]
59. Foo, L.C.; Dougherty, J.D. Aldh1L1 is expressed by postnatal neural stem cells in vivo. *Glia* **2013**, *61*, 1533–1541. [CrossRef] [PubMed]
60. Beyer, F.; Lüdje, W.; Karpf, J.; Saher, G.; Beckervordersandforth, R. Distribution of Aldh1L1-CreER(T2) Recombination in Astrocytes Versus Neural Stem Cells in the Neurogenic Niches of the Adult Mouse Brain. *Front. Neurosci.* **2021**, *15*, 713077. [CrossRef]
61. Machelska, H.; Celik, M. Opioid Receptors in Immune and Glial Cells-Implications for Pain Control. *Front. Immunol.* **2020**, *11*, 300. [CrossRef] [PubMed]

cells

MDPI

Article

Gene Enrichment Analysis of Astrocyte Subtypes in Psychiatric Disorders and Psychotropic Medication Datasets

Xiaolu Zhang [1,†], Alyssa Wolfinger [1,†], Xiaojun Wu [1], Rawan Alnafisah [1], Ali Imami [1], Abdul-rizaq Hamoud [1], Anna Lundh [1], Vladimir Parpura [2], Robert E. McCullumsmith [1,3], Rammohan Shukla [1,*] and Sinead M. O'Donovan [1,*]

1 Department of Neurosciences, University of Toledo, Toledo, OH 43614, USA
2 Department of Neurobiology, The University of Alabama at Birmingham, Birmingham, AL 35294, USA
3 Promedica Neurosciences Institute, Toledo, OH 43606, USA
* Correspondence: rammohan.shukla@utoledo.edu (R.S.); sinead.odonovan@utoledo.edu (S.M.O.)
† These authors contributed equally to this work.

Abstract: Astrocytes have many important functions in the brain, but their roles in psychiatric disorders and their responses to psychotropic medications are still being elucidated. Here, we used gene enrichment analysis to assess the relationships between different astrocyte subtypes, psychiatric diseases, and psychotropic medications (antipsychotics, antidepressants and mood stabilizers). We also carried out qPCR analyses and "look-up" studies to assess the chronic effects of these drugs on astrocyte marker gene expression. Our bioinformatic analysis identified gene enrichment of different astrocyte subtypes in psychiatric disorders. The highest level of enrichment was found in schizophrenia, supporting a role for astrocytes in this disorder. We also found differential enrichment of astrocyte subtypes associated with specific biological processes, highlighting the complex responses of astrocytes under pathological conditions. Enrichment of protein phosphorylation in astrocytes and disease was confirmed by biochemical analysis. Analysis of LINCS chemical perturbagen gene signatures also found that kinase inhibitors were highly discordant with astrocyte-SCZ associated gene signatures. However, we found that common gene enrichment of different psychotropic medications and astrocyte subtypes was limited. These results were confirmed by "look-up" studies and qPCR analysis, which also reported little effect of psychotropic medications on common astrocyte marker gene expression, suggesting that astrocytes are not a primary target of these medications. Conversely, antipsychotic medication does affect astrocyte gene marker expression in postmortem schizophrenia brain tissue, supporting specific astrocyte responses in different pathological conditions. Overall, this study provides a unique view of astrocyte subtypes and the effect of medications on astrocytes in disease, which will contribute to our understanding of their role in psychiatric disorders and offers insights into targeting astrocytes therapeutically.

Keywords: astrocyte subtype; psychotropic medication; schizophrenia; major depression; kinase

Citation: Zhang, X.; Wolfinger, A.; Wu, X.; Alnafisah, R.; Imami, A.; Hamoud, A.-r.; Lundh, A.; Parpura, V.; McCullumsmith, R.E.; Shukla, R.; et al. Gene Enrichment Analysis of Astrocyte Subtypes in Psychiatric Disorders and Psychotropic Medication Datasets. *Cells* **2022**, *11*, 3315. https://doi.org/10.3390/cells11203315

Academic Editor: Daniele Nosi

Received: 3 October 2022
Accepted: 20 October 2022
Published: 21 October 2022

1. Introduction

Astrocytes play many important roles in the brain, from facilitating metabolism and cell signaling [1,2], to responding to injury through regulating oxidative stress [3]. However the role of astrocytes in central nervous system (CNS) disorders is still poorly understood. While studies of postmortem brain broadly support perturbation of astrocytes in neurological and psychiatric disorders, the findings can prove challenging to interpret. For example, there is evidence of both increased [4–8] and decreased [7,9–19] astrocyte marker expression, increased [6,10] and decreased [20–26] astrocyte cell counts, changes in astrocyte morphology [7,27–29], as well as no significant changes in astrocyte expression [27,30–35] in disorders like schizophrenia and major depressive disorder (MDD). These findings also vary depending on the brain region and methodological approach studied [6,7,9,11,21,28,36–39].

A potential cause for these conflicting findings may be due to the complex responses of astrocytes to injury; astrocytes can undergo remodeling that leads to a range of responses [40] from complete atrophy [41,42] to a dynamic state termed "reactive astrogliosis" [43–45].

Reactive astrogliosis describes a spectrum of heterogeneous changes in gene expression [46–48], cell morphology [47–52], and overall function including fluid and ion homeostasis [53,54], oxidative stress response [55–57] and synapse formation [58,59]. These changes can also range from reversible to permanent [60,61]. A diverse set of mechanisms play a role in inducing astrogliosis [62], which can have a protective or pathological effect [63–65], depending on the subtype of astrocyte involved [43,46], and the type of injury sustained [43,49,62]. In mouse models of ischemic injury or lipopolysaccharide (LPS)-induced infection, changes in astrocyte gene expression were identified in a common set of genes and genes that were uniquely regulated by each type of injury [43]. There are also different "types" of astrocytes that can have potentially different responses to injury. These subtypes are classified based on cellular morphology (e.g., protoplasmic, fibrous, bushy), location (e.g., frontal cortex, hippocampus, cerebellum) and primary functions (e.g., metabolic, structural, signaling) [66]. However, our understanding of the different types of astrocytes in humans and their roles in CNS disorders is still limited [66,67]. Although inclusive of this wide spectrum of phenotypic changes, the term "reactive astrogliosis" will be used here as an umbrella term to encompass the range of possible changes that occur in astrocytes as a result of neurological insult.

In addition to disease heterogeneity and methodological differences, psychotropic medications are hypothesized to normalize disease-associated changes in astrocyte expression [27,66,68], and may contribute to the conflicting reports of altered astrocyte expression reported in psychiatric disorders. Antipsychotic treatment is associated with changes in brain volume in schizophrenia patients [69,70] and in nonhuman primates administered antipsychotic drugs [71,72]. Antidepressant use is associated with reduced astrocyte cell counts in rodent brain [73] and reduced astrocyte marker gene expression in postmortem MDD brain tissue studies [15], although other studies report no changes in these variables in postmortem studies [12,23,24,26,74]. Mood stabilizers are also associated with changes in astrocyte morphology and density in rodent brain [75]. Despite the extensive number of studies examining astrogliosis in postmortem and animal studies of neurological and psychiatric disorders, it is still unclear what effect these medications have on astrocyte expression and function in these disorders.

In this study we apply bioinformatic analysis of transcriptomic datasets to assess the relationship between different astrocyte subtypes (mouse astrocyte subtypes, mouse disease model astrocytes, human astrocytes, human SCZ hiPSC-derived astrocytes) and psychiatric disorders, and between astrocytes and psychotropic medications. Our goal is first to explore how different subtypes of astrocytes are implicated in psychiatric disorders and second, to determine whether psychotropic medications are associated with molecular changes in astrocytes in disease.

2. Materials and Methods

2.1. Mouse Astrocyte Subtype Gene-Sets

The gene-sets for each transcriptomically distinct subtype of mouse astrocyte were downloaded from the supplementary data of Batiuk et al. single-cell RNA sequencing astrocyte study [76]. Full gene-sets are listed in Supplementary Table S1. These gene-sets include genes enriched in each of five subtypes of astrocytes (AST1-5) and genes expressed in at least 60% of astrocytes (common) identified in the study. As described in Batiuk et al., AST1-3 appear to be mature astrocyte subtypes. AST1 is found in the subpial layer and hippocampus. AST2 has high expression in cortical layers 2/3 and 5, lower expression in cortical layers 1, 4, and 6, and negligible hippocampal expression. AST3 subtype is expressed throughout cortical layers and hippocampus. AST4 is found predominately in the subgranular zone of the hippocampus and may represent a progenitor or hippocampal neural stem cell population [76]. AST5 subtype expression overlaps substantially with

AST4 and is hypothesized to represent an intermediate state between progenitor and mature astrocyte. AST5 shows enrichment in cortical layer 2/3 and 5, in the stratum lacunosum-moleculare and dentate gyrus of the hippocampus, and subpial layers.

2.2. Mouse Disease Astrocyte Gene-Sets

Astrocytes gene-sets (GSE 35338) from a mouse model of stroke (middle cerebral arterial occlusion (MCAO), 1 day timepoint) and a mouse model of neuroinflammation (lipopolysaccharide (LPS) 1 day timepoint) as described in [43], were obtained. Gene-sets for mouse disease model astrocytes are listed in Supplementary Table S1.

2.3. Human Astrocyte Gene-Sets

FPKM gene expression values were obtained from BrainRNAseq (www.brainrnaseq org, access on 3 October 2022) from enriched human mature astrocytes and whole cortex homogenate samples [67]. Genes with mean FPKM less than 1 across all samples were removed. To identify genes enriched in human astrocytes, the average log fold change (LFC) in gene expression in mature astrocyte samples relative to the total homogenate was calculated. Genes were binned into six different groups based on difference in expression AST_Bin 1: LFC 0-1, AST_Bin 2: LFC 1-1.5, AST_Bin 3: LFC 1.5-2, AST_Bin 4: LFC 2-2.5, AST_Bin 5: LFC 2.5-3, AST_Bin 6: LFC 3-10. Gene-sets for human astrocytes are listed in Supplementary Table S1.

2.4. Schizophrenia hiPSC-Derived Astrocytes

Three gene-sets of human induced pluripotent stem cells (hiPSC) derived into astrocytes were obtained from [77]. The first gene set consists of 1720 differentially expressed genes (adj $p < 0.1$) from subtype (SUB) compared to classical (CLA) hiPSC astrocytes (AST_hiPSC_SUBvCLA). Subtypes astrocytes are defined as GFAP+ and S100B+ cells with a distinct, non-bushy morphology. The second gene-set consists of 504 differentially expressed genes (adj $p < 0.1$) from schizophrenia (SCZ) compared to healthy control (HC) hiPSC astrocytes (AST_hiPSC_SCZvHC). The third gene-set consists of 430 differentially expressed genes ($p < 0.05$) from schizophrenia clozapine responders compared to clozapine non-responders (AST_hiPSC_SCZ-RvSCZ-NR). Gene-sets are listed in Supplementary Table S1.

2.5. Disease-Disease Similarity

Curated disease-associated gene-sets for major psychiatric disorders major depression (MDD), bipolar disorder (BPD), and SCZ were downloaded from DisGeNET [78]. To avoid size related bias and improve the specificity of pathway profiles [79–81], we further restricted our analysis to diseases with gene-set size between 10 to 500 significant differentially expressed genes. Pairwise-overlap between disease-associated gene-sets was calculated using hypergeometric [82] Gene-Overlap R package version 1.26.0 (http://shenlab-sinai github.io/shenlab-sinai/, accessed on 14 September 2022). Disease-specific curated gene-lists were downloaded from DisGenNET (https://www.disgenet.org/downloads, accessed on 14 September 2022), and see Supplementary Table S1.

2.6. Gene Ontology (GO) Analysis

Pathways affected in different astrocyte subtypes and diseases were determined using hypergeometric overlap analysis (HGA) with a background of 21,196 genes (default, GeneOverlap package). The astrocyte-associated gene-sets and disease-associated gene-sets were tested against GO pathways associated with Biological Process (GOBP), Molecular Function (GOMF), and Cellular Component (GOCC). Updated lists of GO-pathways were obtained from the Bader lab (http://download.baderlab.org/EM_Genesets/, accessed on 16 April 2021). To compare enrichment of pathways across different astrocyte subtypes and diseases, the -log10(p-value) was used to generate the heatmap. To better identify the character of biological changes in the overlap results, a focused analysis of forty a priori functional themes was performed. As described in our previous study [83], the pathways were filtered based on the parent–child association between GO-terms in our list of significant pathways (child-pathways) and hand-picked parent-pathways representing the a priori theme from the GO-database. Gene-sets for GO analysis are listed in Supplementary Table S2.

2.7. Density-Index

To quantitatively summarize how common (close to 1) or unique (close to 0) a theme is across different astrocyte subtypes and diseases, we deployed a previously developed density-index [84]. For a given $r \times c$ matrix of -log10(p-value), a density-score is obtained as:

$$Density = 1 - \left(\frac{count\ of\ \ zero\ elements\ in\ the\ matrix}{r \times c}\right)$$

where r and c represent the number of pathways in a theme and number of astrocyte subtypes and diseases, respectively. A zero element represents a non-significant disease-pathway relationship. For density associated with individual pathways and drug signatures, the density-index representing the fraction of non-zero elements in the number of astrocyte subtypes and diseases was used. In such an instance, the numerator of the above density formula represents count of zero elements in a vector. Whereas r and c in the denominator are constant holding a fixed value of 1 and 27, respectively.

2.8. Drug-Target Enrichment Analysis

Enrichment of astrocyte subtypes and drug-induced molecular signatures in the disease-associated gene-sets was calculated using HGA. Drug-specific gene markers were downloaded from the DSigDB library of gene-sets [85]. In order to understand the druggable-mechanism and targets involved, gene-markers of drugs with known modes of action (MOA) and targets were used. Drugs were then further sorted by the typical diseases which the drug is used to treat. DSigDB gene-sets are listed in Supplementary Table S3.

2.9. Rat Studies

All animal studies were carried out in accordance with the IACUC guidelines at the University of Alabama at Birmingham. Male Sprague Dawley rats were housed with a 12-h cycle of light and darkness. They were pair-housed and given access to food and water ad libitum. Rats were administered with either 28.5 mg kg^{-1} haloperidol-decanoate (n = 10) via intramuscular injection every 3 weeks for 9 months or vehicle-treated with sesame oil (n = 10) under the same conditions. At the end of the 9 months, brains were removed and stored at −80 °C.

2.10. qPCR Studies

RNA was extracted (RNeasy Mini Kit, Qiagen, Germantown, MD, USA) from 14 µm thick fresh frozen sections of rat frontal cortex as previously described [86,87]. RNA was reverse-transcribed (High-Capacity cDNA Reverse Transcription Kit, Applied Biosystems, Life Technologies, Carlsbad, CA, USA) at cycling conditions: 1 cycle: 25 °C for 10 min; 2 cycles: 37 °C for 60 min; 1 cycle: 85 °C for 5 min. cDNA was diluted and equalized to 11 ng/ul. Three microliters of cDNA and 17 µL of Taqman master mix (Applied Biosystems, Life Technologies, Carlsbad, CA, USA) containing 1 µL of Taqman primer were loaded per well and ran at the following cycling conditions: 1 cycle: 95 °C for 10 min; 40 cycles: 95 °C for 15 s, 60 °C for 60 s. Taqman primers for rats assayed were: GFAP (Rn01253033_m1), VIM (Rn00667825_m1), and SOX9 (Rn01751070_m1). The samples were run in duplicate. The relative gene expression was calculated by comparing to a standard curve made of a pool of cDNA from all samples. Gene expression was then normalized to the geomean of reference genes PPIA (Rn00690933_m1) and B2M (Rn00560865_m1). Glial fibrillary acidic protein (GFAP), vimentin (VIM) and Sex-determining region Y (SRY)-box 9 (SOX9) are commonly used astrocyte markers. GFAP is highly expressed in many reactive astrocytes and is upregulated in response to neurological disorders [45]. VIM is a cytoskeletal astrocyte marker and is upregulated in response to neurological injury and is also expressed by progenitor astrocytes [45]. SOX9 is a transcription factor whose expression also increases in response to disease or injury to astrocytes. [45]. Data were analyzed for normal distribution (D'Agostino and Pearson test) and equality of variance (F-test). Data were log transformed and analyzed by Student's *t*-test. $\alpha = 0.05$ for statistical tests. Data were analyzed with GraphPad Prism v7.04 (GraphPad Software, San Diego, CA, USA).

2.11. "Look up" Studies

The R shiny application "Kaleidoscope" (https://kalganem.shinyapps.io/BrainDatabases/ accessed on 1 June 2022) [88], was used to search astrocyte subtype gene-sets (Table S4), including expression of GFAP, VIM, SOX9 and excitatory amino acid transporter 2 (EAAT2, gene name SLC1A2) (Table S5) in publicly available transcriptomic datasets. The data is reported from analysis of rodent brain tissue following at least 2 weeks administration with typical or atypical antipsychotics, antidepressants or mood stabilizers. To determine the effects of antipsychotic medications on astrocyte marker gene expression in human brain, gene names GFAP, VIM, SOX9 and SLC1A2 (EAAT2) were searched in the Stanley Medical Research Institute (SMRI) Online Genomics Database repository [89,90]. The SMRI contains meta-analysis of postmortem brain transcriptomic data from different brain regions (BA6, 8/9, 10, 46, cerebellum) from patients diagnosed with schizophrenia and bipolar disorder who were either on or off antipsychotics and mood stabilizers, respectively, at time of death. The fold change and *p*-value of transcript expression in patients on/off medication is reported. No data is available from the SMRI for MDD patients on or off antidepressant medications.

2.12. Identifying Small Molecules to Reverse Astrocyte-Disease Gene Signatures

The DrugFindR R Shiny application was run as described [91] to identify chemical perturbagens that are highly discordant to the AST_hiPSC_SCZvHC dataset. DrugFindR mines the Library of Network Integrated Signatures (iLINCS) chemical perturbagen repository. Antipsychotic drugs enriched in MCF7 or PC3 cell lines in gene enrichment analysis were not included if identified by DrugFindR analysis to prevent redundancy of findings as DSigDB also utilizes a subset of LINCS signatures [85]. The reported Discordance Score is the negative Pearson correlation coefficient between the disease (AST_hiPSC_SCZvHC) signature and the precomputed iLINCS drug signatures. iLINCS identifies candidate drugs with a score range >+0.2 and <−0.2. We present only those drugs identified with a negative (discordance) score. The lower the discordance score, the more effectively a drug reverses gene expression changes associated with the AST_hiPSC_SCZvHC signature.

2.13. Protein kinase Activity Array

Brain tissue from a non-psychiatrically ill control subject (male, age 73, pH 6.4, PMI 17 h, dorsolateral prefrontal cortex) was obtained from the Alabama Brain Collection with consent from next of kin with IRB approval protocols. Approximately 500 mg of brain tissue was homogenized in 5 mL buffer (0.32 M sucrose and 1 mM EDTA, pH 7.4 + HALT Protease Inhibitor) and centrifuged at 1000× *g* for 10 min at 4 °C. The supernatant was layered onto 7.8 mL each of 2%, 6%, 10%, and 20% *v*/*v* Percoll in SEDH buffer (0.32 M sucrose, 1 mM EDTA, 0.25 mM DTT, and 20 mM HEPES, pH 7.4) and centrifuged at 33,500× *g* for 5 min at 4 °C. The turbid layer between the 2% and 6% Percoll layers was collected, resuspended in SEDH buffer, and centrifuged at 1000× *g* for 20 min at 4 °C. The supernatant-containing gliosome fraction was centrifuged at 33,500× *g* for 40 min at 4 °C. The pellet was resuspended and washed ×2 using the same spin conditions. The pellet was resuspended in MPERS buffer. Two micrograms protein from the enriched gliosome fraction and total homogenate was loaded on a Pamgene chip and analyzed using the Pamgene12 (Pamgene International, 's-Hertogenbosch, The Netherlands) as previously described [92–94].

3. Results

3.1. Astrocyte Subtype Enrichment in Psychiatric Disorders

To determine whether different astrocyte subtypes are enriched in psychiatric disorders, we assessed the enrichment of (1) genes representing six mouse astrocyte subtypes (AST1-5 and "common") [76], (2) genes representing astrocytes from two mouse disease models (AST_MCAO and AST_LPS) [43], (3) genes (in 6 bins) representing human astrocytes (AST_Bin 1-6) [67] and (4) genes representing iPSC-derived astrocytes from classical and subtype astrocytes (AST_hiPSC_SUBvCLA), astrocytes from SCZ and HC subjects (AST_hiPSC_SCZvsHC) and astrocytes from SCZ subjects who were clozapine responders and non-responders (AST_hiPSC_SCZ-RvsSCZ-NR) [77]. Astrocyte gene-set enrichment was examined in the DisGeNET gene-sets representing different psychiatric disorders (Figure 1).

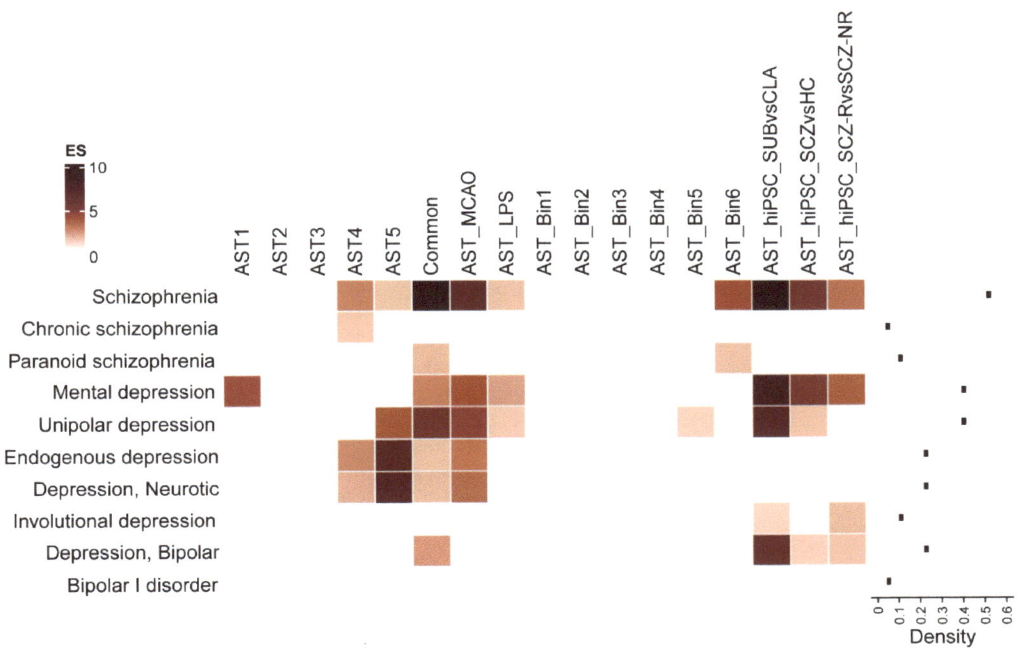

Figure 1. Gene enrichment of mouse astrocyte subtype gene-sets (AST1-5, common), mouse disease model astrocyte gene-sets (AST_MCAO, AST_LPS), and proxy human astrocyte gene-sets (AST_Bin1-6) across neuropsychiatric and neurological disorders (y-axis). Color intensity is proportional to -log10(*p*-value). Density index for all astrocyte gene-sets across all disorders is displayed to right of heatmap. AST astrocyte, CLA astrocyte classical subtype, ES enrichment score, HC healthy control, hiPSC human induced pluripotent stem cell, LPS lipopolysaccharide, MCAO middle cerebral arterial occlusion. SCZ schizophrenia, SCZ_R clozapine responder, SCZ_NR clozapine non-responder, SUB astrocyte subtype.

The set of genes representing the "common" (see details in 2. Materials and Methods) mouse astrocyte subtype were the most frequently enriched, in line with findings that astrocytes play important roles in the pathophysiology of these disorders [66]. The AST4 and AST5 subtypes, proposed to represent progenitor astrocytes and intermediate (progenitor-mature) astrocyte populations, respectively [76], were enriched in a subset of depression and schizophrenia datasets. This may suggest that specific astrocytes are altered in particular disease conditions. The human astrocyte genes contained in bin 6 (AST_Bin 6), which consists of genes with the highest log fold change (LFC 3-10) in expression in human astrocytes relative to total brain, also showed enrichment in schizophrenia but not across the depression datasets. This may be a limitation of the specificity of the "proxy" human astrocyte dataset, as common enrichment was only found with the disease dataset with the highest density index, schizophrenia. Conversely, enrichment of the astrocyte genes associated with the mouse MCAO and LPS disease models and AST_hiPSC genes, in schizophrenia and depression datasets may indicate expression of a common subset of disease–related astrocyte genes whose expression is induced in pathological conditions. Overall, schizophrenia is the disorder with the highest density index (>0.5). As the density index quantitatively compares the significance of disease gene expression to an independent variable (astrocyte population) [84], this suggests greater dysregulation of astrocytes in this disorder compared to other psychiatric disorders, with distinct differences in gene enrichment for certain astrocyte subtypes and in astrocytes in pathological conditions.

3.2. Biological Pathway Enrichment across Astrocyte Subtypes and Neuropsychiatric Disease

Biological pathway enrichment (GOBP, GOMF and GOCC) of astrocyte datasets and psychiatric disorders is shown in Figure 2. Parent–child analysis was applied to cluster pathways [83]. There is enrichment of "immune" related pathways in the astrocyte gene-sets from mouse disease models, SCZ AST_hiPSC (response to cytokine pathways) and psychiatric disorders, but limited enrichment of mouse astrocyte subtypes or human astrocytes, supporting unique changes in astrocyte gene expression associated with pathological conditions. In the metabolism cluster, "carbohydrate metabolism" is enriched in astrocyte subtypes and psychiatric disorders but has limited enrichment in mouse disease models or human hiPSC astrocytes datasets. This may suggest that pathological changes in carbohydrate metabolism are not driven by astrocytes in disease states. "Peptide metabolic process" is highly enriched in AST4 but not other astrocyte subtypes and "response to oxidative stress" is enriched in AST4 and AST5 subtypes and MCAO model but not other subtypes. This supports the idea that different astrocyte subtypes may be implicated in specific biological processes. Conversely, lipid metabolism appears to be an essential function of astrocytes, highly enriched in both the mouse "common" astrocyte subtype and the human AST_Bin 6 dataset. "Proteolyis" pathways are enriched across astrocyte and psychiatric disorder datasets but not in AST_hiPSC datasets which suggests that proteolysis is not perturbed in astrocytes in schizophrenia, or alternatively, may result from differences in gene expression in hiPSC-derived astrocytes compared to astrocytes from in vivo models and human brain tissue.

Signaling cluster pathways, including "modulation of chemical synaptic transmission", "regulation of membrane potential" and "regulation of neurotransmitter levels" are commonly enriched in AST_hiPSC datasets and across psychiatric disorder datasets but have limited enrichment in mouse subtype or human astrocyte datasets, suggesting species and disease specific alteration in astrocyte-related signaling processes. High levels of gene enrichment in the "Synapse" subcluster were largely driven by the hiPSC- derived astrocyte gene-sets and support changes in synaptic transmission and astrocyte sensitivity to synaptic signaling in different human astrocyte subtypes (hIPSC_SUBvCLA), in disease (hIPSC_SCZvHC) and in response to medication (hIPSC_SCZ-NRvSCZ-R) [77]. These findings also implicate altered glutamate system signaling, of which astrocytes are essential regulators via glutamate synthesis and clearance mechanisms [77,95]. "G protein coupled receptor (GPCR) signaling pathways" are highly enriched in psychiatric disorder datasets but have limited enrichment across astrocyte-related datasets. Although astrocytes express GPCRs [96], these results suggest that deficits in GPCR signaling in psychiatric disorders are not related to their astrocyte function. Conversely, "regulation of protein phosphorylation" pathways are enriched across datasets, suggesting a fundamental role for cellular regulatory processes in astrocytes and psychiatric disorders. Overall, pathway analysis identified biological processes that are commonly dysregulated in psychiatric disorders [53–59], and supports a role for astrocytes in a subset of these pathological processes.

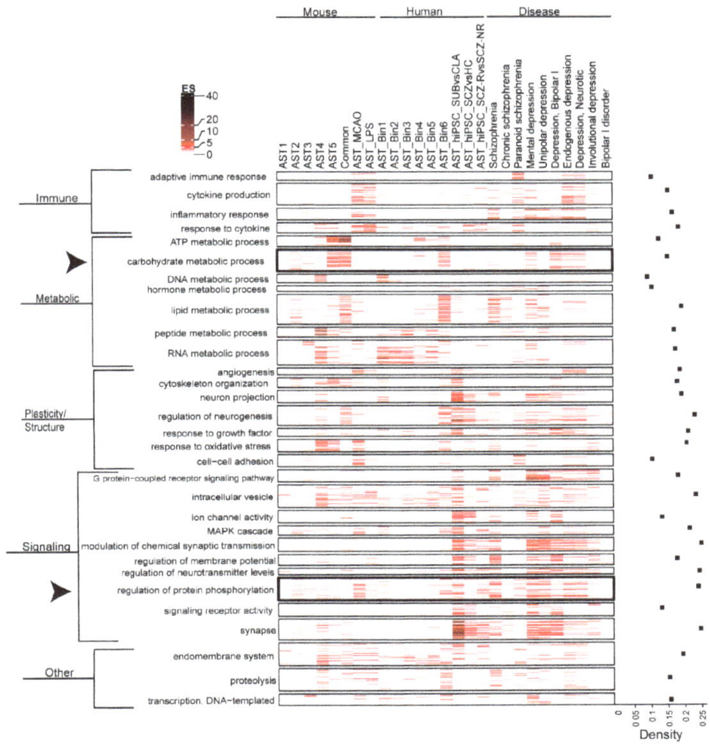

Figure 2. Enrichment of clustered biological pathways across astrocyte gene-sets and neuropsychiatric and neurological disorders. Color intensity is proportional to -log10(p-value). Density index for all pathway clusters across astrocyte groups and disorders is displayed to right of heatmap. Arrowhead highlights pathways "regulation of protein phosphorylation" and "carbohydrate metabolism". AST astrocyte, CLA astrocyte classical subtype, ES enrichment score, HC healthy control, hiPSC human induced pluripotent stem cell, LPS lipopolysaccharide, MCAO middle cerebral arterial occlusion. SCZ schizophrenia, SCZ_R clozapine responder, SCZ_NR clozapine non-responder, SUB astrocyte subtype.

3.3. Psychotropic Medication Effect on Astrocyte Subtypes and Disease

Next, we assessed the gene enrichment of astrocyte subtypes, psychiatric disorders, and the gene-sets representing different types of commonly prescribed psychotropic medications (antipsychotics, antidepressants and mood stabilizers) from the DSigDB library (Figure 3). Overall, gene enrichment of the astrocyte datasets and psychotropic drug datasets is low. In contrast, there are high levels of gene enrichment across psychotropic drug datasets and most psychiatric disorder datasets. All medication classes included at least one medication with a relatively high density index (>0.5). This suggests that psychotropic drugs induce significant changes in gene expression that are common to gene expression associated with psychiatric disorder datasets, but not astrocyte datasets. These findings may also indicate an effect of specific medications on specific astrocyte subtypes. For example, gene expression changes associated with the mood stabilizer lithium chloride are enriched in the mouse AST4 subtype, but not other mouse astrocyte subtypes. Equally, the AST_hiPSC_SUBvsCLA dataset shows low levels of gene enrichment with antidepressant and antipsychotic datasets but moderate levels of gene enrichment with mood stabilizer datasets. These results may suggest that this class of drugs have unique effects on this particular astrocyte subtype. Although drug gene-set enrichment across schizophrenia disease datasets is lower than other disease datasets, there is still common enrichment with specific astrocyte gene-sets and drug datasets (e.g., haloperidol_BOSS, thioridazine_PC3_Up, clozapine_CTD), suggesting that different drugs may target specific subsets of astrocytes and may be a promising therapeutic strategy in this disorder.

Figure 3. Gene enrichment of different classes of psychotropic medications across astrocyte gene-sets and neuropsychiatric and neurological disorders. Color intensity is proportional to -log10(*p*-value). Density index for all medications across astrocyte groups and disorders is displayed on right. AST astrocyte, CLA astrocyte classical subtype, ES enrichment score, HC healthy control, hiPSC human induced pluripotent stem cell, LPS lipopolysaccharide, MCAO middle cerebral arterial occlusion. SCZ schizophrenia, SCZ_R clozapine responder, SCZ_NR clozapine non-responder, SUB astrocyte subtype.

The DSigDB is a collection of drug datasets from different sources and include data generated from varied cell types and tissues that use a range of drug concentrations and dosage time points that may contribute to heterogeneous results. This is reflected in the gene enrichment heatmaps presented here, where common enrichment of astrocyte is highly variable across different drug datasets but is relatively consistent across disease datasets, suggesting robust gene enrichment in disease but not astrocytes.

3.4. Effect of Chronic Medication on Astrocyte Marker Expression in Rat Brain

To further explore the relationship between psychotropic medications and their effects on astrocytes, we carried out a "look-up" study of antipsychotics, antidepressants and mood stabilizers on an extensive range of astrocyte marker genes in rodent frontal cortex. We used those genes defined in the "mouse_common" gene list, which is comprised of genes found in at least 60% of astrocytes analyzed by [76]. Figure 4A and Table S4 shows gene expression (Log2FC) of these astrocyte marker genes in rodent brain following chronic (2 weeks–12 weeks) administration with mood stabilizers, antidepressants or antipsychotics. The largest changes in astrocyte marker expression are induced by mood stabilizers (lithium and valproate) and the weakest effect is seen in antidepressant datasets. Interestingly, gene expression changes following 4 weeks or 12 weeks of antipsychotic (haloperidol and clozapine) administration differ significantly, as previously observed [97]. To further explore differences in the chronic effects of antipsychotics on astrocyte-associated gene expression, we assayed the mRNA levels of three established astrocyte markers, GFAP, VIM, and SOX9, in the frontal cortex of rats administered the typical antipsychotic haloperidol-decanoate for 9 months (36 weeks), as this time-point better models a chronic treatment course in schizophrenia patients. As at the 4- and 12- week time-points (Table S5), no significant differences were detected in the mRNA expression of GFAP, SOX9, or VIM in haloperidol-decanoate treated rats compared to vehicle-treated controls ($p > 0.05$; Figure 4B–D). We have previously reported no significant change in the mRNA expression of the rodent EAAT2 homolog GLT-1, another common astrocyte marker, in the frontal cortex of rats treated chronically with haloperidol-decanoate [86]. These results suggest that chronic haloperidol administration does not significantly alter astrocyte gene marker expression, i.e., it does not make them more reactive. It also highlights the importance of studying the effects of chronic medication treatment in a manner that better recapitulates patient treatment courses. Conversely, GFAP, VIM, and SOX9 mRNA expression were significantly increased in schizophrenia subjects who were "on" antipsychotic medication at time of death (data obtained from SMRI database) but there was no significant difference in astrocyte marker transcript expression in bipolar disorder subjects who were "on" mood stabilizers compared to subjects who were "off" (Figure 4E). These results highlight the importance of conducting translational studies to elucidate the effects of chronic psychotropic medication on the brain of psychiatric patients, as well as studying the effects of chronic medication treatment in a manner that better recapitulates patient treatment courses.

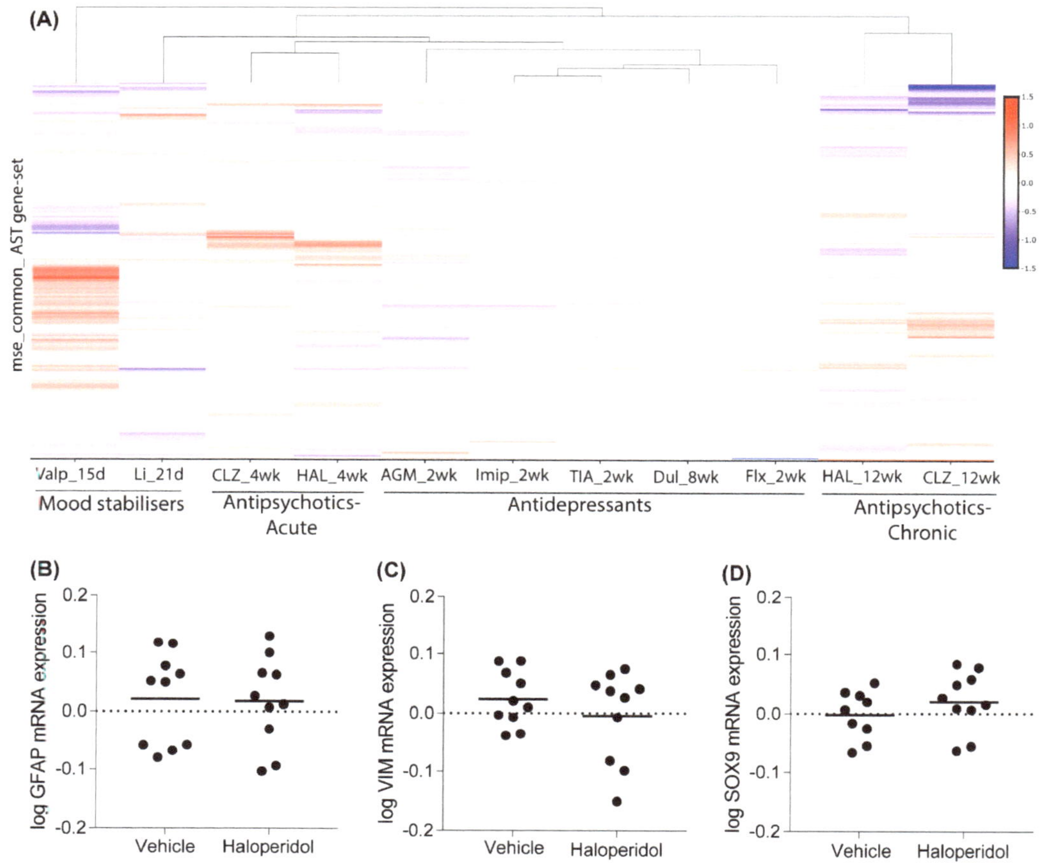

(E)

SMRI	GFAP	VIM	SOX9	EAAT2
On/off antipsychotics	1.15 (p=0.01)*	1.26 (p=0.01)*	1.12 (p=0.002)*	−1.04 (p=0.28)
On/off mood stabilizers	1.05 (p=0.06)	1.06 (p=0.43)	−1.07 (p=0.18)	−1.05 (p= 0.33)

Figure 4. (**A**) Heatmap of log 2 fold change transcript expression of astrocyte marker gene list derived from mouse_common gene-set, in models of psychotropic drug administration. Differences in patterns of gene expression were identified in 12 weeks compared to 4 weeks of antipsychotic administration. In rats treated for 36 weeks (9 months) with the antipsychotic haloperidol decanoate (n = 10/group), frontal cortex mRNA expression of astrocyte markers (**B–D**) GFAP, VIM and SOX9 was not significantly different (Student's t-test, $p > 0.05$) compared to vehicle treated control (n = 10/group). (**E**) Descriptive statistics of gene expression of GFAP, VIM, and SOX9 in the chronic antipsychotic-treated rat study. Data was log-transformed. Data expressed as mean. GFAP glial fibrillary acidic protein, VIM vimentin, SOX9 SRY (sex-determining region Y)-box 9. * $p < 0.05$.

3.5. Exploratory Studies of Common Protein Kinase Activity Profiles in Postmortem Glial Cell Compartments and in Schizophrenia

Protein kinase activity, determined by an increase in phosphorylation signal intensity is higher in a gliosome preparation from postmortem frontal cortex tissue relative to total homogenate from the same sample (Figure 5A,B). This is in line with pathway analysis which suggests that protein phosphorylation activity is significantly enriched across astrocyte populations (see arrowhead, Figure 2). A subset of the kinases identified in the gliosome preparation (Figure 5C) were also identified in our earlier study of kinase activity in schizophrenia postmortem frontal cortex tissue [94], suggesting that changes in astrocyte kinase activity may contribute to signal transduction network perturbations in this disorder. The gliosome preparation represents kinase activity in a single astrocyte compartment and does not exclude a role for dysregulated kinase activity from other astrocyte subcellular compartments in schizophrenia.

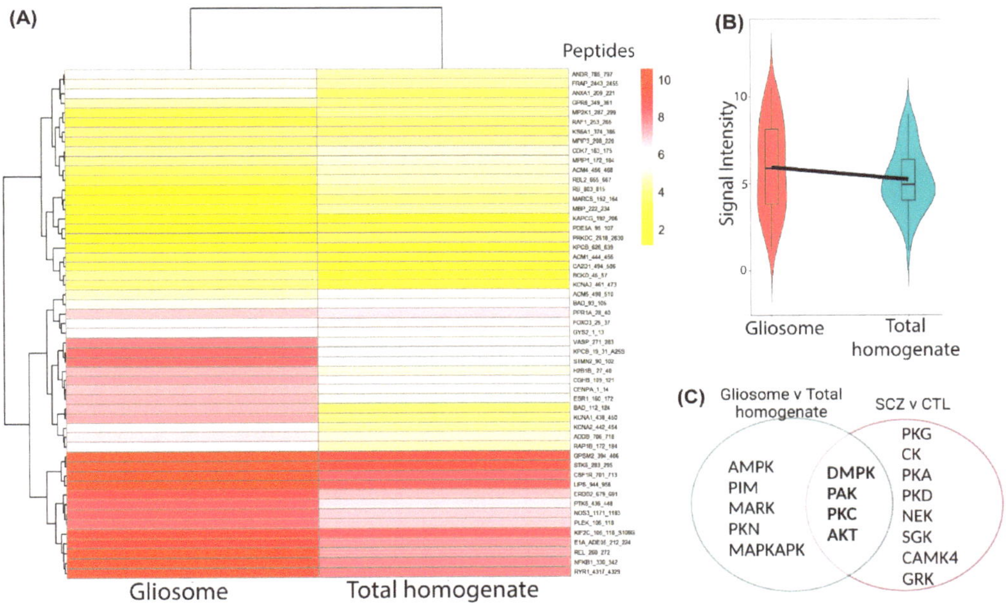

Figure 5. (**A**) Heatmap of Pamgene12 kinome array peptide phosphorylation intensity in a gliosome fraction and total homogenate from postmortem dorsolateral prefrontal cortex. A greater number of peptides are phosphorylated in the gliosome fraction suggesting enrichment of serine/threonine protein kinase activity in this sample. (**B**) Violin plot of peptide phosphorylation signal intensity in gliosome fraction and total homogenate. (**C**) Following peptide mapping, a number of common kinases with differential activity (L2FC +/− 0.2) in the gliosome relative to total homogenate samples and a previous study of postmortem schizophrenia (SCZ) and control (CTL), were identified.

3.6. Chemical Perturbagens That Reverse the hiPSC_SCZvHC Gene Signature

Using DrugFindR, we identified the FDA approved drugs with the greatest discordance (≤ -0.60) with the hiPSC_SCZvHC gene signature (Table 1). This human astrocyte dataset was selected as the best gene-set to identify human disease-relevant discordant chemical perturbagens. Several antipsychotics were identified, although only one, promazine, had a discordance score < -0.60, suggesting that while gene signatures induced by antipsychotics are negatively correlated with the SCZ_AST_hiPSC gene signature, these effects are likely modest, in line with the relatively limited evidence of gene enrichment of astrocyte and psychotropic drug datasets in Figure 3. Interestingly, amongst the other classes of drugs identified in the DrugFindR analysis, several kinase inhibitors (< -0.6) were identified. Pathway analysis suggests that protein phosphorylation activity is dysregulated across psychiatric disorder datasets and different astrocyte populations (Figure 2), and these findings implicate a therapeutic role for modulation of kinases in schizophrenia [94,98].

Table 1. FDA approved drugs that are discordant (discordance score <−0.6) with the AST_hiPSC_SCZvHC gene signature. Drugs were identified using the iLINCS repository of chemical perturbagens. A number of antipsychotics, kinase inhibitors and "other" drugs were identified.

Drug	Score	LINCS ID	Cell Line	Dose	Timepoint	Drug Class
			Antipsychotics			
Promazine	−0.656	LINCSCP_235798	MCF7	10 uM	6 h	antipsychotic
Chlorpromazine	−0.493	LINCSCP_21936	HEPG2	10 uM	6 h	antipsychotic
Trifluoperazine	−0.474	LINCSCP_64508	VCAP	5 uM	24 h	antipsychotic
Phenothiazine	−0.464	LINCSCP_209948	HA1E	10 uM	24 h	antipsychotic
Fluphenazine	−0.460	LINCSCP_22209	HEPG2	10 uM	6 h	antipsychotic
Thioridazine	−0.511	LINCSCP_50729	RKO	10 uM	6 h	antipsychotic
Trifluoperazine	−0.474	LINCSCP_64508	VCAP	5 uM	24 h	antipsychotic
			Kinase inhibitors			
Neratinib	−0.673	LINCSCP_163133	SKBR3	10 uM	3 h	kinase inhibitor
Vemurafenib	−0.658	LINCSCP_162826	SKBR3	1.11 uM	24 h	kinase inhibitor
Tivozanib	−0.653	LINCSCP_128031	HUES3	0.37 uM	24 h	kinase inhibitor
Everolimus	−0.618	LINCSCP_121310	HT29	0.37 uM	24 h	kinase inhibitor
Dacomitinib	−0.602	LINCSCP_84161	ASC	3.33 uM	24 h	kinase inhibitor
			Other drugs			
Oxandrolone	−0.689	LINCSCP_34288	MCF7	10 uM	24 h	Anabolic steroid
Vorinostat	−0.636	LINCSCP_67113	HT29	10 uM	6 h	HDAC inhibitor
Cholic acid	−0.631	LINCSCP_258416	NEU	10 uM	24 h	Bile acid
Aminosalicylic acid	−0.631	LINCSCP_141713	MCF7	10 uM	24 h	Antitubercular agent
Tadalafil	−0.623	LINCSCP_144322	MCF7	0.04 uM	24 h	Phosphodiesterase 5 inhibitor
Omacetaxine mepesuccinate	−0.623	LINCSCP_126799	HT29	0.37 uM	24 h	protein synthesis inhibitor
Abiraterone acetate	−0.613	LINCSCP_153115	PC3	10 uM	24 h	antiandrogen
Mebendazole	−0.611	LINCSCP_27148	HT29	10 uM	6 h	anthelmintic
Bisacodyl	−0.609	LINCSCP_30211	MCF7	10 uM	6 h	stimulant laxative

4. Discussion

Astrocytes play an important, albeit still poorly understood role in many neurological and psychiatric diseases [1–3,66]. Although postmortem studies generally support altered astrocyte expression and morphology in these disorders [6,7,10,20–29], many studies have shown conflicting findings [4–19] or report no changes in astrocytes in disease [27,30–35]. The diversity of astrocyte subtypes [43,46], their responses to different insults [43,49,62] and the potential effects of psychotropic medications on astrocytes [12,15,23–25,69–75] may contribute to challenges in defining and interpreting astrocyte changes in the brain in these disorders.

To improve our understanding of how astrocytes are altered in the brain in disease and the effects of medications on astrocytes, we carried out bioinformatic analysis, taking advantage of emerging single-cell RNAseq datasets that define mouse astrocyte subtypes. As unique molecular signatures for different subtypes of human astrocytes have yet to be elucidated [66,67], we instead used signatures for five different astrocyte subtypes (AST1-5) and a "common" astrocyte signature derived from mouse brain [76]. A proxy human astrocyte gene-set, derived from the Brain-Seq project [67], was included to account for species differences in astrocytes. Finally, to account for changes in astrocyte gene expression due to inherent differences in astrocytes in "healthy" and "pathological" conditions, we also included astrocyte gene-sets derived from two different mouse models of disease (MCAO model and LPS-induced infection model) [43] and from hiPSC derived astrocytes from schizophrenia patients and healthy controls (astrocyte types, disease differences, clozapine responders and non-responders) [77].

4.1. Species, Astrocyte Subtype and Disease-Driven Changes in Gene Enrichment in Psychiatric Disorders

Gene enrichment analysis [84] found that genes associated with different astrocyte datasets were enriched in schizophrenia and depression, supporting a role for astrocytes in these psychiatric disorders. Although the highest level of enrichment across psychiatric disorders was seen with the "common" mouse astrocyte subtype, the AST4-5 subtypes, which are highly expressed in hippocampus and cortical layers II/III and V, and likely represent progenitor or immature astrocytes [76], were also enriched in schizophrenia and depression datasets. Conversely, the human astrocyte gene-sets (AST_Bin 6) were not enriched in depression datasets, but astrocyte gene-sets from the mouse disease models and schizophrenia hiPSC astrocytes were enriched across depression datasets. As changes in astrocyte expression and morphology have been reported in postmortem brain in MDD, including in the hippocampus [11,12,17,19,99–101], these findings highlight the species-specific differences in molecular signatures of astrocyte subtypes [102], and the importance of studying astrocytes under pathological conditions to understand the unique changes in this cell type in different disease states [43,49,62].

There is extensive conservation of gene expression between human and mouse astrocytes [102]. However, whereas genes involved in metabolism are highly enriched in mouse astrocytes, genes with higher enrichment in human astrocytes are implicated in processes described by GO term "defense response", which includes inflammatory responses. Human astrocytes are also more vulnerable to oxidative stress, indicating further differences in mitochondrial metabolism in human and mouse astrocytes [102]. These distinctive physiological responses may contribute to the dissimilarities found in astrocyte gene enrichment in human psychiatric disorders and mouse astrocyte subtypes. Human astrocytes also carry out a wider range of functions including roles in augmenting cognition [103]. Thus, rodent models cannot fully capture the structural, functional and molecular diversity of human astrocyte subtypes, several of which are not found in the rodent brain [66]. Comprehensive molecular profiling of these different, potentially highly specialized cell subtypes in humans, are essential to understand their role in pathology and potential as therapeutic targets.

This is further supported by the diverse patterns of gene enrichment of the mouse disease model astrocyte datasets compared to the astrocyte subtypes from healthy mice. However, differences in astrocyte gene-set enrichment in different disease datasets may also indicate that specific astrocyte subtypes are affected in particular disorders, as seen in mental depression, where mouse AST1, a subtype found in the hippocampus, was uniquely enriched. Overall, our data support a role for astrocytes in psychiatric disorders, but it will be necessary to characterize and study human astrocyte subtypes and their unique responses to different pathological states to fully understand their role in the brain and disease.

Next, we looked at how astrocyte function relates to psychiatric disorders and saw that many of the same biological pathways were enriched in mouse astrocyte subtypes, in the mouse disease model astrocytes, the human astrocytes, and in psychiatric disorders. Pathway analysis identified common enrichment in the "metabolic", "plasticity/structure" and "signaling" pathway clusters, highlighting the role of astrocytes in these processes and the dysregulation of these processes in schizophrenia and depression [66,104,105]. Unsurprisingly, the "immune" related pathways were highly enriched in both mouse disease model astrocyte datasets compared to the other astrocyte subtypes, and in depression and schizophrenia. A subset of pathways, "response to cytokine", were also enriched in AST_hiPSC datasets suggesting a targeted immune response in astrocytes derived from schizophrenia patients. Neuroimmune dysregulation is commonly reported in psychiatric disorders and may contribute to the underlying pathophysiology [13,14,66,104,105]. This finding also highlights the functional differences in astrocytes in disease states compared to healthy conditions.

The "carbohydrate metabolism" pathway cluster is enriched in astrocyte subtypes and psychiatric disorders but has limited enrichment in mouse disease models or AST_hiPSC schizophrenia datasets. Astrocytes play an essential role in ensuring the bioenergetic requirements of neurons are met in the brain [106]. We have previously shown significant deficits in the expression and enzyme activity of hexokinase, a rate-limiting enzyme in glucose metabolism, in pyramidal neurons but not in astrocytes, in the brain in schizophrenia [107]. Thus, this gene enrichment analysis confirms a role for astrocytes in carbohydrate metabolism, but does not support major deficits in these biological processes in astrocytes in disease.

Additionally, "regulation of protein phosphorylation" was enriched across astrocyte and psychiatric disorder datasets, implicating these processes in astrocytes under physiological and pathological conditions. We confirmed enrichment of protein kinase activity in a gliosome preparation relative to total homogenate in human postmortem brain tissue, and found that changes in activity in a subset of these kinases are also implicated in schizophrenia [94,98]. This suggests that the gene enrichment associated with astrocytes and psychiatric disorder datasets generated by bioinformatic analysis reflect biological processes associated with these cells and lends additional support for dysregulation of these processes in disease.

A potentially confounding variable in understanding the role of astrocytes in disease is the effect of psychotropic medications that may be normalizing disease-related changes in astrocytes, or inducing structural, molecular, or functional changes in these cells [23,24,26,73,75,108,109]. Postmortem studies that have considered the effects of psychotropic medications have found changes in astrocyte cell counts [73], morphology [75] and gene expression markers [15]. However, it is not clear if these changes reflect a therapeutic response to medication, an effect of the disease process, or a combination of both. Additionally, if astrocytes are altered in the disease state or as a result of medication, this may support their role as a therapeutic target. To address these questions, we examined enrichment of gene signatures induced by commonly prescribed psychotropic drugs in the different astrocyte and psychiatric disorder datasets. Overall, psychotropic drug gene lists were highly enriched in depression (adding support for off-label use of antipsychotics for the treatment of MDD [110]) and to a lesser extent in schizophrenia, but there was surprisingly limited enrichment across astrocyte datasets. For example, astrocytes express the dopamine D2 receptor [111], the major pharmacological target of currently available antipsychotic medications, and 5-HT2B, a serotonin receptor which is a target of all selective serotonin reuptake inhibitor (SSRI) administration [112,113]. However, our findings suggest that psychotropic drugs do not induce significant changes in astrocyte-associated gene-sets. Higher levels of gene enrichment of the mouse disease model astrocytes and the human astrocyte datasets support the idea that there are inherent differences in the way that psychotropic medications affect astrocytes in different species and in the healthy versus disease state, as discussed [43,49,62]. Notably, we found enrichment of the AST_hiPSC_SCZvsHC gene-set with clozapine drug datasets but not with the AST_hiPSC_SCZ-RvsSCZ-NR (clozapine responder's vs. non-responders) dataset. This may reflect a limitation of the drug datasets available in DSigDB, or that the threshold for identifying significant differentially expressed genes was lower for this dataset.

While the general lack of consistent enrichment of drug-induced genes in astrocyte datasets is somewhat surprising [114,115], our data also suggest that specific astrocyte subtypes may be affected by specific medications, as seen with the unique enrichment of lithium chloride and the mouse AST4 subtype gene-set and the hiPSC_SUBvsCLA dataset. We also found enrichment of mouse MCAO astrocytes, and lithium chloride gene-sets, in line with a recent report showing a neuroprotective effect of lithium in ischemic stroke [116]. This further supports that different astrocytes have diverse responses to different drugs and pathological conditions.

Overall, these bioinformatic findings suggest that astrocytes are not significant primary targets of psychotropic medications used in the treatment of psychiatric disorders but different subtypes of astrocyte may be targets under specific conditions. In support of this, we found that chronic antipsychotic administration did not significantly affect astrocyte marker (GFAP, VIM, SOX9 and EAAT2) transcript expression in the frontal cortex in rodent brain. Future studies will be required to examine marker expression in other brain regions. "Look-up" studies of rodent transcriptomic brain datasets also found few reports of significantly altered astrocyte marker transcript expression changes in response to different psychotropic medications, suggesting that chronic medication administration has a limited effect on transcript expression of established glial markers in rodent models. Importantly, "look-up" studies did find significant increases in expression of astrocyte gene markers in postmortem brain of schizophrenia subjects who were "on" antipsychotics compared to those who were "off" antipsychotics but not in bipolar disorder subjects who were "on" mood stabilizers compared to those who were "off," highlighting the importance of considering specific disease- and drug-interaction effects on astrocytes. Future studies should consider a broader range of astrocyte genes to better capture understand the diversity of astrocyte subtypes and their responses in pathological states.

Finally, we mined the LINCS chemical perturbagen gene signature repository to identify FDA-approved small molecules that could potentially reverse the gene expression changes associated with astrocytes in schizophrenia (AST_hiPSC_SCZvHC). Kinase inhibitors were among the top drug hits identified by this analysis and typically had stronger discordance scores than antipsychotics, suggesting that modulating astrocytic perturbation of kinases may have therapeutic potential in schizophrenia.

4.2. Limitations

A necessary limitation of this study is the use of astrocyte subtype molecular signatures derived from healthy mice for bioinformatic analysis, which may not reflect the subtypes of astrocytes found in human brain [76,102]. Additionally, the gene-sets (Bins 1-6) used as a proxy for a human astrocyte gene signature are derived from the log fold change of gene expression in enriched astrocyte populations relative to the expression in total brain homogenate, and may not represent true human astrocyte molecular signatures. The analyses described here utilize transcript-level data. Many drugs act at the protein level rather than at the transcript level and their effects on astrocytes may not be fully captured by gene enrichment analysis. Finally, qPCR and lookup studies used GFAP, VIM, SOX9 and EAAT2 transcript expression as astrocyte markers. These are commonly used astrocyte markers with conserved expression across species that play unique roles in the reactive astrogliosis response [117–120], but they may not represent all astrocyte subtypes or molecular changes associated with disease. Future studies should also consider how other, lesser studied astrocyte markers are altered in disease states and in response to medication.

5. Conclusions and Future Directions

In summary, this study provides a unique analysis of archived transcriptomic data and emerging molecular characterization of astrocyte subtypes to illuminate the relationship between psychotropic medications and astrocytes in the context of psychiatric disorders. We identified gene enrichment of distinct subtypes of astrocytes in psychiatric disorders but surprisingly, minimal enrichment of psychotropic medication-induced gene expression in astrocyte subtypes. This bioinformatic analysis was supported (generally) by "lookup" studies and qPCR in rodents, suggesting that these medications have limited effects on inducing gene expression changes in astrocytes. However, our results highlight the importance of conducting truly translational studies, including postmortem brain studies that account for the effects of medication, due to the differences in astrocyte gene-sets in different species, cell subtypes and disease states. Interestingly, when we identified the biological processes associated with different astrocyte subtypes, regulation of protein phosphorylation was implicated across astrocyte and disease datasets. In combination with our exploratory studies of gliosome protein kinase activity, and in silico analysis of chemical perturbagens that reverse astrocyte-disease related gene signatures, our findings suggest that dysregulated astrocyte kinase activity is potentially a novel therapeutic target in schizophrenia. However, further work is required to elucidate the role of altered regulation of protein phosphorylation in astrocytes in the brain in psychiatric disorders. These findings serve as a starting point to better understand the complex relationship between psychotropic medications and astrocytes in psychiatric disorders, which could have significant consequences on our understanding of the pathophysiology of disease and potentially offer insight into targeting astrocytes therapeutically.

Supplementary Materials: The following supporting information can be downloaded at: https://www.mdpi.com/article/10.3390/cells11203315/s1, Table S1: Gene lists used for enrichment analysis; Table S2: Pathway enrichment in astrocyte and psychiatric disorder datasets; Table S3: Drug (DSigDB) gene sets; Table S4: Look-up" study of mouse_astrocyte "common" gene-set and psychotropic model datasets (Kaleidoscope); Table S5: GFAP, VIM, SOX9, EAAT2 in rodent psychotropic drug administration "look-up" study.

Author Contributions: R.S., X.Z. and X.W. conducted bioinformatic analyses. X.Z., X.W. and A.I. generated bioinformatic figures. A.W. conducted qPCR and look-up studies and wrote first draft of manuscript. R.A. and A.-r.H. conducted kinome assay and kinome data analysis. A.I. conducted DrugFindR analysis. S.M.O. conceived and directed the project and wrote the final manuscript with input from all co-authors. R.E.M. and V.P. provided expert guidance on data interpretation. All authors have read and agreed to the published version of the manuscript.

Funding: This research was funded by MH107487 and MH121102 (R.E.M.), and GM123971 (V.P.). This project was supported by Grant YIG-1-139-20 awarded to SMOD from the American Foundation for Suicide Prevention. The content is solely the responsibility of the authors and does not necessarily represent the official views of the American Foundation for Suicide Prevention.

Institutional Review Board Statement: Not applicable.

Informed Consent Statement: Not applicable.

Data Availability Statement: Publicly available datasets accessed in this study: DisGenNET (https://www.disgenet.org/downloads); Gene Ontology annotation (http://download.baderlab.org/EM_Genesets/); DrugFindR (https://github.com/CogDisResLab/drugfindR); Kaleidoscope (https://kalganem.shinyapps.io/BrainDatabases/). All other genesets available directly from cited material.

Conflicts of Interest: The authors have declared no competing interest.

References

1. Pekny, M.; Pekna, M.; Messing, A.; Steinhauser, C.; Lee, J.M.; Parpura, V.; Hol, E.M.; Sofroniew, M.V.; Verkhratsky, A. Astrocytes: A central element in neurological diseases. *Acta Neuropathol.* **2016**, *131*, 323–345. [CrossRef] [PubMed]
2. Kettenmann, H.; Ransom, B.R. *Neuroglia*, 3rd ed.; Oxford University Press: New York, NY, USA, 2013; 930p.
3. Chen, Y.; Qin, C.; Huang, J.; Tang, X.; Liu, C.; Huang, K.; Xu, J.; Guo, G.; Tong, A.; Zhou, L. The role of astrocytes in oxidative stress of central nervous system: A mixed blessing. *Cell Prolif.* **2020**, *53*, e12781. [CrossRef] [PubMed]
4. Toro, C.T.; Hallak, J.E.; Dunham, J.S.; Deakin, J.F. Glial fibrillary acidic protein and glutamine synthetase in subregions of prefrontal cortex in schizophrenia and mood disorder. *Neurosci. Lett.* **2006**, *404*, 276–281. [CrossRef]
5. Markova, E.; Markov, I.; Revishchin, A.; Okhotin, V.; Sulimov, G. 3-D Golgi and image analysis of the olfactory tubercle in schizophrenia. *Anal. Quant. Cytol. Histol.* **2000**, *22*, 178–182.
6. Feresten, A.H.; Barakauskas, V.; Ypsilanti, A.; Barr, A.M.; Beasley, C.L. Increased expression of glial fibrillary acidic protein in prefrontal cortex in psychotic illness. *Schizophr. Res.* **2013**, *150*, 252–257. [CrossRef]
7. Qi, X.R.; Kamphuis, W.; Shan, L. Astrocyte Changes in the Prefrontal Cortex from Aged Non-suicidal Depressed Patients. *Front. Cell. Neurosci.* **2019**, *13*, 503. [CrossRef]
8. Rao, J.S.; Harry, G.J.; Rapoport, S.I.; Kim, H.W. Increased excitotoxicity and neuroinflammatory markers in postmortem frontal cortex from bipolar disorder patients. *Mol. Psychiatry* **2010**, *15*, 384–392. [CrossRef]
9. Williams, M.R.; Hampton, T.; Pearce, R.K.; Hirsch, S.R.; Ansorge, O.; Thom, M.; Maier, M. Astrocyte decrease in the subgenual cingulate and callosal genu in schizophrenia. *Eur. Arch. Psychiatry Clin. Neurosci.* **2013**, *263*, 41–52. [CrossRef]

10. Williams, M.; Pearce, R.K.; Hirsch, S.R.; Ansorge, O.; Thom, M.; Maier, M. Fibrillary astrocytes are decreased in the subgenual cingulate in schizophrenia. *Eur. Arch. Psychiatry Clin. Neurosci.* **2014**, *264*, 357–362. [CrossRef]
11. Altshuler, L.L.; Abulseoud, O.A.; Foland-Ross, L.; Bartzokis, G.; Chang, S.; Mintz, J.; Hellemann, G.; Vinters, H.V. Amygdala astrocyte reduction in subjects with major depressive disorder but not bipolar disorder. *Bipolar Disord.* **2010**, *12*, 541–549. [CrossRef]
12. Torres-Platas, S.G.; Nagy, C.; Wakid, M.; Turecki, G.; Mechawar, N. Glial fibrillary acidic protein is differentially expressed across cortical and subcortical regions in healthy brains and downregulated in the thalamus and caudate nucleus of depressed suicides. *Mol. Psychiatry* **2016**, *21*, 509–515. [CrossRef]
13. Choudary, P.V.; Molnar, M.; Evans, S.J.; Tomita, H.; Li, J.Z.; Vawter, M.P.; Myers, R.M.; Bunney, W.E., Jr.; Akil, H.; Watson, S.J.; et al. Altered cortical glutamatergic and GABAergic signal transmission with glial involvement in depression. *Proc. Natl. Acad. Sci. USA* **2005**, *102*, 15653–15658. [CrossRef]
14. Zhao, J.; Verwer, R.W.; van Wamelen, D.J.; Qi, X.R.; Gao, S.F.; Lucassen, P.J.; Swaab, D.F. Prefrontal changes in the glutamate-glutamine cycle and neuronal/glial glutamate transporters in depression with and without suicide. *J. Psychiatr. Res.* **2016**, *82*, 8–15. [CrossRef]
15. Fatemi, S.H.; Laurence, J.A.; Araghi-Niknam, M.; Stary, J.M.; Schulz, S.C.; Lee, S.; Gottesman, I.I. Glial fibrillary acidic protein is reduced in cerebellum of subjects with major depression, but not schizophrenia. *Schizophr. Res.* **2004**, *69*, 317–323. [CrossRef]
16. Arnold, S.E.; Franz, B.R.; Trojanowski, J.Q.; Moberg, P.J.; Gur, R.E. Glial fibrillary acidic protein-immunoreactive astrocytosis in elderly patients with schizophrenia and dementia. *Acta Neuropathol.* **1996**, *91*, 269–277. [CrossRef]
17. Muller, M.B.; Lucassen, P.J.; Yassouridis, A.; Hoogendijk, W.J.; Holsboer, F.; Swaab, D.F. Neither major depression nor glucocorticoid treatment affects the cellular integrity of the human hippocampus. *Eur. J. Neurosci.* **2001**, *14*, 1603–1612. [CrossRef]
18. Hercher, C.; Chopra, V.; Beasley, C.L. Evidence for morphological alterations in prefrontal white matter glia in schizophrenia and bipolar disorder. *J. Psychiatry Neurosci.* **2014**, *39*, 376–385. [CrossRef]
19. Johnston-Wilson, N.L.; Sims, C.D.; Hofmann, J.P.; Anderson, L.; Shore, A.D.; Torrey, E.F.; Yolken, R.H. Disease-specific alterations in frontal cortex brain proteins in schizophrenia, bipolar disorder, and major depressive disorder. The Stanley Neuropathology Consortium. *Mol. Psychiatry* **2000**, *5*, 142–149. [CrossRef]
20. Pakkenberg, B. Pronounced reduction of total neuron number in mediodorsal thalamic nucleus and nucleus accumbens in schizophrenics. *Arch. Gen. Psychiatry* **1990**, *47*, 1023–1028. [CrossRef]
21. Barley, K.; Dracheva, S.; Byne, W. Subcortical oligodendrocyte- and astrocyte-associated gene expression in subjects with schizophrenia, major depression and bipolar disorder. *Schizophr. Res.* **2009**, *112*, 54–64. [CrossRef]
22. Bowley, M.P.; Drevets, W.C.; Ongur, D.; Price, J.L. Low glial numbers in the amygdala in major depressive disorder. *Biol. Psychiatry* **2002**, *52*, 404–412. [CrossRef]
23. Ongur, D.; Drevets, W.C.; Price, J.L. Glial reduction in the subgenual prefrontal cortex in mood disorders. *Proc. Natl. Acad. Sci. USA* **1998**, *95*, 13290–13295. [CrossRef]
24. Rajkowska, G.; Miguel-Hidalgo, J.J.; Wei, J.; Dilley, G.; Pittman, S.D.; Meltzer, H.Y.; Overholser, J.C.; Roth, B.L.; Stockmeier, C.A. Morphometric evidence for neuronal and glial prefrontal cell pathology in major depression. *Biol. Psychiatry* **1999**, *45*, 1085–1098. [CrossRef]
25. Cotter, D.; Mackay, D.; Chana, G.; Beasley, C.; Landau, S.; Everall, I.P. Reduced neuronal size and glial cell density in area 9 of the dorsolateral prefrontal cortex in subjects with major depressive disorder. *Cereb. Cortex* **2002**, *12*, 386–394. [CrossRef]
26. Cotter, D.; Mackay, D.; Landau, S.; Kerwin, R.; Everall, I. Reduced glial cell density and neuronal size in the anterior cingulate cortex in major depressive disorder. *Arch. Gen. Psychiatry* **2001**, *58*, 545–553. [CrossRef] [PubMed]
27. Catts, V.S.; Wong, J.; Fillman, S.G.; Fung, S.J.; Shannon Weickert, C. Increased expression of astrocyte markers in schizophrenia: Association with neuroinflammation. *Aust. N. Z. J. Psychiatry* **2014**, *48*, 722–734. [CrossRef]
28. Torres-Platas, S.G.; Hercher, C.; Davoli, M.A.; Maussion, G.; Labonte, B.; Turecki, G.; Mechawar, N. Astrocytic hypertrophy in anterior cingulate white matter of depressed suicides. *Neuropsychopharmacology* **2011**, *36*, 2650–2658. [CrossRef]
29. Strakowski, S.M.; Adler, C.M.; Almeida, J.; Altshuler, L.L.; Blumberg, H.P.; Chang, K.D.; DelBello, M.P.; Frangou, S.; McIntosh, A.; Phillips, M.L.; et al. The functional neuroanatomy of bipolar disorder: A consensus model. *Bipolar Disord.* **2012**, *14*, 313–325. [CrossRef]
30. Steffek, A.E.; McCullumsmith, R.E.; Haroutunian, V.; Meador-Woodruff, J.H. Cortical expression of glial fibrillary acidic protein and glutamine synthetase is decreased in schizophrenia. *Schizophr. Res.* **2008**, *103*, 71–82. [CrossRef]
31. Trepanier, M.O.; Hopperton, K.E.; Mizrahi, R.; Mechawar, N.; Bazinet, R.P. Postmortem evidence of cerebral inflammation in schizophrenia: A systematic review. *Mol. Psychiatry* **2016**, *21*, 1009–1026. [CrossRef]
32. Pantazopoulos, H.; Woo, T.U.; Lim, M.P.; Lange, N.; Berretta, S. Extracellular matrix-glial abnormalities in the amygdala and entorhinal cortex of subjects diagnosed with schizophrenia. *Arch. Gen. Psychiatry* **2010**, *67*, 155–166. [CrossRef] [PubMed]
33. Damadzic, R.; Bigelow, L.B.; Krimer, L.S.; Goldenson, D.A.; Saunders, R.C.; Kleinman, J.E.; Herman, M.M. A quantitative immunohistochemical study of astrocytes in the entorhinal cortex in schizophrenia, bipolar disorder and major depression: Absence of significant astrocytosis. *Brain Res. Bull.* **2001**, *55*, 611–618. [CrossRef]
34. Dean, B.; Gray, L.; Scarr, E. Regionally specific changes in levels of cortical S100beta in bipolar 1 disorder but not schizophrenia. *Aust. N. Z. J. Psychiatry* **2006**, *40*, 217–224. [CrossRef]

35. Webster, M.J.; Knable, M.B.; Johnston-Wilson, N.; Nagata, K.; Inagaki, M.; Yolken, R.H. Immunohistochemical localization of phosphorylated glial fibrillary acidic protein in the prefrontal cortex and hippocampus from patients with schizophrenia, bipolar disorder, and depression. *Brain Behav. Immun.* **2001**, *15*, 388–400. [CrossRef] [PubMed]
36. Benes, F.M.; Davidson, J.; Bird, E.D. Quantitative cytoarchitectural studies of the cerebral cortex of schizophrenics. *Arch. Gen Psychiatry* **1986**, *43*, 31–35. [CrossRef] [PubMed]
37. Bruton, C.J.; Crow, T.J.; Frith, C.D.; Johnstone, E.C.; Owens, D.G.; Roberts, G.W. Schizophrenia and the brain: A prospective clinico-neuropathological study. *Psychol. Med.* **1990**, *20*, 285–304. [CrossRef] [PubMed]
38. Rajkowska, G.; Miguel-Hidalgo, J.J.; Makkos, Z.; Meltzer, H.; Overholser, J.; Stockmeier, C. Layer-specific reductions in GFAP reactive astroglia in the dorsolateral prefrontal cortex in schizophrenia. *Schizophr. Res.* **2002**, *57*, 127–138. [CrossRef]
39. Schmitt, A.; Steyskal, C.; Bernstein, H.G.; Schneider-Axmann, T.; Parlapani, E.; Schaeffer, E.L.; Gattaz, W.F.; Bogerts, B. Schmitz, C.; Falkai, P. Stereologic investigation of the posterior part of the hippocampus in schizophrenia. *Acta Neuropathol* **2009**, *117*, 395–407. [CrossRef] [PubMed]
40. Verkhratsky, A.; Zorec, R.; Parpura, V. Stratification of astrocytes in healthy and diseased brain. *Brain Pathol.* **2017**, *27*, 629–644 [CrossRef]
41. Verkhratsky, A.; Rodriguez, J.J.; Steardo, L. Astrogliopathology: A central element of neuropsychiatric diseases? *Neuroscientist* **2014**, *20*, 576–588. [CrossRef] [PubMed]
42. Verkhratsky, A.; Augusto-Oliveira, M.; Pivoriunas, A.; Popov, A.; Brazhe, A.; Semyanov, A. Astroglial asthenia and loss of function, rather than reactivity, contribute to the ageing of the brain. *Pflüg. Arch.* **2021**, *473*, 753–774. [CrossRef]
43. Zamanian, J.L.; Xu, L.; Foo, L.C.; Nouri, N.; Zhou, L.; Giffard, R.G.; Barres, B.A. Genomic analysis of reactive astrogliosis. *J Neurosci.* **2012**, *32*, 6391–6410. [CrossRef] [PubMed]
44. Anderson, M.A.; Burda, J.E.; Ren, Y.; Ao, Y.; O'Shea, T.M.; Kawaguchi, R.; Coppola, G.; Khakh, B.S.; Deming, T.J.; Sofroniew, M.V Astrocyte scar formation aids central nervous system axon regeneration. *Nature* **2016**, *532*, 195–200. [CrossRef] [PubMed]
45. Escartin, C.; Galea, E.; Lakatos, A.; O'Callaghan, J.P.; Petzold, G.C.; Serrano-Pozo, A.; Steinhauser, C.; Volterra, A. Carmignoto, G.; Agarwal, A.; et al. Reactive astrocyte nomenclature, definitions, and future directions. *Nat. Neurosci.* **2021**, *24* 312–325.
46. Sofroniew, M.V. Molecular dissection of reactive astrogliosis and glial scar formation. *Trends Neurosci.* **2009**, *32*, 638–647. [CrossRef]
47. Eddleston, M.; Mucke, L. Molecular profile of reactive astrocytes—Implications for their role in neurologic disease. *Neuroscience* **1993**, *54*, 15–36. [CrossRef]
48. Pekny, M.; Nilsson, M. Astrocyte activation and reactive gliosis. *Glia* **2005**, *50*, 427–434. [CrossRef]
49. Wilhelmsson, U.; Bushong, E.A.; Price, D.L.; Smarr, B.L.; Phung, V.; Terada, M.; Ellisman, M.H.; Pekny, M. Redefining the concept of reactive astrocytes as cells that remain within their unique domains upon reaction to injury. *Proc. Natl. Acad. Sci. USA* **2006**, *103*, 17513–17518. [CrossRef]
50. Sofroniew, M.V. Reactive astrocytes in neural repair and protection. *Neuroscientist* **2005**, *11*, 400–407. [CrossRef]
51. Maragakis, N.J.; Rothstein, J.D. Mechanisms of Disease: Astrocytes in neurodegenerative disease. *Nat. Clin. Pract. Neurol* **2006**, *2*, 679–689. [CrossRef]
52. Correa-Cerro, L.S.; Mandell, J.W. Molecular mechanisms of astrogliosis: New approaches with mouse genetics. *J. Neuropathol Exp. Neurol.* **2007**, *66*, 169–176. [CrossRef]
53. Zador, Z.; Stiver, S.; Wang, V.; Manley, G.T. Role of aquaporin-4 in cerebral edema and stroke. *Handb. Exp. Pharmacol* **2009**, *190*, 159–170. [CrossRef]
54. Simard, M.; Nedergaard, M. The neurobiology of glia in the context of water and ion homeostasis. *Neuroscience* **2004**, *129*, 877–896 [CrossRef]
55. Swanson, R.A.; Ying, W.; Kauppinen, T.M. Astrocyte influences on ischemic neuronal death. *Curr. Mol. Med.* **2004**, *4*, 193–205 [CrossRef]
56. Chen, Y.; Vartiainen, N.E.; Ying, W.; Chan, P.H.; Koistinaho, J.; Swanson, R.A. Astrocytes protect neurons from nitric oxide toxicity by a glutathione-dependent mechanism. *J. Neurochem.* **2001**, *77*, 1601–1610. [CrossRef] [PubMed]
57. Hamby, M.E.; Hewett, J.A.; Hewett, S.J. TGF-beta1 potentiates astrocytic nitric oxide production by expanding the population of astrocytes that express NOS-2. *Glia* **2006**, *54*, 566–577. [CrossRef]
58. Christopherson, K.S.; Ullian, E.M.; Stokes, C.C.; Mullowney, C.E.; Hell, J.W.; Agah, A.; Lawler, J.; Mosher, D.F. Bornstein, P.; Barres, B.A. Thrombospondins are astrocyte-secreted proteins that promote CNS synaptogenesis. *Cell* **2005**, *120*, 421–433. [CrossRef] [PubMed]
59. Stevens, B.; Allen, N.J.; Vazquez, L.E.; Howell, G.R.; Christopherson, K.S.; Nouri, N.; Micheva, K.D.; Mehalow, A.K Huberman, A.D.; Stafford, B.; et al. The classical complement cascade mediates CNS synapse elimination. *Cell* **2007**, *131* 1164–1178. [CrossRef] [PubMed]
60. Burda, J.E.; Sofroniew, M.V. Reactive gliosis and the multicellular response to CNS damage and disease. *Neuron* **2014**, *81*, 229–248 [CrossRef]
61. Sofroniew, M.V.; Vinters, H.V. Astrocytes: Biology and pathology. *Acta Neuropathol.* **2010**, *119*, 7–35. [CrossRef]
62. Sofroniew, M.V. Astrogliosis. *Cold Spring Harb. Perspect. Biol.* **2014**, *7*, a020420. [CrossRef]
63. Verkhratsky, A.; Rodrigues, J.J.; Pivoriunas, A.; Zorec, R.; Semyanov, A. Astroglial atrophy in Alzheimer's disease. *Pflüg. Arch* **2019**, *471*, 1247–1261. [CrossRef]

64. Middeldorp, J.; Hol, E.M. GFAP in health and disease. *Prog. Neurobiol.* **2011**, *93*, 421–443. [CrossRef]
65. Norton, W.T.; Aquino, D.A.; Hozumi, I.; Chiu, F.C.; Brosnan, C.F. Quantitative aspects of reactive gliosis: A review. *Neurochem. Res.* **1992**, *17*, 877–885. [CrossRef]
66. Zhang, X.; Alnafisah, R.S.; Hamoud, A.A.; Shukla, R.; Wen, Z.; McCullumsmith, R.E.; O'Donovan, S.M. Role of Astrocytes in Major Neuropsychiatric Disorders. *Neurochem. Res.* **2021**, *46*, 2715–2730. [CrossRef]
67. Zhang, Y.; Sloan, S.A.; Clarke, L.E.; Caneda, C.; Plaza, C.A.; Blumenthal, P.D.; Vogel, H.; Steinberg, G.K.; Edwards, M.S.; Li, G.; et al. Purification and Characterization of Progenitor and Mature Human Astrocytes Reveals Transcriptional and Functional Differences with Mouse. *Neuron* **2016**, *89*, 37–53. [CrossRef] [PubMed]
68. Testen, A.; Kim, R.; Reissner, K.J. High-Resolution Three-Dimensional Imaging of Individual Astrocytes Using Confocal Microscopy. *Curr. Protoc. Neurosci.* **2020**, *91*, e92. [CrossRef]
69. Ho, B.C.; Andreasen, N.C.; Ziebell, S.; Pierson, R.; Magnotta, V. Long-term antipsychotic treatment and brain volumes: A longitudinal study of first-episode schizophrenia. *Arch. Gen. Psychiatry* **2011**, *68*, 128–137. [CrossRef]
70. Lieberman, J.A.; Tollefson, G.D.; Charles, C.; Zipursky, R.; Sharma, T.; Kahn, R.S.; Keefe, R.S.; Green, A.I.; Gur, R.E.; McEvoy, J.; et al. Antipsychotic drug effects on brain morphology in first-episode psychosis. *Arch. Gen. Psychiatry* **2005**, *62*, 361–370. [CrossRef]
71. Konopaske, G.T.; Dorph-Petersen, K.A.; Sweet, R.A.; Pierri, J.N.; Zhang, W.; Sampson, A.R.; Lewis, D.A. Effect of chronic antipsychotic exposure on astrocyte and oligodendrocyte numbers in macaque monkeys. *Biol. Psychiatry* **2008**, *63*, 759–765. [CrossRef]
72. Konopaske, G.T.; Dorph-Petersen, K.A.; Pierri, J.N.; Wu, Q.; Sampson, A.R.; Lewis, D.A. Effect of chronic exposure to antipsychotic medication on cell numbers in the parietal cortex of macaque monkeys. *Neuropsychopharmacology* **2007**, *32*, 1216–1223. [CrossRef]
73. Martin, J.L.; Magistretti, P.J.; Allaman, I. Regulation of neurotrophic factors and energy metabolism by antidepressants in astrocytes. *Curr. Drug Targets* **2013**, *14*, 1308–1321. [CrossRef]
74. Miguel-Hidalgo, J.J.; Waltzer, R.; Whittom, A.A.; Austin, M.C.; Rajkowska, G.; Stockmeier, C.A. Glial and glutamatergic markers in depression, alcoholism, and their comorbidity. *J. Affect. Disord.* **2010**, *127*, 230–240. [CrossRef]
75. Rajkowska, G.; Clarke, G.; Mahajan, G.; Licht, C.M.; van de Werd, H.J.; Yuan, P.; Stockmeier, C.A.; Manji, H.K.; Uylings, H.B. Differential effect of lithium on cell number in the hippocampus and prefrontal cortex in adult mice: A stereological study. *Bipolar Disord.* **2016**, *18*, 41–51. [CrossRef]
76. Batiuk, M.Y.; Martirosyan, A.; Wahis, J.; de Vin, F.; Marneffe, C.; Kusserow, C.; Koeppen, J.; Viana, J.F.; Oliveira, J.F.; Voet, T.; et al. Identification of region-specific astrocyte subtypes at single cell resolution. *Nat. Commun.* **2020**, *11*, 1220. [CrossRef]
77. Akkouh, I.A.; Hribkova, H.; Grabiec, M.; Budinska, E.; Szabo, A.; Kasparek, T.; Andreassen, O.A.; Sun, Y.M.; Djurovic, S. Derivation and Molecular Characterization of a Morphological Subpopulation of Human iPSC Astrocytes Reveal a Potential Role in Schizophrenia and Clozapine Response. *Schizophr. Bull.* **2022**, *48*, 190–198. [CrossRef]
78. Pinero, J.; Ramirez-Anguita, J.M.; Sauch-Pitarch, J.; Ronzano, F.; Centeno, E.; Sanz, F.; Furlong, L.I. The DisGeNET knowledge platform for disease genomics: 2019 update. *Nucleic Acids Res.* **2020**, *48*, D845–D855. [CrossRef]
79. Williams, A.G.; Thomas, S.; Wyman, S.K.; Holloway, A.K. RNA-seq Data: Challenges in and Recommendations for Experimental Design and Analysis. *Curr. Protoc. Hum. Genet.* **2014**, *83*, 11.13.1–11.13.20. [CrossRef]
80. Holmans, P. Statistical methods for pathway analysis of genome-wide data for association with complex genetic traits. *Adv. Genet.* **2010**, *72*, 141–179. [CrossRef] [PubMed]
81. Wang, K.; Li, M.; Hakonarson, H. Analysing biological pathways in genome-wide association studies. *Nat. Rev. Genet.* **2010**, *11*, 843–854. [CrossRef] [PubMed]
82. Fuxman Bass, J.I.; Diallo, A.; Nelson, J.; Soto, J.M.; Myers, C.L.; Walhout, A.J. Using networks to measure similarity between genes: Association index selection. *Nat. Methods* **2013**, *10*, 1169–1176. [CrossRef]
83. Shukla, R.; Prevot, T.D.; French, L.; Isserlin, R.; Rocco, B.R.; Banasr, M.; Bader, G.D.; Sibille, E. The Relative Contributions of Cell-Dependent Cortical Microcircuit Aging to Cognition and Anxiety. *Biol. Psychiatry* **2019**, *85*, 257–267. [CrossRef]
84. Smail, M.A.; Wu, X.; Henkel, N.D.; Eby, H.M.; Herman, J.P.; McCullumsmith, R.E.; Shukla, R. Similarities and dissimilarities between psychiatric cluster disorders. *Mol. Psychiatry* **2021**, *26*, 4853–4863. [CrossRef]
85. Yoo, M.; Shin, J.; Kim, J.; Ryall, K.A.; Lee, K.; Lee, S.; Jeon, M.; Kang, J.; Tan, A.C. DSigDB: Drug signatures database for gene set analysis. *Bioinformatics* **2015**, *31*, 3069–3071. [CrossRef]
86. O'Donovan, S.M.; Hasselfeld, K.; Bauer, D.; Simmons, M.; Roussos, P.; Haroutunian, V.; Meador-Woodruff, J.H.; McCullumsmith, R.E. Glutamate transporter splice variant expression in an enriched pyramidal cell population in schizophrenia. *Transl. Psychiatry* **2015**, *5*, e579. [CrossRef]
87. O'Donovan, S.M.; Sullivan, C.; Koene, R.; Devine, E.; Hasselfeld, K.; Moody, C.L.; McCullumsmith, R.E. Cell-subtype-specific changes in adenosine pathways in schizophrenia. *Neuropsychopharmacology* **2018**, *43*, 1667–1674. [CrossRef]
88. Alganem, K.; Shukla, R.; Eby, H.; Abel, M.; Zhang, X.; McIntyre, W.B.; Lee, J.; Au-Yeung, C.; Asgariroozbehani, R.; Panda, R.; et al. Kaleidoscope: A New Bioinformatics Pipeline Web Application for In Silico Hypothesis Exploration of Omics Signatures. *bioRxiv* **2020**. [CrossRef]
89. Torrey, E.F.; Webster, M.; Knable, M.; Johnston, N.; Yolken, R.H. The stanley foundation brain collection and neuropathology consortium. *Schizophr. Res.* **2000**, *44*, 151–155. [CrossRef]
90. Higgs, B.W.; Elashoff, M.; Richman, S.; Barci, B. An online database for brain disease research. *BMC Genom.* **2006**, *7*, 70. [CrossRef]

91. O'Donovan, S.M.; Imami, A.; Eby, H.; Henkel, N.D.; Creeden, J.F.; Asah, S.; Zhang, X.; Wu, X.; Alnafisah, R.; Taylor, R.T.; et al. Identification of candidate repurposable drugs to combat COVID-19 using a signature-based approach. *Sci. Rep.* **2021**, *11*, 4495. [CrossRef]
92. Chadha, R.; Alganem, K.; McCullumsmith, R.E.; Meador-Woodruff, J.H. mTOR kinase activity disrupts a phosphorylation signaling network in schizophrenia brain. *Mol. Psychiatry* **2021**, *26*, 6868–6879. [CrossRef]
93. Bentea, E.; Depasquale, E.A.K.; O'Donovan, S.M.; Sullivan, C.R.; Simmons, M.; Meador-Woodruff, J.H.; Zhou, Y.; Xu, C.; Bai, B.; Peng, J.; et al. Kinase network dysregulation in a human induced pluripotent stem cell model of DISC1 schizophrenia. *Mol. Omics* **2019**, *15*, 173–188. [CrossRef]
94. McGuire, J.L.; Hammond, J.H.; Yates, S.D.; Chen, D.; Haroutunian, V.; Meador-Woodruff, J.H.; McCullumsmith, R.E. Altered serine/threonine kinase activity in schizophrenia. *Brain Res.* **2014**, *1568*, 42–54. [CrossRef]
95. O'Donovan, S.M.; Sullivan, C.R.; McCullumsmith, R.E. The role of glutamate transporters in the pathophysiology of neuropsychiatric disorders. *NPJ Schizophr.* **2017**, *3*, 32. [CrossRef]
96. Kofuji, P.; Araque, A. G-Protein-Coupled Receptors in Astrocyte-Neuron Communication. *Neuroscience* **2021**, *456*, 71–84. [CrossRef] [PubMed]
97. Moody, C.L.; Funk, A.J.; Devine, E.; Homan, R.C.D.; Boison, D.; McCullumsmith, R.E.; O'Donovan, S.M. Adenosine Kinase Expression in the Frontal Cortex in Schizophrenia. *Schizophr. Bull.* **2020**, *46*, 690–698. [CrossRef]
98. McGuire, J.L.; Depasquale, E.A.; Funk, A.J.; O'Donovan, S.M.; Hasselfeld, K.; Marwaha, S.; Hammond, J.H.; Hartounian, V.; Meador-Woodruff, J.H.; Meller, J.; et al. Abnormalities of signal transduction networks in chronic schizophrenia. *NPJ Schizophr.* **2017**, *3*, 30. [CrossRef]
99. Stockmeier, C.A.; Mahajan, G.J.; Konick, L.C.; Overholser, J.C.; Jurjus, G.J.; Meltzer, H.Y.; Uylings, H.B.; Friedman, L.; Rajkowska, G. Cellular changes in the postmortem hippocampus in major depression. *Biol. Psychiatry* **2004**, *56*, 640–650. [CrossRef] [PubMed]
100. Stockmeier, C.A.; Rajkowska, G. Cellular abnormalities in depression: Evidence from postmortem brain tissue. *Dialogues Clin. Neurosci.* **2004**, *6*, 185–197. [CrossRef]
101. Si, X.; Miguel-Hidalgo, J.J.; O'Dwyer, G.; Stockmeier, C.A.; Rajkowska, G. Age-dependent reductions in the level of glial fibrillary acidic protein in the prefrontal cortex in major depression. *Neuropsychopharmacology* **2004**, *29*, 2088–2096. [CrossRef]
102. Li, J.; Pan, L.; Pembroke, W.G.; Rexach, J.E.; Godoy, M.I.; Condro, M.C.; Alvarado, A.G.; Harteni, M.; Chen, Y.W.; Stiles, L.; et al. Conservation and divergence of vulnerability and responses to stressors between human and mouse astrocytes. *Nat. Commun.* **2021**, *12*, 3958. [CrossRef]
103. Han, X.; Chen, M.; Wang, F.; Windrem, M.; Wang, S.; Shanz, S.; Xu, Q.; Oberheim, N.A.; Bekar, L.; Betstadt, S.; et al. Forebrain engraftment by human glial progenitor cells enhances synaptic plasticity and learning in adult mice. *Cell Stem Cell* **2013**, *12*, 342–353. [CrossRef]
104. Verkhratsky, A.; Parpura, V.; Rodriguez-Arellano, J.J.; Zorec, R. Astroglia in Alzheimer's Disease. *Adv. Exp. Med. Biol.* **2019**, *1175*, 273–324. [CrossRef]
105. Dossi, E.; Vasile, F.; Rouach, N. Human astrocytes in the diseased brain. *Brain Res. Bull.* **2018**, *136*, 139–156. [CrossRef]
106. Sullivan, C.R.; O'Donovan, S.M.; McCullumsmith, R.E.; Ramsey, A. Defects in Bioenergetic Coupling in Schizophrenia. *Biol. Psychiatry* **2018**, *83*, 739–750. [CrossRef] [PubMed]
107. Sullivan, C.R.; Koene, R.H.; Hasselfeld, K.; O'Donovan, S.M.; Ramsey, A.; McCullumsmith, R.E. Neuron-specific deficits of bioenergetic processes in the dorsolateral prefrontal cortex in schizophrenia. *Mol. Psychiatry* **2019**, *24*, 1319–1328. [CrossRef]
108. Peng, L.; Li, B.; Verkhratsky, A. Targeting astrocytes in bipolar disorder. *Expert Rev. Neurother.* **2016**, *16*, 649–657. [CrossRef]
109. Rivera, A.D.; Butt, A.M. Astrocytes are direct cellular targets of lithium treatment: Novel roles for lysyl oxidase and peroxisome proliferator activated receptor-gamma as astroglial targets of lithium. *Transl. Psychiatry* **2019**, *9*, 211. [CrossRef]
110. Ravindran, N.; McKay, M.; Paric, A.; Johnson, S.; Chandrasena, R.; Abraham, G.; Ravindran, A.V. Randomized, Placebo-Controlled Effectiveness Study of Quetiapine XR in Comorbid Depressive and Anxiety Disorders. *J. Clin. Psychiatry* **2022**, *83*, 3. [CrossRef] [PubMed]
111. Shao, W.; Zhang, S.Z.; Tang, M.; Zhang, X.H.; Zhou, Z.; Yin, Y.Q.; Zhou, Q.B.; Huang, Y.Y.; Liu, Y.J.; Wawrousek, E.; et al. Suppression of neuroinflammation by astrocytic dopamine D2 receptors via alphaB-crystallin. *Nature* **2013**, *494*, 90–94. [CrossRef]
112. Zhang, S.; Li, B.; Lovatt, D.; Xu, J.; Song, D.; Goldman, S.A.; Nedergaard, M.; Hertz, L.; Peng, L. 5-HT2B receptors are expressed on astrocytes from brain and in culture and are a chronic target for all five conventional 'serotonin-specific reuptake inhibitors'. *Neuron Glia Biol.* **2010**, *6*, 113–125. [CrossRef]
113. Hertz, L.; Rothman, D.L.; Li, B.; Peng, L. Chronic SSRI stimulation of astrocytic 5-HT2B receptors change multiple gene expressions/editings and metabolism of glutamate, glucose and glycogen: A potential paradigm shift. *Front. Behav. Neurosci.* **2015**, *9*, 25. [CrossRef]
114. Sethi, P.; Virmani, G.; Gupta, K.; Thumu, S.C.R.; Ramanan, N.; Marathe, S. Automated morphometric analysis with SMorph software reveals plasticity induced by antidepressant therapy in hippocampal astrocytes. *J. Cell Sci.* **2021**, *134*, jcs258430. [CrossRef]
115. Bouvier, M.L.; Fehsel, K.; Schmitt, A.; Meisenzahl-Lechner, E.; Gaebel, W.; von Wilmsdorff, M. Sex-dependent alterations of dopamine receptor and glucose transporter density in rat hypothalamus under long-term clozapine and haloperidol medication. *Brain Behav.* **2020**, *10*, e01694. [CrossRef]

116. Chen, B.; Zhang, M.; Ji, M.; Zhang, D.; Chen, B.; Gong, W.; Li, X.; Zhou, Y.; Dong, C.; Wen, G.; et al. The neuroprotective mechanism of lithium after ischaemic stroke. *Commun. Biol.* **2022**, *5*, 105. [CrossRef]
117. O'Leary, L.A.; Davoli, M.A.; Belliveau, C.; Tanti, A.; Ma, J.C.; Farmer, W.T.; Turecki, G.; Murai, K.K.; Mechawar, N. Characterization of Vimentin-Immunoreactive Astrocytes in the Human Brain. *Front. Neuroanat.* **2020**, *14*, 31. [CrossRef]
118. Quist, E.; Ahlenius, H.; Canals, I. Transcription Factor Programming of Human Pluripotent Stem Cells to Functionally Mature Astrocytes for Monocultures and Cocultures with Neurons. *Methods Mol. Biol.* **2021**, *2352*, 133–148. [CrossRef]
119. Balouch, B.; Funnell, J.L.; Ziemba, A.M.; Puhl, D.L.; Lin, K.; Gottipati, M.K.; Gilbert, R.J. Conventional immunomarkers stain a fraction of astrocytes in vitro: A comparison of rat cortical and spinal cord astrocytes in naive and stimulated cultures. *J. Neurosci. Res.* **2021**, *99*, 806–826. [CrossRef]
120. Dragic, M.; Zaric, M.; Mitrovic, N.; Nedeljkovic, N.; Grkovic, I. Two Distinct Hippocampal Astrocyte Morphotypes Reveal Subfield-Different Fate during Neurodegeneration Induced by Trimethyltin Intoxication. *Neuroscience* **2019**, *423*, 38–54. [CrossRef]

cells

MDPI

Article

Astrocytes in the Optic Nerve Are Heterogeneous in Their Reactivity to Glaucomatous Injury

Ying Zhu [1,2,†], Rui Wang [1,3,†], Anthony C. Pappas [1,‡], Philip Seifert [1], Andrej Savol [4,5], Ruslan I. Sadreyev [4,5], Daniel Sun [1] and Tatjana C. Jakobs [1,*]

1 Department of Ophthalmology, Massachusetts Eye and Ear Infirmary/Schepens Eye Research Institute, Harvard Medical School, 20 Staniford Street, Boston, MA 02114, USA
2 Department of Ophthalmology, Stanford University, 1651 Page Mill Road, Palo Alto, CA 94304, USA
3 Department of Ophthalmology, The First Affiliated Hospital of Northwest University, Xi'an 710002, China
4 Department of Molecular Biology, Massachusetts General Hospital, Boston, MA 02114, USA
5 Department of Pathology, Massachusetts General Hospital and Harvard Medical School, 185 Cambridge St., Boston, MA 02114, USA
* Correspondence: tatjana_jakobs@meei.harvard.edu
† These authors contributed equally to this work.
‡ Present address: College of Osteopathic Medicine, Rocky Vista University, 255 E Center Street, Ivins, UT 84738, USA.

Abstract: The optic nerve head is thought to be the site of initial injury to retinal ganglion cell injury in glaucoma. In the initial segment of the optic nerve directly behind the globe, the ganglion cell axons are unmyelinated and come into direct contact to astrocytes, suggesting that astrocytes may play a role in the pathology of glaucoma. As in other parts of the CNS, optic nerve head astrocytes respond to injury by characteristic changes in cell morphology and gene expression profile. Using RNA-sequencing of glaucomatous optic nerve heads, single-cell PCR, and an in-vivo assay, we demonstrate that an up-regulation of astrocytic phagocytosis is an early event after the onset of increased intraocular pressure. We also show that astrocytes in the glial lamina of the optic nerve are apparently functionally heterogeneous. At any time, even in naïve nerves, some of the cells show signs of reactivity—process hypertrophy, high phagocytic activity, and expression of genetic markers of reactivity whereas neighboring cells apparently are inactive. A period of increased intraocular pressure moves more astrocytes towards the reactive phenotype; however, some cells remain unreactive even in glaucomatous nerves.

Keywords: glaucoma; astrocytes; optic nerve head; RNA-sequencing; phagocytosis; mitophagy

Citation: Zhu, Y.; Wang, R.; Pappas, A.C.; Seifert, P.; Savol, A.; Sadreyev, R.I.; Sun, D.; Jakobs, T.C. Astrocytes in the Optic Nerve Are Heterogeneous in Their Reactivity to Glaucomatous Injury. *Cells* **2023**, *12*, 2131. https://doi.org/10.3390/cells12172131

Academic Editors: Vladimir Parpura and Michael Brenner

Received: 31 December 2022
Revised: 14 July 2023
Accepted: 18 August 2023
Published: 23 August 2023

1. Introduction

Glaucoma leads to progressive degeneration and death of retinal ganglion cells and their axons [1,2]. The main risk factors are elevated intraocular pressure (IOP), genetics and a positive family history, and advancing age [3–5]. In spite of many recent advances, it is not completely understood why ganglion cells degenerate in glaucoma.

Retinal ganglion cell axons are unmyelinated in the retina and optic nerve head (ONH) until they have cleared the laminar region, which is made up exclusively of astrocytes in mice ("glial lamina", see Supplementary Figure S1 for a schematic) and of astrocytes supported by a collagenous lamina cribrosa in larger species [6–8]. Due to mechanical stress by elevated IOP, the lack of myelin, and possibly vascular factors, axons may be particularly vulnerable in the nerve head region [9,10]. Several studies have identified morphological and gene expression changes in the glial cells of the ONH early in the course of the disease before the first signs of degeneration of ganglion cells are visible in the retina [11–19]. Given the close proximity of astrocytes to the unmyelinated axons and reactive changes observed in those astrocytes, they have been implicated in the progression of glaucoma [20,21]. Al types of injury to the CNS induce a reactive response in astrocytes and reactive astrocytes can exert many functions, both protective and detrimental to surrounding neurons [22–25]. Recent studies identified gene expression profiles of astrocytes that assumed a neurotoxic (or type A1) state in neuroinflammation, or a neuroprotective (type A2) state that dominates after ischemic injury [26,27]. However, this classification should not be interpreted as denoting fixed cell types but rather as patterns of astrocyte reactivity in response to signals from their environment [28]. Whether astrocytes take a more neuroprotective or neurotoxic route may depend not only on the nature of the injury., but also on the spatial or temporal distance from the point of injury [29–31].

There is evidence that at least in the early stages of glaucoma, the protective role dominates. In the optic nerve (as in other tissues, such as the spinal cord [32,33]), general inhibition of astrocyte reactivity leads to a worse visual outcome and increased ganglion cell death following eye injury [34]. However, it is not yet clear in the context of glaucoma which of the many functions of astrocytes are beneficial for ganglion cell and axon preservation. Optic nerve head astrocytes respond to elevated intraocular pressure with characteristic morphological changes. The processes of astrocytes in uninjured nerves are oriented perpendicularly to the axis of the optic nerve and form glial tubes through which the ganglion cell axons pass as bundles [8]. After elevation of IOP, some astrocyte processes sprout longitudinal processes that often invade the axon bundles and contain degraded mitochondria [14,18]. This behavior is apparently typical of glaucomatous optic nerves; however, in many other ways—axonal swelling, accumulation of abnormal mitochondria, and thickening of blood vessel walls—morphological changes in glaucomatous nerves resemble those that occur during normal aging [35,36]. In the present study, we asked three questions. First, what are the differences between glaucomatous and aged nerves on the level of gene expression? Second, what cellular functions of astrocytes are activated in the optic nerve head after elevation of IOP? And third, do optic nerve head astrocytes behave as a single group, or do individual cells differ in their reactions to elevated IOP? We used a combined approach of RNA sequencing, single-cell quantitative PCR, and morphological analysis of single astrocytes in situ to demonstrate that sprouting of new processes and an up-regulation of phagocytosis are characteristic of glaucomatous, but not aged astrocytes. Furthermore, even though all astrocytes in the optic nerve head apparently belong to a single morphological type [8], their reactive profiles are very heterogeneous.

2. Materials and Methods

2.1. Animal Husbandry

All animal studies were carried out in accordance with the Association for Research in Vision and Ophthalmology (ARVO) guidelines for the Use of Animals in Ophthalmic and Vision Research and approved by the Institutional Animal Care and Use Committee of the Schepens Eye Research Institute. Male and female C57bl/6 mice were obtained from the Jackson Laboratory (stock number 000664). C57bl/6 mice for RNA-sequencing were 13 and 40 weeks (10 months) old. The mouse line B6.*hGFAPpr*::EGFP was bred in our colony. The original strain *hGFAPpr*::EGFP was kindly provided by Helmut Kettenmann (Max Dellbruck Center, Berlin, Germany). This strain expresses EGFP stochastically in individual astrocytes throughout the brain and optic nerve [8,37,38]. Since the strain was originally on the FVB/N background and carried the rd1 mutation, we backcrossed it with C57bl/6 mice for >10 generations. All animals were group housed under 12 h dark and light cycles and received water and food at libitum.

2.2. Microbead Injection and Tissue Preparation

The microbead occlusion model was used to induce elevated IOP [39,40]. Before microbead injection, the baseline IOP was measured with a TonoLab rebound tonometer (ICare, Espoo, Finland) under 1.5% isoflurane anesthesia delivered via a nose cone. Polystyrene microbeads (15 μm diameter, Invitrogen, Carlsbad, CA, USA) were injected into the anterior chamber of the right eye at a concentration of 1.6×10^6 beads in 1 μL sterile PBS using a beveled glass needle (100 μm tip size) attached to a Hamilton syringe. Control groups received an injection of sterile saline solution. All mice were followed by IOP measurements twice weekly for one month. All IOP measurements were taken at the same time of day (mid-morning) to minimize circadian variation. Cumulative IOP (cIOP) was calculated as the area under the curve for IOP over time (in mmHgDays). At the end of the experiment, mice were euthanized by CO_2 inhalation, and retinas and optic nerves were dissected. The unmyelinated segments of the optic nerves were identified under a dissecting microscope and stored in an RNAlater tissue storage agent (Invitrogen). Retinas were whole-mounted on nitrocellulose filters, fixed in 4% (w/v) paraformaldehyde in phosphate-buffered saline (PBS), and stained with a mouse monoclonal antibody against Brn3a (Millipore Sigma, MAB1585, Burlington, MA, USA), a marker for retinal ganglion cells, followed by a fluorescent secondary antibody (donkey anti-mouse IgG, catalog no. 715-586-150, Jackson Immunoresearch, West Grove, PA, USA). For ganglion cell counting, 8 image stacks (2 from each quadrant of the retina) were taken through the ganglion cell layer of each retina at a position equidistant to the border of the retina and optic nerve head. Image stacks were imported into ImageJ and automatically counted by a user-written routine. Ganglion cell densities were averaged for each retina and ganglion cell nuclei per mm^2 were calculated.

2.3. RNA Extraction and RNA-Seq

Unmyelinated portions of 4 right optic nerves (from 4 individual mice) were pooled for each RNA extraction. In all cases, 2 of the nerves were from male, and 2 from female mice. Microbead-injected and saline-injected control nerves were processed in parallel. Total RNA was extracted using RNeasy micro columns (Qiagen, Hilden, Germany) and quantified using a Bio Analyzer 2100 (Agilent, Santa Clara, CA, USA). Only preparations with RNA integrity number > 9 were used for RNA-sequencing (details are summarized in Supplementary Table S1). Four biological replicates were used for each group. RNA sequencing was carried out in the MGH Next Generation core facility (Department of Molecular Biology at MGH). RNA-seq libraries were constructed using the Clontech SMARTer Ultra Low Input RNA Kit for Sequencing v3 (Takara, Kusatsu, Japan) and sequenced on an Illumina HiSeq2500 instrument, resulting in approximately 30 million reads per sample on average. Transcriptome mapping to full genome sequences for *Mus musculus* (UCSC version mm10) was performed with STAR version 2.3.0 [41]. Read counts for individual genes were produced using FeatureCounts [42]. Differential expression analysis was performed using the DESeq2 R package after normalizing read counts and excluding those genes with normalized read counts equal to 0 across all replicates [43]. Differentially expressed genes were defined based on the criteria of a p-value < 0.01 (unpaired t-tests) and false discovery rate (FDR) < 0.05. Ingenuity Pathway Analysis (Qiagen) was used for pathway analysis.

2.4. Dye Injection into the Superior Colliculus

A cohort of 10 mice were injected with microbeads as described above. A control group of 10 mice received saline injections. On day 13 (i.e., four days before the end of the experiment), MitoTracker™ Red FM (Invitrogen, M22425) was injected into the left superior colliculus of each mouse. This time point was chosen because we wanted to maximize the amount of dye in transit for imaging the optic nerve [44,45]. Mice were anesthetized with ketamine and xylazine, and a small hole was drilled in the skull overlying the superior colliculus. About 0.5 μL of a 200 nM solution of MitoTtracker™ Red FM was injected slowly into the tissue with a glass microneedle attached to a Hamilton syringe [46]. After the injection, the hole was filled with Gelfoam (Pfizer) and the skin was closed with a surgical staple.

2.5. Tissue Preparation and Confocal Microscopy

After euthanasia with CO_2 followed by cervical dislocation, the cranium was opened and the bones and brain were removed, exposing the optic nerves including the optic chiasm. The whole skull base was then fixed in 4% paraformaldehyde for one hour. Then the optic nerves were dissected free from the surrounding tissue, postfixed with 4% paraformaldehyde, and embedded in 6% *w/v* agarose in PBS. Serial thick (200 μm) sections of the unmyelinated region of the optic nerve were cut on a Leica VT100S vibratome (Leica Microsystems, Buffalo Grove, IL, USA), mounted in Vectashield (Vector Laboratories, Burlingame, CA, USA), and used immediately for imaging on a Leica SP8 confocal microscope. Image stacks were taken through each optic nerve section using a 63× glycerol immersion objective (numerical aperture 1.4). Image stacks that were to be used for quantification were taken without additional optical zoom at the same laser settings, scan speeds (400 Hz), and pixel size (1024 × 1024 pixels, 0.4 μm/pixel). The z-step size was matched to the pixel size in the x and y planes (0.4 μm/pixel). Some cells were imaged at higher magnification to show fine morphological detail. These image stacks were not used for quantification.

For immunohistochemistry, optic nerves were cryoprotected overnight with 30% *w/v* sucrose in PBS, embedded in OCT compound, and sectioned at 16 μm. Sections were incubated on the slides in primary antibodies (rat monoclonal anti-galectin 3, Cedar Lane CL8942AP; chicken polyclonal anti-GFAP, Abcam, AB4674; and rabbit polyclonal anti-Iba1, Fujifilm-Wako 019-19741). Secondary antibodies (donkey anti-rat, chicken, or rabbit conjugated to FITC or TRITC were obtained from Jackson Immunoresearch).

2.6. Transmission Electron Microscopy (TEM)

The tissue preparation for TEM is described in detail in an earlier publication from our laboratory [35]. Sections from young (3 month), old (10 month), and microbead-injected (1 month after injection) nerves were used to image optic nerve astrocytes. Astrocyte process thickness was measured on TEM images at medium (2900×) magnification. Astrocyte process thickness was measured in young, old, and microbead-injected nerves using ImageJ.

2.7. Image Analysis

Confocal image stacks (8 bit per color channel) were imported into Matlab (The Mathworks, Natick, MA, USA) and split into the red channel (containing the signal from the MitoTracker^{TM} Red FM) and green channel (containing the signal from EGFP-labeled astrocytes). The red channel was converted to black and white, median filtered with a kernel size of 3 × 3 pixels, and mitochondria were identified using the bwlabeln function in Matlab. This function returns matrices of all connected components (i.e., mitochondria). As most of these mitochondria were not inside an EGFP-labeled astrocyte, in the second step the identified mitochondria were compared with the pixel values in the green channel. All labeled mitochondria that were completely surrounded by green signal from astrocytes (defined as a pixel intensity value of >100 in the green channel) were identified by a user-written algorithm. In addition, total astrocyte volume and total mitochondria volume were measured by the same algorithm. The processed image stacks were reconstructed using the Amira software package (Visage Imaging, San Diego, CA, USA). The total number of astrocytes and number of MitoTracker^{TM} Red FM-containing astrocytes in each stack was determined with ImageJ by an observer ignorant of the IOP history of the eye.

2.8. Single-Cell RT-PCR

Naïve 13- and 40-week-old C57bl/6 mice (3 animals per group) were used for the age comparison of astrocyte gene expression. In addition, microbead-injected mice and saline-injected control mice were used (4 animals per group). Unfixed optic nerve heads were dissected free of the surrounding tissue and separated from the myelinated region of the nerve under microscopic control. The tissue was digested with papain and dissociated into single cells as described before [47,48]. Individual astrocytes were collected into thin wall PCR tubes containing the buffer for reverse transcriptase and PCR (Access PCR System, Promega, Madison, WI, USA), primers for phagocytic markers, reactive astrocytes of neuroprotective and neurotoxic types [49], positive and negative controls. All primer sequences were chosen from the Harvard Primer Bank [50] or designed and tested in-house. Primer sequences are given in Supplementary Table S2. Reverse transcription and first-round amplification were carried out for all primers simultaneously using the Access PCR system with 17 cycles. Second-round PCR or qPCR was carried out for all genes separately on a Lightcycler 480 Instrument (Roche Applied Science, Penzberg, Germany). Gapdh and Oaz1 were used as internal controls and cells that did not express detectable levels (defined as a Ct value of >32) of both genes were not analyzed.

2.9. Statistical Analysis

All statistical analysis was carried out in Matlab (The Math Works, Natick, MA, USA) or R (The R foundation). For microbead injection experiments, the sample size needed to detect a 15% loss of ganglion cells at $\alpha = 0.05$ with a power of 0.8 is $n = 6$; however, for RNA-sequencing larger group sizes ($n = 16$) were used because a single optic nerve does not yield sufficient RNA for one biological replicate. For the experiment using MitoTracker^{TM} Red FM, group sizes of $n = 10$ were used. Single-cell PCR has a high variability. We calculated that $n = 12$ cells per condition would be sufficient to detect a 2-fold change in expression levels with $\alpha = 0.05$ at 0.8. Delta Ct values were calculated for every gene (Ct_{target}—Ct_{Gapdh}), and expression levels were calculated as 2^(-ΔCt). Expression levels in microbead-injected versus saline-injected and aged versus young cells were calculated and compared using *t*-tests with Bonferroni correction for multiple comparisons (whole population comparison). Since expression levels were different for different genes, the data was then normalized for further analysis. Delta Ct values were averaged over all cells in the microbead versus saline and aged versus young groups, respectively. Individual Ct values were subtracted from the averages and divided by the standard deviation. The results were used for generating heatmaps with the rows representing genes and the columns representing individual astrocytes. Then, 95% confidence intervals were calculated for cells from microbead-injected eyes and cells from aged eyes, respectively. If the confidence intervals did not include 0, the genes were considered significantly up- or down-regulated in the microbead (or aged) group.

For morphological analysis of astrocytes, all groups of measurements were first tested for normal distribution (Shapiro–Wilk test), and a non-parametric test (Wilcoxon rank sum test) was used if a group did not pass the test at an alpha level of 0.05. ANOVA with Tukey–Kramer post hoc tests was used to compare IOP measurements and measurements of astrocyte process thickness from electron microphotographs. Linear regression was calculated by the least square method. All values are given as mean ± standard deviation unless stated otherwise.

2.10. Data Sharing

RNA sequencing data was deposited to GEO (GSE139605). Original image stacks of MitoTracker^{TM} Red FM-injected optic nerve heads in B6.*hGFAPpr*::EGFP mice were deposited to the Harvard Dataverse and are accessible under https://dataverse.harvard.edu/dataverse/astrocytes2019 (accessed on 21 August 2023).

3. Results

3.1. Gene Expression in Optic Nerve Heads from Eyes with and without Elevated IOP

We profiled the gene expression in the glial lamina, the unmyelinated region of the optic nerve (Supplementary Figure S1) of C57bl/6 mice in four groups: a group of young adult mice (13 weeks, $n = 16$, male:female 1:1) that had experienced one month of unilaterally elevated IOP, an age-matched control group (saline injection, $n = 16$), an age-matched (13 weeks) group of naïve mice, and finally a group of aged (10 month) naïve mice. At 13 weeks (young adulthood), the murine retina and optic nerve are fully mature, whereas a 40-week-old mouse would be considered middle-aged [51,52]. This age was chosen because it corresponds to the age when the incidence of glaucoma in humans starts to rise and corresponds roughly to a human age of about 40 years [3]. Injection of microspheres into the right anterior chamber led to an increase in IOP (Figure 1A–D) and a loss of about 15% of ganglion cells in the injected eye one month later (Figure 1E–G). Injection of saline solution did not change IOP or ganglion cell density in the retina (Figure 1A). Total RNA was isolated from 4 optic nerve heads (originating from 2 male and 2 female mice each, Supplementary Table S1) per each of 4 biological replicates. We did not use the contralateral eyes as controls because IOP manipulation in one eye can lead to glial activation in the untreated contralateral eye [53,54]. A total of 14,926 genes had read counts > 10 in our samples. Principal component analysis showed good clustering for all but 2 outliers (one in the Saline group and one in the 40-week naïve group), which were excluded from further analysis (Supplementary Figure S2). We then performed cluster analysis on all groups with the genes that were differentially expressed ($p < 0.01$, FDR = 0.05) in any of the samples, the 13-week-old naïve mice serving as the baseline. As expected, the young naïve and saline-injected groups were very similar, whereas microbead-injected and old naïve nerves formed their own clusters, but showed overall similarity to each other (Figure 2A).

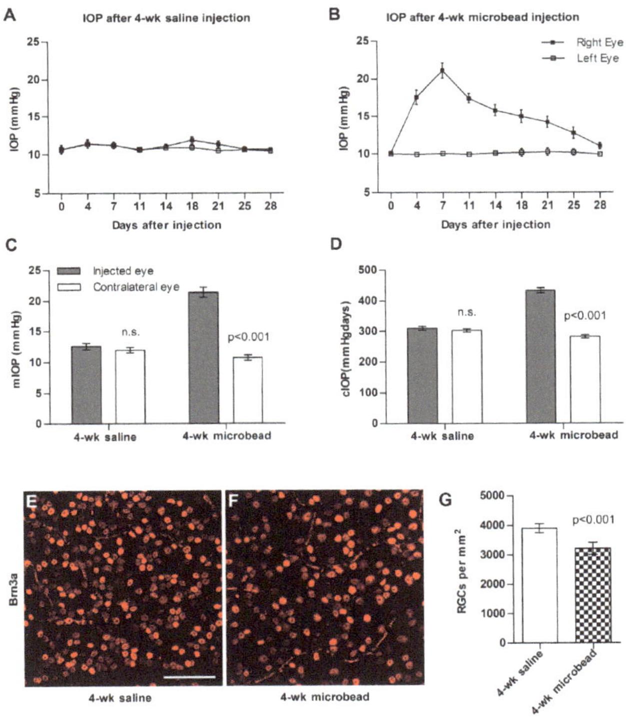

Figure 1. Elevation of IOP and ganglion cell loss in C57bl/6 eyes. (**A**) control animals show no increase in IOP after unilateral injection of sterile saline solution. (**B**) IOP increases after injection of microbeads, peaks at day 7, and gradually returns to baseline. (**C**) Maximum IOP measured during the experimental period in saline- and microbead-injected eyes. (**D**) Cumulative IOP (cIOP, i.e., the area under the curve) of IOP vs. time for saline- and microbead-injected eyes. (**E**) Retinal ganglion cell labeling with Brn3a in a retina 4 weeks after injection of saline solution. (**F**) Retinal ganglion cell labeling with Brn3a in a retina 4 weeks after microbead injection shows marked loss of ganglion cell nuclei. (**G**) 4 weeks of IOP elevation leads to a significant loss of ganglion cells in the injected eyes. Groups were compared using ANOVA with Tukey–Kramer post hoc test. N.s., not significant. Scale bar for panels (**E**,**F**), 25 μm.

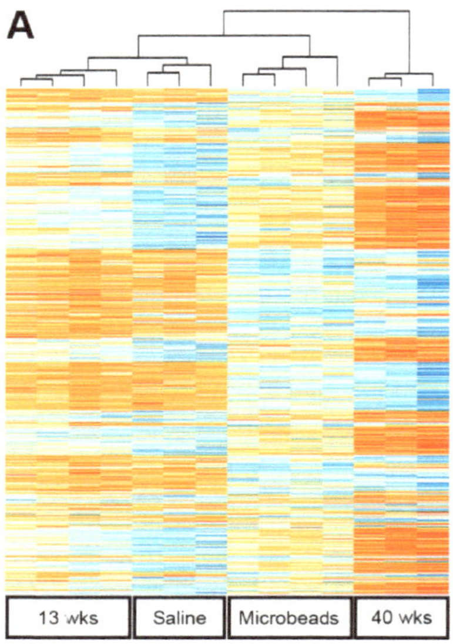

B

	Differentially expressed genes	Up-regulated	Down-regulated
40 wks vs 13 wks (naïve)	3154	1652 (52.4%)	1502 (47.6)
Microbeads vs Saline	1235	962 (77.9%)	273 (22.1%)
Saline vs 13 wks (naïve)	0	n.a.	n.a.

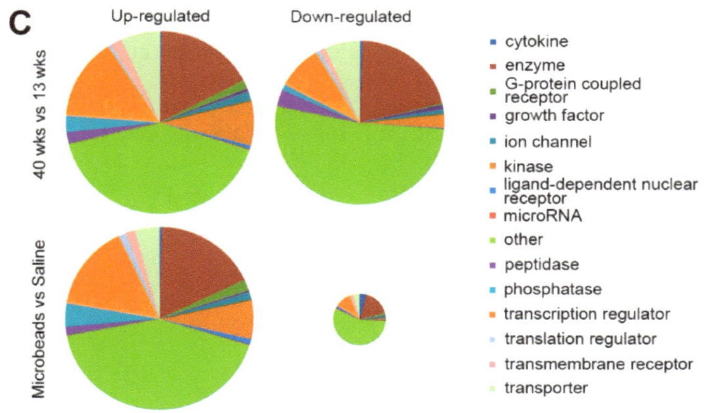

Figure 2. RNA sequencing of optic nerve heads in naïve young, saline-injected (4 weeks after injection), microbead-injected (4 weeks after injection), and naïve aged (40 weeks) mice. (**A**) hierarchical clustering of all differentially expressed genes in the four groups. Differentially expressed genes were defined as genes up- or down-regulated by at least 1.5-fold, $p < 0.01$, FDR = 0.05. Red colors indicate up-regulated genes, blue colors indicate down-regulated genes. The brackets indicate similarities between samples by hierarchical clustering. Saline-injected and young naïve mice cluster together. Microbead-injected and aged nerves form their own clusters, but are overall relatively similar. (**B**) numbers of differentially regulated (DE) genes in microbead (MB) vs. saline-injected nerves and young vs. aged naïve nerves. (**C**) distribution of Ingenuity Pathway Analysis (IPA) annotations for differentially expressed up- or down-regulated genes. The areas of the circles signify the number of total genes in this group. There are roughly the same numbers of up- and down-regulated genes in the aged versus young groups, whereas after microbead injection, about 4 times as many up-regulated genes were detected. n.a., not applicable.

For further analysis of differentially regulated pathways, we directly compared young versus aged naïve and microbead versus saline-injected groups. Differentially expressed genes were defined as genes significantly up- or down-regulated by at least 1.5-fold (2-tailed t-test, $p < 0.01$, FDR= 0.05). Figure 2B,C summarize the numbers of differentially expressed genes for these groups and their annotations, respectively (complete gene lists are available under GEO accession number GSE139605). As expected from the overlap of genes differentially expressed in the glaucoma and aged groups, many canonical pathways predicted to be activated or attenuated in glaucoma or age were similar. This included the endothelin-1, interleukin-6, and Stat3 pathways, whose roles in glaucomatous nerves have been described before [12,13,34]. Neither in the glaucoma nor the aged groups did we find up-regulation of cell cycle or cell proliferation-associated pathways.

The similarity between aged and glaucomatous nerves was also apparent in electron microscopy of the unmyelinated optic nerve (Figure 3). Astrocytes in young nerves show the characteristic arrangement of large processes that delineate axon bundles, and small fine processes that extend into axon bundles (Figure 3A, astrocytes are false-colored in red). Astrocytes in 40-week-old nerves (Figure 3B) or nerves after one month of IOP elevation (Figure 3C) show obvious process hypertrophy. We did not observe more astrocyte nuclei, as an indicator of cell numbers, in old or glaucomatous nerves (4 glaucomatous nerve profiles and 5 age-matched normal nerve profiles were counted).

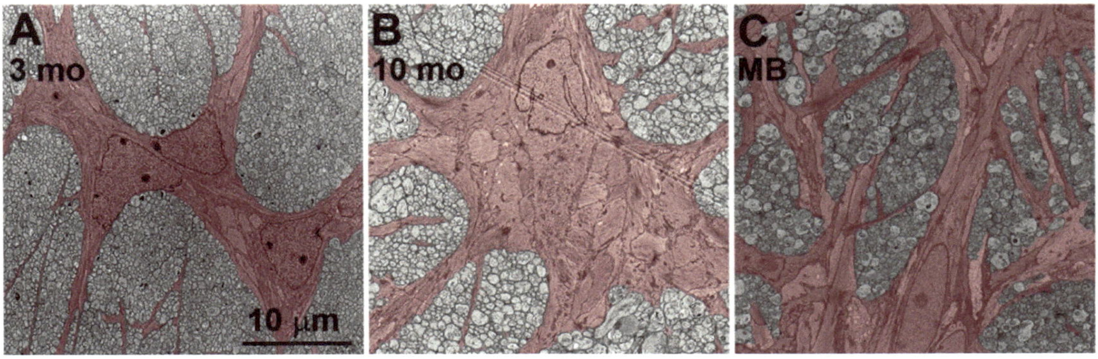

Figure 3. Astrocyte hypertrophy in glaucomatous and aged nerves. (**A**) Transmission electron microphotography of an optic nerve of a 3-month-old naïve C57bl/6 mouse in the coronal plane. The section was taken through the unmyelinated region of the nerve. Astrocyte cell bodies and processes are false-colored in red. (**B**) optic nerve in a 40-week-old naïve mouse. Astrocyte cell bodies and processes are notably enlarged. (**C**) optic nerve in a 3-month-old mouse, 1 month after microbead (MB) injection into the anterior chamber of the eye. Individual processes are 0.97 ± 0.18 μm in 3-month-old nerves, 1.07 ± 0.19 μm in 10-month-old nerves, and 1.16 ± 0.27 μm after microbead injection. Astrocyte processes in aged and glaucomatous mice are significantly different at $p = 0.0014$ and $p < 0.001$ between young and aged, and young and glaucomatous nerves, respectively. ANOVA followed by Tukey-Kramer test.

3.2. Differences between Microbead-Injected and Aged Nerves

We compared genes differentially expressed in glaucomatous versus aged nerves to identify biological functions that are activated specifically in optic nerves after 4 weeks of elevated IOP (Table 1). Several of the pathways predicted to be activated in glaucomatous versus aged nerves are annotated with the formation of protrusions, sprouting, and branching of cells. We therefore asked whether the functional consequence of newly formed astrocyte processes may be an increase in phagocytic activity.

Table 1. Ingenuity Pathway Analysis annotation of the major predicted up- and down-regulated pathways in microbead-injected nerves versus aged nerves.

Annotation	*p* Value	Direction	z Score
Formation of cellular protrusions	1.98×10^{-9}	Increased	2.705
Development of neurons	1.01×10^{-8}	Increased	2.441
Infection of cells	1.08×10^{-6}	Increased	3.768
Sprouting	4.63×10^{-6}	Increased	2.484
Cell movement of tumor cell lines	1.31×10^{-5}	Increased	2.037
Branching of neurites	4.19×10^{-5}	Increased	2.991
Cytostasis	0.000154	Increased	2.411
Inflammatory response	0.000213	Increased	3.093
Cell movement of phagocytes	0.000292	Increased	2.112
Nonhematologic malignant neoplasm	3.18×10^{-33}	Decreased	−2.45
Development of genitourinary system	4.66×10^{-7}	Decreased	−2.435
Neurodegeneration of cerebellum	5.49×10^{-5}	Decreased	−2.8
Neurodegeneration of Purkinje cells	8.68×10^{-5}	Decreased	−2.621
Abnormality of heart ventricle	0.000299	Decreased	−2.126

3.3. Expression of Phagocytic Markers in Individual Astrocytes

Astrocytes are the major cell type in the optic nerve, but microglia, endothelial cells, and fibroblasts from the surrounding pia are also present, and some of the gene expression changes in our study may be due to contributions from these cell types. To validate astrocyte-specific gene expression, we therefore isolated single astrocytes from freshly dissociated optic nerves 2 weeks after microbead injection (high IOP) or saline injection (controls). Aged astrocytes were collected from 40-week-old naïve animals; astrocytes from young naïve nerves were collected as controls. Astrocytes retain their morphology well after dissociation [48], but in every case the cell type was also confirmed by verifying that the cells were positive for the astrocyte marker GFAP and negative for the microglial marker Iba1 and the oligodendrocyte marker MBP by conventional PCR. Phagocytosis-related genes (*Abca1, Abca7, Dock1, Gulp1, Lamp1, Lgals3, Megf10, Mfge8*), genes related to the degradation of mitochondria (*Pink1, Park2*), and markers of astrocyte reactivity (*Fbln5, H2-T23, Psmb8, Srgn*, A1 reactivity; *Clcf1, Emp1, Ptx3, Stat3, Tgm1*, A2 reactivity) were tested using quantitative PCR.

Considering the whole population of astrocytes in our sample, *Gulp1* and *Lgals3* were significantly up-regulated in astrocytes from microbead- versus saline-injected nerves (*Gulp1*, 2.5-fold, $p = 0.00056$ and *Lgals3*, 7.8-fold 7.2×10^{-5}, *t*-test with Bonferroni correction for multiple comparisons). This was also apparent in immunostaining for the gene product of *Lgals3*, galectin-3, which is up-regulated after microbead injection (Figure 4A–C). Galectin-3 co-localized with the astrocyte marker GFAP, but not with the microglia marker IBA1 (Figure 4D,E). Amplification of *Abca1* was weak or below the detection limit for most astrocytes, though *Abca1* is present in RNA samples from the whole optic nerve suggesting that it may localize mainly to cells other than astrocytes. Most of the overall activation of engulfment and phagocytosis genes is driven by a subset of highly active cells, rather than by a shift of the whole population to a more phagocytic state (Figure 5A). There was no clear trend towards an A1 (neurotoxic) or A2 (protective) type of astrocyte reactivity for the genes tested after microbead injection (Figure 5A,C). In contrast, neither phagocytosis-related genes nor markers of reactivity were up-regulated in aged versus young naïve astrocytes (Figure 5B,D).

3.4. Mitochondria Incorporation into Optic Nerve Head Astrocytes

Astrocytes in the unmyelinated segment of the optic nerve are known to incorporate axonal mitochondria [55]. We made use of this property to directly visualize phagocytic activity in optic nerve head astrocytes in vivo. A cohort of B6.*hGFAPpr*::EGFP mice (10 animals) were unilaterally injected with microspheres solution into the anterior chamber (Supplementary Figure S3). In this strain, EGFP is expressed in astrocytes in a stochastic manner so that individual cells can directly be observed microscopically [8,18]. In addition each animal received an injection of MitoTracker™ Red FM into the contralateral superior colliculus on day 13 after microbead injection. Another group of 10 B6.*hGFAPpr*::GFP mice received the MitoTracker™ Red FM injection without prior injection of microbeads. Four days later (day 17) the mice were sacrificed. These timepoints were chosen because the IOP elevation is at a relatively stable plateau between days 11 and 21 (Figure 1B). MitoTracker™ Red FM is taken up by the ganglion cell axonal terminals and labels mitochondria that are transported retrogradely to the cells' somata in the retina. Since there is no other direct connection, all labeled mitochondria in the optic nerve must be axonal in origin. Labeled mitochondria appeared in the retinal ganglion cell bodies (Figure 6A) and axons in the optic nerve (Figure 6B) four days later in the high-IOP and control groups. We tested whether any MitoTracker™ Red FM-labeled axonal mitochondria become incorporated by astrocytes by confocal imaging of EGFP$^+$ cells. A total of 38 image stacks containing 344 unique EGFP$^+$ astrocytes were imaged in the control group, and 34 image stacks containing 339 astrocytes were imaged in the high-IOP group.

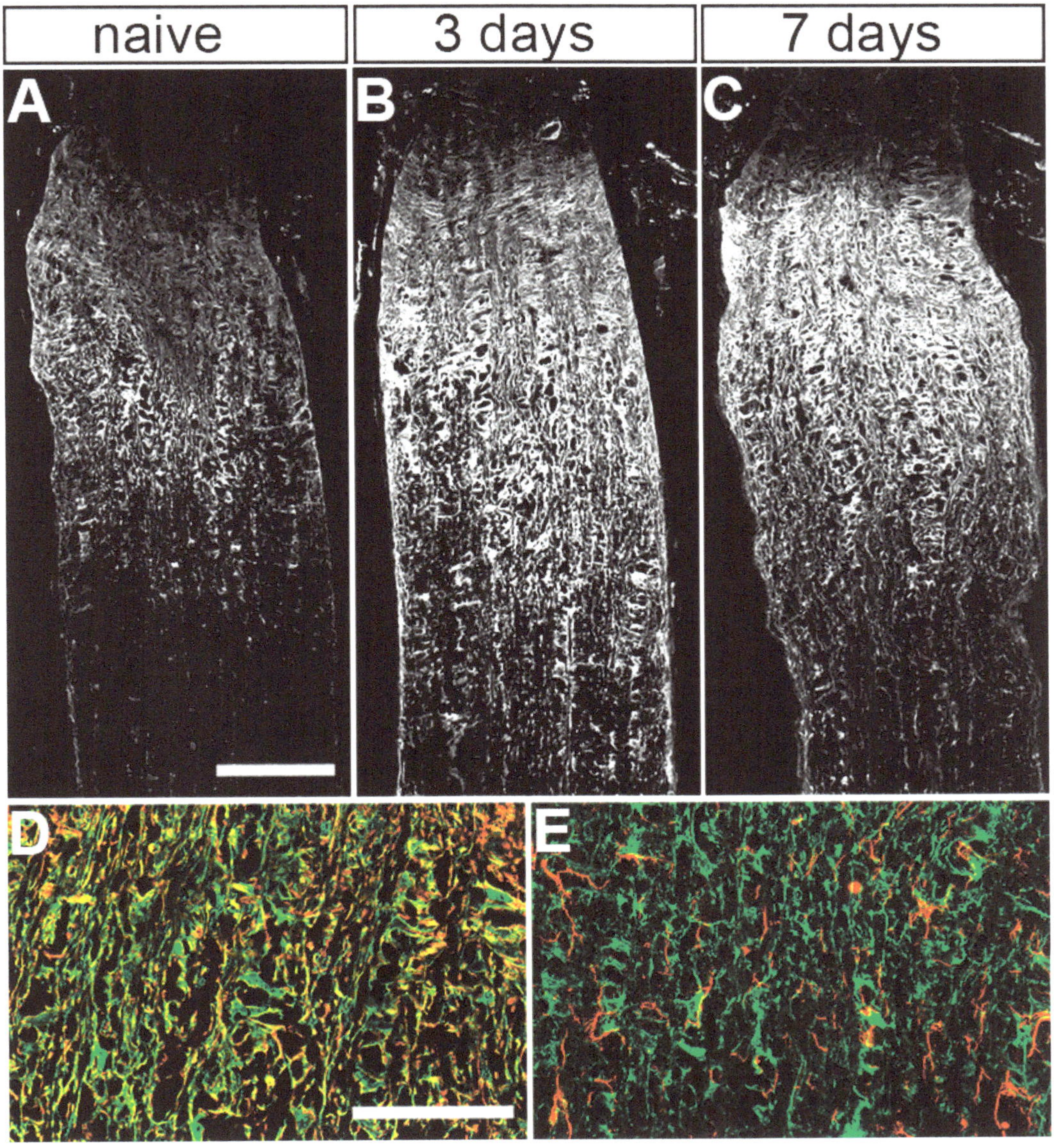

Figure 4. Apparent increase in staining for Galectin-3 after IOP elevation. (**A**) longitudinal section of a naïve optic nerve immunolabeled for Galectin-3. (**B**,**C**) 3 and 7 days after microbead injection. Galectin-3 staining is observed in the naïve control, but qualitatively appears to be up-regulated at 3 and 7 days. Scale bar for A-C, 100 µm. (**D**) higher magnification of an optic nerve 7 days after microbead injection stained for Galectin-3 (green) and the astrocyte marker GFAP (red). (**E**) higher magnification of a nerve 7 days after microbead injection stained for Galectin-3 and the microglia marker IBA1 (red). Scale bar for (**D**,**E**), 50 µm.

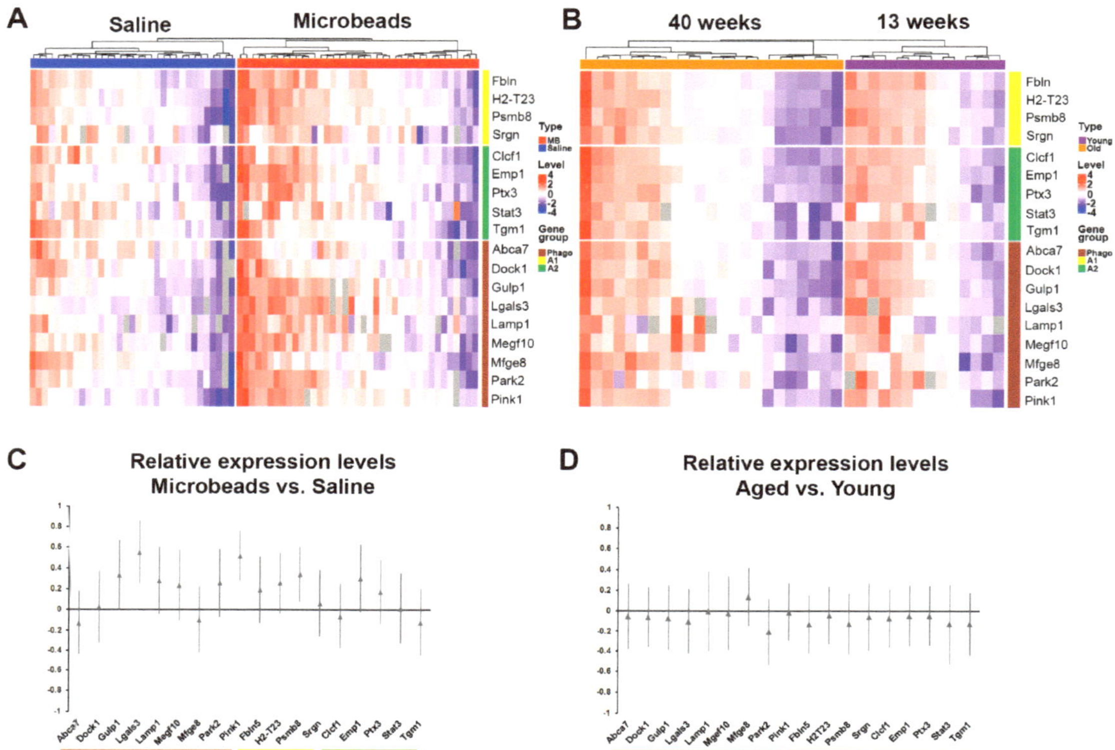

Figure 5. Single-cell qPCR of saline- vs. microbead-injected and young vs. aged astrocytes. (**A**) heatmap of normalized expression values in single, freshly dissociated astrocytes after microbead- or saline injection. Rows represent genes (A1 reactivity markers (yellow bar) *Fbln5*, *H2-T23*, *Psmb8*, *Srgn*; A2 reactivity markers (green bar) *Clcf1*, *Emp1*, *Ptx3*, *Stat3*, *Tgm1*; phagocytosis-related markers (dark red bar) *Abca7*, *Dock1*, *Gulp1*, *Lgals3*, *Lamp1*, *Megf10*, *Mfge8*; mitochondrial degeneration markers *Park2*, *Pink1*). Columns represent individual astrocytes (n = 33 for the saline group (blue bar), n = 39 for the microbead group (red bar)). (**B**) heatmap of normalized expression values in single astrocytes from naïve 13-week (purple bar) or 40-week (orange bar) old C57bl/6 mice. Genes are the same as for A. Columns represent individual astrocytes (n = 14 for the 3-month group, n = 23 for the 9-month group). (**C**) Normalized expression levels for all cells from microbead-injected nerves compared to the population average. A value of 0 would indicate that the distribution of high-expressing and low-expressing cells are identical for the microbead-injected and control (saline) group. A value of 1 would indicate that all microbead-injected cells are in the high expressing group. For most genes, more astrocytes in the microbead-injected group are high-expressing; however, this reached significance (defined as the 95% confidence interval for the gene not containing 0) only for *Gulp1*, *Lgals3*, *Pink1*, and *Psmb8*. Grey bars represent 95% confidence intervals. (**D**) Same analysis as in C for 13-week and 40-week-old naïve astrocytes. No gene is expressed at a significantly higher or lower level in the older group (all confidence intervals include 0).

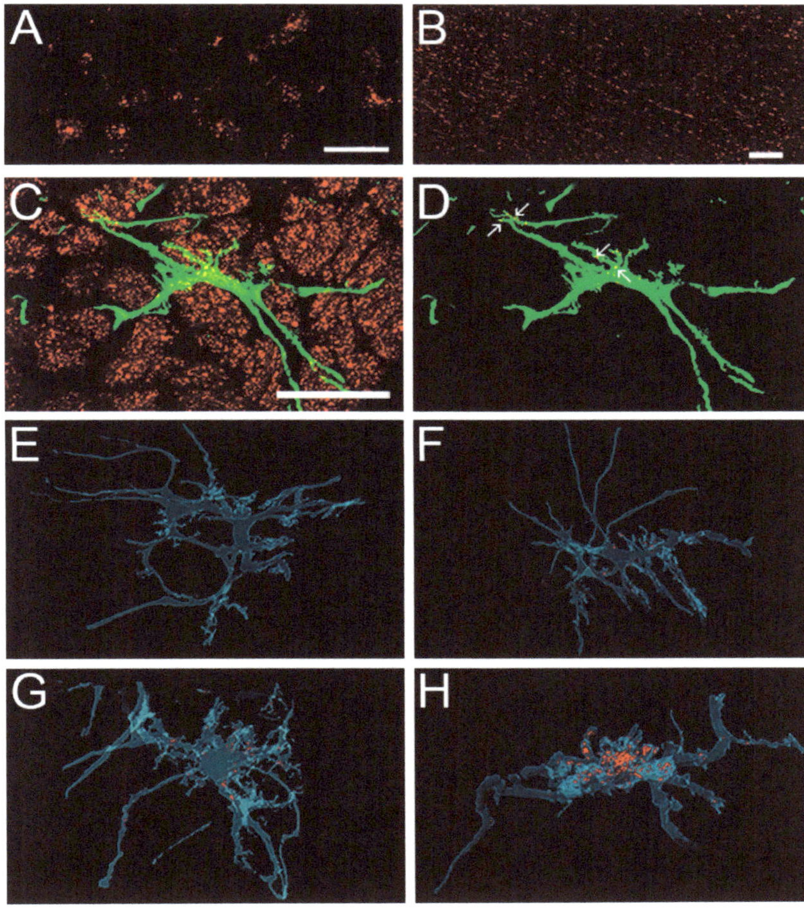

Figure 6. Distribution of axonal mitochondria in surrounding astrocytes. MitoTracker^{TM} Red FM labeling. (**A**) Whole mounted retina 4 days after injection of MitoTracker^{TM} Red FM into the contralateral superior colliculus. Retrograde axonal transported labeled mitochondria appear as puncta in the retina. Scale bar, 20 μm. (**B**) Longitudinal section of the optic nerve 4 days after injection of MitoTracker^{TM} Red FM. Scale bar, 20 μm. (**C**) Cross section of an optic nerve (unmyelinated region) with a single astrocyte expressing EGFP. Axonal mitochondria were labeled by injection of MitoTracker^{TM} Red FM into the contralateral superior colliculus. The majority of mitochondria does not co-localize with the astrocyte. Scale bar, 50 μm. (**D**) Same astrocyte after digitally removing all mitochondria signals that were not enclosed within EGFP$^+$ astrocyte cytoplasm in all 3 dimensions. Several labeled mitochondria (arrows in (**D**)) are visible inside the astrocyte. (**E–H**) representative images of astrocytes with no incorporated mitochondria. (**E**), sporadic (**F**), moderate (**G**) and large amounts of mitochondria incorporation. The images were reconstructed using the Amira software package after digitally removing the mitochondria not incorporated into an EGFP$^+$ astrocyte. The astrocyte is rendered transparent to show the mitochondria inside.

As most of the MitoTracker^{TM} Red FM-labeled mitochondria do not co-localize with EGFP$^+$ astrocytes (Figure 6C), we digitally reconstructed each image stack including only those mitochondria that were surrounded in all three dimensions by astrocytic cytoplasm (Figure 6D). Not all astrocytes incorporated labeled mitochondria (Figure 6E–H). In the control group 146 out of 344 astrocytes contained no MitoTracker^{TM} Red FM (42.4%). This number was lower in the high IOP group (92 out of 339; 27.1%) (Figure 7A). The amount of incorporated MitoTracker^{TM} Red FM per astrocyte differed considerably between individual cells within the same nerve. Astrocytes located between 0–100 μm behind the sclera contained few labeled mitochondria. In addition, incorporated mitochondria were not distributed evenly within individual astrocytes. Most labeled material was observed either in the cell body or in short, tortuous, or longitudinal processes. The long, relatively smooth primary astrocyte processes contained relatively few MitoTracker^{TM} Red FM-labeled mitochondria. Astrocytes that contained labeled mitochondria generally also showed process hypertrophy (Figure 7B). This was the case for astrocytes from both saline-injected nerves and microbead-injected nerves.

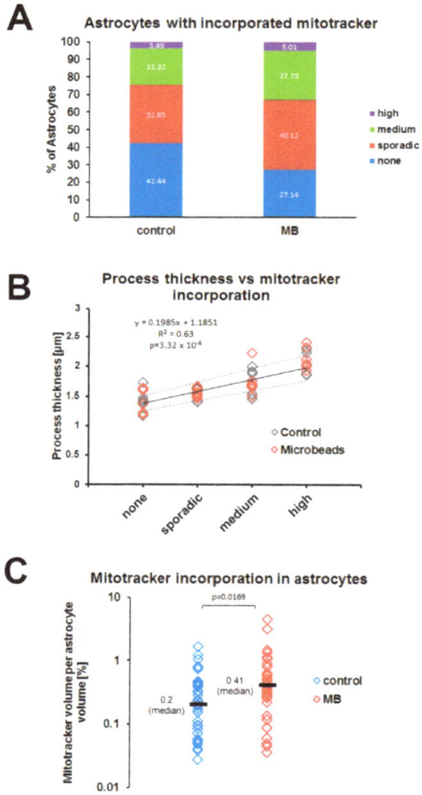

Figure 7. Quantification of mitochondria incorporation into optic nerve astrocytes. (**A**) percentages of astrocytes with no, sporadic, medium, and high incorporation of axonal mitochondria encountered in saline-injected and microbead-injected nerves. (**B**) Process thickness of astrocytes with no, sporadic, medium, and high incorporation of MitoTracker^{TM} Red FM. Process thickness was measured 10 μm distal to the soma in 6 astrocytes from each category in saline-injected (control, grey diamonds) and microbead-injected (red diamonds) nerves; in total n = 24 astrocytes in each group. Solid grey line shows the linear regression line based on the control cells. Dashed grey lines show 95% confidence intervals. There was no significant difference between astrocytes from control and injected nerves. (**C**) total MitoTracker^{TM} Red FM incorporation in astrocytes (in % of the total EGFP+ astrocyte volume) in saline-injected and microbead-injected (MB) nerves. Diamonds indicate individual cell measurement, Wilcoxon rank sum test.

To quantify whether an increase in IOP leads to increased incorporation of mitochondria, we calculated the volume of MitoTracker^{TM} Red FM-labeled mitochondria in astrocytes for every image stack in the control and high-IOP group. Since EGFP expression in the B6.*hGFAPpr*::GFP strain is stochastic, the number of labeled astrocytes is different from nerve to nerve. Therefore, we also determined the total volume of EGFP^{+} astrocytes per stack and calculated the incorporation of labeled mitochondria per astrocyte volume (Figure 7C). The fraction of astrocyte volume occupied by MitoTracker^{TM} Red FM-labeled material was generally small (range 0–1.63% of the volume, median 0.2% for the control group). In the high IOP group, this amount was moderately increased (range 0.036–4.52%, median 0.41%, $p = 0.0151$, Wilcoxon rank sum test).

4. Discussion

4.1. Astrocytes from Aged and Glaucomatous Optic Nerve Heads Appear Similar in Many Respects

Aging has long been recognized as a risk factor for glaucoma and confounding factor in diagnosis since glaucomatous eyes usually are older eyes [56,57]. Age-related changes of course occur in all ocular tissues [58], and though any single one of them may not be clinically relevant by itself, they can jointly contribute to the severity of glaucoma [59]. Most importantly in this context, in all mammalian species where this question has been addressed, ganglion cell axon counts were found to decrease with age [60,61]. Though the rate of axon loss dramatically increases in the presence of elevated IOP [62], similar changes in astrocyte reactions and axon numbers may be expected in glaucomatous and aged optic nerves, albeit on different time scales. This is indeed observed for optic nerve axons and blood vessels, so that glaucomatous nerves appear prematurely aged [35]. Not unexpectedly, the morphological similarity of aged and glaucomatous nerves corresponds to a similar pattern of gene expression changes in both conditions. RNA extracted from the unmyelinated segment of the optic nerve mainly originates from astrocytes, which are the major cell type of this region, whereas microglia and endothelial cells are sparse [6,8]. RNA-seq experiments showed an overall similarity of gene expression profiles in aged and glaucomatous astrocytes (Figure 2A), and especially for up-regulated genes the overlap between both sets was high (77.9% of genes up-regulated after IOP elevation are also up-regulated in aged nerves). We did not find an enrichment of genes associated with cell cycle or proliferation, and counts of astrocyte nuclei in electron microscopy images showed no difference between nerves after one month of IOP elevation and controls. This differs from earlier reports that described astrocyte proliferation in glaucomatous nerves of DBA/2J mice or rats with experimentally induced ocular hypertension [13,63–65]. A possible explanation is that the microbead model used in our study models the early response to a moderate increase in IOP that is not severe or long-lasting enough to induce astrocyte proliferation. However, both in aging and glaucomatous nerves astrocyte processes showed hypertrophy, a sign of a more reactive phenotype [66–68]. Most genes associated with the A1 (neurotoxic) or A2 (neuroprotective) route of astrocyte reactivity were not differentially expressed either in the microbead or the aged groups. This was surprising as a recent study demonstrated elevation of A1 genes in murine optic nerves after IOP elevation [69]. A possible explanation for this discrepancy may be the use of different platforms (microfluidic qPCR versus RNA-sequencing).

4.2. Sprouting of New Astrocytic Processes Is an Early Event in Glaucoma

The main difference between aged and glaucomatous nerve heads was in the up-regulation of pathways annotated with cell sprouting and the formation of processes. This agrees with previous morphological studies in the optic nerves of glaucomatous DBA/2J mice and C57bl/6 mice after one month of IOP elevation [14,18]. Astrocytes in the glial lamina are normally fairly flat and oriented in the coronal plane—i.e., their processes span the diameter of the optic nerve but have only very few and short longitudinal processes [8]. In glaucomatous nerves, the astrocytes extend new processes that run longitudinally in the optic nerve, invade axon bundles, and often contain remnants of mitochondria. However, it was not clear what cell biological function these new processes carried out. Astrocytes are known to be phagocytic in vitro and in vivo [31,70–74]. A particularly intriguing mechanism links the astrocytes of the optic nerve to the degradation of axonal mitochondria. In a process described as "transcellular mitophagy", astrocyte processes enwrap axonal evulsions that contain degenerating mitochondria, and eventually internalize them [55,75]. This process is physiological and presumably helps retinal ganglion cells to dispose of mitochondria that have reached the end of their lifespan, but there is also evidence that axonal evulsions are particularly numerous in the optic nerves of glaucomatous DBA/2J mice [76].

We therefore measured the incorporation of axonal mitochondria into astrocytes in a double-label in vivo assay. Interestingly, both in naïve and glaucomatous nerves, we observed large differences in the amount of material any individual astrocyte incorporated. In all nerves, there were at least some cells that did not contain any labeled material, whereas a small number of cells were highly active. Glaucomatous nerves contained more highly active cells, but even in those nerves some astrocytes remained that had incorporated very little or no axonal mitochondria. Cells that had incorporated many labeled mitochondria also showed hypertrophy of their processes, a finding commonly associated with a more reactive state (Figure 7B). The heterogeneity of astrocytes in normal and glaucomatous nerves was also apparent in single-cell qPCR for phagocytosis- and reactivity-associated genes. Though on the population level the expression of *Lgals3* and *Gulp1* was increased after elevation of IOP, the increase was mainly due to an increase in numbers of highly active cells rather than a shift of the whole population.

4.3. Implications of the Apparent Heterogeneity of Optic Nerve Head Astrocytes

Taken together, our results suggest that at any time a subset of optic nerve head astrocytes is reactive—as indicated by process hypertrophy, high phagocytic activity, and relatively high expression of reactivity genes (however without a clear bias towards the A1 or A2 phenotype described for protoplasmic astrocytes in other regions of the brain). An episode of IOP elevation leads to an increase of astrocyte reactivity on the population level. However, we did not find that the astrocytes move in unison towards a more reactive phenotype. Neither were reactive astrocytes in glaucomatous nerves obviously different from their counterparts in naïve nerves—actually, they were indistinguishable by all parameters we tested. The only difference was that the number of reactive astrocytes was higher in the glaucomatous nerve heads. For example, the percentage of astrocytes that had not incorporated any axonal mitochondria was 42.44% in control nerves and decreased to 27.14% in nerves after IOP elevation (Figure 7).

There are several possible interpretations of this observation. Perhaps the easiest is that there are (at least) two different types of astrocyte present in the optic nerve head, only one of which is highly phagocytic and prone to becoming reactive. However, there is no clear evidence on the anatomical level that the astrocytes of the glial lamina belong to different types. Rather, their shape is fairly uniform [8,77]; and without the clear morphological distinctions that usually form the basis of type definitions in other systems [78]. In addition, all astrocytes in our single cell PCR expressed most (usually all) phagocytic and reactivity markers tested, though not necessarily at the same levels. This suggests that differences between individual astrocytes in the optic nerve head are a matter of degree, not of clearly identifiable types.

Another explanation would be that astrocytes react to local signals in their vicinity, such as "find me" signals like ATP or UDP leaking out of injured axons, or microglia-derived factors [49,79]. Such signals might instruct an astrocyte to extend new processes towards the origin of the signal or increase its phagocytic activity. Yet, the anatomy of the optic nerve head argues against this explanation. In contrast to protoplasmic astrocytes in the grey matter of the CNS, which occupy spatially separated domains [80,81], the astrocytes in the optic nerve head often span the whole diameter of the nerve and their processes overlap considerably forming the sieve-like meshwork of the glial lamina [8

It is hard to imagine how a localized signal would affect only a subset of the closely intermingled astrocyte processes.

An alternative interpretation of our observations would be that astrocytes cycle through periods of high phagocytic activity followed by a period of relative inactivity. Astrocytes are not professional phagocytes and are relatively slow in degrading and eliminating phagocytized material [82], which may explain why a resting period is needed. Since it is at present not possible to follow individual optic nerve astrocytes in vivo in the intact optic nerve, our experiments necessarily show only a static snapshot of what is most likely a continuous physiological process. Cells that were not active during the period the MitoTracker^{TM} Red FM was present would have remained unlabeled, and hence invisible to our observations. The implication would be that optic nerve head astrocytes are always at least somewhat reactive. This may be caused by the constant mechanical strain they are exposed to as the IOP fluctuates due to circadian rhythm, eye movements, and posture even in healthy individuals [83–85]. Astrocytes express several mechanosensitive channels and extend long processes that insert into the pial sheet around the nerve [8,48,86–89]. They are therefore in an ideal anatomical position to sense stretch on the optic nerve head imparted through IOP fluctuations. In the presence of elevated IOP, more astrocytes might transition to the reactive state, or alternatively, activated cells are prevented from returning to the resting level.

5. Conclusions

Life cell imaging and gene expression surveys have expanded our understanding of astrocytes and their reaction to injury. Experimental evidence indicates that reactive astrocytes perform many functions that are beneficial for neuronal survival, especially in the early stages of glaucoma. Hopefully, stimulating the beneficial (and inhibiting the harmful) activities of reactive astrocytes can be used as a neuroprotective approach to glaucoma therapy.

Supplementary Materials: The following supporting information can be downloaded at: https://www.mdpi.com/article/10.3390/cells12172131/s1, Table S1: Group Summaries for RNA-sequencing; Table S2: Primer Sequences; Figure S1: Schematic of the Optic Nerve; Figure S2: Principal Component Analysis; Figure S3: Intraocular Pressure.

Author Contributions: T.C.J., R.W. and D.S. conceived the study. Y.Z., R.W., A.C.P., D.S. and T.C.J. performed experiments on live mice, immunohistochemistry, and confocal microscopy. Y.Z. performed quantitative RT-PCR on astrocytes. A.C.P., P.S. and T.C.J. did electron microscopy. Y.Z., A.S. and R.I.S. performed RNA-sequencing. A.S., R.I.S. and Y.Z. analyzed RNA-sequencing data. Y.Z. and T.C.J. analyzed qPCR data. T.C.J. wrote the Matlab scripts to quantify mitochondria incorporation by astrocytes. T.C.J., D.S., R.W. and R.I.S. acquired funding. T.C.J. wrote the original draft of the manuscript. Y.Z., D.S. and T.C.J. prepared the figures, and all authors reviewed the final version. All authors have read and agreed to the published version of the manuscript.

Funding: This research was funded by the NIH grants R01EY019703 (TCJ) and R01EY029268 (DS), the NIH Core grants for vision research P30EY003790 and P30EY014104, NIH NIDDK Grant P30DK040561 (RIS); grants from the Chinese Scholarship Council, the National Natural Science Foundation of China grant 81902014 (RW), The Massachusetts Lions Eye Research Fund, the Bright Focus Foundation, and Research to Prevent Blindness.

Institutional Review Board Statement: The animal study protocol was approved by the Institutional Review Board of Schepens Eye Research Institute (protocol code 2021N000075, last approval date: 10 April 2022).

Informed Consent Statement: Not applicable.

Data Availability Statement: Confocal image stacks are available at https://dataverse.harvard.edu/dataverse/astrocytes2019. RNA-sequencing data has been submitted to GEO (Accession number GSE139605).

Conflicts of Interest: The authors declare no conflict of interest.

References

1. Nickells, R.W. Ganglion cell death in glaucoma: From mice to men. *Vet. Ophthalmol.* **2007**, *10*, 88–94. [CrossRef] [PubMed]
2. Quigley, H.A. Glaucoma. *Lancet* **2011**, *377*, 1367–1377. [CrossRef] [PubMed]
3. Tham, Y.C.; Li, X.; Wong, T.Y.; Quigley, H.A.; Aung, T.; Cheng, C.Y. Global prevalence of glaucoma and projections of glaucoma burden through 2040: A systematic review and meta-analysis. *Ophthalmology* **2014**, *121*, 2081–2090. [CrossRef] [PubMed]
4. Wiggs, J.L. The Cell and Molecular Biology of Complex Forms of Glaucoma: Updates on Genetic, Environmental, and Epigenetic Risk Factors. *Investig. Ophthalmol. Vis. Sci.* **2012**, *53*, 2467–2469. [CrossRef] [PubMed]
5. Jonas, J.B.; Aung, T.; Bourne, R.R.; Bron, A.M.; Ritch, R.; Panda-Jonas, S. Glaucoma. *Lancet* **2017**, *390*, 2183–2193. [CrossRef] [PubMed]
6. Morcos, Y.; Chan-Ling, T. Concentration of astrocytic filaments at the retinal optic nerve junction is coincident with the absence of intra-retinal myelination: Comparative and developmental evidence. *J. Neurocytol.* **2000**, *29*, 665–678. [CrossRef] [PubMed]
7. May, C.A.; Lütjen-Drecoll, E. Morphology of the murine optic nerve. *Investig. Ophthalmol. Vis. Sci.* **2002**, *43*, 2206–2212.
8. Sun, D.; Lye-Barthel, M.; Masland, R.H.; Jakobs, T.C. The morphology and spatial arrangement of astrocytes in the optic nerve head of the mouse. *J. Comp. Neurol.* **2009**, *516*, 1–19. [CrossRef]
9. Quigley, H.A.; Addicks, E.M.; Green, W.R.; Maumenee, A.E. Optic nerve damage in human glaucoma. II. The site of injury and susceptibility to damage. *Arch. Ophthalmol.* **1981**, *99*, 635–649. [CrossRef]
10. Howell, G.R.; Libby, R.T.; Jakobs, T.C.; Smith, R.S.; Phalan, F.C.; Barter, J.W.; Barbay, J.M.; Marchant, J.K.; Mahesh, N Porciatti, V.; et al. Axons of retinal ganglion cells are insulted in the optic nerve early in DBA/2J glaucoma. *J. Cell Biol.* **2007**, *179*, 1523–1537. [CrossRef]
11. Balaratnasingam, C.; Morgan, W.H.; Bass, L.; Ye, L.; McKnight, C.; Cringle, S.J.; Yu, D.-Y. Elevated pressure induced astrocyte damage in the optic nerve. *Brain Res.* **2008**, *1244*, 142–154. [CrossRef] [PubMed]
12. Howell, G.R.; Macalinao, D.G.; Sousa, G.L.; Walden, M.; Soto, I.; Kneeland, S.C.; Barbay, J.M.; King, B.L.; Marchant, J.K. Hibbs, M.; et al. Molecular clustering identifies complement and endothelin induction as early events in a mouse model of glaucoma. *J. Clin. Investig.* **2011**, *121*, 1429–1444. [CrossRef] [PubMed]
13. Johnson, E.C.; Doser, T.A.; Cepurna, W.O.; Dyck, J.A.; Jia, L.; Guo, Y.; Lambert, W.S.; Morrison, J.C. Cell Proliferation and Interleukin-6–Type Cytokine Signaling Are Implicated by Gene Expression Responses in Early Optic Nerve Head Injury in Rat Glaucoma. *Investig. Ophthalmol. Vis. Sci.* **2011**, *52*, 504–518. [CrossRef] [PubMed]
14. Lye-Barthel, M.; Sun, D.; Jakobs, T.C. Morphology of Astrocytes in a Glaucomatous Optic Nerve. *Investig. Ophthalmol. Vis. Sci.* **2013**, *54*, 909–917. [CrossRef] [PubMed]
15. Tehrani, S.; Johnson, E.C.; Cepurna, W.O.; Morrison, J.C. Astrocyte Processes Label for Filamentous Actin and Reorient Early within the Optic Nerve Head in a Rat Glaucoma Model. *Investig. Ophthalmol. Vis. Sci.* **2014**, *55*, 6945–6952. [CrossRef] [PubMed]
16. Tehrani, S.; Davis, L.; Cepurna, W.O.; Choe, T.E.; Lozano, D.C.; Monfared, A.; Cooper, L.; Cheng, J.; Johnson, E.C.; Morrison, J.C. Astrocyte Structural and Molecular Response to Elevated Intraocular Pressure Occurs Rapidly and Precedes Axonal Tubulin Rearrangement within the Optic Nerve Head in a Rat Model. *PLoS ONE* **2016**, *11*, e0167364. [CrossRef] [PubMed]
17. Cooper, M.L.; Crish, S.D.; Inman, D.M.; Horner, P.J.; Calkins, D.J. Early astrocyte redistribution in the optic nerve precedes axonopathy in the DBA/2J mouse model of glaucoma. *Exp. Eye Res.* **2016**, *150*, 22–33. [CrossRef]
18. Wang, R.; Seifert, P.; Jakobs, T.C. Astrocytes in the Optic Nerve Head of Glaucomatous Mice Display a Characteristic Reactive Phenotype. *Investig. Ophthalmol. Vis. Sci.* **2017**, *58*, 924–932. [CrossRef]
19. Tehrani, S.; Davis, L.; Cepurna, W.O.; Delf, R.K.; Lozano, D.C.; Choe, T.E.; Johnson, E.C.; Morrison, J.C. Optic Nerve Head Astrocytes Display Axon-Dependent and -Independent Reactivity in Response to Acutely Elevated Intraocular Pressure. *Investig. Ophthalmol. Vis. Sci.* **2019**, *60*, 312–321. [CrossRef]
20. Neufeld, A.H.; Liu, B. Glaucomatous Optic Neuropathy: When Glia Misbehave. *Neurosci.* **2003**, *9*, 485–495. [CrossRef]
21. Hernandez, M.R.; Miao, H.; Lukas, T. Astrocytes in glaucomatous optic neuropathy. *Prog. Brain Res.* **2008**, *173*, 353–373. [CrossRef] [PubMed]
22. Pekny, M.; Nilsson, M. Astrocyte activation and reactive gliosis. *Glia* **2005**, *50*, 427–434. [CrossRef] [PubMed]
23. Sofroniew, M.V. Reactive Astrocytes in Neural Repair and Protection. *Neurosci.* **2005**, *11*, 400–407. [CrossRef] [PubMed]
24. Sofroniew, M.V.; Vinters, H.V. Astrocytes: Biology and pathology. *Acta Neuropathol.* **2010**, *119*, 7–35. [CrossRef] [PubMed]
25. Sun, D.; Jakobs, T.C. Structural Remodeling of Astrocytes in the Injured CNS. *Neuroscientist* **2012**, *18*, 567–588. [CrossRef] [PubMed]
26. Zamanian, J.L.; Xu, L.; Foo, L.C.; Nouri, N.; Zhou, L.; Giffard, R.G.; Barres, B.A. Genomic Analysis of Reactive Astrogliosis. *J. Neurosci.* **2012**, *32*, 6391–6410. [CrossRef] [PubMed]
27. Liddelow, S.A.; Barres, B.A. Reactive Astrocytes: Production, Function, and Therapeutic Potential. *Immunity* **2017**, *46*, 957–967. [CrossRef] [PubMed]
28. Escartin, C.; Galea, E.; Lakatos, A.; O'Callaghan, J.P.; Petzold, G.C.; Serrano-Pozo, A.; Steinhäuser, C.; Volterra, A.; Carmignoto, G. Agarwal, A.; et al. Reactive astrocyte nomenclature, definitions, and future directions. *Nat. Neurosci.* **2021**, *24*, 312–335. [CrossRef]
29. Oberheim, N.A.; Tian, G.-F.; Han, X.; Peng, W.; Takano, T.; Ransom, B.; Nedergaard, M. Loss of Astrocytic Domain Organization in the Epileptic Brain. *J. Neurosci.* **2008**, *28*, 3264–3276. [CrossRef]

30. Wanner, I.B.; Anderson, M.A.; Song, B.; Levine, J.; Fernandez, A.; Gray-Thompson, Z.; Ao, Y.; Sofroniew, M.V. Glial Scar Borders Are Formed by Newly Proliferated, Elongated Astrocytes That Interact to Corral Inflammatory and Fibrotic Cells via STAT3-Dependent Mechanisms after Spinal Cord Injury. *J. Neurosci.* **2013**, *33*, 12870–12886. [CrossRef]
31. Morizawa, Y.M.; Hirayama, Y.; Ohno, N.; Shibata, S.; Shigetomi, E.; Sui, Y.; Nabekura, J.; Sato, K.; Okajima, F.; Takebayashi, H.; et al. Reactive astrocytes function as phagocytes after brain ischemia via ABCA1-mediated pathway. *Nat. Commun.* **2017**, *8*, 28. [CrossRef] [PubMed]
32. Bush, T.G.; Puvanachandra, N.; Horner, C.H.; Polito, A.; Ostenfeld, T.; Svendsen, C.N.; Mucke, L.; Johnson, M.H.; Sofroniew, M.V. Leukocyte Infiltration, Neuronal Degeneration, and Neurite Outgrowth after Ablation of Scar-Forming, Reactive Astrocytes in Adult Transgenic Mice. *Neuron* **1999**, *23*, 297–308. [CrossRef] [PubMed]
33. Herrmann, J.E.; Imura, T.; Song, B.; Qi, J.; Ao, Y.; Nguyen, T.K.; Korsak, R.A.; Takeda, K.; Akira, S.; Sofroniew, M.V. STAT3 is a Critical Regulator of Astrogliosis and Scar Formation after Spinal Cord Injury. *J. Neurosci.* **2008**, *28*, 7231–7243. [CrossRef] [PubMed]
34. Sun, D.; Moore, S.; Jakobs, T.C. Optic nerve astrocyte reactivity protects function in experimental glaucoma and other nerve injuries. *J. Exp. Med.* **2017**, *214*, 1411–1430. [CrossRef] [PubMed]
35. Zhu, Y.; Pappas, A.C.; Wang, R.; Seifert, P.; Sun, D.; Jakobs, T.C. Ultrastructural Morphology of the Optic Nerve Head in Aged and Glaucomatous Mice. *Investig. Ophthalmol. Vis. Sci.* **2018**, *59*, 3984–3996. [CrossRef] [PubMed]
36. Xu, Q.; Rydz, C.; Huu, V.A.N.; Rocha, L.; La Torre, C.P.; Lee, I.; Cho, W.; Jabari, M.; Donello, J.; Lyon, D.C.; et al. Stress induced aging in mouse eye. *Aging Cell* **2022**, *21*, e13737. [CrossRef]
37. Nolte, C.; Matyash, M.; Pivneva, T.; Schipke, C.G.; Ohlemeyer, C.; Hanisch, U.K.; Kirchhoff, F.; Kettenmann, H. GFAP promoter-controlled EGFP-expressing transgenic mice: A tool to visualize astrocytes and astrogliosis in living brain tissue. *Glia* **2001**, *33*, 72–86. [CrossRef]
38. Emsley, J.G.; Macklis, J.D. Astroglial heterogeneity closely reflects the neuronal-defined anatomy of the adult murine CNS. *Neuron Glia Biol.* **2006**, *2*, 175–186. [CrossRef]
39. Sappington, R.M.; Carlson, B.J.; Crish, S.D.; Calkins, D.J. The Microbead Occlusion Model: A Paradigm for Induced Ocular Hypertension in Rats and Mice. *Investig. Ophthalmol. Vis. Sci.* **2010**, *51*, 207–216. [CrossRef]
40. Chen, H.; Wei, X.; Cho, K.-S.; Chen, G.; Sappington, R.; Calkins, D.J.; Chen, D.F. Optic Neuropathy Due to Microbead-Induced Elevated Intraocular Pressure in the Mouse. *Investig. Ophthalmol. Vis. Sci.* **2011**, *52*, 36–44. [CrossRef]
41. Dobin, A.; Davis, C.A.; Schlesinger, F.; Drenkow, J.; Zaleski, C.; Jha, S.; Batut, P.; Chaisson, M.; Gingeras, T.R. STAR: Ultrafast universal RNA-seq aligner. *Bioinformatics* **2013**, *29*, 15–21. [CrossRef] [PubMed]
42. Liao, Y.; Smyth, G.K.; Shi, W. feature Counts: An efficient general purpose program for assigning sequence reads to genomic features. *Bioinformatics* **2014**, *30*, 923–930. [CrossRef] [PubMed]
43. Anders, S.; McCarthy, D.J.; Chen, Y.; Okoniewski, M.; Smyth, G.K.; Huber, W.; Robinson, M.D. Count-based differential expression analysis of RNA sequencing data using R and Bioconductor. *Nat. Protoc.* **2013**, *8*, 1765–1786. [CrossRef] [PubMed]
44. Dacey, D.M.; Peterson, B.B.; Robinson, F.R.; Gamlin, P.D. Fireworks in the Primate Retina: In Vitro Photodynamics Reveals Diverse LGN-Projecting Ganglion Cell Types. *Neuron* **2003**, *37*, 15–27. [CrossRef] [PubMed]
45. Hu, B.; Huang, Y.; Jakobs, T.C.; Kang, Q.; Lv, Z.; Liu, W.; Wang, R. Viability of mitochondria-labeled retinal ganglion cells in organotypic retinal explant cultures by two methods. *Exp. Eye Res.* **2022**, *226*, 109311. [CrossRef] [PubMed]
46. Choi, H.J.; Wang, R.; Jakobs, T.C. Single-Cell Dissociation and Characterization in the Murine Retina and Optic Nerve. In *Glaucoma Methods and Protocols*; Springer: Berlin/Heidelberg, Germany, 2018; Volume 1695, pp. 311–334. [CrossRef]
47. Choi, H.J.; Sun, D.; Jakobs, T.C. Isolation of intact astrocytes from the optic nerve head of adult mice. *Exp. Eye Res.* **2015**, *137*, 103–110. [CrossRef] [PubMed]
48. Choi, H.J.; Sun, D.; Jakobs, T.C. Astrocytes in the optic nerve head express putative mechanosensitive channels. *Mol. Vis.* **2015**, *21*, 749–766. [PubMed]
49. Liddelow, S.A.; Guttenplan, K.A.; Clarke, L.E.; Bennett, F.C.; Bohlen, C.J.; Schirmer, L.; Bennett, M.L.; Münch, A.E.; Chung, W.-S.; Peterson, T.C.; et al. Neurotoxic reactive astrocytes are induced by activated microglia. *Nature* **2017**, *541*, 481–487. [CrossRef]
50. Spandidos, A.; Wang, X.; Wang, H.; Seed, B. PrimerBank: A resource of human and mouse PCR primer pairs for gene expression detection and quantification. *Nucleic Acids Res.* **2010**, *38*, D792–D799. [CrossRef]
51. Dutta, S.; Sengupta, P. Men and mice: Relating their ages. *Life Sci.* **2016**, *152*, 244–248. [CrossRef]
52. Flurkey, K.; Currer, J.M.; Harrison, D.E. Mouse models in aging research. In *The Mouse in Biomedical Research*; Fox, J.G., Davisson, M.T., Quimby, F.W., Barthold, S.W., Newcomer, C.E., Smith, A.L., Eds.; Elsevier: Burlington, MA, USA, 2007; Volume 3, pp. 637–672.
53. Ramírez, A.I.; Salazar, J.J.; de Hoz, R.; Rojas, B.; Gallego, B.I.; Salinas-Navarro, M.; Alarcón-Martínez, L.; Ortín-Martínez, A.; Avilés-Trigueros, M.; Vidal-Sanz, M.; et al. Quantification of the Effect of Different Levels of IOP in the Astroglia of the Rat Retina Ipsilateral and Contralateral to Experimental Glaucoma. *Investig. Ophthalmol. Vis. Sci.* **2010**, *51*, 5690–5696. [CrossRef] [PubMed]
54. Gallego, B.I.; Salazar, J.J.; De Hoz, R.; Rojas, B.; Ramírez, A.I.; Navarro, M.S.; Ortín-Martínez, A.; Soriano, F.J.V.; Avilés-Trigueros, M.; Pérez, M.P.V.; et al. IOP induces upregulation of GFAP and MHC-II and microglia reactivity in mice retina contralateral to experimental glaucoma. *J. Neuroinflamm.* **2012**, *9*, 92. [CrossRef] [PubMed]
55. Davis, C.-H.O.; Kim, K.-Y.; Bushong, E.A.; Mills, E.A.; Boassa, D.; Shih, T.; Kinebuchi, M.; Phan, S.; Zhou, Y.; Bihlmeyer, N.A.; et al. Transcellular degradation of axonal mitochondria. *Proc. Natl. Acad. Sci. USA* **2014**, *111*, 9633–9638. [CrossRef] [PubMed]

56. Vianna, J.R.; Danthurebandara, V.M.; Sharpe, G.P.; Hutchison, D.M.; Belliveau, A.C.; Shuba, L.M.; Nicolela, M.T.; Chauhan, B.C. Importance of Normal Aging in Estimating the Rate of Glaucomatous Neuroretinal Rim and Retinal Nerve Fiber Layer Loss. *Ophthalmology* **2015**, *122*, 2392–2398. [CrossRef]
57. Wu, Z.; Saunders, L.J.; Zangwill, L.M.; Daga, F.B.; Crowston, J.G.; Medeiros, F.A. Impact of Normal Aging and Progression Definitions on the Specificity of Detecting Retinal Nerve Fiber Layer Thinning. *Am. J. Ophthalmol.* **2017**, *181*, 106–113. [CrossRef] [PubMed]
58. Liu, B.; McNally, S.; Kilpatrick, J.I.; Jarvis, S.P.; O'Brien, C.J. Aging and ocular tissue stiffness in glaucoma. *Surv. Ophthalmol.* **2018**, *63*, 56–74. [CrossRef] [PubMed]
59. Guedes, G.; Tsai, J.C.; Loewen, N.A. Glaucoma and aging. *Curr. Aging Sci.* **2011**, *4*, 110–117. [CrossRef] [PubMed]
60. Neufeld, A.H.; Gachie, E.N. The inherent, age-dependent loss of retinal ganglion cells is related to the lifespan of the species. *Neurobiol. Aging* **2003**, *24*, 167–172. [CrossRef]
61. Calkins, D.J. Age-Related Changes in the Visual Pathways: Blame It on the Axon. *Investig. Ophthalmol. Vis. Sci.* **2013**, *54*, ORSF37–ORSF41. [CrossRef]
62. McKinnon, S.J.; Schlamp, C.L.; Nickells, R.W. Mouse models of retinal ganglion cell death and glaucoma. *Exp. Eye Res.* **2009**, *88*, 816–824. [CrossRef]
63. Johnson, E.C.; Jia, L.; Cepurna, W.O.; Doser, T.A.; Morrison, J.C. Global Changes in Optic Nerve Head Gene Expression after Exposure to Elevated Intraocular Pressure in a Rat Glaucoma Model. *Investig. Ophthalmol. Vis. Sci.* **2007**, *48*, 3161–3177. [CrossRef] [PubMed]
64. Son, J.L.; Soto, I.; Oglesby, E.; Lopez-Roca, T.; Pease, M.E.; Quigley, H.A.; Marsh-Armstrong, N. Glaucomatous optic nerve injury involves early astrocyte reactivity and late oligodendrocyte loss. *Glia* **2010**, *58*, 780–789. [CrossRef] [PubMed]
65. Lozano, D.C.; Choe, T.E.; Cepurna, W.O.; Morrison, J.C.; Johnson, E.C. Early Optic Nerve Head Glial Proliferation and Jak-Stat Pathway Activation in Chronic Experimental Glaucoma. *Investig. Ophthalmol. Vis. Sci.* **2019**, *60*, 921–932. [CrossRef] [PubMed]
66. Jyothi, H.; Vidyadhara, D.; Mahadevan, A.; Philip, M.; Parmar, S.K.; Manohari, S.G.; Shankar, S.; Raju, T.R.; Alladi, P.A. Aging causes morphological alterations in astrocytes and microglia in human substantia nigra pars compacta. *Neurobiol. Aging* **2015**, *36*, 3321–3333. [CrossRef] [PubMed]
67. Robillard, K.N.; Lee, K.M.; Chiu, K.B.; MacLean, A.G. Glial cell morphological and density changes through the lifespan of rhesus macaques. *Brain Behav. Immun.* **2016**, *55*, 60–69. [CrossRef] [PubMed]
68. Palmer, A.L.; Ousman, S.S. Astrocytes and Aging. *Front. Aging Neurosci.* **2018**, *10*, 337. [CrossRef] [PubMed]
69. Guttenplan, K.A.; Stafford, B.K.; El-Danaf, R.N.; Adler, D.I.; Münch, A.E.; Weigel, M.K.; Huberman, A.D.; Liddelow, S.A. Neurotoxic Reactive Astrocytes Drive Neuronal Death after Retinal Injury. *Cell Rep.* **2020**, *31*, 107776. [CrossRef]
70. Shuangshoti, S.; Samranvej, P.; Netsky, M.G. Phagocytic astrocytes and neurons in old encephalomalacia. *J. Neuropathol. Exp. Neurol.* **1979**, *38*, 235–241. [CrossRef]
71. Kalmár, B.; Kittel, Á.; Lemmens, R.; Környei, Z.; Madarász, E. Cultured astrocytes react to LPS with increased cyclooxygenase activity and phagocytosis. *Neurochem. Int.* **2001**, *38*, 453–461. [CrossRef]
72. Nguyen, J.V.; Soto, I.; Kim, K.-Y.; Bushong, E.A.; Oglesby, E.; Valiente-Soriano, F.J.; Yang, Z.; Davis, C.-H.O.; Bedont, J.L.; Son, J.L.; et al. Myelination transition zone astrocytes are constitutively phagocytic and have synuclein dependent reactivity in glaucoma. *Proc. Natl. Acad. Sci. USA* **2011**, *108*, 1176–1181. [CrossRef]
73. Mills, E.A.; Davis, C.-H.O.; Bushong, E.A.; Boassa, D.; Kim, K.-Y.; Ellisman, M.H.; Marsh-Armstrong, N. Astrocytes phagocytose focal dystrophies from shortening myelin segments in the optic nerve of *Xenopus laevis* at metamorphosis. *Proc. Natl. Acad. Sci. USA* **2015**, *112*, 10509–10514. [CrossRef] [PubMed]
74. Gomez-Arboledas, A.; Davila, J.C.; Sanchez-Mejias, E.; Navarro, V.; Nunez-Diaz, C.; Sanchez-Varo, R.; Sanchez-Mico, M.V.; Trujillo-Estrada, L.; Fernandez-Valenzuela, J.J.; Vizuete, M.; et al. Phagocytic clearance of presynaptic dystrophies by reactive astrocytes in Alzheimer's disease. *Glia* **2018**, *66*, 637–653. [CrossRef] [PubMed]
75. Davis, C.-H.O.; Marsh-Armstrong, N. Discovery and implications of transcellular mitophagy. *Autophagy* **2014**, *10*, 2383–2384. [CrossRef] [PubMed]
76. Kim, K.-Y.; Perkins, G.A.; Shim, M.S.; Bushong, E.; Alcasid, N.; Ju, S.; Ellisman, M.H.; Weinreb, R.N.; Ju, W.-K. DRP1 inhibition rescues retinal ganglion cells and their axons by preserving mitochondrial integrity in a mouse model of glaucoma. *Cell Death Dis* **2015**, *6*, e1839. [CrossRef] [PubMed]
77. Butt, A.M.; Colquhoun, K.; Tutton, M.; Berry, M. Three-dimensional morphology of astrocytes and oligodendrocytes in the intact mouse optic nerve. *J. Neurocytol.* **1994**, *23*, 469–485. [CrossRef]
78. Masland, R.H. Neuronal cell types. *Curr. Biol.* **2004**, *14*, R497–R500. [CrossRef] [PubMed]
79. Li, W. Eat-me signals: Keys to molecular phagocyte biology and "Appetite" control. *J. Cell. Physiol.* **2012**, *227*, 1291–1297. [CrossRef]
80. Bushong, E.A.; Martone, M.E.; Jones, Y.Z.; Ellisman, M.H. Protoplasmic Astrocytes in CA1 Stratum Radiatum Occupy Separate Anatomical Domains. *J. Neurosci.* **2002**, *22*, 183–192. [CrossRef]
81. Wilhelmsson, U.; Bushong, E.A.; Price, D.L.; Smarr, B.L.; Phung, V.; Terada, M.; Ellisman, M.H.; Pekny, M. Redefining the concept of reactive astrocytes as cells that remain within their unique domains upon reaction to injury. *Proc. Natl. Acad. Sci. USA* **2006**, *103*, 17513–17518. [CrossRef]

82. Lööv, C.; Hillered, L.; Ebendal, T.; Erlandsson, A. Engulfing Astrocytes Protect Neurons from Contact-Induced Apoptosis following Injury. *PLoS ONE* **2012**, *7*, e33090. [CrossRef]
83. Wilensky, J.T. Diurnal variations in intraocular pressure. *Trans. Am. Ophthalmol. Soc.* **1991**, *89*, 757–790. [PubMed]
84. Downs, J.C.; Burgoyne, C.F.; Seigfreid, W.P.; Reynaud, J.F.; Strouthidis, N.G.; Sallee, V. 24-Hour IOP Telemetry in the Nonhuman Primate: Implant System Performance and Initial Characterization of IOP at Multiple Timescales. *Investig. Ophthalmol. Vis. Sci.* **2011**, *52*, 7365–7375. [CrossRef] [PubMed]
85. Gautam, N.; Kaur, S.; Kaushik, S.; Raj, S.; Pandav, S.S. Postural and diurnal fluctuations in intraocular pressure across the spectrum of glaucoma. *Br. J. Ophthalmol.* **2016**, *100*, 537–541. [CrossRef]
86. Ho, K.W.; Lambert, W.S.; Calkins, D.J. Activation of the TRPV1 cation channel contributes to stress-induced astrocyte migration. *Glia* **2014**, *62*, 1435–1451. [CrossRef] [PubMed]
87. Ho, K.W.; Sidorova, T.N.; Lambert, W.S.; Calkins, D.J. Age and elevated pressure increase astrocyte expression of TRPV1 in retina and optic nerve of the DBA/2J mouse model of glaucoma. *Investig. Ophthalmol. Vis. Sci.* **2011**, *52*, 2446.
88. Dai, C.; Khaw, P.T.; Yin, Z.Q.; Li, D.; Raisman, G.; Li, Y. Structural basis of glaucoma: The fortified astrocytes of the optic nerve head are the target of raised intraocular pressure. *Glia* **2012**, *60*, 13–28. [CrossRef] [PubMed]
89. Li, Y.; Li, D.; Ying, X.; Khaw, P.T.; Raisman, G. An energy theory of glaucoma. *Glia* **2015**, *63*, 1537–1552. [CrossRef]

MDPI AG
Grosspeteranlage 5
4052 Basel
Switzerland
Tel.: +41 61 683 77 34

Cells Editorial Office
E-mail: cells@mdpi.com
www.mdpi.com/journal/cells

www.ingramcontent.com/pod-product-compliance
Lightning Source LLC
Chambersburg PA
CBHW041430210926
51850CB00013B/3392